Physical and Chemical Hydrogeology

Physical and Chemical Hydrogeology

Second Edition

Patrick A. Domenico

David B. Harris Professor of Geology
Texas A&M University

Franklin W. Schwartz

Ohio Eminent Scholar in Hydrogeology
The Ohio State University

John Wiley & Sons, Inc.

New York   Chichester   Weinheim   Brisbane   Toronto   Singapore

To Diane and Cynthia and
the Memory of Lucy, Phil, and Daniel

ACQUISITIONS EDITOR Cliff Mills
MARKETING MANAGER Kimberly Manzi
PRODUCTION EDITOR Deborah Herbert
DESIGNER Ann Marie Renzi
ILLUSTRATION EDITOR Edward Starr
COVER Designed by Carolyn Joseph; Photo: Tony Stone Worldwide/David Carriere

This book was set in 10/12 ITC Garamond Book by Bi-Comp, Inc. and printed and bound by Courier Companies Inc, (Westford). The cover was printed by Phoenix Color Corp.

This book is printed on acid-free paper. ∞

The paper in this book was manufactured by a mill whose forest management programs include sustained yield harvesting of its timberlands. Sustained yield harvesting principles ensure that the numbers of trees cut each year does not exceed the amount of new growth.

Library of Congress Cataloging in Publication Data

Domenico, P. A. (Patrick A.)
   Physical and chemical hydrogeology / Patrick A. Domenico, Franklin
W. Schwartz. — 2nd ed.
      p.   cm.
   Includes bibliographical references and index.
   ISBN 0-471-59762-7 (cloth : alk. paper)
   1. Hydrogeology.   I. Schwartz, F. W. (Franklin W.)   II. Title.
GB1003.2.D66   1997
551.49—dc21   97-21776
10 9 8 7 6 5 4   CIP

Preface

One main objective in producing this Second Edition of *Physical and Chemical Hydrogeology* was to incorporate the new, broadly-based scientific advances that have been evident in the field. Most noteworthy in this respect is new knowledge on ground-water microbiology, theoretical and practical knowledge related to contamination by NAPLs and DNAPLs and multiphase fluids in general, new strategies for site clean ups, and risk assessment as a tool for making decisions about contaminated sites. However, we have not introduced these new concepts at the expense of other essential material—both traditional and modern—that makes up the essence of hydrogeology. As with the previous edition, the transport of fluid, energy, and mass in porous media remains the guiding theme throughout the book. Additionally, the Second Edition preserves the process oriented focus of the original book and continues to emphasize the relationship between theory and practice. We have again attempted to retain an understandable style while explaining complex hydrogeological matters. To this end, as in the previous edition, we include a significant number of worked examples and a problem set for most of the chapters. The book is still intended for students at the advanced undergraduate or beginning graduate level.

The Second Edition also reflects thoughtful suggestions and criticisms of colleagues who made clear what elements of the original book worked or didn't work. Clearly some reorganization was called for with an attempt to make each chapter as self sufficient as possible. This Edition, like the previous one, deliberately contains more material than can be covered in a single course so that choices have to be made by the instructor. We hope that this reorganization makes the task of choice easier.

We acknowledge the special contribution of Dr. Stephen Worthington of McMaster University, and the late Jim Quinlan. They helped by writing sections dealing with karst in Chapters 2 and 16. Like all professors, we benefit from the continuing help and assistance of present and former students. Dr. Hubao Zhang worked to expand our treatment of well hydraulics through the addition of the code, WELLz, which is included on disk at the back of the book. Dr. Dea-Ha Lee and Dr. Abe Springer commented on Chapter 7, with Dr. Springer providing MODFLOW data for his study site in Ohio. Dr. Rob Schincariol kindly provided computer generated hydraulic conductivity fields in Chapter 10. Dr. Alan Fryar contributed section 16.4 that deals with self-organizing systems. Gordon McClymont inspired the sections on risk assessment, and he and Hubao Zhang helped in reviewing this material. Our editors at John Wiley & Sons have been extremely supportive and worked to modernize the layout of the book.

Patrick A. Domenico
Franklin W. Schwartz

Contents

Chapter 1

Introduction

1.1 **What Is Hydrogeology?**
1.2 **The Relationship Between Hydrogeology and Other Fields of Geology**
1.3 **The Hydrologic Cycle**

This book was written from the perspective that the reader is interested in becoming a hydrogeologist. We recognize that this is not likely to be the case generally. However, for us to take a different perspective would require a different kind of book, one that is perhaps not as detailed or as rigorous. And this would be unfair to all students in several different ways. First, the student interested in hydrogeology as a career is entitled to know the breadth of the field and the fact that there is much to study, so much in fact that it cannot all be learned in one course. Additional courses will be required, not only in hydrogeology but in supporting sciences such as soil physics, soil mechanics, geochemistry, and numerical methods. Graduate training is an essential requirement. Additionally, all students are entitled to know that hydrogeology contains not only geology but a heavy dose of physics and chemistry as well, and its main language is mathematics. Thus, the entry level to the field is high, but so are the intellectual and practical rewards. For earth scientists not interested in hydrogeology as a career, let us state outright that virtually every activity in the earth sciences requires some knowledge of subsurface fluids, and rock-water interactions in particular, and that is what hydrogeology and this book are about.

Last, from our own biased perspective, we think it would be difficult to find a more rewarding introduction to the concepts of science than those offered by the study of hydrogeology. The discipline blends field, experimental, and theoretical activities. Sometimes the experimental activities take place in the laboratory; at other times they take place in the field, where the comfort of laboratory control is lost. Additionally, the field and experimental activities often play a major role in the formulation of reliable theoretical models of processes and events. In increasingly more cases, the methods of hydrogeology give us the wherewithal to study and perhaps quantify some of nature's experiments. These same methods help us study, quantify, and sometimes rectify some unfortunate "experiments" of an industrialized society. All of this, and much more, is hydrogeology.

1.1 What Is Hydrogeology?

We have chosen to call this book *Physical and Chemical Hydrogeology* for two reasons. First, this is what the book and the field are about. Second, most other potential candidate titles have been preempted. The simple title *Ground Water* is appealing and was used by Tolman in 1937 and by Freeze and Cherry in 1979. The books are quite different in scope because during the period 1935 to the present, the field of knowledge was undergoing tremendous expansion and this expansion is expertly captured by Freeze and Cherry (1979). Lamarck wrote a book entitled *Hydrogeology* in 1802, as did Davis and DeWiest in 1966. Obviously, due to the times, there is no similarity between the subject matter of these books. A book entitled *Geohydrology* was prepared by DeWiest in 1965 and one entitled *Groundwater Hydrology* was written by Todd, first in 1959 with a later edition appearing in 1980.

Some people remain uncomfortable with the situation where the subject matter of subsurface fluids can be organized under a variety of titles and taught under a variety of disciplines. The meaning of the terms *hydrogeology* and *geohydrology* in particular have caused some debate. Frequently it is stated that the former deals with the geologic aspects of ground water whereas the latter places more emphasis on hydraulics and fluid flow. This arbitrary division is no longer taken seriously by most people in the field. In fact, the term *hydrogeology* was defined long before the modern era in hydrogeology, which differs markedly from its early beginnings, and both definitions likely reflect the special interest of their promulgators. In 1919, Mead published a book on hydrology where he defined hydrogeology as the study of the laws of occurrence and movement of subterranean water. There is nothing wrong with this definition. However, Mead stressed the importance of ground water as a geologic agent, especially as it contributes to an understanding of rivers and drainage systems. As a hydrologist interested largely in surface phenomena, this emphasis was well suited to Mead's interests. In 1942, Meinzer edited a book called *Hydrology,* which he defined within the context of the hydrologic cycle, that is, the march of events marking the progress of a particle of water from the ocean basins to the atmosphere and land masses and back to the ocean basins. He divided this science into surface hydrology and subterranean hydrology, or "geohydrology." Meinzer had an illustrious career devoted almost exclusively to the study of ground water as a water supply. In 1923, he published his famous volume on the occurrence of ground water in the United States, which essentially brought to a close the exploration period that started before the turn of the century. Indeed, because of this volume and some modern supplements, we are no longer exploring for ground water in North America and have not been for several decades. Shortly after this publication, Meinzer turned to inventorying the resource, for example, measuring or estimating the inflows and outflows and the changes in storage over time within ground-water bodies (Meinzer and Stearns, 1928; Meinzer, 1932). Such studies require detailed information on the interrelationships between subterranean water and other components of the hydrologic cycle to which it is connected. Thus, Meinzer's definition of geohydrology as the subsurface component of the hydro-

logic cycle suited his interests well. It still remains a good definition, but it does not go far enough. We will offer a definition after we develop an understanding of what the field is today and how it got that way, and who the major players were. Only then will we become aware of the scope of hydrogeology.

Physical Hydrogeology Before the Early 1940s

The turn of the century was an exciting time for hydrogeologists, especially those inclined to the rigors of fieldwork. Their main tools were rock hammers, compasses, and some crude water-level or fluid-pressure measuring devices. These hydrogeologists were likely aware of two important findings of the previous century. First was the experimental work of Henry Darcy in 1856 providing a law that described the motion of ground water, and second was some work by T. C. Chamberlin in 1885 that described water occurrence and flow under "artesian" conditions. Armed with these details, these workers were busy delineating the major water-yielding formations in North America and making important measurements of the distribution of hydraulic head within them. It was a good time to be a field hydrogeologist, with the emphasis being on exploration and understanding the occurrence of ground water and its interrelationship with other components of the hydrologic cycle.

In addition, there was some abstract thinking occurring at this time, but it had little or no impact over the following three or four decades. Slichter (1899) in particular did original theoretical work on the flow of ground water, but he was several decades ahead of his time. King (1899), too, attempted to provide some calculations to support his field findings.

The culmination of this era probably occurred in 1923, when Meinzer published his book on the occurrence of ground water in the United States. The major water-yielding formations were described, and water supply was the order of the day. As we shall learn later, there are generally four stages in the ordered utilization of ground water for water supply: exploration, development, inventory, and management. With the exploration stage completed and the resource already undergoing development, much of the effort of the U.S. Geological Survey turned toward inventory. However, at least one major theoretical finding resulted from this work, and this was provided by Meinzer in 1928 as a result of his study of the Dakota sandstone (Meinzer and Hard, 1925). In his inventory it seemed that more water was pumped from a region than could be accounted for; that is, the inventory (which is really a water balance) could not be closed. Meinzer concluded that the water-bearing formation possessed some elastic behavior and that this elastic behavior played an important role in the manner in which water is removed from storage. Although nothing was made of these ideas for another seven years, this was the start of something that would change the complexion of hydrogeology for at least two decades, if not forever.

That "something" occurred in 1935, when C. V. Theis, with the help of a mathematician named C. I. Lubin, recognized an analogy between the flow of heat and the flow of water (this analogy was also recognized earlier by Slichter as well as by others). The significance of this finding was that heat flow was already mathematically sophisticated whereas hydrogeology was not. However, by analogy, a solution to a heat flow problem, of which there were several, could be used to provide a solution to a fluid-flow analogy. Thus, Theis presented a solution that described the transient behavior of water levels in the vicinity of a pumping well. Imagine the significance of such a finding to a group of scientists who were totally committed to studying the response of ground-water basins to pumping for water supply. For this work, Theis was awarded the treasured Horton medal almost 50 years later (1984).

Two additional major contributions came about in 1940. The first was by Hubbert (1940), who published his detailed work on the theory of ground-water flow. In this work Hubbert was not interested in small-scale induced transients in the vicinity of pumping wells, but in the natural flow of ground water in large geologic basins. The hydrogeologic community at that time was far too preoccupied with transient flow to wells, so this contribution did not become part of the mainstream of hydrogeology until the early 1960s.

The second finding, and we think the most important of the first half of the twentieth century, was provided by Jacob (1940). Now let us fully appreciate the situation at the time. The Theis solution was provided by analogy from some heat-flow solution that was rigorously derived from some equation describing heat flow. Is there not some equation describing the flow of fluid from which it (as well as other solutions) could be derived directly? In 1940, Jacob derived this equation, and, significantly, it incorporated the elastic behavior of porous rocks described by Meinzer some 12 years earlier. Now, why all the fuss about something as abstract as a differential equation? A differential equation establishes a relation between the increments in certain quantities and the quantities themselves. This property allows us to say something about the relation between one state of nature and a neighboring state, both in time and in space. As such, it is an expression for the principle of causality, or the relation between cause and effect, which is the cornerstone of natural science. Thus, the differential equation is an expression of a law of nature.

Following Theis, Jacob, Hantush (one of Jacob's students), and several of Hantush's students, the transient flow of water to wells occupied center stage for about two decades. Training in well hydraulics today is as necessary to a hydrogeologist as training in the proper use of a Brunton compass is to a field geologist. Today, Meinzer, Theis, Jacob, and Hantush are no longer with us. However, of all the hydrogeologists who ever lived, over 95% are still alive and still working. In spite of this enormous pool, it is unlikely that such an era as occurred from 1935 to about 1960 will ever repeat itself.

Chemical Hydrogeology Before the Early 1960s

It would be pleasing indeed if we could trace a bit of historical development in chemical hydrogeology that parallels the common interests demonstrated on the physical side of things. However, no such organized effort materialized, largely because there was no comparable early guidance such as that provided by Darcy, Chamberlin, and the early fieldworkers. Back and Freeze (1983) recognize an "evolutionary" phase in chemical hydrogeology that started near the turn of the century and ended sometime during the late 1950s. By evolutionary, we mean a conglomeration of ideas by different workers interested in a wide variety of hydrochemical aspects. Several good ideas came out of this period, including certain graphical procedures that are still used to interpret water analyses (Piper, 1944; Stiff, 1951) and a particularly good treatment of sodium bicarbonate water by Foster (1950) that is still highly referenced today. In addition to these studies, two others stand out from the perspective of the chemical evolution of ground water with position in ground-water flow (Chebotarev, 1955; Back, 1960). For the most part, however, the early effort on the chemical side of things was directed at determining water quality and fitness for use for municipal and agricultural purposes.

The first extensive guidance to the working chemical hydrogeologist was provided by Hem's (1959) treatment on the study and interpretation of the chemical character of natural waters. This work demonstrated, among other things, that most of the important reactions were known. The majority of the contributions to this knowledge come from chemists, sanitary engineers, biologists, and limnologists. Hem's volume served as the bible for the working hydrogeologist for several years.

The early parts of the 1960 decade marked some unification of the divergent interests of the previous 50 years. Garrels' (1960) book in particular focused on the equilibrium approach in chemical thermodynamics. From this point on, a main focus of much of the work in chemical hydrogeology shifted to understanding regional geochemical processes in carefully conducted field investigations. Interestingly, at about that same time, hydrogeologists became interested in regional ground-water flow as described by Hubbert (1940) some 20 years earlier. Thus, the timing was perfect for a serious bonding of regional hydrogeology and the chemical evolution of formation waters. Today, some 30 years later, training in regional ground-water flow and the chemical evolution of ground-water basins is an essential part of hydrogeology.

Post-1960 Hydrogeology

We have already given the reader a glimpse of post-1960 hydrogeology by making reference to the marriage of physical and chemical hydrogeology by our contemporaries well aware of the pioneering work of Hubbert (1940) and Garrels (1960). Basically, we may isolate three contributing factors that are more or less responsible for additional developments beyond the early 1960s, one technological and two institutionally motivated. The technological factor was the development and eventual accessibility of high-speed computers. Problems that could be solved tediously by analytical mathematics (or those that could not be solved at all) are readily handled by numerical methods requiring computer devices. Thus, numerical methods in hydrogeology is a core course in any serious curriculum in hydrogeology.

Two institutionally motivating factors initiated the interest of hydrogeologists in transport processes, or processes where some physical entity such as heat or chemical mass is moved from one point to another in ground-water flow. Interest in heat flow and geothermal energy in particular was at its peak in the early 1970s in response to an effort to develop alternative energy sources due to the oil embargo. With a return to pre-1970 prices for oil and gas, the interest in geothermal resources diminished considerably. However, the groundwork was laid for a continued interest in heat transport by ground-water flow, especially with regard to the thermal evolution of sedimentary basins, low-temperature mineralization, the dissipation of frictional heat following earthquake generation, thermal pollution, and the fluid and thermo-mechanical response of rocks used for repositories to store high-level nuclear waste.

A second institutionally motivating factor occurred in the mid-1970s in the form of a federal environmental law. Environmental laws were quite common during the 1960s and 1970s: the Clean Water Act, the Clean Air Act, the Clean Drinking Water Act, the Surface Mining Act. Many of these dealt largely with surface water and were relatively simple to initiate and enforce. The Clean Water Act, for example, made funds available for treatment of sewage disposed of in surface waters, thereby contributing to the cleanup of polluted rivers. In 1976, the environmental loop was closed with the introduction of the Resource Conservation and Recovery Act (RCRA), which is fundamentally a ground-water protection act. The purpose of RCRA is to manage solid hazardous waste from the time it is produced to its ultimate disposal. The act had provisions for forced monitoring of waste disposal facilities, which eventually revealed the extent of ground-water contamination throughout the land. This led to the Comprehensive Environmental Response, Compensation, and Liability Act (CERCLA) in 1980, more commonly known as the "superfund" for cleanup. RCRA does not address radioactive or mining wastes. Nor until recently did it address leaky storage tanks. All wastes will eventually be regulated. Hydrogeologists will be involved with the fate and transport of contaminants in the subsurface for many years to come.

Contaminant transport is merely a special application of mass transport in ground-water flow. The transport processes are essentially physical phenomena, for example, the movement of dissolved mass by a moving ground water. This movement is frequently accompanied by the transfer of mass from one phase (liquid or solid) to another. Mass transfer from one phase to another proceeds at a decreasing rate until the two phases come to equilibrium with each other. Thermodynamic information regarding this equilibrium is basic to an understanding of mass transfer, but a more immediate concern is the rate at which the transfer takes place. This is a kinetic, as opposed to an equilibrium, concern, which in many cases is influenced by the rate of ground-water flow. Here again the timing is perfect for a serious bonding of physical and chemical hydrogeology. Training in mass transport and reactions in ground-water flow is an essential part of modern hydrogeology.

Given this brief description of what hydrogeology is and how it got that way, we are now in a better position to attempt a partial description of the field. We should still like to hedge a bit here by making a distinction between the principles of the science and their various applications. The principles of the science reside in the principles of fluid, mass, and energy transport in geologic formations. The applications of the science are not so easily enunciated, but so far we have emphasized problems in human affairs such as water supply, contamination, and energy resources. There is, however, a historical connotation to hydrogeology, and this is taken up in the next section.

1.2   The Relationship Between Hydrogeology and Other Fields of Geology

The cornerstone of geology resides in four areas: mineralogy, petrology, stratigraphy, and structure. The role of hydrogeology in geologic studies can best be appreciated if, when making observations, we ask ourselves, "How did it get this way?" This is an historical question that is fair in this case because geology is an historical science studied largely from the perspective of observation. Let us remember further that the rock mass we are examining consists of a stationary phase and a mobile phase. The terms "stationary" and "mobile" refer to the solids making up the rock body and the fluids contained within the rock, respectively. This is essentially a long-term "through-flow system," with the fluids continually replaced over geologic time frames. Thus, the ability of a moving ground water to dissolve rocks and minerals and to redistribute large quantities of dissolved mass has important implications in chemical diagenesis, economic mineralization, and geologic work in general. Additionally, it has long been understood that the fluids need not reside in a passive porous solid, but instead there frequently exists a complex coupling among stress, strain, and the pore fluids themselves. This coupling is often complicated by heat sources in crustal portions of the earth. The net effect of this coupling is fully demonstrated in several modern environments throughout the world where "abnormally" high fluid pressures reside in deforming rocks. Ultimately, these pressures can lead to rock fracture and further deformations. Is it possible that such environments were equally widespread throughout geologic time, and what consequences can be attributed to them?

Even metamorphic processes cannot be fully examined in the absence of a fluid phase. According to Yoder (1955), water is an essential constituent of many minerals common to metamorphic rocks, it is the primary catalyst in reaction and recrystallization of existing minerals, and it is one of the chief transport agents of material and possibly heat in metasomatic processes. As early as 1909, Munn recognized the role of ground water in hydrocarbon migration and entrapment. To discuss these aspects further would preempt some of the main considerations in this book. However, it should be clear from the perspective of processes that one cannot ignore the role of ground water in performing geologic work.

Bearing in mind all the factors discussed in the preceding paragraphs, we venture the following definition of hydrogeology. Hydrogeology is the study of the laws governing the movement of subterranean water, the mechanical, chemical, and thermal interaction of this water with the porous solid, and the transport of energy, chemical constituents, and particulate matter by the flow.

_1.3_ _The Hydrologic Cycle_

This book is concerned solely with ground water. Because ground water is one component of the hydrologic cycle, some preliminary information on the hydrologic cycle must be introduced to set the proper stage for things to come.

Components of the Hydrologic Cycle

Schematic presentations of the hydrologic cycle such as Figure 1.1 often lump the ocean, atmosphere, and land

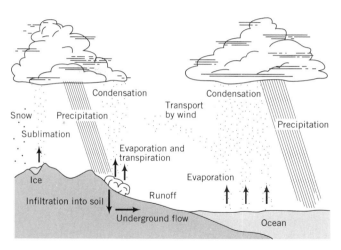

Figure 1.1 Schematic representation of the hydrologic cycle.

areas into single components. Yet another presentation of the hydrologic cycle is one that portrays the various moisture inputs and outputs on a basin scale. This is shown in Figure 1.2, where precipitation is taken as input and evaporation and transpiration (referred to as evapotranspiration) along with stream runoff are outputs. The stream runoff component, referred to as overland flow, can be augmented by interflow, a process that operates below the surface but above the zone where rocks are saturated with water, and by base flow, a direct component of discharge to streams from the saturated portion of the system. Infiltration of water into the subsurface is the ultimate source of interflow and recharge to the ground water.

The ground-water component is more clearly demonstrated in Figure 1.3. In this water profile, the vadose zone corresponds to the unsaturated zone, whereas the phreatic zone corresponds to the saturated zone. The so-called intermediate zone separates the saturated phreatic zone from soil water. It can be absent in areas of high precipitation and hundreds of feet thick in arid areas. The water table marks the bottom of capillary water and the beginning of the saturated zone.

The terms _saturated_ and _unsaturated_ require some clarification. Given a unit volume of soil or rock material, designated as V_T, the total volume consists of both solids (V_s) and voids (V_v). Only the voids are capable of containing a fluid, either air or water. The degree of saturation is defined as the ratio of the water volume to the void volume, V_w/V_v expressed as a percentage. For a fully saturated medium, the ratio is one (or 100%). A degree of saturation less than 100% indicates that air occupies some of the voids. Another term in common usage is moisture content, θ, which is defined as the volume of water divided by the total volume (V_w/V_T), expressed as a percentage.

At the contact between the lower part of a dry porous

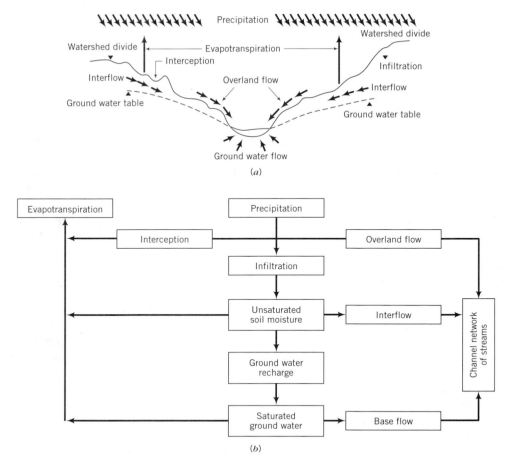

Figure 1.2 The basin hydrologic cycle.

material and a saturated material, the water rises to a certain height above the top of the saturated material. This gives rise to the capillary water zone of Figure 1.3. The driving force responsible for the rise is termed surface tension, a force acting parallel to the surface of

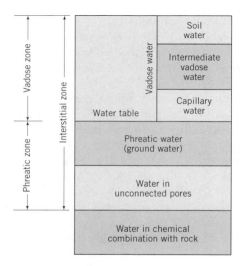

Figure 1.3 The water profile.

the water in all directions because of an unbalanced molecular attraction of the water at the boundary. The tensional nature of these forces can be compared with those set up in a stretched membrane.

Capillarity results from a combination of the surface tension of a liquid and the ability of certain liquids to wet the surfaces with which they come in contact. This wetting (or wetability) causes a curvature of the liquid surface, giving a contact angle between liquid and solid different from 90°. The idealized system commonly used to examine the phenomenon is a water-containing vessel and a capillary tube. When the tube is inserted in the water, the water rises to a height h_c (Figure 1.4). The meniscus, or curved surface at the top of the tube, is in contact with the walls of the tube at some angle α, the value of the contact angle depending on the wall material and the liquid. For a water–glass system, α is taken as zero.

Points A and C in Figure 1.4 are at atmospheric pressure (1.013×10^5 newtons per meter squared (N/m^2) or 14.7 lb per square inch (psi) or approximately 1 bar). A fundamental law of hydrostatics states that the pressure intensity of a fluid at rest can vary in the vertical direction only. Hence, point B is also at atmospheric pressure, and

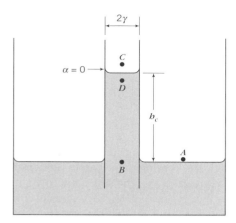

Figure 1.4 Capillary rise in a glass tube.

the pressure at point D must be less than atmospheric by an amount dictated by the pressure exerted by the length of the water column above B, or $h_c\gamma_w$, where γ_w is the unit weight of water. At a height of 10.3 m, the value of $h_c\gamma_w$ is just equal to atmospheric pressure. Hence, 10.3 m is a theoretic upper bound for the height of capillary rise, at least in glass tubes. For pure water at 20°C, the height of capillary rise in a glass tube is expressed as

$$h_c = \frac{0.153}{r} \qquad (1.1)$$

where h_c and r are expressed in centimeters and r is taken as the radius of the tube. In real systems, r is taken as the radius of passage.

Of further significance to this discussion is now an absolute definition of the water table. From Figure 1.4 we recognize that water above the water table is at a pressure below atmospheric while water below the water table is at a pressure greater than atmospheric. The water table is the underground water surface at which the pressure is exactly equal to atmospheric.

The intermediate zone lies above the capillary fringe and consists of water in the form of thin films adhering to the pore linings. This water is free to drain downward under the forces of gravity. The soil, intermediate, and capillary zones are collectively taken as the unsaturated or vadose zone (Figure 1.3).

Evapotranspiration and Potential Evapotranspiration

Evaporation can take place from both soil and free water surfaces. Evaporation from plants is called transpiration, and the combined process is often referred to as evapotranspiration.

The concept of potential evapotranspiration (or the potential for evapotranspiration) and methods for calcu-

lating it were introduced by Thornthwaite (1948). Potential evapotranspiration is defined as "the amount of water that would evaporate or transpire from a surface if water was available to that surface in unlimited supply." There is thus a clear distinction between what is actual and what is potential evapotranspiration. For example, the actual evapotranspiration off the surface of Hoover Reservoir very likely equals the potential rate. However, during most of the year, the actual evapotranspiration in regions adjacent to the reservoir is generally less than the potential rate, simply because of the lack of unlimited water. The actual evaporation off the Sahara desert is nil whereas the potential evapotranspiration (or potential for evapotranspiration) is quite high. In fact, if we wished to irrigate parts of the Sahara, we would have to supply irrigation water at about the potential rate.

From this definition and the few examples stated earlier, potential evapotranspiration is a maximum water loss (or upper limit of actual evapotranspiration), is a temperature-dependent quantity, and is a measure of the moisture demand for a region. The last point stated is better demonstrated by the ratio of precipitation to potential evapotranspiration. In a desert or tundra region, this ratio may be less than 0.1. For these regions, precipitation is not sufficient to meet the water demand so that very little grows naturally. For a successful irrigation scheme in such a region, water must be supplied at the potential rate. The ratio of precipitation to potential evapotranspiration in parts of Montana and North Dakota ranges from 0.2 to 0.6, again indicating a need for irrigation water in crop production. The range in eastern United States is from 0.8 to 1.6, indicating a rather well-balanced situation and, in some cases, a water surplus. The vegetation (or lack thereof) in a region is thus a reflection of the precipitation–potential evapotranspiration ratio. This is obvious when comparing the Sahara desert with the Amazon basin, both of which have excessive potential evapotranspiration, but only one of which is characterized by an equally excessive amount of precipitation.

Figure 1.5 is a qualitative demonstration of the natural water demand and water supply in a region as measured by the potential evapotranspiration and the precipitation, respectively. From these curves, we note three important time periods:

1. Precipitation equals potential evapotranspiration.
2. Precipitation is less than potential evapotranspiration.
3. Precipitation is greater than potential evapotranspiration.

In formulating some rules (laws) that govern the behavior of water in the hydrologic cycle, we first stipulate that the demands of potential evapotranspiration must be met

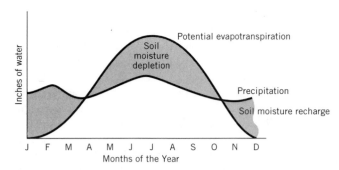

Figure 1.5 Relationship between precipitation and potential evapotranspiration.

(if at all possible) before water is permanently allocated to other parts of the cycle. Thus, when precipitation is equal to potential evapotranspiration, the surplus is theoretically zero, and all the rainfall for that period is available to satisfy the evaporation needs. During such time periods the actual evapotranspiration equals the potential evapotranspiration. This does not mean that infiltration cannot take place. It means simply that any infiltrated water will be available in the soil moisture for use by plants in the transpiration process. In addition, any time the rate of precipitation exceeds the rate of infiltration, there will be some water available for overland runoff so that this suggested balance of precipitation and potential evapotranspiration is not exactly true. At any rate, the surplus after all demands of potential evapotranspiration are met will be zero.

When precipitation is less than potential evapotranspiration, all the precipitation is available to partially satisfy potential evapotranspiration. During such time periods, the actual evapotranspiration would appear to be less than the potential amount. Such is not the case, however, as the part of the demand not met by precipitation may be met by drawing on whatever moisture is in the soil zone. So here again the actual rate can equal the potential rate. If this situation continues over a prolonged period, the soil becomes depleted of its moisture (Figure 1.5), and the actual rate of evapotranspiration will fall below the potential rate. Such periods are normally labeled droughts.

Finally, when temperature drops sufficiently so that precipitation exceeds potential evapotranspiration, a water surplus is realized. This surplus immediately goes into rebuilding the soil moisture component. This rate of infiltration will generally decrease with time as the soil zone becomes more saturated with water. It is during these periods that the underlying ground water becomes recharged and, in response to a decreased infiltration rate, overland flows become more pronounced. It is thus seen that the soil zone is a buffer in the workings of the basin hydrologic cycle, taking in moisture during surplus periods and releasing moisture during deficit periods.

Infiltration and Recharge

If water is applied to a soil surface at an increasing rate, eventually the rate of supply has to exceed the rate of entry, and excess water will accumulate. Not so obvious is the fact that if water is supplied to a soil surface at a constant rate, the water may at first enter the soil quite readily, but eventually will infiltrate at a rate that decreases with time. Horton (1933, 1940) was the first to point out that the maximum permissible infiltration rate decreases with increasing time. This limiting rate is called the infiltration capacity of the soil, an unfortunate choice of words in that capacity generally refers to a volume or amount whereas the infiltration capacity of Horton (1933) is actually a rate.

The concept of field capacity is another useful idea in soil science to denote an upper limit of moisture content in soils. At field capacity the soil is holding all the water it can under the pull of gravity. In a conceptual sense, both interflow and ground-water recharge can commence when the moisture content exceeds field capacity. The proportioning of the available moisture to interflow or ground-water recharge is dependent to a large extent on the nature of the soil zone. If the soil zone is thin and underlain by rather impervious rock, the interflow component may be the dominant one. For thicker pervious soils, downward migration may dominate, with most of the moisture being allocated to ground-water recharge.

Base Flow

The term _recession_ refers to the decline of natural output in the absence of input and is assumed or known from experience to follow an exponential decay law. The baseflow component of streams represents the withdrawal of ground water from storage and is termed a groundwater recession. As this recession is generally determined from stream hydrographs, the hydrograph must be separated into its component parts, which normally consist of overland flow, interflow, and base flow. Frequently, the interflow component is ignored so that the hydrologist generally deals with an assumed two-component system.

Whatever the assumptions, methods employed in hydrograph separation are wrought with difficulties. One of the most acceptable techniques is illustrated in Figure 1.6, which portrays an ideal stream hydrograph that plots the discharge of a river, usually in cubic feet per second (cfs) at a single point in the watershed as a function of time. According to Linsley and others (1958), the direct runoff component is terminated after some fixed time T^* after the peak of the hydrograph, where this time T^* in days is expressed

$$T^* = A^{0.2} \tag{1.2}$$

where A is the drainage area in square miles and 0.2 is some empirical constant. Normally, the recession ex-

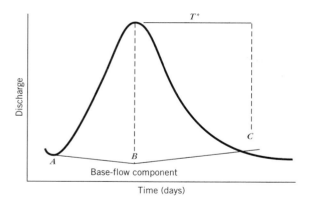

Figure 1.6 Determining base-flow component from stream hydrograph.

isting prior to the storm is extended to a point directly under the peak (*A-B*, Figure 1.6) and then extended again to the point T^* after the peak (*B-C*, Figure 1.6). The depressed base-flow component under the peak is justified because high water levels in the stream have the tendency to cut off or retard the ground-water discharge component.

Consider now the point on the hydrograph T^* days after the peak. From this point onward, the total flow is the base-flow component derived from ground-water discharge into the stream, at least until the next storm. This recession starting at time T^* can be described as

$$Q = Q_o e^{-kt} \qquad (1.3)$$

where Q_o is the discharge at time T^*, Q is the discharge at any time, k is a recession constant, and time t is the time since the recession began. In this equation, the time t varies from zero, which corresponds to the time (T^*), to infinity. Thus, at the beginning of the recession $t = 0$ and $Q = Q_o$. For later times, Q decreases and follows the exponential decay law of Eq. 1.3.

From Eq. 1.3, the recession constant can be expressed

$$k = -\left(\frac{1}{t}\right)\ln\left(\frac{Q}{Q_o}\right) \qquad (1.4)$$

where, as described, Q_o is the base flow at time $t = 0$ and Q is the base flow t time units later. This provides a means for determining the recession constant. Clearly this constant will be some number less than one and will be large (approach one) for flat recessions and small (approach zero) for steep recessions.

The influence of geology on the shape of the recession curve has been discussed and demonstrated by Farvolden (1964). Streams in limestone regions are characterized by flat recessions, a reflection of the fact that most of the drainage takes place in the subsurface. Granitic and other low-permeability regions typically have rather steep base-flow recessions.

A plot of discharge versus time on semilogarithmic paper will yield a straight line, the slope of which defines the recession constant (Figure 1.7). For this case the recession expression becomes

$$Q = \frac{Q_o}{10^{t/t_1}} \qquad (1.5)$$

where Q_o is the discharge at time $t = 0$, t_1 is the time 1 log cycle later, and t equals any time of interest for which the value of Q is desired. For example, for $t = t_1$

$$Q = \frac{Q_o}{10} \qquad (1.6)$$

The total volume of base-flow discharge corresponding to a given recession is found by integrating Eq. 1.5 over the times of interest

$$\text{Vol} = \int_{t_0}^{t} Q \, dt = -\frac{Q_o t_1/2.3}{10^{t/t_1}} \Big|_{t_0}^{t} \qquad (1.7)$$

where t_0 is the starting time of interest. Meyboom (1961) has employed Eq. 1.7 to determine ground-water recharge between recessions. For example, if t_o equals zero and t equals infinity

$$\text{Vol} = \frac{Q_o t_1}{2.3} \qquad (1.8)$$

which is termed the total potential ground-water discharge. This volume is defined as the total volume of

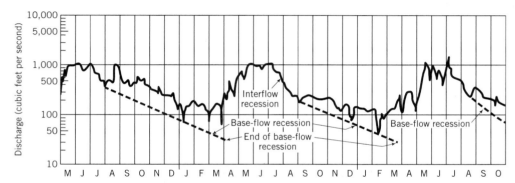

Figure 1.7 Stream hydrograph showing base-flow recession (from Meyboom, J. Geophys. Res., v. 66, p. 1203–1214, 1961. Copyright by Amer. Geophys. Union).

ground water that would be discharged by base flow during an entire recession if complete depletion takes place uninterruptedly. It follows that this volume describes the total volume of water in storage at the beginning of the recession. The difference between the remaining potential ground-water discharge at the end of a given recession and the total potential ground-water discharge at the beginning of the next recession is a measure of the recharge that takes place between recessions (Meyboom, 1961).

Example 1.1

Determine the approximate recharge volume between the first two recessions of Figure 1.7.

The first recession has an initial value of 500 ft³/s and takes about 7.5 months to complete a log cycle of discharge. The total recession takes about 8 months to complete. Total potential discharge is calculated from Eq. 1.8:

$$Q_{tp} = \frac{Q_o t_1}{2.3}$$

$$= \frac{500 \text{ ft}^3/\text{s} \times 7.5 \text{ months} \times 30 \text{ days/month} \times 1440 \text{ min/day} \times 60 \text{ s/min}}{2.3}$$

$$= 4222 \times 10^6 \text{ ft}^3$$

The ground-water volume discharged through the total recession lasting approximately 8 months is determined by evaluating Eq. 1.7 over the limits t equals zero to t equals 8 months

$$\frac{Q_o t_1}{2.3} - \frac{Q_o t_1/2.3}{10^{t/t_1}} = 4222 \times 10^6 \text{ ft}^3 - \frac{4222 \times 10^6 \text{ ft}^3}{10^{8/7.5}}$$

or about 3800×10^6 ft³. Base-flow storage still remaining at the end of the recession can be determined by merely subtracting actual ground-water discharge from total potential discharge, which gives 422×10^6 ft³.

The second recession has an initial value of about 200 ft³/s and takes about 7.5 months to complete a log cycle of discharge. Total potential discharge is calculated

$$Q_{tp} = \frac{Q_o t_1}{2.3}$$

$$= \frac{200 \text{ ft}^3/\text{s} \times 7.5 \text{ months} \times 30 \text{ days/month} \times 1440 \text{ min/day} \times 60 \text{ s/min}}{2.3}$$

or about 1400×10^6 ft³. The recharge that takes place between recessions is the difference between this value and remaining ground-water potential of the previous recession, or 978×10^6 ft³.

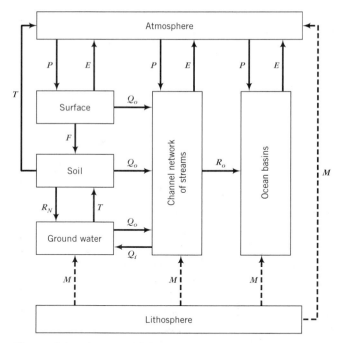

Figure 1.8 Elements of the global hydrologic cycle.

Hydrologic Equation

The hydrologic cycle as shown in Figure 1.2 and described throughout this chapter is a network of inflows and outflows that may be conveniently expressed as

$$\text{input} - \text{output} = \text{change in storage} \qquad (1.9)$$

The word equation given as Eq. 1.9 is a conservation statement and assures us that all the water is accounted for; that is, we can neither gain nor lose water. The interconnections between the components of the hydrologic cycle can be demonstrated on a global scale (Figure 1.8) or on a basin scale (Figure 1.9). On a global scale, the atmosphere gains moisture from the oceans and land areas E and releases it back in the form of precipitation P. Precipitation is disposed of by evaporation to the atmosphere E, overland flow to the channel network of streams Q_o, and infiltration through the soil F. Water in

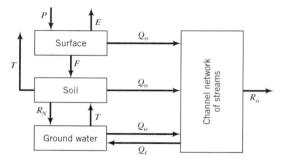

Figure 1.9 Elements of the basin hydrologic cycle.

the soil is subject to transpiration T, outflow to the channel network (the interflow component) Q_o, and recharge to the ground water R_N. The ground-water reservoir may receive water Q_i and release water Q_o to the channel network of streams and the atmosphere. Streams that receive water from the ground-water reservoir by the base-flow component Q_o are termed effluent or gaining streams. This occurs when the ground water table in the vicinity of the stream is above the base of the stream. Streams that lose water to the ground-water reservoir as designated by the Q_i component are termed influent, or losing, streams. This occurs when the ground water table is well below the base of the stream. Also noted on Figure 1.8 is the potential contribution of water of volcanic or magmatic origin.

Figure 1.9 illustrates that isolation of the basin scale hydrologic subsystem cannot exclude the lines of moisture transport connecting this subsystem to the global cycle. This connection is accomplished by accounting for precipitation derived from the global system and total runoff R_o to the global system. From this diagram we reformulate our word equation given earlier as

$$P - E - T - R_o = \Delta S \qquad (1.10)$$

where ΔS is the lumped change in all subsurface water. Each item in the equation has the units of a discharge, or volume per unit time.

The hydrologic equation (Eq. 1.10) must balance for every time period to which it is applied. This balancing will occur provided the accounting of the inflow and the outflow is done over a common period of time. An exact balance is an unreasonable goal due to poor instrumentation, lack of data, and the assumptions that are generally applied.

Equation 1.10 may be expanded or abbreviated, depending on what part of the cycle the hydrologist is interested in and, of course, depending on the available data base over the period of record. For example, for the ground-water component, another form of Eq. 1.10, is arrived at from Figure 1.9:

$$R_N + Q_i - T - Q_o = \Delta S \qquad (1.11)$$

Over long periods of time, providing the basin is in its natural state and no ground-water pumping is taking place, the natural inputs R_N and Q_i will be balanced by the natural outputs T and Q_o so that the change in storage will be zero. This gives

$$R_N + Q_i = T + Q_o \qquad (1.12)$$

or, stated simply, input equals output. This means that the ground-water component in the basin is hydrologically in a steady state, and the variables (or, more properly, the averaged values of the variables) have not changed over the time period over which the averaging took place. On the other hand, if a pumpage withdrawal is included,

Eq. 1.12 becomes

$$R_N + Q_i - T - Q_o - Q_p = \Delta S \qquad (1.13)$$

where Q_p is the added pumping withdrawal. As pumpage is new output from the system, the water levels in the basin will decline in response to withdrawals from ground-water storage. The stream will eventually be converted to a totally effluent one. In addition, with lower water levels, the transpiration component will start to decline and eventually approach zero. If potential recharge to the system was formerly rejected due to a water table at or near land surface, the drop in water levels will permit R_N to increase over its steady value. Thus, at some time after pumping starts, Eq. 1.13 becomes

$$R_N + Q_i - Q_p = \Delta S \qquad (1.14)$$

A new steady state can, at least theoretically, be achieved if the pumping withdrawal does not exceed the inputs R_N and Q_i. If pumping continually exceeds these input values, water is continually removed from storage and water levels will continue to fall over time. Here, the steady state has been replaced by a transient, or unsteady, state, where some parameter (in this case, ground-water storage) continually changes over time. Not only is ground-water storage being depleted, but some of the surface flow has been lost from the stream.

The term *inventory* is generally reserved for investigations in which a detailed accounting of inflow, outflow, and changes in storage is attempted for time intervals, such as years or other units of time, during a period of observation (Meinzer, 1932). In the Pomperaug Basin, Connecticut, for example, available data were organized on a monthly basis for a three-year period of study (Meinzer and Stearns, 1928). The hydrologic equation is the basis for such studies.

Last, our hydrologic equation may be expressed in the form of an ordinary differential equation

$$I(t) - O(t) = \frac{dS}{dt} \qquad (1.15)$$

where I is input and O is output, both taken as a function of time. Equation 1.15 is also referred to as a lumped-parameter equation, a term that is reserved for any equation or analysis wherein spatial variations in the parameters are not considered.

Example 1.2

The following is a simplified inventory based on the hydrologic equation

{recharge from direct precipitation plus
recharge from stream flow}

{minus discharge by pumping minus
discharge by evapotranspiration}
{equals change in ground-water storage}

With regard to this inventory, the following are worth noting:

1. Previous to pumping outputs, natural recharge was equal to natural discharge and the ground-water basin was in a steady state.

2. With the addition of pumpage and in the course of withdrawals from storage, net recharge from stream flow tends to increase and reaches some maximum value, whereas discharge by evapotranspiration seems to decrease and approach some minimum value.

3. During the course of these withdrawals, the basin is in a transient state where water is continually being withdrawn from storage. Although not shown, this results in a continual decline in water levels. A new steady state can be achieved by reducing pumping to about 38×10^6 m³/yr.

Table E1.2

Time Year	Recharge from Direct Precipitation m³	Net Recharge from Stream Flow m³	Discharge by Pumping m³	Discharge by Evapotranspiration m³	Change in Ground-Water Storage m³
1	3×10^7	0	0	3×10^7	0
2	3×10^7	6×10^5	1×10^7	3×10^7	94×10^5
⋮	⋮	⋮	⋮	⋮	⋮
7	3×10^7	1×10^6	3×10^7	9×10^6	8×10^6
8	2.8×10^7	2×10^6	3.5×10^7	5×10^6	10×10^6
9	2.5×10^7	3×10^6	3.5×10^7	3×10^6	10×10^6
10	3.5×10^7	4×10^6	4×10^7	1×10^6	2×10^6
11	3.5×10^7	4×10^6	4.2×10^7	1×10^6	3×10^6
12	3.5×10^7	4×10^6	4×10^7	1×10^6	2×10^6

Problems

1. Calculate the base-flow recession constant for the first recession on Figure 1.7.

2. The diagram below illustrates a hydrograph determined from flow measurements on a spring located in carbonate rocks. The variations in spring flow are due to seasonal changes in rainfall (recharge). Calculate the recession constant and the recharge that takes place between recessions.

3. If the spring in the figure below represented a water supply and a minimum of 10 ft³/s was required by the users, what length of recession (in months) would be required to reach this minimum flow?

4. Ten-year averages (1981–91) for a ground-water inventory for the San Fernando Basin are as follows: Recharge from precipitation and other sources, 79.3×10^6 m³, recharge from injection wells 38×10^6 m³, ground-water inflow, 1.01×10^6 m³, ground-water pumping, 120.55×10^6 m³, ground-water outflow 0.521×10^6 m³, and base flow to streams, 3.8×10^6 m³. Perform a water balance for the ground-water component based on these 10-year averages.

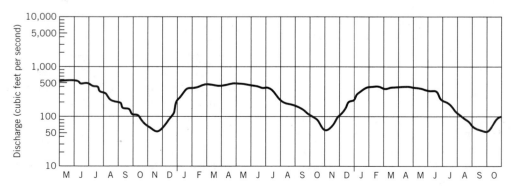

Problem 2.

Chapter 2

The Origin of Porosity and Permeability

A porous body is a solid that contains holes. All rocks are considered porous to some extent, with the pores containing one or more fluids—air, water, or some minority fluid such as a hydrocarbon. The holes may be connected or disconnected, normally or randomly distributed, interstitial or planar cracklike features. The degree of connectivity of the pores dictates the permeability of the rock, that is, the ease with which fluid can move through the rock body. There is nothing of more vital interest to those concerned with the subsurface movement of fluids than permeability, its creation and destruction, and its distribution throughout a rock body. Hence, this introductory chapter is intended to be as complete as possible—given our limited state of knowledge—in describing those processes that result in the formation of porous bodies along with other processes that ultimately modify the rock body, the connectivity of its pore space, and its contained fluids. To help us along in these tasks we call upon simple concepts known to most students in the earth sciences. Thus, we approach this topic from the simple rock cycle to describe the origin and formation of porous bodies, and from tectonic and chemical processes to ascertain whatever mechanical or chemical alterations rocks have been subjected to at some point in their history.

2.1 Porosity and Permeability

Porosity and Effective Porosity

Total porosity is defined as the part of rock that is void space, expressed as a percentage

$$n = \frac{V_v}{V_T} \qquad (2.1)$$

where V_v is the void volume and V_T is the total volume. A related parameter is termed the void ratio, designated as e, and stated as

$$e = \frac{V_v}{V_s} \qquad (2.2)$$

expressed as a fraction, where V_s is the solid volume. As total volume is the sum of the void and solid volume, the following relationships can be derived:

$$e = \frac{n}{1-n} \qquad or \qquad n = \frac{e}{1+e} \qquad (2.3)$$

Figure 2.1 shows some typical kinds of porosity associated with various rocks. The term *primary porosity* is reserved for interstitial porosity (Figures 2.1a through d), and the term *secondary* is used for fracture or solution porosity (Figures 2.1e and f). Interstitial porosity has

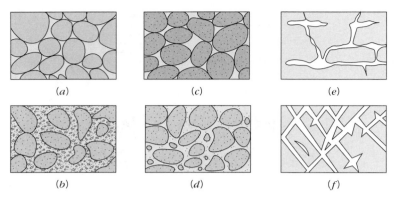

Figure 2.1 Relation between texture and porosity. (*a*) Well-sorted sedimentary deposit having high porosity; (*b*) poorly sorted sedimentary deposit having low porosity; (*c*) well-sorted sedimentary deposit consisting of pebbles that are themselves porous, so that the deposit as a whole has a very high porosity; (*d*) well-sorted sedimentary deposit whose porosity has been diminished by the deposition of mineral matter in the interstices; (*e*) rock rendered porous by solution; (*f*) rock rendered porous by fracturing (from Meinzer, 1923).

been investigated by Graton and Fraser (1935), who demonstrated that its value can range from about 26 to 47% through different arrangements and packing of ideal spheres. In actuality, the porosity of a sedimentary rock will depend not only on particle shape and arrangement, but on a host of diagenetic features that have affected the rock since deposition.

Porosity can range from zero or near zero to more than 60% (Table 2.1). The latter value is reflective of

Table 2.1 Range in Values of Porosity

Material	Porosity (%)
SEDIMENTARY	
Gravel, coarse	24–36
Gravel, fine	25–38
Sand, coarse	31–46
Sand, fine	26–53
Silt	34–61
Clay	34–60
SEDIMENTARY ROCKS	
Sandstone	5–30
Siltstone	21–41
Limestone, dolomite	0–40
Karst limestone	0–40
Shale	0–10
CRYSTALLINE ROCKS	
Fractured crystalline rocks	0–10
Dense crystalline rocks	0–5
Basalt	3–35
Weathered granite	34–57
Weathered gabbro	42–45

In part from Davis (1969) and Johnson and Morris (1962).

recently deposited sediments whereas the former value is for dense crystalline rocks or highly compacted soft rocks such as shales. In general, for sedimentary materials, the smaller the particle size, the higher the porosity. This is best demonstrated by comparing the porosity of coarse gravels with fines, and the total gravel assemblage with silts and clays.

An important distinction is the difference between total porosity, which does not require pore connections, and effective porosity, which is defined as the percentage of interconnected pore space. Many rocks, crystallines in particular, have a high total porosity, most of which may be unconnected. Effective porosity implies some connectivity through the solid medium and is more closely related to permeability than is total porosity. Some data on effective porosity are shown in Table 2.2. As noted, effective porosity can be over one order of magnitude smaller than total porosity, with the greatest difference occurring for fractured rocks.

Heath (1982) recognizes five types of porosity in dominant water-bearing bodies at or near the Earth's surface in the United States and attempts to map their distribution (Figure 2.2). There are some difficulties with this map because of the necessity of mapping a single type of opening in areas where two or more types are present. However, this is a useful presentation and one to which we will refer frequently in this chapter. Each pattern on Figure 2.2 is associated with one or more major water-yielding formations in the United States. Thus, the solution-enlarged openings in carbonate rocks that make up the Florida peninsula are known as the Floridian system; the sands and gravels stretching from New Jersey into Texas are sediments of the Atlantic and Gulf coastal plains; the sand and gravel in the Midwest represents glacial deposits; the sandstones in the northern mid-conti-

Table 2.2   **Range in Values of Total Porosity and Effective Porosity**

	Total Porosity (%)	Effective Porosity (%)
Anhydrite[1]	0.5–5	0.05–0.5
Chalk[1]	5–40	0.05–2
Limestone, dolomite[1]	0–40	0.1–5
Sandstone[1]	5–15	0.5–10
Shale[1]	1–10	0.5–5
Salt[1]	0.5	0.1
Granite[2]	0.1	0.0005
Fractured crystalline rock[2]	—	0.00005–0.01

[1] Data from Croff and others (1985).
[2] Data from Norton and Knapp (1977).

nent represent several formations, including the Dakota Sandstone and the Cambrian-Ordovician system; the sands and gravels in the western part of the nation occupy alluvial basins, whereas the occurrence of these same sediments in the central United States represents the remnant of a giant alluvial apron that formed on the eastern slope of the Rocky Mountains; and the basalt in the western United States occupies the Columbia River Plateau.

Permeability

Permeability may be described in qualitative terms as the ease with which fluid can move through a porous rock

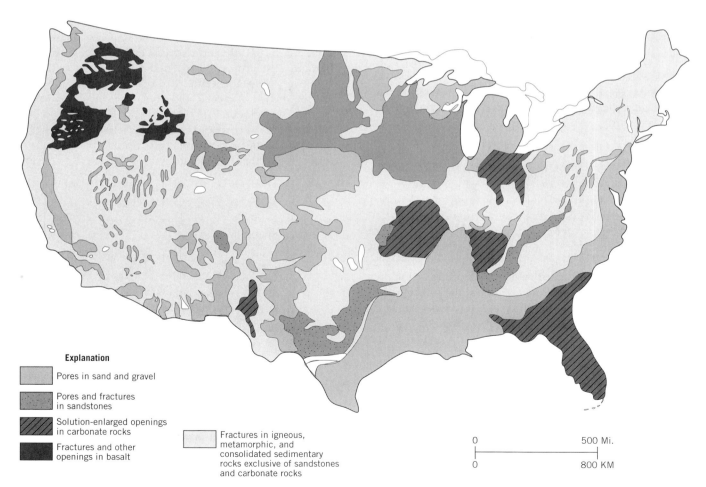

Explanation

- Pores in sand and gravel
- Pores and fractures in sandstones
- Solution-enlarged openings in carbonate rocks
- Fractures and other openings in basalt
- Fractures in igneous, metamorphic, and consolidated sedimentary rocks exclusive of sandstones and carbonate rocks

0   500 Mi.

0   800 KM

Figure 2.2   Map of conterminous United States showing types of water-bearing openings in the dominant aquifers (from Heath, 1982). Reprinted by permission of Ground Water. © 1982. All rights reserved.

and is measured by the rate of flow in suitable units. This qualitative definition will suffice for this section, although later the topic is taken up in quantitative terms.

Contrasts in permeability from one rock to another and, less frequently, within a given rock unit have given rise to a variety of terms and definitions in hydrogeology, most of which pertain to water supply. A rock unit that is sufficiently permeable so as to supply water to wells is termed an aquifer. Major aquifers are referred to by their stratigraphic names, such as the Patuxent Formation in Maryland, the Ogallala Formation throughout most of the High Plains, and the Dakota Sandstone. These aquifers may be considered "commercial" in the sense that they supply sufficient quantities of water for large-scale irrigation or municipal usage.

Aquitards have been defined as beds of lower permeability in the stratigraphic sequence that contain water but do not readily yield water to pumping wells. Major aquitards are generally considered to be low-permeability formations that overlie major aquifers. Examples include the Pierre Shale overlying the Dakota Sandstone, the Maquoketa Shale overlying high-permeability rocks in Illinois, and the Hawthorn Formation overlying the high-permeability Floridian aquifer. Rocks considered to be aquitards in one region may serve as aquifers in others. For example, water supplies are obtained from many low-permeability materials throughout North America, such as glacial tills in the Midwest and in Canada, and fractured crystallines and shales in many parts of the continent. These materials are not able to supply sufficient quantities of water for municipal and irrigation use, but frequently are adequate for domestic or farm usage in rural areas. Thus, the terms *aquifer* and *aquitard* are ambiguous.

More recently, the term *hydrostratigraphic unit* has been employed rather extensively. A hydrostratigraphic unit is a formation, part of a formation, or a group of formations in which there are similar hydrologic characteristics that allow for a grouping into aquifers and associated confining layers. Thus, the Dakota Sandstone and various other aquifers in combination with the Pierre Shale confining unit constitute a hydrostratigraphic unit, as do the Floridian carbonates in combination with the Hawthorn Formation, and several Cambrian-Ordovician formations overlain by the Maquoketa Shale.

The field occurrence of permeability is well documented by the fact that about one-half of the United States is underlain by aquifers capable of yielding moderate to large quantities of water to pumping wells. Unconsolidated deposits such as sand and gravel constitute the largest and most productive aquifers in terms of the volumes of water produced annually. Of the consolidated rocks, sandstones are the most important from the perspective of annual withdrawals, followed by carbonate and volcanic rocks. Sandstones and carbonates are virtually ubiquitous in sedimentary terranes. Volcanics, as the

fourth major aquifers in the United States, depend on fractures for the transmission of water.

2.2 Continental Environments

Weathering

Rocks at the Earth's surface are subjected to either physical disintegration or chemical decomposition. The weathering products can accumulate in place to form a soil or can be transported by wind or water to accumulate elsewhere as sedimentary materials. Weathering is also responsible for much of the secondary porosity illustrated in Figure 2.1. This is especially so with the chemical decomposition of carbonate rocks, where sinkholes and crevices can readily form due to dissolution along preexisting fractures or other pathways of fluid movement (Figure 2.3). A similar type of process occurs in less soluble igneous and metamorphic rocks, where chemical weathering tends to open preexisting fractures, contributing to more pervious pathways. An excellent example of this occurs in the Columbia Plateau, which is characterized by numerous individual basaltic flows, along with occasional sand or gravel units that mark periods of flooding (Figure 2.4). After a particular flow is extruded, it undergoes a cooling period accompanied by the formation of cooling joints, fractures, and vesicles. These rocks then become subjected to weathering, a process that is immediately terminated when another flow is extruded. However, the weathering features within the flow tops are retained in the rock and constitute major pathways for fluid movement. The pathways are largely horizontal, reflecting a widespread surficial weathering process, with the depth of weathering restricted to a few tens of feet, depending on the length of time before the volcanic

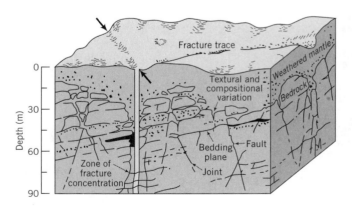

Figure 2.3 Occurrence of permeability zones in fractured carbonate rock. Highest well yields occur in fracture intersection zones (from Lattman and Parizek. 1964). Reprinted from J. Hydrol., v. 2, L. A. Lattman and R. R. Parizek, "Relationship between fracture traces and the occurrence of groundwater in carbonate rocks," p. 73–91, 1964, with kind permission from Elsevier Science, NL, Sara Burgerhartstraat 25, 1055 KV Amsterdam, The Netherlands.

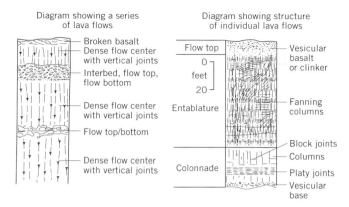

Figure 2.4 Flow tops and flow interiors, basalt, Hanford Reservation (after Vaccaro, 1986).

rock becomes inundated in another sea of lava. Below the weathering zone, the rocks remain rather dense and are characterized by poorly connected cooling joints. Hence, this weathering pattern produces a predictable layercake type of hydrology, with some of the permeable flow tops extending over great distances in the plateau.

Recognizing that rocks are exposed at the Earth's surface for long periods of time, chemical weathering can operate for millions of years, contributing to pervious pathways in even the most insoluble of rocks. Figure 2.5 is a conceptual presentation of the occurrence of fractures in crystalline rocks in the Piedmont region of the Appalachian Mountains. The higher permeability resides where the fractures are connected. In these rocks, the enhanced fracture or joint system frequently becomes tighter with depth. The data of Figure 2.6 were collected from this region and show that the average yield of wells per foot of depth below the position where the rocks are saturated with water decreases with increasing depth.

Thus, either the connectivity is decreasing with depth, or the fractures are getting tighter due to limitations on the depth of weathering, or both. It is noted that even the most productive wells in this region have very low yields, commonly less than 25 gallons per minute (1.58 liters/s). However, in the Piedmont, much of the rural domestic supply is furnished by these fractured low-permeability rocks. Once more, the weathering pattern produces a predictable type of hydrology.

Weathering can take place both above and below the water table but is generally slower in the saturated environment. In addition, a primary mineral subjected to weathering may simply dissolve or a portion of it may reprecipitate to form secondary minerals. The weatherability of igneous silicate minerals is directly proportional to their temperature of formation from molten materials, that is, the high-temperature-phase olivine weathers faster than do the various plagioclase families. One of the more common minerals, quartz, weathers the most slowly. On the other hand, halite, gypsum, pyrite, calcite, dolomite, and volcanic glass weather faster than do the common silicate minerals.

Erosion, Transportation, and Deposition

The main surficial process in continental environments is erosion of the landscape. There are several agents of erosion, but rivers are the most dominant. They carve their own valleys, transporting the eroded material downstream to lakes, rivers, or oceans. Rivers also receive the products of weathering, frequently by an overland flow process.

Fluvial Deposits
Fluvial deposits are generated by the action of streams and rivers. Boggs (1987) recognizes three broad environ-

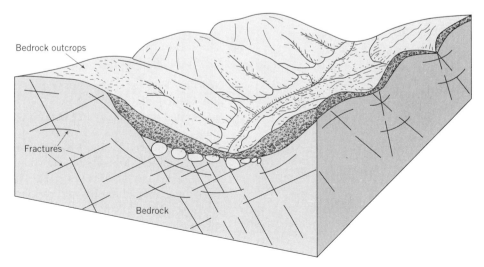

Figure 2.5 Topographic and geologic features of the Piedmont and Blue Ridge region showing fracture interconnections (from Heath, 1984).

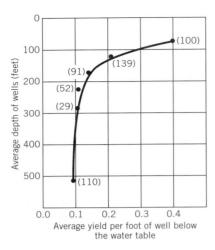

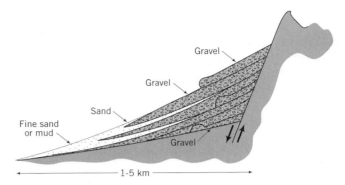

Figure 2.8 Diagrammatic cross section of an alluvial fan (from Rust and Koster, 1984). Reprinted with permission of the Geol. Assoc. of Canada.

Figure 2.6 Decrease in well yields (gpm/ft of well below the water table) with depth in crystalline rocks of the Statesville area, North Carolina. Numbers near points indicate the number of wells used to obtain the average values that define the curve (from LeGrand, 1954).

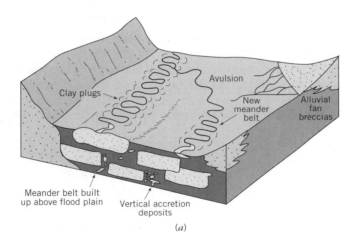

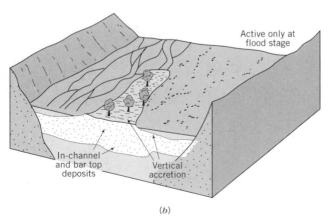

Figure 2.7 Contrasting the geometry of (*a*) meandering and (*b*) braided rivers (from Walker and Cant, 1984). Reprinted with permission of the Geol. Assoc. of Canada.

mental settings for fluvial systems: braided rivers, meandering rivers, and alluvial fans. Figure 2.7 contrasts the geometry of meandering and braided streams. According to Cant (1982), meandering streams generally produce linear shoestring sand bodies that are aligned parallel to the river course, and these are normally bounded below and on both sides by finer materials. The shoestring sands are many times wider than they are thick. The braided river, on the other hand, frequently produces sheetlike sands that contain beds of clay enclosed within them. The reason postulated by Cant (1982) lies in the meander width, with the braided streams capable of extensive lateral migration, whereas the meandering streams are considered to be more rigorously confined in narrow channels. Braided rivers are characterized by many channels, separated by islands or bars. Meandering streams have a greater sinuosity and finer sediment load.

Alluvial fans form at the base of mountains where erosion provides a supply of sediment (Figure 2.8). Fans can occur in arid areas, such as those in Death Valley, California, and in humid areas as well. The upper part of the alluvial fan is characterized by coarse sediment due to confinement of flow to one or a few channels. The toe of the fan has the finest sediment, where more than one channel persisted over long periods of time.

Alluvial Valleys The most important alluvial valleys in the United States are shown in Figure 2.9. The sand and gravel deposits are in hydraulic communication with their stream systems, which provides for their continual replenishment. The more permeable material occurs in clearly defined deposits that normally do not extend beyond the flood plain. Figure 2.9 does not differentiate between those channels cut by glacial melt water, such as those in the Midwest, and those stream channels not affected by glacial melt water, such as those in the southern Appalachians. Nor is there any distinction between meandering and braided patterns. Alluvial valleys are frequently underlain by silt and clay deposits and are among the most productive aquifers in the United States. Heath

Figure 2.9 Alluvial valleys (from Heath, 1984).

(1982), in his modern classification of ground-water regions in the United States, established the alluvial valley system as a separate entity, independent of their geographic occurrence.

As rivers become rejuvenated from time to time, they cut through their own deposits, with the margins of the original valley floor left as terraces (Figure 2.10). These terraces also contain permeable materials and are considered part of the alluvial valley system. In Figure 2.10, the current flood plain alluvium is the youngest deposit and the terrace occupying the highest ground is the oldest and was part of a flood plain prior to rejuvenation.

Alluvial Basins Alluvial basins in the United States occupy a discontinuous region in excess of 1×10^6 km² extending from the state of Washington to the western tip of Texas. They are demonstrated on Figure 2.2 as the patterns of pores in sand and gravel. The material filling most of the western basins was derived by erosion of the adjacent mountains, with alluvial fan development

extending to the basin bottoms. The basins themselves are characterized by block faulting, volcanism and intrusion, high average elevation, and high average heat flow.

Also noted on Figure 2.2 is an extensive sand and gravel region in the central United States extending from the southern part of South Dakota into Texas. This is part of an extensive alluvial apron deposited on the sloping bedrock surface extending from the foot of the Rocky Mountains eastward. Although much of the apron has been removed by erosion or otherwise dissected, the remaining part is known today as the Ogallala Formation. Pumping from the Ogallala for irrigation purposes is the one factor accounting for the agricultural prosperity of the High Plains. Unfortunately, this prosperity is threatened because of depletion of the ground-water resource.

Other than the Ogallala, which was deposited during Late Tertiary time, alluvial fans are noted in stratigraphic sequences of earlier ages. The Van Horn Sandstone of West Texas is an example of a Precambrian fan system, consisting of massive conglomerates that grade into thin-

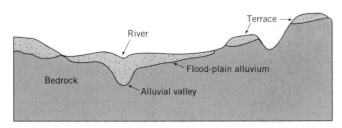

Figure 2.10 Cross section showing associations between alluvial valley and terrace development.

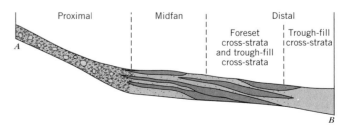

Figure 2.11 Facies and sedimentary structures in Van Horn Sandstone (from McGowen and Groat, 1971).

ner pebbly sandstones, and finally into mudstones (Figure 2.11). Triassic sediments on the margins of the Newark Basin are also made up of the remnants of alluvial fans and were noted by Meinzer (1923) for their water-yielding characteristics.

Eolian Deposits

Eolian is a term used in reference to wind erosion or deposition. There are two types of wind deposits: loess, which is an unstratified deposit composed of uniform grains in the silt-size range, and dunes or drifts composed of sand. Loess deposits are frequently associated with glacial terrane, whereas dunes and drifts are associated with deserts. The porosity of loess is very high, 40 to 50%, but it is not a good transmitter of water because of the poor connectivity of the pore space. The best examples of ancient eolian deposits are in the Colorado Plateau and include the Jurassic Navajo Sandstone and the Entrada, Wingate, White Rim, Coconino, and Lyons Sandstones. Both the Navajo and the Wingate form what is known as the La Plata Group, noted by Meinzer (1923) for its water supply. Other eolian deposits made up of dunes and drifts are expected to have a reasonably high permeability.

Lacustrine Deposits

The term *lacustrine* pertains to lakes, which currently form about 1% of the Earth's surface. Ancient lake deposits may include rock types ranging from conglomerates to mudstones, as well as carbonates and evaporites. Lacustrine deposits formed during the glacial period are generally characterized by low-permeability silts and clays, with an occasional high-permeability delta associated with the terminus of rivers. Ancient lacustrine deposits include the Morrison Shale in the Colorado Plateau, the Chugwater Series of Wyoming, and the Green River Formation in Utah, Wyoming, and Colorado. These are all low-permeability rocks.

Glacial Deposits

The glacial environment is a composite one and incorporates fluvial, lacustrine, and eolian environments (Figure 2.12). The high permeability resides in the glaciofluvial depositional environments, including the alluvial valleys, the alluvial deposits in buried bedrock valleys, and the

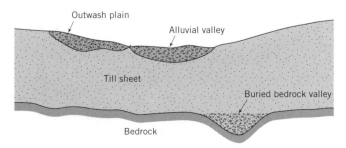

Figure 2.12   Glaciofluvial deposits in glaciated terrain.

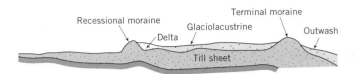

Figure 2.13   Deposits associated with a hypothetical advance and retreat of a glacier.

well-sorted sand and gravel resulting from glacial melt water (outwash plains). Buried bedrock valleys are valleys that are no longer occupied by the streams that cut them. The till sheets are the most extensive deposit and consist of an assortment of grain sizes ranging from boulders to clay and are normally of low permeability. Frequently, the till is fractured and contains a higher permeability than was present originally.

The advance and retreat of continental glaciers generally results in a complex vertical stratigraphy. This is demonstrated on Figure 2.13 for a hypothetical case. The basal till sheet marks the advance of a glacier across a region. The terminal moraine marks the farthest advance, with subsequent melting forming the glaciofluvial deposits downstream from the moraine. We also note the presence of lacustrine deposits associated with a glacial lake that was trapped between the retreating glacier and the end moraine complex. The recessional moraine marks the position where the retreating glacier was temporarily halted, with its melt water forming the lake and associated delta deposits. This is only one simple example of a complex stratigraphy.

2.3   The Boundary Between Continental and Marine Environments

The environments that occupy the boundary between the continents and the ocean have been referred to as marginal marine (Boggs, 1987). Within this boundary, the following environments may be identified: deltas or deltaic systems, beach and barrier island systems, estuarine and lagoonal systems, and tidal flats. A delta is a fluvial deposit that is built into a standing body of water. Deltas consist of an upper deltaic plain, with characteristic braided or meandering patterns above the influence of tides; the lower deltaic plain, which extends from the low-tide shoreline landward to the uppermost tidal influence; and the subaqueous delta plain, which is submerged (Coleman and Prior, 1982). The finer-grained material is normally transported seaward in the subaqueous region. Ancient deltaic systems have been identified in the Illinois Basin and in Gulf Coast sediments. In the Gulf Coast region, both the Frio and the Wilcox are well-known deltaic aquifers. The features and occurrence of ancient deltas are discussed by Eliott (1978).

The beach and barrier island environment consists of offshore bars or barrier islands with shallow lagoons between the islands and the mainland (Figure 2.14). Such

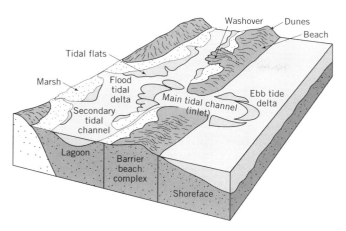

Figure 2.14 Subenvironments of a barrier island system (from Reinson, 1984). Reprinted with permission of the Geol. Assoc. of Canada.

coasts normally form where the offshore terrain is smooth and the seas shallow. The beach and barrier island system results in a narrow body of sediments parallel to the shoreline. Numerous sandstones have been recognized as beach and barrier island sytems, including the Gallup and La Ventana Sandstone in New Mexico, the Muddy Sandstone in Wyoming, and the Mission Canyon Sandstone in the Williston Basin. These are permeable formations, with the Mission Canyon being part of the Madison Group, an important aquifer system in the Williston Basin.

Estuarine and lagoonal environments consist of the estuary, which is the lower part of a river open to the sea, and lagoons, which are shallow seawater bodies such as sounds or bays. Tidal flats are the regions affected by the rise and fall of tides. One of the better documented tidal flat environments is the Upper Dakota Group along the Front range in Colorado. The dominant stratigraphy consists of fine-grained tidal flat deposits overlying thicker coarse-grained channel deposits. The depositional environment for the Dakota Group in general ranged from fluvial to marine in a marginal marine environment, with the greatest amount of sediment deposited as deltas in southward-spreading seas.

2.4 Marine Environments

The marine environment extends from the continental shelf to the oceanic ridges. Shelf sediments normally consist of sands or, where tropical to subtropical climates persist, carbonates. Ancient shelf deposits are well known in the geologic record. Examples of modern carbonate platforms include the Yucatan shelf and the eastern Gulf of Mexico off Florida. The lower part of the Edwards Formation in Texas has been cited as a possible carbonate shelf deposit (Wilson and Jordan, 1983). Deep-water sediments can be of various kinds, but generally are muds of one sort or another, along with carbonates.

Lateral and Vertical Succession of Strata

The term *facies* is defined as any areally restricted part of a designated stratigraphic unit that exhibits a character significantly different from those of other parts of the unit. Thus, a lithofacies is a distinct unit, that is, sand, that may grade into finer material (silt and clay) as one moves from the shoreline toward the deeper parts of the ocean. The concept of a facies is one that relates to the nature of the depositional environment. As shorelines shift over geologic time, the deposits of one environment will eventually underlie the deposits of yet a younger and different environment. With a transgressing sea, near-shore deposits become progressively overlain by deeper-water environments (Figure 2.15*a*). With a regressing sea, the order is reversed. Figure 2.15*b* shows a complete transgressive–regressive sequence. Such sequences are common in many parts of central and southern United States (Driscoll, 1986). Frequently, periods of erosion following the regression will remove one or more of the layers. Shifts in the shoreline bring different major environments into contact. A seaward shift of the shoreline (regression) can result in the establishment of lagoons and beach-barrier systems on older finer-grained, deeper-water sediments; a landward shift of the shoreline (transgression) can result in the formation of beach deposits on finer-grained sediments associated with lagoons and marshes.

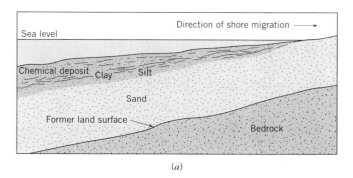

(*a*)

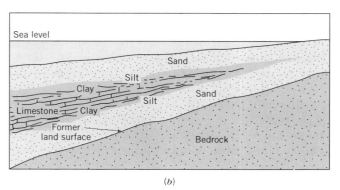

(*b*)

Figure 2.15 (*a*) Cross section of sediments associated with a transgressing sea and (*b*) transgressive–regressive sedimentary sequence (from Driscoll, 1986). Reprinted with permission from *Groundwater and Wells,* 2nd Ed., 1986.

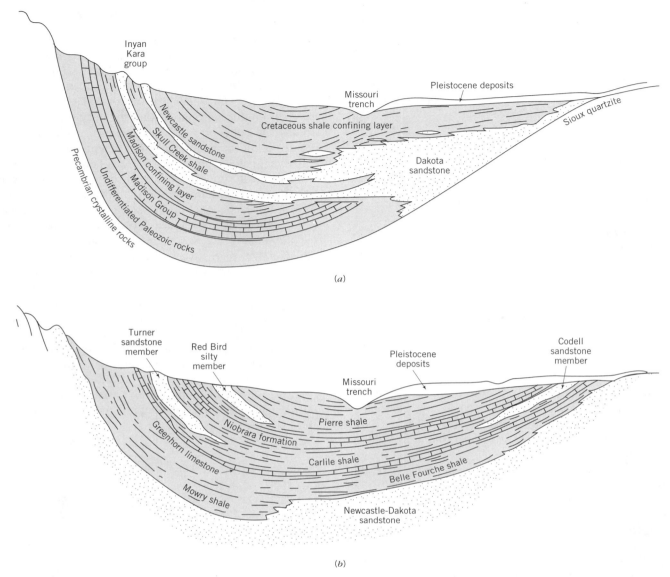

Figure 2.16 (*a*) Dakota aquifer system and (*b*) associated confining layers (after Bredehoeft, J. D., C. E. Neuzil, and P. C. D. Milly, 1983).

The points discussed above are perhaps best demonstrated by the Dakota aquifer system and associated confining layers (Figure 2.16*a*). Here we note three major aquifers, Mississippian carbonate rocks collectively referred to as the Madison Group, the Inyan Kara Group Sandstones, and the Newcastle Sandstone. All of these aquifers crop out on the eastern flank of the Black Hills. The confining layers separating these aquifers are largely shales and include the Madison Group confining layer, the Skull Creek Shale, and the Cretaceous Shale confining layer (Figure 2.16*b*). The Skull Creek Shale thins eastward and eventually pinches out, permitting the Inyan Kara and Newcastle Sandstones to merge and form the Dakota Sandstone of eastern South Dakota. The Madison confining layer also pinches out to the east, permitting carbon-

ate Madison Group waters to enter the basal sands of the Dakota Sandstone. Swenson (1968) considers this a major source of recharge to the Dakota Sandstone. The Cretaceous Shale confining layer, which contains not only shale but some minor chalk and limestone aquifers, is made up largely of three Cretaceous Shales: the Mowry-Belle Fourche, Carlile, and Pierre.

Ancestral Seas and Their Deposits

We have already seen Figure 2.2, which was presented as a map showing the major types of openings in dominant aquifers across the United States. Let us supplement Figure 2.2 with yet another map that gives the ages of the

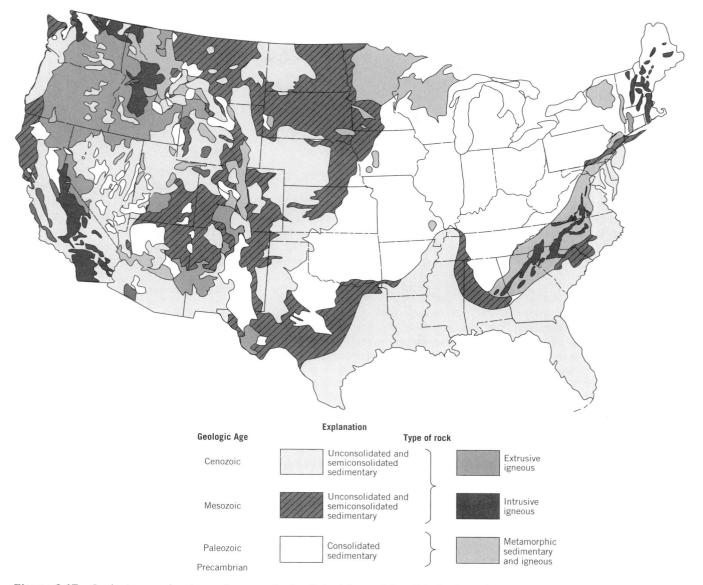

Explanation

Geologic Age	Type of rock

Cenozoic — Unconsolidated and semiconsolidated sedimentary

Mesozoic — Unconsolidated and semiconsolidated sedimentary

Paleozoic — Consolidated sedimentary

Precambrian

Extrusive igneous

Intrusive igneous

Metamorphic sedimentary and igneous

Figure 2.17 Geologic age of major rock groups in the United States (after U.S. Geological Survey, 1970).

major rock groups associated with the porosity patterns (Figure 2.17). Excluding the environments previously discussed, three major groups remain:

1. The Paleozoic consolidated sedimentary group occupying a large southwest-northeast trend across the United States, with a few areally restricted occurrences elsewhere.

2. A rather dispersed pattern of unconsolidated and semiconsolidated Mesozoic rocks.

3. A continuous pattern of unconsolidated and semiconsolidated sedimentary rocks of Cenozoic age extending along the eastern coastline from New Jersey to Texas along with a few areally restricted occurrences elsewhere.

Virtually all these rocks are marine or marginal marine in origin and are related to ancestral seas that inundated parts of the continent during periods of geologic time. We cannot discuss all the formations in this grouping, but we can discuss some of the major ones.

The Paleozoic Rock Group

The paleogeographic maps of Dunbar and Waage (1969) show that seas occupied virtually all the southwest-northeast region of the United States from the Cambrian period through the Carboniferous (sometimes divided into the Mississippian and Pennsylvanian periods). The resulting deposits consist chiefly of sandstone, shale, limestone, and dolomite, with an aggregate thickness of several thousand feet. During this period of time there were several regressions and transgressions with a substantial amount of erosion. Except for those sediments along the eastern margins (the Appalachian Geosyncline region), the formations are nearly horizontal. It was during this period that the Cambrian-Ordovician aquifer was deposited, extending from the Northern Great Plains (Williston Basin) eastward. This formation occurs near the land surface in the Appalachian Mountains and in the Midwest. Cambrian sandstones in the Midwest include the Mount Simon and Eau Claire, whereas Ordovician sandstones include the St. Peter. Important carbonate rocks of Ordovician age include the Plattville and Galena Formations. The Maquoketa Shale marks the top of the Ordovician in Wisconsin and Illinois. Thus, the Cambrian-Ordovician-Maquoketa Shale sequence constitutes a well-defined hydrostratigraphic unit. Other important aquifers above the Cambrian-Ordovician Group include the Silurian age Monroe (restricted to Ohio) and the Niagara Groups, both of which are carbonates. There are relatively few good water-producing rocks in the Devonian and Carboniferous. Extensive deposits of Permian age occur only in the southwest part of this region and are poor water producers. Thus, from the perspective of water supply, the Cambrian-Ordovician and alluvial valley systems in the Midwest are of considerable importance.

The Mesozoic Rock Group

The paleogeographic maps of Dunbar and Waage (1969) for the Mesozoic era show a rather dispersed seaway pattern, which accounts for the dispersed pattern of Mesozoic rocks shown in Figure 2.17. Seaways were very restricted from the Triassic period through the Early Jurassic, except for what is now the west coast of the continent. During the Late Jurassic period, the region that now constitutes the Rocky Mountains was inundated along with some parts of Texas and Louisiana. During Early Cretaceous a major seaway inundated most of Texas and Louisiana and the northern Rocky Mountain states. During Late Cretaceous a major seaway extended from Mexico through the entire Rocky Mountain region, in both the United States and Canada.

The marine formations associated with the Triassic seas in the western United States produced few aquifers worth noting. Some of the Jurassic rocks in the Colorado Plateau are good water producers, for example, the La Plata Group, but these have already been described as eolian deposits. By far, it is the Cretaceous period with its extensive seaway that gave rise to the major water producers in the western United States. The Early Cretaceous produced the Trinity Group in Texas, the Dakota Sandstone in the Black Hills, the Cloverly Sandstone in Wyoming, and the Kootenai in Montana. The Upper Cretaceous produced the Woodbine in Texas and the Dakota Group in the Northern Great Plains, both of which are noted in Figure 2.17, and the Milk River Sandstone in Alberta, Canada.

The Cenozoic Rock Group

The Cenozoic era is noted for the rise of the Rocky Mountains, the extrusion of lava beds in the Columbia River Plateau, block faulting in the basin and range region, and the advance of glaciers across the northern part of the continent. Our interest here, however, is restricted to the Cenozoic band of rocks that bound the eastern coast of the United States, extending around most of the Gulf of Mexico. These are marine deposits and, once again, the result of continental inundation by seas.

Seas covered the eastern continental shelf from Newfoundland to the Gulf Coast of Mexico throughout the Cenozoic and, on occasion, spread over the coastal plains. Deposits from these seas have a minimal thickness at their western-most outcrop and thicken toward the Atlantic. The deposits were progressively tilted seaward. Carbonate facies are not abundant in the northern parts of the region and are restricted to Georgia and Florida. The Mississippi River was formed by Miocene time, carrying large volumes of sediments to the Gulf of Mexico, building the Mississippi Delta. Thus, Cenozoic sediments in the Gulf of Mexico region are largely of the clastic variety.

Within this geologic setting, numerous sandstones were deposited in the Atlantic coastal plain region, giving way largely to carbonate groups to the south. This fact motivated Heath (1984) to break from precedent in his modern classification of ground-water regions to establish the Atlantic and Gulf coastal plain region, which he describes as interbedded sand, silt, and clay, and a southeast coastal plain region, which he describes as layers of sand and silt over carbonate rocks.

Diagenesis in Marine Environments

The progressive burial of sediments in depositional environments is accompanied by physical and chemical changes that affect the sediments. These physical and chemical changes occur because of increases in the overburden pressure, increased temperature, and the chemical interactions between minerals and migrating pore water. From a hydrologic perspective, diagenesis is important because of its effect on pore space. Some processes act to reduce the existing pore space, and others act to enhance it. An end result of these processes is the lithification of the sedimentary pile, that is, the conversion of sediments to consolidated sedimentary rocks.

Porosity Reduction: Compaction and Pressure Solution

The term *compaction* refers to a diagenetic process where the weight of the overburden contributes to porosity reduction from some initial higher value. Compaction can take place by changing either the arrangement of the grains or their shape. In a fluid-saturated environment, a

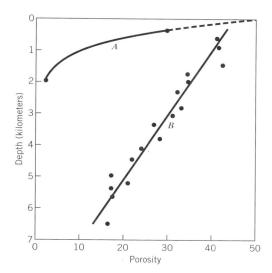

Figure 2.18 Porosity versus depth curves. Curve *A* from Athy (1930) for shales; curve *B* from Blatt (1979) for sandstones. Data for Blatt's curve represent 1000-ft averages of 17,367 porosity measurements (from an unpublished manuscript by Atwater and Miller).

decrease in porosity requires an expulsion of the pore fluids from the sediments, with the total volume of fluid expelled balanced by the total volume of porosity loss. The term *normal compaction* is reserved for the condition where the fluid expulsion takes place more or less concurrently with the deformation such that there is no apparent increase in the pressure of the fluid. Disequilibrium compaction implies a time lag between geologic loading of the sediments and the expulsion of their pore waters, meaning that some of the load is carried by the pore water, resulting in fluid pressures higher than would normally be expected in a hydrostatic (oceanic) environment. As a general rule, the finer grained the sediment, the greater its initial porosity, and the greater the potential porosity decrease. Hence, clays will be affected by compaction to a greater extent than sands, and carbonate muds may not be materially affected if crystallization occurs early in the depositional history.

Measured porosity versus depth curves for shales and sandstones are shown in Figure 2.18. The curve for shale was produced by Athy (1930) from a study of Paleozoic shales in Oklahoma. Athy proposed the relationship

$$n = n_o e^{-az} \tag{2.4}$$

where n is porosity, n_o is the average porosity of near surface clays, z is the depth below surface, and a is an empirical constant, equal to 1.42×10^{-3} m^{-1} for his data. The data of Athy suggest that given a porosity of surface clays on the order of 0.4 to 0.5, the porosity of compacted clay (shale) is reduced to about 0.03 at 2000 m. The curve for sandstone (Blatt, 1979) is based on some 17,000 measurements of Late Tertiary sandstones in Louisiana.

Below 350 m, the relationship is a linear one, with porosity decreasing from 38 to about 18% at 6000 m.

The decrease in porosity versus depth for sandstones shown in Figure 2.18 does not necessarily prove that grain rearrangement in closer packings is the responsible mechanism for porosity reduction. Another candidate is pressure solution, where grains dissolve at grain-to-grain contacts where the stress is greatest. If the material so dissolved is precipitated locally on the unstressed surface of the solid exposed to the pore space, the bulk volume is reduced at the expense of the pore volume. A conceptual visualization of this process given by Weyl (1959) demonstrates that the reduction in bulk volume is compensated by a commensurate decrease in pore volume (Figure 2.19).

The deformation resulting from pressure solution is directly proportional to the grain-to-grain stress established in the rock. Thus, we expect it accompanies basin loading in depositional environments and is driven by the same stress as that responsible for a closer packing of the grains. There is an abundance of literature on pressure solution in rocks, including overviews by Rutter (1983) and DeBoer (1977). However, there still remains some controversy regarding the driving forces. Bosworth (1981) argues for a plastic deformation mechanism, whereas Green (1984) suggests that volume transfer creep during phase transformations has many characteristics similar to that of pressure solution. Whatever the mechanism, the reduction of porosity causes a rather large reduction in permeability. Hsu (1977), for example, determined that permeability of sandstones in the Ventura Basin in California decreased by three orders of magnitude over 3000 meters of depth.

Palciauskas and Domenico (1989) present a simple constitutive law for porosity reduction based on interpenetration of spheres and the melting heat at points of contact. Figure 2.20 shows their theoretical curve for porosity reduction with depth for sandstone. Based on this model, they concluded that the pore volume change for sandstone subjected to this inelastic deformation is

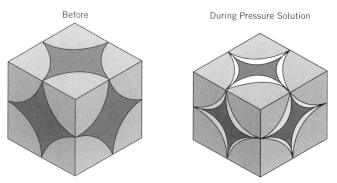

Figure 2.19 Pressure solution of identical spheres of simple cubic packing (from Weyl, J. Geophys. Res., v. 64, p. 2001–2025, 1959. Copyright by Amer. Geophys. Union).

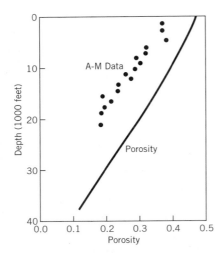

Figure 2.20 Theoretical upper bounds of porosity variation with depth in sandstone. The 17 data points represent 1000-ft averages of 17,367 porosity measurements of Late Tertiary sandstones from an unpublished manuscript by Atwater and Miller, as presented in the work by Blatt (1979) (from Palciauskas and Domenico, Water Resources Res., v. 25, p. 203–213, 1989. Copyright by Amer. Geophys. Union).

substantially larger than the pore volume changes associated with grain rearrangement. Also shown on this figure are the data of Blatt (1979) for sandstones. Since the sandstone cores contain unspecified amounts of pore filling clays and cements, the measured porosity should be lower than the theoretical curve, but ideally should have a similar slope with depth. The theoretical curve given in Figure 2.20 is in good agreement with observations by Maxwell (1964), who stated that porosity of sandstones diminished with depth to less than 5% porosity below 10 km.

Chemical Rock–Water Interactions: Secondary Porosity in Sandstones

From a strictly chemical perspective, porosity can be reduced by cementation, replacement of one mineral by another, and recrystallization. Conversely, porosity can be enhanced by dissolution of grains and cements. A very large percentage of existing porosity in sandstones is not original intergranular porosity, but a secondary or solution-type porosity created by some dissolution process. The following statements by Hayes (1977) sum up much of modern thinking on porosity development in sandstones:

1. Primary intergranular porosity is subject to almost total destruction by several mechanical and chemical processes early in the diagenetic period.

2. Later in diagenesis, secondary porosity can be produced by dissolution.

3. Chemical diagenesis of sandstones is a kinetic process involving reactions between minerals and a moving aqueous solution. The main source of water in the depositional environment is from dewatering of shales interbedded with sandstone.

The first two statements appear to be amply supported by thin section and field studies (see, for example, Schmidt and McDonald, 1977). The third statement is a hypothesis, albeit a reasonable one. Before accepting any of these statements, there are several questions that must be addressed, including the responsible driving forces for water movement from shale into sandstone in an oceanic (hydrostatic) environment, the mass transport mechanisms, and the reactions that contribute to the diagenesis. These are major topics that will be covered elsewhere in this text. However, at this time a short statement on the driving forces for fluid movement may be in order in that these driving forces relate to the equilibrium and disequilibrium concepts introduced earlier, as well as to the process of pressure solution.

Simply, fluid moves from the shales into the sands because the shales are in a greater state of disequilibrium compaction than the sands. Indeed, if sands are continuous to the overlying oceanic waters, they are likely undergoing equilibrium compaction. Given this statement, there are two questions that may be pursued further: (1) For a given thickness of overburden and in the absence of fluid flow, will the fluid in all rocks be equally pressurized? (2) What is responsible for maintaining high fluid pressures in certain rocks and not in others over long periods of geologic time; that is, what contributes to disequilibrium compaction?

The first of these can be addressed immediately with a statement describing the fluid pressure production in the absence of fluid flow

$$dP = \chi \, d\sigma \qquad (2.5)$$

where P is fluid pressure, σ is stress, and the factor χ is called the pore pressure coefficient. If $\chi = 0$, $dP = 0$, and if $\chi = 1$, fluid pressure change equals the change in overburden pressure $d\sigma$. The pore pressure coefficient expresses the change in fluid pressure with changes in stress in the absence of fluid flow; that is,

$$\chi = \left(\frac{dP}{d\sigma}\right)_M \qquad (2.6)$$

where the subscript M means at constant fluid mass, that is, no loss or gain of mass due to fluid flow. If the rocks and the water contained within them are totally incompressible, the pore pressure coefficient is zero, and the fluid pressure would not be affected by the stress changes. Here the total load is carried by the rigid matrix. On the other hand, if virtually all the overburden is carried by the fluid and none by the matrix, the pore pressure coefficient is close to one, and the fluid pressure starts to approach the overburden pressure. As the individual grains, the pores, and the fluids within the pores all exhibit compressibility to some extent, the pore pressure coefficient will be greater than zero but less than one. Work by Palciauskas and Domenico (1989) suggests that the pore pressure coefficient is on the order of 0.99 for clays, 0.83 for mudstones, 0.67 for sandstones, and 0.25 for limestones. Thus we note that the more rigid rocks (that is, limestones and, to a lesser degree, sandstones) are not capable of being excessively pressurized even in the absence of fluid flow. On the other hand, the fluid pressure in clays and mudstones can approach the overburden pressure because most of the incremental load is carried by the water in the pores. Hence, depending on the magnitude of the pore pressure coefficient, it may be difficult to generate high fluid pressures in some rocks relative to others, irrespective of the rate of loading and limitations on the rate of fluid flow.

Low-permeability sediments, such as shales, are those in which the fluid pressures will be maintained the longest because they do not drain efficiently. This statement is only partially correct, but it is sufficient for our purposes here. It follows that the driving force for fluid movement is the pressure differences between the fluid in the various rocks in the depositional environment, and this pressure difference is established because all rocks cannot be equally pressurized due to compressibility factors. Those that are pressurized the most maintain their pressures longer due to permeability factors.

Boggs (1987) cites seven major diagenetic processes other than compaction, at least four of which have the potential for effecting porosity and permeability in depositional environments. These are replacement, where one mineral is replaced by another; inversion, which is the replacement of a mineral by its polymorph (polymorphs are minerals having the same chemical composition but different crystal forms); recrystallization, or a change in crystal size and shape, generally resulting in an increase in grain size; and dissolution, a selective process that removes the most soluble minerals for a given set of environmental conditions.

2.5 Uplift, Diagenesis, and Erosion

With uplift and erosion, the sedimentary sequence we have been following has gone through one complete cycle. Over all geologic time, weathering and erosion act to modify any elevated landscape, with numerous rock bodies entering the sedimentary cycle time and again. From the perspective of permeability development, there are several important factors associated with the uplift of sedimentary rocks from their marine or marginal marine environment. First, the rocks are thrust into a meteoric regime so that fresh water can continually enter their outcrop areas. The entering meteoric water mixes with and displaces the original formation waters of the deposi-

tional environment. In some places, the displacement has been quite effective, with the original formation waters being completely displaced from the formation, at least in their current continental positions. In other places, the displacement process has not been effective. At any rate, the rocks apparently undergo yet another stage of diagenesis, rather unevenly perhaps, this time meteorically driven. With regard to porosity development, which is our main concern here, sandstone diagenesis merely means yet another generation of secondary porosity development by dissolution of those mineral phases that are not stable in the meteoric regime. For carbonates, again from the perspective of porosity development, our main interest here is the dissolution of mature carbonate terrain uplifted into a ground-water circulation system.

The Style of Formations Associated with Uplift

Figure 2.21 shows three styles of development for formations associated with uplift. Figure 2.21*a* shows the geologic features associated with the Rocky Mountain development throughout the Cenozoic era. This style of development gives rise to exposed outcrops in the uplifted area, with the Dakota Group referred to earlier providing a classic example. Several of the beds pinch out with depth and distance eastward, whereas others continue across the basin and outcrop or subcrop in structurally undisturbed areas. The Dakota Sandstone, for example, outcrops along the Black Hills in South Dakota and underlies about 90% of Nebraska, and both Iowa and Minnesota to some extent. The Dakota also occurs in the Williston Basin in North Dakota and in the Denver Basin, where it contains salt water.

Figure 2.21*b* shows those features associated with the uplift of the Colorado Plateau, which was raised vertically along with the Front Range of the Rocky Mountains. The classic basin-type structure is evident, with outcrops or subcrops circumscribing the outer rim of the basins. It is noted that water can enter this type of basin along its entire circumscribed outcrop area.

Yet a third type of association of formations with mountains is shown in Figure 2.21*c*, but the relationship here is a passive one. This setting is supposed diagrammatically to represent the Atlantic coastal plain, where the sediments were derived largely from the Appalachians and were not "uplifted" with them. As mentioned previously, the sediments thicken toward the sea, and, although there is evidence of some eastward tilting, the association of these sediments with tectonics is quite different from that mentioned earlier for the western United States. However, because of their topographic relationships, water can enter the formations in the outcrop areas and move eastward. Relatively freshwater zones extending several miles offshore into the Atlantic Ocean are not uncommon.

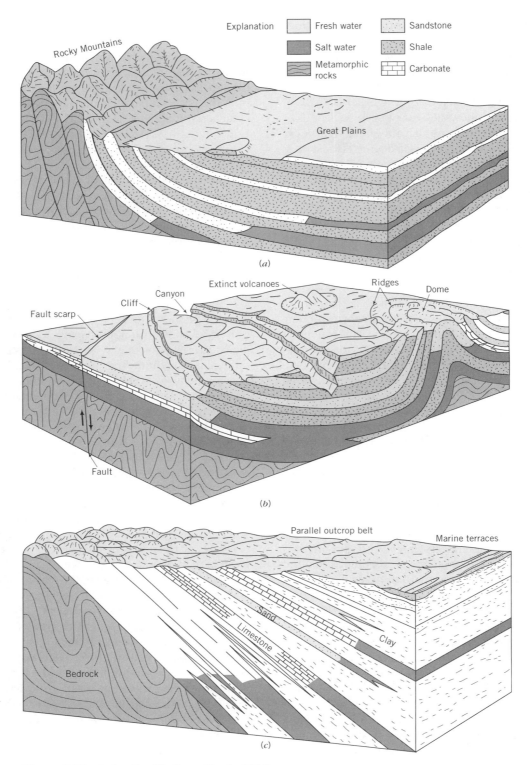

Figure 2.21 Style of uplift (from Heath, 1984).

Following Heath (1984), we have attempted to show the occurrence of salty or brackish waters in association with the uplift patterns of Figure 2.21. In these diagrams, the transition from fresh water to salt water is portrayed as an abrupt interface. As noted, the transition is a pro-

gressive one that can occur over many miles. However, the diagrams do provide a background for a general description of our observations, several of which can be made here.

One notable feature of Figure 2.21 is that the younger

permeable units can be more or less completely flushed of their original waters, while the deeper (older) ones are only partly flushed, with the degree of flushing getting poorer with depth. One possible reason for this is that the younger beds may have had more recent regressions and the formations may have been subjected to long-term meteoric flushing in the depositional environment prior to uplift. Further, they are closer to the surface and may be replenished from meteoric sources other than those available at the outcrop. A second feature, as shown in Figure 2.21b, is the effect of basin perimeter recharge. Down-dip flushing in this case has the effect of concentrating the original formation water and can be effective only for short distances from the intake areas. As the formation fluids become more concentrated with depth, there is a marked density difference between the incoming meteoric waters and the original fluids, and density stratification occurs. For these cases, the lighter fluid will actually attempt to move above or around the more dense one if at all possible, or actually discharge cross-formationally where further movement is impeded by the more dense fluid. It is difficult to conceive of any manner in which flushing can be effective in such environments.

Secondary Porosity Enhancement in Carbonate Rocks

According to Stringfield and LeGrand (1966), if the deposits overlying a carbonate terrain are of low permeability and are thick, and if the carbonate was never elevated into a ground-water circulation system, then little secondary porosity will develop. On the other hand, elevation into a ground-water circulation system at either an early stage or a late stage will lead to the development of secondary porosity and permeability. Hence, if a carbonate is unconformably overlain by a low-permeability rock such as shale, there is little question that the permeability of the carbonate rock has been enhanced by solution. The same can be stated for carbonate rocks in direct contact with glacial till laid down during the Pleistocene, for such carbonates represent long-time erosional surfaces that were characterized by paleohydrologic drainage systems. On the other hand, secondary permeability development cannot be assured when limestones are in conformable contact with overlying marine shales. Thus, the significance of the unconformity should not go unnoticed in the geologic record, especially when the concern is with secondary porosity and permeability development in carbonates.

Major solutional enhancement of permeability in carbonate aquifers is dependent on the circulation of ground water. Thus, the capability must exist for (1) chemically aggressive ground water to recharge the system, (2) the rock unit to transmit water through the fractures, and (3) ground water to drain from the system (Stringfield and LeGrand, 1966). These three conditions are best satisfied in unconfined carbonate aquifers, which readily allow for recharge to and discharge from the aquifer. Older carbonate rocks have low porosities, serving to concentrate flow along the fractures. A positive feedback loop of dissolution and transport is the logical result of these processes and results in a self-organized pattern of solutionally enlarged fractures. These have been modeled using both physical models (Ewers, 1982) and mathematical models (Dreybrodt, 1990; Palmer, 1991; Howard and Groves, 1995).

The threshold, above which there is a flow in interconnected solutionally enlarged fractures, appears to be a few tens of meters at most. At a scale of hundreds of meters the patterning results in conduits, which are interconnected solutionally enlarged fractures wider than 1 cm, and in which turbulent flow is likely. Finally where flow paths are kilometers in length, then patterning results in caves (conduits accessible to humans, thus generally larger than 1 m in diameter). The vast speleological literature is largely concerned with this largest scale.

There has been some investigation of techniques to use in wells to identify the extent and size of enlarged, interconnected fractures (Hickey, 1984; Price, 1994; Sauter, 1992; Worthington and Ford, 1995). The only definitive test for interconnectivity is the direct measurement of flow velocity by point-to-point tracer testing. Many thousands of such tests have been successfully completed in carbonate terrains and almost always have revealed ground-water velocities >100 m/day.

Extensive near-surface solution in carbonate rocks results in a karst landscape. Sinkholes are the primary landform associated with mature karst. They can be classed generally into four types: (1) the classical solution sink, which involves a closed depression in the bedrock with or without a soil cover, (2) a cavern collapse sink that develops when the roof falls in on a near surface conduit, (3) subsidence sinks caused by upward stoping [sic] of a collapsed solution cavity through substantial thicknesses of bedrock, and (4) soil piping sinkholes, which represent the abrupt collapse of a soil arch on mechanically sound bedrock through soil loss into solution cavities (White and White, 1987). Sinkholes are a serious geologic hazard. Considerable damage has resulted from their formation beneath highways, railroads, dams, reservoirs, pipelines and vehicles.

2.6 Tectonism and the Formation of Fractures

A fractured rock may be regarded as intact rock bodies separated by discontinuities. The rock bodies are generally considered to be impermeable, but frequently some permeability has to be attributed to these intact rock blocks. From the perspective of ground-water flow, a fluid-conducting medium can be described as fractured if the majority of the flow takes place through discrete fracture channels that in some fashion form an integrated, interconnected system.

Van Golf-Racht (1982) defines a fracture as the surface along which a loss in cohesion takes place. A fault is a fracture where displacement has occurred, whereas a joint is a fracture along which no notable displacement has occurred. From a purely geologic perspective, these definitions are adequate; from a hydrologic perspective, the most important factors are fracture properties irrespective of displacements, including orientation, density, aperture opening, smoothness of fracture walls, and—possibly above all else—the degree of connectivity. For example, given a set of fractures in a rock, none of which extend over the scale of the formation and none of which are interconnected, the rock cannot transmit water via the fracture network.

Style of Fracturing

Fracture style is closely related to stress history and rock type. Brittle rocks of low porosity are most susceptible to fractures, given the proper tectonic setting. Van Golf-Racht (1982) cites three cases where stress-related fractures may occur:

1. In response to folding and faulting.

2. Deep erosion of the overburden, which will produce differential stresses that can cause fractures.

3. Rock volume shrinkage (shrinkage cracks) where water is lost, say, for example, in shales or shaley sands.

Frequently, the terms *microfracture* and *macrofracture* are applied to describe the magnitude of fracture. These terms are imprecise, but a microfracture is on the order of the grain size or smaller and is not readily detected by the naked eye. Microfractures are frequently caused by differential thermal expansion of individual minerals. A macrofracture is a fracture that is readily detected without a microscope and can range from a simple joint to a throughgoing discontinuity.

Stearns and Friedman (1972) use the following descriptions in the study of fractures:

1. Conjugate fractures, or fractures related to a common origin that form under a single state of stress. Conjugate fractures are intersecting fractures, with the angle of intersection commonly found to be about 60°.

2. Orthogonal fractures, which are normal to each other and are not related to a single stress state.

3. Regional fractures, which extend over large areas and apparently are unrelated to local structures.

Orthogonal fractures are independent of structure and are generally well defined, even in flat lying rocks. Structurally related fractures are associated with specific struc-

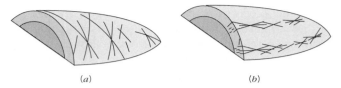

Figure 2.22 Schematic illustration of most common fractures associated with a fold: (*a*) type I, (*b*) type II. Both types maintain a consistent relation to bedding, but not to folding (from Stearns and Friedman, 1972). Reprinted by permission.

tures such as faults and folds. Fractures associated with faults are thought to be caused by the same stresses that caused the faulting, and their strike will in general parallel that of the fault. Stearns (1964, 1967) reported five types of fracture patterns commonly associated with folding, only two of which produced a significant fracture density. As noted in Figure 2.22, both patterns consist of two conjugate shear fractures and an extension fracture. The pattern designated type I is thought to be earlier in origin than type II and is common even on low-dipping folds where type II is absent. The two patterns occur in the same bed and represent two different stress states. Pattern type I has been described as single zones of parallel fractures across the entire structure whereas type II fractures may range in length from a few inches to several tens of feet. Type II fractures, however, consist of a fracture zone with three different orientations, whereas type I fractures appear to have a single orientation and are not an assemblance of all three orientations. Hence, Stearns and Friedman (1972) postulate that the type II fractures, although smaller in length and aperture, have a greater degree of connectivity and may be the better fluid conductors. On the other hand, because type I fractures are continuous over large regions, they may be an important regional pathway for fluid movement.

In comparing rock deformation under the same environmental influences, Stearns (1967) notes that the density of fractures is dependent on the lithology of the rock (Figure 2.23). This is in agreement with DeSitter (1956).

Several other studies attempt to relate fractures to lithology or to regional and local structures. Harris and others (1960) conclude that fractures associated with compressional deformation have certain features in common, such as repetition and continuity of trend. The influence of lithology is again pointed out along with the fact that thinner sedimentary rocks are more susceptible to fracturing than are thicker ones. Regan and Hughes (1949) discuss the productivity of fractured shales in California, where the fractures occur in chert zones and in zones of calcareous and platy siliceous shale.

Other types of fractures include those produced by volume shrinkage, caused by cooling in igneous rocks such as basalts, and desiccation in sedimentary rocks.

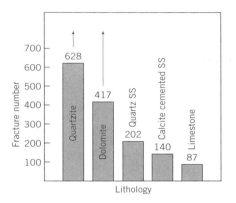

Figure 2.23 Average fracture density for several common rock types naturally deformed in the same physical environment (from Stearns, 1967).

Landes and others (1960) conclude that water-bearing fractures are a general rule in crystalline rocks between the depths of 600 and 1000 ft below the surface.

Weathered till is yet another material that shows a high fracture density. According to Barari and Hedges (1985), studies in the Midwest and Canada show that the upper part of most glacial tills has been weathered and contains a dense fracture pattern. Few if any fractures persist in the lower unweathered portions of till sheets. The permeability for weathered tills is three to four orders of magnitude larger than that for the unweathered tills.

Fluid Pressure and Porosity

The in situ porosity and permeability of fractured rocks are generally considered to be a function of fluid pressure. This statement holds for intergranular material as well, but to a lesser extent. In discussing these concepts, Snow (1968) provides abundant evidence that fractures "breathe," or open and close in response to changes in fluid pressure. As there is a direct relationship between fluid pressure and aperture opening (and, consequently, permeability), there follows some relationship between fracture strength, frictional resistance to sliding, and fluid pressure. Brace and others (1968) discovered an increase in permeability in fractured granites due to an increase in fluid pressure, which caused a decrease in the stress acting on the rocks. Serafin and del Campo (1965) suggested that the percolation of reservoir water beneath dams caused a widening of the fracture openings with a consequent weakening of the structure. A series of field experiments by Jouanna (1972) revealed that as the normal stress increases, the permeability decreases, and this process is irreversible. According to Sharp and Maini (1972), the relationship between stress and permeability is nonlinear because the normal stiffness of the fracture increases with decreasing aperture width. Thus, decreases in stress in fractured rock caused by increases in fluid pressure have a significant effect on rock permeabil-

ity. Witherspoon and Gale (1977) review the relationship between fracture permeability and fluid pressure. Quantitative evaluations have been provided by Gangi (1978) and Walsh (1978, 1981).

Connectivity

Consider a rock mass with a relatively impermeable matrix that contains several fractures. This is shown in Figure 2.24 for two cases with the same fracture orientation and density. Assuming that all other parameters that control flow are held constant, the only difference between these cases is the degree of interconnection between individual fractures. In Figure 2.24*a*, a condition is shown in which none of the fracture clusters are connected to other clusters. This prevents flow across the sample, indicating that the unconnected clusters add nothing to the permeability of the medium. Figure 2.24*b* shows clusters that are interconnected across the sample, suggesting that flow can be achieved across the region.

Percolation theory deals with the idea of connectivity (Shante and Kirkpatrick, 1971; Dienes 1982). A percolation threshold is defined as the density of fractures that intersect sufficiently to promote flow. A finite set of fracture clusters can be interconnected, but the connections between various finite clusters may not exist. Above the percolation threshold, one cluster becomes "infinite" so that flow can take place through the medium. However, many other clusters may not be connected to the infinite one. As the fracture density increases, isolated clusters become more rare. De Marsily (1985) gives several examples where the percolation threshold has been related to the average number of intersections of a single fracture with all other fractures, as well as to the density and length of the fractures. Robinson's (1982) results are based on the latter and indicate that when the product of the fracture density and the square of their average length equals 1.5, the percolation threshold is achieved and flow commences. Below this value, the fractures are insufficiently connected to promote flow.

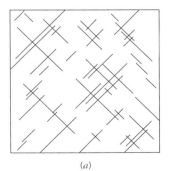 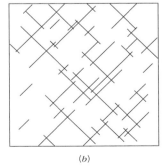

(*a*)   (*b*)

Figure 2.24 Representations of a conceptual rock mass illustrating the importance of connectivity. (*a*) Unconnected clusters, (*b*) Most clusters connected.

These results at this stage of the research are the consequence of numerical experiments wherein fracture patterns are randomly generated with a computer and the individual intersections noted. Key questions and difficulties remain in identifying the percolation threshold and obtaining field data on actual fracture densities. Three-dimensional data on fracture density and volumes (Charlarx and others, 1984) are obviously even more difficult to obtain. Further, from our discussion of fracture style, it is clear that there exists in nature different classes of fractures and that connectivity studies for the individual classes as well as any possible assemblage of classes would give results that might be less abstract and more pertinent to the problems of flow in real rocks. That is to say, if classes of fractures can be identified from field mapping, perhaps something can be said of a qualitative or semiquantitative nature about the potential for connectivity and the degree of connectivity that might exist on the basis of theoretical studies.

Chapter 3

Ground-Water Movement

Considering the fact that the flow of fluid can be observed in "real time" in laboratory experiments, it is not surprising that laws have been developed to describe some of the macroscopic details of the motion. Because "motion" involves measurements of time and distance, the experimental observations are quantitative ones, so that the laws take on a mathematical form. Our main law may be stated outright as

The velocity, or distance traveled over some time interval, is proportional to some driving force.

The surprising feature of this statement is that—provided certain conditions are met—it holds irrespective of the complex details of the connected pore space described in Chapter 2.

3.1   Darcy's Experimental Law and Field Extensions

Henry Darcy was a civil engineer concerned with the public water supply of Dijon, France, in particular the acquisition of data that would improve the design of filter sands for water purification. In search of this information, Darcy set out "to determine the laws of flow of water through sand." His method was experimental, the results of which were published in 1856.

An experimental apparatus similar to that employed by Darcy is shown in Figure 3.1. It consists of a cylinder having a known cross-sectional area A (L^2) that contains a porous medium (e.g., sand) and includes appropriate plumbing to flow water through the column. The cylinder contains two manometers, whose intakes are separated from one another by a distance Δl (L). The manometers are nothing more than small open tubes. As will become clear shortly, the elevation to which the water level rises in a manometer is a measure of the energy that the ground water possesses at the inlet of the manometer. In an actual experiment, water flows into and out of the cylinder at a known volumetric flow rate Q (L^3/T), and the elevation of water levels in the manometers, h_1 and h_2 (L), are measured relative to a local datum (Figure 3.1).

The parameters of this experiment may be combined to yield two key master variables. The specific discharge, q, has the units of velocity (L/T) and represents the volumetric flow rate per unit surface area of the cylinder. It is determined from the experiment as Q/A. The dimensionless hydraulic gradient, $(h_1 - h_2)/\Delta l$, represents the change in water-level elevation in the manometers separated by Δl. Over small distances, the gradient is expressed in familiar calculus terms as $\partial h/\partial l$.

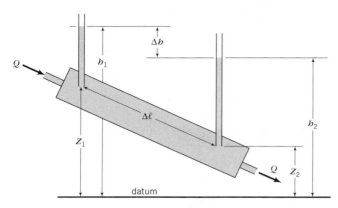

Figure 3.1 Laboratory apparatus to demonstrate Darcy's law.

In the terms of the variables just given, Darcy's law is expressed

$$\frac{Q}{A} = q = -K\frac{(b_1 - b_2)}{l} = -K\frac{\partial b}{\partial l} \qquad (3.1)$$

where q is the volumetric flow rate per unit surface area, with units of velocity, and K is a constant of proportionality. Because the gradient is a dimensionless quantity, the proportionality constant K has units of velocity. The minus sign is a convention that is employed because flow is in the direction of a decrease in water levels in the manometers, that is, from where b is high to where b is low (Figure 3.1). Darcy's equation may be stated in words as

The velocity of flow is proportional to the hydraulic gradient.

Darcy's law is valid for flow through most granular material. The law suggests a linear relationship between the specific discharge and the hydraulic gradient. This relationship holds as long as the flow is laminar. Under conditions of turbulent flow, the water particles take more circuitous paths. At the other extreme, for very-low-permeability materials, a minimum threshold gradient may be required before flow takes place (Bolt and Groenevelt, 1969).

Four aspects of Darcy's law require clarification from a field perspective: the specific discharge q, the water-level measurements, the gradient, and the proportionality constant K.

The Nature of Darcy's Velocity

As defined, the specific discharge q is a volumetric flow rate per unit surface area of sample. Because water moves only through the pore openings making up the surface area, Darcy's q is a "superficial" velocity. We thus define a more realistic velocity v that is a volumetric flow rate

per unit area of connected pore space. The expression for v comes directly from Eq. 3.1

$$\frac{Q}{n_e A} = \frac{q}{n_e} = v = -\left(\frac{K}{n_e}\right)\frac{\partial b}{\partial l} \qquad (3.2)$$

where $n_e A$ is the effective area of flow and n_e has been introduced as the effective porosity. Here, the quantity v is referred to as the linear or pore velocity of ground water. The linear velocity v will always be larger than the superficial velocity (specific discharge) and increases with decreasing effective porosity. Finally, it is convenient to designate the gradient $\partial b/\partial l$ as i so that Darcy's law may be expressed in the following forms:

$$q = Ki \qquad (3.3a)$$

$$Q = KiA \qquad (3.3b)$$

$$v = Ki/n_e \qquad (3.3c)$$

$$v = q/n_e \qquad (3.3d)$$

The quantity q (as well as v) has both a magnitude and a direction, the latter being toward decreasing elevation of water in the manometers. This makes q and v vector quantities, a fact that shall be exploited in later chapters.

Hydraulic Head: Hubbert's Force Potential

In Figure 3.1, the water-level elevations in the manometers are measured with reference to a common datum, taken as a horizontal plane beneath the flow apparatus. Thus, the absolute value of the water-level elevations was of no concern to Darcy, only the differences between them. In this section, we are not interested in these differences but in what the actual water-level measurements mean. For this purpose, it is convenient to introduce a counterpart of the laboratory manometer, known as a piezometer.

The piezometer is a tube or pipe used to measure water-level elevations in field situations. It is open at the top where measurements are taken and open at the bottom to facilitate the entrance of water. A simplified version of this device is shown in Figure 3.2, where the common datum is taken as sea level (zero elevation). As demonstrated by Hubbert (1940), the terms *elevation, pressure,* and *total head* on Figure 3.2 can be explained in terms of the conventional Bernoulli equation. This equation states that under conditions of steady flow, the total energy of an incompressible fluid is constant at all positions along a flow path in a closed system. This may be written as

$$gz + \frac{P}{\rho_w} + \frac{v^2}{2} = \text{constant} \qquad (3.4)$$

where g is the acceleration due to gravity, z is the elevation of the base of the piezometer, P is the pressure exerted by the water column, ρ_w is the fluid density, and

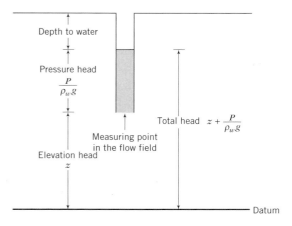

Figure 3.2 Diagram showing elevation, pressure, and total head for a point in the flow field.

v is velocity. Dividing through by *g*, Eq. 3.4 becomes

$$z + \frac{P}{\rho_w g} + \frac{v^2}{2g} = \text{constant} \quad (3.5)$$

where the quantity $\rho_w g$ equals the unit weight of water. Equations 3.4 and 3.5 describe the total energy contained by the fluid, where the first term is the energy of position, the second term is the energy due to sustained fluid pressure, and the third term is the energy due to fluid movement. The dimensions of Eq. 3.4 are L^2/T^2, or energy per unit mass. Equation 3.5 is expressed as energy per unit weight, say, foot-lb per pound, which reduces simply to feet. In SI units, the dimensions are newton-meters per newton, or simply meters.

The three terms of Bernoulli's equation, when expressed as energy per unit weight, are referred to as elevation head, pressure head, and velocity head, respectively. The term *z* in Eq. 3.5 is the elevation head and represents the elevation at the base of the piezometer. In a theoretical sense it represents the work required to increase the elevation of a unit weight of water from datum to height *z* (Figure 3.2). Stated another way, every body at the surface of the earth has a gravitational attraction toward the earth's center. To raise this body counter to this attraction requires work, and this work is stored in the form of a potential energy. The quantity $P/\rho_w g$ represents the length of the water column in the piezometer. It represents the work that a fluid is capable of doing because of its sustained pressure. The sum of these terms is called the potential energy of the fluid. The third term, $v^2/2g$, is the kinetic energy, or energy due to the fluid motion. As the velocity of ground water is slow, this term is ignored. Thus, the sum of the elevation head *z* and the pressure head $P/\rho_w g$ represents the total head *h* in the system; that is,

$$h = z + \frac{P}{\rho_w g} \quad (3.6)$$

Stated simply, the total head *h* is the sum of the elevation of the base of the piezometer plus the length of the water column in the piezometer. Hence, the total head at a point is found by measuring the elevation of the water level in a piezometer (Figure 3.2). However, the point to which this head refers is not the water level, but the point at the terminus of the piezometer at elevation *z*. The total head is sometimes referred to as the energy of position and is a scalar quantity.

In SI units, *h* in Eq. 3.6 is meters (m), *z* is meters (m) above the datum (usually sea level), *P* is Pascals (Pa), ρ_w is kg/m³, and *g* is m/s². Expressed in basic SI units, a Pascal is kg/m·s². The density ρ_w varies as a function of temperature and chemical composition, with fresh water at 15.5°C having a density of 1000 kg/m³. The gravitational constant, *g*, is 9.81 m/s². The following example illustrates the relationship among the various parameters.

***

Example 3.1

With reference to Figure 3.2, assume that the ground surface has an elevation of 1000 m above sea level, the depth to water is 25 m, the total depth of the piezometer is 50 m, and the water has a density of 1000 kg/m³. What is (a) the total hydraulic head at the measurement point, (b) the pressure head, and (c) the pressure?

(a) *h* = elevation of water in the piezometer
    = surface elevation − depth to water
    = 1000 − 25 = 975 m
(b) From Eq. 3.6, $P/\rho_w g = h − z$
    = 975 − 950 = 25 m
(c) From Eq. 3.6, $P = (h − z) \cdot \rho_w g = 25 \times 1000 \times 9.81 = 2.45 \times 10^5$ Pa or 0.245 MPa where the designation M means mega or 10^6.

***

An additional expression for the energy of position of a fluid can be obtained from the Bernoulli expression of Eq. 3.4 where, ignoring the velocity term

$$\phi = gz + \frac{P}{\rho_w} = gh \quad (3.7)$$

where ϕ is referred to as the hydraulic potential, in units of energy/mass. Defining potential in this way provides yet another form of Darcy's law commonly found in the literature. From the statement $\phi = gh$, it follows that

$$\frac{\partial h}{\partial l} = \left(\frac{1}{g}\right)\frac{\partial \phi}{\partial l} \quad (3.8)$$

Substituting this result in Eq. 3.1 gives

$$q = -\left(\frac{K}{g}\right)\frac{\partial \phi}{\partial l} \quad (3.9)$$

where $\partial\phi/\partial l$ is the potential gradient (Hubbert, 1940).

The Gradient and Ground-Water Flow

Darcy's law shows that for flow to occur there must be differences in hydraulic head, providing what has been referred to as a hydraulic gradient. From a field perspective, it follows that a value for hydraulic head can be defined and measured at every point within some region. The term *field* is generally used to describe a region where some physical quantity can be described in terms of a space coordinate system and time. In this case, the quantity is a scalar, giving rise to a scalar field, more commonly referred to as a potential field. There are numerous quantities that satisfy this definition, including temperature $T(x, y, z, t)$ and the concentration of some substance $C(x, y, z, t)$. Here the Cartesian coordinate system is employed to identify the value of the stated scalar quantity at a point of interest at a given time.

The potential field in any region can be described by measuring the hydraulic head at a large number of points and contouring the resulting set of data. Starting at any one of the piezometers, it is likely that the head will increase in some directions and decrease in others. The gradient in x can be written as $\partial h/\partial x$, the partial derivative indicating how h changes in x irrespective of changes in y and z. That is, $\partial h/\partial x$ is the change in head as we move along the x axis. Further, the gradient in y becomes $\partial h/\partial y$, which describes the change in head along the y axis, and the gradient in z is $\partial h/\partial z$. Thus, Darcy's law expressed as

$$q_x = -K_x \frac{\partial h}{\partial x} \qquad (3.10a)$$

$$q_y = -K_y \frac{\partial h}{\partial y} \qquad (3.10b)$$

$$q_z = -K_z \frac{\partial h}{\partial z} \qquad (3.10c)$$

describes the flow of fluid along the x, y, and z axes, where the material properties in x, y, and z are different; that is, $K_x \neq K_y \neq K_z$.

As it is rather limiting to restrict ourselves to one of three directions, we may ask for the rate of change in head in any direction and, most important, the direction of maximum change. For this purpose, the gradient of a scalar field h can be defined as

$$\text{grad } h = \nabla h = \mathbf{i} \frac{\partial h}{\partial x} + \mathbf{j} \frac{\partial h}{\partial y} + \mathbf{k} \frac{\partial h}{\partial z} \qquad (3.11)$$

where the vectors $\mathbf{i}$, $\mathbf{j}$, $\mathbf{k}$ are unit vectors in the x, y, and z directions, respectively, and ∇h (which is pronounced del h and is an upside-down "delta" to remind us of differentials) is the abbreviation for grad h. Thus, in vectorial notation, for $K_x = K_y = K_z$, Darcy's law becomes

$$\mathbf{q} = -K \text{ grad } h = -K \nabla h \qquad (3.12)$$

where grad h can be read "gradient of h." Equation 3.12 is a vectorial equation where the x, y, and z components are given by Eq. 3.11.

The gradient of h or ∇h is a vector that represents the spatial rate of change of hydraulic head (Eq. 3.11). It consists of three components x, y, and z, each of which represents how fast the head changes in that respective direction. The direction of ∇h is that which coincides with the direction in which the head changes the fastest. For the condition $K_x = K_y = K_z$, this direction is perpendicular to lines of equal head. This direction is of particular interest in that it coincides with the direction of ground-water flow.

Example 3.2 helps to illustrate the concept of a gradient in a two-dimensional flow field.

Example 3.2

The measured hydraulic head at piezometers A, B, and C is 150 m, 140 m, and 130 m, respectively. Assuming that the potential field is two dimensional in x and y, it can be represented by a dipping plane in space. The direction of flow, which is depicted on the figure, is normal to the contour lines, or, in other words, the direction of steepest decrease in head. The gradient calculation is shown on the figure

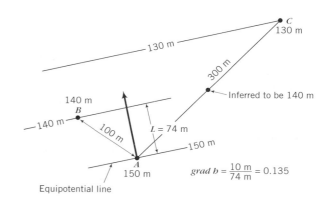

Figure 3.E2

Physical Interpretation of Darcy's Proportionality Constant

By experimentally varying fluid density, viscosity, and the geometrical properties of sands, Hubbert (1956) reported that Darcy's proportionality constant K varied in the following manner

$$K \propto \rho_w$$

$$K \propto 1/\mu$$

$$K \propto d^2$$

where μ is viscosity and d is the mean grain diameter of the sand. This can be expressed as

$$K = K^* \frac{\rho_w d^2}{\mu} \qquad (3.13)$$

where K^* is yet another constant of proportionality containing variables not yet evaluated. It is obvious at this point that Darcy's K is a function of both properties of the medium and properties of the fluid.

Insight into the parameter K^* may be obtained by comparing Darcy's law with the Hagen-Poiseuille equation long known to govern laminar flow through small-diameter passages. Incorporating Eq. 3.13 into Darcy's law gives

$$\mathbf{q} = \frac{-K^* \rho_w d^2}{\mu} \operatorname{grad} h \qquad (3.14)$$

whereas the Hagen-Poiseuille equation is

$$\mathbf{q} = \frac{-N \rho_w g R^2}{\mu} \operatorname{grad} h \qquad (3.15)$$

where N is a dimensionless shape factor relating to the geometry of passage and R is the diameter of passage. Equations 3.14 and 3.15 are perfectly equivalent if the diameter of passage R is equivalent to the mean grain diameter so that the constant K^* is equal to the product of the dimensionless shape factor N and the acceleration due to gravity g. Making this analogy, Darcy's constant of proportionality is expressed as

$$K = \frac{N \rho_w g d^2}{\mu} = \frac{k \rho_w g}{\mu} \qquad (3.16)$$

where Nd^2 characterizes the properties of the medium and ρ_w/μ the properties of the fluid. The parameter K is referred to as the hydraulic conductivity and contains properties of both the medium and the fluid, and the parameter k is referred to as the intrinsic permeability (equal to Nd^2) and contains properties of the medium only. The hydraulic conductivity with units of velocity characterizes the capacity of a medium to transmit water, whereas the permeability with units L^2 characterizes the capacity of the medium to transmit any fluid.

The term *hydraulic conductivity* is used most frequently in ground-water literature when dealing with the single water phase. The term *permeability* is used in the petroleum industry where the fluids of interest are oil, gas, and water. Typical values of permeability in square centimeters or square meters are quite small so that the "darcy" is commonly defined as a unit of permeability. For a material of 1-darcy permeability, a pressure differential of one atmosphere will produce a flow rate of 1 cc (cubic centimeter) per second for a fluid with 1-centipoise viscosity through a cube having sides 1 cm in length. One darcy is approximately equal to 10^{-8} cm^2 or, for water of normal density and viscosity, 10^{-3} cm/s. For tight materials, the millidarcy (md) is used, where 1 md equals 0.001 darcys, or approximately 10^{-6} cm/s.

It is now possible to state yet another form of Darcy's law commonly used in the petroleum industry. Incorporating our definition of the constant of proportionality

K, the hydraulic head h, and the gradient of the head h, Darcy's law becomes

$$\mathbf{q} = \frac{N \rho_w g d^2}{\mu} \operatorname{grad}\left(z + \frac{P}{\rho_w g}\right) \qquad (3.17)$$

For those cases where all piezometers are bottomed at the same elevation in a flat lying bed, z is a constant so its derivative goes to zero. Taking both the fluid density and the acceleration of gravity as constants, Eq. 3.17 becomes

$$\mathbf{q} = -\frac{Nd^2}{\mu} \operatorname{grad} P = -\frac{k}{\mu} \operatorname{grad} P \qquad (3.18)$$

This expression is quite convenient when dealing with multiphase fluid systems where the permeability k by definition is invariant with respect to whatever fluid is being considered.

Units and Dimensions

Various units and dimensions are given in Table 3.1 in both the foot-pound-second system with its FLT base (force, length, and time) and the International System (SI) with its MLT base (mass, length, and time).

3.2 Hydraulic Conductivity and Permeability of Geologic Materials

Observed Range in Hydraulic Conductivity Values

Virtually thousands of measurements of hydraulic conductivity and permeability have been obtained in both the laboratory and in the field. Davis (1969) provides the best summary of these data. Table 3.2 is taken largely from this review along with additional input from Johnson and Morris (1962) and Croff and others (1985). The values cited are given in meters per second, but conversions can be made to centimeters per second, feet per second, feet per year, or gallons per day per square foot, the last referred to as a meinzer unit in honor of O. E. Meinzer. The meinzer unit is defined as the flow of water in gallons per day through an opening of 1 ft^2, under a unit hydraulic gradient at a temperature of 60°F. Conversion to permeability (square centimeters, square feet, or darcys) is readily accomplished with the cited conversion factors.

As noted in Table 3.2, hydraulic conductivity can range in value over about 12 orders of magnitude, with the lowest values for unfractured igneous and metamorphic rocks and the highest values for gravels and some karstic or reef limestones and permeable basalts. The range in hydraulic conductivity within a given rock type is greatest for the crystalline rocks and smallest for the sedimentary material. In general, a hydraulic conductivity

Table 3.1   **Dimensions and Common Units for Flow Parameters**

Parameter	Symbol	ft-lb-sec System Dimension	Units	Conversion Factor Multiply by	SI Dimension	Units
Hydraulic head	h	L	ft	3.048×10^{-1}	L	m
Elevation head	z	L	ft		L	m
Pressure head	—	L	ft		L	m
Fluid pressure	P	F/L^2	lb/ft^2	4.788×10	M/LT^2	N/m^2 or Pa
Fluid potential	ϕ	L^2/T^2	ft^2/s^2	9.29×10^{-2}	L^2/T^2	m^2/s^2
Density of water	ρ_w	—	—		M/L^3	kg/m^3
Gravitational constant	g	L/T^2	ft/s^2	3.048×10^{-1}	L/T^2	m/s^2
Unit weight of water	$\gamma_w = \rho_w g$	F/L^3	lb/ft^3		—	—
Volumetric flow rate	Q	L^3/T	ft^3/s	2.832×10^{-2}	L^3/T	m^3/s
Specific discharge	q	L/T	ft/s	3.048×10^{-1}	L/T	m/s
Hydraulic conductivity	K	L/T	ft/s		L/T	m/s

approaching 10^{-9} m/s and smaller can be characterized as low-permeability material. Clay, shale, chalk, and unfractured igneous and metamorphic rocks fall within this category. However, if these rocks or sedimentary accumulations are fractured, the conductivity can easily exceed this limiting value by two or three orders of magnitude.

Character of Hydraulic Conductivity Distribution

Because of the way in which geological deposits are formed, hydraulic conductivity values for a particular unit can vary from place to place. This scatter in values can be depicted graphically using a histogram or represented statistically by fitting some kind of statistical distribution. Let us explore these ideas with a set of 32 measured hydraulic-conductivity values for a hypothetical geologic unit (Table 3.3).

Figure 3.3*a* illustrates the frequency histogram for these data. Like thousands of histograms constructed from various data (Law, 1944; Bennion and Griffith, 1966), there is a strong positive skew in the distribution. Statistically, this distribution is complex and quite unlike the Gaussian or normal frequency distribution, which characterizes many distributions. We can transform the hydraulic conductivity data in Table 3.3 by taking the logarithm. For example, $\log(4.0 \times 10^{-8}) = -7.4$. Interestingly, the histogram (Figure 3.3*b*) shows the log-transformed data to be approximately normally distributed. Such a distribution can be characterized in terms of a sample mean and standard deviation, which for the distribution in Figure 3.3*b* are -6.99 and 0.76, respectively.

In many studies, hydraulic conductivity has been shown to be a log-normally distributed geological parameter. In practice, data are transformed using either base 10 logarithms or natural logarithms. However, there is no hard and fast rule that hydraulic conductivity is log-

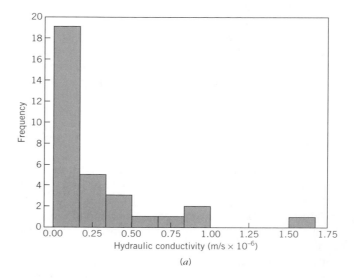

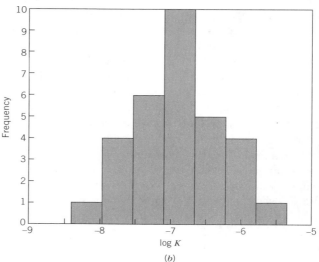

Figure 3.3   (*a*) Frequency distribution histogram of the original 32 values of hydraulic conductivity, (*b*) histogram of the log-transformed hydraulic conductivity data.

Table 3.2 **Representative Values of Hydraulic Conductivity for Various Rock Types**

Material	Hydraulic Conductivity (m/s)
SEDIMENTARY	
Gravel	3×10^{-4}–3×10^{-2}
Coarse sand	9×10^{-7}–6×10^{-3}
Medium sand	9×10^{-7}–5×10^{-4}
Fine sand	2×10^{-7}–2×10^{-4}
Silt, loess	1×10^{-9}–2×10^{-5}
Till	1×10^{-12}–2×10^{-6}
Clay	1×10^{-11}–4.7×10^{-9}
Unweathered marine clay	8×10^{-13}–2×10^{-9}
SEDIMENTARY ROCKS	
Karst and reef limestone	1×10^{-6}–2×10^{-2}
Limestone, dolomite	1×10^{-9}–6×10^{-6}
Sandstone	3×10^{-10}–6×10^{-6}
Siltstone	1×10^{-11}–1.4×10^{-8}
Salt	1×10^{-12}–1×10^{-10}
Anhydrite	4×10^{-13}–2×10^{-8}
Shale	1×10^{-13}–2×10^{-9}
CRYSTALLINE ROCKS	
Permeable basalt	4×10^{-7}–2×10^{-2}
Fractured igneous and metamorphic rock	8×10^{-9}–3×10^{-4}
Weathered granite	3.3×10^{-6}–5.2×10^{-5}
Weathered gabbro	5.5×10^{-7}–3.8×10^{-6}
Basalt	2×10^{-11}–4.2×10^{-7}
Unfractured igneous and metamorphic rocks	3×10^{-14}–2×10^{-10}

To convert meters per second to	Multiply by
cm/s	10^2
(gal/day)/ft^2	2.12×10^6
ft/s	3.28
ft/yr	1×10^8
darcy	1.04×10^5
ft^2	1.1×10^{-6}
cm^2	1×10^{-3}
To convert any of the above to meters per second →	Divide by the appropriate number above

normally distributed, or for that matter are other flow parameters. For example, Woodbury and Sudicky (1991) show that other distributions may be appropriate for hydraulic conductivity data as well. In the case of porosity, studies show that distributions from a single formation typically have normal rather than log-normal distributions (Davis, 1969).

Anisotropicity and Heterogeneity Within Units

Most rocks have a directional quality to their overall structure. Metamorphic rocks are noted for their schistosity, most sediments for their horizontal stratification, and basalts for their preferred orientation of shrinkage cracks when the cooling period is rather short (and a sparsity of such cracks when the cooling history is long). For such materials, the hydraulic conductivity, as measured from some representative sample, will not be equal in all directions. The term *anisotrophic* is used to describe materials where the permeability or conductivity at a point has a directional dependency. When permeability is the same in all directions, the material at that point is isotropic. Davis (1969) cites several cases for bedded sediments in particular where the permeability is greater in the direction of the stratification and smaller perpendicular to the stratification. Although supporting data are sparse, similar statements can be made for maximum directional permeability associated with metamorphic structures such as schistosity. Table 3.4 provides a summary of information on the anisotropic nature of some sedimentary materials as determined from core samples. In these tabulations, the horizontal conductivity is taken in the direction of the structural features, such as stratification, and the vertical conductivity is taken at right angles to the stratification.

The classical definition considers a unit to be homogeneous if the permeability in a given direction is the same from point to point in a geologic unit. Materials that do not conform with this condition are heterogeneous. A variety of modern studies has shown that there is no possibility of ever finding a truly homogeneous unit in nature. Hydraulic conductivity distributions in reality

Table 3.3 **Hydraulic Conductivity Values Measured in a Hypothetical Geologic Unit**

		K (m/s)		
4.0×10^{-8}	1.5×10^{-6}	4.0×10^{-7}	2.0×10^{-7}	8.0×10^{-8}
1.0×10^{-7}	8.0×10^{-9}	5.0×10^{-7}	2.5×10^{-8}	4.0×10^{-8}
2.0×10^{-8}	1.0×10^{-7}	2.5×10^{-8}	1.0×10^{-7}	3.0×10^{-7}
4.0×10^{-7}	1.0×10^{-7}	2.0×10^{-7}	1.0×10^{-6}	6.0×10^{-7}
1.0×10^{-6}	1.0×10^{-8}	2.0×10^{-7}	2.0×10^{-10}	2.5×10^{-7}
1.5×10^{-7}	8.0×10^{-7}	5.0×10^{-8}	1.0×10^{-7}	4.0×10^{-8}
6.0×10^{-8}	5.0×10^{-8}			

Table 3.4 **The Anisotropic Character of Some Rocks**

Material	Horizontal Conductivity (m/s)	Vertical Conductivity (m/s)
Anhydrite	10^{14}–10^{-12}	10^{-15}–10^{-13}
Chalk	10^{-10}–10^{-8}	5×10^{-11}–5×10^{-9}
Limestone, dolomite	10^{-9}–10^{-7}	5×10^{-10}–5×10^{-8}
Sandstone	5×10^{-13}–10^{-10}	2.5×10^{-13}–5×10^{-11}
Shale	10^{-14}–10^{-12}	10^{-15}–10^{-13}
Salt	10^{-14}	10^{-14}

look like the one shown on Figure 3.4, which is from the shallow, unconfined aquifer at Canadian Forces Base Borden (Sudicky, 1986). The contoured field is based on 720 permeameter measurements along a single cross section. Compared to other carefully studied aquifers, the Borden aquifer is "mildly" heterogeneous.

Geostatistical concepts have proven useful in representing the heterogeneous character of a hydraulic conductivity distribution. One can represent the variability of the parameter in terms of the variance and spatial autocovariance in the measured data. The variance for a normal distribution represents the spread in measured hydraulic conductivity around the mean. The autocovariance represents the spatial continuity in the hydraulic conductivity distribution. What is meant by spatial continuity? As shown on Figure 3.4, hydraulic conductivity values are not random, but are correlated such that two measurements taken close together are quite similar, but measurements separated by a larger distance are much less similar. Effectively, there is a decay in the correlation as a function of separation distance.

These geostatistical concepts have much in common with the classical definitions of isotropicity and homogeneity. For example, depending upon how the geological materials are formed, the expected autocovariance at any point in vertical direction (e.g., across bedding) may be the same or may be different than that in the horizontal direction. This dependency determines whether the me-

dium has a statistically isotropic or a statistically anisotropic correlation structure. Similarly, we can examine the point-to-point variability in the expected values of mean, variance, and autocovariance. When these are constant, the medium is considered to be stationary or statistically homogeneous. This geostatistical concept of homogeneity is useful because it represents the reality of natural hydraulic conductivity distributions within the context of the classical definitions. A distribution in which there is spatial dependence is considered to be statistically heterogeneous. Chapter 10 will include a quantitative approach to characterize the correlation structure of a hydraulic conductivity distribution.

One source of statistical heterogeneity is variability in the expected mean hydraulic conductivity as a function of position in the field. An example of this kind of conductivity field is the Milk River sandstone in Alberta, Canada. The Milk River sandstone has been interpreted as the seaward margin of a littoral environment, some of the sand having been deposited by streams or currents running parallel to the coastline. This depositional environment has resulted in a northwest-southeast linear trend of sand lenses (Meyboom, 1960). The trend observed in Figure 3.5 indicates that the largest mean conductivity values are associated with thick clean sands in the southern and central parts of the formation, with a progressive decrease northward and to the west and east as well in response to greater proportions of fine-grained material

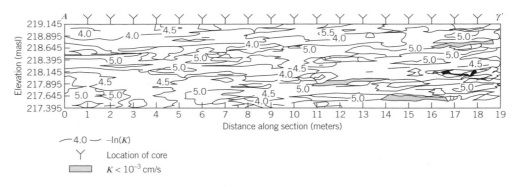

Figure 3.4 Hydraulic conductivity distribution at Canadian Forces Base Borden (from Sudicky, E. A., Water Resources Res., v. 22, p. 2069–2082, 1986. Copyright by American Geophysical Union).

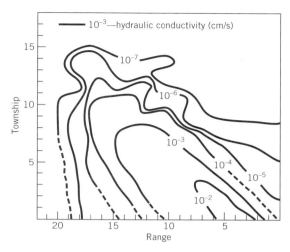

Figure 3.5 Map of the Milk River sandstone showing hydraulic conductivity distribution (from Schwartz and others, 1982). Reprinted from Symposium on Geochemistry of Groundwater, F. W. Schwartz, K. Muehlenbachs, and D. W. Chorley, "Flow system controls on the chemical evolution of groundwater," 1982 with kind permission from Elsevier Science, NL, Sara Burgerhartstraat 25, 1055 KV Amsterdam, The Netherlands.

in the formation. Drift in hydraulic conductivity in sedimentary accumulations is often controlled by the early geologic history as established within the depositional environment.

Research is continuing to develop alternative conceptualizations of permeability distributions. One direction this work is moving in is the application of fractal methods (e.g., Hewitt, 1986; Wheatcraft and Tyler, 1988). Another is the integration of statistical and sedimentological models, where the "alluvial architecture" of a deposit (e.g., as is shown in Figure 2.7) provides yet additional information with which to characterize the hydraulic conductivity distribution (Phillips and others, 1989). Information on the alluvial architecture could come from conventional mapping approaches in the field or through the application of models capable of simulating complex depositional histories (Koltermann and Gorelick, 1992).

Heterogeneity Among Units and the Classification of Aquifers

Hydrogeologists are often concerned with examining hydraulic conductivity distributions at a large scale that could encompass several different stratigraphic units. At this scale, any description of hydraulic conductivity naturally takes into account the organization imposed on the field by the geology. One simple example relates to stratigraphic layering where the statistical description of the field is based on individual geologic units that commonly have very different hydraulic conductivities.

This model of layered heterogeneity is historically im-

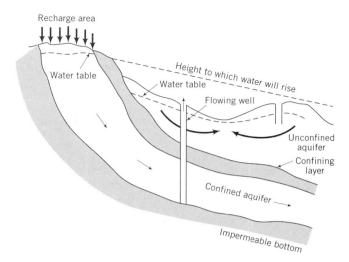

Figure 3.6 Schematic cross section illustrating the difference between a confined and an unconfined aquifer.

portant in categorizing the occurrence of ground water in the hydrologic cycle. If a homogeneous rock of high permeability exists continuously from land surface to some great depth, the water in this aquifer will occur exclusively under unconfined conditions. An unconfined aquifer is one in which the water table forms an upper boundary and is at atmospheric pressure (Figure 3.6). In some instances, where low-permeability materials are interbedded with higher-permeability units, downward percolating water in the unsaturated zone may become "perched" on the low-permeability units. Thus, a localized zone of saturation could form above the low-permeability unit. This condition is referred to as a "perched water table," which is a local zone of saturation completely surrounded by unsaturated conditions (Figure 3.7).

Given the layered heterogeneity demonstrated in Figure 3.6, a confined aquifer can occur as a high-permeability unit between two low-permeability units, or aquitards. The aquitards in this case are referred to as confining layers. The confined high-permeability unit frequently contains water under pressure due to the elevation of the intake area in dipping beds. The water level registered in wells tapping such a confined aquifer can be above

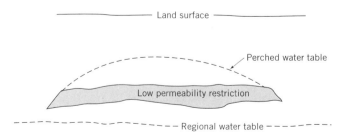

Figure 3.7 Schematic presentation of a perched water table.

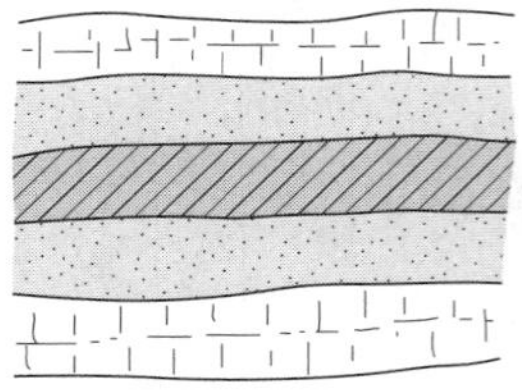

Figure 3.8 Layered heterogeneity.

or below the regional water table in the overlying unconfined aquifer (Figure 3.6). In some cases, the wells may even flow at the surface, in which case they are referred to as artesian wells.

Creating Hydraulic Conductivity Averages

In many applications, it is advantageous to represent the local scale heterogeneity in hydraulic conductivity by some "average" hydraulic conductivity value. One case that has been analyzed extensively is that caused by layering of high- and low-permeability units (Figure 3.8). If the conductivity variation within a unit is much smaller than the conductivity contrasts between beds, as might be the case for interlayered sandstone and shale, it can be assumed that each layer is homogeneous and isotropic. Each layer, however, is characterized by a different hydraulic conductivity, rendering the sequence as a whole heterogeneous.

According to Leonards (1962), an equivalent horizontal hydraulic conductivity in the horizontal or x-direction is

$$K_x = \frac{\sum (m_i K_i)}{\sum m_i} \tag{3.19}$$

where K_x is the equivalent horizontal conductivity, K_i is the homogeneous conductivity of an individual layer, and m_i is the thickness of the layer. For the direction at right angles to the stratification

$$K_z = \frac{\sum m_i}{\sum (m_i/K_i)} \tag{3.20}$$

where K_z is an equivalent vertical conductivity for the layered system.

These equations give the hydraulic conductivity values parallel and vertical to the stratification for a single homogeneous, anisotropic rock that is the equivalent of the layered system shown in Figure 3.8.

Example 3.3 illustrates the use of Eqs. 3.19 and 3.20 in calculating an average hydraulic conductivity for a layered sequence of rocks.

Example 3.3

Consider a 300-m sequence of interbedded sandstone and shale that is 75% sandstone. The sandstone has a horizontal and vertical hydraulic conductivity of 10^{-5} m/s, and the shale has a horizontal and vertical hydraulic conductivity of 1.92×10^{-12} m/s. From equation 3.19

$$K_x = \frac{225\ \text{m} \cdot 1 \times 10^{-5}\ \text{m/s} + 75\ \text{m} \cdot 1.92 \times 10^{-12}\ \text{m/s}}{300\ \text{m}}$$

$$K_x = 7.5 \times 10^{-6}\ \text{m/s}$$

From Eq. 3.26

$$K_z = \frac{300\ \text{m}}{\dfrac{225\ \text{m}}{1 \times 10^{-5}\ \text{m/s}} + \dfrac{75\ \text{m}}{1.92 \times 10^{-12}\ \text{m/s}}}$$

$$K_z = 7.7 \times 10^{-12}\ \text{m/s}$$

Thus, for horizontal flow the most permeable units dominate the system; for vertical flow the least permeable units dominate the system. Under the same hydraulic gradient, horizontal flow is on the order of six orders of magnitude faster than vertical flow. For this example, horizontal flow through 300 m of sediments with a uniform hydraulic conductivity of 7.5×10^{-6} m/s would be the same as the sum of the flows through each of the individual layers in the example. That is, they are "equivalent."

For more complex media, probability theory provides a straightforward approach to averaging. For example, Matheron (1967) shows that for uniform flow (i.e., parallel flow lines) in a medium with a statistically isotropic correlation structure the average hydraulic conductivity for a system of any dimension (e.g., 1-D, 2-D, or 3-D) is between the harmonic mean and the arithmetic mean of the set of local values. This result holds irrespective of the spatial correlation. The harmonic mean of a set of numbers is given as

$$H = \frac{N}{\sum \left(\dfrac{1}{X}\right)} \tag{3.21}$$

where H is the harmonic mean, N is the number of values in the sample, and X is the individual values. As an illustration, the harmonic mean of the numbers 2, 4, and 8 is 3.43 whereas the arithmetic mean for the same set is 4.66. Matheron's theoretical result is also demonstrated by measurements of actual distributions (Cardwell and Parsons, 1945; Warren and Price, 1961; and Bennion and Griffiths, 1966).

For some media, the average hydraulic conductivity is a function of the geometric mean, which for a series of numbers is given as

$$G = (X_1 X_2 X_3 \ldots X_N)^{1/N} \tag{3.22}$$

which is the Nth root of a product of N numbers. For our simple example, the geometric mean of 2, 4, and 8 is 4. Hence, the geometric mean is intermediate between the harmonic and arithmetic means. It turns out for log-normally distributed hydraulic conductivity data and two-dimensional uniform flow, the average hydraulic conductivity is exactly equal to the geometric mean. Similar scaling laws for uniform flow in one- and three-dimensional systems with a statistically isotropic correlation structure are given by Gutjahr and others (1978). Dagan (1989) develops more general relationships for statistically anisotropic media similar to that depicted in Figure 3.4.

All of these scaling laws for heterogeneous fields require that flow be uniform. If flow lines are not parallel (e.g., converging flow to a well), one cannot derive an average hydraulic conductivity from a collection of local measurements.

Darcy's Law for Anisotropic Material

In most of the discussions of Darcy's law, it has been tacitly assumed that conductivity was independent of direction. For this assumption, Darcy's law was expressed as

$$\mathbf{q} = -K \operatorname{grad} h \tag{3.23}$$

which is correct only when $K_x = K_y = K_z$, that is, the material is isotropic. For anisotropic material, the following forms of Darcy's law were given as

$$q_x = -K_x \frac{\partial h}{\partial x} \tag{3.24a}$$

$$q_y = -K_y \frac{\partial h}{\partial y} \tag{3.24b}$$

$$q_z = -K_z \frac{\partial h}{\partial z} \tag{3.24c}$$

which describes the flow of fluid along the x, y, and z axes where the material properties are different, that is, anisotropic. However, Eqs. 3.24 are simplifications of yet a more complex form of Darcy's law for anisotropic material. This more complex form can be examined by considering the fact that the velocity in each of the expressions of Eqs. 3.24 is a vector and can be resolved into components parallel to the x, y, and z axes. Consider that these x, y, and z axes form the edges of a small cube through which the fluid is flowing (Figure 3.9). We may now condense and express the information in Figure 3.9 in terms of q_{ij}, where the first subscript indicates the direction perpendicular to the plane upon which the velocity vector acts and the second subscript indicates the direction of the velocity vector in that plane. That is to say, q_x given in Eq. 3.24a describes the flow of fluid along the x axis. Thus, the components of the velocity vector q_x on the plane of the cube normal to the x axes become q_{xx}, q_{xy}, and q_{xz}. Here q_{xz} is a normal velocity component in that it acts along the normal to the desig-

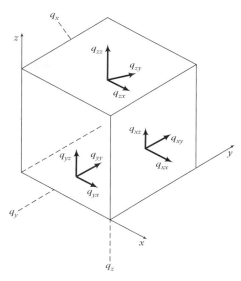

Figure 3.9 Array of nine components of three velocity vectors acting at a point on the three faces of a cube.

nated plane, whereas q_{xy} and q_{xz} act tangential to the plane in question. Thus, Eq. 3.24a can be expressed

fluid flow in the plane normal to $x = q_x = -K_{xx}\frac{\partial h}{\partial x} - K_{xy}\frac{\partial h}{\partial y} - K_{xz}\frac{\partial h}{\partial z}$

in the direction of x axis y axis z axis

(3.25)

Thus q_x depends not only on gradients in x, but on gradients in y and z as well.

In a similar fashion, the velocity vector q_y of Eq. 3.24 can be resolved into three components, all of which act on a plane normal to the y axis (Figure 3.9). One of these will be a normal velocity component (q_{yy}) in that it acts in the direction normal to the designated plane whereas the others will be tangential. The same line of reasoning is followed for the resolution of q_z into three components. Hence, Eqs. 3.24 can be expressed as

fluid flow

in the plane normal to $x = q_x$

$$= -K_{xx}\frac{\partial h}{\partial x} - K_{xy}\frac{\partial h}{\partial y} - K_{xz}\frac{\partial h}{\partial z}$$

in the plane normal to $y = q_y$

$$= -K_{yx}\frac{\partial h}{\partial x} - K_{yy}\frac{\partial h}{\partial y} - K_{yz}\frac{\partial h}{\partial z}$$

in the plane normal to $z = q_z$

$$= -K_{zx}\frac{\partial h}{\partial x} - K_{zy}\frac{\partial h}{\partial y} - K_{zz}\frac{\partial h}{\partial z}$$

in the direction of x axis y axis z axis

In this form it is seen that there are nine components of the hydraulic conductivity in anisotropic material. These components can be placed in matrix form to give what is known as the hydraulic conductivity tensor

$$
\begin{matrix}
K_{xx} & K_{xy} & K_{xz} \\
K_{yx} & K_{yy} & K_{yz} \\
K_{zx} & K_{zy} & K_{zz}
\end{matrix}
$$

This is a second-order symmetric tensor that has the property $K_{ij} = K_{ji}$; that is, $K_{xy} = K_{yx}$, $K_{xz} = K_{zx}$, and so on. For the special case where the principal directions of anisotropy coincide with x, y, and z directions of the coordinate axes, the six components K_{xy}, K_{xz}, K_{yx}, K_{yz}, K_{zx}, and K_{zy} are all equal to zero. In this case, the x, y, and z axes are called the principal axes of the porous medium. The conductivity tensor for the principal axes then becomes

$$
\begin{matrix}
K_{xx} & 0 & 0 \\
0 & K_{yy} & 0 \\
0 & 0 & K_{zz}
\end{matrix}
$$

Stated another way, when the coordinate axes are oriented parallel to the principal axes of the porous medium, we recover the original expressions for Darcy's law for anisotropic material (Eqs. 3.24) when x, y, and z are principal axes and $K_x \neq K_y \neq K_z$. For the very special condition where the material properties do not differ with direction, we obtain the isotropic form of Darcy's law (Eq. 3.23), where the hydraulic conductivity is taken as a scalar quantity.

Measurement of Hydraulic Conductivity

Hydraulic conductivity can be measured in a variety of ways. In practice, only three common methods are employed: field tests, laboratory tests, and empirical or semi-empirical methods based on grain diameter and grain-size distributions, or simple hydraulic models. Field tests are by far the most reliable for they permit the testing of large volumes of rock with one pumping well and one or more observation wells. Pumping-in tests have a considerably smaller area of influence but are important in low-permeability rocks that do not readily yield water to pumping wells. Field testing is an extensive topic and will be covered in Chapter 6.

Laboratory Testing

Permeameters for measuring conductivity are shown in Figure 3.10. In the constant head test, a valve at the base of the sample is opened and the water starts to flow. After a sufficient volume of water is collected over the time of the test, the volumetric flow rate Q is ascertained.

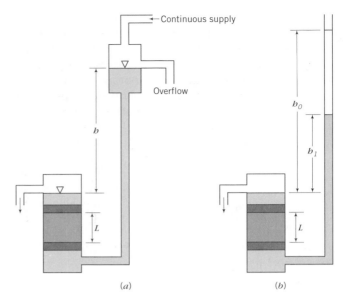

Figure 3.10 (*a*) Constant head and (*b*) falling head permeameters.

Hydraulic conductivity is then determined with Darcy's law of the form

$$
K = \frac{QL}{Ab} \tag{3.26}
$$

where L is the length of the sample, A is the cross-sectional area of the sample, and b is the constant head shown on Figure 3.10*a*.

In the falling head test, the head is measured in the standpipe of Figure 3.10*b*, along with the time of measurement. For a sample of length L and a cross-sectional area A, the conductivity is determined by

$$
K = 2.3 \frac{aL}{A(t_1 - t_0)} \log_{10} \frac{b_0}{b_1} \tag{3.27}
$$

where a is the cross-sectional area of the standpipe and $(t_1 - t_0)$ is the elapsed time required for the head to fall from b_0 to b_1. Some permeameters are designed so the sample can be brought to the original stress condition at the depth it was collected. This is important for deep samples where the measured conductivity changes as a function of the applied stress.

The Search for Empirical Correlations

As hydraulic conductivity can be readily measured in the laboratory, there have been numerous attempts to relate the measured values to various properties of a porous medium. Table 3.5 summarizes some commonly accepted relationships. The first three equations relate hydraulic conductivity or permeability to grain size or grain-size distributions. The remaining two relationships are more hydraulically based.

In general, the permeability of porous rocks appears

Table 3.5 **Examples of Empirical Relationships for Estimating Hydraulic Conductivity or Permeability Values**

Source	Equation	Parameters
Hazen (1911)	$K = Cd_{10}^2$	K = hydraulic conductivity (cm/s)
		C = constant 100 to 150 (cm/s)$^{-1}$ for loose sand
		d_{10} = effective grain size cm (10% particles are finer, 90% coarser)
Harleman et al. (1963)	$k = (6.54 \times 10^{-4})d_{10}^2$	k = permeability (cm^2)
Krumbein and Monk (1943)	$k = 760\, d^2 e^{-1.31\sigma}$	k = permeability (darcys)
		d = geometric mean grain diameter (mm)
		σ = log standard deviation of the size distribution
Kozeny (1927)	$k = Cn^3/S^{*2}$	C = dimensionless constant: 0.5, 0.562, and 0.597 for circular, square, and equilateral triangle pore openings
		k = permeability (L^2)
		n = porosity
		S^* = specific surface–interstitial surface areas of pores per unit bulk volume of the medium
Kozeny-Carmen Bear (1972)	$K = \left(\dfrac{\rho_w g}{\mu}\right)\dfrac{n^3}{(1-n)^2}\left(\dfrac{d_m^2}{180}\right)$	K = hydraulic conductivity
		ρ_w = fluid density
		μ = fluid viscosity
		g = gravitational constant
		d_m = any representative grain size
		n = porosity

to be proportional to some mean grain diameter squared, which reflects the size of a pore, along with the spread or distribution of the grain sizes (pores). Collins (1961) points out that the grain- (pore-) size distribution also influences the magnitude of specific surface, which increases with decreasing grain size. Consequently, pore-size distributions are indirectly incorporated in the Kozeny-type formulations.

Hydraulic arguments have also been applied to fractured rocks. For a parallel array of planar joints of aperture width b, with N joints per unit distance across the rock face, the permeability may be described by (Snow, 1968)

$$k = Cnb^2 = CNbb^2 \qquad (3.28)$$

where k is the permeability with dimensions of length squared, C is a dimensionless constant, in this case (1/12), and the porosity n is taken to be a planar type of porosity equal to Nb. Here, N has the units L^{-1}.

3.3 *Mapping Flow in Geological Systems*

Section 3.1 showed how the spatial variation in hydraulic head determined the hydraulic gradient, and ultimately the direction of ground-water flow. This concept forms the basis of a practical technique for mapping the pattern of ground-water flow in the field. Hydraulic heads are determined by measuring water levels in wells or piezometers installed at different locations and depths. Normally, the piezometers are located within a study area to sample different topographical regimes in the area of interest. Often, at each drilling location, several piezometers are completed to sample hydraulic head at different depths

in the same unit, or in various units (Figure 3.11). Such a collection of piezometers is referred to as a piezometer nest.

Hydraulic head distributions vary in three spatial directions and time. The time element can be removed by assuming all the head measurements are made at the same time (i.e., a "snapshot" of hydraulic head conditions at one point in time). The remaining difficulty in representing a three-dimensional field is overcome by placing piezometers in the flow field where the flow can be interpreted as largely two dimensional. In other words, one can construct two-dimensional planes (e.g., a cross section or a map view) where the flow is essentially two dimensional. However, the resulting cross section or map must be viewed as a two-dimensional projection of a three-dimensional field (Hubbert, 1940). Some projec-

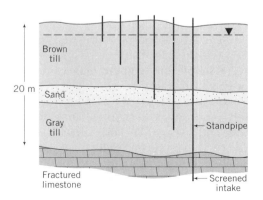

Figure 3.11 Illustration of a piezometer nest including a water-table observation well. Piezometers are emplaced in both high- and low-permeability units.

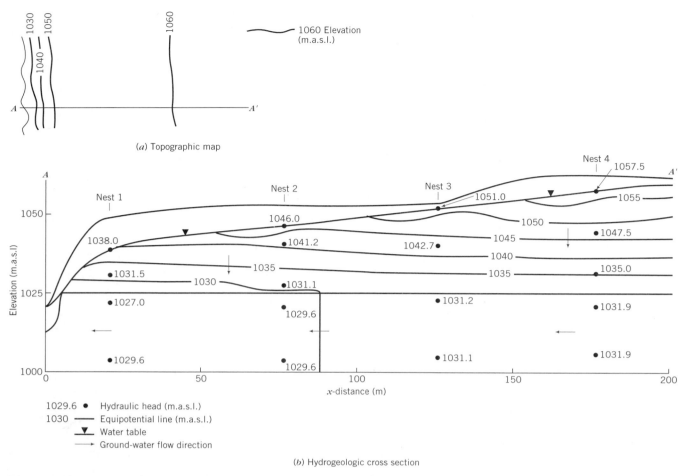

(a) Topographic map

(b) Hydrogeologic cross section

Figure 3.12 Panel *(a)* shows the orientation of the cross section in relation to a hill slope. Panel *(b)* is an example of a hydrogeologic cross section describing the pattern of ground-water flow.

tions will contain most of the variations in hydraulic head that have to be addressed.

Hydrogeological Cross Sections

Hydrogeological cross sections are vertical sections through a three-dimensional flow region. By aligning the section parallel to the direction of mean ground-water flow, flow conditions can be represented accurately in terms of a two-dimensional cross section. Figure 3.12*a* illustrates a simple topographic map encompassing an upland area and an adjacent valley. The hydrogeological cross section *A-A'* is located parallel to the direction of mean flow. Normally, the section includes basic information about the stratigraphy and variation in hydraulic conductivity, as well as hydraulic head data from nests of piezometers located along the section (Figure 3.12*b*). A convenient way of presenting these data is to represent the measurement point (i.e., intake of the piezometer or elevation of the water table) by a dot and noting the measured value of hydraulic head. Contouring these head

data defines the spatial variation in hydraulic head. For this simple example, ground-water flow as represented by the arrows is normal to the equipotential lines.

To maximize the quantity of information along a given hydrogeological cross section, the choice is often made in a field investigation to install piezometer nests along lines in the field. To select the orientation of this line before installing the piezometers, one can examine other existing hydraulic head data or, in the absence of such data, infer that the direction of ground-water flow follows the regional gradient in the surface topography (e.g., Figure 3.12*a*).

Let us return to the hydrogeological cross section in Figure 3.12*b* because it helps to illustrate yet another two-dimensional plane that can be constructed through a complex hydraulic head field. In the more permeable lower unit, the equipotential lines are vertical, and flow in the aquifer is essentially horizontal. Within this unit, there is an absence of vertical gradients, which suggests that the gradient in hydraulic head varies only in the plane of an aquifer (here *x-y* space).

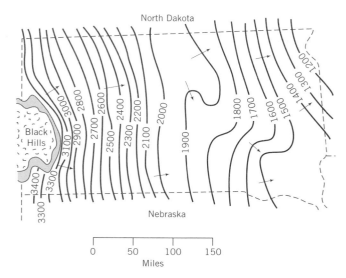

North Dakota

Black Hills

Nebraska

0   50   100   150
Miles

Figure 3.13 Potentiometric surface of the Dakota Sandstone, contour interval 100 ft (from Darton, 1909).

Potentiometric Surface and Water-Table Maps

As our example suggests, measurements of hydraulic head obtained from water wells or piezometers in a confined aquifer can be contoured on a map. Such a map of hydraulic heads is referred to as a potentiometric surface map, defined by Meinzer (1923) as an imaginary surface that everywhere coincides with the level of water in the aquifer. If the aquifer is unconfined, the contoured surface is referred to as a map of the water table.

Figure 3.13 shows the potentiometric surface for the Dakota Sandstone. This artesian aquifer crops out along the eastern flanks of the Black Hills in South Dakota and dips eastward. Water enters this unit at its elevated intake areas and moves downdip in an easterly direction. The water moves from where the head is high to where the head is low, with each of the lines presumably connecting points of equal head.

In working with these kinds of maps, be aware of these important points. First, a potentiometric map must be related to a single aquifer. Other aquifers deeper or shallower in the section will have different potentiometric surfaces that may exhibit heads that are higher or lower than the one of immediate concern. For example, the Madison Formation underlies the Dakota Sandstone throughout much of South Dakota and has its own potentiometric surface showing more or less the same slope and same direction of flow, but higher heads than those encountered in the Dakota. Second, it is assumed that flow in the aquifer is horizontal, that is, parallel to upper and lower confining layers. If a piezometer is placed in such an aquifer, and a hydraulic head is noted, the hydraulic head is presumed not to change with increasing depth

of piezometer penetration. Hence the potentiometric surface is in reality a projection of vertical equipotential lines into the horizontal plane. Lastly, head losses between adjacent pairs of equipotential lines are equal, and the hydraulic gradient varies inversely with distance between lines of equal head. This statement is simply a necessary convention in the construction of potentiometric maps where the contour interval is constant, for example, 100 ft in Figure 3.13. This accomplished, the second statement follows automatically. From Figure 3.13, the contours are closely spaced and the gradient is steep near the Black Hills. Farther east, the contours are farther spaced, and the gradient is flattened out somewhat.

Example 3.4 illustrates the calculations one can make from potentiometric surfaces.

Example 3.4

The potentiometric map and intersecting flow lines given below are for a nonhomogeneous but isotropic aquifer. It is assumed further that there are no gains or losses in the total flow so that the volumetric flow rate passing one equipotential line must equal the flow rate passing an adjacent equipotential line, that is, there can be no gaps in the fluid. This means that

$$Q_1 = K_1 \frac{\Delta b}{L_1} A_1 = Q_2 = K_2 \frac{\Delta b}{L_2} A_2$$

where K_i is the hydraulic conductivity of region i and A_i is the area of flow between adjacent flow lines. Recognizing that $A_1 = A_2$ for a uniformly thick aquifer and $\Delta b_1 = \Delta b_2$ by construction

$$\frac{K_1}{K_2} = \frac{L_1}{L_2}$$

or the ratio of the conductivities equals the ratios of the lengths between equipotential lines. Thus, if the hydraulic conductivity is known in one region, it is known for the other regions.

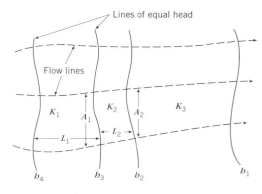

Figure 3.E4

Closing Statements

There is a theoretical point to be made in closing this section. To help this point along, let us briefly review a few statements from the preceding pages. First, the term *field* is used whenever some quantity can be specified at every point in space and time. A scalar field was defined as a property that takes on a value (number) at any point within a specific region of space and time. Clearly, the hydrogeological cross sections discussed above (Figure 3.12b) as well as the potentiometric surface (Figure 3.13) satisfies this description, with the scalar property being hydraulic head. As mentioned previously, the time element has been removed by assuming that all the head measurements are made at the same time. Further, the gradient was introduced as consisting of components in x, y, and z, each representing how fast the head changes in that respective direction. It was also stated that for the condition $K_x = K_y = K_z$, the gradient of the head is a vector perpendicular to the lines of equal head. Thus, as the flow is presumed to be at right angles to the lines of equal head, the gradient of the head, or grad h, is colinear with the flow. The gradient of the head is thus seen to be the force that drives the flow, and the direction of this driving force and the flow is at right angles to the lines of equal head. The velocity of this flow, so directed, is given by Darcy's law. The connection between what Darcy observed and calculated at the experimental scale and what can be mapped and calculated at the field scale is thus complete.

The following example shows the variety of hydraulic calculations that can be made on the field scale.

Example 3.5

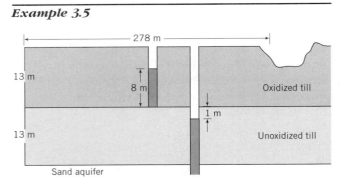

Figure 3.E5

A waste disposal facility is to be constructed in the materials given above. The facility will occupy a region 278 m long (in the direction of ground-water flow) by 200 m wide. The trenches will extend about 13 m into the oxidized till, which has a water table about 5 m below land surface, and a measured hydraulic conductivity of 10^{-7} m/s. The underlying unoxidized till has a hydraulic conductivity of 10^{-8} m/s. A water-table map has been

prepared for the oxidized till layer and demonstrates an average gradient of about 0.072 across the site. The materials underlying the small stream on the right-hand side of the diagram are to be excavated to the top of the unoxidized zone. This trench will be backfilled with gravel with some sort of collection system. Assume an effective porosity of 0.1 for the entire till layer.

1. Horizontal velocity in oxidized till layer
$$v = \frac{K}{n_e}\left(\frac{\Delta h}{\Delta x}\right) = \left(\frac{10^{-7}}{0.1}\,\text{m/s}\right)7.2 \times 10^{-2}$$
$$= 7.2 \times 10^{-8}\,\text{m/s} = 2.3\,\text{m/yr}$$

2. Volumetric flow into trench
$$Q = KiA = (10^{-7}\,\text{m/s})(7.2 \times 10^{-2})(4572\,\text{m}^2)$$
$$= 3.3 \times 10^{-5}\,\text{m}^3/\text{s}$$

3. Vertical velocity through unoxidized till layer
$$v = \frac{K_z}{n_e}\left(\frac{\Delta h}{\Delta z}\right) = \left(\frac{10^{-8}\,\text{m/s}}{0.1}\right)(1)$$
$$= 10^{-7}\,\text{m/s} = 3.15\,\text{m/yr}$$

4. Travel time for waste to reach sand aquifer
$$\text{time } t = L/v = 13\,\text{m}/3.15\,\text{m/yr} = 4\,\text{yr}$$

3.4 Flow in Fractured Rocks

In fractured rocks, the interconnected discontinuities are considered to be the main passages for fluid flow, with the solid rock blocks considered to be impermeable. Thus, on the scale of the field problem, one of two approaches might be followed when dealing with the flow of fluids in fractured rock: continuum or discontinuum (discrete). The continuum approach assumes that the fractured mass is hydraulically equivalent to a porous medium. The obvious advantage of treating fractured rocks as a continuum is that Darcy's law as developed can be applied so that no new theories are involved. If the conditions for a continuum do not exist, the flow must be described in relation to individual fractures or fracture sets.

Continuum Approach to Fluid Flow

Applying a continuum-based approach to fluid flow in fractures first requires an understanding of some of the basic assumptions implicit in a model for an intergranular porous medium. This understanding is essential in establishing to what extent these conditions can be achieved in some fractured rocks. Let us state at the outset that the conditions for a continuum approach to fluid flow in intergranular porous media are seldom, if ever, challenged. Such is not the case for fractured rocks.

Intergranular Porous Rocks

In Darcy's own words, he set out "to determine the laws of flow of water through sand." And this he did, exclusively at the experimental scale. There are, of course, other scales from which it is possible to study certain phenomena. At the molecular scale, for example, one is interested in the behavior of molecules. This scale of behavior probably never crossed Darcy's mind because it is patently impossible to fully understand the behavior of fluids at this level. Besides, Darcy, like most of the rest of us, was more interested in the manifestations of molecular motion, or things that could be expressed in measurable terms, such as viscosity, density, temperature, concentration, and, of course, velocity.

If we step up a scale, we find ourselves at the pore scale, referred to as the microscopic level. The microscopic scale was clearly not for Darcy because nothing could be observed within the individual pore in 1856. Darcy was interested in a macroscopic law, and his only recourse was to develop it on a macroscopic scale. That is, he experimented with a volume of sand that was large with respect to a single pore but small with respect to the space within which significant variations of the macroscopic properties may be anticipated. That is to say, he sought out the smallest possible sample that exhibited an acceptable level of homogeneity. We call this the macroscopic approach, more frequently referred to as a continuum approach.

In this regard, it is convenient to introduce a macroscopic control volume that is large with respect to an individual pore, but small with respect to the space within which significant variations of the macroscopic properties may be anticipated. Hubbert (1956) emphasized this point with a diagram such as Figure 3.14, which is a plot of porosity versus the volume of the sample in which it is measured at some point. At the microscopic level, the value of porosity varies widely. For example, the sample may be all solid or all pore. As the volume increases, a statistical average smooths out the microscopic variations, and we find there are no longer any variations with the size of the sample. Bear (1972) defines this limit as the representative elementary volume, defined as a volume of sufficient size such that there are no longer any significant statistical variations in the value of a particular property with the size of the element.

As noted in Figure 3.14 for a reasonably homogeneous medium, the property of interest (porosity) is adequately represented by some statistical average with little or no variance at any scale larger than the representative elementary volume. At larger scales in real rocks, a greater number of heterogeneities may be encountered, mostly as a result of the regional stratigraphic framework that includes several facies changes and resulting changes in the material properties. An average value can, of course, be obtained at any scale regardless of the degree of heterogeneity, but the variance about the mean will increase

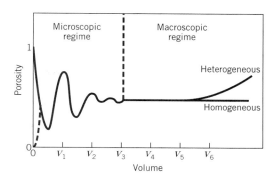

Figure 3.14 Diagram illustrating the representative elementary volume (from Hubbert, 1956). Reprinted with permission of the Amer. Inst. Mining, Met. and Petrol. Engrs.

with the scale of the problem. This larger scale may be referred to as megascopic. It thus follows that in both the microscopic and megascopic regimes, there may be no single value that can be assigned to represent faithfully any one of our material properties. The continuum approach, then, is restricted exclusively to the macroscopic regime where properties are only a function of position, as defined by some appropriate coordinate system, and time, and do not vary with the size of the field.

Fractured Rocks

As with porous intergranular media, a small control volume of fractured rock may be filled entirely with openings or with solid rock. A larger volume will include cracks and solid rock, but the proportion of each will change as different points are sampled within the same small-scale feature. This variation presumably gets smaller as the volume increases so that the final representative volume is achieved. For many fractured rocks, it is likely that as the volume of rock increases beyond some representative volume, the parameters will start to vary again before becoming constant once more. Thus, the representative volume may exist on several scales. This hierarchy of scales is demonstrated in Figure 3.15, where a rock mass may be homogeneous on several scales. At small scales, say, the individual fracture scale of *A*, testing could encounter the intact rock, or the crack openings, with large variations in the measured parameter. As larger-scale sampling is conducted over the entire basalt interior, some representative volume may be achieved in that increasing the volume of the unit tested has no further effect on the value of the average property. This behavior is demonstrated at the *B* scale of Figure 3.15. Including a few permeable flow tops in the testing scheme results in an increase in the average value of the permeability (*C*). Including many flow tops provides for homogeneity at yet a larger scale. If the rock mass was characterized by highly permeable vertical discontinuities, the rock again would exhibit nonhomogeneity until the volume was increased to include several such discon-

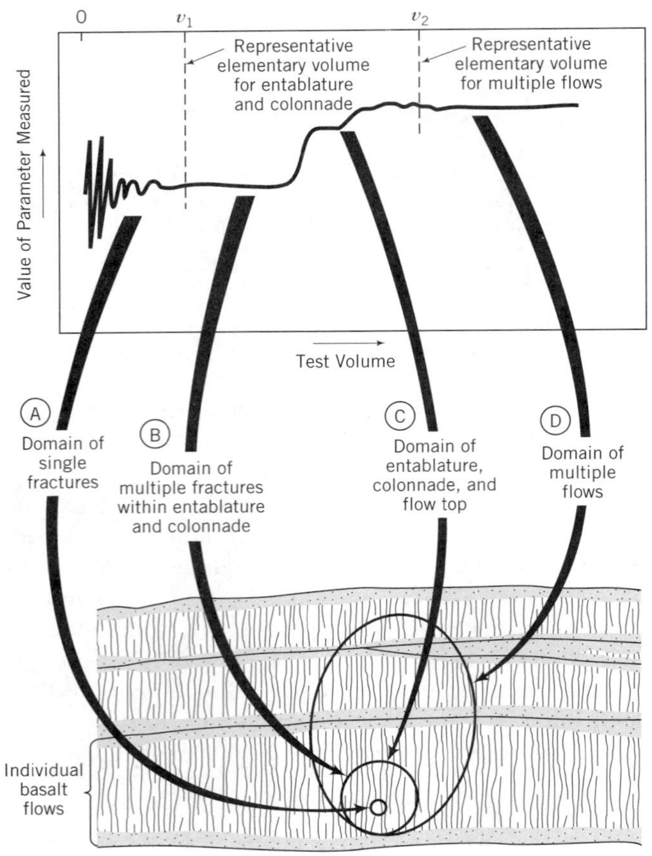

Figure 3.15 Schematic definition of a representative elementary volume as applied to basalt (from Department of Energy, 1986).

tinuities. The average permeability would thus increase again.

There are a few points to be made here. If one wishes to apply continuum-based models to a large-scale problem, the data employed in that analysis should come from a testing program that is also large scale. This applies to all problems in ground water, not just in fractured rocks. Virtually all the important parameters such as permeability, dispersivity, and even rock compressibility are significantly smaller on the laboratory scale than on the field scale. Hence, representative volumes for both porous and fractured mediums can exist on several scales.

The second point deals with the question of when can a fractured medium be treated within the continuum approach. This particular question has been the subject of continued research. First, it should be emphasized that many models of fractured rock have been employed in certain flow problems without directly demonstrating either the size or the presence of a representative volume. One practical approach is to test the rock properties over the largest practical rock volume. This approach certainly does not guarantee success. Some theoretical papers dealing with determination of the representative volume are given by Endoe (1984) and Long and others (1982).

One last issue to be raised is whether representative elementary volumes even exist for some types of fractured rocks. Some theorists believe that as the scale of testing increases, the scales of heterogeneity expand right along with it. In other words, every block of rock will contain a few significant heterogeneities that are not representative of the fracturing found over the rest of the volume. If ongoing research indeed could support this idea that representative elemental volumes may not exist, support for continuum approaches will diminish.

If the condition for a continuum does not exist, the flow must be described through the individual fractures or fracture sets. This gives rise to several problems. First, to characterize the flow, information is required about the orientation, fracture density, degree of connectivity, aperture opening, and smoothness of fractures. Second, if the apertures are large, the flow may be turbulent rather than laminar, and Darcy's law no longer applies. Third, the values of hydraulic conductivity will vary with changes in the three-dimensional stress field and the fluid pressure.

Equivalent Hydraulic Conductivity An equivalent hydraulic conductivity or permeability may be calculated for a set of planar fractures with equations developed by Snow (1968)

$$K = \frac{\rho_w g N b^3}{12\,\mu} \qquad k = \frac{N b^3}{12} \tag{3.29}$$

where K is hydraulic conductivity, k is permeability, b is the aperture, and N is the number of joints per unit distance across the face of the rock (L^{-1}). For this case, the product Nb is the planar porosity. The expression for hydraulic conductivity given has been plotted for pure water at 20°C by Hoek and Bray (1981), where the equivalent conductivity is taken as parallel to the array (Figure 3.16). As an example, for a spacing of 1 joint per meter and an opening of 0.1 cm, the equivalent hydraulic conductivity is 8.1×10^{-2} cm/s. This means that 1 m^2 of porous material with a hydraulic conductivity of 8.1×10^{-2} cm/s will conduct just as much water as will 1 joint per meter with an opening of 0.1 cm under identical hydraulic gradients.

The Cubic Law The cubic law states that for a given gradient in head, flow through a fracture is proportional to the cube of the fracture aperture. For laminar flow between two smooth parallel plates, the volumetric flow rate can be expressed (Romm, 1966) as

$$Q = \frac{\rho_w g b^2}{12\,\mu}(bw)\frac{\partial h}{\partial L} \tag{3.30}$$

where Q is a volumetric flow rate, ρ_w is the density of water, g is the gravitational acceleration, μ is viscosity, b is the aperture opening, w is the fracture width perpendicular to the flow direction, and $\partial h/\partial L$ is the gradient

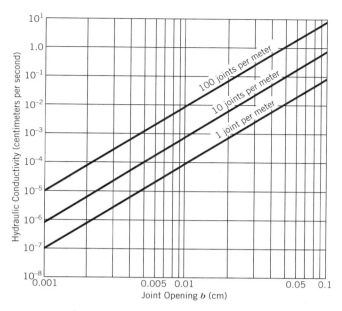

Figure 3.16 Influence of joint opening b and joint spacing on the hydraulic conductivity in the direction of a set of smooth, parallel joints in a rock mass (from Hoek and Bray, 1981). Reprinted with permission of the Institution of Mining and Metallurgy, London.

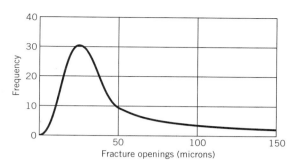

Figure 3.17 Statistical frequency curve of opening width (from Van Golf-Racht, 1982). Reprinted from T. D. Van Golf, 1982. Fundamentals from Fractured Reservoir Engineering, with kind permission from Elsevier Science, NL, Sara Berger-hartstraat 25, 1055 KV Amsterdam, The Netherlands.

in the flow direction. This equation is of the form $Q = KiA$, where i is the gradient and the area A is (bw). Hence, the flow rate is related to the cube of the fracture aperture. The hydraulic conductivity for this parallel plate model is

$$K = \frac{\rho_w g b^2}{12\,\mu} \qquad (3.31)$$

The plates employed in experimental work confirming Eq. 3.30 were smooth optical glass. Gale and others (1985) review the several attempts to incorporate fracture roughness in the experiments. We can think of fracture roughness as the local or point-to-point variability in aperture along a fracture. Roughness, among other things, reduces the aperture openings, and most of the expressions for conductivity in rough channels are of the form

$$K = \frac{\rho_w g b^2}{12\,\mu[1 + C(x)^n]} \qquad (3.32)$$

where C is some constant larger than one, x is a group of variables that describe the roughness, and n is some power greater than one. Hence, roughness causes a decrease in hydraulic conductivity.

It thus follows that the influence of aperture opening is of most importance in the discrete flow of fluids in throughgoing fractures. The frequency of such openings in a given rock, like permeability, appears to be skewed to the right (Figure 3.17). The validity of the cubic law

has been demonstrated in several studies (Huitt, 1956; Gale, 1975; Witherspoon and others, 1980) that conclude that the law is valid where fluid pressure effects are not important.

3.5 Flow in the Unsaturated Zone

The vadose or unsaturated zone occurs between the water table and the ground surface (Figure 1.3). With the exception of parts of the capillary fringe, the pores contain both water and soil gases. The quantity of water in a partially saturated medium can be represented in terms of the volumetric water content (θ), which is defined as

$$\theta = \frac{V_w}{V_T} \qquad (3.33)$$

where V_T is some unit volume of soil or rock, and V_w is the volume of water. If the pore is completely saturated, the volumetric water content is equal to the porosity (n). Therefore, in the unsaturated zone, water contents vary over the range, $0 \leq \theta \leq n$.

Hydraulic and Pressure Heads

In the zone of saturation, the driving force for ground water was demonstrated to be the hydraulic head, defined mathematically as

$$b = z + \psi \qquad (3.34)$$

where z is the elevation head and ψ is the pressure head $(P/\rho_w g)$. Darcy's law describes this flow. These same concepts apply to the movement of water in the unsaturated zone. There are, however, several differences and complications that have to be considered. The most important difference is that pressure heads in the unsaturated zone are less than atmospheric. We see this pattern in Figure 3.18, which illustrates measurements of pressure heads below the water table ($\psi > 0$), at the water

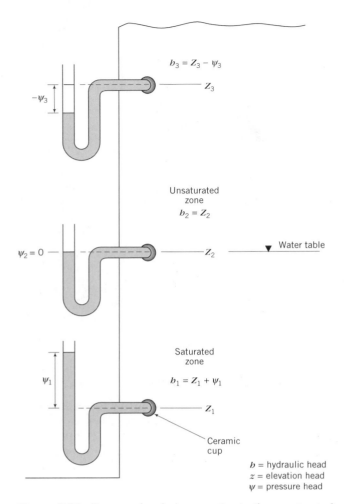

$b_3 = z_3 - \psi_3$

z_3

$-\psi_3$

Unsaturated
zone
$b_2 = z_2$

$\psi_2 = 0$

z_2 ▼ Water table

Saturated
zone
$b_1 = z_1 + \psi_1$

ψ_1

z_1

Ceramic
cup

b = hydraulic head
z = elevation head
ψ = pressure head

Figure 3.18 Pressure heads are negative in the unsaturated zone, zero along the water table, and positive in the saturated zone. As shown, the total hydraulic head is the algebraic sum of the elevation head and the pressure head.

table ($\psi = 0$), and above the water table ($\psi < 0$). Pressure heads less than atmospheric are also referred to as tension heads, or suction heads acknowledging the capillary forces that bind water to solids in the unsaturated zone. It is this "negative" pressure head in the unsaturated zone that accounts for the fact that water present in partially saturated soils cannot flow into a borehole (pressures less than atmospheric are considered negative). That is to say, the pressure in boreholes is atmospheric and the pressure in the surrounding rock is less than atmospheric, and water does not move from low pressure to high pressure.

A device known as a tensiometer is used to measure pressure heads in the unsaturated zone. In its simplest form, it consists of a porous ceramic cup connected by a water column to a manometer. The very fine pores of the cup remain filled with water that provides a hydraulic connection between the soil water and the water column. As the pressure heads change in the soil, water

flows into or out of the tensiometer to maintain hydraulic equilibrium. Tensiometers used in field applications are somewhat more sophisticated, but the operating principles remain the same.

Although the water table usually serves as a major boundary between the disciplines of hydrogeology and soil physics, it is much less significant from a process point of view. The total potential of the system as indicated by hydraulic head is continuous across the water table, and flow always proceeds in the direction of decreasing hydraulic head. The only concession to the negative pressure heads above the water table is that the hydraulic head is determined as the algebraic sum of ψ and z.

This point can be illustrated with the example represented in Figure 3.19a. With a constant and continuous rainfall on the soil surface, a flow of water will occur through the porous medium. By making measurements with tensiometers at a large number of points and contouring these data, one could discover the pressure head

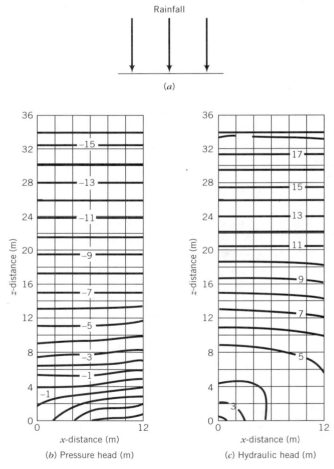

Figure 3.19 Example of saturated–unsaturated flow in a vertical cross section (*a*) under a constant rainfall of 4.0×10^{-8} m/s. Panels (*b*) and (*c*) illustrate the pressure head and hydraulic head fields, respectively (modified from Sullivan and Suen, 1989).

distribution in Figure 3.19*b*. Near the bottom of this field is the water table, which is defined by the $\psi = 0$ contour. As expected, the pressure heads above the water table are negative, and those below are positive. The hydraulic head distribution for this example is shown in Figure 3.19*c*. It is obviously continuous across the entire domain, and it is not really particularly obvious where the water table actually is. The gradient of hydraulic head is generally downward, reflecting the predominantly downward flow of water in the medium.

Water-Retention Curves

Generally, as the water content of a soil decreases, the pressure head becomes more negative or, alternatively, the capillary pressure increases. The capillary pressure increases because the water that remains as the soil becomes drier finds itself in smaller and smaller voids. This is the same behavior we noted in Eq. 1.1, which showed that capillary pressure increased as the diameter of a capillary tube decreased. The relationship between pressure head and water content is determined experimentally and given in Figure 3.20. This so-called water retention curve is nonlinear whether pressure head is plotted on an arithmetic or logarithmic scale. Typically, at both large and small water contents, small changes in water content are accompanied by extremely large changes in pressure head. The behavior at low water content reflects the fact that soils never completely lose all of their water. This lower limit in water content is termed the "residual volumetric water content."

In modeling applications, one commonly fits these experimental data by mathematical relationships. Although many different fitting equations have been derived, one of the most common is that proposed by van Genuchten (1980)

$$\theta = \theta_r + \frac{\theta_s - \theta_r}{(1 + |\alpha\psi|^n)^m} \qquad (3.35)$$

where θ_r and θ_s are the residual and saturated volumetric water contents, respectively, ψ is the pressure head, and α, m, and n are empirical constants determined by nonlinear regression. Usually, it is assumed that $m = 1 - 1/n$.

The van Genuchten equation fits the pressure head-saturation curve in Figure 3.20 with $\theta_r = 0.034$; $\theta_s = 0.391$; $\alpha = 0.0298$; $n = 2.513$. We can check this relationship by calculating the volumetric water content for Figure 3.20 that coincides with a pressure head of -100 cm

$$\theta = 0.034 + \frac{0.391 - 0.034}{(1 + |0.0298 \cdot -100|^{2.513})^{0.603}} \qquad (3.36)$$

$$\theta = 0.098 \approx 0.1$$

A quick check of the figure confirms the fit.

The actual shape of the water-retention curves depends on several factors with pore-size distribution probably being the most important. Figure 3.21 illustrates curves for sand, fine sand, and silt loam. The sand has the most uniform distribution of large pores, while the silt loam with a broad, grain-size distribution contains small pores, which will facilitate much smaller pressure heads. The shape of the water-retention curves changes depending upon whether the soil is drying out or wetting. In other words, the relationship is hysteretic. A detailed discussion of this phenomenon is beyond the scope of

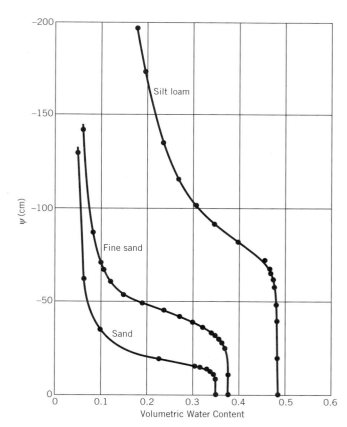

Figure 3.21 Water retention curves for sand, fine sand, and silt loam (from Brooks and Corey, 1966).

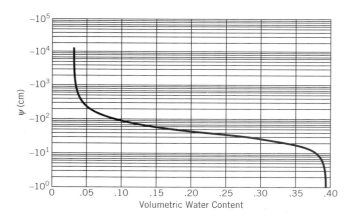

Figure 3.20 Water retention curve for the Berino fine sandy loam (modified from Wierenga and others, 1986).

this introductory chapter. Readers can refer to more advanced textbooks like Greenkorn (1983).

Darcy's Law for Variably Saturated Flow

Yet another complication with the unsaturated zone is that the hydraulic conductivity is strongly dependent in a nonlinear fashion upon the degree of saturation and, consequently, pressure head. When the soil is near saturation with a pressure head close to zero, the hydraulic conductivity takes on its maximum value. As the pores become filled with air, the saturation decreases, the pressure head becomes more negative, and the hydraulic conductivity decreases. This relationship is demonstrated in Figure 3.22. It thus follows that the form of Darcy's law applicable to the unsaturated zone requires that the hydraulic conductivity be expressed either as a function of moisture content or pressure (suction) head. From Hillel (1971)

$$\mathbf{q} = -K(\psi)\, \text{grad}\, H \tag{3.37}$$

where H may contain both suction and gravitational components, or

$$\mathbf{q} = -K(\theta)\, \text{grad}\, H \tag{3.38}$$

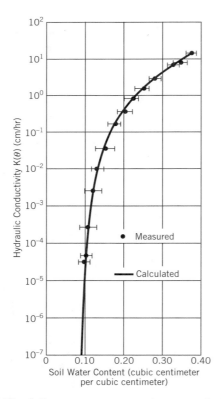

Figure 3.22 Soil water content–suction curve for Lakeland sand and calculated and experimental values of hydraulic conductivity (from Elzeftawy, Mansell, and Selim, 1976. Distribution of water and herbicide in Lakeland Pond during initial stages of infiltration. Soil Science, v. 122, p. 297–307. © by Williams and Wilkins, 1976).

Detailed information on flow in the unsaturated zone can be found in Richards (1931) and Kirkham and Powers (1972).

Unsaturated Flow in Fractured Rocks

The movement of fluid in unsaturated fractured rock is of some interest to hydrogeologists concerned with hazardous and radioactive waste transport. Of immediate concern, for example, is the proposal for storage of high-level nuclear waste in unsaturated fractured rocks in Nevada. Although this is a problem of recent interest, some progress has been made. Of major concern for flow in such rocks is the conceptual model from which the physics of the flow process must be deduced. If the conceptual model is correct, mathematical solutions may capture the essence of the flow. If the conceptual model is incorrect or incomplete, the mathematical models may be misleading.

The presence of surface-connected root channels and worm holes has historically served as an analog for fractures, prompting both theoretical and experimental work. Bevin and Germann (1982) show that these channels conduct water only during rainfall events where the infiltration rate through the soil matrix is less than the precipitation rate at the surface. The water enters the vertical cracks and can laterally infiltrate the matrix at depth. Ponding of water can occur in vertically discontinuous fractures. Davidson (1984, 1985) has modeled such a system, and Bouma and Dekker (1978) have performed experimental work using dyes. Bevin and Germann (1982) provide an excellent review.

When the infiltration rate at the surface exceeds the rainfall rate, the water at the surface is taken in by the pores in the rock matrix. The conceptual model of Wang and Narasimhan (1985) attempts to describe the flow process by a capillary theory that recognizes that large pores (fractures) desaturate first during the drainage process and small pores (matrix) desaturate last. Thus, fractures will tend to remain dry when the aperture is large, with some water films or continuous water contained in the micro aperture portions of the fracture. A dry fracture has a hydraulic conductivity smaller than the conductivity of the partially saturated matrix. The continuous air phase thus produces an infinite resistance to flow across the fracture except for those places that contain water in the micro apertures. The water in micro aperture parts of the fracture constitute pipelines for the transport of water from one matrix block to another. Thus, the flow lines avoid the dry portions of the fracture (Figure 3.23), which act as barriers to flow both along and normal to the fracture.

Peters and Klavetter (1988) recognize the relationships among vertical flux, hydraulic conductivity, and matrix-fracture flow. If the vertical flux in an unsaturated fractured rock is less than the saturated hydraulic conductivity of the matrix, the flow will tend to remain in the

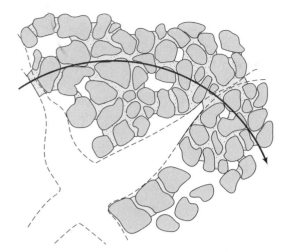

Figure 3.23 Conceptual model of partially saturated, fractured, porous medium showing schematically the flow lines moving around the dry portions of the fractures (from Wang and Narasimhan, Water Resources Res., v. 21, p. 1861–1874, 1985. Copyright by Amer. Geophys. Union).

matrix. For this case, Wang and Narasimhan (1985) suggest that the steady-state flow field of a partially saturated fractured rock can be understood without detailed knowledge of the discrete fracture network. If the flux is greater than the saturated hydraulic conductivity of the matrix, the matrix will saturate and fractures or open fault zones will then accept the flow. Although this statement appears reasonable from the theory discussed thus far, field evidence suggests that fracture flow will commence at some critical matrix saturation less than full saturation (Rasmussen and others, 1989).

Yet another conceptual model of fracture flow in arid zones has been presented by Nativ and others (1995). This model is based on field observations at an industrial complex site in Israel, where a variety of chemicals apparently bypassed the low-permeability matrix, contaminating the lower aquifer via rapid fracture flow. The authors proposed a conceptual model in which a small portion of the rainwater percolates downward through the matrix while a larger percentage of the percolating water moves through the preferential fracture pathways. In arid regions, during periods of zero or low precipitation, the fractures drain rapidly due to gravity and/or imbibition into the rock matrix. Most of the matrix remains at near-saturation because the time it takes to drain the low-permeability pores is large compared to the frequency of recharge events from storms or, in this case, industrial effluent. Thus, to a large degree most of the matrix water moves slowly and appears not to play an active role in short-time scale hydrology, short time again referring to significant precipitation events. The larger the storm event or sequence of events, the more water will enter the fracture and penetrate to greater depths. During small precipitation events, significant fracture flow does not occur, and infiltration takes place slowly through the matrix. This conceptual model is more in tune with observations made in tunnels and underground mines following precipitation events.

Problems

1. Calculate the specific discharge in m/s for a hydraulic conductivity of 10^{-6} m/s and a hydraulic gradient of 0.019.

2. Calculate the actual velocity in m/s from the information given above for an effective porosity of 0.1 and 0.0001.

3. What is one likely reason for the large differences in effective porosity cited above?

4. Prove by dimensional analysis, using Table 3.1, that pressure head has the units of length. For pressure use F/L^2, for acceleration of gravity use L/T^2, and for density use M/L^3.

5. Prove, by dimensional analysis, that hydraulic head has the units of energy per unit fluid weight (L) and hydraulic potential has the units of energy per unit fluid mass (L^2/T^2).

6. Determine both the direction of flow and the gradient of flow from the following three points as measured in the field.

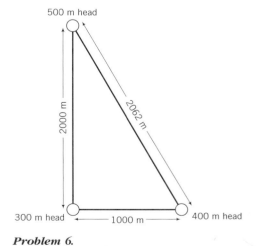

Problem 6.

7. In the following diagram, assume a hydraulic gradient of 0.019 with the water level at point *A* being at an elevation of 300 m above sea level. Assume further that the water level in all the piezometers is at the top of the

---

piezometers. The piezometers are located 5000 m apart. Calculate the following:

Piezometer	A	B	C	D
Depth of piezometer, m	200	190	50	8
Total head, m	300	—	—	—
Pressure head, m	—	—	—	—
Elevation head, m	—	—	—	—

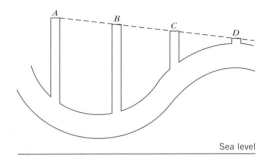

Problem 7.

8. A geologic formation has a hydraulic conductivity of 10^{-6} m/s.

a. What is the conductivity in feet per second?
b. What is the conductivity in gallons per day per ft²?
c. What is the permeability in square centimeters?
d. What is the permeability in darcys?

9. Draw or demonstrate the conditions under which Darcy's law can be expressed as $q_x = (k/\mu)(\partial P/\partial x)$, where k is the permeability, μ is viscosity, and P is the fluid pressure as measured at the bottom of a borehole.

10. Which of the following statements for Darcy's law are incorrect (if any)?

a. $Q = KiA$
b. $\mathbf{q} = K$ grad h
c. $\mathbf{q} = (K/g)$ grad ϕ
d. $\mathbf{q} = \mathbf{v}/n_e$
e. $\mathbf{v} = (K/n_e)$ grad h
f. $\mathbf{v} = (k\rho g/\mu n_e)$ grad h
g. $\mathbf{q} = (k\rho g/\mu)$ grad h

11. Three horizontal, homogeneous, and isotropic formations overlie one another, each of which is 20 m thick, with horizontal conductivities of 10^{-6}, 10^{-7}, and 10^{-8} m/s. Compute the horizontal and vertical components of hydraulic conductivity for an equivalent homogeneous, anisotropic formation.

12. A sample is subjected to a hydraulic gradient of $\partial h/\partial x = 0.1$ cm/cm in the x direction and a flux $\mathbf{q}_y$ of 0.001 cm/s is measured in the y direction in response to this gradient (there is no gradient in the y direction). The same sample is then subjected to a hydraulic gradient $\partial h/\partial y$ of 0.01 cm/cm in the y direction and a flux $\mathbf{q}_x$ is measured in the x direction in response to this hydraulic gradient (there is no gradient in the x direction). What is the magnitude of the flux $\mathbf{q}_x$ in response to $\partial h/\partial y$?

13. Consider the diagram in Example 3.4. Assume that no flow is added to or lost from the system, that is, inflow = outflow.

a. Which area has the highest hydraulic conductivity?

b. Which area has the lowest hydraulic conductivity?
c. Approximately how many times higher is the conductivity for the region identified in (*a*) than for the region identified in (*b*)?

14. Consider the potentiometric surface for a homogeneous, isotropic aquifer that, for inflow equal to outflow, would be characterized by a uniform spacing of the equipotential lines across the flow field.

a. Describe the equipotential spacing in the direction of flow if the area was subjected to uniform recharge across the flow field (*hint:* inflow not equal to outflow).
b. Describe the equipotential spacing in the direction of flow if the area was subjected to uniform natural discharge across the flow field.

15. The aquifer in Example 3.4 is underlain by a uniformly thick homogeneous clay layer. Beneath this clay layer is another aquifer. The head in this aquifer near b_4 (which is assumed to be 70 m) is 60 m, or 10 m lower than the measured value for b_4. In the vicinity of b_1, which is assumed to have a measured head of 30 m, the head in the lower aquifer is on the order of 25 m, or 5 m lower than the measured 30 m equipotential. In which of these two regions is the velocity of vertical movement across the clay layer greatest and why? Which way is the flow directed, upward or downward?

16. Calculate the specific discharge across the clay layer in cm/s where the vertical hydraulic conductivity of the clay layer is 10^{-7} cm/s. In which direction is the flow?

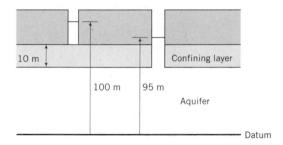

Problem 16.

17. Consider an upper confined aquifer overlying a lower aquifer, separated by a low-permeability confining

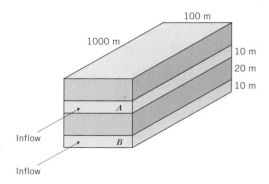

Problem 17.

layer. Other than the information in the diagram, you are given the following:

    (1) The hydraulic conductivity of both the upper and lower aquifers is on the order of 10^{-6} m/s.

    (2) The vertical conductivity of the low-permeability confining layer is on the order of 10^{-9} m/s.

    (3) The heads in the lower aquifer average about 10 m higher than those in the upper aquifer.

    (4) The hydraulic gradient in the near vicinity of the inflow areas of both aquifers is on the order of 10^{-1} m/m.

a. Calculate the inflow Q (m³/s) into both aquifers.

b. Calculate the outflow Q (m³/s) out of both aquifers.

c. Calculate the hydraulic gradient in the vicinity of the outflow areas of both aquifers.

d. Sketch the general form for the piezometric surfaces of the upper and lower aquifers.

Chapter 4

Main Equations of Flow, Boundary Conditions, and Flow Nets

From a field or laboratory perspective, we have pursued the concepts of fluid movement in porous rocks about as far as they can be carried. We have some idea about the range in magnitude of porosity and permeability in geologic materials, the meaning and measurement of hydraulic head, and the determination of flow rates and direction of ground-water movement. Thus if we can map the hydraulic head, we can perform many useful calculations. To go further than this requires new information. This chapter is intended to provide this new information, and the next three chapters will put it into practice.

4.1   Organizing the Study of Ground-Water Flow Equations

What sets hydrogeology apart from many of the other geosciences is the emphasis on treating problems quantitatively. These problems might include predicting how much hydraulic head at a point will decline by pumping a nearby well for some specified time, or how the concentration of a contaminant at a point will change in response to some proposed remedial scheme. The quantitative approaches also provide the basis for interpreting the results of various tests (e.g., slug tests; aquifer tests) that hydrogeologists routinely undertake in the field.

The essence of the quantitative approach involves representing information about a problem in mathematical terms and eventually solving the resulting equation. Let us illustrate this idea with the simple ground-water flow problem shown in Figure 4.1a. The question we would like to solve is given the pattern of geological layering (and the resulting hydraulic conductivity distribution) and the time invariant configuration of the water table, what does the pattern of ground-water flow look like? The conceptualization of the problem mathematically (Figure 4.1b) requires (1) finding the appropriate equation to describe the flow of ground water, (2) establishing a domain or region where the equation is to be solved, and (3) defining the conditions along the boundary (i.e., boundary conditions). The solution of the governing equation establishes the hydraulic head at specified (x, z) positions within the domain. These hydraulic head values can be contoured to provide equipotential distributions from which we can deduce the patterns of flow (Figure 4.1c).

This simple example is designed to highlight some of the new knowledge that is required for the quantitative treatment of ground-water flow. The starting place is in understanding the form and function of ground-water flow equations and how they are derived. There is a need to learn about boundary conditions and how they

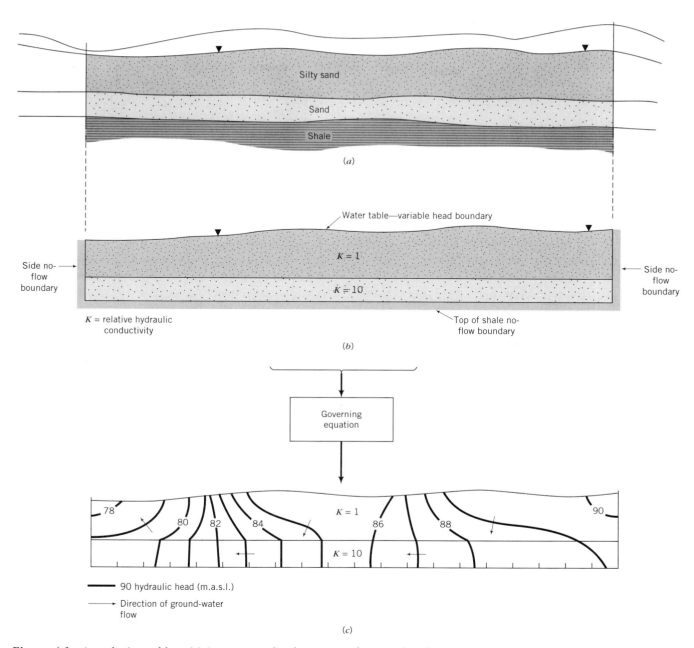

Figure 4.1 A geologic problem (*a*) is conceptualized onto a mathematical problem (*b*), which is ultimately solved in terms of the hydraulic head distribution (*c*).

represent the influence of the world beyond the region of interest on the specific flow problem. The development of equations that apply to flow in aquifers makes it clear that there are important hydraulic parameters that remain to be explained in detail. Finally, we need to make a start on actually solving ground-water flow equations through the use of flow nets.

4.2 *Conservation of Fluid Mass*

The mathematical treatment of ground-water flow through a porous medium depends upon an equation

that captures the essence of the physics of flow. The basis for developing such an equation is a conservation statement that balances the inflow, outflow, and change in water mass within a representative volume of porous medium.

In other words, a conservation of fluid mass statement may be given as

$$\text{mass inflow rate} - \text{mass outflow rate} = \\ \text{change in mass storage with time}$$

in units of mass per time. In general, this statement may be applied to a domain of any size. Consider this state-

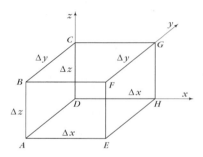

Figure 4.2 Representative volume.

ment as it applies to the small cube of porous material of unit volume where $\Delta x\, \Delta y\, \Delta z$ = unity (Figure 4.2). This box serves as a representative elementary volume. From this figure

mass inflow rate through the face $ABCD = \rho_w q_x\, \Delta y\, \Delta z$

$$(4.1)$$

The density ρ_w has the units M/L^3 whereas the specific discharge q_x is a velocity L/T so that $\rho_w q_x\, \Delta y\, \Delta z$ has the units of mass per time.

In general, the mass outflow rate can be different than the input rate and is given as

$$\begin{matrix} \text{mass outflow rate} \\ \text{through the face } EFGH \end{matrix} = \left[\rho_w q_x + \frac{\partial(\rho_w q_x)\, \Delta x}{\partial x} \right] \Delta y\, \Delta z$$

$$(4.2)$$

The net outflow rate is the difference between the inflow and the outflow or, subtracting Eq. 4.2 from 4.1

$$\text{net outflow rate through } EFGH = -\frac{\partial(\rho_w q_x)\, \Delta x\, \Delta y\, \Delta z}{\partial x}$$

$$(4.3)$$

Making similar calculations for the remainder of the cube

$$\begin{matrix} \text{net outflow rate through} \\ \text{the face } CDHG \end{matrix} = -\frac{\partial(\rho_w q_y)\, \Delta x\, \Delta y\, \Delta z}{\partial y} \quad (4.4a)$$

$$\begin{matrix} \text{net outflow rate through} \\ \text{the face } BCGF \end{matrix} = -\frac{\partial(\rho_w q_z)\, \Delta x\, \Delta y\, \Delta z}{\partial z} \quad (4.4b)$$

Adding these results gives

$$\text{net outflow rate through all the faces } =$$

$$-\left[\frac{\partial(\rho_w q_x)}{\partial x} + \frac{\partial(\rho_w q_y)}{\partial y} + \frac{\partial(\rho_w q_z)}{\partial z} \right] \Delta x\, \Delta y\, \Delta z$$

$$(4.5)$$

in units of mass per time.

The right-hand side of the conservation statement is merely a change in mass storage with respect to time. For the unit volume presented earlier, this becomes

$$\frac{\partial(\rho_w n)}{\partial t} \Delta x\, \Delta y\, \Delta z \quad (4.6)$$

again in units of mass per time. Collecting Eqs. 4.5 and

4.6 and dividing both sides by $\Delta x\, \Delta y\, \Delta z$ gives

$$-\left[\frac{\partial(\rho_w q_x)}{\partial x} + \frac{\partial(\rho_w q_y)}{\partial y} + \frac{\partial(\rho_w q_z)}{\partial z} \right] = \frac{\partial(\rho_w n)}{\partial t} \quad (4.7)$$

which states that the net outflow rate per unit volume equals the time rate of change of fluid mass per unit volume. By making a further assumption that the density of the fluid does not vary spatially, the density term on the left-hand side can be taken out as a constant so that Eq. 4.7 becomes

$$-\left[\frac{\partial q_x}{\partial x} + \frac{\partial q_y}{\partial y} + \frac{\partial q_z}{\partial z} \right] = \frac{1}{\rho_w} \frac{\partial(\rho_w n)}{\partial t} \quad (4.8)$$

With the simple transformation from Eq. 4.7 to Eq. 4.8 our equation now deals with volumes of fluid per unit volume instead of mass per unit volume, where the two are related through $\rho_w = M/V_w$. For the left-hand side of Eq. 4.8, q_x is merely Q_x/A, which expresses a volumetric flow of fluid per unit time per unit area. Thus, the left-hand side of Eq. 4.8 expresses the net fluid outflow rate per unit volume. This being so, the right-hand side of Eq. 4.8 must describe the time rate of change of fluid volume within the unit volume, that is, $\partial(V_w/V_T)/\partial t$. Thus Eq. 4.8 states that the net fluid outflow rate for the unit volume equals the time rate of change of fluid volume within the unit volume.

Main Equations of Flow

As the q's on the left-hand side of Eq. 4.8 represent Darcy's specific discharge, we may directly substitute Darcy's law for anisotropic material, giving

$$\frac{\partial}{\partial x}\left(K_x \frac{\partial h}{\partial x} \right) + \frac{\partial}{\partial y}\left(K_y \frac{\partial h}{\partial y} \right) + \frac{\partial}{\partial z}\left(K_z \frac{\partial h}{\partial z} \right)$$

This expression has a positive sign because the q's are negative. Assuming that the material is isotropic and homogeneous

$$K\left[\frac{\partial}{\partial x}\left(\frac{\partial h}{\partial x} \right) + \frac{\partial}{\partial y}\left(\frac{\partial h}{\partial y} \right) + \frac{\partial}{\partial z}\left(\frac{\partial h}{\partial z} \right) \right]$$

$$= K\left[\frac{\partial^2 h}{\partial x^2} + \frac{\partial^2 h}{\partial y^2} + \frac{\partial^2 h}{\partial z^2} \right]$$

Some clarification is required here. The term $(\partial/\partial x)(\partial h/\partial x)$ represents a space rate of change in the gradient across the unit volume. Accordingly, there must follow velocity variations in the three component directions. If these velocity variations cancel each other, for example, increases in x are compensated by decreases in y, and so forth, the fluid mass per unit volume is not changing with time. For this condition, the right-hand side of Eq. 4.8 is zero, which gives

$$\frac{\partial^2 h}{\partial x^2} + \frac{\partial^2 h}{\partial y^2} + \frac{\partial^2 h}{\partial z^2} = 0 \quad (4.9)$$

Equation 4.9 is Laplace's equation, one of the most useful field equations employed in hydrogeology. The solution to this equation describes the value of the hydraulic head at any point in a three-dimensional flow field.

Yet another possibility concerning the right-hand side of Eq. 4.8 is the case where the unit volume has some storage qualities. We can go no further here unless something can be said about how this storage takes place. Let us assume that the gains or losses in fluid volume within the unit volume are proportional to changes in hydraulic head, which turns out to be a reasonably good assumption. That is to say, an increase in head suggests that water has gone into storage, and a decrease in head suggests just the opposite, that is, a removal of water from storage. To account for these gains and losses, the right-hand side of Eq. 4.8 must be of the form

$$\frac{1}{\rho_w} \frac{\partial(\rho_w n)}{\partial t} = S_s \frac{\partial h}{\partial t} \qquad (4.10)$$

where S_s is some proportionality constant. Our reasoning here is that with fluid flow, the substance of interest is the gains or losses in fluid volume, not hydraulic head, so that the constant S_s is needed to convert head changes to the amount of fluid added to or removed from storage in the unit volume. As Eq. 4.10 describes the time rate of change of fluid volume within the unit volume, S_s as a proportionality constant must be a measure of the volume of water withdrawn from or added to the unit volume when the head changes a unit amount. In this form, nothing can be ascertained about the physics of the proportionality constant S_s, but it obviously must have the units L^{-1} to make the equality of Eq. 4.10 dimensionally correct. We call S_s the specific storage, and we will have more to say about it later.

With this assumption, the conservation statement becomes

$$\frac{\partial^2 h}{\partial x^2} + \frac{\partial^2 h}{\partial y^2} + \frac{\partial^2 h}{\partial z^2} = \frac{S_s}{K} \frac{\partial h}{\partial t} \qquad (4.11)$$

which is called the diffusion equation. The quantity K/S_s is called the hydraulic diffusivity and has the units L^2/T. Equation 4.11 is another important field equation in hydrogeology and is used to describe unsteady or transient flow problems. The solution to this equation describes the value of the hydraulic head at any point in a three-dimensional flow field at any given time or, more precisely, how the head is changing with time.

There are now a few more, perhaps subtle, observations that can be made. For openers, let us compare the left-hand side of Eq. 4.7 with the expression for the gradient

$$\mathbf{i} \frac{\partial h}{\partial x} + \mathbf{j} \frac{\partial h}{\partial y} + \mathbf{k} \frac{\partial h}{\partial z} \qquad (4.12a)$$

$$\frac{\partial(\rho_w q_x)}{\partial x} + \frac{\partial(\rho_w q_y)}{\partial y} + \frac{\partial(\rho_w q_z)}{\partial z} \qquad (4.12b)$$

Both expressions describe the space rate of change of some quantity, in one case the hydraulic head (or a scalar) and in the other case a mass flux, $M/L^2 T$, which is a vector. The abbreviation grad h or ∇h has been used for the gradient. To cover both cases, it is convenient to introduce the notation that is common to both

$$\frac{\partial(\ )}{\partial x} + \frac{\partial(\ )}{\partial y} + \frac{\partial(\ )}{\partial z}$$

which, of course, means nothing until something is put into the parentheses signs. If we put in hydraulic head (or a scalar), this becomes the gradient, which has already been defined as a vector perpendicular to the equipotential lines, and is given the abbreviation grad h or ∇h. If we put in a mass flux (or a vector) this is called a divergence and is given the abbreviation div $(\rho_w q)$ or $\nabla \cdot \rho_w q$ so that we have

$$-\nabla \cdot (\rho_w \mathbf{q}) = -\text{div}(\rho_w \mathbf{q})$$

$$= -\left[\frac{\partial(\rho_w q_x)}{\partial x} + \frac{\partial(\rho_w q_y)}{\partial y} + \frac{\partial(\rho_w q_z)}{\partial z} \right]$$

$$(4.13)$$

which is read "divergence of $\rho_w \mathbf{q}$." Now, why the term *divergence*? Any useful dictionary will define divergence as a difference or deviation. And that is exactly what it is, a difference between the mass inflow rate and the mass outflow rate for the unit volume. Thus, divergence has the physical meaning of net outflow rate per unit volume.

The left-hand side of our conservation equation is, in mathematical terms, a divergence, that is, a net outflow rate per unit volume. With this foresight our main equation could have been immediately stated as

$$-\text{div}(\rho_w \mathbf{q}) = \frac{\partial \rho_w n}{\partial t} \qquad (4.14)$$

which is perfectly equivalent to Eq. 4.7 and which means the net outflow rate per unit volume equals the time rate of change of mass per unit volume. In addition, Eq. 4.8 could have been deduced immediately

$$-\text{div}\,\mathbf{q} = \frac{1}{\rho_w} \frac{\partial \rho_w n}{\partial t} \qquad (4.15)$$

which states that the net fluid outflow rate equals the time rate of change of fluid volume.

So we have a simple notation that not only is efficient in expressing long, sometimes cumbersome equations but contains physical meaning. Now, as long as we are on the topic of operators, there is yet another one that can be used for Laplace's equation (Eq. 4.9) and the left-hand side of the diffusion equation (Eq. 4.11). For these expressions we define

$$\nabla^2(\ ) = \frac{\partial^2(\ )}{\partial x^2} + \frac{\partial^2(\ )}{\partial y^2} + \frac{\partial^2(\ )}{\partial z^2}$$

where ∇^2 is a new operator, called a Laplacian. Again, this operator means nothing until something is placed in the parentheses. Thus, Laplace's equation and the diffusion equation can be neatly and compactly expressed as

$$\nabla^2 h = 0 \qquad (4.16a)$$

$$\nabla^2 h = \frac{S_s}{K} \frac{\partial h}{\partial t} \qquad (4.16b)$$

respectively.

We have covered much ground in this section. We not only have learned about the conservation of mass, but have reformulated this conservation theorem in terms of the important field equations referred to as the diffusion equation and Laplace's equation. We have also introduced a shorthand set of operators that are useful to describe the three-dimensional problem. These mathematical ideas are among the most abstract that we will encounter in hydrogeology. Nevertheless, they can be understood conceptually, and the student is advised to study them well because they will appear time and again.

4.3 The Storage Properties of Porous Media

The specific storage was defined as a proportionality constant relating the volumetric changes in fluid volume per unit volume to the time rate of change in hydraulic head. This concept is easy to grasp for the unconfined condition of flow such as depicted in Figure 4.3a, where drainage release from storage is accompanied by a drop in the water table, or hydraulic head. The amount of

water obtained per unit volume drained is rather substantial and is obviously equal to the volume of pore space actually drained. For the confined case of Figure 4.3b, however, a drop in head is not accompanied by drainage from storage as the aquifer remains fully saturated at all times. Further, the amount of water obtained in response to the unit head drop is a small fraction of that obtained in the unconfined case. An immediate question arises as to how a fully saturated confined aquifer can release water from storage or, equally perplexing, how it can take in additional water in response to head increases. In what manner is space created to accommodate this added water?

A similar set of questions was pondered in 1925 by Meinzer (Meinzer and Hard, 1925) as a result of his study of the Dakota Sandstone. Meinzer concluded that both the water and the porous structure are elastically compressible, and changes in head are accompanied by changes in both water and pore volume. His calculations for the Dakota Sandstone indicated that over a 38-year pumping period, more water was pumped and removed from the aquifer than the total recharge volume over that time span. In addition, this volumetric withdrawal could not be accounted for by the slight expansion of water when the pressure is lowered. Meinzer (1928) concluded that the interstitial pore space of the sandstone was reduced to the extent of the unaccountable volumetric withdrawals from storage. His problem, then, was to demonstrate that aquifers compressed when the fluid pressure was reduced and expanded when the fluid pressure was increased. Several lines of evidence were explored, including the response of water levels in wells to ocean tides and passing trains, and subsidence of the

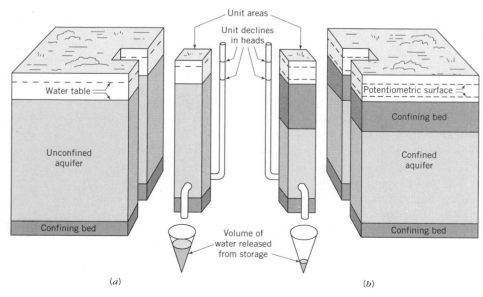

Figure 4.3 Diagrams illustrating the concept of storativity in (*a*) an unconfined aquifer and (*b*) a confined aquifer (from Heath, 1982).

land surface. These investigations established the ground-work for later work by Theis (1940) on the source of water derived from wells and by Jacob (1940), who developed most of the flow equations discussed in this section.

Compressibility of Water and Its Relation to Specific Storage for Confined Aquifers

The isothermal compressibility of water may be defined as

$$\beta_w = \frac{1}{K_w} = -\frac{1}{V_w}\left(\frac{\partial V_w}{\partial P}\right)_{T,M} \quad (4.17)$$

where β_w is the fluid compressibility, with units of pressure^{-1}, K_w is the bulk modulus of compression for the fluid, V_w is the bulk fluid volume, and P is pressure. Here, β_w reflects the bulk fluid volume response to changes in fluid pressure at constant temperature T and mass M. The minus sign is used because the fluid volume decreases as the pressure increases. The isothermal compressibility β_w can be taken as 4.8×10^{-10} m^2/N (2.3×10^{-8} ft^2/lb) for ground water at 25°C.

As mass conservation requirements demand that the product of fluid density and fluid volume remain constant, that is, $M = \rho_w V_w$, a decrease in fluid density in response to a decrease in pressure is accompanied by an increase in fluid volume. That is, as $\rho_w V_w$ equals a constant, $d(\rho_w V_w) = 0$, or

$$\rho_w dV_w + V_w d\rho_w = 0 \quad (4.18)$$

From this equation and Eq. 4.17,

$$dV_w = -\frac{V_w d\rho_w}{\rho_w} \quad (4.19a)$$

$$dV_w = -V_w \beta_w dP \quad (4.19b)$$

Equating the right-hand side of these statements gives

$$\frac{\partial \rho_w}{\partial t} = \beta_w \rho_w \frac{\partial P}{\partial t} = \beta_w \rho_w^2 g \frac{\partial h}{\partial t} \quad (4.20)$$

The last term in this expression is arrived at by recognizing that the pressure head is only one component of total head $h = z + P/\rho_w g$ so that if the elevation head z is invariant with respect to time, $\partial h/\partial t = (1/\rho_w g)\partial P/\partial t$.

Equation 4.20 states that fluid density decreases with decreasing pressure (the water expands) and increases with increasing pressure (the water contracts). Thus, a confined saturated volume will release water from storage when the pressure head is lowered. This idea is incorporated in the concept of specific storage where, from Eq. 4.10

$$\frac{1}{\rho_w}\frac{\partial(\rho_w n)}{\partial t} = \frac{1}{\rho_w}\left[n\frac{\partial \rho_w}{\partial t} + \rho_w\frac{\partial n}{\partial t}\right] = S_s\frac{\partial h}{\partial t} \quad (4.21)$$

Assuming that the matrix is incompressible, that is, the porosity does not change with time, substitution of Eq.

4.20 into this result gives

$$\frac{1}{\rho_w}\left[n\frac{\partial \rho_w}{\partial t}\right] = [\rho_w g\, n\beta_w]\frac{\partial h}{\partial t} \quad (4.22)$$

The bracketed quantity $[\rho_w g\, n\beta_w]$ is the specific storage of a rock with an incompressible matrix, defined as the volume of water released from or taken into a unit volume due exclusively to expansion or contraction of the water when the pressure head changes by a unit amount. For a porosity of 20% and the value of water compressibility cited, this amounts to about 9×10^{-7} m^3 of water released from each cubic meter of saturated sediment when the pressure head declines 1 m. This is not much, but this is the nature of confined aquifers.

Example 4.1

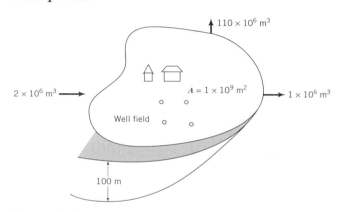

Figure 4.E1

The specific storage due exclusively to expansion or contraction of the water has been determined to be on the order of 9×10^{-7} m^{-1} for a sandstone with a porosity of 20%. How much of the total pumpage of 110×10^6 m^3 has been supplied by expansion of the water? Assume that the total head drop over the area was on the order of 100 m.

According to the definition of specific storage, there are 9×10^{-7} m^3 of water removed per cubic meter of rock when the head is lowered 1 m. The total volume of water removed from the part of the aquifer shown due to water expansion per unit decline in head is readily calculated:

total volume/unit head decline
 $= S_s \times$ area $\times$ formation thickness
 $= 9 \times 10^{-7}$ m$^{-1} \times 1 \times 10^9$ m$^2 \times 1 \times 10^2$ m
 $= 9 \times 10^4$ m^3/m

For a 100-m head drop, the total volume supplied by expansion of the water is

$$9 \times 10^4 \text{ m}^3/\text{m} \times 1 \times 10^2 \text{ m} = 9 \times 10^6 \text{ m}^3$$

This leaves 100×10^6 m^3 unaccounted for, which was the essence of Meinzer's (1928) dilemma.

Compressibility of the Rock Matrix: Effective Stress Concept

For a porous rock to undergo compression, there must be an increase in the grain-to-grain pressures within the matrix; oppositely, for it to expand, there must be a decrease in grain-to-grain pressure. Without such pressure changes, no volumetric changes can occur. A simple analogy of this process has been given in several places (Terzaghi and Peck, 1948), one modification of which incorporates a spring, a watertight piston, and a cylinder.

In Figure 4.4*a*, a spring under a load σ has a characteristic length z. If the spring and piston are placed in a watertight cylinder filled with water below the base of the piston, the spring supports the load σ and the water is under the pressure of its own weight, as demonstrated by the imaginary manometer tube in Figure 4.4*b*. In Figure 4.4*c*, an additional load $\Delta\sigma$ is placed on the system. Because water cannot escape from the cylinder, the spring cannot compress, and the additional load must be borne by the water. This is again demonstrated by the imaginary manometer, which shows the fluid pressure in excess of hydrostatic pressure. The term *excess pressure* is used in the sense that the pressure exceeds the original hydrostatic pressure.

As conservation laws of fluid mass state that no change in fluid pressure can occur except by loss (or gain) of water, a hermetically sealed cylinder will maintain an excess fluid pressure indefinitely. If some of the water is allowed to escape, the water pressure is lowered, and the spring compresses in response to the additional load it must support (Figure 4.4*d*). Hence, there has been a transfer of stress from the fluid to the spring. When the excess pressure is completely dissipated, hydrostatic conditions once more prevail, and the stress transfer to the spring is complete (Figure 4.4*e*).

The analogy to be made is that in any porous water-filled sediment, there are pressures in the solid phase by virtue of the points of contact, the resilient grain structure represented by the spring, and there are pressures due to the contained water, represented by the watertight cylinder. The former are referred to as intergranular pressures, or effective stresses, and the latter as pore-water pressures, or neutral stresses. The total vertical stress

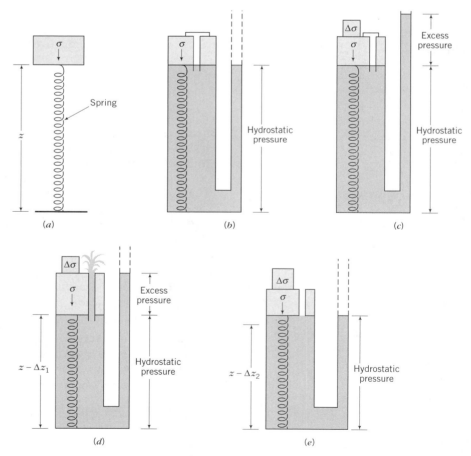

Figure 4.4   Piston and spring analogy showing the transfer of the support for the added load from water pressure to the spring.

acting on a horizontal plane at any depth is resolved into these neutral and effective components

$$\sigma = \bar{\sigma} + P \qquad (4.23)$$

where σ is the total vertical stress, $\bar{\sigma}$ is effective stress, and P is neutral stress. For the analogy cited, $\Delta\sigma$ equals ΔP at the initial instant of loading, and $\Delta\sigma$ equals $\Delta\bar{\sigma}$ at the terminal condition. Intermediate between these extremes, total vertical stress is always in balance with the sum of the effective and neutral stresses. Neutral stresses, whatever their magnitude, act on all sides of the granular particles, but do not cause the particles to press against each other. All measurable effects, such as compression, distortion, and a change in shearing resistance, are due exclusively to changes in effective stress.

A simple demonstration of the effective stress concept is shown in Figure 4.5. This figure shows an observation well in the vicinity of a railroad station. As a train approaches the station, the load of the train is added to the total stress as expressed on the left-hand side of Eq. 4.23. This added stress appears to be carried at least in part by the ground water, as witnessed by the rise in water levels. The water level in this confined aquifer then declines to a new position as the fluid pressure is dissipated by diffusion of the pore fluids to areas of lower pressure. The added weight of the train then appears to be carried entirely by the aquifer skeleton. Hence, there has been a rather rapid stress transfer from the water to the solids with a commensurate decrease in porosity. When the train leaves the area, the effective stress is decreased immediately, which now causes a vertical expansion of the aquifer; that is, the porosity has been recovered. Owing to this sudden increase in pore volume, the water level in the well declines sharply, but shortly returns to its original position.

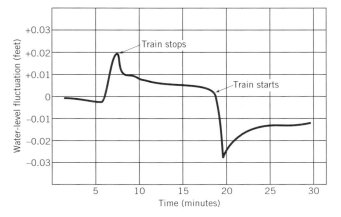

Figure 4.5 Water-level fluctuations in an observation well near a railroad station (from Jacob, Trans. Amer. Geophys. Union, v. 20, p. 666–674, 1939. Copyright by Amer. Geophys. Union).

To grasp the full significance of the effective stress concept, it is important to remember that any increases to the left-hand side of Eq. 4.23 are compensated by corresponding increases on the right-hand side. Witness the train example. On the other hand, a decrease in pressure caused by pumping a confined aquifer has no effect on the total stress. Instead, that part of the load carried by the grain structure must increase in proportion to the decrease in fluid pressure. Expressed mathematically, σ is constant and, from Eq. 4.23,

$$d\bar{\sigma} = -dP \qquad (4.24)$$

the negative sign indicating that a decrease in fluid pressure is accompanied by an increase in intergranular pressure.

Matrix Compressibility and Its Relation to Specific Storage of Confined Aquifers

Earlier in this chapter, a relationship was developed between the compressibility of the fluid and the specific storage, or at least that part of the specific storage dealing with the fluid component. A similar approach may be taken for the component of storage due to pore volume changes. That is, we should like to state that the change in pore volume is proportional to the change in effective stress. If such is the case, effective stress can be expressed in terms of fluid pressure and, ultimately, in terms of total head. Accepting such an effective stress–pore volume relationship requires another proportionality constant, which in this case is an elastic property of the porous medium. This property is a material compressibility of some sort, and it is important to ascertain exactly what compressibility we require. For example, the compressibility of water as defined was expressed for the condition of constant mass, a volumetric dilation of water where the mass remained constant throughout. The definition we seek must be intimately related to the manner in which pore volume changes are related to changes in effective stress.

From a physical perspective, a change in effective stress as defined so far merely means that the individual grains (which are assumed to be incompressible) have been pushed closer together; that is, the pore volume has been reduced due to grain rearrangement. For our purposes here, water must be permitted to drain freely while the sample is being compressed. In addition, as the stress giving rise to this reduction in pore volume is a vertical stress, the compression is assumed to take place only in the vertical direction. These conditions dictate the nature of the proportionality constant relating pore volume changes to changes in effective stress. The compressibility we seek must be described at constant pressure; that is, the water must be permitted to drain freely out of the deforming sample so that fluid mass is lost.

Other constraints include incompressible grains and one-dimensional (vertical) compression. Within these constraints, the vertical reduction of the pore volume is exactly equal to the volume of pore fluid expelled. The required expression is

$$\beta_b = \frac{1}{K_b} = -\frac{1}{V_b}\left(\frac{\partial V_b}{\partial \bar{\sigma}}\right)_{P,T} = -\frac{1}{V_b}\left(\frac{\partial V_p}{\partial \bar{\sigma}}\right)_{P,T} = \beta_p = \frac{1}{H_p}$$

(4.25)

where $\Delta V_b = \Delta V_p$ for the special case of incompressible grains. In this formulation, β_b is the bulk (total) compressibility in units of pressure^{-1}, K_b is a bulk modulus of compression, β_p is the vertical compressibility, H_p is a modulus of vertical compression referring to the pores only, V_b is the bulk volume, V_p is the pore volume, and $\bar{\sigma}$ is effective stress. Here, β_p reflects the pore volume decrease in response to changes in stress at constant temperature and pressure. The negative sign is a convention employed to account for a volume decrease with an increase in stress. Typical values for the vertical compressibility are given in Table 4.1. Note that the compressibility of water is of the same order of magnitude as the compressibility of sound rock.

As noted, the compressibility β_p is perfectly equivalent to bulk (total) compressibility β_b, that is, $\partial V_p/\partial \bar{\sigma} = \partial V_b/\partial \bar{\sigma}$. When the individual grains are compressible, the pore volume change is no longer equal to the bulk volume change.

Following a development similar to that given by Freeze and Cherry (1979), we now present an expression for the specific storage of a rock containing an incompressible fluid. For the case of incompressible grains, the change in total volume dV_b is equal to the change in water volume dV_w. From equation 4.25

$$dV_b = dV_w = -\beta_p V_b d\bar{\sigma} = \beta_p V_b dP$$

(4.26)

where $d\bar{\sigma} = -dP$ (Eq. 4.24). From the expression for total head $h = z + \dfrac{P}{\rho_w g}$, consider the elevation head to

remain constant so that all changes in head are due to changes in pressure head. This means

$$dP = \rho_w g \, dh$$

(4.27)

Equation 4.26 thus becomes

$$dV_w = \beta_p V_b \rho_w g \, dh$$

(4.28)

Dividing both sides by V_b and dh and considering a unit volume ($V_b = 1$) and a unit pressure head decline ($dh = 1$) gives

$$dV_w^* = \rho_w g \beta_p$$

(4.29)

The quantity dV_w^* is an expression for a volume of water removed from a unit volume of rock when the pressure head declines one unit, and has the units of length^{-1}. The quantity on the right-hand side is thus defined as the specific storage of a rock containing an incompressible fluid, or the volume of water released from or taken into storage in a unit volume due exclusively to compression or expansion of the matrix when the pressure head changes a unit amount. For a dense, sandy gravel with a vertical compressibility of 10^{-9} m^2/N (Table 4.1), the component of storage released due to compression of the matrix amounts to about 1×10^{-5} m^3 of water for each cubic meter of sediment when the pressure head declines 1 m. This volume is equivalent to the pore volume decrease. As with the volumes associated with fluid expansion, the amount of water produced is quite small.

From the developments above, and Eq. 4.22, the specific storage S_s is given by

$$S_s = \rho_w g(\beta_p + n\beta_w)$$

(4.30)

The derivations above are sufficient to provide a physical understanding of the process of storage release in elastic media but have been simplified considerably. Actually, it is not possible to obtain a correct expression for the component of storage due to matrix compression from Eq. 4.7 alone in that both the fluid and the solids are in motion in a deforming medium. To conserve the

Table 4.1 **Vertical Compressibility**

Material	Coefficient of Vertical Compressibility		
	ft^2/lb	m^2/N	bars^{-1}
Plastic clay	1×10^{-4}–1.25×10^{-5}	2×10^{-6}–2.6×10^{-7}	2.12×10^{-1}–2.65×10^{-2}
Stiff clay	1.25×10^{-5}–6.25×10^{-6}	2.6×10^{-7}–1.3×10^{-7}	2.65×10^{-2}–1.29×10^{-2}
Medium-hard clay	6.25×10^{-6}–3.3×10^{-6}	1.3×10^{-7}–6.9×10^{-8}	1.29×10^{-2}–7.05×10^{-3}
Loose sand	5×10^{-6}–2.5×10^{-6}	1×10^{-7}–5.2×10^{-8}	1.06×10^{-2}–5.3×10^{-3}
Dense sand	1×10^{-6}–6.25×10^{-7}	2×10^{-8}–1.3×10^{-8}	2.12×10^{-3}–1.32×10^{-3}
Dense, sandy gravel	5×10^{-7}–2.5×10^{-7}	1×10^{-8}–5.2×10^{-9}	1.06×10^{-3}–5.3×10^{-4}
Rock, fissured	3.3×10^{-7}–1.6×10^{-8}	6.9×10^{-10}–3.3×10^{-10}	7.05×10^{-4}–3.24×10^{-5}
Rock, sound	less than 1.6×10^{-8}	less than 3.3×10^{-10}	less than 3.24×10^{-5}
Water at 25°C	2.3×10^{-8}	4.8×10^{-10}	5×10^{-5}

Modified from Domenico and Mifflin (1965), Water Resources Res. 4, p. 563–576. Copyright by Amer. Geophys. Union.

total mass, two conservation statements are required, one for the fluids and one for the solids. This derivation is algebraically messy and has been relegated to appendix 1A.

Example 4.2

Returning to Example 4.1 determine how much of the pumpage is supplied by compression of the matrix for $S_s = 10^{-5}$ m^{-1} if one ignores expansion of the water.

total volume/unit head decline
$$= S_s \times \text{area} \times \text{formation thickness}$$
$$= 10^{-5} \text{ m}^{-1} \times 1 \times 10^9 \text{ m}^2 \times 1 \times 10^2 \text{ m}$$
$$= 10^6 \text{ m}^3/\text{m}$$

For a 100-m drop in head,

$$10^6 \text{ m}^3/\text{m} \times 1 \times 10^2 \text{ m} = 100 \times 10^6 \text{ m}^3$$

Thus the problem of Meinzer (1928) is resolved.

Equation for Confined Flow in an Aquifer

From developments given, the diffusion equation is expressed as

$$\nabla^2 h = \frac{\rho_w g (\beta_p + n\beta_w)}{K} \frac{\partial h}{\partial t} = \frac{S_s}{K} \frac{\partial h}{\partial t} \qquad (4.31)$$

The conditions for steady flow are more readily seen from this expression. When both the fluid and the matrix are incompressible ($\beta_w = \beta_p = 0$), the right-hand side of Eq. 4.31 becomes zero and the steady-state Laplace equation is obtained. In addition, both S_s and K may be multiplied by the formation thickness m, giving

$$\nabla^2 h = \frac{S_s m}{K m} \frac{\partial h}{\partial t} = \frac{S}{T} \frac{\partial h}{\partial t} \qquad (4.32)$$

Here the product of S_s and m is called the storativity S and is dimensionless, whereas the product of the hydraulic conductivity and formation thickness is called transmissivity T, with units of L^2/T. The storativity, or coefficient of storage, is defined as the volume of water an aquifer releases from or takes into storage per unit surface area of aquifer per unit change in the component of pressure head normal to that surface. Stated in another way, the storativity is equal to the volume of water removed from each vertical column of aquifer of height m and unit basal area when the head declines by one unit. Figure 4.6 illustrates the field concept of storativity for confined aquifers.

Transmissivity, or the coefficient of transmissibility, is defined as the rate of flow of water at the prevailing temperature through a vertical strip of aquifer one unit wide, extending the full saturated thickness of the aquifer, under a unit hydraulic gradient. In Figure 4.7, it is the quantity of water flowing through opening B. It may

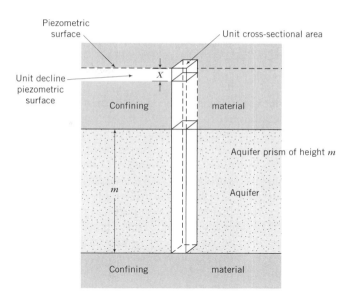

Figure 4.6 Diagram illustrating the storativity for confined conditions (from Ferris and others, 1962).

be useful to note that the storativity is the volume of water removed from a prism of unit basal area extending the full saturated thickness of the aquifer under a unit head decline, whereas transmissivity is the volume of water flowing through one face of that same prism in a unit time under a unit hydraulic gradient. In this sense, both storativity and transmissivity are terms exclusively defined for field conditions in that they relate to the storage and transmissive properties of geologic formations of a specified thickness. From Figure 4.7, hydraulic conductivity is the quantity of water flowing in one unit time through a face of unit area (opening A) under a driving force of one unit of hydraulic head change per

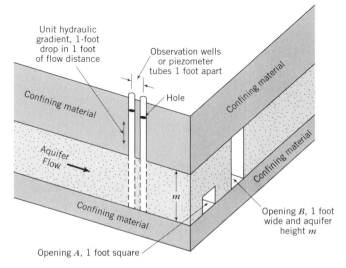

Figure 4.7 Diagram illustrating hydraulic conductivity and transmissivity (from Ferris and others, 1962).

unit length. Hence, transmissivity is the product of the formation thickness and the hydraulic conductivity.

Specific Yield of Aquifers

The discussions thus far have been concerned exclusively with confined aquifers. Upon the lowering of head in such aquifers, they remain fully saturated so that no dewatering occurs, and the water released is volumetrically equivalent to the volumetric expansion of the water and contraction of the pore space. These processes also occur in unconfined aquifers, but the water volumes associated with them are negligibly small compared to the volumes obtained from drainage of the pores. Figure 4.8 shows the concept of storativity associated with the unconfined condition, which is the volume of water drained from the *x* portion of the aquifer. The storativity under unconfined conditions is referred to as the specific yield, defined as the ratio of the volume of water that drains by gravity to the total volume of rock. Specific retention, on the other hand, is the ratio of the volume of water the rock retains against the force of gravity to the total rock volume. These relationships may be expressed

$$S_y = \frac{V_{wd}}{V_T} \tag{4.33a}$$

$$S_r = \frac{V_{wr}}{V_T} \tag{4.33b}$$

where S_y is the specific yield, V_{wd} is the volume of water drained, V_T is the total rock volume, S_r is the specific retention, and V_{wr} is the volume of water retained against the force of gravity. It follows that the sum of V_{wd} and V_{wr} represents the total water volume contained within the interconnected pore space within a unit volume of saturated material. Total porosity is then equal to the sum of specific yield, specific retention, and the ratio of the volume of water contained in the unconnected pore

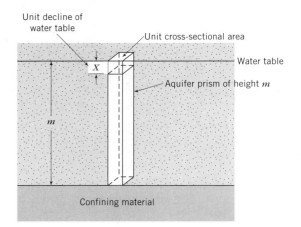

Figure 4.8   Diagram illustrating the storativity (specific yield) for unconfined conditions (from Ferris and others, 1962).

Table 4.2   Values of Specific Yield for Various Geologic Materials

Material	Specific Yield (%)
Gravel, coarse	23
Gravel, medium	24
Gravel, fine	25
Sand, coarse	27
Sand, medium	28
Sand, fine	23
Silt	8
Clay	3
Sandstone, fine-grained	21
Sandstone, medium-grained	27
Limestone	14
Dune sand	38
Loess	18
Peat	44
Schist	26
Siltstone	12
Till, predominantly silt	6
Till, predominantly sand	16
Till, predominantly gravel	16
Tuff	21

From Johnson (1967).

space to the total volume. Johnson (1967) has summarized values of specific yield for a variety of rock types and sediments (Table 4.2). The coarser the material, the more closely the specific yield approaches total porosity.

4.4   *Boundary Conditions and Flow Nets*

Given the differential equations described in the previous section, there now remains one additional problem to be concerned with, namely, their solution or, more specifically, some additional information required in their solution. To appreciate this new information, let us first put down all the essentials of our flow theory:

1.   A potential field is presumed to exist; that is, $h(x, y, z, t)$ is a well-defined scalar quantity in the field of interest.

2.   The potential $h(x, y, z, t)$ changes over space and time. The mapping of this potential for one point in time constitutes the potentiometric map.

3.   Because the potential changes over space, a gradient exists. This gradient has been demonstrated to be a vector perpendicular to the equipotential lines; that is, it is colinear with the flow for an isotropic porous medium.

4.   Provided with this gradient, the water is in continuous motion and is completely described by Darcy's law. With such motion, the fluid must be conserved. Thus we define the divergence, which is a net outflow rate per unit volume. If the net outflow rate is zero

$$\nabla^2 h = 0 \text{ (steady flow)}$$

If the net outflow rate is not zero

$$\nabla^2 h = \frac{S_s}{K}\frac{\partial h}{\partial t} \text{ (unsteady flow)}$$

These statements represent what this chapter has been mostly about. In their present form, the steady and unsteady flow equations constitute conservation principles. By themselves they mean little until they are solved. To this end we add a few more statements.

5. If the flow is steady, given the head or the gradient of head on the total boundary of the region of interest, it is possible to calculate $h(x, y, z)$ for the interior.

6. If the flow is unsteady, given the head or the gradient of the head along the total boundary of the region, along with additional information on the initial head before transients take place, it is possible to calculate $h(x, y, z, t)$ for the interior.

Statements 5 and 6 represent some important new information. To put this information in the proper context, it will be recalled that one way to obtain the head distribution for a region is to map it. We now learn that it can also be calculated if certain information is known along the boundaries of the flow region. Slichter first stated these ideas in a hydrogeologic context in 1899, but the impact of his effort was practically nil until the late 1930s. It was during that 25-year period from the late 1930s to the mid-1960s that hydrogeology was transformed from largely a descriptive practice to a highly quantitative and even predictive one.

Basically, there are two kinds of boundary conditions to deal with. The first, as discussed, is a specified head along a given boundary. An important special case for this boundary condition is where the specified head is constant, termed a constant head boundary. Any line along which the head is constant must be an equipotential line. This means that there can be no flow along the equipotential line, but instead, for isotropic material, the flow must be at right angles to the equipotential line. We conclude that for any region bounded in part by a constant head, the flow must be directed at right angles either away from or toward that boundary, and the boundary is an equipotential line.

A second type of boundary is where the normal component of the gradient is specified. Given the normal component of a gradient along a boundary, there must be a flux across the boundary. Thus, we use the term *flux boundary* or, more commonly, *flow boundary*, where flow either enters or leaves a region across a boundary. A special type of flow boundary is a no-flow boundary, that is, a region across which no flow is permitted to

leave or enter a region. Expressed mathematically

$$\frac{\partial \phi}{\partial N} = 0 \qquad (4.34)$$

where ϕ is the potential and N is the normal to boundary. This means there is no gradient normal to the boundary so that there is obviously no flow across the boundary. If there is no flow across a boundary in a two-dimensional flow region, then the flow must be tangential to the boundary. We conclude that a no-flow boundary must be aligned at right angles to the equipotential lines and must therefore be colinear with the lines of flow. No-flow boundaries are called flow lines, and they intersect the equipotential lines at right angles. Thus, the solution to our main differential equations is simply two intersecting families of curves. In isotropic media, the lines are mutually perpendicular, and one family of curves determines the other. Either may be taken as the solution. Such intersecting families of curves are called flow nets, with the boundaries as well as the interior consisting of flow lines and equipotential lines. For steady flow, the curves do not change over time; for unsteady flow, any given configuration represents an instantaneous condition. Unsteady flow is best viewed as a continuous succession of steady states, each of which is "steady" for short periods of time.

A few examples of flow nets are given in Figure 4.9. Figure 4.9a portrays flow in the x-z plane in a pervious rock unit beneath the base of a dam. This flow is estab-

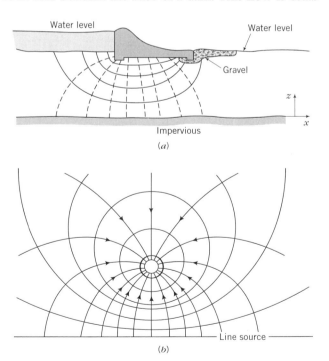

Figure 4.9 Examples of flow nets. Panel (a) represents flow in the x-z plane of a pervious stratum underlying a dam. Panel (b) shows flow toward a discharging well in the x-y plane as influenced by a line source (constant head boundary).

lished by the head of water behind the dam, the base of which can be taken as an equipotential line, that is, a constant head boundary across which the flow is directed downward. The contact between the impermeable and permeable rock unit is another boundary, in this case a no-flow boundary. Hence, this contact is a flow line. The base of the dam also functions as a flow line, whereas the stream below the dam that receives the ground-water discharge functions as an equipotential line. As the flow pattern forms a system of squares or curvilinear squares, the material is assumed to be isotropic and homogeneous. Given this information along with the boundaries, the interior net has been sketched and represents a graphical solution to Laplace's equation.

Figure 4.9b is a set of flow lines and equipotential lines in plan view around a discharging well bordered by an infinite line source. The line source represents a constant head boundary so that the flow lines intersect it at right angles. This flow field is depicted in the x-y plane and represents a mathematical solution to the diffusion equation.

Figure 4.10 is from the classic study by Bennett and Meyer (1952). This flow net is for the Patuxent Formation where the heads are below sea level (thus, when water is moving from the 20-ft equipotential to the 30-ft equipotential, it is moving in the direction of decreasing head). The numerous piezometric lows depicted on the map were caused by extensive pumping. Note that the surface forms a set of squares or curvilinear squares so it is represented as isotropic and homogeneous. However, the hydraulic conductivity for the various pumping regions differs in accordance with the differences in the size of the squares. In this projection, the head distribution was first mapped and the flow lines then superposed. Figure 4.10

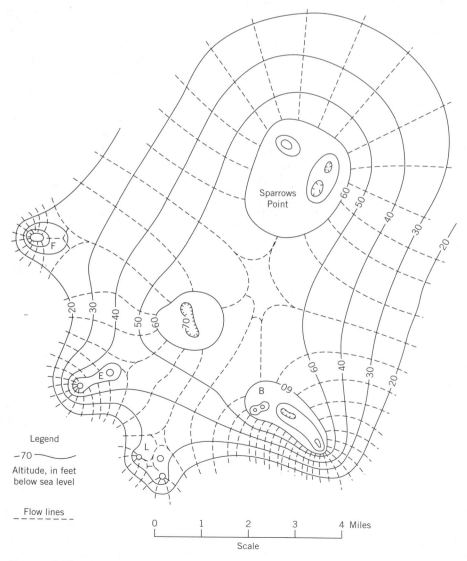

Legend

—70— Altitude, in feet below sea level

- - - - Flow lines

0   1   2   3   4 Miles
Scale

Figure 4.10 Flow net for the Patuxent Formation (from Bennett and Meyer, 1952).

thus represents a field or "mapped" solution to Laplace's equation for whatever boundary conditions are prevalent in this area.

The volumetric flow rate within flow nets can be determined by summing the individual Q's in each of the flow tubes of Figure 4.10. Another common approach is derived from Darcy's law expressed for one section of a flow channel

$$\Delta Q = K \frac{\Delta b}{L} a \qquad (4.35)$$

where ΔQ is the flow in one flow channel, Δb is the head drop across a pair of equipotential lines, L is the distance over which the head drop takes place, and a is the distance between adjacent flow lines. For the homogeneous, isotropic system and its characteristic curvilinear squares, $L = a$. As the total Q in the system is the ΔQ times the number of flow channels (n_f), and the total head drop across the flow system (ΔH) is the Δb times the number of head drops (n_d), the total flow is expressed

$$Q = \frac{n_f}{n_d} K \, \Delta H \qquad (4.36)$$

The flow rate Q is expressed as the volumetric flow rate per unit time per unit thickness of the flow field. For Figure 4.9a this flow must be multiplied by the length of the dam (measured perpendicular to the page) to give the total flow in the previous statement. For Figure 4.10, the Q must be multiplied by the saturated aquifer thickness to give the total flow in the aquifer.

Example 4.3

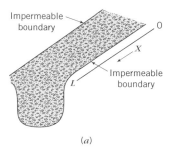

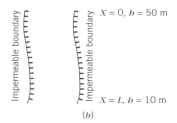

(a)

(b)

Figure 4.E3

Consider the flow in a channel filled with gravel, cut into an impermeable rock. Panel (b) shows a plan view of the aquifer.

1. Given the head at $x = 0$ of 50 m and the head at $x = L$ of 10 m, the equipotential lines can be drawn for a homogeneous aquifer. Thus, at $x = L/2$, $b = 30$ m, at $x = 3L/4$, the head is 20 m, and so on. By superposing flow lines on the equipotential lines, it is noted that the impermeable boundaries are flow lines.

2. For the simple situation given, Laplace's equation is expressed

$$\frac{\partial^2 b}{\partial x^2} = 0$$

which is integrated to give

$$b = A_1 x + A_2$$

where A_1 and A_2 are constants that have to be evaluated for the boundary conditions. Assume $b = b_o$ at $x = 0$ and $b = b_L$ at $x = L$. Thus, from this equation, $A_2 = b_o$ and $A_1 = (b_L - b_o)/L$. The head distribution is thus expressed as

$$b(x) = (b_L - b_o)\frac{x}{L} + b_o$$

Thus, for panel b, $b(x) = b_o$ at $x = 0$, and $b(x) = b_L$ at $x = L$. Given numerical values for the head at these boundaries, the interior may be calculated.

Example 4.4

In Figure 4.10, pumpage from the Sparrows Point District has been measured at 1.0×10^6 ft³ per day. The potentiometric surface is assumed to be steady; that is, the amount of water passing through the formation to supply the pumping wells is equal to the pumping rate. It is noted that there are 15 flow channels associated with the Sparrows Point pumping. For a total head drop ΔH of 30 ft, the number of drops is three so that $\Delta H/n_d = 10$. For a total head drop ΔH of 40 ft, the number of potential drops is 4 so that $\Delta H/n_d$ is again 10. Thus, $\Delta H/n_d$ will always be a constant for a region. The regional transmissivity is determined from Eq. 4.36

$$T = Km = \frac{1.0 \times 10^6 \text{ ft}^3/\text{day}}{(15)(10) \text{ ft}} = 6666 \frac{\text{ft}^2}{\text{day}}$$

$$= 7.16 \times 10^{-3} \frac{\text{m}^2}{\text{s}}$$

Graphical Flow Net Construction

Guidelines for flow net construction have been provided by various authors for over 50 years, with the treatment by Cedergren (1967) being among the most thoughtful. Perhaps no better tool is available for an appreciation of the nature of flow patterns in porous media. The first part of this discussion will focus on the nonhomogeneous but isotropic conditions assumed for most aquifers. It follows immediately that, because of isotropicity, flow lines will intersect equipotential lines at right angles. For every flow net considered, some equipotential lines and some flow lines will be established by the boundary conditions. When the boundary of a region is of the no-flow variety (impermeable), it is represented by a flow line and equipotential lines must intersect it at right angles. When the boundary of a region is of the constant head variety, it is an equipotential line, and flow lines must intersect it at right angles. The same quantity of flow moves between adjacent pairs of flow lines, provided no flow enters or leaves the region in the internal part of the net. It follows further that the number of flow channels must remain constant throughout the net.

The guidelines stated above should all be reasonably apparent from our previous discussions on boundary conditions and flow nets. Some "graphical" guidelines might include selecting a practical scale for the net and a practical number of flow channels so as not to clutter up the pattern.

Not discussed in this section but demonstrated on Figure 4.1 is the refraction of flow lines across the boundary of two formations of different hydraulic conductivity. This will be discussed in detail in Chapter 5.

In anisotropic material, flow net construction is impeded by the fact that flow lines and equipotential lines no longer intersect at right angles. The steady-state equation for this flow condition is

$$K_b \frac{\partial^2 b}{\partial x^2} + K_v \frac{\partial^2 b}{\partial z^2} = 0 \qquad (4.37)$$

where K_b and K_v mean horizontal and vertical hydraulic conductivity, respectively. If we keep in mind that a flow net is a graphical solution to Laplace's equation, a coordinate system transformation will yield the appropriate equation, where we define $\bar{x} = x/(K_b)^{1/2}$ and $\bar{y} = y/(K_v)^{1/2}$ so that

$$\frac{\partial^2 b}{\partial \bar{x}^2} + \frac{\partial^2 b}{\partial \bar{z}^2} = 0 \qquad (4.38)$$

From a graphical perspective a flow net can always be constructed by multiplying the original horizontal distances by $(K_v/K_b)^{1/2}$. The horizontal dimensions of the transformed section will always be shorter or longer than the original section, depending on whether K_b is larger or smaller than K_v (Figure 4.11).

Cedergren (1967) discusses the procedures to follow for obtaining a flow net for anisotropic conditions. First, the cross section is redrawn to the appropriate reduced or extended horizontal scale as shown in Figures 4.11b and 4.11d. This is referred to as the transformed section. The flow net is then constructed on the transformed scale. Following this, the flow net is redrawn on the original sections (Figures 4.11a and 4.11c). The resulting

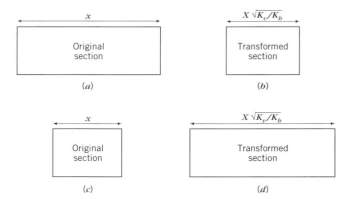

Figure 4.11 Transformation for flow nets in anisotropic material. In panel (b) $K_b > K_v$.

flow net on the original cross section will be composed of rectangles elongated in the direction of greater hydraulic conductivity.

4.5 Dimensional Analysis

Solutions to the diffusion equation will in general encompass a dimensionless group of terms that either dictate the nature of the diffusion process or demonstrate the competition between two rate processes. One of the more important uses of dimensional analysis is that it permits one to recognize the importance of parameters or, more precisely, groups of parameters in a fluid flow problem. To arrive at this dimensionless group for the diffusion equation, we first establish the following dimensionless variables that are noted with a superscript "+"

$$x^+ = x/L$$
$$y^+ = y/L$$
$$z^+ = z/L$$
$$t^+ = t/t_e$$
$$b^+ = b/b_e$$
$$\nabla^+ = L\nabla$$
$$\nabla^{2+} = L^2\nabla^2$$

where L is some characteristic length, t_e is some characteristic time, and b_e is some characteristic head. Note that the operators are also written in dimensionless form. With these dimensionless quantities, the diffusion equation for fluid flow expressed as

$$\nabla^2 b = \frac{S_s}{K} \frac{\partial b}{\partial t} \qquad (4.39)$$

becomes, in dimensionless form,

$$\nabla^{2+} b^+ = \left(\frac{S_s L^2}{K t_e}\right) \frac{\partial b^+}{\partial t^+} \qquad (4.40)$$

where all terms are dimensionless, including the quantity in the brackets. This bracketed quantity is known as the Fourier number, N_{FO}, and occurs in all unsteady flow problems.

The Fourier number is more readily expressed as

$$N_{FO}^{-1} = \frac{K/S_s}{L^2/t_e} \qquad (4.41)$$

where the numerator has already been introduced as the hydraulic diffusivity. The quantity $1/u$ that determines the "well function" $W(u)$ used in well hydraulics is of this form (Chapter 6). Another useful form for the inverse of this dimensionless group is

$$N_{FO} = \frac{S_s L^2/K}{t_e} = \frac{T^*}{t_e} \qquad (4.42)$$

where T^* is the time constant for a basin. If the time t_e at which we wish to observe some transient is significantly larger than T^*, the transient will not be observable, and the basin would appear to be in some steady state. If the time of observation is less than the characteristic time T^*, the transient is readily observable. Note that the characteristic time T^* gets large for large values of S_s and L^2 and small values of K. This form for the dimensionless group is useful in studying problems in disequilibrium compaction (Chapter 8).

Problems

1. Address the following problems as they pertain to Eq. 4.31.

a. Express this equation for one-dimensional flow.

b. Express this equation for a porous medium in which the fluid can be considered to be incompressible.

c. Express this equation for a porous medium in which both the fluid and the matrix are considered to be incompressible.

d. In what type of material (clay, siltstone, sandstone, gravel, limestone, granite) would you consider ignoring the term $\rho_w g n \beta_w$?

2. The storativity of an elastic aquifer is 3×10^{-4}. The porosity of the aquifer is 0.3, the compressibility of water is 4.8×10^{-10} m²/N, and the thickness of the aquifer is about 60 m.

a. How many cubic meters of water are removed from storage under an area of 1×10^6 m² when the head declines one meter?

b. Ascertain the storativity if the aquifer matrix is incompressible.

c. What is the value for the coefficient of vertical compressibility of the aquifer?

3. Consider a well pumping from storage at a constant rate from an extensive elastic aquifer.

a. The rate of head decline times the storativity summed up over the area over which the head decline is effective equals _____?

b. At any given time, the total head decline, summed up over the area over which the head change is effective, times the storativity equals _____?

4. A clay layer extending from ground surface to a depth of 50 ft is underlain by an artesian sand. A piezometer drilled to the top of the aquifer registers a head of 20 ft above land surface. The saturated unit weight of the clay is 125 lb/ft³. Assume a water table at land surface.

a. What are the values of total, neutral, and effective pressures at the bottom of the clay layer?

b. If a trench is dug to a depth of 20 ft, what are the total, neutral, and effective pressures at the bottom of the clay layer beneath the trench? Assume the trench is filled with water to land surface.

c. Is it possible to maintain the stability of the bottom of the trench in *b* above if the trench is dewatered?

d. What is the approximate minimum depth of a trench excavated in the clay layer that will cause a quick condition or failure at its bottom? Assume the trench to be filled with water.

5. If the total withdrawal from pumping center *B* of Figure 4.10 is 3×10^4 m³/day, compute the average transmissivity of the aquifer in the vicinity of pumping center *B*.

6. Designate whether the following are equipotential lines or flow lines and if they are constant head or no-flow boundaries: *AB, BC, CD, DE, FG*.

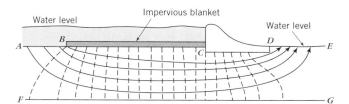

Problem 6.

7. Redraw the flow net of Problem 6 for (*a*) an impervious blanket one half as long, and (*b*) in the absence of an impervious blanket. For a pool height of 20 m and a hydraulic conductivity of 5×10^{-4} m/s, calculate the volumetric flow for the redrawn flow nets above and for the flow net of Problem 6.

8. Consider a vertical cross section *ABCD* where *AB* is the upper boundary, *DC* is the lower boundary, *AD* is the left-hand side boundary, and *BC* is the right-hand side boundary. Let the side boundaries be one half the length of the upper and lower boundaries. Draw flow nets for the following conditions:

a. *AB* is a constant head of 60 m, *AD* and *BC* are impermeable, and *DC* is divided into two equal sections, with the left-hand part being a constant head of 10 m and the right-hand side being impermeable.

b. *AB* is as in (*a*), *DC* is impermeable, and the side boundaries are divided into two equal segments, with the upper segments being impermeable and the lower segments being constant heads of 10 m.

9. Alter your vertical section of Problem 8 by making the right-hand boundary *BC* one and one-half times as long as side boundary *AD*. Here the bottom boundary *DC* remains the same whereas the upper boundary has a slope. Consider this upper boundary to be a sloping water table, and draw a flow net in the region assuming *DC*, *AD*, and *BC* are all no-flow boundaries, and the water table (*AB*) is a variable head boundary.

10. Repeat Problem 9 for anisotropic material, where the horizontal hydraulic conductivity is five times larger than the vertical hydraulic conductivity.

Chapter 5

Ground Water in the Basin Hydrologic Cycle

5.1 *Topographic Driving Forces*
5.2 *Surface Features of Ground-Water Flow*
5.3 *Some Engineering and Geologic Implications of Topographic Drive Systems*

The driving forces responsible for ground-water movement were known before the turn of the century and were categorized by King (1899) as gravitational (topographic), thermal, and capillary. King did not mention the importance of tectonic strain as a driving force, although he clearly demonstrated an up-dip migration of water in response to the compaction of sediments in active depositional environments. Our interest in this chapter is on how driving forces can be manifested in the basin hydrologic cycle, and what physical consequences result from these manifestations. Such consequences are important in a variety of engineering projects and in understanding a host of processes operating near the Earth's surface.

5.1 *Topographic Driving Forces*

The Early Field Studies

"The basal principles of artesian wells are simple. The schoolboy reckons himself as their master." This is the opening statement in Chamberlin's (1885) classic treatment on the requisites and qualifying conditions for the occurrence of artesian wells in pervious rocks. The pervious stratum discussed by Chamberlin has been defined as an aquifer, and the watertight or impervious beds are known as confining beds, commonly referred to as aquitards. Chamberlin noted further that no stratum is entirely impervious so that the aquifer can progressively discharge its fluids down dip. Of special importance here is the fact that the elevation differences of the outcrops, or the outcrop and the subcrop, provide the topographic drive for ground-water movement, hydraulically depicted by the potentiometric surface (Figure 5.1).

Other configurations of ground-water flow were also being recognized at about this same time. This is documented in the United States Geological Survey's Nineteenth Annual Report (1899), part two of which is entitled "Papers Chiefly of a Theoretical Nature." Included in this volume is the classic work of Slichter (1899) on the principles and conditions for the movement of ground water, and a theoretical-field analysis by King (1899). Of importance here is King's observations regarding the influence of gravity on shallow ground-water flow:

The contours of the ground water level show that the (water) surface presents the features of the hills and valleys approximately conformable with the relief forms of the surface above, the water being low where the surface of the ground is low, and higher where the surface of the ground is high.

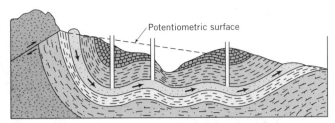

Figure 5.1 Schematic cross section showing trace of the potentiometric surface and areas of flowing and nonflowing wells (from Hubbert, 1953). Reprinted by permission.

In short, King stated that the water table was everywhere a subdued replica of the topography, and water moves from topographically high areas to topographically low areas. Figure 5.2, taken directly from King, diagrammatically illustrates these facts. As with confined systems, the topographic drive is provided (and limited) by differences in surface elevation.

The field evidence offered by Chamberlin (1885) and King (1899) still provides the basis for the categorization of ground-water flow under confined and unconfined conditions. At one extreme is Chamberlin's concept of the confined condition, exemplified by dipping aquifers bounded by low but finite permeability units with continuous replenishment at the outcrop areas. If a well is placed in such an aquifer, the water will rise in the well to the elevation of the water table in the intake area, minus any head losses incurred from the point of intake to the point of measurement. The pressure-producing mechanism is the hydrostatic weight of the body of water extending down dip from the water table in the outcrop area. The dynamic mechanism required to maintain the high-pressure system is the continuous replenishment by precipitation. At the other extreme is the unconfined condition discussed by King (1899), where the water table is a subdued replica of topography. The dynamic mechanism required to maintain this flow is likewise a continuous replenishment by precipitation. Indeed, as water offers no resistance to deformation, its movement from high elevations to low elevations would, in the absence of recharge, result in drainage of the water contained in the topographic highs, thereby producing a flat

surface of minimum potential energy (the hydrostatic or nonmoving condition). This tendency is opposed by continual replenishment. Elevation differences in the outcrops areas for the confined case, or in the land surface in the unconfined case, provide the topographic drive mechanism for movement of the contained fluids. Continual replenishment assures that this drive persists indefinitely.

Conceptual, Graphical, and Mathematical Models of Unconfined Flow

From what we have learned from the field studies discussed in the previous section, the question arises as to the possibility of establishing at least some features of a conceptual model that will always be valid. One such feature, for example, might be that the water table is a subdued replica of topography, at least for unconfined flow systems that are undisturbed by pumping. Thus, given information on the topography of a region—even if nothing else is known—it is at least possible to say something about the uppermost part of the saturated zone. In addition, following King's portrayal of flow, the flow appears to "branch" at the topographic divides, with the flow going in divergent directions. Thus, topographic divides appear to be ground-water divides as well, this being the obvious result when the water table is a subdued replica of the topography. Again, in concert with King, streams appear to receive all the ground-water discharge from the adjoining parts of the valley, with none of the flow passing underneath the stream. The streams, or at least the major ones occupying the major topographic lows, are ground-water divides in that they prevent the intermingling of ground water from opposite flanks of the same basin. This, too, is inevitable for the lowest topographic points in a region where the water table reflects the topography.

It follows that a useful conceptual model of a flow field should incorporate those facts that have been learned from field investigations. Such a model was provided by Hubbert (1940) and is shown in Figure 5.3. Notice in this model that King's measurements on the relationship between the water table and the land surface

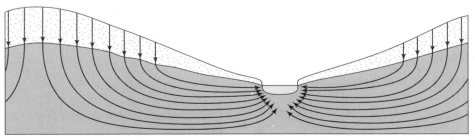

Figure 5.2 Diagrammatic section illustrating ground-water flow in a watershed (from King, 1899).

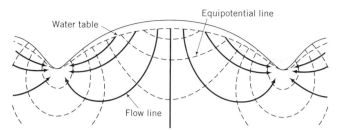

Figure 5.3 Topographically controlled flow pattern (from Hubbert, 1940). Reprinted by permission of the Journal of Geology, University of Chicago Press. Copyright © 1940.

are faithfully reproduced. In addition we find that King's observations at the topographic highs and lows are likewise incorporated in this model. This has the effect of setting up individual flow cells in each of the basins shown in Figure 5.3, with the flow in one basin being more or less independent of the flow in adjacent basins. It is noted also that the symmetrical portrayal of the land surface has the effect of producing a mirror image flow pattern for the flow in adjacent flanks of a given basin.

There are, of course, some differences between the presentations of King (1899) and those of Hubbert (1940). In Hubbert's conceptual diagram, the flow is presented by intersecting flow lines and equipotential lines. It is noted that the water table has a point of intersection with each of these equipotential lines, with the value of the hydraulic head anywhere on a given equipotential line taking on this water table elevation at its point of intersection with the equipotential line. This is an important detail and provides information that cannot be obtained from King's diagram. For example, consider a piezometer emplacement somewhere in the vicinity of the topographic high. After passing through the unsaturated zone, the piezometer will encounter the water table and the water will stand in the well at the water table elevation. The piezometer is then set deeper, intersecting one of the equipotential lines. This means that the water will now stand in the well at an elevation equal to the elevation of the water table where that equipotential intersects the water table. Thus, the water level in the well will be at a lower elevation than the water table at the point of piezometer emplacement. With continued depth of emplacement, the water level in the well will continually occupy a lower elevation. The reason for this is obvious from Figure 5.3, which shows that in the vicinity of topographic highs, the potential (head) decreases with increasing depth as the flow is directed downward. Regions where the flow of water is directed downward with respect to the water table are called recharge areas (Tóth, 1963). The incorporation of recharge areas is thus part of our conceptual model; the delineation of such areas is a field problem, but we expect them to coincide with topographic highs.

Focus now on the topographic lows, where a piezome-

ter emplacement encounters higher hydraulic heads with depth so that the water level in the well will actually be above the water table adjacent to the well. Hubbert (1940) noted that it is possible to obtain a flowing well for such conditions in that all that is required is the intersection of the piezometer with an equipotential line that has a larger value than the land surface at that point. Again, the reason for this behavior is shown in Figure 5.3, where the head increases with increasing depth so that flow is directed upward. Regions where the flow of water is directed upward with respect to the water table are called discharge areas, and the line that separates the recharge area from the discharge area is called a hinge line (Tóth, 1963). Clearly, the concept of a discharge area must be incorporated in our conceptual model; the field problem is to delineate such areas, and we expect them to coincide with topographically low areas.

Now let's stretch our basin out a bit, giving it a flatter water table (topographic) expression (Figure 5.4). We note that the recharge and discharge areas are diminished somewhat, and we notice a large region where the equipotential lines are vertical or near vertical. If a piezometer is emplaced in this region, it will follow a single equipotential line. Thus, once the piezometer encounters the water table, the water level will not change with increasing depth. It will be noticed here that this is the only set of conditions in the flow field where a piezometer emplaced below the water table actually provides a direct measurement of the water table elevation. Such a region can be called a region of lateral flow, which corresponds to Tóth's (1963) hinge line.

As long as the water table is not fluctuating too severely in response to recharge and discharge, the flow depicted in Figures 5.3 and 5.4 is steady, or nearly so. It follows that these figures represent graphical solutions to Laplace's equation in the two-dimensional region depicted. Thus, they are in fact mathematical models of the flow field. A question now arises as to the nature of the boundary conditions so that it is possible to reproduce other con-

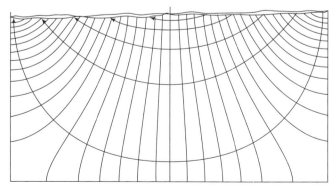

Figure 5.4 Topographically controlled flow pattern (from Tóth, J. Geophys. Res., v. 67, p. 4375–4387, 1962. Copyright by Amer. Geophys. Union).

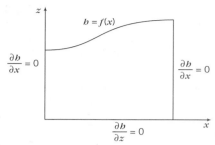

Figure 5.5 Two-dimensional region demonstrating typical boundary conditions for regional flow.

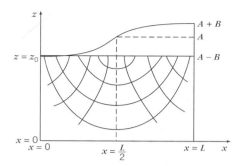

Figure 5.E1 Two-dimensional flow region.

figurations of flow for other water table configurations and basin depth to basin length ratios. Because of the symmetry demonstrated on either side of a stream discharge area, one limb of a basin will serve these purposes. Because of the diverging flow at the topographic high and the confluence of the flow at the topographic low, these lateral boundaries are obviously of the no-flow type. That is to say, no flow crosses the topographic high into or out of the region, or crosses the topographic low into or out of the region. As no-flow boundaries, these may be represented by flow lines indicating that the flow is tangential to them. Provided we assume there is a real impermeable layer at some depth in the region, this impermeable layer is likewise a no-flow boundary. It, too, must be represented by a flow line. These boundaries can be placed in the general category; that is, they will always be present for the major topographic low and the major topographic high of a region, and for a specified depth. The upper surface or water table is neither a flow line nor an equipotential line, and is a specified head boundary where the head varies as a function of space in the same fashion that the elevation of the land surface varies as a function of space. Clearly, this boundary condition will be specific for a given topographic expression. If this head varies as a function of time as well, the problem is no longer one of steady flow, but of unsteady flow. The boundary conditions so described are given in Figure 5.5. An analytical solution to this problem for three fixed boundary conditions and a variable water table configuration is given in Example 5.1.

Example 5.1 (From Domenico and Palciauskas, 1973)

A solution for Laplace's equation in the region of Figure 5E.1 where the upper boundary is not specified but all other boundaries are of the no-flow variety is

$$\phi(x, z) = a_o + \sum_{n=1}^{\infty} a_n \cosh \frac{n\pi x}{L} \cos \frac{n\pi x}{L}$$

where ϕ is the hydraulic potential and the coefficients a_o and a_n are determined from the equation of the water table for various special cases. The coefficients are deter-

mined from

$$a_0 = \frac{1}{L} \int_0^L \phi(x, z_o) \, dx$$

and

$$a_n = \frac{2}{L \cosh (n\pi z_o/L)} \int_0^L \phi(x, z_o) \cos \frac{n\pi x}{L} \, dx$$

The equations given are general in that they apply to any water table configuration. For a specific case, assume that the equation of the water table $\phi(x, z_o) = A - B \cos \pi x/L$. The coefficients then become

$$a_o = A; \qquad a_1 = \frac{-B}{\cosh (\pi z_o/L)}; \qquad \text{other } a\text{'s} = 0$$

The solution for Laplace's equation then becomes

$$\phi(x, z) = A - \left[\frac{B \cosh (\pi z/L)}{\cosh (\pi z_o/L)} \right] \cos \frac{\pi x}{L}$$

This case permits a good understanding of the geometrical controls on the spatial distribution of the potential. It is noted that at the points $z = z_o$ for $x = 0$, $x = L$, and $x = L/2$, the value of the potential is $A - B$, $A + B$, A, respectively. Further, at $x = L/2$ for all z, the potential equals A; that is, a vertical equipotential line exists at $x = L/2$. This vertical equipotential corresponds to the midline of flow.

Effects of Basin Geometry on Ground-Water Flow
The mathematical description of the boundary conditions given in Figure 5.5 was first stated by Tóth (1962, 1963) in two modern-day classic studies. This work was the first significant extension of Hubbert's (1940) ideas in over 20 years. Tóth's work is based on the following assumptions:

1. The medium is isotropic and homogeneous to a specified depth, below which there exists an impermeable basement.

2. Flow is restricted to a two-dimensional vertical section. The topography can be approximated by simple

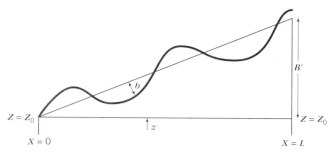

Figure 5.6 Sinusoidal water table with a regional slope.

curves, such as straight lines or sine waves, and the water table is a subdued replica of the topography.

3. The upper boundary of the two-dimensional section is the water table, the lower boundary is the impervious basement, and the lateral boundaries are groundwater divides.

Tóth was not content to investigate the individual flow cells demonstrated in Figure 5.3. Instead he considered a sinusoidal water table with a regional slope of the form (Figure 5.6)

$$h(x, z_o) = \left[z + \frac{B'x}{L} + b \sin \frac{2\pi x}{\lambda} \right] \qquad (5.1)$$

where the second term on the right-hand side corresponds to the regional slope and the third term to the local relief superposed on the regional slope. In this third term, b is the amplitude of the sine wave and λ is the number of oscillations, so that L/λ is the number of flow cells. As noted in Figure 5.6, B' is the height of the major topographic high above the major topographic low and x is a variable ranging from zero to the basin length L. From this analysis, Tóth (1963) identified a local system, which has its recharge area at a topographic high and its discharge area at a topographic low that are adjacent to each other; an intermediate system, which is characterized by one or more topographic highs and lows located between its recharge and discharge areas; and a regional system, which has its recharge area at the major topographic high and its discharge area at the major topographic low (Figure 5.7). The important conclusions from this study are as follows:

1. If local relief is negligible ($b \sin 2\pi x/\lambda = 0$, Eq. 5.1) and there is a general slope of the topography, only regional systems will develop where ground water moves from the major topographic high to the major topographic low.

2. If regional slope is negligible ($B'x/L = 0$, Eq. 5.1), only local systems will develop. The greater the relief the deeper the local systems that develop.

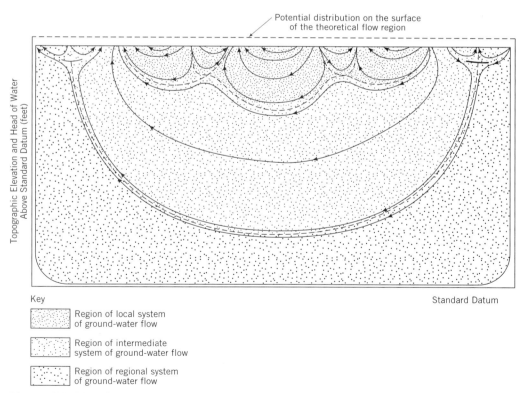

Potential distribution on the surface of the theoretical flow region

Topographic Elevation and Head of Water Above Standard Datum (feet)

Standard Datum

Key

Region of local system of ground-water flow

Region of intermediate system of ground-water flow

Region of regional system of ground-water flow

Figure 5.7 Two-dimensional isotropic flow model showing the distribution of local, intermediate, and regional ground-water flow systems (from Tóth, J. Geophys. Res., v. 68, p. 4795–4812, 1963. Copyright by Amer. Geophys. Union).

3. If both local relief and a regional slope are negligible, neither regional nor local systems will develop. Waterlogged areas will be common, and the ground water may discharge by evapotranspiration in response to a flat water table near the land surface.

4. Given both local relief and a regional slope, local, intermediate, and regional systems will develop.

Effects of Basin Geology on Ground-Water Flow

The work of Tóth demonstrates the importance of basin geometry on ground-water flow. By geometry, we mean the ratio of basin depth to length and the relief or topography making up the surface. As the medium was assumed isotropic and homogeneous, little can be inferred about the influence of basin geology on ground-water flow, that is, the influence of the stratigraphic framework with its attendant permeability variations. This problem was addressed by Freeze and Witherspoon (1966, 1967) within the framework of the Tóth model. In their approach, n homogeneous and isotropic interlayered formations were treated within the two-dimensional vertical section (Figure 5.8). The water table is incorporated with a series of straight-line segments so that any configuration can be approximated as a function of the space variables.

The boundary value problem solved by Freeze and Witherspoon is actually n interrelated problems. When n equals 3, Figure 5.8 applies. The condition that $b_n = b_{n+1}$ ensures a continuous potential across the contact of two layers. A second condition is required so that fluid mass is conserved when flow crosses a boundary between adjacent strata of differing permeability. This is referred to as the tangent refraction law (Figure 5.9). Analytically, a refraction of the flow lines occurs such that the permeability ratio of the two units equals the

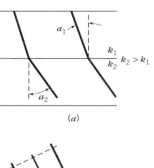

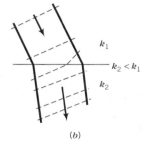

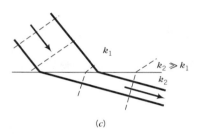

(c)

Figure 5.9 Diagrams of flow-line refraction and conditions at the boundaries between materials of differing permeability.

ratios of the tangents of the angles the flow lines make with the normal to the boundary.

$$\frac{k_1}{k_2} = \frac{\tan a_1}{\tan a_2} \qquad (5.2)$$

In Figure 5.9*b*, the hydraulic gradient in the lower-permeability unit is steepened to accommodate the flow crossing the boundary from the unit of higher conductivity; in Figure 5.9*c*, both the hydraulic gradient and the cross-sectional flow area in the high-permeability unit are decreased to accommodate the flow crossing the boundary from the low-permeability unit.

Freeze and Witherspoon (1966, 1967) employed numerical methods to solve the flow problem discussed earlier. With analytical methods, both the problem and the solution are expressed in terms of parameters. This was the method used by Tóth (1962, 1963). With numerical solutions, both the problem and the solution are expressed in terms of numbers. Hence, with the latter method, each case is quite specific, and several cases must be solved to obtain some general information on the subject.

The main conclusions of Tóth (1963) regarding control of topography on the ground-water flow were verified

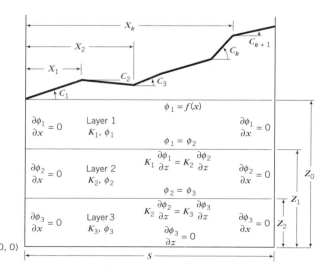

Figure 5.8 Two-dimensional, unconfined model with n homogeneous, isotropic layers (from Freeze and Witherspoon, Water Resources Res., v. 2, p. 641–656, 1966. Copyright by Amer. Geophys. Union).

with the modeling effort of Freeze and Witherspoon. This is demonstrated in the sequence of figures given as Figure 5.10. Figure 5.10*a* shows that a lack of local relief results in development of a single regional system. Figure 5.10*b* demonstrates the effect of adding a slight amount of topographic relief, and Figure 5.10*c* demonstrates the local, intermediate, and regional flow system development when the relief is altered to that characteristic of hummocky terrain.

The primary effort of Freeze and Witherspoon (1967) dealt with determining the role of permeability contrasts in influencing various degrees of flow-line refraction at the permeability boundaries between adjacent strata as well as the influence of this contrast in controlling the resulting equipotential and flow-line distributions throughout the formations. These effects are shown in Figure 5.11. Figures 5.11*a* and 5.11*b* show the effects of a high-permeability layer underlying a layer of lower permeability. As the permeability of the lower layer increases, it acts like a pirating agent for the flow, forcing near-vertical flow through the uppermost unit. This is best demonstrated by the orientation of the equipotential lines in the lower-permeability unit for a 100-to-1 permeability contrast. Figures 5.11*c* and 5.11*d* show the effects of a high-permeability lens in the flow field. If the lens is in the recharge area, it too acts like a pirating agent, forcing vertical flow through the overlying unit and creat-

ing a discharge scenario at its terminus that would not be predicted on the basis of topography. Shifting the high-permeability lens to the discharge end of the system has similar effects on the overlying units.

The sloping stratigraphy case is demonstrated in Figures 5.12*a* and 5.12*b*. If the dip of the bed is in the direction of topographic drive (Figure 5.12*a*), the flow is down dip. With the hydraulic conductivity contrasts demonstrated, the behavior here is identical to what we would expect from the Chamberlin model. If the dip of the bed is opposite the topographic drive, the flow is up dip, with the large conductivity contrasts causing downward flow in the overlying low-permeability unit. Such a system has been postulated for part of the flow field in southern Maryland (Figure 5.13).

The points just discussed become important when ascertaining the influence of structure on ground-water flow. For example, most diagrams illustrating flow in folded beds consisting of anticlines and synclines show the flow directed at right angles to the synclinal and anticlinal axes, that is, up and down the structural limbs (see, for example, Figure 5.1). This, of course, is appropriate for such folded rocks that obtain their topographic drive from the elevated outcrop areas. If, however, a series of anticlines and synclines have been uplifted into a meteoric recharge area and possess a distinct plunge to lower elevations, the flow will be parallel to the structural

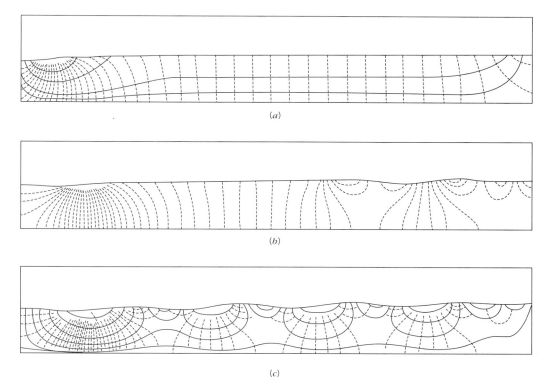

(*a*)

(*b*)

(*c*)

Figure 5.10 Effect of water table configuration on regional ground-water flow through homogeneous isotropic mediums (from Freeze and Witherspoon, Water Resources Res. v. 3, p. 623–634, 1967. Copyright by Amer. Geophys. Union).

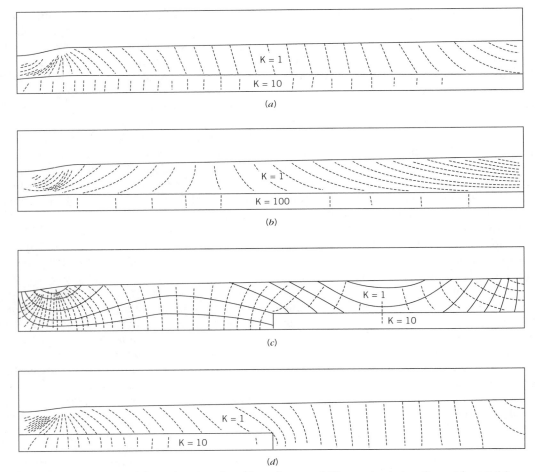

Figure 5.11 Regional flow showing the effect of permeability contrasts in adjacent layers (*a*) and (*b*) and the effect of a high permeability lens in the flow field of *c* and *d* (from Freeze and Witherspoon, Water Resources Res., v. 3, p. 623–634, 1967. Copyright by Amer. Geophys. Union).

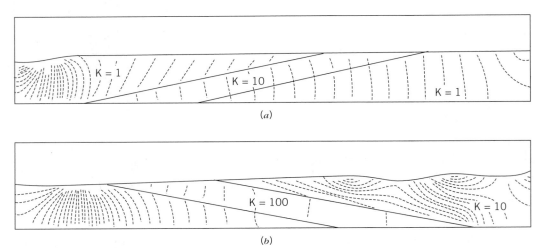

Figure 5.12 Regional flow in sloping topography (from Freeze and Witherspoon, Water Resources Res., v. 3, p. 623–634, 1967. Copyright by Amer. Geophys. Union).

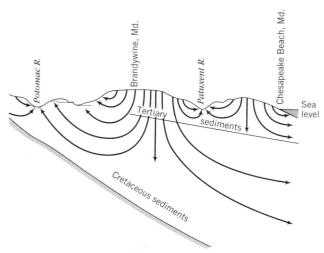

Figure 5.13 Diagrammatic cross section through southern Maryland showing the lines of ground-water flow (from Back, 1960).

axes in the direction of the plunge. In this case, the elevated rocks in the recharge area provide the topographic drive, and the anticlinal axes serve as ground-water divides separating the flow systems in adjoining synclines.

Based on the information presented thus far, it is possible to obtain a picture of regional flow for any basin for which the pertinent data are available. Data requirements include permeability distribution and geometry of the basin boundaries. Although the ideal homogeneous case is contrary to field observations, it is merely a condition in the mathematical development, not an assumption necessary to the general validity of the theory. Further, the steady-state solutions do not deny the existence of a fluctuating upper boundary of flow. The main argument is that the effect on flow patterns will be small if (1) the fluctuating zone is small compared with total saturated thickness and (2) the relative configuration of the water table is unchanged throughout the cycles of fluctuation. Given that these conditions are reasonably satisfied, the value of the information provided by the flow net can best be appreciated if one is interested in the distribution of recharge and discharge areas along the water table, the depth and lateral extent of local systems in hummocky terrain, and the degree to which the zones of high permeability act as major conductors of water and their overall influence on the flow pattern. In summarizing the factors that influence these items, Freeze and Witherspoon (1967) cite the following as the most important:

1. The three most influential factors affecting the potential distribution are
 a. The ratio of basin depth to lateral extent.
 b. The configuration of the water table.
 c. Variations in permeability.

2. A major valley will tend to concentrate discharge in the valley. Where the regional water table slope is uniform, the entire upland is a recharge area. In hummocky terrain, numerous subbasins will be superimposed on the regional system.

3. A buried aquifer of significant permeability will act as a conduit that transmits water to principal discharge areas, and it will thus affect the magnitude of recharge and the position of the recharge area.

4. Stratigraphic pinchouts at depth can create recharge or discharge areas where they would not be anticipated on the basis of water table configuration.

A field application of the ideas just presented is shown in Figure 5.14 for the Palo Duro Basin in Texas. The Palo Duro has been studied as a potential repository for the disposal of high-level radioactive waste by the U.S. Department of Energy. Figure 5.14*a* is a geologic cross section demonstrating the various geologic units in the section, where the salt beds have been considered to serve as the repository unit. Figure 5.14*b* shows the flow patterns as measured or inferred in the field, with the large hydraulic gradient across the salt beds reflecting their low permeability. Figure 5.14*c* is a numerical model of the flow field. Note that the downgradient lateral boundary is depicted as a constant head in that the flow presumably continues across this area to discharge in parts of the basin not depicted in this flow diagram.

Ground Water in Mountainous Terrain

Mountainous terrain occupies about 20% of the Earth's land surface and presents some special problems in hydrologic analysis, mostly because hydraulic head data are seldom available to confirm the suggested conformation between the land surface elevation and elevation of the water table. Further, mountainous regions promote deep circulation of ground water, denying access to ground water outcrops that aid in regional investigation. A further complication arises because mountainous regions are frequently fractured and may be in an active state of compression or extension, suggesting that fracture apertures may be functionally related to the state of stress in the Earth's crust. Thus, the elevation of the water table may be intimately related to a changing hydraulic conductivity of the region and the variable climatic factors that influence infiltration. For a given hydraulic conductivity distribution, the lower the infiltration rate, as controlled by climatic factors, the deeper the water table.

It follows that the simple relations given by King (1899) and Hubbert (1940) and substantiated in numerous studies in low-lying sediments may not apply to mountainous terrain characterized by a fracture permeability. In such cases, the water table may be considered a free surface whose depth and configuration depend on the interplay between infiltration and permeability distribution.

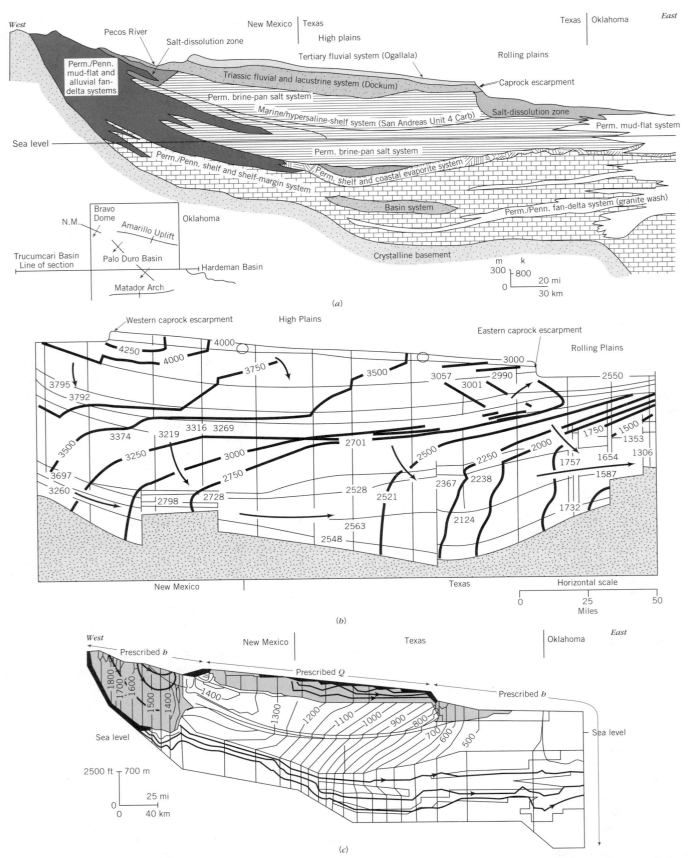

Figure 5.14 *Geology, regional flow, and simulation of flow in the Palo Duro Basin. (a) Geology and (c) simulation of flow, from Senger and others (1987); (b) regional flow, from Bair (1987). Figure (b) reprinted by permission of* **Ground Water.** *Copyright © 1987. All rights reserved.*

84

Jamieson and Freeze (1983) were among the first to use a free surface technique as applied to mountainous terrain in British Columbia. They used a fixed infiltration rate to estimate the range of hydraulic conductivity that might be expected to produce a water table at a given depth. Each hydraulic conductivity pattern resulted in a different elevation of the free water surface. This study has been expanded by Forster and Smith (1988) to include the effects of varying infiltration rates, surface topography, topographic symmetry, and permeable frac-

ture zones. They conclude that permeability has the greatest impact on the mountainous flow system. Asymmetry can cause the displacement of the ground-water divides from the topographic divides and a relatively small increase in the vertical permeability of fractures relative to the horizontal permeability causes significant declines in the water table elevation (on the order of 400 m). The authors also state that high relief can promote deep ground-water circulation to elevated temperature regions, requiring a modeling of both the fluid flow

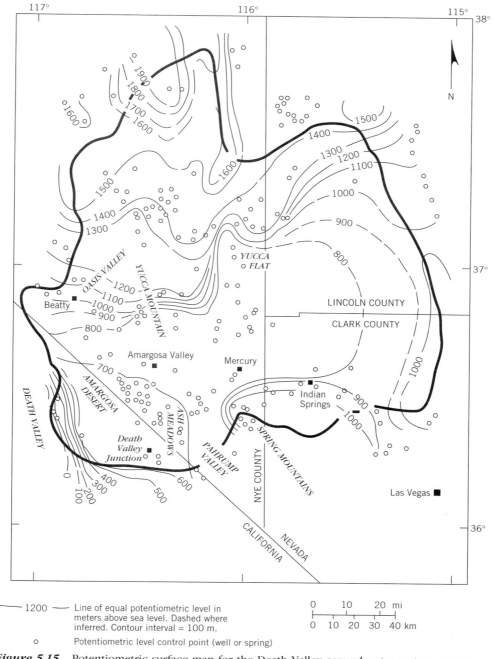

---- 1200 ---- Line of equal potentiometric level in meters above sea level. Dashed where inferred. Contour interval = 100 m.

o   Potentiometric level control point (well or spring)

Figure 5.15 Potentiometric surface map for the Death Valley ground-water system encompassing Yucca Mountain (modified from DOE, 1988b).

and the thermal regime. The thermal aspects of mountainous regions will be discussed in a later chapter.

One mountainous region that does not qualify as "data poor" is Yucca Mountain in southern Nevada, a volcanic tuff pile located north of Las Vegas. Yucca Mountain is currently under study as the first site in the United States for disposal of commercial nuclear waste. As part of the southern Great Basin, a distinct feature of the Yucca Mountain region is crustal extension, which suggests that the mountain has become "wider" during Late Miocene–

Late Quaternary time, with the "widening" occurring in the form of vertical fracture development. The mountain is characterized by extremely low infiltration (less than 1 mm/yr) and a deep (600 m), rather contorted water table configuration. Winograd and Thordarson (1975) have shown that the water table throughout the Death Valley flow system (which includes Yucca Mountain) is a series of "plateaus" separated by sharp steps and ranges in altitude from over 1900 m to below sea level in Death Valley (Figure 5.15). On the "plateaus," the gradients are

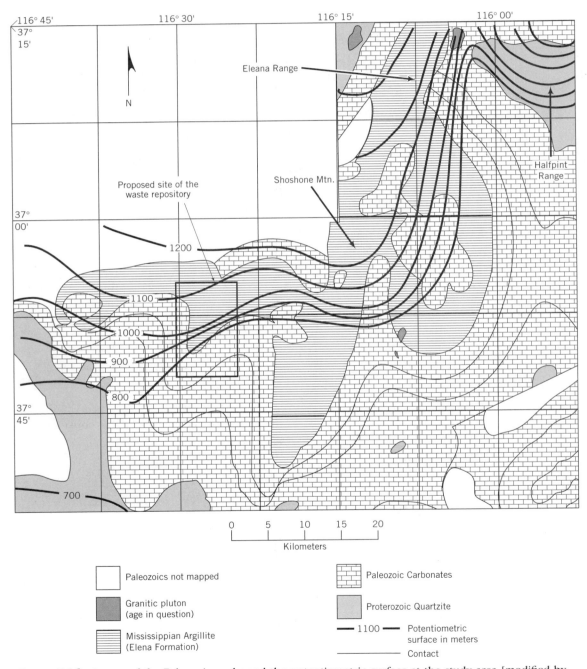

Figure 5.16 A map of the Paleozoic rocks and the potentiometric surface at the study area [modified by Strom (1993) from Robinson (1985)].

small, measured in tens of meters per kilometer. Between the "plateaus," slopes as large as 10 and 30% are present. Szymanski (1989) relates this water table configuration to the hydraulic conductivity structure, with steep slopes occurring where the upper transmissive part is missing or thin, and small gradients in areas characterized by a thick upper transmissive unit. There are, of course, alternative causes such as faulted barriers to flow or highly transmissive zones conducting a limited flux. Indeed, the geology underlying the volcanic tuffs may be the primary control on steep and shallow gradients throughout the region (Fridrich and others, 1991, 1993). Note on Figure 5.16 that the steep gradient in the volcanic rocks roughly coincides with those areas where the Eleana aquitard no longer overlies the Paleozoic high-permeability carbonates (Fridrich and others, 1991, 1993). Thus water may be "perched" in the volcanics where the Eleana is present, only to cascade into the carbonates at the contact. In the northeastern part of the region, the low-permeability Proterozoic quartzite serves as the perching layer. The flat gradient automatically follows in that much of the water in the volcanics has been siphoned off to the directly underlying carbonates. The geologic map shown is rather crude, and more information will be required to fully understand this phenomenon.

Lastly, it is important to recognize the water table–topographic coincidence in the model studies discussed thus far is part of the boundary conditions. The free surface analysis suggests that the depth and configuration of the water table will depend on the interplay between infiltration and permeability distribution. A subsurface topographic high on an aquitard, for example, may result in a ground-water divide in an area of uniform precipitation and recharge. In Figure 5.17, the aquitard acts as a perching layer and its topographic high supports a ground-water divide that would not be suspected from surface topography.

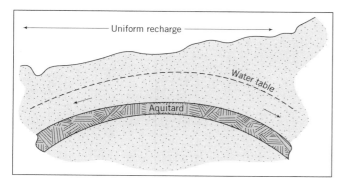

Figure 5.17 Subsurface topographic high on perching layer supporting ground-water divide.

Ground Water in Carbonate Terrain

Carbonate terrains occupy about 10% of the Earth's surface and supply ground water to about a quarter of the global population (Ford and Williams, 1989). Carbonates provide a challenge to the hydrogeologist because the rocks are soluble and thus susceptible to the formation of self-organized networks of solutionally enlarged fractures. The result is a double-porosity aquifer. Most of the water moves slowly through the matrix and little-enlarged fractures and is the fraction predominantly sampled in wells. The network of enlarged fractures, or conduits, makes up the second fraction. These commonly make up a tiny fraction of aquifer volume and thus are unlikely to be intercepted by wells. Nevertheless, the conduits will transmit solutes rapidly. For instance, in a limestone aquifer in England, Atkinson (1977) recorded linear velocities of 520 m to 21,200 m/day using tracers in 38 tests over distances of up to 11 km. Calculations showed that the conduits made up a mere 0.03% of aquifer volume, showing that almost all of the specific yield of 0.92% is from the matrix and from little-enlarged fractures through which water moves very slowly.

In Central Kentucky flow in the unconfined Mississippian aquifer is toward the Green River. Core samples indicate that the limestone has a primary porosity of 3.3% and a hydraulic conductivity of 2×10^{-11} m/s (Brown and Lambert, 1963). Solutional enlargement of fractures has been extensive, resulting in integrated conduit networks and caves. Quinlan and Ray (1981) used water levels from 1400 wells to produce a water table map, part of which is shown in Figure 5.18. The water table is marked by a number of troughs, and each trough terminates in a downgradient direction at one or more springs along the Green River. More than 500 tracer tests were conducted to these springs to identify flow directions, over distances up to 22 km. The tracer tests showed that flow routes converge on the water level troughs and that this convergence may occur at considerable distances from the springs. Thus tributary flow occurs throughout the aquifer. Linear velocities along the conduits, as demonstrated by the tracer tests, varies from less than 200 m/day to more than 10 km/day.

For more than 20 years sewage effluent rich in heavy metals was discharged into the ground about 2 km from the axis of one major water table trough. The effluent was identified in a number of springs along the Green River, but in none of the 22 sampled wells between the sewage-treatment plant and the river (Quinlan and Rowe, 1977). This demonstrates that springs can be effective monitoring points in carbonate rocks, since they may discharge the integrated flow from considerable areas. Conversely, wells that are "downgradient" from a facility are not reliable unless they can be shown by tracer tests to lie on the flow path from that facility (Quinlan, 1990).

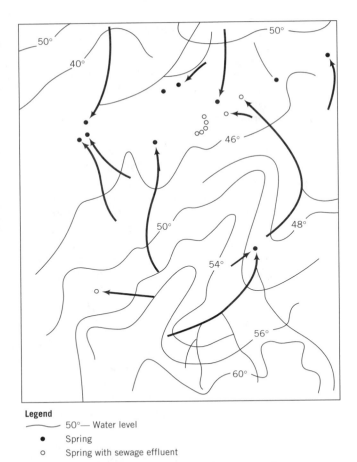

Legend

⎯⎯⎯ 50°⎯ Water level
● Spring
○ Spring with sewage effluent

Figure 5.18 Water table map and flow routes from tracer tests.

Ground Water in Coastal Regions

Although frequently of limited relief, topographic drive systems play a major role in coastal areas where mixing of fresh water and salt water occurs. Under natural conditions, it is the flow of fresh water toward the sea that limits the landward encroachment of seawater. With the development of ground-water supplies and subsequent lowering of the water table or piezometric surface, the dynamic balance between fresh and seawater is disturbed, permitting the seawater to intrude usable parts of the aquifer (Figure 5.19). This phenomenon has been reported in several parts of the world, including the Netherlands (Ernest, 1969) and Israel (Schmorak, 1967). The best documented cases in the United States include Long Island (Lusczynski and Swarzenski, 1966), Miami (Kohout, 1961), and many parts of California (California Department of Water Resources, 1958). Saltwater intrusion is a special case of ground-water contamination and is yet another phenomenon subject to numerical simulation (Pinder and Cooper, 1970; Segol and Pinder, 1976).

Banks and Rictor (1953) suggest five methods of controlling saltwater intrusion, four of which require control of the water table or piezometric surface at or near the coast. The five methods include artificial recharge, a reduction or rearrangement of the pumping wells, establishing a pumping trough along the coast, thereby limiting the area of intrusion to the trough, formation of a pressure ridge along the coast, and installation of a subsurface barrier. These methods have been discussed at length by Todd (1980).

Field observations in coastal areas suggest the presence of a mixing zone where miscible sea- and fresh water interfuse about their boundary. This phenomenon is best documented in the Biscayne aquifer near Miami, Florida (Kohout, 1960; Kohout and Klein, 1967). There is ample evidence to suggest that the mixing zone between the fresh and salt water moves seaward during periods of heavy recharge and inland during periods of low freshwater head. During the periods of inland flow, the cyclic flow of salt water in the mixing zone tends to limit the extent to which seawater invades the aquifer. That is to say, salinity differences establish density differences that drive the cyclic motion. A typical cross section of the mixing zone is given in Figure 5.20, and a generalized theory of the phenomenon is given by Bear and Bachmat (1967).

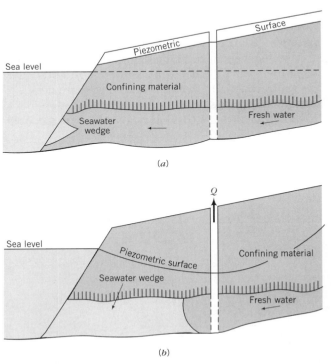

Figure 5.19 Hydraulic conditions near a coastline (*a*) not subject to seawater intrusion and (*b*) subject to seawater intrusion with an advancing seawater wedge.

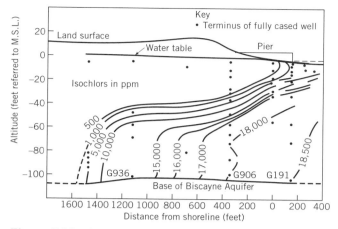

Figure 5.20 Cross section of a mixing zone near Miami, Florida (from Kohout and Klein, 1967). Reprinted with permission of Intl. Assoc. of Hydrol. Sci.

The Freshwater–Saltwater Interface in Coastal Regions

There are many studies that attempt to ascertain the position of the freshwater–saltwater interface, both in coastal areas and underlying oceanic islands, and for conditions of flow as might be induced in the vicinity of pumping wells. Many of these studies have one major assumption in common, namely, that the two fluids—salt water and fresh water—are immiscible and are separated by a rigid interface. That is to say, the mixing zone discussed above is replaced by a line (or, more appropriately, a plane) across which no flow can occur. The interface is common to both fluids and in some cases only one fluid is permitted to move and in others both fluids may move. We will discuss these theories here as applied to coastal regions, focusing on their applications and limitations.

The Ghyben–Herzberg Relation

At the turn of the century, it was generally thought that salt water in coastal areas occurred at a depth approximating sea level. Two investigators, working independently, demonstrated that seawater actually occurred at depths below sea level equivalent to approximately 40 times the height of fresh water above sea level (Ghyben, 1889; Herzberg, 1901). The analytical explanation of this phenomenon is referred to as the Ghyben–Herzberg formula and is derived through simple hydrostatics. For two segregated fluids with a common interface, the weight of a column of fresh water extending from the water table to the interface is balanced by the weight of a column of seawater extending from sea level to the same depth as the point on the interface; that is, the weight of the column of fresh water of length $b_f + z$ equals the weight of the column of salt water of length z (Figure 5.21).

By designating ρ_f and ρ_s as the densities of fresh and salt water, respectively, the condition of hydrostatic balance is expressed

$$\rho_s g z = \rho_f g (b_f + z) \tag{5.3}$$

or

$$z = \frac{\rho_f}{\rho_s - \rho_f} b_f \tag{5.4}$$

where z is the depth below sea level to a point on the interface. If the density of fresh water is taken as 1.0 and seawater as 1.025

$$z = 40 b_f \tag{5.5}$$

as confirmed by Ghyben and Herzberg by observation. This means, where this condition is approximately correct, a freshwater level of 20 ft above sea level corresponds to 800 ft of fresh water below sea level. Stated in another way, a lowering of the water table by 5 ft will cause a 200-ft rise of salt water. It is further noted that the slope of the interface is 40 times greater than the slope of the water table.

The length of the prevailing saltwater wedge under the hydrostatic conditions of the Ghyben–Herzberg formula may be established for the simple geometry of Figure 5.22. From Darcy's law of the form $Q = KiA = KimY$, where i is the hydraulic gradient, m is thickness, and Y is the length of the shoreline

$$Q' = \frac{Q}{Y} = K \frac{(b_f - 0)}{L} m \tag{5.6}$$

where Q' is the discharge per unit length of shoreline, and $(b_f - 0)/L$ is the hydraulic gradient where b_f is taken at the distance L from the shoreline. The Ghyben–

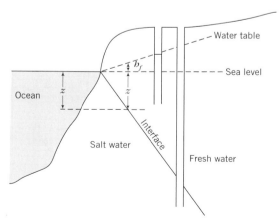

Figure 5.21 Hydrostatic conditions of the Ghyben–Herzberg relation.

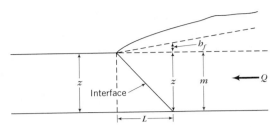

Figure 5.22 Idealized geometry to calculate the length of the saltwater wedge from the Ghyben–Herzberg relation.

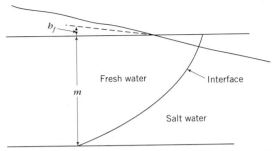

Figure 5.24 Idealized geometry to calculate the length of the saltwater wedge.

Herzberg relation predicts that at a distance L from the shoreline

$$b_f = z \frac{(\rho_s - \rho_f)}{\rho_f} = m \frac{(\rho_s - \rho_f)}{\rho_f} \qquad (5.7)$$

Solving Eq. 5.6 for b_f and equating this result to the right-hand side of Eq. 5.7 gives

$$L = \frac{(\rho_s - \rho_f)Km^2}{\rho_f Q'} \qquad (5.8)$$

Thus, the length of protrusion of salt water under natural conditions is directly proportional to the hydraulic conductivity and thickness squared and inversely proportional to the flow of fresh water to the sea. As the flow of fresh water to the sea is reduced, say by pumping along coastal reaches, the length of the intruded saltwater wedge increases.

The Shape of the Interface with a Submerged Seepage Surface The relationships just cited are only approximately correct because of the inherent hydrostatic assumptions. Further, the coincidence at the coastline of a zero freshwater and saltwater head, along with the assumption of a rigid interface across which no flow can occur, effectively closes the system at its discharge point. Another approach to this problem is demonstrated on Figure 5.23, which gives the results of an analysis by

Glover (1964) where the total discharge takes place below sea level. The position of the freshwater–seawater interface is determined from

$$z^2 = \frac{2Q'x\rho_f}{K(\rho_s - \rho_f)} + \left[\frac{Q'\rho_f}{K(\rho_s - \rho_f)} \right]^2 \qquad (5.9)$$

where z is the depth below sea level to the interface, x is the distance measured positive inland from the shoreline, and Q' is the freshwater flow per unit length of shoreline. The width of the freshwater discharge gap is determined where $z = 0$

$$x_o = - \frac{Q'\rho_f}{2K(\rho_s - \rho_f)} \qquad (5.10)$$

whereas the depth to the interface below sea level at the coastline ($x = 0$) is given as

$$z_o = \frac{\rho_f Q'}{(\rho_s - \rho_f) K} \qquad (5.11)$$

Last, the height of the water table at any distance x is given as

$$b_f = \frac{2Q'x(\rho_s - \rho_f)}{\rho_f K} \qquad (5.12)$$

so that at $x = 0$, $b_f = 0$. Thus, the greater the flow to the sea, the deeper the interface, and the larger the required freshwater discharge gap.

For the geometry of Figure 5.24, the landward protrusion of the saltwater wedge is determined directly from Eq. 5.9

$$L = \frac{1}{2} \left[\frac{m^2 K(\rho_s - \rho_f)}{Q'\rho_f} - \frac{\rho_f Q'}{(\rho_s - \rho_f) K} \right] \qquad (5.13)$$

It is noted that the first term in the brackets is identical to Eq. 5.8 based on the hydrostatic condition, whereas the second term is equal to the depth to the interface at $x = 0$ (Eq. 5.11). The role of fresh water flow to the sea as the major control on retarding the advance of salt water is clearly shown in this development.

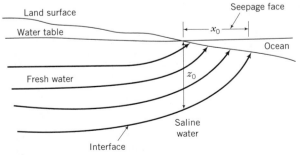

Figure 5.23 Flow pattern near the coast (modified from Glover, 1964).

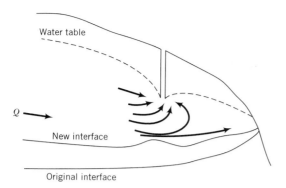

Figure 5.25 Freshwater flow to a well pumping above the interface.

Upconing of the Interface Caused by Pumping Wells

The Ghyben–Herzberg relationship indicates that any lowering of head results in a rise in the interface. When the lowering takes place by pumping wells that withdraw water from above the interface, the interface can rise, a phenomenon referred to as upconing (Figure 5.25). An approximate analytical solution for this upconing has been given by Schmorak and Mercado (1969). Their solution gives the new equilibrium elevation to an interface in direct response to pumping

$$z = \frac{Q\rho_f}{2\pi dK(\rho_s - \rho_f)} \qquad (5.14)$$

where z is the new equilibrium elevation, Q is the pumping rate, and d is the distance from the base of the well to the original (prepumping) interface (Figure 5.26). Both laboratory and field observations suggest that this relationship holds only for very small rises in the interface and there exists a critical elevation at which the interface is no longer stable and salt water flows to the well (Schmorak and Mercado, 1969).

Dagan and Bear (1968) suggest that the interface will be stable for upconed heights that do not exceed one-third of d given in Figure 5.26. Thus, if z is taken as $0.3d$,

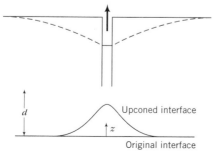

Figure 5.26 Upconing of interface in response to pumping.

the maximum permitted pumping rate should not exceed

$$Q_{max} \leq 0.6\pi\, d^2 K \left(\frac{\rho_s - \rho_f}{\rho_f} \right)$$

Dagan and Bear (1968) present other relationships dealing with upconing of the interface as well as estimates of maximum discharge to assure stability.

5.2 Surface Features of Ground-Water Flow

The saturated ground-water system discussed thus far has been isolated from the basin hydrologic cycle by introducing the effect of its environment through boundary conditions. In the real world there is a recognizable interaction between ground water and the rest of the basin hydrologic cycle. Some of the manifestations of this interaction are readily seen at the surface, whereas other types are best examined with models of some sort.

Surface features of ground-water flow include all observations that can be used to ascertain the occurrence of ground water, including springs, seeps, saline soils, permanent or ephemeral streams, ponds, or bogs in hydraulic connection with underground water. The areal occurrence of these features is invariably restricted to areas of ground-water discharge, and their comprehension requires some knowledge of the nature of ground-water outcrops (Meyboom, 1966b). The significance of these features may best be understood by examining a few studies that focus on their interpretation.

Recharge–Discharge Relations

The prairie profile (Figure 5.27), consisting of a central topographic high bounded on both sides by areas of major natural discharge, has been offered as a model of ground-water flow to which all observable ground-water phenomena in a prairie environment can be related (Meyboom, 1962, 1966b). Geologically, the profile consists of two layers, the uppermost being the least permeable, with a steady-state flow of ground water toward the discharge areas. The unconfined flow pattern has been substantiated by numerous borings for both small-scale systems, say, a typical knob and adjacent kettle common to rolling prairie topography, and for a scale of the magnitude of the prairie profile. Recharge and discharge areas have been delineated and are characterized by a decrease in head with increased depth and by an increase in head with increased depth, respectively. The occurrence of flowing wells is noted in parts of the discharge area.

As it is apparent that most of the natural discharge occurs by evapotranspiration, considerable attention has been given to this phenomenon and to the surface features observed where it occurs. Included among these

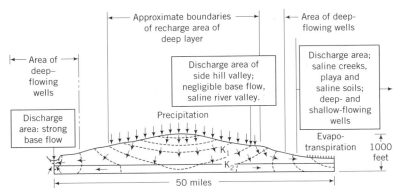

Figure 5.27 The prairie profile (from Meyboom, 1966b: Ground-water studies in the Assiniboine River drainage, Pt. I, The evaluation of a flow system in southern central Saskatchewan Geol. Survey Canada Bull. 139. Energy, Mines, and Resources Canada. Reproduced by permission of Supply and Services, Canada, 1990).

observations are (1) the occurrence of willow rings and the chemical character of water bodies centered within them, (2) the distribution and types of vegetation of saline soils with respect to the occurrence of local and regional flow systems, and (3) the location of ponds and bogs with respect to ground-water flow.

With regard to (1), Meyboom (1966a, 1966b) noted that most of the willows are located in the higher areas, which suggests that the recharge area is covered by numerous discharging points, each of which is characterized by a willow ring. Willows are phreatophytes, defined by Meinzer (1927) as plants whose roots extend to the water table so that they obtain water for transpiration directly from the saturated zone. As willows have a low tolerance for saline (alkali) conditions, their occurrence is associated with waters that have not moved very far in the system. Although their occurrence in the higher watershed areas is widespread, they do occur elsewhere in the basin. These other occurrences are regarded as possible manifestations of local flow systems superimposed on the regional ground-water flow.

With regard to (2), extensive areas of saline soil occur in areas of regional ground-water discharge, where a net upward movement of mineralized ground water takes place. A consistent transition noted is from willow vegetation on the watershed areas to halophytic plant communities within the discharge areas. A halophyte is a phreatophyte that thrives on saline waters. A related transition is noted for local flow systems that receive water from highly saline formations. Where the local system is not replenished by saline water formations, the ground water is relatively fresh, and saline zones fail to develop. In the latter case, freshwater phreatophytes occupy the discharge area and make possible its delineation as an end point of a local system. Tóth (1966) reported the results of a similar surface mapping method based on

observations in the prairie environment of Alberta, Canada.

The mapping techniques discussed here are only of value in arid and semiarid regions, where surface water is not sufficiently abundant to mask or conceal the surface effects of ground-water flow.

Investigations of the type discussed have not been restricted to the prairie environment. Extensive investigations in Nevada, starting perhaps with Meinzer (1922), have focused on surface features as they relate to the occurrence of ground water. In discussing basins in the Great Basin and the lakes that occupied them during the Pleistocene epoch, Meinzer (1922) recognized three types: (1) those in which lakes still exist, (2) those that no longer have lakes and are discharging ground water from the subterranean reservoirs into the atmosphere by evaporation and transpiration, and (3) those that do not have lakes and in which the water table is everywhere so deep that they do not discharge ground water except by subterranean leakage out of the basin.

The basin of type 2 is exemplified in early studies of Big Smokey Valley (Meinzer, 1917). Water enters the alluvial basin by influent seepage of streams of the alluvial fan areas, and at the contact between the surrounding mountain ranges and the alluvium. Ground water is discharged by evaporation and transpiration in the valley lowlands. Studies of the Big Smokey Valley and similar environments provided the first understanding of the surficial manifestations of ground-water flow, including the distribution of soluble salts in discharge areas (Meinzer, 1927). The concentric arrangement of phreatophytes was first noted in these early studies, with salt grass occupying the inner belt where the water table is near land surface. Other species, such as greasewood in northern Nevada and mesquite in southern Nevada, occupy the outer belt, where the water table is farther below land surface (Meinzer, 1927). This type of basin is exemplified in Figure 5.28, where the basin alluvium is underlain by low-permeability rock.

The basins of type 3 have been investigated by Winograd (1962), Eakin (1966), and Mifflin (1968), resulting in the delineation of two regional flow systems in carbonate terrain in southern and eastern Nevada. These systems are typified by large drainage areas encompassing several topographic basins, relatively long flow paths, and large spring areas of invariant discharge, where water temperature is several degrees higher than mean air temperature. In Figure 5.28 these basins are exemplified where the basin alluvium is underlain by highly permeable carbonate rock that siphons off the flow so as to eliminate ground-water discharge at the surface.

One of these regional flow systems occurs in eastern Nevada in the White River area (Figure 5.29). The major discharge points occur as springs in White River Valley, Pahranagat Valley, and Moapa Valley, the last identified as the terminal point of the flow system. Boundaries to this flow system, at least to the south, appear to be controlled by thick clastic rocks whose permeability is considerably less than that of the carbonates. Eakin (1964) obtained closure of the water balance within a 13-valley area, with 78% of the recharge estimated as occurring in the four northern valleys, and 62% of the discharge estimated to be from spring areas in Pahranagat and Moapa Valley. A tentative analysis suggests a 15- to 20-year lag in spring discharge response to recharge from precipitation in the recharge areas (Eakin, 1964). This is demonstrated on Figure 5.30, where precipitation about 100 miles north of the springs in Moapa Valley is plotted against spring discharge. The two curves match for the sharp rise in precipitation during the period 1935–1941 and the rise in the discharge graph during the period 1956–1960. The above-average precipitation for that

time period occurred regionally throughout eastern Nevada and western Utah, thus suggesting that above-average precipitation was widespread throughout the drainage area.

Most of Winograd's (1962) conclusions are based on the results of an extensive drilling program within the Nevada test site and adjacent areas. Hydraulic head data suggest that ground water moves from at least as far north as Yucca Flat to Ash Meadows, without regard to topographic divides (Figure 5.31). The northern part of the system is a recharge area, receiving water from the overlying alluvial reservoirs. Discharge through spring areas occurs in Ash Meadows, the terminal point of the system. Boundaries to the system, at least to the south, appear to be controlled by thick sequences of lower-permeability clastic rock. Winograd and Thordarson (1975) discuss the hydraulic details of flow in the South Central Great Basin, with special reference to this carbonate flow system.

The examples just cited deal with discharge phenomena in conspicuous topographic lows. Such areas reflect the observable manifestations of the theories of ground-water flow and are logical starting places for study of the system of flow contributing to large spring areas in volcanic and carbonate terrain.

Ground-Water–Lake Interactions

In performing the simulations leading to his conclusions on regional ground-water flow, Tóth (1963) recognized the presence of stagnation zones at the juncture of flow systems. A stagnation point is a point of minimum head along a subsurface divide separating one flow system from another. This is shown diagrammatically in Figure 5.32a, where the subsurface divide separates a local flow system associated with a lake from a regional system.

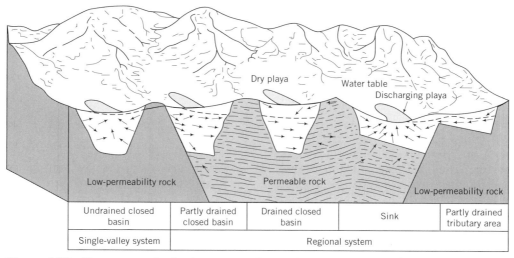

Figure 5.28 Flow systems in the Great Basin (from Eakin and others, 1976).

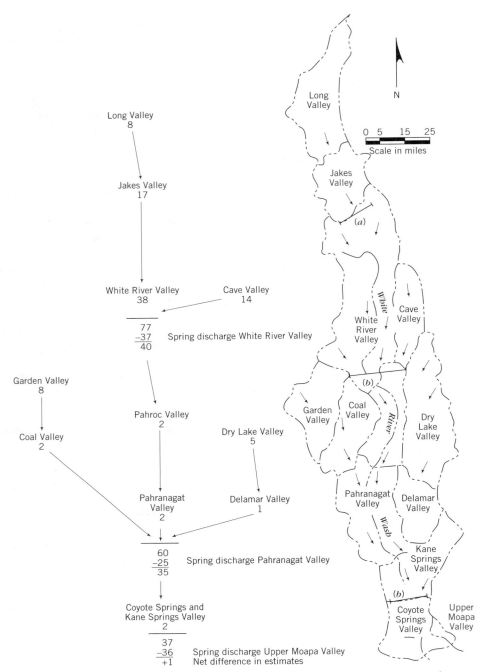

Figure 5.29   Estimated average annual recharge to and discharge from a regional ground-water system in eastern Nevada (from Eakin, Water Resources Res., v. 2, p. 251–271, 1966. Copyright by Amer. Geophys. Union).

The minimum head at the stagnation point is greater than the head in the lake. Thus, the lake of Figure 5.32a is termed a discharge lake in that it receives ground-water discharge from the aquifer. A recharge lake is one that recharges the ground-water flow, and a through-flow lake (Figure 5.32b) is one that both receives and releases water to the ground-water flow (Born and others, 1974, 1979). For a 63-lake sample studied by Born and others

(1974), most of which were in the Midwest region, 24 were discharge lakes, 23 were through-flow lakes, 6 were recharge lakes, 4 were placed in some combined category, and 6 changed from one type to another in response to seasonal changes (Anderson and Munter, 1981).

The search for stagnation points is aided by model studies of the type initiated by Tóth (1963) and Freeze and Witherspoon (1967). One example is provided by

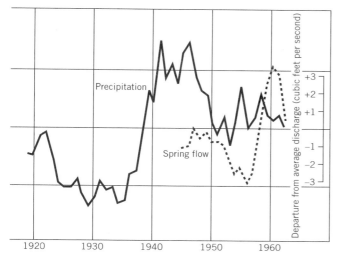

Figure 5.30 Cumulative departure from average precipitation 100 miles north of Moapa Valley versus cumulative departure from average spring flow from Moapa Valley (from Eakin, 1964).

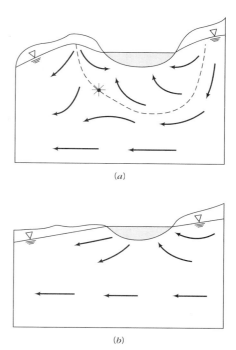

(a)

(b)

Figure 5.32 Ground-water flow paths in the vicinity of lakes. Panel (*a*) is a discharge lake, where the asterisk shows the location of a stagnation point. Panel (*b*) is a flow-through lake (from Anderson and Munter, Water Resources Res., v. 17, p. 1139–1150, 1981. Copyright by Amer. Geophys. Union).

Winter (1976, 1978). Figure 5.33*a* gives results typical of these studies. In Figure 5.33*a*, a stagnation zone exists on the downgradient side of the single lake. The lake is thus classified as a discharge lake and receives water from the flow local system. If a high-conductivity layer is introduced at depth (Figure 5.33*b*), the stagnation point is eliminated and the pirating effect of the high-conductivity layer converts the lake into a recharge lake. Multiple-lake systems on a regional slope have also been

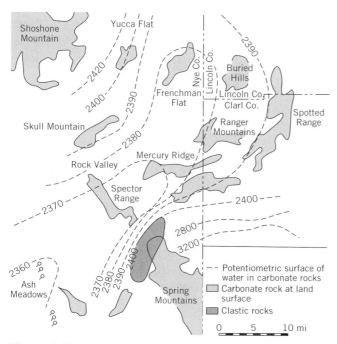

Figure 5.31 Regional flow systems in rocks underlying the Nevada test site and adjoining area (from Winograd, 1962).

investigated by Winter (1976) and can be recharge, discharge, or through-flow depending on basin geometry and geology. Steady-state models have commonly been used to describe such behavior in the vicinity of lakes (McBride and Pfannkuch, 1975; Larson and others, 1976). In addition to these model studies, other investigators (Meyboom, 1966a, 1967; Anderson and Munter, 1981) have conducted field investigations in the vicinity of lakes for the purpose of determining seasonal changes in the flow pattern.

Ground-Water–Surface-Water Interactions

Flow studies of ground-water basins may be usefully applied to gain an understanding of the interrelations between the processes of infiltration and recharge at topographically high parts of the basin, and of ground-water discharge via evapotranspiration and base flow at topographically low points. Such studies permit a spatial evaluation of the flow of ground water in the hydrologic cycle. For example, at least some of the water derived from precipitation that enters the ground in recharge areas will be transmitted to distant discharge points, and so cause a relative moisture deficiency in soils overlying recharge areas. Water that enters the ground in discharge areas cannot overcome the upward potential gradient,

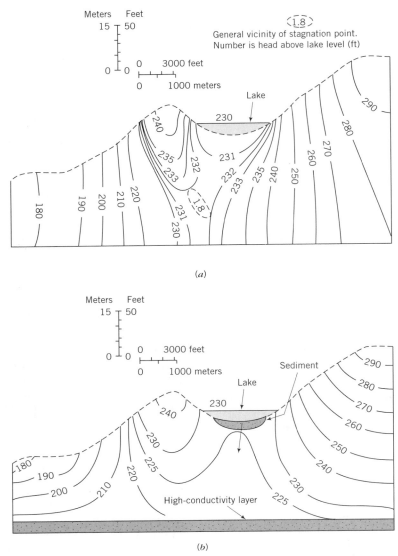

Figure 5.33 Flow conditions in the vicinity of a lake demonstrating the effect of a high-permeability layer at depth (from Winter, 1976).

and therefore becomes subject to evapotranspiration in the vicinity of its point of entry. The hinge line separating areas of upward and downward flow may thus serve as a boundary common to areas of relative soil moisture surplus and deficiency (Tóth, 1966). In nonirrigated agricultural areas, this may be reflected by variations in crop yield. Further, the ramifications of human activities in discharge areas are immediately apparent. Some of these include waterlogging problems associated with surface-water irrigation of lowlands, or due to the destruction of phreatophytes.

The spatial distribution of flow systems will also influence the intensity of natural ground-water discharge. The main stream of a basin may receive ground water from the area immediately within the nearest topographic high, and possibly from more distant areas if the region is characterized by both regional and local systems. If base-flow calculations are used as indicators of average recharge, significant error may be introduced because base flow may represent only a small part of the total discharge occurring downgradient from the hinge line.

Other interesting aspects resulting from such studies involving the basin hydrologic cycle deal with determining the actual amount of ground water that effectively participates in the hydrologic cycle. Tóth (1963) calculated that about 90% of the recharge never penetrates deeper than 75 to 90 m. Similar conclusions have been arrived at by tritium dating studies, which demonstrated a stratification of tritium and tritium-free waters (Carlston and others, 1960).

As might be expected, the majority of surface-water–ground-water studies have been conducted with model calculations. Transient flows of water in the unsaturated zone have been examined by Rubin (1968), Hornberger and others (1969), and Freeze (1971). The models couple saturated–unsaturated conditions. The ideas of infiltration capacity have likewise been studied in a transient mode with models that incorporate coupled unsaturated and saturated behavior (Freeze, 1969). Models that couple surface and subsurface flow have been used to study the runoff components of the hydrologic cycle (Smith and Woolhiser, 1971) as well as aquifer–stream connections (Hornberger and others, 1970).

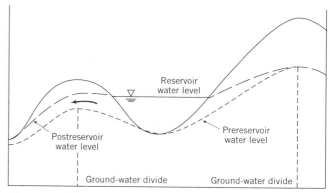

Figure 5.34 Changes in water levels due to construction of reservoirs (from Bryan, 1925).

5.3 Some Engineering and Geologic Implications of Topographic Drive Systems

Large Reservoir Impoundments

Dams, with their attendant surface-water reservoirs, are generally constructed in areas of ground-water discharge. If a reservoir is to occupy a deeply incised stream, the water level in the reservoir can be higher than the regional ground-water levels underlying the stream reach and in the adjacent flood plains that are inundated. With such an imposed head due to the filling of the reservoir, ground-water discharge to the stream and flood plain in the entire region that is inundated is no longer possible. This has little or no influence on the rate of recharge taking place at adjacent topographic highs where the ground-water levels are higher than the highest reservoir level. A situation is thus created where more water is entering the system than can be discharged from the system under the prevailing hydrologic regime. The prevailing hydrologic regime, however, does not prevail long. What was likely a steady state in preconstruction days now becomes a transient one, where the incoming water goes into ground-water storage, everywhere raising the water table (Figure 5.34). Eventually, the total discharge from the system must once more equal the recharge. Normally this will take place after the water table has risen sufficiently so that new discharge regions are created to take the place of those that can no longer function properly. Cady (1941) describes such effects on ground-water levels in response to surface storage in the vicinity of Flathead Lake, Montana, making note of the potential impact on agricultural lands where the water table is not far below the surface.

In association with an impoundment and subsequent rising of the water table, a reversal in flow directions in the subsurface is a distinct possibility. This has been

documented by Van Everdingen (1972), who studied the changes in the ground-water regime in the vicinity of Lake Diefenbacker on the South Saskatchewan River. Piezometric levels were obtained along a line crossing the Saskatchewan River both before and after construction of the reservoir, with the piezometers placed in the permeable sand units that are part of the Bearpaw shales. The prereservoir flow system was exactly what one would predict, with flow directed toward the Saskatchewan River, where it eventually ends up as ground-water discharge. After construction, there were some notable changes. In the deeper unit, the transversal flow was maintained toward the river valley, but at much lower gradients. In the intermediate unit, the transversal flow was reversed away from the river valley. In the uppermost unit, all the transversal flow was unidirectionally away from the site.

Frequently, an increase in spring discharge has been observed in the vicinity of new reservoirs. This has been reported by Gupta and Suknija (1974), who measured increments in spring flow on the order of 0.145 liters/s in springs some 15 miles from a dam in India. The dam is constructed on basalt. Gupta and Rastogi (1970) report fluctuations in four piezometers installed in limestone near a reservoir in Greece that were identical to those associated with fluctuations in the lake level. Snow (1972) has also reported increases in spring discharge in the vicinity of the same reservoir.

The impounding of reservoirs can also lead to some stability problems. Underneath the reservoir area itself, the increased fluid pressures as shown in Figure 5.35 are partially compensated by the increased weight of the water that must be added to the total stress. Downstream from the reservoir, however, no such compensation occurs. In Figure 5.35b, two principal effects are noted. These include increased deformation of the valley bottom and increased deformation and landslides in the valley walls.

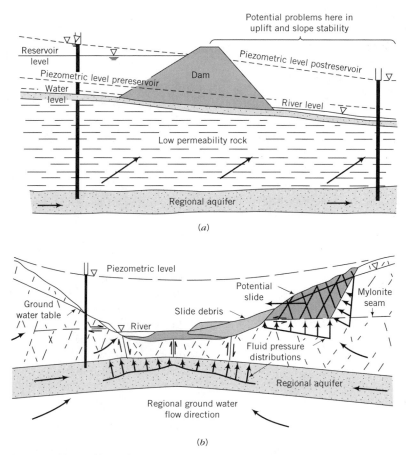

(a)

(b)

Figure 5.35 Effect of dams on water levels showing (*a*) possible stability problems created by reservoir blocking regional ground-water discharge and (*b*) effects of high fluid pressures on valley bottom and wells downstream from the reservoir (from Patton and Hendron, 1974).

Excavations: Inflows and Stability

The Sea-Level Canal

By their very nature, sea-level canals are built in coastal areas of low topographic relief, such as the Florida and Panama peninsulas. The natural equilibrium position of the freshwater–saltwater interface is already at a fairly high elevation, although still below sea level. Excavation below the water table permits the canal to act as a natural drain, causing water levels to decline with an attendant rise of the interface (Figure 5.36). In a preliminary investigation of a sea-level canal across Florida, Paige (1936) contended that a new ground-water table would evolve and be steepest near the canal and approach the older one asymptotically at a distance not in excess of 15 miles. As the canal would be cut approximately 10 m below sea level, it was taken as nothing more than an extension of the existing coast line of the Gulf of Mexico. The argument here was that fresh water is currently present in coastal wells along the Gulf of Mexico, and similar conditions will prevail along the canal cut. These conclu-sions were criticized by Brown (1937) and other notable experts in the profession at that time, including Thompson, Meinzer, and Stringfield (1938).

In his investigation, Paige noted the general relation-

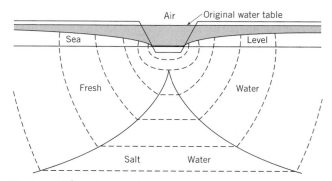

Figure 5.36 Saltwater-freshwater relationship due to excavation along coastal areas (from Hubbert, 1940). Reprinted by permission of the Journal of Geology, University of Chicago Press. Copyright © 1940.

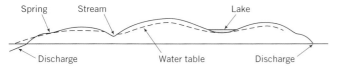

Figure 5.37 General relation of ground water to topography (from Paige, 1936). Reprinted from Economic Geology, 1936, v. 31, p. 537–570.

ship of ground water to topography, focusing in particular on the various water table highs associated with topographic highs and the fact that the outcrop areas of the formation to be cut (Ocala limestone) actually served as discharge areas (Figure 5.37). He noted topographically high areas both south and north of the canal, between which there was generally lower ground along the canal route. Based on this information along with various borings and detailed investigations, Paige concluded that the canal route traversed the most important discharge area of the Floridian plateau. In his reply to his critics, Paige (1938) stated that this discharge area played an important part in the selection of the canal route, as opposed to many others examined by the Army Corps of Engineers. This was likely one of the first large-scale engineering projects where the ground-water flow system with its attendant recharge and discharge areas actually figured in the decision process on site location.

One of the important issues was the potential effect on Silver Springs, the largest limestone spring in the United States and an important tourist attraction. Silver Springs is 3 miles north of the canal route and about 12 m above sea level. Paige argued that Silver Springs was an outlet of flow systems draining southward from the north so that the canal should not affect the source of water supplied to the springs. In the final analysis, Paige (1936) recognized that there would be a substantial effect on water levels near the canal, but these effects would be of little consequence in terms of damage. As it turned out, the potential impact of the canal was probably not as severe as the impact of ground-water development that took place in later years. Observations by Brown and Parker (1945) along a completed portion of the canal showed that saltwater encroachment had been facilitated by hydraulic gradients created not by the canal but by pumping.

From this brief discussion, it is clear that most of the arguments for or against the canal were largely qualitative, based exclusively on the geologic controls of the hydrology. In view of the times it could be no other way. If such an event were to repeat itself today, the debate would not go on in a scientific journal, but in the courts, with a host of lawyers and their various consultants and more than likely more than one mathematical model that

purportedly completely describes the projected response.

Ground-Water Inflows into Excavations

Ibrahim and Brutsaert (1965) provide a simple analytical solution for predicting inflows and a projected water-level decline in the vicinity of excavations such as the sea-level canal just discussed. The case they considered is shown in Figure 5.38 as a two-dimensional homogeneous isotropic region. The transient response of the water table is presented by the dimensionless graph of Figure 5.38b, where

$$N_{FO} = t \left/ \frac{S_y L^2}{KH} = \frac{t}{T^*} \right. \tag{5.15}$$

where S_y is the specific yield and the product KH is the transmissivity, here obviously assumed not to be affected by the water-level decline, and T^* is the time constant for the basin. The dimensionless discharge χ is given in Figure 5.38c and is expressed

$$\chi = \frac{S_y L q'}{KH^2} = \frac{T^* q'}{HL} \tag{5.16}$$

where q' is the flow rate into the excavation per unit length of excavation and has the units L^2/T.

To apply this method, information is required for K, S_y, H, and L. Verma and Brutsaert (1970, 1971) provide numerical models for similar problems.

Example 5.2

Assume that the basin parameters S_y, L^2, K, and H are such that $T^* = 10^2$, in days. For a time period $t = 100$ days, $t/T^* = 1$. From Figure 5.38b, this means that the ground-water head at the face of the excavation ($x = 0$) is about 0.1 of the original saturated thickness H, and at the ground-water divide ($x = L$) it is about 0.5 of the original saturated thickness. The position of the water table may be readily determined at other points. For $t/T^* = 1$ and Figure 5.38c, the dimensionless discharge of Eq. 5.16 is determined to be about 0.25, from which q' can be determined. Note that for longer times, t/T^* increases and the discharge rate q' declines in response to the flatter hydraulic gradients.

The Stability of Excavations in Ground-Water Discharge Areas

One of the most common problems in engineering construction is the bottom heaving of trenches or other excavations constructed in areas of ground-water discharge. Several case histories are available, but we shall focus on only one to bring out the principles involved. This particular situation occurred during a project where cuts up to 20 m were required in glacial materials. The

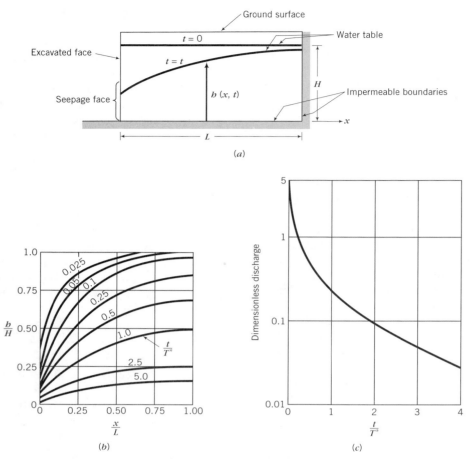

Figure 5.38   Prediction of ground-water inflows into an excavation (from Ibrahim and Brutsaert, 1965). Reprinted by permission of the American Society of Civil Engineers.

major drainage in the region adjacent to the proposed excavations occupied an intermoranic sag; such sags are noted for their heterogeneity, variation in grain size, and general lenticular character of the sands and gravels. The sands and gravels that do occur in these geologic environments generally have a high permeability in that they have been well sorted by glacial melt water.

A lake occupying a gravel pit was located adjacent to the construction area. The lake was pumped prior to construction for a period of several months, lowering the lake level about 15 m. Five observation wells indicated changes of 2.5 to 5 m over this period. Four of the wells were bottomed in the underlying limestone, which is highly fractured with excellent hydraulic communication throughout. The one well in the glacial material showed a head drop of about 2.5 m, and this well was closer to the gravel pit than some of the deeper wells that penetrated the limestone, which showed a greater response to pumping. It is likely, then, that pumping from the pit did not materially affect the water levels in the glacial

materials where the overall heterogeneity prevented a good hydraulic communication between the permeable lenses. The response in the underlying limestone was likely the result of pressure relief associated with pumping the lake.

Contract documents required the contractor to maintain the lake level about 10 m below its natural level with the idea that this would sufficiently reduce pressures in the glacial material so that stability problems would not be encountered. However, during construction, the bottom of the excavations heaved, and the construction had to be stopped until dewatering wells were put in at an additional cost.

One interesting point in all this is that the position of the water table was entirely overlooked in these investigations. All the observation wells in the area indicated water levels that were at or a few feet below land surface. Indeed, one of the piezometers was flowing. This was termed the "water table" in the preconstruction investigations. However, backhoe data from several different locations indicated that the sediments were dry from land surface to in excess of 4.6 m in depth; in none of these excavations was the water table encountered. Hence, we reemphasize the fact that water levels in piezometers that penetrate below the top of the zone of saturation do not indicate the position of the water table, but indicate a head or potential that may be greater than the water table, or less than the water table, depending upon where one is measuring in the flow system. In this case, as in all discharge areas, the former is the actual situation.

A simple calculation could have removed any doubt as to the need of dewatering or pressure relief prior to construction. Assume that the water table is at 15 ft below the surface, the dry unit weight of the material is on the order of 100 lb/ft^3, and the saturated unit weight is 110 lb/ft^3. The total vertical stress at the base of the glacial material (100 ft) is thus 10,850 lb/ft^2. If the water level in a piezometer finished at the base of the glacial material registered a head at land surface, the neutral stress is calculated to be about 6240 lb/ft^2. Hence, after removal of about 43 ft of material the effective stress at the base of the glacial material is reduced to zero. It would appear that cuts approaching 60 ft in depth are not likely to be stable without some pressure relief provided by pumping wells.

Landslides and Slope Stability

In his classic treatment of landslide development, Terzaghi (1950) noted the following causes for the relatively rapid movement of slopes: (1) external changes, (2) earthquake shocks, (3) lubrication by water, and (4) ground-water level rises. External changes refer to man-made or natural activities that disturb the stability of an existing slope, such as undercutting the toe, say, by meandering rivers or excavations. Earthquake shocks increase the shearing stresses relative to the shearing resistance as accelerations associated with earthquakes represent horizontal forces. The lubrication by water was dismissed by Terzaghi because water actually acts as an antilubricant in that there is more friction between "wet" mineral grains than between dry ones. The main emphasis of Terzaghi's paper is the effect of water-level changes in the interpretation of the Mohr–Coulomb equation, stated here as

$$\tau = \tau_o + (\sigma - P)\tan\phi \qquad (5.17)$$

where τ is the shearing resistance per unit area, τ_o is the cohesion per unit area, σ is the total normal stress at a point due to the weight of the solids and fluids, P is the fluid pressure at the point, and ϕ is the angle of sliding friction. The greater the fluid pressure P, the greater is the portion of the overburden that is carried by the water;

when $\sigma - P = 0$, the overburden is in a state of floatation (Terzaghi, 1950), and the resistance to shear is reduced to zero for cohesionless material, and to the cohesive strength for materials that possess cohesive properties.

Terzaghi (1950) argued that seasonal variations in rainfall can give rise to seasonal variations in the fluid pressure, thereby reducing the shearing resistance independent of any effect on the angle of sliding friction. Thus, during periods of heavy or prolonged rainfall, slopes become more susceptible to failure because of the attendant increases in water levels and decreases in effective stress. This sometimes leads to a definite periodicity in slope failures. Seepage from sources of water such as newly built reservoirs or unlined canals can have similar results. Hence, if a slope starts to move, the means for stopping the movement must be adapted to the processes that started it (Terzaghi, 1950). As most slides are related to increases in fluid pressures, this means that both surface and subsurface drainage or dewatering is called for.

As with most original developments in science, once the preliminary physics of a problem are brought out in theory, others immediately rush in to embellish the original ideas. By embellish we mean to improve by adding detail. The detail added in this case is of a hydrologic and geologic nature. Deere and Patton (1967) put Eq. 5.17 in a hydrologic perspective, recognizing that slopes are part of a regional flow system that is characterized by elevation differences in water levels and in land surface. They recognize that small topographic deviations within a regional system produce local flow systems that are often the most critical with regard to fluid pressures and slope stability, especially in areas of ground-water discharge. In addition, these authors demonstrated sequences of dipping beds susceptible to high pressures, the susceptibility of weathered shales to landslides when the slope is in a discharge area, and the pore water pressures that can develop at the base of low-permeability materials overlying more permeable rocks in a ground-water discharge area. These same ideas are carried over by Patton and Hendron (1974).

Fractured rock masses are treated more or less along the same lines. There are, however, some differences, again, purely geological. In that the porosity of fractured rock may be two or three orders of magnitude smaller than that of sedimentary rock, a typical rainfall event should result in a relatively larger water-level response. Further, because of the prevailing shear stresses in a body of rock forming a slope at some inclination to the horizontal, the joints may be wider and more numerous than those located in otherwise similar, flat-lying rock. Hence the quantity of water that can enter the rock mass is greater in the vicinity of the slope.

Numerous references are available on ground-water and slope stability. These include Cedergren (1967), Morgenstern (1970), Kiersch (1973), and Royster (1979).

Problems

1. Briefly discuss the meaning of the boundary conditions given on Figure 5.5. What boundary conditions are included in Figure 5.8 that are not in Figure 5.5? Why?

2. Construct flow nets in the *x-z* plane for Figure 5.4 where the left-hand boundary is a flow (as opposed to a no-flow) boundary, and for Figure 5.27 for the case where $K_1 = K_2$.

3. Set up a problem similar to the one discussed for the sea-level canal. Assume a flat water table 12 m above sea level, a canal that extends 9 m below sea level, and a ground-water divide (springs) approximately 3 miles from the canal. Assume a hydraulic conductivity of 8.35 cm/day and a specific yield of 0.01.

a. Calculate the head *h* at the canal and at the divide for times of 10, 20, 40, and 100 years.

b. Calculate the flow rate q' into the canal for 10 years and 100 years.

4. Perform the calculations for the case history given in the section dealing with stability of excavations.

5. A saltwater interface is stable at a position of about 40 ft beneath the base of a well in a formation with a hydraulic conductivity of 1×10^{-1} cm/s. Calculate the maximum pumping rate so as not to cause upconing.

6. The flow of fresh water toward the sea is 2×10^{-4} ft³/s per foot of coastline in an aquifer that is 100 ft thick with a hydraulic conductivity of 3.3×10^{-3} ft/s. Calculate the inland distance from the shoreline to the toe of the saltwater wedge. What is the new position from the shoreline if the seaward flow is reduced by one-half?

Chapter 6

Hydraulic Testing:
Models, Methods, and Applications

Hydraulic testing as used in this book is a description of the field tests and testing procedures required to obtain certain hydraulic properties, pressure measurements, or indirect determinations of ground-water velocity in rocks of various kinds. Such testing procedures were born out of the practical necessity of evaluating ground water as a water supply. Thus the early beginnings in this field dealt exclusively with pumping tests designed for permeable formations that were important from the perspective of water supply. In more recent times, due largely to concerns of hazardous waste migration and nuclear waste burial, interest has shifted to testing procedures suited to low-permeability rocks or rocks with a fracture-type permeability. For simplicity we classify the various tests as conventional or specialized. By conventional we mean those tests that virtually all hydrologists conduct in a routine fashion in relatively simple geologic environments. Specialized testing procedures are often required in low-permeability rocks and especially in fractured rocks.

6.1 Prototype Geologic Models in Hydraulic Testing

Every hydrologic model employed in hydraulic testing is patterned after some specific or prototype geologic environment. Three such environments are shown in Figure 6.1. Figure 6.1*a* shows a typical confined aquifer of large areal extent, assumed to be isotropic and homogeneous, overlain by a confining layer over which there is a water table. There are three variations of this prototype model. First, if the confining layer is impermeable and contains an incompressible fluid within an incompressible matrix ($S'_s = 0$), all the water removed from the aquifer will come from storage within the aquifer, and the water-level change (drawdown) versus time response at some observation point will be exponential, that is, will plot as a straight line on semilogarithmic paper (curve *A*, Figure 6.2). For reasons that will be clear later, this is referred to as the nonleaky response. If the confining layer has some finite permeability but still contains an incompressible fluid within an incompressible matrix, it may be possible to invoke the transfer of water across the confining layer. The time–drawdown observation will reflect this additional water source with an upward inflection at some point in time (curve *B*, Figure 6.2). This is referred to as the leaky response. Two things are worth noting here. First, because of the time delay for water to enter the pumped aquifer, an aquifer may appear nonleaky over several hours or days of pumping and will be so classified if the test is terminated before the leakage has been detected. This, however, will not affect the

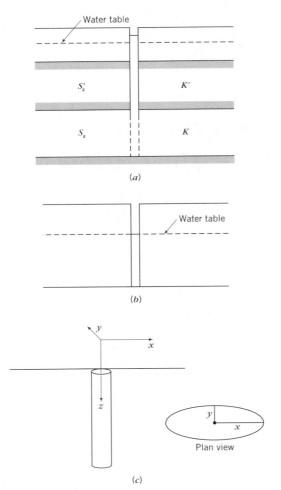

Figure 6.1   Prototype geologic models.

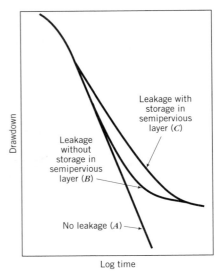

Figure 6.2   Comparison of time–drawdown response for geologic prototypes of Figure 6.1*a* (from Hantush, J. Geophys. Res., v. 65, p. 3713–3725, 1960. Copyright by Amer. Geophys. Union).

analysis in that the leaky and nonleaky curves remain coincident or nearly so until the inflection point is realized (Figure 6.2). Second, leakage across a confining layer may not be the only cause for the upward inflection. If pumping is near a stream or a lake and surface water is brought into the aquifer due to the established hydraulic gradients, a similar inflection will occur.

The last version of Figure 6.1*a* is for a permeable confining layer with a finite specific storage. In this case, the confining layer itself can contribute water to the aquifer, resulting once more in an upward inflection (Figure 6.2*c*). This is referred to as leakage with storage in the confining layer. It is noted now that the time delay is considerably shortened (Figure 6.2). For long times, the leakage with storage case is coincident with the leaky case, but this coincidence evolves through a different time–drawdown pathway.

Yet another geologic prototype is the aquifer that is considered homogeneous but anisotropic. In Figure 6.1*c* the aquifer is considered to be horizontal and one of the principal directions of the hydraulic conductivity tensor is vertical, that is, parallel to the well. In this case, the

horizontal conductivities in a horizontal plane (x and y) are not equal. The response of such a system is shown in the plan view given in Figure 6.3 where the transmissivity in the x direction is larger than the transmissivity in the y direction. Variations of this model include situations where the principal directions of the hydraulic conductivity tensor are neither vertical nor horizontal. Two- and three-dimensional anisotropic models may be useful in the analysis of fractured rock.

This brief discussion does not exhaust the prototype geologic model guiding the hydraulic testing of real rock. Others include a double-porosity concept where water can be removed from both the fractures and matrix of a fractured porous medium and discrete flow in individual fractures.

Operational hydraulic equations have been derived for the ideal prototype models discussed earlier. The solutions have been obtained from some form of the diffusion

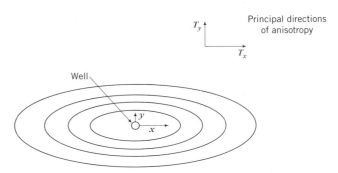

Figure 6.3   Plan view of drawdown response in an anisotropic aquifer.

equation, where time is a parameter. The essence of hydraulic testing calls for the matching of real response data with the theoretical response expected from the prototype condition. In most of the conventional methods, this matching is performed with a curve-fitting procedure or by some simple graphical technique. With some of the specialized testing procedures, the matching is frequently conducted with the aid of a computer.

6.2   Conventional Hydraulic Test Procedures and Analysis

The term *conventional* is used here in reference to hydraulic testing procedures that are most commonly used in formations of moderately high permeability with the stated purpose of determining the transmissive and storage properties of aquifers. In this section we will restrict our discussions to tests that require one pumping well and one or more observation wells or piezometers in which it is possible to measure the response to pumping. From an historic perspective, these procedures originated with the work of C. V. Theis in 1935.

The Theis Nonequilibrium Pumping Test Method

The mathematical expression for removal of heat at a constant rate from a homogeneous, infinite slab has provided a useful analogy for study of ground-water flow to a pumping well. It is assumed, initially, that the slab is at some uniform temperature. An infinitesimal rod of lower temperature parallel to the z axis is then allowed to draw off the heat (Figure 6.4a). In mathematical terminology, the rod represents a continuous line sink. The temperature change at some distance r from the rod is a function of the rate at which heat is withdrawn, the properties of the slab, and time. By analogy, the infinite, homogeneous slab is replaced by an extensive, homogeneous aquifer, and the rod by a well of infinitesimal diameter (Figure 6.4b). Similarly, the rate of pumping is analogous to the rate of heat withdrawal, and water-level change at any distance r from the pumping well is a

function of the pumping rate, the properties of the aquifer, and time. A solution to the heat-flow problem just stated is given by Carslaw and Jaeger (1959). The hydrologic analog was given by Theis (1935) and is of the form

$$h_o - h = s = \frac{Q}{4\pi T} \int_{r^2 S/4Tt}^{\infty} \frac{e^{-z}}{z} dz \qquad (6.1)$$

where h_o is the original head at any distance r from a fully penetrating well at time t equals zero, h is the head at some later time t, s is the difference between h_o and h and is called the drawdown, Q is a steady pumping rate, T is the transmissivity, S is the storativity, and the exponential integral is a well-known tabulated function, the value of which is given by the infinite series

$$\int_u^{\infty} \frac{e^{-z}}{z} dz = -0.577216 - \ln u + u$$

$$- \frac{u^2}{2 \cdot 2!} + \frac{u^3}{3 \cdot 3!} - \frac{u^4}{4 \cdot 4!} + \cdots \qquad (6.2)$$

where

$$u = \frac{r^2 S}{4Tt} \qquad (6.3)$$

Equation 6.1 is a solution to the polar coordinate form of the diffusion equation, or

$$\frac{\partial^2 h}{\partial r^2} + \frac{1}{r} \frac{\partial h}{\partial r} = \frac{S}{T} \frac{\partial h}{\partial t} \qquad (6.4)$$

for the initial and boundary conditions

$$h(r, 0) = h_o$$

$$h(\infty, t) = h_o$$

$$\lim_{r \to 0} \left(r \frac{\partial h}{\partial r} \right) = \frac{Q}{2\pi T} \qquad \text{for } t > 0$$

The first two of these are read: "The head at some radius r from the well at time zero equals the initial head h_o, and the head at an infinite distance at any time t equals the initial head h_o." The second condition stipulates that water withdrawal may affect distant parts of the aquifer but the exterior boundary is never encountered (as expected for an aquifer of infinite extent). The third condition provides for a constant withdrawal rate at r_w, which is the radius of the well, which, in turn, is taken as infinitesimal.

Equation 6.1 is referred to as the nonequilibrium equation. It is sometimes referred to as the nonleaky well flow equation in that all the water pumped is removed from a single aquifer with no contributions from other beds either above or below the pumped aquifer.

A verbal interpretation of the parameters incorporated in the nonequilibrium equation can provide insight into the shape and growth of a cone of depression in an ideal aquifer. It is first noted that the exponential integral of

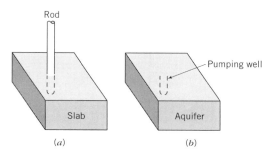

Figure 6.4   Schematic diagram of (*a*) an infinite slab and (*b*) the well flow analogy.

Table 6.1   Values of $W(u)$ for Values of u

u	1.0	2.0	3.0	4.0	5.0	6.0	7.0	8.0	9.0
$\times 1$	0.219	0.049	0.013	0.0038	0.0011	0.00036	0.00012	0.000038	0.000012
$\times 10^{-1}$	1.82	1.22	0.91	0.70	0.56	0.45	0.37	0.31	0.26
$\times 10^{-2}$	4.04	3.35	2.96	2.68	2.47	2.30	2.15	2.03	1.92
$\times 10^{-3}$	6.33	5.64	5.23	4.95	4.73	4.54	4.39	4.26	4.14
$\times 10^{-4}$	8.63	7.94	7.53	7.25	7.02	6.84	6.69	6.55	6.44
$\times 10^{-5}$	10.94	10.24	9.84	9.55	9.33	9.14	8.99	8.86	8.74
$\times 10^{-6}$	13.24	12.55	12.14	11.85	11.63	11.45	11.29	11.16	11.04
$\times 10^{-7}$	15.54	14.85	14.44	14.15	13.93	13.75	13.60	13.46	13.34
$\times 10^{-8}$	17.84	17.15	16.74	16.46	16.23	16.05	15.90	15.76	15.65
$\times 10^{-9}$	20.15	19.45	19.05	18.76	18.54	18.35	18.20	18.07	17.95
$\times 10^{-10}$	22.45	21.76	21.35	21.06	20.84	20.66	20.50	20.37	20.25
$\times 10^{-11}$	24.75	24.06	23.65	23.36	23.14	22.96	22.81	22.67	22.55
$\times 10^{-12}$	27.05	26.36	25.96	25.67	25.44	25.26	25.11	24.97	24.86
$\times 10^{-13}$	29.36	28.66	28.26	27.97	27.75	27.56	27.41	27.28	27.16
$\times 10^{-14}$	31.66	30.97	30.56	30.27	30.05	29.87	29.71	29.58	29.46
$\times 10^{-15}$	33.96	33.27	32.86	32.58	32.35	32.17	32.02	31.88	31.76

From Wenzel (1942).

Eq. 6.2 is a function only of the lower limit of integration so that Eq. 6.1 can be written

$$s = \frac{Q}{4\pi T} W(u) \qquad (6.5)$$

where u is defined in Eq. 6.3. The term $W(u)$ is termed the well function of u, which indicates its dependence on values of u expressed in Eq. 6.3. It follows that the value of the exponential integral may be ascertained and tabulated for each of several values of u. Such a tabulation is shown in Table 6.1 for values of u ranging from 10^{-15} to 9.0. If u equals 4.0×10^{-10}, then $W(u)$ equals 21.06, which is the value of the series of Eq. 6.2 for this particular value of u.

Of the variables comprising u, the storativity and transmissivity may be considered constant for a given set of conditions, and distance and time considered variables. Values of $W(u)$ may be plotted against values of $1/u$ in such a way that for any fixed distance r, the plotted curve reveals the exact shape of the drawdown curve as a function of time t (Figure 6.5). Curves showing the relation between $W(u)$ and u are termed "type" curves.

In that the value of u depends on time, distance, transmissivity, and storativity, u determines the radius of a cone of depression (Theis, 1940). The radius not only increases with increasing time but, for a given time, is larger for decreasing values of storativity and increasing values of transmissivity. By examining the complete statement for drawdown (Eq. 6.5), drawdown at any point for a given time is proportional to discharge and inversely proportional to transmissivity. The lateral extent of a cone of depression at any given time and its rate of growth are independent of the pumping rate.

As a final point, the dimensionless variable u is of the form of an inverse Fourier number introduced in Chapter 4

$$u = \frac{r^2/t}{4(T/S)} = \frac{Sr^2/4T}{t} = \frac{T^*}{t} \qquad (6.6)$$

where $Sr^2/4T$ has the units of time and may be referred to as a time constant T^*. Thus, if the time constant T^* is small compared to time t, u becomes small, $W(u)$ becomes large, and drawdowns become large (Figure 6.2). Small values for the time constant are associated with small values of storativity, say, for confined conditions, small distances to the point of observation, or large values of transmissivity. If the time constant $Sr^2/4T$ becomes large with respect to a proposed time of observation t, the expected drawdown at the point of observation r becomes imperceptibly small. Large values for the time

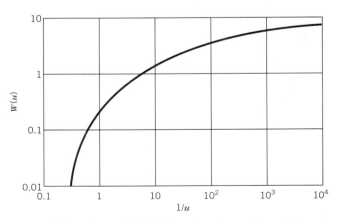

Figure 6.5   Values of $W(u)$ plotted against values of $1/u$.

constant are frequently associated with large values for the storativity, as expected for unconfined conditions.

The Curve-Matching Procedure

The Theis (or nonequilibrium) equation is used extensively to determine the hydraulic properties of aquifers. For these purposes, it will be recalled that for a fixed distance r, drawdown versus time is of the form of the curve in Figure 6.5. If drawdown s can be measured at one point r for several values of time t, and if the discharge is steady and is known, the coefficients of transmissivity and storativity can be determined by a graphical method of superposition. Field data composed of drawdown versus time collected at a nonpumping observation well at a known distance r from a pumped well are plotted on logarithmic paper of the same scale as the type curve (Figure 6.6a). The field curve is superimposed on the type curve, with the coordinate axes of the two curves kept parallel while matching field data to the type curve (Figure 6.6b). Any point on the overlapping sheets is selected arbitrarily (the point need not be on the matched curves). The selected point is defined by four coordinate values: $W(u)$ and s, and $1/u$ and t (Figure 6.6b). Equation

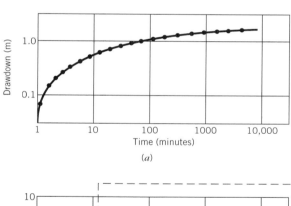

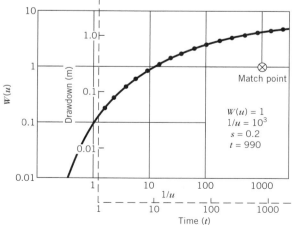

Figure 6.6 Graphs of (a) field data and (b) superposition of the field data on the type curve to obtain the formation parameters.

6.5 can be solved for transmissivity by using the match-point coordinates s and $W(u)$ and the discharge Q. Equation 6.3 can be solved for storativity by using the match-point coordinates $1/u$ and t, the distance r from the pumped well to the observation well, and the value of transmissivity as determined earlier.

Example 6.1

The data of Figure 6.6 were collected at a distance of 150 m from a well pumped at a rate of 5.43×10^3 m³/day. From Figure 6.6

$$W(u) = 1 \qquad s = 0.2 \text{ m}$$
$$\frac{1}{u} = 10^3 \qquad t = 990 \text{ min (about 0.7 day)}$$

Transmissivity is determined from Eq. 6.5,

$$T = \frac{Q}{4\pi s} W(u) = \frac{5.43 \times 10^3 \text{ m}^3/\text{day} \times 1}{4 \times 3.14 \times 0.2 \text{ m}}$$
$$= 2.2 \times 10^3 \text{ m}^2/\text{day}$$
$$= 2.5 \times 10^{-2} \text{ m}^2/\text{s}$$

Storativity is determined from Eq. 6.3,

$$S = \frac{4uTt}{r^2}$$
$$= \frac{4 \times 1 \times 10^{-3} \times 2.2 \times 10^3 \text{ m}^2/\text{day} \times 7 \times 10^{-1} \text{ day}}{225 \times 10^2 \text{ m}^2}$$
$$= 2.7 \times 10^{-4}$$

Assumptions and Interpretations

The operational hydraulic equations were derived for an ideal model aquifer, whereas time–drawdown data represent real aquifer response. The procedure requires a matching of real response data with the response expected under ideal conditions. The most conspicuous assumption is that the aquifer is infinite in areal extent. This means the following:

1. The cone of depression will never intersect a boundary to the system.

2. An infinite amount of water is stored in the aquifer.

Condition 1 is likely to be met for very short times, or for longer times in extensive aquifers of uniform material. In that the cone of depression is expanding with time, any geologic boundary within a short distance from the pumping well will immediately disrupt the postulated behavior. According to assumption 2, water levels will eventually return to their prepumping level once the wells are shut down.

In addition, it is assumed that the well is of infinitesimal diameter and fully penetrates the aquifer. This means that storage in the well can be ignored and that the well receives water from the entire thickness of the aquifer. Methods of treatment for deviations from full penetration have been presented in a number of papers (Muskat, 1937; Hantush and Jacob, 1955; Hantush, 1957). A solution for drawdown in large-diameter wells that takes into consideration the storage within the well has been presented by Papadopulos and Cooper (1967). Yet another assumption is that water is released instantaneously with decline in head in the aquifer. This assumption may be met when water is released by compression of the aquifer and expansion of the water, but fails to describe adequately the gravity flow system of the unconfined case.

Homogeneity and isotropicity are major assumptions in the development of the descriptive differential equation of which the well flow equation is a solution. This means that the transmissivity and storativity are assumed to be constants, both in space and in time. Hence, the geologic medium is assumed to be the simplest type conceivable.

Modifications of the Nonequilibrium Equation

At least two important modifications of the nonequilibrium equation can be traced to a very simple observation made by Cooper and Jacob (1946), namely, that the sum of the series of Eq. 6.2 beyond ln u becomes negligible when u becomes small. This occurs for large values of time t or small value of distance r. By neglecting the series beyond ln u, Eq. 6.1 can be expressed

$$s = \frac{Q}{4\pi T}\left(-0.5772 - \ln \frac{r^2 S}{4Tt}\right) \qquad (6.7)$$

or

$$s = \frac{Q}{4\pi T}\left(\ln \frac{4Tt}{r^2 S} - 0.5772\right) \qquad (6.8)$$

As 0.5772 equals ln 1.78 and ln x equals 2.3 log x, Eq. 6.8 becomes

$$s = \frac{2.3Q}{4\pi T}\log \frac{2.25Tt}{r^2 S} \qquad (6.9)$$

which is presented as the modified nonequilibrium equation. This equation may be applied to

1. Time–drawdown observations made in a single observation well, as in the original nonequilibrium method (modified nonequilibrium method).

2. Drawdown observations made in different wells at the same time (distance-drawdown method).

Time–Drawdown Method

If drawdown observations are made in a single well for various times, a plot of drawdown versus the logarithm

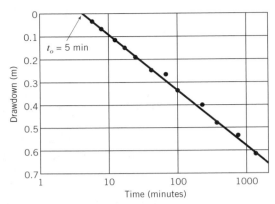

Figure 6.7 Semilogarithmic plot of drawdown versus time in an observation well.

of time will yield a straight line (Figure 6.7). At time t_1, the drawdown s_1 is expressed from Eq. 6.9

$$s_1 = \frac{2.3Q}{4\pi T}\log \frac{2.25Tt_1}{r^2 S} \qquad (6.10)$$

At time t_2, the drawdown s_2 will be

$$s_2 = \frac{2.3Q}{4\pi T}\log \frac{2.25Tt_2}{r^2 S} \qquad (6.11)$$

It follows that

$$s_2 - s_1 = \frac{2.3Q}{4\pi T}\log \frac{t_2}{t_1} \qquad (6.12)$$

If t_1 and t_2 are selected one log cycle apart, log $(t_2/t_1) = 1$. Equation 6.12 then becomes

$$\Delta s = \frac{2.3Q}{4\pi T} \qquad (6.13)$$

where Δs is the drawdown per log cycle. The value of T can be obtained from Eq. 6.13.

The storativity can be obtained by selecting any drawdown on Figure 6.7 for a given time and substituting this value in Eq. 6.9. For convenience, s equals zero is selected, so that Eq. 6.9 becomes

$$s = 0 = \frac{2.3Q}{4\pi T}\log \frac{2.25Tt}{r^2 S} \qquad (6.14)$$

This requires that $2.25Tt_o/r^2 S = 1$ so that

$$S = \frac{2.25Tt_o}{r^2} \qquad (6.15)$$

where t_o is the time intercept where the drawdown line intercepts the zero drawdown axis.

Example 6.2

The data of Figure 6.7 were collected at an observation well 305 m from a well pumping at a rate of 5.43×10^3 m³/day. Drawdown per log cycle is determined from the graph to be 0.24 m. Solving for T and S,

$$T = \frac{2.3Q}{4\pi\Delta s} = \frac{2.3 \times 5.43 \times 10^3 \text{ m}^3/\text{day}}{4 \times 3.14 \times 0.24 \text{ m}}$$

$$= 4.1 \times 10^3 \text{ m}^2/\text{day}$$

$$= 4.8 \times 10^{-2} \text{ m}^2/\text{s}$$

$$S = \frac{2.25Tt_o}{r^2} = \frac{2.25 \times 4.1 \times 10^3 \text{ m}^2/\text{day} \times 5 \text{ min}}{93025 \text{ m}^2 \times 1440 \text{ min}/\text{day}}$$

$$= 3.4 \times 10^{-4}$$

Distance–Drawdown Method

The time–drawdown analysis just described requires one pumping well and one observation well, where a plot of drawdown versus time on semilogarithmic paper yields a straight line. This straight line demonstrates the exponential decline of water levels at a point in a single aquifer. It is also possible to obtain the hydraulic properties by examining drawdown at two or more points at one instant of time. At time t, the drawdown s_1 at a distance r_1 is, from Eq. 6.9,

$$s_1 = \frac{2.3Q}{4\pi T} \log \frac{2.25Tt}{r_1^2 S} \qquad (6.16)$$

and the drawdown s_2 at a distance r_2 is

$$s_2 = \frac{2.3Q}{4\pi T} \log \frac{2.25Tt}{r_2^2 S} \qquad (6.17)$$

It follows that

$$s_1 - s_2 = \frac{2.3Q}{4\pi T} \log \frac{r_2^2}{r_1^2} \qquad (6.18)$$

Recognizing that $\log(r_2^2/r_1^2) = \log(1/r_1^2) - \log(1/r_2^2)$ and that $\log(1/r^2) = 2\log(1/r)$, Eq. 6.18 becomes

$$s_1 - s_2 = \frac{2.3Q}{2\pi T} \log \frac{r_2}{r_1} \qquad (6.19)$$

The graphic procedure calls for the plotting of drawdowns observed at the end of a particular pumping period in two or more observation wells at different distances from the pumped well against the logarithms of the respective distances (Fig. 6.8). By considering drawdown per log cycle, Eq. 6.19 becomes

$$\Delta s = \frac{2.3Q}{2\pi T} \qquad (6.20)$$

By extrapolating the distance–drawdown curve to its intersection with the zero drawdown axis, the storativity can be determined in the same manner as described for the modified nonequilibrium method,

$$S = \frac{2.25Tt}{r_o^2} \qquad (6.21)$$

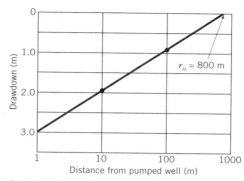

Figure 6.8 Semilogarithmic plot of drawdown versus distance.

where r_o is the intersection of the straight-line slope with the zero drawdown axis.

Example 6.3

Consider the following data for Figure 6.8. Observation wells are placed 10 m and 100 m, respectively, from a pumping well. The well is pumped at a rate of 5.43×10^3 m^3/day. At the end of 200 min (1.39×10^{-1} days), the drawdown per log cycle is noted to be 1.1 m. From Eqs. 6.20 and 6.21

$$T = \frac{2.3Q}{2\pi \Delta s} = \frac{2.3 \times 5.43 \times 10^3 \text{ m}^3/\text{day}}{2 \times 3.14 \times 1.1 \text{ m}}$$

$$= 1.8 \times 10^3 \text{ m}^2/\text{day}$$

$$= 2.1 \times 10^{-2} \text{ m}^2/\text{s}$$

$$S = \frac{2.25Tt}{r_o^2}$$

$$= \frac{2.25 \times 1.8 \times 10^3 \text{ m}^2/\text{day} \times 1.39 \times 10^{-1} \text{ days}}{64 \times 10^4 \text{ m}^2}$$

$$= 8.8 \times 10^{-4}$$

Steady-State Behavior as a Terminal Case of the Transient Response

The straight line in Figure 6.8 represents two points on a profile of a cone of depression for a given instant of time. At some time later yet another curve can be constructed, and it will parallel the curve shown in the figure, with the same drawdown per log cycle. For very long periods of time, the drawdown in the observation wells will have approached their steady-state value and no longer decline with time. A plot of this drawdown on Figure 6.8 will again form a straight line, parallel to the other lines, with the same drawdown per log cycle. This problem involving steady-state drawdowns at two observation points was solved by Thiem (1906) completely independent of any transient flow considerations. The solution for Thiem's confined condition is identical to that given in Eqs. 6.19 and 6.20. However, as time is not a factor in Thiem's steady-state derivation, the storativity cannot be determined for the steady drawdown case. If the flow takes place under unconfined conditions, Thiem's solution changes somewhat and may be expressed as

$$s_1^2 - s_2^2 = \frac{2.3Q}{\pi K} \log \frac{r_1}{r_2} \quad (6.22)$$

As in the application of the confined case, any pair of distances may be employed. This equation, along with its steady-state counterpart for confined conditions, is frequently referred to as the equilibrium, or Thiem equation.

The Hantush–Jacob Leaky Aquifer Method

One major assumption of the Theis nonequilibrium solution is that all the water pumped is removed from storage within the aquifer. There are several ways in which this condition may be compromised in field conditions: direct recharge to the aquifer from streams, direct recharge across bounding low-permeability materials, and release of water from bounding low-permeability rocks that possess their own hydraulic characteristics (i.e., specific storage and hydraulic conductivity). The problem of leakage has been extensively investigated by Hantush and Jacob (1955) and Hantush (1956, 1960, 1964). In this section we will be concerned only with their earliest contribution, which provides methods most commonly used in field problems.

Figure 6.1a will help develop the pertinent ideas. If the artesian sand in the aquifer is pumped, and the bounding low-permeability rocks are in fact "impermeable," the anticipated response at an observation well can be described completely by the Theis equation. However, if water is entering the aquifer, say, by either leakage across the upper low-permeability unit or by release of water from storage in the low-permeability unit, we would anticipate some deviation from the Theis response. If we assume that the low-permeability unit has a negligible specific storage (that is, is incompressible) but a finite permeability, water can be transmitted across it from the overlying water table. Figure 6.9 shows a typical response at an observation well in such an aquifer. Note that the time–drawdown curve departs from the expected nonequilibrium behavior, resulting in less drawdown per log cycle of time.

There are two ways to analyze drawdown curves of the type shown on Figure 6.9, depending on how much information is desired. The simplest method recognizes that it takes some length of time for water to permeate through the upper low-permeability material, and the upward inflection from the straight-line behavior of the

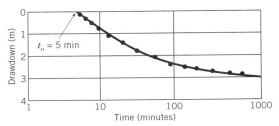

Figure 6.9   Time–drawdown response of a leaky aquifer.

Theis solution provides some estimate of this required time period. Hence, for all practical purposes, the time–drawdown behavior previous to the inflection represents a withdrawal of water from storage from the aquifer, no part of which was contributed from other sources. This part of the curve, along with its straight-line extension, may be analyzed with methods previously discussed for time–drawdown behavior. That is to say, over this period of the pumping test, the assumption that all the water pumped comes from storage in the aquifer is a reasonably valid one.

The second method requires a curve-matching procedure where the drawdown for all time is given as (Hantush and Jacob, 1955; Hantush, 1956)

$$s = \frac{Q}{4\pi T} W\left(u, \frac{r}{B}\right) \quad (6.23)$$

where $W(u, r/B)$ is the tabulated well function for leaky aquifers. It is thus recognized that short-term drawdowns previous to leakage are described by the Theis nonequilibrium equation and the well function $W(u)$ whereas long-term drawdowns are described by the well function $W(u, r/B)$. Values of $W(u, r/B)$ are given in Table 6.2, and a type curve is presented as Figure 6.10. Note that the family of curves converges on the nonleaky curve for small values of $1/u$ (which correspond to small values of time) and small values of r/B. For small values of time, $W(u, r/B) \approx W(u)$.

The graphic procedure for solving for the formation constants is similar to that described previously. Early drawdown data from the field curve are matched with the nonleaky part of the curve, but they soon deviate and follow one of the leaky r/B curves (Figure 6.11). The match point yields values of $W(u, r/B)$, $1/u$, t, and s. In addition, the r/B curve followed by the field data is noted. Transmissivity and storativity are readily determined from

$$T = \frac{Q}{4\pi s} W\left(u, \frac{r}{B}\right) \quad (6.24)$$

and

$$S = \frac{4uTt}{r^2} \quad (6.25)$$

Table 6.2   Values of $W(u\ r/B)$

u \ r/B	0.01	0.015	0.03	0.05	0.075	0.10	0.15	0.2	0.3	0.4
0.000001										
0.000005	9.4413									
0.00001	9.4176	8.6313								
0.00005	8.8827	8.4533	7.2450							
0.0001	8.3983	8.1414	7.2122	6.2282	5.4228					
0.0005	6.9750	6.9152	6.6219	6.0821	5.4062	4.8530				
0.001	6.3069	6.2765	6.1202	5.7965	5.3078	4.8292	4.0595	3.5054		
0.005	4.7212	4.7152	4.6829	4.6084	4.4713	4.2960	3.8821	3.4567	2.7428	2.2290
0.01	4.0356	4.0326	4.0167	3.9795	3.9091	3.8150	3.5725	3.2875	2.7104	2.2253
0.05	2.4675	2.4670	2.4642	2.4576	2.4448	2.4271	2.3776	2.3110	1.9283	1.7075
0.1	1.8227	1.8225	1.8213	1.8184	1.8128	1.8050	1.7829	1.7527	1.6704	1.5644
0.5	0.5598	0.5597	0.5596	0.5594	0.5588	0.5581	0.5561	0.5532	0.5453	0.5344
1.0	0.2194	0.2194	0.2193	0.2193	0.2191	0.2190	0.2186	0.2179	0.2161	0.2135
5.0	0.0011	0.0011	0.0011	0.0011	0.0011	0.0011	0.0011	0.0011	0.0011	0.0011

u \ r/B	0.5	0.6	0.7	0.8	0.9	1.0	1.5	2.0	2.5
0.000001									
0.000005									
0.00001									
0.00005									
0.0001									
0.0005									
0.001									
0.005									
0.01	1.8486	1.5550	1.3210	0.1307					
0.05	1.4927	1.2955	1.2955	1.1210	0.9700	0.8409			
0.1	1.4422	1.3115	1.1791	1.0505	0.9297	0.8190	0.4271	0.2278	
0.5	0.5206	0.5044	0.4860	0.4658	0.4440	0.4210	0.3007	0.1944	0.1174
1.0	0.2103	0.2065	0.2020	0.1970	0.1914	0.1855	0.1509	0.1139	0.0803
5.0	0.0011	0.0011	0.0011	0.0011	0.0011	0.0011	0.0010	0.0010	0.0009

Trans. Amer. Geophys. Union, 37, p. 702–714. Copyright by Amer. Geophys. Union.
After Hantush (1956).

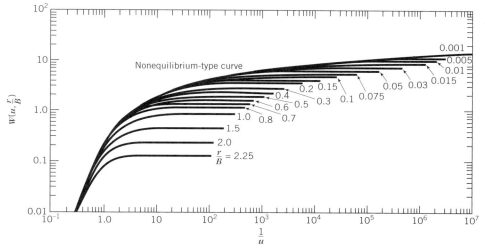

Figure 6.10   Type curve for leaky aquifers.

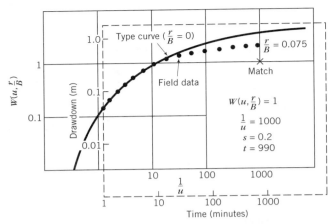

Figure 6.11 Graphic procedure for solving for formation constants for leaky aquifers.

In addition, the coefficient of vertical hydraulic conductivity can be determined from the relationship

$$\frac{1}{B} = \left(\frac{k'/m'}{T}\right)^{1/2} \tag{6.26}$$

where $1/B$ is determined from the match, m' is the thickness of the bounding low-permeability formation, and k' is the vertical hydraulic conductivity.

Example 6.4

The data in Figure 6.11 are similar to those in Figure 6.6 except for the leakage. Assuming the data were obtained from an observation well located 150 m from a well pumped at a rate of 5.43×10^3 m^3/day, the coefficient of transmissivity is calculated

$$T = \frac{Q}{4\pi s} W\left(u, \frac{r}{B}\right) = \frac{5.43 \times 10^3 \text{ m}^3/\text{day} \times 1}{4 \times 3.14 \times 0.2 \text{ m}}$$

$$= 2.2 \times 10^3 \text{ m}^2/\text{day}$$

$$= 2.5 \times 10^{-2} \text{ m}^2/\text{s}$$

The coefficient of storage is

$$S = \frac{4uTt}{r^2}$$

$$= \frac{4 \times 1 \times 10^{-3} \times 2.2 \times 10^3 \text{ m}^2/\text{day} \times 7 \times 10^{-1} \text{ days}}{22,500 \text{ m}^2}$$

$$= 2.7 \times 10^{-4}$$

Both results are identical with the numerical values obtained for Figure 6.6. Assuming the semipervious layer is 3 m in thickness,

$$k' = \frac{Tm'(r/B)^2}{r^2}$$

$$= \frac{2.2 \times 10^3 \text{ m}^2/\text{day} \times 3 \text{ m} \times 5.6 \times 10^{-3}}{22,500 \text{ m}^2}$$

$$= 1.6 \times 10^{-3} \text{ m/day}$$

Water Table Aquifers

The tests just described are designed to test the confined flow condition. This means that during the testing, the aquifer undergoes no dewatering, provided the heads are not lowered below the uppermost confining layer. For this case water is released from storage due to elastic compression of the matrix and expansion of the water itself. For a water table, or unconfined aquifer, the lowering of water levels will actually cause some dewatering of the aquifer, and the value of the storativity should increase by a few orders of magnitude over its confined counterpart. In Chapter 4, the storativity for such cases was referred to as the specific yield, S_y.

Two major concerns are involved in the analysis of water table aquifers. First, dewatering of the aquifer can actually lead to a decrease in transmissivity during pumping in that $T = Km$, where m is the saturated thickness of the aquifer. The second problem relates to the assumption of the Theis equation, where "water is released from storage instantaneously with decline in head." If this condition is not satisfied, the time–drawdown data will deviate from the expected response for the confined condition. The mathematical description of this response has been the subject of numerous papers starting with Boulton (1954, 1955, 1963), extended by Prickett (1965) based on the Boulton analysis, and advanced significantly by Dagan (1967), Streltsova (1972) and Steltsova and Rushton (1973), and Neuman (1972, 1975).

In view of the attention given to this subject, it is informative now to obtain a semiquantitative perspective of a typical time–drawdown response in an unconfined aquifer. Actually, there are three distinct parts of the time–drawdown curve as obtained in the vicinity of a pumped well. The very early time–drawdown data should correspond to that predicted by the Theis equation where water is released from storage due to elastic compression and expansion of the water. Thus the time–drawdown data should follow the Theis curve where the storativity is on the order expected for the confined condition. With increased time, the effects of gravity drainage take over, with vertical flow components in the vicinity of the pumped well. During this period, the time–drawdown response will be a function of the ratio of horizontal to vertical conductivity, the thickness of the aquifer, and the distance to the pumped well. If departures from the Theis-type curve do occur, they will be similar to those obtained when the slope decreases in the leaky aquifer test. Although this is occurring in response to vertical flow components, one may view this in a conceptual sense as an increase in the "storativity" over its confined value. For later time, the time–drawdown data will again follow some Theis-type curve, with the storativity now reflecting the specific yield S_y; that is, the "storativity" is no longer increasing with time.

On a logarithmic plot the curve has the shape of an

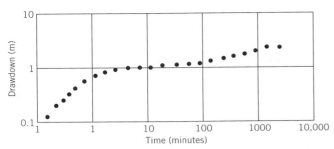

Figure 6.12 Time-drawdown response of an unconfined aquifer.

elongated letter "s" (Figure 6.12). Given this information, there are two approaches to the interpretation of pumping tests performed in unconfined aquifers. First, let us recognize that previous to Boulton's contribution (1954), no distinction was made between the confined and unconfined response. Indeed, virtually hundreds of tests were conducted on unconfined systems using the Theis analysis with no apparent reference to the need for modifications. Jacob (1950) recognized the difficulties with unconfined flow to wells and recommended a procedure to account for an apparent increase in storativity with time. In this procedure, the transmissivity is first determined from the early time–drawdown data by the modified nonequilibrium method (Eq. 6.13). Field values of drawdown s for later time t are then introduced into Eq. 6.9 and the equation solved for the specific yield. If the time–drawdown curve actually contains three segments, as in Figure 6.12, the ultimate value of S_y can be readily determined. If only two segments are discernible, an intermediate value will be obtained.

A more exact analysis requires a curve-matching procedure. According to Neuman (1975) the solution to the flow problem is

$$s = \frac{Q}{4\pi T} W(u_A, u_B, \eta) \qquad (6.27)$$

where $W(u_A, u_B, \eta)$ is the well function for the unconfined aquifer. The arguments of the well function are described as

$$u_A = \frac{r^2 S}{4Tt} \qquad \text{(for early time–drawdown data)} \qquad (6.28a)$$

$$u_B = \frac{r^2 S_y}{4Tt} \qquad \text{(for late time–drawdown data)} \qquad (6.28b)$$

$$\eta = \frac{r^2 K'}{m^2 K} \qquad (6.28c)$$

where m is the initial saturated thickness, K is the horizontal conductivity, and K' is the vertical conductivity. Figure 6.13 shows the type curve employed in the matching procedure where the type A curves merge out of the Theis curve shown on the left and the type B curves merge into the Theis-type curve shown on the right. The following method is recommended for the analysis. First, the early time–drawdown data should be matched to the type A curves and any match point selected. This match point will have the coordinates $W(u_A, \eta)$, $1/u_A$, t, and s. The match will provide a value for η. The transmissivity is then determined with Eq. 6.27 and the storativity with Eq. 6.28a. The later time–drawdown data are then matched to the specific type B curve of the same η value determined for the A match. The match point yields

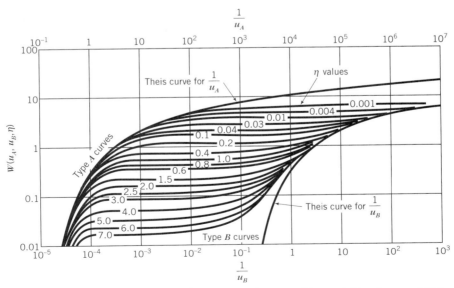

Figure 6.13 Theoretical curves of $W(u_A, u_B, \eta)$ versus $1/u_A$ and $1/u_B$ for an unconfined aquifer (from Neuman, Water Resources Res., v. 11, p. 329–342, 1975. Copyright by Amer. Geophys. Union).

values of $W(u_B, \eta)$, $1/u_B$, s, and t. Transmissivity is determined from Eq. 6.27, which should provide a value similar to that previously determined. The storativity can be determined with Eq. 6.28b and should yield a value somewhat larger than previously determined. As $K = T/m$ Eq. 6.28c can be used to determine the vertical hydraulic conductivity. Neuman (1975) provides tables for the function $W(u_A, u_B, \eta)$.

6.3 Single-Borehole Tests

The one feature shared by the methods described in the previous section is the necessity of one pumping well and one or more observation wells. Single-borehole tests offer some advantages in economy, frequently with a certain loss of information. In spite of this, the single-borehole test is becoming increasingly more common in hydraulic testing associated with waste disposal sites and in low-permeability material. As in the previous section, we will be concerned here mainly with the conventional or common methods of analysis.

Recovery in a Pumped Well

If a well is pumped for a given period of time t and then shut down, the residual drawdown (the original prepumping water level minus the water level at any time after shutdown) can be approximated as the numerical difference between the drawdown in the well if the discharge had continued and the recovery of the well in response to an imaginary recharge well, of the same flow rate, superimposed on the discharging well at the time it is shut down (Figure 6.14). Designating original head as h_o and the recovered head at any time as h', residual drawdown is expressed from the modified nonequilib-

rium Eq. 6.9

$$h_o - h' = \Delta s' = \frac{2.3Q}{4\pi T}\left(\log\frac{2.25Tt}{r^2S} - \log\frac{2.25Tt'}{r^2S}\right)$$

(6.29)

where $\Delta s'$ is residual drawdown, t is the time since pumping started, and t' is the time since pumping stopped. This equation reduces to

$$\Delta s' = \frac{2.3Q}{4\pi T}\log\frac{t}{t'}$$

(6.30)

Field procedure requires a drawdown measurement at the end of the pumping period (t) and recovery measurements during the recovery period (t'). The graphic procedure is to plot residual drawdown on the arithmetic scale and the value of t/t' on the logarithmic scale. The time t includes the interval over the pumping plus recovery period whereas the time t' includes the recovery interval only. If calculations are made over one log cycle of t/t',

$$\Delta s' = \frac{2.3Q}{4\pi T}$$

(6.31)

where $\Delta s'$ is the residual drawdown per log cycle. The storativity is not determined directly with this method.

The Drill Stem Test

The drill stem test used in petroleum engineering is the equivalent of the recovery test. Figure 6.15 shows the main components of a typical drill stem test. The packer assembly is set at the isolated stratigraphic interval of interest. The bypass valve is initially opened and the formation flows under its head. This is equivalent to the pumping period of the recovery test. The valve is then closed, shutting in the formation pressure. The ensuing period is analogous to the recovery period of the single-borehole test previously discussed. The method of Horner (1951) is generally used to analyze the data

$$P_w - P_o = \frac{2.3Q\mu}{4\pi km}\log\frac{t}{t'}$$

(6.32)

where P_w is the pressure in the well bore, P_o is the undisturbed formation pressure, Q is the rate of production, μ is the viscosity of the fluid, k is the permeability, m is the formation thickness, t is the time since the test began (the sum of the production and the recovery time), and t' is the time since the recovery portion of the test started.

A typical pressure buildup plot is shown in Figure 6.16. By considering the pressure change over one log cycle, Eq. 6.32 reduces to

$$\frac{km}{\mu} = \frac{2.3Q}{4\pi\,\Delta P}$$

(6.33)

where ΔP is the change in pressure over one log cycle.

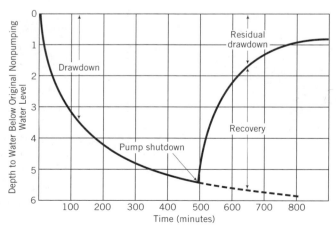

Figure 6.14 Arithmetic plot of drawdown and recovery curve versus time.

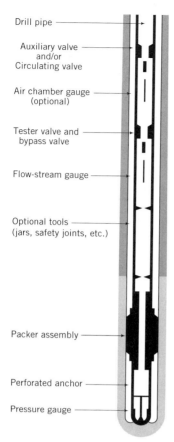

Figure 6.15 Main components of a typical drill-stem test string (from Bredehoeft, 1965). Reprinted by permission of Ground Water. © 1965. All rights reserved.

By extrapolating the pressure curve to $t/t' = 1$, the undisturbed formation pressure P_o is determined; that is, in Eq. 6.32, log (1) = 0.

Pressure buildup tests in single boreholes have long been of interest to petroleum engineers, with the monographs of Matthews and Russel (1967) and Earlougher (1977) providing good reviews.

Slug Injection or Withdrawal Tests

The pressure recovery in a borehole after withdrawal of a known volume of water (slug), or the pressure decline after injection of a known volume of water (slug), is termed the slug test. The slug test is one of the more common methods employed in field practice in that it can be used in low- to moderately high-conductivity materials and requires no pumping apparatus. Thus, small-diameter boreholes normally used in geologic exploration may serve double duty in a hydraulic testing program.

The analysis of water level versus time data in a single borehole in response to a slug injection or withdrawal has been pioneered by Hvorslev (1951) and Cooper and

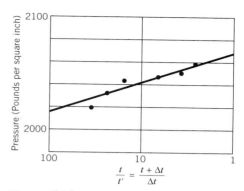

Figure 6.16 Drill-stem test result.

others (1967). Whatever method is being employed, the procedures for the analysis are reasonably straightforward. The original head in the borehole is first noted and recorded. For the slug injection, water is instantaneously added to the borehole, raising the original head above its static level. This is termed h_o, that is, the height of the slug at time equal to zero, say, for example, 2 or 3 ft. The slug will then start to decay as water enters the formation. This change in head is noted over time, where h is designated the height of the slug for any given time. Thus, at time equal to zero, $h/h_o = 1$, and as time gets large, h/h_o approaches zero, that is, the slug has been completely dissipated (Figure 6.17) If the well or borehole is screened above the water table, as is common for investigating oil spills or leaks, a slug withdrawal procedure is necessary.

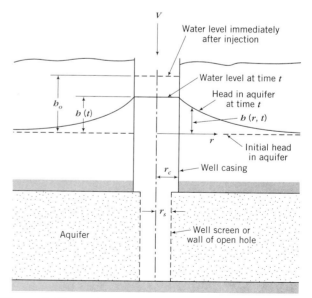

Figure 6.17 Well into which a volume V of water is suddenly injected for a slug test of a confined aquifer (from Cooper and others, Water Resources Res, v. 3, p. 263–269, 1967. Copyright by Amer. Geophys. Union).

Cooper and others (1967) present a type curve solution for the interpretation of slug tests. Papadopulos and others (1973) extended the range of the type curves for greater use in practice. However, it has been noted that it is difficult to achieve a definitive fit in the matching procedure in that field data frequently overlap or fit parts of several of the type curves. Consequently, the method is infrequently used.

The method of Hvorslev (1951) is probably the most widely used in field practice. Field data such as given in Figure 6.18 are plotted on semilogarithmic paper with b/b_o on the log scale and time t on the arithmetic scale (Figure 6.19). The hydraulic conductivity may be determined from the relationship (Cedergren, 1967)

$$K = \frac{A}{F(t_1 - t_2)} \log_e \frac{b_1}{b_2} \qquad (6.34)$$

where A is the area of the borehole and F is a shape factor that depends on the size and shape of the intake area. Hvorslev (1951) has evaluated several values of the shape factor F for a variety of conditions, as has the U.S. Navy Bureau of Yards and Docks (1961) (see, for example, Cedergren, 1967). For a cased, uncased, or perforated extension into an aquifer of finite thickness for $L/r > 8$,

$$F = \frac{2\pi L}{\ln(L/r)} \qquad (6.35)$$

where L is the length of the intake area and r is the radius of the borehole. For a borehole area $= \pi r^2$, Eq. 6.34 becomes

$$K = \frac{r^2 \ln(L/r)}{2L} \frac{\log_e(b_1/b_2)}{(t_1 - t_2)} \qquad (6.36)$$

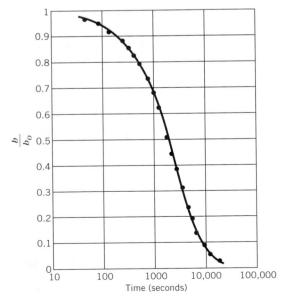

Figure 6.18 Field response of slug test.

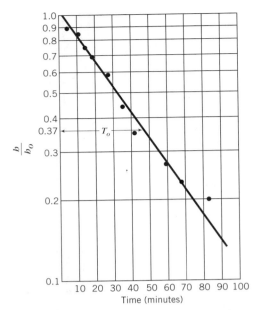

Figure 6.19 Field response of slug test.

It is convenient now to take $b_1 = b_o$ at $t = 0$ and $b_2 = 0.37b_o$ so that

$$\log_e\left(\frac{b_1}{b_2}\right) = \log_e\left(\frac{b_o}{0.37b_o}\right) = \log_e 2.7 = 1.0$$

Equation 6.36 then becomes

$$K = \frac{r^2 \ln(L/r)}{2LT_o} \qquad (6.37)$$

where T_o is the time intercept on the field curve where $b/b_o = 0.37$ (Figure 6.19). T_o is termed the basic time lag and is determined where the head ratio $= 0.37$. The time lag refers to the fact that the transients in a well seldom correspond exactly to those occurring in the adjacent formation so that flows are associated with various time lags that depend on the shape factor F, the hydraulic conductivity, and the radius of the borehole.

Example 6.5

The field data of Figure 6.18 are plotted on Figure 6.19. For an observed value of $T_o = 44$ minutes, a casing radius of 2.54 cm, and a well screen length of 3.8 m, the hydraulic conductivity is determined from Eq. 6.37:

$$K = \frac{r^2 \ln(L/r)}{2LT_o} = \frac{(2.54)^2 \ln(380/2.54)}{(2)(380)(44)(60)} = 1.61 \times 10^{-5} \text{ cm/s}$$

Response at the Pumped Well: Specific Capacity and Well Efficiency

Well capacity, or discharge per unit time, is often used as a measure of well yield. When comparing the strength of one well versus another, it is better to express well capacity in relation to some common standard. The accepted standard of comparison is the unit drawdown. Thus, dividing the pumping rate by the total drawdown in a well gives the specific capacity, that is, cubic meters per day per meter of drawdown.

The drawdown observed in a pumping well is composed of two parts: (1) drawdown due to laminar flow of water in the aquifer toward the well, referred to as formation loss s and calculated with the well flow equations discussed in this chapter, and (2) drawdown resulting from the turbulent flow of water in the immediate vicinity of the well, through the well screen or casing openings and in the well casing, referred to as well loss s_w. Jacob (1950) stated that well loss is proportional to some power of the discharge exceeding the first power and approaching the second. As an approximation,

$$s_w = CQ^2 \qquad (6.38)$$

where s_w is well loss (L), C is the well loss constant, and Q is the pumping rate. Total drawdown in a pumping well, s_t, is then expressed

$$s_t = s + s_w = \frac{Q}{4\pi T} W(u) + CQ^2 \qquad (6.39)$$

The well loss constant C may be determined by comparing the actual drawdown in a pumping well with that predicted for a 100% efficient well ($s_w = 0$) as based on the Theis equation, where r is taken as the radius of the well r_w. The efficiency of the well is taken as the ratio

of the predicted drawdown, where $s_w = 0$, to the actual measured drawdown s_t.

A theoretical specific capacity may be defined as the specific capacity that would be achieved for a 100% efficient well, that is, one in which the well loss $s_w = 0$. Theis and others (1963) demonstrated that the theoretical specific capacity of a well can be determined from the abbreviated nonequilibrium equation

$$T = \frac{Q}{4\pi s}\left(-0.5772 - \ln\frac{r^2 S}{4Tt}\right) \qquad (6.40)$$

or, solving for Q/s

$$\frac{Q}{s} = \frac{4\pi T}{\ln\dfrac{4Tt}{r_w^2 S} - 0.5772} \qquad (6.41)$$

The assumption here is that well loss is zero.

From this equation, the theoretical specific capacity of a well is directly proportional to T and inversely proportional to $\ln t$, $\ln 1/r_w^2$, and $\ln 1/S$. Hence, large changes in T cause correspondingly large changes in specific capacity. Large changes in t, r_w, and S cause comparatively small changes in specific capacity. Given the actual specific capacity of a pumped well, the radius of the well, and the duration of pumping, and assuming some value for S, the transmissivity can be estimated with Eq. 6.41. Figure 6.20 has been prepared to aid in this determination. In this figure the pumping period has been taken as one day and the other pertinent parameters are given in so-called American practical hydrology units. Here spe-

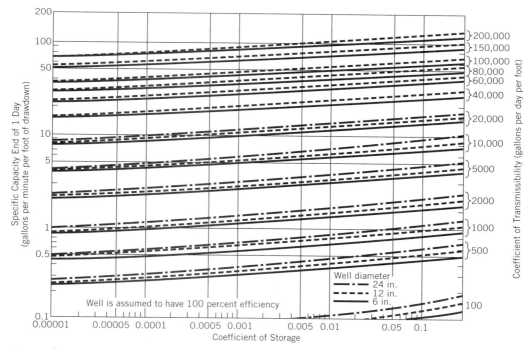

Figure 6.20 Graph showing interrelation of well diameter, specific capacity, transmissivity, and storativity (from Theis and others, 1963).

cific capacity is given in gallons per day per foot of drawdown, transmissivity is given in gallons per day per foot, and the storativity remains a dimensionless constant. To use the figure, measured specific capacity (left-hand axis) is followed across to an intersection with an estimated storativity, and then the appropriate curve for the well diameter is followed to the right, yielding an approximate value for the transmissivity. If well losses are involved in the measured specific capacity, which is inevitable, the transmissivity so determined represents an underestimated value. That is to say, the actual formation transmissivity is higher than the calculated one. Gallons per foot per day can be readily converted to m²/day by multiplying by 1.242×10^{-2}.

Example 6.6

An aquifer has a transmissivity of 125 m²/day and a storativity of 1×10^{-4}. A 60-cm diameter well in this aquifer has a drawdown of 91 m after 24 hours of pumping at a rate of 5440 m³/day. The efficiency of the well is determined as follows:

For drawdown in the pumping well in the absence of well loss

$$s = \frac{Q}{4\pi T} W(u) = \frac{5440 \, W(u)}{(4)(3.14)(125)} = 3.5 \, W(u)$$

The value of u is determined for r equal to the well radius

$$u = \frac{r^2 S}{4Tt} = \frac{(0.3)^2(1 \times 10^{-4})}{(4)(125)(1)} = 1.8 \times 10^{-8}$$

From Table 6.1, $W(u) = 17.2$ so that $s = (3.5)(17.2) = 60$ m, and well efficiency is determined as $E = 60/91 = 66\%$.

The well loss constant C is determined from $s_w = CQ$

$$C = \frac{s_w}{Q^2}$$

$$= 31 \, \text{m}/2.96 \times 10^7 \, \text{m}^6/\text{day}^2 = 1.05 \times 10^{-6} \, \text{day}^2/\text{m}^5$$

$$= 7.8 \times 10^3 \, \text{s}^2/\text{m}^5$$

6.4 Partial Penetration, Superposition, and Bounded Aquifers

Partial Penetration

When the screened or open section of a well casing does not coincide with the full thickness of the aquifer it penetrates, the well is referred to as partially penetrating. With existing wells, this is the rule rather than the exception. Under such conditions the flow toward the pumping well (or observation point) will be three dimensional because of vertical flow components (Figure 6.21). Our

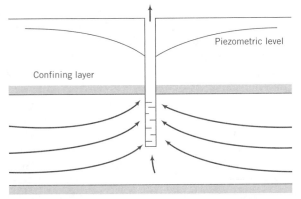

Figure 6.21 The effect of partial penetration on flow to a well.

experience with vertical flow components in the unconfined case previously discussed suggests upward inflection points in the time–drawdown response.

The topic of partial penetration has been the subject of numerous papers (Muskat, 1937; Hantush, 1961, 1964; Neuman, 1974). Two aspects of the problem are of interest here. First, solutions for partially penetrating wells have been derived and are discussed by Hantush (1961, 1964) and Walton (1970). These solutions may be applied directly if partial penetration is a factor affecting the response. However, with such solutions, it is necessary to know the actual degree of penetration of the aquifer, which may not be known in many cases. A second point is the effect of partial penetration on the water-level response. Hantush (1964) has conducted the most work in this area and provides some general guidelines for confined aquifers. The effects of partial penetration will not normally affect the pumping test result if the observation well is located at some distance $r > 1.5m \, (K/K')^{1/2}$ from the pumped well, where K is the horizontal conductivity, K' is the vertical conductivity, and m is the aquifer thickness. If the conductivities K and K' are of the same order of magnitude, the condition $r > 1.5m$ is sufficient. If this condition is not satisfied, there will be an upward inflection in the response, similar to that obtained in the leaky method or for some sort of recharge boundary. In some cases, depending on the relative positions of the screens, more distant observation points can actually show more drawdown than closer ones. For long periods of time $t > Sm/2K'$, the time–drawdown slope of a partially penetrating response will be the same as expected if the well completely penetrated the aquifer. The reader is referred to Hantush (1961) or Walton (1970) for the methodology required to treat partial penetration.

Principle of Superposition

The Theis nonequilibrium equation represents a solution to the diffusion equation for a prescribed set of boundary and initial conditions. The pumping test procedure is relatively straightforward where a transient response to a steady pumping rate is observed over a limited period of time (say, 24 hours) at some known distance from the pumping well. Given the transmissive and storage properties as determined over this short pumping period, it is possible to predict the water level at later times and at other distances in response to any steady pumping rate Q. This assumes, of course, that the storativity and transmissivity do not change with time, whereas the postulated time of pumping t, the distance r, and the pumping rate Q may be taken as variables.

The Theis equation (for that matter the Hantush–Jacob equation) may also be used to obtain a theoretical drawdown for any time t at various distances r in response to a battery of wells pumping at various rates. This is because the diffusion equation is linear, that is, consists of a sum of linear terms where a linear term is first degree in the dependent variables and their derivatives. Linearity in differential equations is synonymous with the principle of superposition, which states that the derivative of a sum of terms is equal to the sum of the derivatives of the individual terms. Expressed in terms of the pumping response, the total effect resulting from several wells pumping simultaneously is equal to the sum of the individual effect caused by each of the wells acting separately.

Example 6.7

The principle of superposition is an important one in the application of hydraulic methods to real-world problems. The purpose of this example is to demonstrate an efficient method with which well test data are organized to examine well interference. We assume that the data of Figure 6.1 apply; that is, a well is pumped for a relatively short period at a rate of 5.43×10^3 m³/day and a transmissivity of 2.2×10^3 m²/day and a storativity of 2.7×10^{-4} is determined. We assume these properties are constant over a large region. The table below demonstrates a procedure for tabulating values for the drawdown for times ranging from 1 to 100 days and over distances up to 100 m. It is noted that these calculations are for the specific pumping rate of 5.43×10^3 m³/day from a single well.

time = 1 day

$$u = \frac{r^2 S}{4Tt} = \frac{r^2 \times 2.7 \times 10^{-4}}{4 \times 2.2 \times 10^3 \times 1} = 3.07 \times 10^{-8}\, r^2$$

r, m	u	W(u)	s, m
0.1	3.07×10^{-10}	21.33	4.18
1	3.07×10^{-8}	16.72	3.28
10	3.07×10^{-6}	12.12	2.38
100	3.07×10^{-4}	7.51	1.47

time = 10 days

$$u = \frac{r^2 S}{4Tt} = \frac{r^2 \times 2.7 \times 10^{-4}}{4 \times 2.2 \times 10^3 \times 10} = 3.07 \times 10^{-9}\, r^2$$

r, m	u	W(u)	s, m
0.1	3.07×10^{-11}	23.63	4.64
1	3.07×10^{-9}	19.02	3.74
10	3.07×10^{-7}	14.42	2.83
100	3.07×10^{-5}	9.81	1.93

time = 100 days

$$u = \frac{r^2 S}{4Tt} = \frac{r^2 \times 2.7 \times 10^{-4}}{4 \times 2.2 \times 10^3 \times 100} = 3.07 \times 10^{-10}\, r^2$$

r, m	u	W(u)	s, m
0.1	3.07×10^{-12}	25.93	5.09
1	3.07×10^{-10}	21.33	4.19
10	3.07×10^{-8}	16.72	3.28
100	3.07×10^{-6}	12.12	2.38

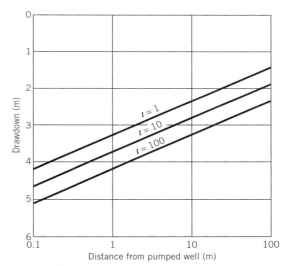

Figure 6.E7

The graph may be referred to as a mathematical model for the determination of the response to pumping in this region of constant transmissivity and storativity. We note that the drawdown some 100 m from the pumping well at the end of 100 days of pumping will be on the order of 2.4 m, whereas at 10 m it will be 3.3 m. On the other hand, if we are interested in the water-level response at one point 100 m distant from a pumping well and 10 m distant from another pumping well, the principle of superposition states that we require the sum 2.4 + 3.3 = 5.7 m. These calculations are only valid for the "design" pumping rate of 5.43×10^3 m³/day. However, we note that the pumping rate Q is not one of the variables making up the variable u and only occurs in the drawdown statement $s = (Q/4\pi T)W(u)$. Hence, the magnitude of drawdown is proportional to the pumping rate, and so if we double the rate in the examples, we double the drawdown, and if we pump half as much from each well, we halve the drawdown.

With these points in mind, consider the following example. Two wells are pumping for 10 days at rates of 10.86×10^3 and 2.72×10^3 m^3/day. Five days after these wells start pumping, a third well turns on at a rate of 4.07×10^3 m^3/day. The pumping wells are at a distance of 100, 10, and 1 m, respectively, from an observation well. At the end of this 10-day period, the drawdown at the point of observation is determined from the preceding figure for the postulated pumping rates

$$s_1 = 3.9 \qquad s_2 = 1.4 \qquad s_3 = 2.8$$

giving a total drawdown of 8.1 m.

Suppose now that the observation well is itself a pumping well pumping at a rate of 5.43×10^3 m^3/day for this same 10-day period. If the diameter of the well is about 0.6 m, its self-induced drawdown, ignoring well loss, is about 4.0 m. Hence, the total drawdown at this point is on the order of 12.1 m.

___

Bounded Aquifers

An assumption employed throughout this chapter is that the aquifer is infinite in areal extent. Geologic boundaries limit the extent of real aquifers and serve to distort the calculated cones of depression forming around pumping wells. The method of images, which plays an important role in the mathematical theory of electricity and is employed in the solution of some geophysical problems, aids in the evaluation of the influence of aquifer boundaries on well flow. This theory, as described by Ferris (1959), permits treatment of the aquifer limited in one or more directions. However, the additional assumption of straight-line boundaries has been added. This gives aquifers of rather simple geometric form.

The image well theory can be explained as follows (Stallman, 1952): formation A is bounded by the relatively impermeable formation B, the boundary between the two located at a variable distance r from a pumping well (Figure 6.22a). As formation B is relatively impermeable, no flow can occur from it toward the pumping well, and the boundary is a no-flow (barrier) boundary. The effect of a barrier boundary is to increase the drawdown in a well. The problem now is to duplicate this physical situation by substituting a hydraulic entity that serves this purpose. As no flow occurs across a ground-water divide, the barrier boundary is simulated by the supposition that formation A is infinite in areal extent and that an imaginary well is located across the real boundary, on a line

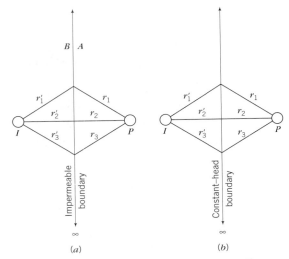

Figure 6.22 Simple two-well system for an aquifer bounded by (a) a no-flow boundary and (b) a constant-head boundary (from Stallman, 1952).

at right angles thereto, and at the same distance from the boundary as a real pumping well. If the imaginary well is assumed to start pumping at the same time and at the same rate as the real well, the boundary will be transformed into a ground-water divide.

Similarly, if the aquifer is bounded by a stream that provides recharge to the aquifer, the effect is to decrease the drawdown in a well. A zero-drawdown (constant-head) boundary can be simulated by an imaginary well, located as earlier, with the exception that the imaginary well must recharge water at the same rate as the pumping well (Figure 6.22b).

The drawdown at any point in the real aquifer or at the boundary for the simple two-well system is the sum of the effects of the real and imaginary wells operating simultaneously. From the principle of superposition

$$s = s_p \pm s_i = \frac{Q}{4\pi T}[W(u)_p \pm W(u)_i]$$

$$= \frac{Q}{4\pi T}\Sigma\, W(u) \qquad (6.42)$$

where s is the observed drawdown at any point, consisting of the sum of the effects of the real well s_p and the imaginary well s_i, the well function for the real well is $W(u)_p$, and $W(u)_i$ is the well function for the imaginary well. If the imaginary well is a recharging well, the negative sign is used.

By similar reasoning,

$$u_p = \frac{r_p^2 S}{4Tt} \qquad u_i = \frac{r_i^2 S}{4Tt} \qquad (6.43)$$

where r_p is the distance from the pumping well to the observation point and r_i is the distance from the imaginary well to the observation point. From the equality

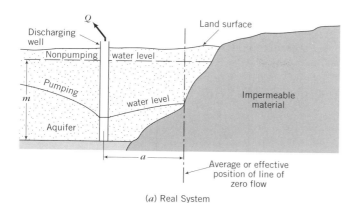

(a) Real System

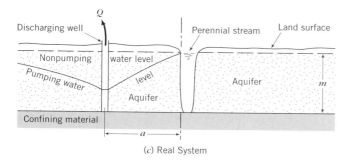

Note: Aquifer thickness m should be very large compared to resultant drawdown near real well.

(b) Hydraulic Counterpart of Real System

(c) Real System

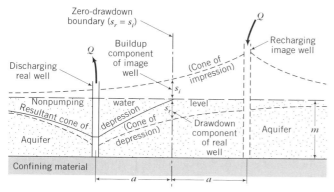

Note: Aquifer thickness m should be very large compared to resultant drawdown near real well.

(d) Hydraulic Counterpart of Real System

expressed in Eq. 6.43

$$u_i = \left(\frac{r_i}{r_p}\right)^2 u_p \qquad (6.44)$$

or

$$u_i = \overline{K}^2 u_p \qquad (6.45)$$

where $\overline{K}$ equals the constant r_i/r_p. For any point on the boundary, r_p equals r_i ($\overline{K}$ equals 1), and u_i equals u_p. It follows that $W(u)_i$ equals $W(u)_p$ (Eq. 6.42), and the drawdown is either zero (constant-head boundary) or twice the effect of one well pumping in an infinite aquifer.

When a well in a bounded aquifer is pumped, water levels in observation wells will initially decline under the influence of the pumping well only. When the cone of depression reaches an exterior boundary, deviations from the ideal response will be noted in the observation well. Under these conditions, the early part of the time–drawdown curve unaffected by the boundary behaves as if the aquifer were infinite in areal extent, and this part of the curve can be used to determine the hydraulic properties.

Figure 6.23 shows the effects of pumping near a barrier and recharge boundary and the role of image wells in simulating this pumpage.

Aquifers are often bounded by two or more boundaries, and time–drawdown data will respond accordingly. As mentioned, the resulting geometry must be rather simple in order to apply the image theory. Hence, two converging boundaries delineate a wedge-shaped aquifer, two parallel boundaries intersected at right angles by a third boundary forms a semiinfinite strip, and four intersecting right-angle boundaries form a rectangular aquifer.

A number of image wells associated with each pumping well characterizes a multiple-boundary system. Clearly, primary image wells placed across a boundary will balance the effect of a pumping well at that boundary, but will cause an unbalanced effect at the opposite boundary in violation of the no-flow or constant-head requirement. It is then necessary to add secondary image wells at appropriate distances to satisfy the conditions of no flow or zero drawdown. Figure 6.24 shows some plan views of image well systems for some simple aquifers.

Once the boundaries of a finite aquifer have been simulated by means of image wells, analysis of drawdown effects can proceed as if the aquifer were infinite in areal

Figure 6.23 Diagrams of the effect of pumping near barrier and constant-head boundaries and appropriate hydraulic counterparts for image well theory (from Ferris and others, 1962). Note: Aquifer thickness m should be very large compared to resultant drawdown near real well.

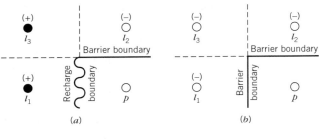

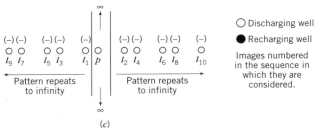

Figure 6.24 Image well systems for (*a*) wedge-shaped aquifer with both barrier and recharge boundaries, (*b*) wedge-shaped aquifer with barrier boundaries, and (*c*) infinite strip.

extent. In other words, graphic models such as given in Example 6.7 can be used to find the drawdown in response to one or several wells pumping simultaneously. Once the distance from real and imaginary wells to the point of interest has been determined, drawdown (or recovery) may be obtained from the graphs. For example, the drawdown at any point in the real aquifer of Figure 6.24 is the algebraic sum of the effect of one real well, one discharging imaginary well, and two recharging imaginary wells. For the infinite array (Figure 6.24c), image wells are added until the most remote pair has negligible influence on water-level response.

Example 6.8

A 60-cm-diameter well is located in an aquifer with a transmissivity of 125 m²/day and a storativity of 5×10^{-2}. A fault (barrier) is located 300 m from the well. The drawdown at the midpoint between the fault and the barrier at the end of one year's pumpage at a rate of 2700 m³/day is determined

$$s = s_p + s_i = \frac{Q}{4\pi T}[W(u)_p + W(u)_i]$$

$$(u)_p = \frac{r_p^2 S}{4Tt} = \frac{(150\ \text{m})^2\, 5 \times 10^{-2}}{(4)(125)\ \text{m}^2/\text{day}\,(365)\ \text{day}} = 6.16 \times 10^{-3}$$

$$(u)_i = \frac{r_i^2 S}{4Tt} = \frac{(450\ \text{m})^2\, 5 \times 10^{-2}}{(4)(125)\ \text{m}^2/\text{day}\,(365)\ \text{day}} = 5.55 \times 10^{-2}$$

From Table 6.1, $W(u)_p = 4.52$, $W(u)_i = 2.37$.

$$s = \frac{2700\ \text{m}^3/\text{day}}{4 \times 3.14 \times 125\ \text{m}^2/\text{day}}(6.89) = 11.85\ \text{m}$$

Example 6.9

The accompanying figure depicts a time–drawdown response in an observation well affected by one pumping well and a barrier boundary. By construction, $s_1 = s_2$. It thus follows that $W(u)_1 = W(u)_2$ and

$$u_1 = \frac{r_1^2 S}{4Tt_1} \qquad \text{and} \qquad u_2 = \frac{r_2^2 S}{4Tt_2}$$

where $u_1 = u_2$. Thus

$$r_2 = r_1 \left(\frac{t_2}{t_1}\right)^{1/2}$$

where r_1 is the distance from the pumping well to the observation well, which is known, and r_2 is the distance from the observation well to the imaginary pumping well causing the inflection, which is readily calculated. However, because the exact location of the barrier is not known, r_2 merely defines a set of points that form a circle of radius r_2 with the observation well located at its center.

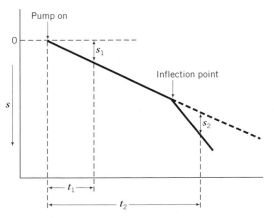

Figure 6.E9

6.5 Hydraulic Testing in Fractured or Low-Permeability Rocks

Advances in hydraulic testing over the past decade have been born out of the practical necessity of evaluating rocks as hosts for chemical or nuclear wastes. In some cases, this requires testing in deep boreholes up to a few thousand feet in depth, often in rocks that are of a low permeability, and frequently in rock that is fractured. In many cases, the emphasis has been on the single-borehole test. These tests generally require special equipment such as packers to isolate testing zones, feed in apparatus whereby a slug of water may be removed from the isolated zone or the zone quickly pressurized, and pressure transducers to measure the response. With deep boreholes, temperature or salinity effects on fluid density must sometimes be accounted for. Hydraulic testing in this modern era is thus a field within itself, generally requiring a large degree of specialization. In this section, we will consider only a few of the methods in detail.

Single-Borehole Tests

The slug test discussed earlier is well suited for low-permeability testing. The pressure recovery or decline in a borehole can be measured after the removal or injection of a volume of water. The pulse test is a modification of the slug test whereby a testing interval within a single borehole is instantaneously under- or overpressured by removing or adding water. The time response of the pressure buildup or decay is then observed, with the analysis being similar to that used in the slug test.

Wang and others (1978) extended the methods developed by Cooper and others (1967) to pulse testing in a single fracture. The method is based on the parallel plate model and assumes that the aperture width is independent of fluid pressure and that no flow is derived from the matrix of the fractured medium. Barker and Black (1983) extend this method to examine the effect of flow in the matrix as well as in the fractures. Their aquifer model is identical to that employed by de Swaan (1976) and Boulton and Streltsova (1977).

The determination of vertical permeability of low-permeability rocks in a single borehole has also been addressed in a number of papers (Burns, 1969; Pratts, 1970; Hirasaka, 1974). The monograph of Earlougher (1977) also discusses the available methodology and technology.

Multiple-Borehole Tests

Of the numerous multiple-borehole testing procedures available, only two will be discussed in this section: the method of Hantush (1960, 1964), which provides information on the vertical permeability and storativity of confining layers, and a modification of this method by Neuman and Witherspoon (1969a, 1969b, 1972).

The method of Hantush (1960, 1964) requires a pumping test procedure identical to the conventional pumping well–observation well setup discussed previously. The parameters obtained from the test include the storativity and transmissivity of the aquifer and the storativity and vertical permeability of the confining layer. The solution is an asymptotic one where the water-level response in the observation well is analyzed over a short time period and over a long time period. A curve-fitting procedure is involved for each of the designated time periods. For the case corresponding to Figure 6.1a, the short-term drawdown can be expressed as

$$s = \left(\frac{Q}{4\pi T}\right) H(u, \beta) \qquad (6.46)$$

where u is defined as in all previous cases as $r^2 S/4Tt$ and β is defined as

$$\beta = \left(\frac{r}{4B}\right)\left(\frac{S'}{S}\right)^{1/2} \qquad (6.47)$$

where S' is the storativity of the confining layer and $1/B$ has already been defined as $(k'/m'T)^{1/2}$. By short times we mean times of observation less than $m'S'/10k'$. The quantity $H(u, \beta)$ is an infinite integral whose value is approximated as

$$H(u, \beta) = W(u) - \frac{4\beta}{(\pi u)^{1/2}}\left[0.2577 + 0.6931 \exp\left(\frac{-u}{2}\right)\right] \qquad (6.48)$$

where $W(u)$ is the familiar well function of u. Some typical curves for the well function $H(u, \beta)$ are shown in Figure 6.25. Hantush (1960) has tabulated relevant values for the function $H(u, \beta)$ that may be adequate in applications. Note from Eq. 6.48 that this function approaches $W(u)$ as β approaches zero.

The long-time unsteady drawdown is expressed as

$$s = \left(\frac{Q}{4\pi T}\right) W\left(u\delta, \frac{r}{B}\right) \qquad (6.49)$$

where

$$\delta = 1 + \left(\frac{S'}{3S}\right)$$

Note that as S' approaches zero, we recover the Hantush–Jacob leaky formula.

The procedure for analysis is as follows.

1. Early time–drawdown data are fitted to the family of curves shown on Figure 6.25 with a match point producing values of $H(u, \beta)$, u, s, and t. The value of β is also obtained as β is a parameter of the type

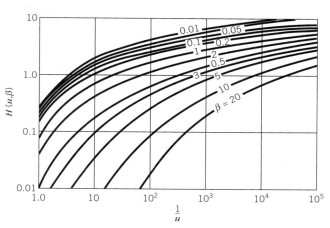

Figure 6.25 Type curves for a well in an aquifer confined by a leaky layer that releases water from storage (with permission from William Walton).

curve in the same manner that r/B is a parameter of the leaky curves. The transmissivity is determined from Eq. 6.46 and the storativity S is determined from the relationship $u = r^2S/4Tt$.

2. The usual procedures for leaky aquifers are then followed for the long-term data with the type curve of Figure 6.10. The match point will produce values of $W(u\delta, r/B)$, δu, s, and t. The value of r/B is again noted. This will produce values of transmissivity with Eq. 6.49 and the vertical hydraulic conductivity from the determined B value. As both β and B are now known, Eq. 6.47 may be solved for the storativity of the confining layer or, conversely, with known values of δS and S, the storativity of the confining layer may be determined. Details of analysis for other geometric configurations of leakage are given by Domenico (1972).

Interestingly, the solutions just described recover the leaky aquifer formula as the storativity of the confining layer approaches zero, as well as the Theis solution where both the storativity and permeability of the confining layer approach zero. Unfortunately, because of the similarity in the shape of the family of curves of Figure 6.25 used in the analysis, it is difficult to achieve a definitive match. Figure 6.2 shows the expected time–drawdown response where leakage from storage is involved and compares this response with the conventional leaky and nonleaky drawdown curves.

The asymptotic short-term–long-term solution of Hantush (1960) has been solved over continuous time by Neuman and Witherspoon (1969a, 1969b). These results were organized into a field testing procedure whereby the hydraulic properties of the aquifer and bounding low-permeability material may be determined (Neuman and Witherspoon, 1972). The test, referred to as the "ratio test," requires an observation well in the confining layer. Testing the hydraulic response of a confining layer is not a new development and was part of the requirements of earlier work dealing with low-permeability response (Wolff, 1970; Wolff and Papadopulos, 1972). However, the so-called ratio method represents a sophisticated approach to testing low-permeability rock where low-permeability zones bound higher-permeability aquifers that can be pumped over an extended period of time.

Many fractured rocks qualify as low-permeability media, and the testing of such rocks has been long viewed with skepticism. Much of the recent research was conducted by the staff of the University of Arizona, where a field test facility was established in Precambrian granite near Oracle, Arizona. A complete description of the site and the geologic, geophysical, and hydrologic investigations are given in Jones and others (1985). The hydrologic test theory has been given by Hsieh and Neuman (1985) and Hsieh and others (1985).

The method of Hsieh and others consists of injecting fluid into a number of isolated zones in boreholes and measuring the response in the same isolated zones in adjacent boreholes. The response is then fitted to one of several possible analytical solutions, generally with the aid of a computer. A simple curve-matching procedure is given by Hsieh and others (1985).

The hydraulic testing of fractured rock is a special application of flow in homogeneous but anisotropic material; that is, the hydraulic conductivity changes with direction. With the method of Hsieh and others, the principal planes of anisotropicity need not be established in advance. Other pumping test methods designed for anisotropic horizontal aquifers in which one of the principal directions is vertical have been given by Papadopulos (1965), Hantush (1966a, 1966b), and Hantush and Thomas (1966). These models correspond to the geologic prototype given as Figure 6.1c. Much of the recent research refers to the contribution of Papadopulos (1965) as one of the first original contributions in this field, including Way and McKee (1982), Loo and others (1984), and Hsieh and others (1985). A method of determining vertical conductivities has been given by Weeks (1969) and Way and McKee (1982).

6.6 Some Applications to Hydraulic Problems

There are various solutions or partial solutions to hydrologic problems that require either the results obtained from a hydrologic test or the utilization of the mathematical model upon which the formation response was predicated. For instance, any application of Darcy's law requires a priori knowledge of hydraulic conductivity. Hence, if we wish to make some calculations of flow rates or velocity of movement, the hydraulic conductivity must be known. If we wish to say something about travel time for a contaminant that moves at the speed of ground water, some determination of hydraulic conductivity is required. At yet another level, if we wish to apply some numerical model to ascertain the long-term response of a basin to prolonged pumping, information is required on the spatial distribution of hydraulic conductivity and storativity. Hence, the necessity of hydraulic testing.

On the other hand, there is also a need for the mathematical models employed in well testing. For example, a short-term pumping test (say, 24 hours) will yield the pertinent formation parameters T and S. As time and distance are variables in the mathematical model, it is possible to say something about what the drawdowns will be at later times and other distances. Due to the principle of superposition, it is possible to do the foregoing for a whole battery of wells pumping at different rates. Such predictions are invaluable in the design of dewatering operations, in water well interference considerations, in the design of well interceptors to contain

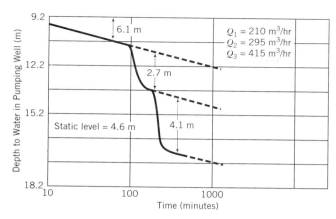

Figure 6.26 Semilogarithmic plot of time–drawdown curve obtained during a step–drawdown test.

contaminant migration, and for predicting pressure response and contaminant migration due to deep-well injection. As no new principles are involved in these applications, these points are best demonstrated with a few examples, most of which are based on actual case histories. That is to say, we already know the theory; now we need some rules for applying it.

Screen Diameter and Pumping Rates

The term *screen* is used to signify the diameter of a well that is open to the flow of water. It may be a perforated casing or some manufactured steel case apparatus. The larger the pumping rate we wish to employ, the larger the pumping apparatus required so that some minimum well diameter should be associated with a given pumping rate. The following is a rule of thumb guideline. For a desired pumping rate of 28 m³/hr, the minimum well diameter should be 15 cm. The following pairs are recommended: 68 m³/hr, 20 cm; 136 m³/hr, 25 cm; 272 m³/hr, 30 cm; 454 m³/hr, 35 cm; 682 m³/hr, 40 cm; 1136 m³/hr, 45 cm. It follows that using a 20-cm-diameter well will not assure a well yield of 68 m³/hr; the actual yield will be controlled by the transmissivity. On the other hand, a 20-cm-diameter well will not produce 682 m³/hr from a formation that is capable of delivering such amounts because of limitations placed on the pumping equipment.

Well Yield: The Step–Drawdown Test

A step–drawdown test is one in which the well is operated during successive periods at constant fractions of its full capacity (Figure 6.26). Such tests are useful for determining well yield and for establishing the depth for the pump setting. The specific capacity may be determined for each of the steps, that is, from Figure 6.26, S.C. = 210/6.0 = 35 m²/hr, S.C. = 295/8.8 = 33.5 m²/hr, and so forth. The specific capacity decreases with increasing pumping rates due to increased well loss.

Two useful kinds of information may be obtained from step–drawdown tests: pump setting requirements and the well loss constant. Consider that the well in question penetrates 15 m below the water table. The depth of penetration below the water table is called the maximum available drawdown. Thus at a pumping rate of 210 m³/hr or 295 m³/hr, only 6.1 m or 8.8 m of the available drawdown will be utilized. With a pump setting near the bottom of the well, either of these rates may be used in a long-term pumping test. At 415 m³/hr, approximately 13 m of the maximum available drawdown will be utilized; that is only 2 m of water remains in the well after a few hours of pumping. We thus establish a reasonable upper bound for the pumping rate on the order of 340 m³/hr.

The well loss constant C may be determined directly from equations presented by Jacob (1950) where

$$C = \frac{\Delta s_i/\Delta Q_i - \Delta s_{i-1}/\Delta Q_{i-1}}{\Delta Q_{i-1} + \Delta Q_i} \quad (6.50)$$

or, for steps 1 and 2 of Figure 6.26

$$C = \frac{\Delta s_2/\Delta Q_2 - \Delta s_1/\Delta Q_1}{\Delta Q_1 + \Delta Q_2} = \frac{(2.7/85) - (6.1/210)}{210 + 85}$$

$$= 9.2 \times 10^{-6} \text{ m/(h}^3\text{/hr)}^2$$

and for steps 2 and 3,

$$C = \frac{\Delta s_3/\Delta Q_3 - \Delta s_2/\Delta Q_2}{\Delta Q_2 + \Delta Q_3} = \frac{(4.1/120) - (2.7/85)}{85 + 120}$$

$$= 1.2 \times 10^{-5} \text{ m/(m}^3\text{/hr)}^2$$

for an average of 1.06×10^{-5} m/(m³/hr)². Thus the well loss associated with each of the pumping rates used in the step–drawdown test is readily determined from the relationship $s_w = CQ^2$. For each of the pumping rates employed, the well loss is 0.47 m, 0.92 m, and 1.83 m, respectively.

A Problem in Dewatering

Figure 6.27 shows a situation where a tunnel must be constructed some 18 m below a water table. To facilitate construction, the water table must be lowered below the tunnel during the construction period. Several questions must be addressed:

1. What pumping rates will be employed?
2. What are the depth and diameter of the pumping wells?
3. What is the duration of pumping?
4. How many wells will be required, and how should they be spaced?

Many of these questions are interdependent, and some have to be answered on the basis of value judgment. The values for transmissivity, storativity, and specific capacity

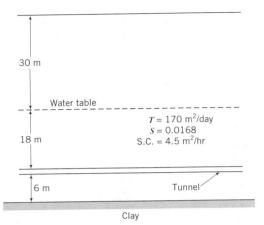

Figure 6.27 Schematic cross section of a proposed tunnel through an unconfined aquifer.

are noted in the diagram and were obtained by hydraulic testing.

The first thing to note is that the maximum available drawdown for wells drilled to the top of the clay layer is on the order of 24 m. Thus, with a measured specific capacity of 4.5 m²/hr, the maximum possible pumping rate is calculated to be 108 m³/hr. This, however, is somewhat unrealistic in that such a pumping rate would readily remove all the water from the well. Restricting the drawdown in a single pumping well to 21 m, the design pumping rate is calculated to be 95 m³/hr. From information provided in the previous example, this requires a design diameter of 20 cm for the dewatering wells. Thus, we have determined that the wells will be 55 m deep, their diameter will be 20 cm, and their design pumping rate will be 95 m³/hr. All that remains is the determination of the pumping duration and well spacing.

Figure 6.28 shows some theoretical distance–drawdown curves for pumping durations of 10 and 100 days. Note that the difference in drawdown between the two curves is only about 2 m. The longer-duration

pumping period has the benefit of a greater spacing between the wells to obtain the desired drawdown. However, the shorter-duration period is the better one here for several reasons. First, in most dewatering operations, only a few tens of meters along the tunnel alignment are dewatered during construction. Once the construction activities are completed in a certain reach, the pumps are moved farther along the alignment. Thus, the whole alignment need not be dewatered in advance. In addition, a 100-day period would require the disposal of greater quantities of water. Thus, the 10-day design period is the better one.

Figure 6.29 shows a plan view of one feasible design. The 10 m offset from the centerline of the tunnel was a contractual agreement. The water levels must be lowered some 20 m all along the centerline. As a first guess, assume the wells in each row are 30 m apart. The least drawdown will occur at points 1 and 3, and the most at point 2. Using superposition with the aid of Figure 6.28, the drawdown at points 1 and 3 is calculated to be on the order of 30 m. Clearly, we do not have 30 m of available drawdown, but the model calculation is not constrained and can produce this nonsensical result. For a 60-m spacing, the drawdown is 24 m at points 1 and 3 and 29 m at point 2. For a 90-m spacing the drawdown is approaching 23 m at points 1 and 3 and is slightly more at point 2. Thus it appears we are slowly moving toward an optimal spacing, and this spacing is in excess of 60 m. However, there is need for some caution here in that the calculated spacings have been based on a design pumping rate of 95 m³/hr, and drawdown is directly proportional to the pumping rate. With five interfering wells, this pumping rate cannot be maintained for the target 10-day period. For the 90-m spacing, a decrease in the effective pumping rate from 95 m³/hr to 83 m³/hr will produce a 20-m drawdown at points 1 and 3. For the 60-m spacing, a decrease in the pumping rate to 78 m³/hr will likewise produce a drawdown of 20 m

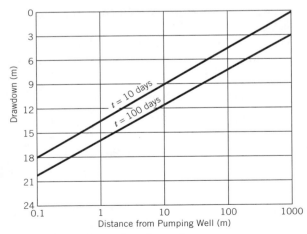

Figure 6.28 Time–drawdown data.

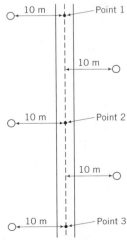

Figure 6.29 Plan view of well alignment for dewatering.

at points 1 and 3. Thus the need for value judgments, with the 60-m spacing being the more prudent one, at least for the preliminary design. This means that the design pumping rate of 95 m³/hr can fall off to as little as 78 m³/hr over the 10-day pumping period and the dewatering can still be accomplished.

A Problem in Water Supply

A power company plans to build a facility in the southwestern desert of the United States and requires a water supply for cooling purposes. The plant is to be constructed a few miles from a known ground-water source, and the plan is to deliver the pumped water to the plant, where it will be consumed in the cooling process (no return flows). The plant is to be built in three units, phased in every five years, with each unit requiring a continuous flow of water amounting to about 477 m³/hr. A cross section of the water supply is shown in Figure 6.30, which is a rather narrow alluvial valley cut into low-permeability sediments. The company has access to property that extends across the entire valley (about 4570 m) and about 3050 m along the valley reach. Three wells are shown in the plan view of this property, and their locations have been idealized to facilitate the required computations (Figure 6.31).

There are a few facts and a few questions of importance here. First, because of the geologic configuration of the alluvial valley, the problem should be treated as an infinite strip with image wells repeating to infinity (Figure 6.24c). Second, drawdowns in any given well should be limited to about 30 m or the subsequent dewatering of the aquifer will destroy its transmissive properties. This means that the pumping should be distributed as equally as possible. With these constraints, can the power company obtain sufficient water for each of the units, each of which is phased in every 5 years? If not, can they supply one unit for the life of the plant (50 years)? If not, how long can they supply one unit without excessive dewatering of the aquifer? The transmissivity has been determined to be 497 m²/day, the storativity equals 10⁻³, the well diameters are 36 cm, and the actual specific capacity of the wells is about 11 m²/hr.

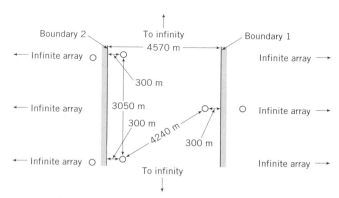

Figure 6.31 Plan view of well location relative to boundaries.

It is not intended to treat this entire complex problem, but merely to show some elements of the use of the pertinent hydraulic equations in addressing quantitative questions. Figure 6.32 shows the familiar distance–drawdown plot for an infinite aquifer with a design pumping rate of 160 m³/hr. Let us focus on the centermost well. With a specific capacity of 11 m²/hr and a pumping rate of 160 m³/hr, the drawdown is anticipated to be about 14 m. This will certainly increase with time, but let us not dwell on this detail. The influence from the two neighboring real wells at the end of a five-year period will be on the order of 4 m, giving a total drawdown of 18 m. Consider now only four image wells associated with the boundaries affecting this well. These will be located 300 m and 8840 m from boundary 1 and 4270 m and 4870 m from boundary 2. These images are thus located the following distances from the pumped well: 600 m, 9140 m, 8590 m, and 9140 m. Their cumulative effect on the pumping well is about 16 m giving a total drawdown of 34 m. Yet to be considered are the other images associated with this well, which have a measurable influence from as far away as 27,420 m (Figure 6.31),

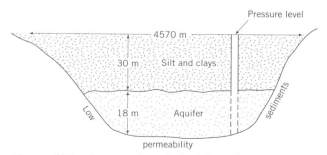

Figure 6.30 Cross section of alluvial aquifer.

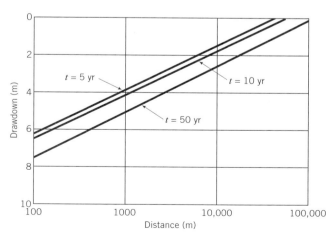

Figure 6.32 Time–drawdown response of an infinite aquifer.

plus the images associated with the other two pumping wells. Based on this preliminary analysis, an attempt to develop sufficient water for one unit over a five-year period will cause substantial dewatering of the aquifer. This conclusion must be tempered by other considerations, for example, the possibility of leakage from the overlying clays and silts (which was not detected over a mere 24-hour pumping period, but which is virtually assured over pumping periods measured in terms of years), and the possibility of some recharge to the aquifer over extended periods of pumping. Given that these other sources of water act to limit the drawdown, it can still be concluded that a water supply for more than one unit is totally out of the question. The life of the water supply for a single unit will depend on recharge, which may not be substantial in a desert environment, and the rate of leakage from the overlying sediments.

6.7 Computer-Based Calculations

This section describes the operation and potential applications of an interactive computer code, WELLz, for estimating drawdown(s) due to one or more pumping and/or injection wells. The present version of the code accommodates calculations involving a fully penetrating well in a confined, leaky, or water-table aquifer. The equations for drawdown for a confined and a leaky confined aquifer can be described by

$$\text{Confined Aquifer: } s = \frac{Q}{4\pi T} W(u)$$

$$\text{where} \quad u = \frac{r^2 S}{4Tt} \quad (6.51a)$$

$$\text{Leaky Aquifer: } s = \frac{Q}{4\pi T} W\left(u, \frac{r}{B}\right)$$

$$\text{where} \quad u = \frac{r^2 S}{4Tt}, \quad \text{and} \quad \frac{1}{B} = \left(\frac{k'/m'}{T}\right)^{1/2} \quad (6.51b)$$

According to Jacob (1944), the drawdown s for an unconfined aquifer can be expressed as

$$s = m - (m^2 - 2s'm)^{1/2} \quad (6.52)$$

where s' is the drawdown for a confined aquifer of the same hydraulic properties, which is calculated from Eq. 6.52, and m is the thickness of the aquifer, before pumping.

The main difficulty in implementing these or similar well-hydraulics solutions in the code comes in evaluating the well functions $\{W(u)$; and $W(u, r/B)\}$ for the specified T, S, and Q. Fortunately, approximations are available to most of the frequently used functions in analytic groundwater equations.

WELLz 1.0 is programmed in C++ and runs under Microsoft Windows. The code is provided on the disk included at the back of the book and instructions for loading are included in Appendix B. When the code is run from Windows, the initial dialog box has three main pull-down menus—"File," "Editing," and "Options." Files–New is used when the user wishes to work a new problem. A set of default data appears in subsequent dialog boxes, which the user edits. Files–Open provides the opportunity to load a preexisting data set saved from an earlier run. The user selects the particular file—saved with a *.wzf extension—and the appropriate data appear in the "Input Menu" dialog box and subsequent dialog boxes (Figure 6.33). When a preexisting data set is loaded, the first screen that is opened is the contour plot for the problem. To edit aquifer or well parameters for a subsequent run, one would use the Edit menu. To return to the Input Menu or first screen, choose Files–New. In effect, if changes are required to the Input Menu, the simulation data will likely require major changes, making it a new trial.

Once the Input–Menu dialog box appears, one selects the proper unit parameters, defines the size of the region over which drawdowns are to be contoured, selects the particular aquifer type of interest, and defines the total simulation time (Figure 6.33a). For contouring, drawdowns are calculated at equally spaced data points across the region. Normally, users are advised to keep the number of rows and columns in the calculational grid large. The default 50, 50 is recommended, providing drawdown calculations at 2500 separate points. In any case, the number of rows and columns each cannot exceed 60. Note that the simulation region identifies a region in infinite space where drawdowns are to be calculated and contoured. Defining this region has no impact on the calculation, as is the case with numerical model. Thus, it is possible to define pumping/injection wells outside of the region and their impacts may still be felt within the simulation region. Similarly, drawdown may exist outside of the calculational region. The well coordinates need to be defined so that all values are greater than zero.

The next dialog box serves to input the relevant aquifer parameters. The particular dialog box that appears depends upon the aquifer type. For example, Figure 6.33b illustrates the dialog box for a confined aquifer. The next dialog box (Figure 6.33c) contains information on various pumping and observation wells. One uses the mouse to add or edit well information. Observation wells are identified by a pumping rate of 0.0. After all of the necessary parameters are provided through the dialog boxes, a window provides the contoured drawdown over the region of interest (Figure 6.33d). The drawdown is automatically provided at the designated observation wells. By selecting "Options—Label Contours," one can use the mouse to label the contour lines, or display drawdowns at any designed point on the screen. Simply move the arrow to the position where you wish the drawdown to be shown and click the mouse. At this point, "File" options

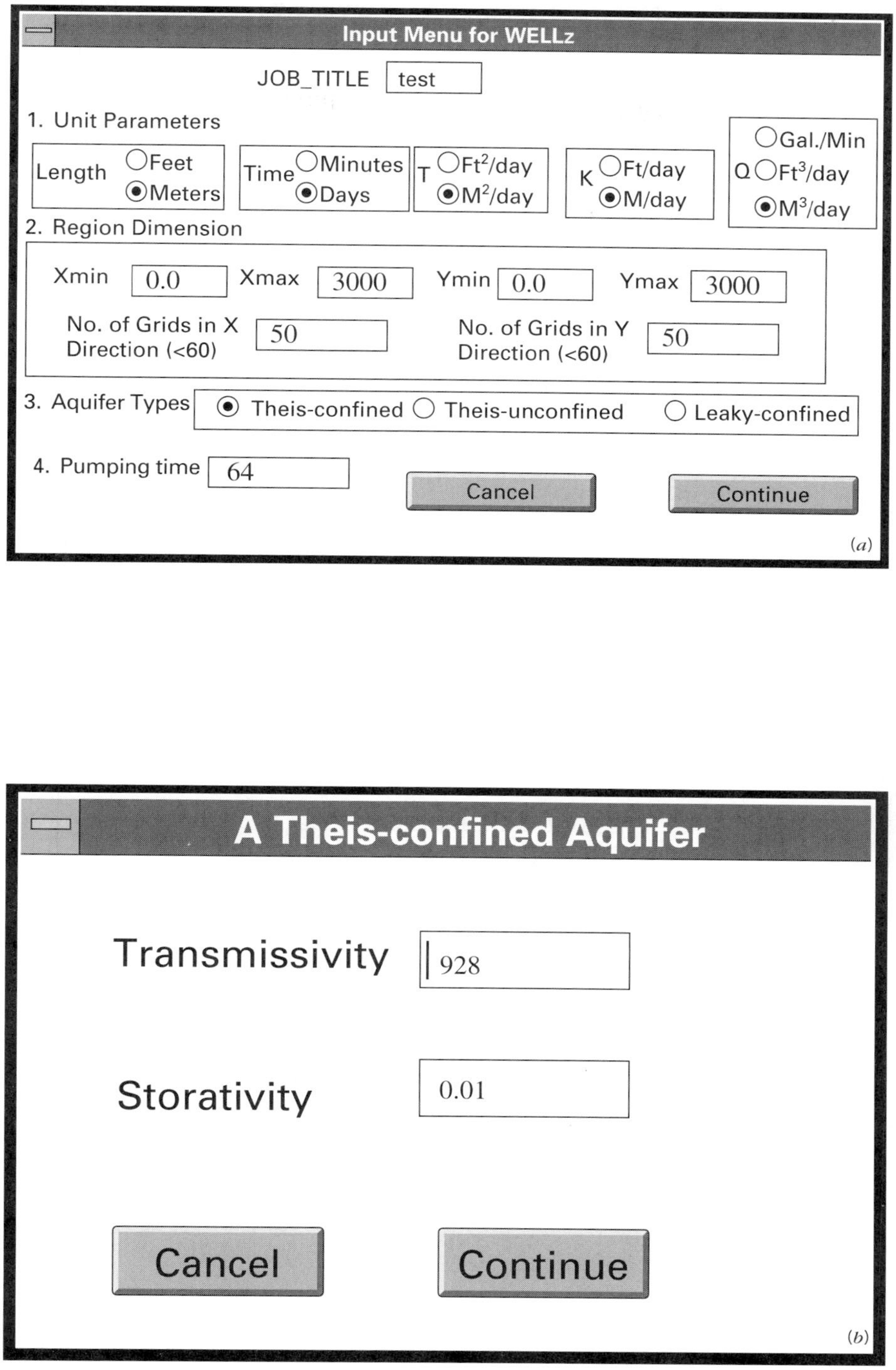

(a)

(b)

Figure 6.33   Example of the sequence of dialog boxes appearing in WELLz (Panels *a*, *b*, and *c*) along with the printed drawdown map for the problem (Panel *d*).

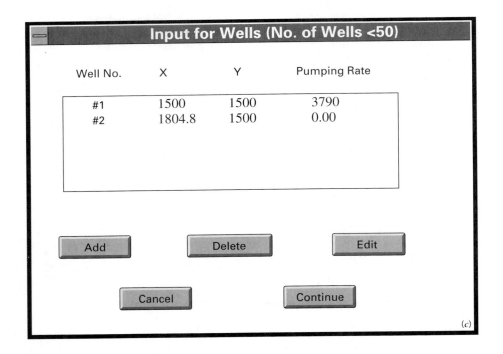

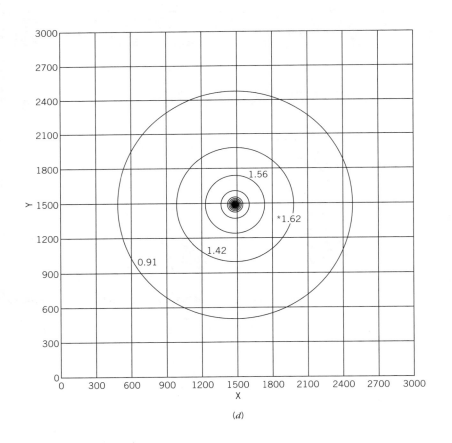

Figure 6.33 **Continued**

similar to those used previously are provided as follows:

- New—rerun the simulation from the beginning
- Open—open a previously saved data file with a *.wzf extension
- Save As—save the data file with a *.wzf extension
- Print—print the contour map
- Save Report—export the drawdowns to a file for plotting, and create a report of the run to be stored in a file
- Exit—quit the code

The number of contour lines and contour intervals are either fixed or specified by the user by selecting Options—Contour values. Contour lines are defined by simply dividing the fixed minimum and maximum drawdowns in the field into equal intervals to produce 10 contour lines. The export capabilities as File—Save Report facilitate the creation of a more polished drawdown map using other plotting packages. Provision is also included with Export to save a summary report of the run in another file.

Code Demonstration

The following simple well-hydraulic problem illustrates how the code is used.

Example 6.10

A well that is pumped at a constant rate of 3790 m³/day fully penetrates a confined artesian aquifer. The aquifer can be assumed infinite with a transmissivity and storativity of 928 m²/day and 1.0×10^{-2}, respectively. Determine the drawdown in an observation well located 304.8 m from the pumping well at 64 days.

We solve the problem using WELLz, as follows.

1. Select the WELLz icon from the Windows Programs Manager by double clicking the mouse.

2. Select "File–New," because the data set does not already exist.

3. Edit the information that appears in the dialog box as shown in Figure 6.33*a*. The unit parameters for this problem are all meters and days, as shown. The region is set up with the pumping well in the middle having coordinates (1500, 1500) (see Fig. 6.34). The calculations will be carried out within a region 3000 m by 3000 m. Note that as required by the code all wells have coordinates greater than zero. For this problem, "Theis confined aquifer" is selected. Having completed editing the form, we move to the next dialog box by clicking on "Continue."

4. Edit values of *T* and *S* on the next form (Fig. 6.33*b*), and continue.

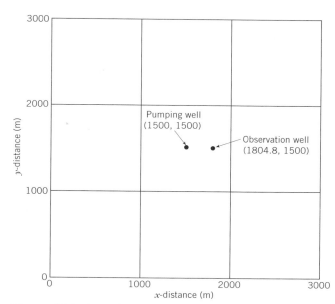

Figure 6.34 Map showing the location of wells within the calculation region.

5. Add information on the pumping well and the observation well (Fig. 6.33*c*). The coordinates of the observation well are (1804.8, 1500). Click on "Continue," and move to the final dialog box.

6. The drawdown due to the well is contoured in Figure 6.33*d*. At the observation well, the drawdown is indicated to be 1.6 m. By selecting "Option—Label–Contours," one can label the contour lines and other points of interest.

7. Finally, we save the data and exit the code.

Bounded Aquifers Revisited

WELLz and similar codes are particularly useful in applications to bounded aquifer systems, where the wells and image wells coupled with a relatively large number of observation points make drawdown calculations tedious. Example 6.11 provides another, more realistic look at image-well theory and the application of WELLz to solve a problem complicated by a relatively large number of wells.

Example 6.11 (based on Walton, 1962)

Walton (1962) illustrated the application of image-well theory in the interpretation of an aquifer test by Mikels (1952) at Zion, Illinois. The test involved a single pumping well (A-P), pumped at 99 gpm for 3180 minutes and six observation wells (Figure 6.35). The unconfined gravelly sand aquifer had a saturated thickness of approximately 20 ft. Walton (1962) interpreted the asymmetry in the cone of depression and similarities in water-level responses in the wells and Lake Michigan as indications of good hydraulic connection and a recharge boundary. He estimated the hydraulic conductivity to be 387.7 ft/day the specific yield to be 0.01, and the distance to the effective line of recharge to be 206 ft.

Use WELLz to simulate the 3180-minute test at Zion. See how well the simulated drawdowns match the observed drawdowns shown in Figure 6.35.

1. Let us begin by formulating the problem. First, we define a region over which the calculations are to be conducted and a coordinate system for defining the location of the wells (Fig. 6.36). The exact location of the origin doesn't matter except that the region should be located appropriately. The simulation region we have selected stops at the recharge boundary 306 ft from the origin. Nevertheless, the image (injection) well is included as part of the input data and its influence is felt within the simulation domain.

2. We run the code as before, entering the appropriate information. For those more curious, the data set is

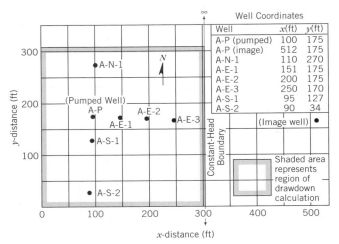

Figure 6.36 Formulation of the problem for simulation. The table lists the coordinates of each of the wells (modified from Walton, 1962, and Mikels, 1952).

included as "EX6_11.wzf" on disk and can be loaded through File–Open.

3. The simulation results at 3180 minutes are plotted in Figure 6.37. Note how the presence of the recharge boundary leads to reduced drawdowns between the pumping well and Lake Michigan. The simulated drawdowns at the observation wells match the observed drawdowns very well.

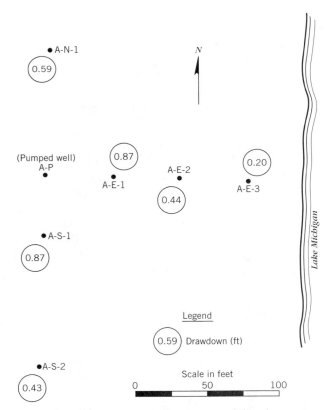

Figure 6.35 Map showing the location of the observation wells for the aquifer test at Zion, Illinois. The drawdown in feet at each of the observation wells at 3180 minutes is circled (modified from Walton, 1962, and Mikels, 1952).

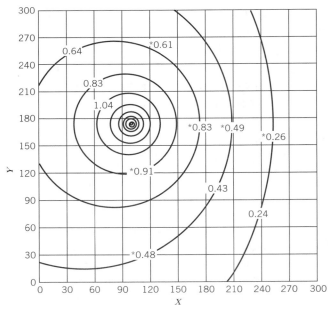

Figure 6.37 Output from WELLz of the calculated drawdown due to pumping A-P.

Problems

1. After 24 hours of pumping a confined aquifer, the drawdown in an observation well at a distance of 97.5 m is 0.6 m and the drawdown in an observation well at a distance of 33.5 m is 1.1 m. The pumping rate is 5.43×10^3 m³/day. Find the transmissivity.

2. Time–drawdown data collected at a distance of 30.5 m from a well pumping at a rate of 5.43×10^3 m³/day are as follows.

t, min	s, m
1	1.2
2	1.6
3	1.9
4	2.1
5	2.3
6	2.5
7	2.7
10	3.1
20	3.7
40	4.3
80	4.8
100	5.0
300	5.8
500	6.2
1000	6.6

Calculate the transmissivity and storativity.

3. A well 76 m deep is planned in an aquifer with a transmissivity of 125 m²/day and a storativity of 1×10^{-2}. The well is expected to yield 2720 m³/day and will be 30 cm in diameter. If the nonpumping water level is 15 m below land surface, estimate the depth to water after one year's operation and after three years' operation.

4. A 46-cm-diameter well within an aquifer with a transmissivity of 100 m²/day and a storativity of 7×10^{-2} is to be pumped continuously. What pumping rate should be used so that the maximum drawdown after two years will not exceed 6 m?

5. You are asked to design a pumping test for a confined aquifer in which the transmissivity is estimated to be about 745 m²/day and the storativity about 1×10^{-4}. What pumping rate would you recommend for the test if it is desired that there be a drawdown of about 2 m in the first five hours of the test at an observation well 30 m from the pumping well?

6. Replace the fault in Example 6.8 with a fully penetrating stream, and calculate the drawdown at the stream and midpoint between the well and the stream for the same pumping rate and duration of pumping.

7. An aquifer has a transmissivity of 93 m²/day and a storativity of 1×10^{-4}. A 60-cm pumping well has a drawdown of 72 m at the end of one day's pumping at a rate of 2800 m³/day. The efficiency of the well is

determined as 56%. What is the efficiency when the well is pumped at 5600 m³/day for a one-day pumping period?

8.

a. List three reasons that might explain an upward inflection of a semilogarithmic time–drawdown plot.

b. List three reasons that might explain a downward inflection of a semilogarithmic time–drawdown plot.

9. Suppose the only type curve available to you was a plot of $W(u)$ versus u. How would you plot time–drawdown data obtained from the field in order to use this curve in the matching procedure?

10. A water well is located 300 m from a waste interceptor trench that acts to intercept a plume emitting from a uranium tailings pile. The well is to be pumped for 100 consecutive days at a rate of 5.43×10^3 m³/day. The transmissivity of the formation is 497 m²/day and the storativity is 0.1. At the end of the 100-day pumping period, will the cone of depression reach the interceptor trench and what will be the drawdown?

11. Use the figure in Example 6.7 to answer the following:

a. A 60-cm-diameter well is located about 3 m from a fault (barrier) and is pumped at a rate of 5.43×10^3 m³/day for a 10-day period. What is the approximate drawdown in the well at the end of the 10-day period (ignore well losses).

b. What sort of shift would you expect in the curves given in Example 6.7 if the storativity was much greater, say, 1×10^{-1}?

12. A disposal well for liquid waste injection commences operation in a horizontal confined aquifer that has the following characteristics: thickness = 9 m, porosity = 10%, hydraulic conductivity = 81 m/day, specific storage = 3.3×10^{-6} m⁻¹, injection rate 5660 m³/day. The disposal well has a radius of 0.3 m

a. To what distance from the well will the front of the cone of impression have extended after 10 days' injection. (Assume, for all practical purposes, that 0.03 m of head above the regional preinjection piezometric head marks the front of the cone of impression.)

b. Approximately how far will the injected contaminant move by the end of the 10-day injection period? Assume that the contaminant moves at the same speed as the ground water and that the induced hydraulic gradient due to the injection is imposed instantaneously and is linear.

13. Consider the enclosed figure to represent the pumping history of a single well over a six-year period. This pumping caused some drawdown in a nearby observation well as shown in the companion diagram. Explain very carefully how you would use superposition in this problem.

Start out with "I would pump the pumping well at a rate of _____ for _____ years to obtain a response; then I would pump it at a rate of _____ for _____ years, and (add) (subtract) the results to or from the previous response; then I would pump the pumping well at a rate of _____ for _____ years, and (add) (subtract) the results to or from the previous sum."

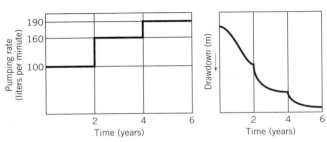

Problem 13

14. The accompanying diagram gives some pumping test results as measured in an observation well. You are given the following:

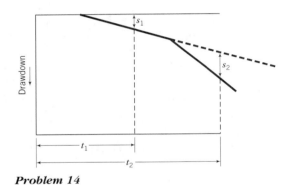

Problem 14

transmissivity = 930 m²/day

storativity = 1×10^{-4}

the distance from the observation well to the pumping well (r_1) is 300 m

the time (t_1) at which s_1 is measured is 0.1 day

the time (t_2) at which the drawdown s_2 is measured is 10 days

the drawdown s_2 is two times the drawdown s_1

Find the distance from this observation well to the image well responsible for this behavior.

15. An injection test in a single well produced the curve-fitting parameters for the transmissivity T, storativity S, vertical permeability k_v, and the storativity S' for the geometry shown. As no other facts are given, you

are required to compute the recovery in the well in response to the injection. Calculate the short-term recovery in the injection well of radius 0.15 m at the end of 0.22 days of injection at a rate of 28 m³/day. (Hint: Figure 6.25 is required to obtain the result.)

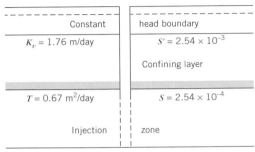

Problem 15

16. During a field investigation of a leaky underground storage tank for oil, a group of small-diameter observation wells is emplaced. In the interest of interception of some of the oil product floating on the water table, the perforations are placed both above and below the water table. Later, it is decided that a value of the hydraulic conductivity of the saturated zone is necessary for further calculations. Explain how the Hvorslev test for a withdrawal procedure would work. Use the data of Figure 6.19 to obtain a value for the hydraulic conductivity as has been done in Example 6.5, reinterpreting the data for a withdrawal as opposed to an injection test.

17.

a. Design a dewatering scheme for the following situation. A tunnel is to be constructed 17 m below land surface. The water table is at a depth of 12 m, and the tunnel will reside 5 m below the water table. The available saturated thickness from the water table to a clay layer is 12 m. The transmissivity of the material to be dewatered is 3730 m²/day, and the storativity is 0.01. The actual specific capacity of the well used to conduct the hydraulic testing has been measured at 48 m²/hr. The dewatering wells will be finished 6 m below the tunnel, and they must be offset 5 m from the centerline of the tunnel. The wells will be placed alternately on each side of the tunnel (see Figure 6.29). Consider four wells pumping simultaneously over a 10-day period at a pumping rate of 270 m³/hr. Determine the well diameter and well spacing required to lower the water table by 6 m along the tunnel alignment.

b. Repeat the exercise for paired wells, that is, four wells set in groups of two on the same line perpendicular to the tunnel alignment. Comment on the "best" design in terms of well spacing.

c. Your assignment is to type a report not to exceed one page on your selected design. Include two figures, one that shows your mathematical model such as depicted in Figure 6.28 and one that shows the optimal design (Figure 6.29).

18. Extend the boundaries of Figure 6.31 an additional 5330 m on both sides and analyze this problem for the centermost well and the pumping demand of 480 m³/ hr. The wells must remain in their current position; only the boundaries have been moved. Consider the effect of three real wells pumping and four image wells associated with each of the pumping wells. Include a sketch of your real and image-well location and comment on the

possibility of obtaining 480 m³/hr for one unit for a 5-, 10-, and 50-year period.

19. A well was pumped at 30 gallons per minute. The drawdown response was characteristic of a leaky artesian aquifer. The transmissivity of the aquifer is 200 ft²/day, the storativity is 0.0004, the thickness of the confining bed is 15 ft, and hydraulic conductivity of the confining bed is 0.015 ft/day. Calculate the drawdown at three observation wells using WELLz after 2000 minutes of pumping. The pumping well has coordinates (100 ft, 100 ft), and the coordinates for the observation wells are OB1 (100, 100), OB2 (78, 5), and OB3 (340, 177).

Chapter 7

Ground Water as a Resource

Thus far we have learned about aquifers and how water contained within them can be developed by wells. This chapter extends these ideas in addressing questions of how a ground-water resource as a whole is managed and some of the important tools and strategies. One cannot discuss issues of regional-scale aquifer development and management without talking about simulation models that are now at the core of most basin-wide assessments. What has made aquifer simulation models indispensable for ground-water studies is the power to integrate the complexities of hydrogeologic settings, hydrogeologic processes, and water utilization.

7.1 Development of Ground-Water Resources

A resource represents a supply of something that can be drawn upon for use. Like petroleum, water is a resource that can be readily transported to achieve a better balance between the location of its supply and the demand for its use. Unlike petroleum, ground water is not a minority fluid in the subsurface environment and its value is not normally determined by the marketplace. However, ground water can also be nonrenewable, at least when viewed within a human time frame, and its exploitation is subject to supply and demand. In areas with

abundant surface water, ground water is frequently an underexploited resource. Conversely, in areas without surface-water supplies, ground water is almost always overexploited. In areas between these extremes, the development of the total water resource depends on the demand for water. How the total water-resource system is developed and the manner in which it is operated depends not only on availability of supply but on legal, political, and socioeconomic precedents and constraints.

The Response of Aquifers to Pumping

Prior to the initiation of pumping, ground-water recharge is balanced by the natural discharge of water to springs, creeks, rivers, and lakes. Withdrawal of water by wells is an additional stress put on the ground-water system that must be balanced by (1) an increase in recharge, (2) a decrease in natural discharge, (3) a loss of ground-water storage, or (4) a combination of these factors. Theis characterized this pattern of change as a "state of dynamic equilibrium" whereby a new balance between inputs and outputs might be achieved at a new but lower level of ground-water storage. As the system adjustment often requires time, the basin can remain in a long-term transient state, with water levels falling in response to the pumping. Thus, in Theis's words, ground water may be classified as a renewable resource, but there are instances where it may not be so within a human time frame.

Yield Analysis

Management of ground-water basins requires some kind of yield analysis to determine how much ground water is available for pumping. Frequently, the management strategy is to obtain the maximum possible pumping compatible with the stability of the supply. The term *safe yield* as an indicator of this maximum use rate has had an interesting evolution since first introduced by Lee (1915). Lee defined safe yield as

The limit to the quantity of water which can be withdrawn regularly and permanently without dangerous depletion of the storage reserve.

This definition was expanded by Meinzer (1923), who defined safe yield as

The rate at which water can be withdrawn from an aquifer for human use without depleting the supply to the extent that withdrawal at this rate is no longer economically feasible.

Thus, the "dangerous depletion" of Lee is described in economic terms by Meinzer, and both speak of permanency of withdrawals.

As a philosophical concept, the definition of Lee is a good one. However, no guidance is provided on how this rate or rates may be determined and the concept remains far too ambiguous for practical use. Conkling (1946) attempted to make the concept less ambiguous by specifying the conditions that constitute a safe yield. He described safe yield as an annual extraction of water that does not:

1. Exceed average annual recharge.
2. Lower the water table so that the permissible cost of pumping is exceeded.
3. Lower the water table so as to permit intrusion of water of undesirable quality.

A fourth condition, the protection of existing water rights, was added by Banks (1953).

The single-valued concept of safe yield as proposed by Conkling and modified by Banks encompasses hydrologic, economic, quality, and legal considerations. This overspecification of the term is not likely what Lee or Meinzer had in mind. The controversial nature of the concept as defined by Conkling is clearly demonstrated in 43 pages of discussion of his original 28-page paper by no fewer than 10 authorities. In practice, safe yield has no unique or constant value, its value at any time depending on the spacing and location of the wells and their influence on the dynamics of interchange between ground water and other elements of the hydrologic cycle.

In addition, it is not possible to develop a safe yield in the absence of initial developmental overdraft, a necessary first stage in ground-water development where withdrawals cause a lowering of the water table in areas of natural discharge and recharge. Further, there are seasonal or cyclical overdrafts where water levels eventually return to their original levels during periods of limited withdrawals. Thomas (1951) and Kazmann (1956) have suggested abandonment of the term because of its indefiniteness. Freeze (1971) introduced a concept of maximum stable basin yield, determined from a three-dimensional saturated–unsaturated numerical model. Although the concept is a good one, determining its value is not a simple task.

Case Study: The Upper Los Angeles River Area

There are ground-water basins, especially in the arid southwestern part of the United States, where a basin-wide response to pumping on a massive scale is evident. A case in point is the Upper Los Angeles River area (ULARA) in southern California, which is comprised of one large ground-water basin, the San Fernando Basin, and three other smaller basins—the Sylmar Basin, the Verdugo Basin, and the Eagle Rock Basin (Figure 7.1). The San Fernando Valley area includes the cities of Los Angeles, Burbank, and Glendale and is home to various Hollywood studios and industrial companies such as, Lockheed, Rockwell, and 3M.

The ULARA is bounded by various mountain ranges and hills (Figure 7.1). Ground water is extracted by wells from the thick valley-fill aquifer across much of the San Fernando Valley. While the valley-fill aquifer is locally productive, rates of recharge are relatively low due mainly to the arid climate and seasonal nature of precipitation. Over the past 100 years, the valley floor received an average of about 14 inches (35.6 cm) of rainfall, most of which falls from December to March (ULARA Watermaster Report, 1994).

As the population of the San Fernando Valley increased from the 1930s onward, it became inevitable that the valley-fill aquifer could not provide an inexhaustible supply of water. Lost initially was the natural discharge to the rivers and a loss of storage in the ground-water system itself. Shown in Figure 7.2 are two observation wells (3700A, located at 2; and 3914H, located at 5). The hydrographs for these wells depict the historical decline of water levels in the valley-fill aquifer from the late 1930s on to 1968. As suggested by the hydrographs, water-level declines of the order of 100 to 200 feet (30.4 to 60.8 m) were not uncommon.

Ground-water withdrawals in the basin were limited by court actions in 1968 to approximately 104,000 acre-feet per water year (ULARA Watermaster Report, 1994).

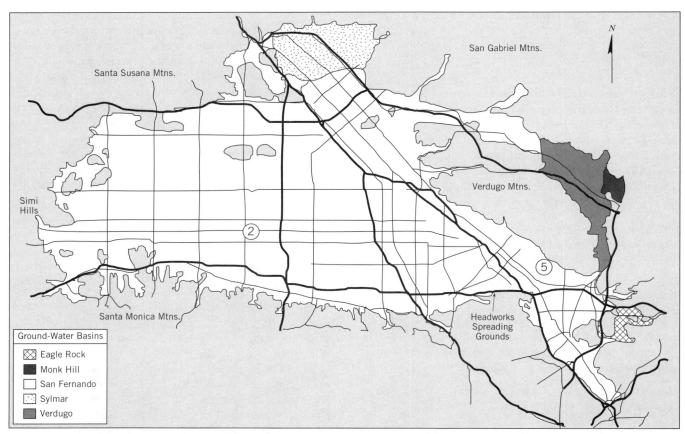

Figure 7.1 Location of ground-water basins of the Upper Los Angeles River Area. Shown in circles are the location of observation wells referred to in the text (modified from Watermaster Report, 1994).

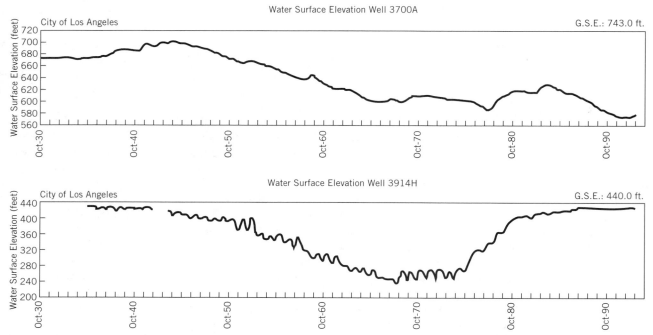

Figure 7.2 Hydrographs of two wells located in the San Fernando Basin (modified from the Watermaster Report, 1994).

This rate of withdrawal was about two thirds of the average of the previous six years. This decrease is reflected in the behavior of the hydrographs, which subsequently leveled out or recovered (Figure 7.2). Overall, the basin is managed to keep total ground-water withdrawals within what is referred to as the safe yield of the basin.

Safe-yield concepts were applied to the San Fernando Basin to help allocate the available water. The safe yield of the basin was estimated to be 43,660 acre-feet per year. The entire amount is available to the City of Los Angeles for extraction. This yield represents the long-term average recharge from precipitation. Some of the water that is imported to the basin from other sources ends up as additional sources of recharge to the valley-fill aquifer. This water is available as well water for pumping by the cities of Los Angeles, Burbank, and Glendale. In the 1993–94 water year, some 43,900 acre-feet of this imported water was available for pumping (ULARA Watermaster Report, 1994). Across the San Fernando Basin actually more water ends up in the ground from recharge of imported water than from recharge due to natural precipitation on the basin.

The management scheme for the ULARA also provides the capabilities for Los Angeles, Glendale, and Burbank to store water and extract equivalent amounts in the future. In effect, in years when the pumping of the cities is less than their allocated amount, the water is deemed to be saved and is available for future use in years of high demand.

Historically, water-management activities like those in the ULARA relied on water-balance calculations for broad, basin-scale estimates of inflow and outflow. In the future, there will be more reliance on calibrated mathematical simulation models at the basin scale that more rigorously account for the inherent variability in material properties, recharge rates, and ground-water withdrawals. The topic of numerical simulation is taken up in Section 7.2.

Management Strategies

The previous discussions provide the context for why ground-water resources require management and some of the tools for yield analysis to establish quantities of water available for pumping. This section explores in more detail some of the issues related to the management of ground-water basins. The California State Department of Water Resources (1980) describes "ground water basin management" in the following terms:

Ground water basin management includes planned use of the ground water basin yield, storage space, transmission capability, and water in storage. It includes 1. protection of natural recharge and use of artificial recharge; 2. planned variation in amount and location of pumping over time; 3. use of ground water storage conjunctively with surface water from local

and imported sources; and 4. protection and planned maintenance of ground-water quality.

As this definition implies, the most important element in any water-management scheme is a well-defined limit on the quantity of water that can be pumped. In its simplest form, this limit may be related to the safe yield of the basin as determined by the quantity of recharge from natural sources. However, most management schemes are more complicated given sources of water other than ground water and the possibilities for various recharge schemes. Thus, in the definition above, item 1 makes reference to artificial recharge, whereas item 3 refers to conjunctive use. The use of artificial recharge implies that both surface water and ground water are being used conjunctively, although all conjunctive-use systems do not rely on artificial recharge as the means of augmenting a ground-water supply.

Artificial Recharge

Artificial recharge is defined by Todd (1980) as augmenting the natural infiltration of precipitation or surface water into the ground by some method of construction, spreading of water, or a change in natural conditions. Artificial recharge is often used to (1) replenish depleted supplies, (2) prevent or retard saltwater intrusion, or (3) store water underground where surface-storage facilities are inadequate to supply seasonal demands. California has long been the leader in artificial-recharge operations, generally for the reasons cited in (1) and (2) and, in more recent years, for reason (3).

Several recharge schemes have been developed, including recharge pits, ponds, wells, and water-spreading grounds (Figure 7.3). Literature in this field is voluminous and includes the publications of the Ground Water Recharge Center in California, which deal primarily with infiltration and water spreading (Schiff, 1955; Behnke and Bianchi, 1965); the extensive work of Baumann, dealing primarily with the theoretical aspects of recharge through wells (1963; 1965); the experience on Long Island (Brashears, 1946; Johnson 1948; Cohen and others, 1968; Seaburn, 1970); and comprehensive annotated bibliographies prepared by Todd for the United States Geological Survey (1959a) and Signor and others (1970). The potential of ground-water basins as storage facilities is demonstrated in the San Joaquin Valley, where the ground-water storage capacity has been estimated to be nine times the capacity of surface-water reservoirs associated with the California Water Plan (Davis and others, 1959).

There are at least five interrelated questions associated with artificial recharge operations:

1. What is the nature of the rechargeable water source?

2. What system of recharge will be used?

3. What are the expected injection rates?

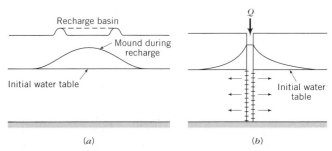

Figure 7.3 Schematic illustration of recharge through basins and wells.

4. How will the system respond hydraulically to injection?

5. How will the injected water be managed as part of the total water-resource system?

There is considerable interdependency between the first two of these questions. For example, with water-spreading methods, water is generally diverted directly from surface-water sources to topographically lower areas. On Long Island, New York, abandoned gravel pits are used to collect storm runoff that previously discharged to the ocean. Moreover, in this same region, more than 1000 recharge wells are in operation in response to legislation that requires direct recycling of ground water pumped for air-conditioning and industrial cooling purposes. In the Santa Clara River Valley, flood waters are stored in reservoirs and later released at low rates so as to enhance their infiltration into the natural streambed. Alternately, municipal waste water subjected to secondary treatment is generally recharged by irrigation or spreading methods (Todd, 1980). Recharge wells for waste water must be accompanied by tertiary treatment.

Within the San Fernando Basin, the Watermaster Report (1994) outlines plans to enhance the water available for pumping by artificial recharge schemes. The proposed East Valley Water Reclamation Project involves the use of up to 40 million gallons per day of treated sewage from the Donald C. Tillman Water Reclamation Plant to provide water for various purposes (Figure 7.4). A significant proportion of this water will be used for ground-water recharge via two spreading grounds.

Pilot studies have been under way in the San Fernando Basin to examine the water-quality implications of recharging treated sewage. At the Headworks Spreading Ground (see Figure 7.1), water from the Los Angeles River, whose low flow is mainly treated reclaimed water from the Tillman Plant, was spread over an area of approximately 30 acres. The spread water was removed from the ground by a well 1000 feet away. The water removed by pumping was approximately 45% reclaimed water and 55% native ground water. Preliminary indications are that transport through the ground improved the water quality. Reductions were apparent in coliforms,

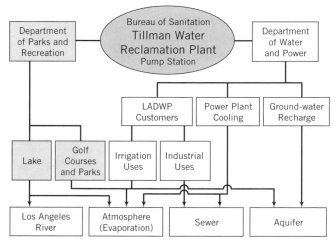

Figure 7.4 Flow diagram illustrating ways proposed for the use of reclaimed water from the Tillman Plant (modified from Watermaster Report, 1994).

total organic carbon, biological oxygen demand, nitrite, ammonia, and turbidity levels.

The most important design issues concerned with artificial recharge are developing and maintaining appropriately high inflow rates (Todd, 1959b). Table 7.1 gives some representative spreading basin recharge rates and some average well recharge rates. Recharge rates associated with spreading grounds vary from 15 m^3/m^2/day (m/day) for some gravels to as little as 0.5 m/day in sand and silt (Bear, 1979). Typically, the infiltration rates fall off with time due to swelling of the soil after wetting (Figure 7.5). There is some threshold value of infiltration where continued recharge is no longer economical so that the spreading ground is either abandoned or some scraping and cleaning process must be put into operation. Decreases in well recharge rates can occur in response to silt introduced in the recharge water, dissolved air in the recharge water, which tends to decrease the permeability in the vicinity of the screen openings, and bacteria-induced chemical growth on the well screen.

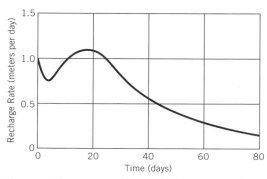

Figure 7.5 Time variation of recharge rate for water spreading on undisturbed soil (from Muckel, 1959).

Table 7.1 **Representative Artificial Recharge Rates**

Spreading Basins		Recharge Wells	
Location	Rate (m/day)	Location	Rate (m³/day)
California		California	
Los Angeles	0.7-1.9	Fresno	500-2200
Madera	0.3-1.2	Los Angeles	2900
San Gabriel River	0.6-1.6	Manhattan Beach	1000-2400
San Joaquin Valley	0.1-0.5	Orange Cove	1700-2200
Santa Ana River	0.5-2.9	San Fernando Valley	700
Santa Clara Valley	0.4-2.2	Tulare County	300
Tulare County	0.1		
Ventura County	0.4-0.5		
New Jersey		New Jersey	
East Orange	0.1	Newark	1500
Princeton	<0.1		
New York		Texas	
Long Island	0.2-0.9	El Paso	5600
		High Plains	700-2700
Iowa		New York	
Des Moines	0.5	Long Island	500-5400
Washington		Florida	
Richland	2.3	Orlando	500-51,000
Massachusetts		Idaho	
Newton	1.3	Mud Lake	500-2400

From Todd (1959b). Reprinted from *Groundwater Hydrology.* Copyright © 1959. John Wiley & Sons, Inc. Reprinted by permission of John Wiley & Sons, Inc.

Conjunctive Use

Conjunctive use involves the coordinated use of surface and ground water to meet some specified water demand in a given area. The following questions from Buras (1966) provide insight into the questions associated with conjunctive use:

1. What system has to be built to minimize the discrepancy (in time, space, and quality) between the natural supply of water and the demand for it?

2. To what extent should the water-resource system be developed, and how extensive should the region serviced be?

3. How should the system be operated so as to achieve a given set of objectives in the best possible way?

Conjunctive operations may be of various kinds. The most common type that is of interest because of the hydraulics is the interconnected stream–aquifer system, where the development of one affects the other. This is the type of system addressed by Bredehoeft and Young (1970) and Young and Bredehoeft (1972). An overall objective of such studies might be to use the total resource in such a way as to maximize benefits. Here, the hydraulic connection between the two sources plays a major role. Yet another type of operation has been described by Chun and others (1964) for the coastal plain of

Los Angeles. This study proposes full use of underground supplies and storage capacity together with local and imported supplies where artificial recharge plays a sizable role in the project. The plan of basin operation is to provide water to the consumer at the lowest possible cost. Yet a third type of operation is where surface water is imported and is used directly to supplement the ground-water supply with little or no connection between the two sources, other than their availability.

Cochran (1968) describes such a situation in Nevada, where the objective was to minimize the cost of operation. For this case, Domenico (1972) derived a mathematical decision rule for the timing of surface-water importation to supplement an overdeveloped ground-water supply. As a marginal value rule, this decision rule states that the importation should take place when the cost of importation equals the cost of mining. The cost of importation includes the initial investment plus operating costs, whereas the cost of mining includes current pumping charges as well as the capitalized cost of all future pumping charges associated with a lowered ground-water storage level.

In recent years, the concept of conjunctive use has been extended to issues of riparian protection. In the arid southwestern states, plants and animals thrive in fragile ecosystems developed along the perennial streams. These systems are particularly at risk when the overdevelopment of ground-water resources lowers water tables in the riparian zones, or results in significant water-table fluctuations (Arizona Department of Water Resources, 1994). The Santa Cruz River, near Tucson, Arizona, is one of several rivers where changes in the riparian ecosystems have raised concerns. Along the lower reaches of this river, the extensive pumping of ground water for irrigation and public supply wells have eliminated riparian vegetation. Lower ground-water levels along the lower Santa Cruz were a major cause of the destruction of mesquite woodlands (Arizona Department of Water Resources, 1994). The State of Arizona is pursuing active programs to maintain and improve the ecological health of riparian ecosystems through the management of ground-water and surface-water resources.

7.2 Introduction to Ground-Water Flow Simulation

For more than 50 years, the quantitative orientation of hydrogeology has differentiated this field from other geosciences. A fundamental tenet holds that properly designed and constructed mathematical models can form the basis for accurate future predictions. Fundamental beliefs in the inherent usefulness of models have established the direction for academic research, as well as the nature of industrial practice. Numerical approaches let hydrogeologists leave behind the analytical approaches of well hydraulics and their limitations. They provide tools to treat hydrologic systems as they occur in nature—where transmissive properties of aquifers are not constant, where recharge is spatially distributed, and where many wells can operate at the same time with differing rates.

Generalized Modeling Approach

A variety of different tasks are required in the construction of a ground-water model. As Figure 7.6 illustrates, the main steps include: the development of a conceptual model, creation of the model and execution of various trials, evaluation of the model results, and either the compilation of new data or validation testing depending on the success of the calibration. Once a model is successfully validated, it can be used predictively.

Conceptual Model

A conceptual model is created through the evaluation of the hydrogeologic data to provide a picture of the hydrogeologic setting over some region of interest. One element of the conceptual model is the hydrogeologic framework, which encompasses the main features of the hydrogeology normally determined from geologic investigations and hydraulic testing. Normally, information is

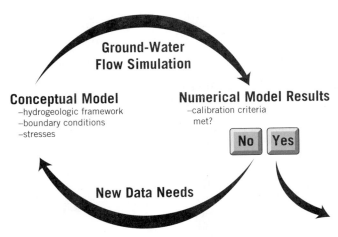

Figure 7.6 Synthesis of the modeling process.

synthesized on the shape, thickness, and hydraulic properties of the major geologic units (for example, hydraulic conductivity and storativity), the distribution of hydraulic head within the major hydrogeologic units, and the distribution and rates of ground-water recharge.

Other elements of the conceptual model are the boundary conditions and the distribution of stresses on the aquifer due to pumping. Boundary conditions are defined along the edges of the simulation domain, including the top and bottom. Their main function is to separate the model region from the rest of the world. In other words, the boundary conditions account for the influence of flow conditions outside of the simulation domain. For example, in the construction of flow nets, we saw how boundary conditions influenced the resulting pattern of flow. Boundary conditions, thus, are required for the solution of any ground-water flow equation. In general, there are two commonly used boundary conditions: (1) specified-hydraulic-head boundaries and (2) specified-flow boundaries. A no-flow boundary is a special case of a specified-flow boundary, and a constant-head boundary is a special case of a specified-head boundary.

Boundary conditions are often difficult to define along the edges of the domain because hydraulic heads or inflow–outflow rates can be poorly defined. One strategy to overcome this problem is to place model boundaries along "natural" hydrogeologic boundaries, or parallel to pathlines. An example of a natural hydrogeologic boundary is a river defined by constant hydraulic-head values equivalent to the stage of the river. Locally, such a boundary condition would provide a zone of recharge or discharge. Another example would be the selection of the major watershed divide as a no-flow boundary. The boundary is no-flow because the gradient is such that flow moves away from a divide. Features of the hydrogeologic setting also help in assigning boundaries. For example, a thick, low hydraulic conductivity unit is selected as the bottom of the simulation domain. At some depth in almost any system, it should be possible to define a boundary of no flow, thus implying that deeper circulation is minimal.

Defining boundary conditions along the top of the simulation region is more difficult because often there are active inflows and outflows of water due to recharge and discharge. Experience shows that recharge–discharge rates vary in time and space and are difficult hydrogeologic parameters to measure. An example is provided later that illustrates how boundary conditions are defined as part of a model conceptualization.

With transient models (that is, models where hydraulic head varies as a function of time), one also needs to define the initial condition for the simulation. The initial condition is the hydraulic-head distribution at time zero. In other words, it reflects the state of the system before the simulation begins.

Ground-Water Flow Simulation

The response of an aquifer system is calculated by the numerical solution to a ground-water flow equation. The term *numerical* implies that numerical methods, as opposed to analytical mathematical approaches, are used to solve the governing partial-differential equation. In the ground-water flow equation, hydraulic head is the unknown. The main numerical approaches used in practice today for solving ground-water flow equations are finite-difference and finite-elements methods. Both are sophisticated methods that in different ways replace the governing differential equation for ground-water flow by a system of algebraic equations.

Both methods require that the region of interest be subdivided using a mesh or grid network. Sometimes, the cumbersome term *discretization* is used to describe this process. The finite-difference approaches use a regular discretization, where an aquifer is subdivided into a series of rectangular grid blocks. In a two-dimensional model, each model cell is assumed to have a thickness, m. Thus, each of the grid blocks represents a volume of the aquifer, $m\, \Delta x\, \Delta y$. In a three-dimensional model, consisting of aquifers and aquitards, individual units are subdivided vertically into cells of a specified thickness.

Operationally, the size of a grid block in the *x-y* plane (that is, a map view) is usually kept small relative to the overall extent of the aquifer. The spacings between the rows and columns vary, but for simplicity the example grids (Figure 7.7*a,b*) assume a constant spacing (that is, $\Delta x = \Delta y$). Associated with the grid blocks are nodes that represent the points where the unknown hydraulic head is calculated. Depending on the formulation of the numerical model, node points can be either mesh centered (Figure 7.7*a*) or block centered (Figure 7.7*b*).

The finite-element method permits a much more general arrangement of node points. The discretization with triangular elements (Figure 7.7*c*) illustrates how easy it is to define the boundaries of irregularly shaped aquifers

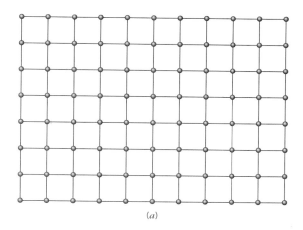

(a)

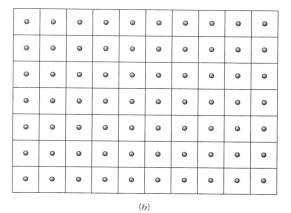

(b)

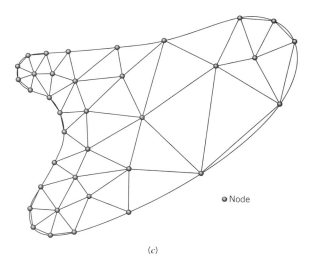

(c)

Figure 7.7 Examples of (*a*) mesh-centered and (*b*) block-centered finite-difference grids. Panel (*c*) illustrates the discretization of an irregularly shaped aquifer with linear triangular elements.

and to ensure that node points coincide with monitoring wells or with geographic features (for example, a river).

In this and following sections, we develop preliminary concepts of finite-difference modeling in relation to the industry-standard code, MODFLOW (McDonald and Harbaugh, 1988). We will leave the detailed discussion of modeling methods to other, more specialized texts like Zheng and Bennett (1995) and Huyakorn and Pinder (1983).

One of the important features of the numerical approaches is that hydraulic heads are calculated only at the node points. This aspect of the formulation will become clear in Section 7.3, which examines the development of the finite-difference form of the ground-water flow equation.

Applying a model requires the construction of a data set that replicates the most important features of the actual system. The basic data for this purpose come from the conceptual model. Exactly how the data are organized depends on the particular model. Generally, every node or cell must be supplied with information on hydraulic conductivity or transmissivity, storativity, and fluxes due to sources and sinks (that is, recharge, pumping, and evaporation). The model also needs to be supplied with information on boundary and initial conditions.

Evaluation of Model Results

The hydraulic-head prediction that comes from a flow model is commonly used as the basis for model calibration. Calibration is a process of selecting model parameters to achieve a good match between the predicted and measured hydraulic heads, or other relevant hydrogeologic data like streamflow changes between gauging stations. Most commonly, calibration is accomplished by a trial-and-error adjustment of model parameters. This procedure involves the systematic variation of model parameters like hydraulic conductivity, storativity, flows, or boundary conditions. Automated inverse procedures, like those in MODFLOWP (Hill, 1992) may speed up calibration. Calibration or model tuning is required because ground-water systems are so poorly known.

Each model run provides a set of predictions about hydraulic-head distributions or other flow rates that can be compared to existing field measurements. Commonly, the comparison is not favorable—suggesting problems in the system conceptualization or inappropriate parameter values. The lack of calibration may force a reexamination of how the model is constructed or in some cases prompt additional data-collection studies. After many iterations around the circle in Figure 7.6, the predictions of hydraulic heads and various flow rates should match the measured values.

Common practice is to set calibration criteria in advance of the calibration exercise. There are no hard and fast rules for what constitutes a good calibration except that errors should be small relative to the total hydraulic head. Criteria may be based on (1) the mean error, (2) the mean absolute error, or (3) the root-mean squared error (Anderson and Woessner, 1992). These measures of the difference between measured and simulated hy-

draulic-head data are given as

$$\text{Mean Error:} \quad \frac{1}{n} \sum_{i=1}^{n} (b_m - b_s)_i$$

$$\text{Mean Absolute Error:} \quad \frac{1}{n} \sum_{i=1}^{n} |(b_m - b_s)_i|$$

$$\text{Root-Mean Squared Error:} \quad \left[\frac{1}{n} \sum_{i=1}^{n} (b_m - b_s)_i^2 \right]^{0.5}$$

where n is the number of points where comparisons are made, b_m is the measured hydraulic head at some point i, and b_s is simulated hydraulic head at the same point (Anderson and Woessner, 1992). Of the three error estimates, Anderson and Woessner (1992) point to the root-mean squared error as the best quantitative measure if the errors are normally distributed. The mean error is not preferred because large positive and negative errors can cancel each other out. A small error estimate, thus, may hide a poor model calibration. Other requirements are sometimes applied in addition to error estimates, such as quantitatively correct flow directions and flow gradients.

Model Verification

Once calibration is complete, a verification test is commonly added to check that the model is a valid representation of the hydrogeologic system. Commonly, model verification involves using the calibrated model to simulate a hydrologic response that is known. For example, one might hold back results from one or more large-scale aquifer tests and examine how well the calibrated model simulates the test. Again, the errors between the observed and simulated hydraulic-head values can be quantified in terms of the error measures. If the model successfully passes this last test, then it can be used for predictive analyses.

A Note of Caution

Predictions made with simulation models must be interpreted with caution. The "aura of correctness" (Bredehoeft and Konikow, 1993) attached to model calculations often exerts much more influence than is reasonable, given the typically uncertain data on which models are built. Oreskes and others (1994) believe that "models are representations, useful for guiding further study but not susceptible to proof."

Predictability becomes a problem because ground-water systems are often so poorly characterized. It is usually unrealistic to find data sufficient to describe hydrologic processes in space and time. Thus, the model design depends significantly on the "informed judgment" of its builder rather than real information. This uncertainty does not disappear simply because a model is constructed.

The calibration–verification process does not lead to a unique description of a hydrogeologic system. For poorly known systems, a very large number of different models can be developed without knowing which, if any, is correct. Stated another way, different model developers, given the same hydrogeologic data, will probably develop different conceptualizations of the same system, each of which can be calibrated and verified. An example of the difficulty of calibrating a ground-water flow model is discussed by Freyberg (1988). Different groups using the same set of synthetic data developed very different predictions concerning system behavior. Sources of variability in the calibration and the resulting prediction were related to (1) the lack of hydrogeologic data, (2) the use of different measures of success in calibration by each group, and (3) differing strategies for calibration (for example, changing local transmissivity values around wells versus changing values over large areas).

The success in model predictions can be examined with post-audits. This term describes the process of checking a prediction made by a ground-water model. For example, if one makes a prediction about the behavior of a system 10 years from now, one could return after 10 years, make the necessary hydraulic-head measurements, and check whether the original model can be validated. Bredehoeft and Konikow (1993) commenting on results from the few available post-audits indicate that "extrapolations into the future were rarely very accurate." They identified the following problems with the models: "the period of history match (that is, calibration) was too short to capture an important element of the model, or the conceptual model was incomplete, or the parameters were not well defined." They concluded that the record of "validating" models was "not encouraging."

If model predictions are suspect, what use are they? In addressing this issue, Oreskes and others (1994) pointed out:

models can corroborate a hypothesis by offering evidence to strengthen what may be already partly established through other means. Models can elucidate discrepancies in other models. Models can also be used for sensitivity analysis—for exploring "what if" questions—thereby illuminating which aspects of the system are most in need of further study, and where more empirical data are needed.

In summary, one must use the power provided by computer models carefully. These are useful tools that should be used with full knowledge of their limitations.

7.3 Formulating a Finite-Difference Equation for Flow

This section develops finite-difference equations for aquifer simulations. The theory here relates specifically to the industry-standard code, MODFLOW. In the finite-difference method, the governing differential equation for ground-water flow is replaced by a difference equation that embodies conservation principles of the original differential equation. Three-dimensional flow is described by the following equation:

$$\frac{\partial}{\partial x}\left(K_{xx}\frac{\partial b}{\partial x}\right) + \frac{\partial}{\partial y}\left(K_{yy}\frac{\partial b}{\partial y}\right) + \frac{\partial}{\partial z}\left(K_{zz}\frac{\partial b}{\partial z}\right) - W = S_s\frac{\partial b}{\partial t}$$

(7.1)

where K_{xx}, K_{yy}, and K_{zz} are values of hydraulic conductivity along the x, y, and z coordinate axes, b is hydraulic head; W is a flux term that accounts for pumping, recharge, or other sources and sinks; S_s is the specific storage; x, y, and z are coordinate directions; and t is time. The form of the equation implies that the principal directions of the hydraulic conductivity ellipse coincide with the coordinate axes.

Description of the Finite-Difference Grid

The finite-difference solution requires that the domain be discretized by a grid. With MODFLOW, the cells are brick shaped (Figure 7.8). The grid system is referenced in terms of a row, column, and layer-numbering scheme with block-centered nodes (Figure 7.8). As the notation implies (Figure 7.8), the dimensions of each cell can be varied. Thus, a dense system of nodes can be provided around features of interest and a sparse system in areas of lesser concern. Overall, the variable grid minimizes the number of nodes in a simulation. However, care must

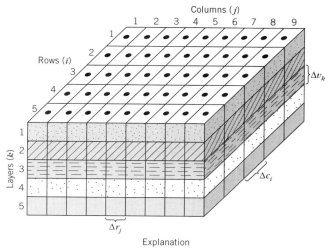

Explanation

• Node

Δr_j Dimension of cell along the row direction. Subscript (j) indicates the number of the column

Δc_i Dimension of cell along the column direction. Subscript (i) indicates the number of the row

Δv_k Dimension of the cell along the vertical direction. Subscript (k) indicates the number of the layer

Figure 7.8 Discretization of a three-dimensional system (modified from McDonald and Harbaugh, 1988).

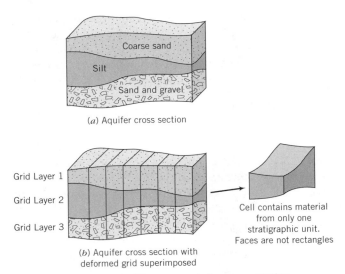

(a) Aquifer cross section

Grid Layer 1
Grid Layer 2
Grid Layer 3

Cell contains material from only one stratigraphic unit. Faces are not rectangles

(b) Aquifer cross section with deformed grid superimposed

Figure 7.9 One possible approach for representing (a) complex layering by (b) a deformed grid (modified from McDonald and Harbaugh, 1988).

be taken to change cell sizes gradually. The rule of thumb is that the dimensions of adjacent cells in a given direction should not differ by more than a factor of 1.5.

The cell sizes in the row and column directions are specified explicitly in the input data. The vertical dimension is specified implicitly by specifying transmissivity values for individual cells. McDonald and Harbaugh (1988) discuss the variety of ways of representing the vertical layering in relation to the hydrogeologic units. It is beyond the scope of this brief overview to discuss all these options. The most common strategy has layers conform with readily identifiable geologic units. Consider the simple example depicted in Figure 7.9a with a coarse sand, a silt, and a sand and gravel layer. Vertical discretization is accomplished using three model layers that coincide with the three stratigraphic units (Figure 7.9b). The changing vertical thickness of individual cells, for example within layer 1, would be accommodated by changing the transmissivity of each cell.

With the model layers so defined, individual cells are no longer bricks. They have an irregular shape where the cell faces are no longer rectangles. Although this representation of a cell gives rise to errors, they are usually insignificant in relation to errors caused by poor estimates of transmissivity, storativity, and recharge rates.

Derivation of the Finite-Difference Equation

The development of the finite-difference flow equation is based on the same ideas of continuity that were developed in Chapter 4. In words, the equation of continuity states that the sum of flows into and out of any cell is equal to the time rate of storage plus or minus additions of water from sources or sinks. It is written mathemati-

cally as

$$\sum Q_i = S_s \frac{\Delta b}{\Delta t} \Delta V \qquad (7.2)$$

where Q_i's account for flow into the cell from adjacent cells (for example, see Figure 7.10) through six sides, and for water added or withdrawn (for example, recharge pumping), S_s is the specific storage, Δb is the change in head over a time interval Δt, and ΔV is the volume of the cell. The LHS of Eq. 7.2 can be expanded in terms of flows through the six sides of the cell, the sources and sinks, and the RHS in terms of the cell dimensions to give

$$Q_{i,j-1/2,k} + Q_{i,j+1/2,k} + Q_{i-1/2,j,k} + Q_{i+1/2,j,k} + Q_{i,j,k-1/2}$$

$$+ Q_{i,j,k+1/2} + QS_{i,j,k} = S_{s_{i,j,k}} \frac{\Delta b_{i,j,k}}{\Delta t} \Delta r_j \Delta c_i \Delta v_k \qquad (7.3)$$

where $Q_{i,j-1/2,k}$ is the volumetric fluid discharge through the face between cells i, j, k; and $i, j - 1, k$, $QS_{i,j,k}$ is the sum of all other inflows to and outflows from the cell; and Δc_i, Δv_k, Δr_j are the dimensions of the cell i, j, k (Figure 7.10).

Inflow through the cell faces can be expanded with Darcy's equation as

$$Q_{i,j-1/2,k} = KR_{i,j-1/2,k} \Delta c_i \Delta v_k \frac{(b_{i,j-1,k} - b_{i,j,k})}{\Delta r_{j-1/2}} \qquad (7.4)$$

where $b_{i,j,k}$ is the hydraulic head at node i, j, k; $b_{i,j-1,k}$ is the hydraulic head at node $i, j - 1, k$; $KR_{i,j-1/2,k}$ is the hydraulic conductivity along the row between nodes i, j, k and $i, j - 1, k$; $\Delta c_i \Delta v_k$ is the area of the cell face normal to the row direction; and $\Delta r_{j-1/2}$ is the distance between nodes i, j, k and $i, j - 1, k$ (McDonald and Harbaugh, 1988). The minus sign is neglected to simplify the development. The form of this and similar equations for the other faces can be simplified by defining a "hydraulic conductance," in this case CR where

$$CR_{i,j-1/2,k} = KR_{i,j-1/2,k} \frac{\Delta c_i \Delta v_k}{r_{j-1/2}} \qquad (7.5)$$

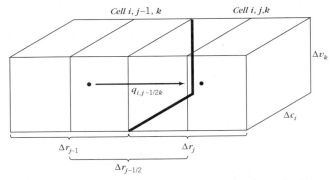

Figure 7.10 Example of inflow through the face of cell i, j, k from cell $i, j - 1, k$. The dimensions of cell i, j, k are Δc_i, Δr_j, and Δv_k (from McDonald and Harbaugh, 1988).

Making this substitution, the six inflow equations can be written as

$$Q_{i,j-1/2,k} = CR_{i,j-1/2,k}(h_{i,j-1,k} - h_{i,j,k}) \quad (7.6)$$

$$Q_{i,j+1/2,k} = CR_{i,j+1/2,k}(h_{i,j+1,k} - h_{i,j,k}) \quad (7.7)$$

$$Q_{i-1/2,j,k} = CC_{i-1/2,j,k}(h_{i-1,j,k} - h_{i,j,k}) \quad (7.8)$$

$$Q_{i+1/2,j,k} = CC_{i,+1/2,j,k}(h_{i+1,j,k} - h_{i,j,k}) \quad (7.9)$$

$$Q_{i,j,k-1/2} = CV_{i,j,k-1/2}(h_{i,j,k-1} - h_{i,j,k}) \quad (7.10)$$

$$Q_{i,j,k+1/2} = CV_{i,j,k+1/2}(h_{i,j,k+1} - h_{i,j,k}) \quad (7.11)$$

where the hydraulic conductances in the column and layer directions are given by CC and CV, respectively (McDonald and Harbaugh, 1988).

Following the notation of McDonald and Harbaugh (1988), the combined flows due to N sources or sinks is given as $QS_{i,j,k}$ where

$$QS_{i,j,k} = P_{i,j,k}h_{i,j,k} + Q_{i,j,k} \quad (7.12)$$

where $P_{i,j,k}$ and $Q_{i,j,k}$ represent the sum of constants related to N different inflow or outflow processes like pumping, induced recharge, etc.

Combining these equations yields the close to final form of the finite difference equation or

$$CR_{i,j-1/2,k}(h_{i,j-1,k} - h_{i,j,k}) + CR_{i,j+1/2,k}(h_{i,j+1,k} - h_{i,j,k})$$
$$+ CC_{i-1/2,j,k}(h_{i-1,j,k} - h_{i,j,k}) + CC_{i+1/2,j,k}(h_{i+1,j,k} - h_{i,j,k})$$
$$+ CV_{i,j,k-1/2}(h_{i,j,k-1} - h_{i,j,k}) + CV_{i,j,k+1/2}(h_{i,j,k+1} - h_{i,j,k})$$
$$+ P_{i,j,k}h_{i,j,k} + Q_{i,j,k} = SS_{i,j,k}(\Delta r_j \Delta c_i \Delta v_k)\frac{\Delta h_{i,j,k}}{\Delta t} \quad (7.13)$$

One final step in the derivation replaces the $\Delta h/\Delta t$ term on the RHS of Eq. 7.13 by a backward time difference

$$\frac{\Delta h_{i,j,k}}{\Delta t} = \frac{h_{i,j,k}^m - h_{i,j,k}^{m-1}}{t_m - t_{m-1}} \quad (7.14)$$

where m designates the present time at which the heads are unknown and $m - 1$ designates the previous time step. All of the h terms on the LHS of Eq. 7.13 would carry the superscript "m." Essentially, the head terms carrying the superscript "m" are unknowns, and those with the superscript "$m - 1$" are known from the previous time step.

The final form of the finite difference equation can be expressed

$$CV_{i,j,k-1/2}h_{i,j,k-1}^m + CC_{i-1/2,j,k}h_{i-1,j,k}^m + CR_{i,j-1/2,k}h_{i,j-1,k}^m$$
$$+ (-CV_{i,j,k-1/2} - CC_{i-1/2,j,k} - CR_{i,j-1/2,k} - CR_{i,j+1/2,k}$$
$$- CC_{i+1/2,j,k} - CV_{i,j,k+1/2} + HCOF_{i,j,k})h_{i,j,k}^m \quad (7.15)$$
$$+ CR_{i,j+1/2,k}h_{i,j+1,k}^m + CC_{i+1/2,j,k}h_{i+1,j,k}^m$$
$$+ CV_{i,j,k+1/2}h_{i,j,k+1}^m = RHS_{i,j,k}$$

where

$$HCOF_{i,j,k} = P_{i,j,k} - \frac{SC1_{i,j,k}}{t_m - t_{m-1}};$$

$$RHS_{i,j,k} = -Q_{i,j,k} - \frac{SC1_{i,j,k}h_{i,j,k}^{m-1}}{t_m - t_{m-1}}; \text{ and}$$

$$SC1_{i,j,k} = S_{si,j,k}\Delta r_j \Delta c_i \Delta v_k$$

Writing one of these equations for each of the nodes in the system yields a system of equations

$$[A]\{h\} = \{q\} \quad (7.16)$$

where $[A]$ is the coefficient matrix, $\{h\}$ is the vector of unknown head values, and $\{q\}$ is a vector of constant-head terms. The mathematical solution of this system of equations provides the hydraulic head for the given time step. Given the complexity of these approaches relative to the introductory character of our book, we must leave readers on their own to examine the details.

7.4 *The MODFLOW Family of Codes*

MODFLOW has emerged as the de facto standard code for aquifer simulation. The original code was developed with an extensive User's Guide and released in 1984. This version was superseded in 1988 (McDonald and Harbaugh, 1988). The popularity of this code can be attributed to four factors. First, the code has proved to be a powerful, robust, and well-crafted product. Early users had great success in applying the code to a wide variety of practical problems. Second, the User's Guide is extremely detailed and provides clear descriptions of how various code options are used. Third, the code has been supported strongly by the U.S. Geological Survey and is readily available. In recent years, more specialized versions of the code have been marketed by several private companies. Finally, the success of the original code has spawned an extensive array of training courses and a large number of related products and calculational modules. Several companies market software that assists in the preparation of MODFLOW data sets, such as, Geraghty & Miller's ModelCad product, Waterloo Hydrogeologic's Visual MODFLOW, and the Department of Defense's Groundwater Modeling System. Other utilities facilitate the contouring of hydraulic-head fields (SURFER; Golden Software, Inc.). The examples presented later assume readers have access to one of these versions of MODFLOW.

There also has been a variety of follow-on products related to the code. Pollock (1989) has developed MOD-PATH—a post-processing package that takes output from steady-state simulations with MODFLOW and computes three-dimensional pathlines. This package has found applications in the simple modeling of contaminant transport and in the location of wells for pump-and-treat systems for the recovery of dissolved contaminants. Hill (1992) has developed MODFLOWP, a code that includes capabilities for estimating various parameters required in a MODFLOW simulation. This code provides a tool for automatic calibration. Other enhancements include the ability to simulate ground-water and surface-water interactions (Prudic, 1989), a new equation solver (Hill, 1990), approaches to accommodate the rewetting of cells in the model that have become dry (McDonald and others, 1991), the ability to model aquifer compaction (Leake and Prudic, 1991), techniques to represent a transmissivity field that is smoothly varying (Goode and Appel, 1992), and a capability to treat narrow horizontal barriers (e.g., faults) that may impede ground-water flow (Hsieh and Freckleton, 1993). These new features make the MODFLOW code tremendously powerful and capable of handling a variety of ground-water conditions.

Solving Systems of Finite-Difference Equations

Procedures for solving systems of algebraic equations can be broadly categorized as direct and iterative. Direct approaches involve rearranging the system of equations to a form that can easily be solved. The methods that most of us learned in high school to solve simple systems of equations are examples of the direct methods. The iterative approaches involve making some initial guess at the unknowns and refining these guesses through a series of repeated calculations until an accurate solution is obtained.

The original MODFLOW contained two iterative schemes. The simpler is Slice-Successive Overrelaxation (SSOR). Instead of solving the entire system of unknowns at the same time, the equations are formulated for a two-dimensional slice (Figure 7.11) with the assumption that the heads in the adjacent two slices are known. The resulting system of equations (actually formulated as the change in hydraulic head) is solved by Gaussian elimination. The slice will usually contain a relatively small num-

ber of nodes because in most cases the number of model layers is small. One iteration is complete when all of the slices are processed. After a large number of iterations, the solution converges.

The Strongly Implicit Procedure (SIP) is a more complicated approach. This procedure involves solving the unknowns for the entire grid simultaneously (McDonald and Harbaugh, 1988). More recently, Hill (1990) implemented a preconditioned conjugate gradient procedure for use with MODFLOW. In general, this and similar solution techniques are extremely fast and robust and, for these reasons, are now often used in solving systems of linear equations.

Modular Program Structure

MODFLOW is built with a modular design that consists of a main program and "packages." The packages are groups of independent subroutines that carry out specific simulation tasks such as accounting for flow into or out of a river, adding recharge at the top surface, or invoking a specific calculational procedure for solving the finite-difference flow equations like slice-successive overrelaxation (McDonald and Harbaugh, 1988).

This modular design is useful in several ways. It provides a logical basis for organizing the actual code with similar program elements or functions grouped together. Such a structure facilitates the integration of new packages to enhance the code's capabilities. The modular structure also provides a convenient way to organize the user's manual (McDonald and Harbaugh, 1988). Finally, the modular structure simplifies the preparation of data for a simulation trial. Typically, data are read by a subroutine within a package. Thus, for simple problems, where only a few packages are used, the input data will be modest.

Table 7.2 lists the packages in the basic code (McDonald and Harbaugh, 1988) along with a brief description of their function.

Illustrative Example

Let us consider again Example 6.10, which was solved analytically. A well fully penetrates a confined and infinite artesian aquifer having a transmissivity of 928 m^2/day and a storativity of 1.0×10^{-2}. The initial head in the aquifer is 0.0 m. The well is pumped at a constant rate of 3790 m^3/day. Determine the drawdown versus time for 64 days at an observation well located 304.8 m from the pumping well. Determine the drawdown at this node at 64 days.

The first step in solving this problem is to construct a grid. To keep the problem simple, we will use a grid with fixed spacings. MODFLOW provides for variable grid spacing to increase the density of nodes around wells. Let us begin by placing a node to represent an

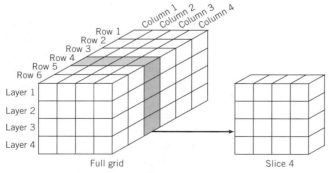

Figure 7.11 Division of the three-dimensional model array into vertical slices for processing in the SSOR package (from McDonald and Harbaugh, 1988).

Table 7.2 **A Summary of the Packages Contained in MODFOW**

Package Name	Package Description
Basic (*.BAS)[1]	Handles those tasks that are part of the model as a whole. Among those tasks are specification of boundaries, determination of time-step length, establishment of initial conditions, calculating a water budget, and printing of results.
Block-Centered Flow (*.BCF)	Required in all model calculations. Reads basic information on the grid spacing, pattern of layering, aquifer type, and hydraulic parameters to calculate terms of finite-difference equations that represent flow within porous medium, specifically flow from cell to cell and flow into storage.
Well (*.WEL)	Designed to add effects of injection or withdrawal wells. Reads information on the well numbers, their location, and pumping schedules. The package adds terms representing flow to wells to the finite-difference equations.
Recharge (*.RCH)	This package is designed to simulate the addition of natural recharge from precipitation to the model. Adds terms representing distributed recharge to the finite-difference equations.
River (*.RIV)	This package provides the capability of modeling ground-water, surface-water interactions. Accounts for both the size of the stream and the possibility of a discrete low-permeability streambed. Adds terms representing flow to rivers to the finite-difference equations.
Drain (*.DRN)	Similar to the river package except considers only inflow to the drain. The package adds terms representing flow to drains to the finite-difference equations.
Evapotranspiration (*.EVT)	Adds terms representing evapotranspiration to the finite-difference equations. Accounts for the effects of plant evapotranspiration and direct evaporation of ground water. Rate varies as a function of the water-table depth.
General-Head Boundaries (*.GHB)	Provides the possibility of variable flow across the model boundaries. The extent of flow depends on the head difference between the cell and some constant head at a source away from the boundary. Adds terms representing general-head boundaries to the finite-difference equations.
Strongly Implicit Procedure (*.SIP)	Iteratively solves the system of finite-difference equations using the Strongly Implicit Procedure.
Slice-Successive Overrelaxation (*.SOR)	Iteratively solves the system of finite-difference equations using Slice-Successive Overrelaxation.
Output Control (*.OC)	Provides user with the ability to control what calculations are printed as output and at what time steps.

[1] Shown in parentheses are the commonly used extensions for naming files on a PC. For example, EX1.BAS is the file containing the basic input data. Modified from McDonald and Harbaugh (1988).

observation well 304.8 m from the pumped well. For the regular grid of this problem, the grid spacing must be 304.8 m or some fraction thereof (for example, 1/2 of 304.8, 1/3 of 304.8, etc.). Second, the aquifer must be infinite, which from a modeling point of view means that the boundaries should be located far from the well. This consideration with constant grid spacings pushes us to select the largest grid spacing or 304.8 m. With a variable grid, it would be no problem to locate the edges of the model far from the well. By default, the four sides of the models are no-flow or impermeable boundaries. Thus, during the 64-day simulation, as long as the cone of depression does not reach the boundaries, the model will represent an infinite aquifer condition. If significant drawdown occurs along the boundaries, the aquifer is no longer infinite. This detail should be checked once preliminary runs are completed.

Another question is how many grid blocks are required in the row and column directions to simulate an infinite aquifer? There are no hard-and-fast guidelines in this respect except that small grids (for example, 15 rows and 15 columns) appear poorly resolved. Extremely large grids (for example, 200 × 200) impose a significant computational burden and usually require graphics software to interpret the results. A recommended starting number of rows and columns would be between 25 and 50. The simulation region in this case is subdivided by 30 grid blocks along the row and column directions (Figure 7.12). The numbering scheme for the rows and columns is always the same as shown in Figure 7.12, with column 1 and row 1 located at the top left corner of the grid. The pumping well is located in the middle of the grid (row 15, column 15). The observation well is located one node or 304.8 m away. For this symmetrical problem, the actual direction does not matter. In real problems, the actual locations of the observation wells will be known. This simple problem involves a single confined aquifer having a constant transmissivity. It is represented in MODFLOW as a single layer.

Running this problem requires four packages—basic, block-centered flow, well, and strongly implicit procedure. The first two packages provide basic information

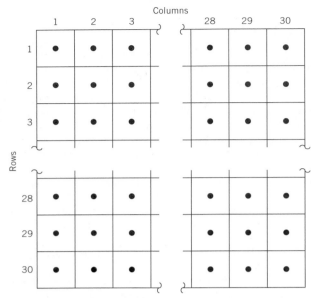

Figure 7.12 MODFLOW grid for the calculation involving the illustrative Example.

about the problem, the grid design, and time stepping. The well package gives details concerning the well. Finally, the strongly implicit package gives information for the iterative solver that is required. Detailed specifications on what information needs to be provided in each of these files and what format is required is provided in the MODFLOW User's Guide (McDonald and Harbaugh, 1988). Operationally, the data for each one of these modules is included in a separate file with information pro-

vided in the basic package that tells the code to prompt for the particular four files. The input data for each of the four packages are listed in Table 7.3, along with explanatory comments.

Having prepared the necessary data files, the user runs MODFLOW. With this output, one will notice that the code echoes back most of the data provided in the various input files. Incidentally, when debugging a new set of data, errors in the input data are found by scanning these output summaries. In looking at the predicted head calculations, you will see that only results from time step 18 or the last step are written out. This style of output is the default. To obtain a more complete set of data one would need to include the "output control," which gives the user extensive control over how the model data are written.

One check on the success of the model calculation is to examine the mass balance results on the output. The results are shown in Table 7.4.

The left-hand column summarizes the cumulative inflows to and outflows from the modeled region over all 18 time steps. The right-hand column is the water balance for time step 18. In both cases, inflows match outflows well, and from a calculational point of view the model seems to have run appropriately. Normally, the percent discrepancies should be less than 1%. The percent discrepancies can be lowered by reducing the convergence parameter in EX1.SIP. In some cases, data errors may prevent the solution from converging. Beyond its diagnostic uses, the mass balance in complicated problems provides information on how much water comes from a particular source. The calculated hydraulic heads at time step 18 in the output are similar to those calculated

---

Table 7.3 **Annotated Summary of Data Required for MODFLOW to Run the Example Problem**

File: EX1.BAS—Basic Package

RERUN OF EXAMPLE 6.10 WITH MODFLOW					
					title up to 80 characters
					title continued, 80 characters, here blank
	1	30	30	1	4 no. layers, rows, columns, and stress periods, unit, see Note 1
11 12 0 0 0 0 0 0 19 0 0 0					switch to turn on indicated packages, see Note 2
	0	0			memory allocation, mostly used as shown
	0	1			description of boundary conditions, see Note 3
999.99					value of head assigned to inactive cells, see Note 4
	0	0.			initial head value, 0.0 assigned to each node, see Note 5
64.058		18	1.2		length of stress period, number of steps in the stress period, time step multiplier 1.2, see Note 6.

Note 1: A stress period defines a specified period of time when sources and sinks (e.g., well withdrawals, river leakage, evaporation, etc.) are defined by a fixed set of parameters. It is this feature that enables one to adjust model parameters with time. For example, when a new well starts up, one would begin a new stress period. The well in this problem is pumped at a constant rate; therefore, only 1 stress period is required.

The last number on this line specifies the consistent time unit for the input data. For this parameter, the following codes apply 0—undefined; 1—seconds; 2—minutes; 3—hours; 4—days; 5—years. We have selected days as the time units, which means all time units will be days (for example, transmissivity is m^2/day).

Note 2:   The row of numbers on this line coincides with the following packages; BCF, WEL, DRN, RIV, EVT, XXX, GHB, RCH, SIP, XXX, SOR, OC. The XXX are packages that remain to be defined. The other package names are defined in Table 7.2. To turn on a package requires that a unique nonzero parameter be included. Here, we have turned on the BCF, WEL, and SIP. When the code is run, it will prompt the user for the corresponding file names for all of these files. Following common practice, each file is named so as to make the particular package name clear (for example, EX1.BCF).

Note 3:   This line applies the boundary condition to every node on a given layer of the model. The 0 indicates that the same boundary condition is assigned to every node on the given layer. Providing a second number greater than zero means the head is unknown and must be determined, as is the case here with 1. A zero means that the cell is inactive (that is, no-flow). A value less than zero applies a constant-head boundary. We are fortunate with our simple problem that the boundary array is filled with 1's and can be accomplished in the shorthand manner shown in the data set. When values are not the same for all nodes, the user would need to specify every value in the grid (see User's Guide). For our problem, letting the values be different would mean an array of integers 30 × 30.

Note 4:   On occasion, hydraulic heads are not calculated for some cells. When model results are presented, the head value of 999.99 will be printed for such no-flow cells—essentially making it obvious that the head has not been calculated for these cells. Any number could be used for this purpose.

Note 5:   This section of the input data file provides the initial value of hydraulic head for all the nodes in a given layer. Each layer would need a separate set of input data. Again, because all nodes in layer 1 will receive the same initial value for hydraulic head (that is, 0) we can use the handy feature where the first value of 0 tells the code to fill the 30 × 30 array of nodes with the second value or 0. If values had varied, one would input the entire matrix according to the instructions in the User's Guide.

Note 6:   One of these lines is provided for each of the stress periods. For this exercise, there is only one stress period that is 64.058 days long. We ask there to be 18 time steps with a time step multiplier of 1.2.

File: EX1.BSF—Block-Centered Flow Package

0	0	0 = transient simulation, 0 = no cell-by-cell terms printed
0		layer type = 0 or confined; see Note 7
0	1.	anisotropy factor each layer; 1.0 here means isotropic
0	304.8	cell width along rows; notation here makes each cell 304.8 m
0	304.8	cell width along the columns, again 304.8; see Note 8
0	.01	storage coefficient, the value 0.01 is assigned to all nodes
0	928.	transmissivity along the rows; 928 m²/day is assigned to all nodes; see Note 9.

Note 7:   MODFLOW supports a variety of different aquifer types. Briefly, 0 = confined aquifer, 1 = unconfined aquifer, 2 = confined–unconfined constant transmissivity, and 3 = confined–unconfined variable transmissivity. One value is provided for each layer (40I2 format).

Note 8:   The anisotropy factor is the ratio of transmissivity or hydraulic conductivity along a column to transmissivity or hydraulic conductivity along a row. Read one value per layer.

Note 9:   This simplified input format works because each row dimension is the same. For a variable grid, the input data would be more complicated.

Note 10:   The transmissivity along rows is provided. The same value is assigned here to every node. The code determines the column values by multiplying this transmissivity value by the anisotropy factor. Note that the sequence of required input data with respect to transmissivity and storativity changes depending on the type of aquifer. For example, with an unconfined aquifer, it is necessary to specify the elevation of the bottom of the aquifer.

File EX1.WEL—Well Package

1	0			max number of wells at any time, print no cell-by-cell terms
1				for each stress period—the number of wells active, here 1
1	15	15	3790.	for each well—location—layer, row, column, and withdrawal rate (negative for withdrawal)

File EX1.SIP—Solver for Finite Difference Equations

50	5				max no. iterations, no. of iteration parameters,
1.	.001	0	.001	1	acceleration parameter (usually 1.), convergence criterion for iteration (.001), and other default parameters, see Note 11.

Note 11:   It is generally difficult for the casual user to decide what parameters to use in this subroutine. As an initial estimate, we could simply set the number of iterations around 50, and convergence criteria of .001 or less, and accept the values suggested in the User's Guide for the rest of the parameters.

Table 7.4

VOLUMETRIC BUDGET FOR ENTIRE MODEL AT END OF TIME STEP 18 IN STRESS PERIOD 1	
CUMULATIVE VOLUMES L**3	RATES FOR THIS TIME STEP L**3/T
IN:	**IN:**
STORAGE = 0.00000	STORAGE = 0.00000
CONSTANT HEAD = 0.00000	CONSTANT HEAD = 0.00000
WELLS = 0.24278E+06	WELLS = 3790.0
TOTAL IN = 0.24278E+06	TOTAL IN = 3790.0
OUT:	**OUT:**
STORAGE = 0.24274E+06	STORAGE = 3788.1
CONSTANT HEAD = 0.00000	CONSTANT HEAD = 0.00000
WELLS = 0.00000	WELLS = 0.00000
TOTAL OUT = 0.24274E+06	TOTAL OUT = 3788.1
IN − OUT = 43.359	IN − OUT = 1.9360
PERCENT DISCREPANCY = 0.02	PERCENT DISCREPANCY = 0.05

earlier with the analytical model. At 304.8 m away from the pumping well, the calculated drawdown is 1.641 m.

Operational Issues

Time-Step Size

Normally, care must be exercised in selecting a time-step size to avoid errors that can occur due to large time steps. These errors are most pronounced in the first few time steps when pumping begins or pumping rates change. The most preferred way of controlling errors is to use relatively small time-step sizes. Overall, it is prudent to place less confidence in results coming from the first few time steps after a significant change in withdrawal or injection rates. MODFLOW lets the user increase the size of the time step as the simulation proceeds. The given time-step size is multiplied by a number greater than 1 between time steps. A time-step multiplier of 1.2 works well in most cases. Thus, an initial time step of 1 day becomes 1.2 days for the second time step, and 1.2 × 1.2 or 1.44 days for the third time step, and so on. A variable time-step size will reduce the computational effort compared to a constant step size and yet preserve the accuracy of the calculation.

Drawdowns at "Pumping" Nodes

In numerical models, predicting drawdowns at "pumping" nodes is usually a problem. The head or drawdown at a pumping node does not represent drawdown in the well because the hydraulic head is a "cell average." This value is not the same as that calculated with an analytical model (for example, the Theis equation), which is the exact drawdown at the point of interest.

To illustrate this point, we will consider the example

problem in Section 7.4. Now, however, the drawdowns at the pumping well will be compared. For the analytic calculation, the observation point is assumed to be 0.1 m away (that is, the radius of the well). As expected, the drawdown determined analytically is much greater than that predicted by the model (Figure 7.13).

Techniques are available to provide better estimates of the actual drawdown that can be expected at a node representing a pumping well. One of the simplest involves a variant on the distance–drawdown theory. In effect, one takes relatively accurate model predictions of drawdown at nodes away from the well, plots them on a semilog distance versus drawdown plot, and obtains the required estimate by extrapolation of the curve. In the illustrative example at 64.1 days (9.23×10^4 minutes), the calculated drawdown at points 304.8 m and 3048 m

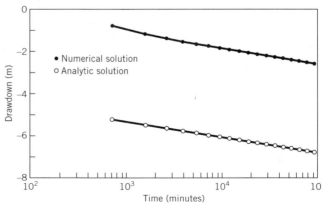

Figure 7.13 Comparison of time–drawdown curves at the pumping well computed using the numerical code and the analytical solution.

away from the well are -1.64 m and -0.26 m, respectively. Having created the appropriate distance–drawdown plot (Figure 7.14), the projection of the straight line provides an estimate of the drawdown 0.1 m away. The estimated value of -6.4 m compares to the analytic value of -6.8 m.

This method is applicable only to confined aquifer with no local changes in the transmissivity. It is not generally useful for the full range of cases that might be encountered in a numerical modeling exercise. Fortunately, there are a variety of other approaches in the literature that can be used to make these estimates.

Water-Table Conditions

With unconfined aquifers, the transmissivity changes as a function of drawdown in the vicinity of the well. For example, examine Figure 7.15, where drawdown in an unconfined aquifer causes the water table to fall. The decrease in the aquifer thickness causes the transmissivity to decrease as well.

The way the numerical methods accommodate the varying transmissivity is to recalculate the transmissivity at the beginning of each iteration cycle within each time step. Thus, every time that the hydraulic head at a node changes, the transmissivity is recalculated as a product of the hydraulic conductivity and the saturated thickness. Within a time step, as the change in hydraulic conductivity becomes smaller after successive iterations, the changes in transmissivity are also reduced. When the iterations are complete, both hydraulic head and transmissivity values will have converged to correct values.

In MODFLOW, a comparison is made at the start of each iteration to determine the magnitude of the hydraulic head relative to elevations of the top ($TOP_{i,j,k}$) and bottom ($BOT_{i,j,k}$) of the unit of interest. Transmissivity for the cell is calculated using one of the three following equations:

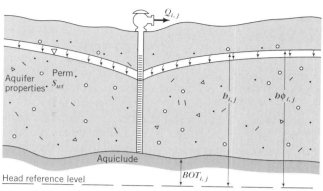

Figure 7.15 Parameters for the problem of flow to a well with water-table conditions (from Prickett and Lonnquist, 1971).

if $\quad HNEW_{i,j,k} \geq TOP_{i,j,k} \quad$ (aquifer is confined)

then $\quad TR_{i,j,k} = (TOP_{i,j,k} - BOT_{i,j,k})\, HYR_{i,j,k}$;

if $\quad TOP_{i,j,k} > HNEW_{i,j,k} > BOT_{i,j,k}$,

$\quad\quad\quad$ (aquifer is unconfined)

then $\quad TR_{i,j,k} = (HNEW_{i,j,k} - BOT_{i,j,k})\, HYR_{i,j,k}$;

if $\quad HNEW_{i,j,k} \leq BOT_{i,j,k}, \quad$ (aquifer has drained)

then $\quad TR_{i,j,k} = 0$

where $HYR_{i,j,k}$ is the hydraulic conductivity of cell i, j, k in the row direction, $TOP_{i,j,k}$ is the elevation of the top of the cell i, j, k, and $BOT_{i,j,k}$ is the elevation of the bottom of cell i, j, k.

Boundary Conditions

The six boundaries of the three-dimensional domain in MODFLOW (that is, top, bottom, and sides) are no-flow boundaries by default. Creating flow across these boundaries requires that a different type of boundary condition be created. For example, one could make one of the side boundaries a constant-head boundary or a flux boundary by adding water to particular nodes along the boundary. There is a place in the Basic Input package to designate specific nodes as fixed head or inactive. A constant-flux boundary may be simulated by simply adding or withdrawing water from boundary nodes as if a well is present. These boundaries are specified with the Basic Input and/or Well Packages.

Fluxes into or out of the top and other layers of the model could occur due to the presence of lakes or rivers. In the model region, these features can be represented by constant-head boundaries or variable-flux boundaries using the River Package. Similarly, recharge from precipitation can be added by applying a small flux at all nodes present in recharge areas. It is this ability to adjust the boundary conditions and treat various groups of internal nodes that make numerical models so powerful in representing complex hydrogeologic systems.

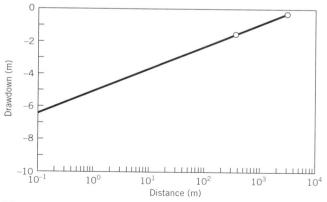

Figure 7.14 Application of a distance–drawdown extrapolation to estimate the drawdown at a "pumping" node.

7.5 Case Study in the Application of MODFLOW

Springer (1990) and Springer and Bair (1992) have applied MODFLOW to a buried valley aquifer complex near Wooster, Ohio. Figure 7.16 shows the geometry of the buried valley, Killbuck Creek, and its tributaries Little Killbuck and Apple Creeks, as well as the location of two major municipal wellfields. From 1984 to 1988 the North and South Wellfields produced from 3.5 to 4.4 million gallons per day. More recently, both wellfields may be abandoned due to contamination.

Cross section A-A' shows the bedrock valley and the distribution of geologic units. Immediately overlying bedrock is glacial till. This unit is overlain in turn by sand and gravel outwash, which comprise the aquifer. The outwash unit may be separated by a silt unit up to 21 feet thick (Figure 7.17a). The silt unit is relatively discontinuous and, for example, does not occur in the area of the South Wellfield. The uppermost geologic unit is comprised of silty-clay lacustrine and floodplain deposits ranging in thickness from 10 to 30 feet. Shown on the map (Figure 7.16) is the location of alluvial fans along Little Killbuck and Clear and Apple Creeks that directly overlie the outwash aquifer. A variety of hydrogeologic studies and careful stream-discharge measurements have shown that the alluvial fans provide the most important hydraulic connection between surface water and ground water (Springer and Bair, 1992; Breen and others, 1995).

Through the years, a great deal of hydrogeologic data have been collected in relation to this aquifer system (see Springer, 1990). The data base includes extensive stream-discharge measurements, hydraulic conductivity measurements of the streambed, lithologic logs from various test holes, hydraulic information from aquifer tests with a variety of wells, production rates for municipal and industrial wells, and measured hydraulic head data. The model is supported by extensive hydrogeologic data, which are discussed in detail by Springer (1990). See Breen and others (1995) for a more recent USGS modeling study of the same system.

Model Development

The finite difference model is composed of 28 rows, 46 columns, and 4 layers. The active portion of the grid (Figure 7.18) is determined by the buried valley and is composed of cells having dimensions $\Delta r_j = 250$ ft and Δc_i ranging between 250 and 500 ft. Figure 7.17b shows how the observed stratigraphy at the site relates to the 4 layers of the grid. The silty-clay unit that subdivides the outwash along A-A' is not explicitly included as a separate layer. Its hydraulic effects, however, are included by lowering the vertical conductivity between units 2 and 3. For multilayer problems, it is necessary in the Block-Centered Flow Package to input a matrix called Vcont that reflects the vertical hydraulic conductivity

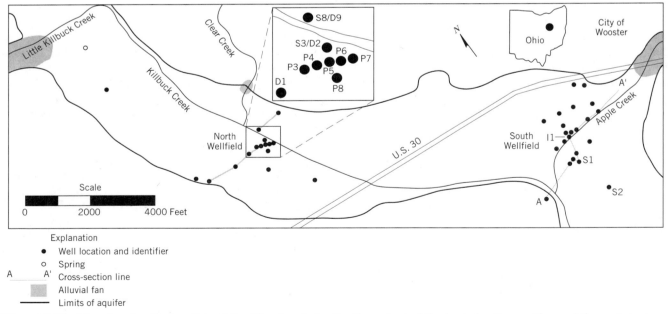

Figure 7.16 Map showing the location of the Wooster study site, the extent of the buried valley aquifer, and the region encompassed by the MODFLOW model (modified from Springer and Bair, 1992). Reprinted by permission of Ground Water. Copyright © 1992. All rights reserved.

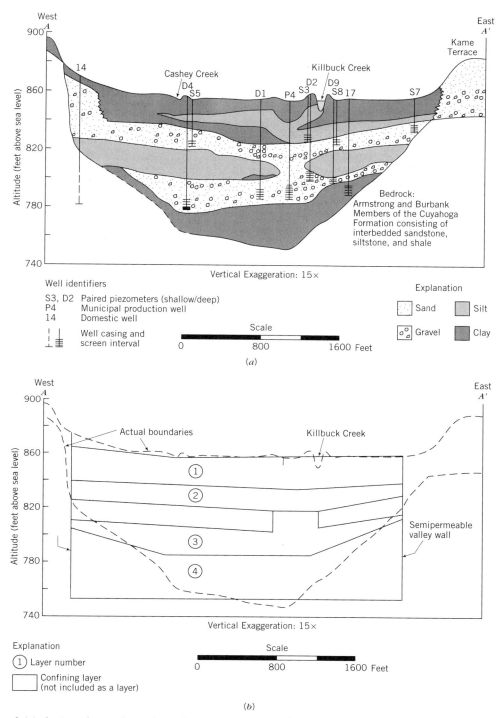

Figure 7.17 Panel (*a*) depicts the geology along the cross section *A-A'* (from Springer and Bair, 1992). Reprinted by permission of Ground Water. Copyright © 1992. All rights reserved. Panel (*b*) illustrates how the various units are conceptualized for MODFLOW.

between layers divided by the distance between nodes. McDonald and Harbaugh (1988) explain in detail how Vcont is calculated when a confining layer of some thickness and hydraulic conductivity is present.

Figure 7.18 shows the boundary conditions for layer 1. The various streams are represented with the River Package. Thus, with pumping, induced infiltration from the creeks occurs with the quantity of infiltration deter-

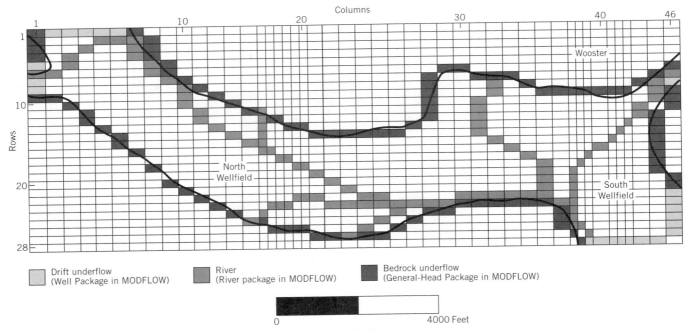

Figure 7.18 Location of active grid cells and a summary of boundary conditions for layer 1 (modified from Springer, 1990).

mined by the head gradient between the stream and the aquifer, as well as the area and hydraulic conductivity of the streambed. The relatively low hydraulic conductivity of the lacustrine and floodplain deposits restrict the quantity of induced infiltration. The alluvial fans are important because much more induced infiltration is possible through their higher permeability streambeds as the streams pass over them. Inflows to the alluvial aquifer farther up or down valley (that is, drift underflow) are represented by constant fluxes created by adding wells along the end boundaries of the model using the Well Package (Figure 7.18). Inflows to the active cells due to flow through the low-permeability valley walls (that is, bedrock underflow) is simulated using the General-Head Package (Figure 7.18). In effect, the General-Head Package is a variable-flux boundary condition that lets the inflow through the boundaries increase as the drawdown in the aquifer increases. By being able to specify a conductance (proportional to hydraulic conductivity) between the aquifer and the boundary, the inflow through the boundary can be controlled. In the Wooster example, 29% of the water pumped from the aquifer is thought to originate from the inflow of bedrock water through the side boundaries (Breen and others, 1995). The boundary conditions for the deeper units (that is, layers 2, 3, and 4) are essentially the same except that the streams and recharge are not included. There are also minor changes around the alluvial fans, but generally the boundary conditions are set up to provide bedrock and drift underflows as was the case with layer 1.

The Recharge Package provides a component of inflow

due to precipitation and water from melting snow. In general, the recharge rates are not constant everywhere. Recharge rates on the alluvial fan are greater than those for the middle part of the valley, where lower permeability materials and agricultural drainage limit infiltration (Springer and Bair, 1992).

This conceptualization directly determines what MOD-FLOW packages would need to be included in the model. Beyond those packages required for all simulations—Basic, Block-Centered Flow, and a solver (here the Strongly Implicit Procedure)—the Well Package accounts for the municipal and industrial withdrawal wells and some boundary fluxes; the Recharge Package adds natural recharge to the system; the River Package provides for the possibility of induced infiltration from the streams; the General-Head Boundary Package represents inflow from the bedrock laterally adjacent to the various layers; and the Output Control specifies in detail how the results are to be presented.

Overall, the potential sources of inflow to the aquifers are more than capable of balancing the water withdrawn by pumping. Therefore, the flow system will be at steady state. The simulation model then is configured to provide a steady-state result. The drawdown around the North and South Wellfields is sufficiently large that layer 1 will desaturate locally and portions of layer 2 will behave as an unconfined aquifer. Thus, layer 1 is considered to be an unconfined aquifer, layer 2 is considered capable of being converted from a confined to an unconfined aquifer, and layers 3 and 4 are considered to be confined.

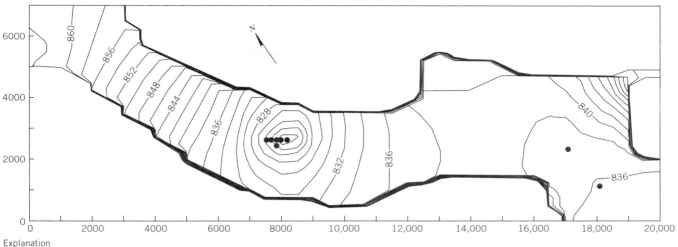

Explanation
Scales in feet
● Pumping well location
—836— Contour of equal hydraulic head (ft asl)
Contour interval: 2 feet

Figure 7.19 Map of the simulated distribution of hydraulic head for layer 3 (from Springer, 1990).

Data Preparation and Model Calibration

In the Wooster case study and most every model study, there will never be sufficient data to provide necessary parameters (for example, transmissivity, storativity, recharge rates) for every node in every layer. When hydrogeologic data are available, they form a basis for estimating model parameters required for every node. For example, Springer (1990), through a detailed interpretation of geologic data, was able to develop isopack maps that together with lithologic and hydraulic test data provided a detailed description of transmissivity distributions for the layers representing the outwash aquifer. Thus, transmissivity values for layer 2 varied from 500 to 6000 ft²/day and in layer 3 from 20 and 90,000 ft²/day. The transmissivity distributions for the lower hydraulic conductivity units (layers 1 and 4) are less well known and are not represented in detail. Transmissivities of layer 1 range between 0.01 to 200 ft²/day (Springer and Bair, 1992). Layer 4 is assigned a constant transmissivity of 0.0005 ft²/day. Streambed conductance values are based on measured hydraulic conductivities for various reaches of the streambed (Springer, 1990). Recharge rates of 4 in./yr for the valley and 10 in./yr are assigned to alluvial fans. These values are based on estimates of the U.S. Geological Survey (Breen and others, 1995). Pumping rates for the individual wells are quantified through a synthesis of existing records.

Calibration involved making minor adjustments to the transmissivity field, vertical conductance values (between vertical nodes), streambed-hydraulic conductivity values, and bedrock leakage rates until the steady-state model was successfully calibrated. The calibration criteria involved (1) a good visual comparison between the measured and simulated potentiometric surfaces, (2) a root-mean square error between measured and simulated heads of less than 2.5 feet, (3) simulated water losses from streams that compared favorably to measured losses, and (4) simulated leakage from bedrock that agreed with estimates from chemical mixing and isotopic studies (Springer and Bair, 1992).

The enclosed computer disk contains the complete set of files required to run the Wooster case study {WOO. BAS, WOO.BCF, etc.). We would encourage readers to examine these data files to see just how complex the data files can be for a realistic MODFLOW problem. Shown in Figure 7.19 is the simulated, steady-state potentiometric surface for layer 3 under the imposed conditions of pumping. The cone of depression due to large withdrawals at the North and South Wellfields is obvious. The model was validated by removing the pumping stress and successfully simulating the historic, nonpumping potentiometric surface (Springer and Bair, 1992).

Example 7.1

One of the main benefits of using a code like MODFLOW is that the mass balance calculations provide a very useful way to examine the source of water provided to a system of pumping wells. With the help of the MODFLOW code and the data files supplied for the Wooster case study, determine the proportion of water pumped from the system due to natural recharge, leakage from the creeks, drift underflow, and bedrock underflow. To obtain the answer to the problem, it is necessary to plow through the output to find the mass balance summary.

Problem

1. The purpose of this problem is to illustrate the impact of boundary conditions on a simulation. Reconsider the illustrative example in Section 7.3 and assume that the two different sets of boundary conditions shown in the figures apply, that is, no-flow boundaries on four sides, and no-flow boundaries on three sides and a constant-head boundary along the top. Use MODFLOW and run simulations for 1474 days (35 time steps with a 1.2 time-step multiplier) with the two different boundary conditions to compare the drawdowns at a point 304.8 m away from the withdrawal well.

Hint: The original problem was actually set up with four no-flow boundaries (see Figure 7.20a).

Thus, the first part of this problem is solved simply by running the simulation for another 17 time steps (35 in total) to produce the drawdown shown on the figure. A constant-head boundary is simulated in MODFLOW by adjusting the information provided on boundary conditions that are provided in EX1.BAS. The existing line describing the boundary conditions is {9 spaces} 0 {9 spaces} 1. The 0 tells the code that the array describing the boundary conditions has a constant value. The 1 is that constant value. Adding a constant-head edge requires that the constant-head nodes be designated by −1. Thus, it becomes necessary to include the complete boundary condition array. The first few lines of this piece become

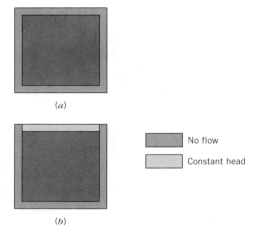

(a)

(b)

No flow

Constant head

Problem 1

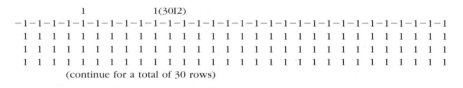

```
        1              1(30I2)
-1-1-1-1-1-1-1-1-1-1-1-1-1-1-1-1-1-1-1-1-1-1-1-1-1-1-1-1-1-1
 1  1  1  1  1  1  1  1  1  1  1  1  1  1  1  1  1  1  1  1  1  1  1  1  1  1  1  1  1  1
 1  1  1  1  1  1  1  1  1  1  1  1  1  1  1  1  1  1  1  1  1  1  1  1  1  1  1  1  1  1
 1  1  1  1  1  1  1  1  1  1  1  1  1  1  1  1  1  1  1  1  1  1  1  1  1  1  1  1  1  1
        (continue for a total of 30 rows)
```

The 1 next to the format (30I2) is a multiplier for all the integers in the following array.

Chapter 8

Stress, Strain, and Pore Fluids

8.1 *Deformable Porous Media*
8.2 *Abnormal Fluid Pressures in Active Depositional Environments*
8.3 *Pore Fluids in Tectonic Processes*

As fluids are ubiquitous within the crust, more and greater emphasis has been placed on the physics that describe fluid pressures in deformable porous rocks. Outstanding problems include the generation and dissipation of abnormally high fluid pressures in areas of active sediment deposition, the generation of earthquakes due to fluid injections and in the vicinity of man-made reservoirs, landslides and slope failures caused by high fluid pressure, and the role of fluid pressures in faulting and fracturing. Thus in this chapter we are obliged to learn about driving forces for fluid movement other than those associated with topographic relief. To gain an understanding of such forces we start with the premise that water does not reside in a passive porous solid, but instead there exists a complex coupling between the moving fluid, entities that might be carried by the fluid, and the solid matrix itself. In this chapter we focus on the couplings among stress, strain, and pore fluids. The moving entities (heat energy and mass) are discussed in later chapters.

8.1 Deformable Porous Media

One-Dimensional Consolidation

The concept of disequilibrium compaction was discussed in Chapter 2 and was described in terms of a time lag between the loading of a sedimentary pile and the dissipation of the fluid pressure caused by the loading. This type of transient behavior is frequently associated with clays and other low-permeability sediments that have a relatively high specific storage and low hydraulic conductivity; that is, the hydraulic diffusivity is small. In addition to these fluid pressure transients, there are a host of others associated with high-permeability materials subjected to natural loading, such as earth and oceanic tides, earthquakes, and even changes in atmospheric pressure. To examine the nature of these transients, it is necessary to reexamine Darcy's law and the concept of effective stress and to reformulate these ideas into a different form of a diffusion-type equation.

Development of the Flow Equation

An excess fluid pressure has been defined as a pressure in excess of some preexisting hydrostatic value. Within the concept of the effective stress statement, this may be stated as

$$\sigma + \Delta\sigma = \bar{\sigma} + (P_s + P_{ex}) \qquad (8.1)$$

where the total pore water pressure consists of two parts, P_s and P_{ex}, where P_s is a hydrostatic pressure and P_{ex} is a transient pore water pressure in excess of the hydrostatic value, and is presumed to be caused by the increment in total stress $\Delta\sigma$. In a similar fashion, total head may be restated as

$$b = z + \frac{P_s}{\rho_w g} + \frac{P_{ex}}{\rho_w g} \qquad (8.2)$$

For one-dimensional vertical flow, Darcy's law becomes

$$q_z = -K_z \frac{\partial b}{\partial z} = -K_z \frac{\partial}{\partial z}\left(z + \frac{P_s}{\rho_w g} + \frac{P_{ex}}{\rho_w g}\right) \qquad (8.3)$$

For the problem under consideration, the flow occurs in response to a gradient in excess pressure, not total head, giving

$$q_z = -\frac{K_z}{\rho_w g}\frac{\partial P_{ex}}{\partial z} \qquad (8.4)$$

as the appropriate form of Darcy's law.

The conservation equation required for further analysis has already been given as Eq. A.6 in Appendix A, reexpressed here for one-dimensional vertical flow of a homogeneous fluid

$$-\frac{\partial q_z}{\partial z} = \frac{n}{\rho_w}\frac{\partial \rho_w}{\partial t} + \frac{1}{(1-n)}\frac{\partial n}{\partial t} \qquad (8.5)$$

where

$$\frac{1}{\rho_w}\frac{\partial \rho_w}{\partial t} = \beta_w \frac{\partial P}{\partial t} \qquad \frac{1}{(1-n)}\frac{\partial n}{\partial t} = \beta_p\left(\frac{\partial P_{ex}}{\partial t} - \frac{\partial \sigma}{\partial t}\right) \qquad (8.6)$$

In this development, $\partial \sigma / \partial t$ is taken as the vertical stress change that gives rise to the excess pressure development. It is noted that such stress changes were not required in our previous development of the flow equation for confined aquifers. Substituting Darcy's law along with Eq. 8.6 into the conservation equation gives the one-dimensional consolidation equation

$$K_z \frac{\partial^2 P_{ex}}{\partial z^2} = \rho_w g(n\beta_w + \beta_p)\frac{\partial P_{ex}}{\partial t} - (\rho_w g \beta_p)\frac{\partial \sigma}{\partial t} \qquad (8.7)$$

Except for the one-dimensional flow condition and the inclusion of the stress term that gives rise to the fluid pressure production, the assumptions here are identical to those employed in the previous development of the diffusion equation. The quantity $\rho_w g(n\beta_w + \beta_p)$ is the specific storage, whereas the quantity $\rho_w g \beta_p$ is the component of specific storage due exclusively to pore compressibility.

Equation 8.7 incorporates the time rate of change of vertical stress as a pressure-producing mechanism, which is of some importance to many loading problems in hydrogeology. As there is a relationship between stress and strain, one would expect that there is a companion equation that incorporates the time rate of change of pore volume strain as a pressure-producing mechanism. For the simple case we are examining, where the fluids are compressible but the individual solid grains are not,

this companion equation may be stated as

$$K_z \frac{\partial^2 P_{ex}}{\partial z^2} = (\rho_w g n \beta_w)\frac{\partial P_{ex}}{\partial t} + (\rho_w g)\frac{\partial \varepsilon}{\partial t} \qquad (8.8)$$

where $\rho_w g n \beta_w$ is the specific storage due exclusively to expansion of the water and ε is the one-dimensional volume strain, or dilation, taken as the pressure-producing mechanism.

The Undrained Response of Water Levels to Natural Loading Events

Numerous cases of water-level fluctuations in wells in response to atmospheric pressure, ocean and earth tides, earthquakes, and passing trains have been reported (Jacob, 1939; Robinson, 1939; Parker and Stringfield, 1950; Bredehoeft, 1967; van der Kamp, 1972; van der Kamp and Gale, 1983). These fluctuations are generally accepted as evidence that confined aquifers are not rigid bodies, but are elastically compressible, an idea fostered by Meinzer (1928) and advanced by Jacob (1940). Currently, it is a routine exercise to compare the water-level response to the loads that cause them to calculate certain hydraulic properties of the medium.

In discussing and interpreting these fluctuations, two factors must be considered. First, there are loads that act not only on the rock matrix and its contained water, but also on the water level in an open observation well, and there are loads that act only on the rock matrix and its contained water. Examples of the former include changes in atmospheric pressure, and for the latter there are earth and oceanic tides. Second, it is important to consider the boundary conditions under which the loading takes place. Two types are possible. For a field deformation that is slower than the characteristic times for the diffusion of the pore fluid, the fluid pressure may rise above its ambient value and then dissipate rather quickly. Hence, for all practical purposes, the fluid pressure remains constant during the deformation, and such boundary conditions are called constant pressure, or drained. This behavior is an expected one in high-permeability materials. The effect of passing trains on water levels as demonstrated in Figure 4.5 is an example of a drained response, and the parameter β_p given in Table 4.1 provides an example of a drained compressibility. Conversely, when the load variations are rapid in comparison to the diffusion time for the pore fluid, the local fluid mass remains essentially constant during deformation. This response thus occurs at constant mass, or under undrained conditions. The undrained deformation is elastically stiffer than its drained counterpart so that the respective deformations are characterized by different coefficients. Hence, when a coefficient is sought to characterize a prescribed type of response, it is important to know the boundary conditions under which that response occurs.

All water-level fluctuations as observed at a point that

are caused by rapid loading phenomena (tidal and atmospheric pressure variations) are treated as constant mass phenomena. As constant mass implies an absence of fluid flow, all the information we require about these deformations may be obtained by setting the fluid flow term in Eqs. 8.7 and 8.8 equal to zero, which gives, after minor rearrangement

$$\left(\frac{\partial P}{\partial \sigma}\right)_M = \frac{\beta_p}{\beta_p + n\beta_w} \quad (8.9a)$$

$$\left(\frac{\partial P}{\partial \varepsilon}\right)_M = \frac{1}{n\beta_w} \quad (8.9b)$$

$$\left(\frac{\partial \varepsilon}{\partial \sigma}\right)_M = \frac{\beta_p n\beta_w}{\beta_p + n\beta_w} \quad (8.9c)$$

In these developments, the minus sign has been ignored because we are interested only in absolute values.

Equation 8.9a is termed the pore pressure coefficient and is defined as the change in fluid pressure at constant fluid mass as the stress increases. Interpreted literally, this coefficient describes the percentage of an incremental load that is carried by the pore fluid provided the pore fluid is not permitted to drain. The pore pressure coefficient has been referred to as the tidal efficiency by Jacob (1940), an unfortunate choice of terms in that this equation describes the change in fluid pressure at constant mass in response to any blanket load, not just ocean tides. As used by Jacob (1940), tidal efficiency (T.E.) is defined as the ratio of a piezometric level amplitude as measured in a well to the oceanic tidal amplitude, or

$$\text{T.E.} = \frac{dP}{\gamma_w dH} = \frac{\beta_p}{\beta_p + n\beta_w} \quad (8.10)$$

where H is the height of the tide. Figure 8.1 shows an example of the correspondence between ocean tides and water levels in wells located in aquifers with an oceanic suboutcrop.

Equation 8.9b has been used by Bredehoeft (1967) to examine the effect of earth tides on water levels in wells.

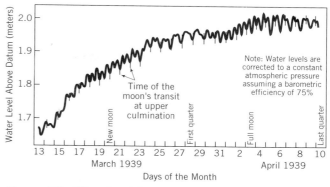

Figure 8.2 Water-level fluctuations in a well caused by earth tides (from Robinson, Trans. Amer. Geophys. Union, v. 20, p. 656–666, 1939. Copyright by Amer. Geophys. Union).

This dilation can cause water-level changes of about 1 to 2 cm. As noted in this expression, the volume change of a dilated rock–water system is represented by a volume change of water in the pores, where $n\beta_w = (S_s - \rho_w g\beta_p)/\rho_w g$. Figure 8.2 demonstrates the effect of earth tides on water levels.

Barometric pressure has an inverse relationship with water levels in wells (Figure 8.3). The barometric efficiency (B.E.) is used to describe how faithfully a water level responds to changes in atmosphere pressure, and is given as

$$\text{B.E.} = \frac{\gamma_w db}{dP_a} \quad (8.11)$$

where P_a is atmospheric pressure. The response of water levels to changes in atmospheric pressure may be reasoned as follows. Incremental increases or decreases in atmospheric pressure acting on a column of water in a well are, respectively, added to or subtracted from the pressure of the water in the well. These same stresses acting on any part of the confined aquifer, however, are

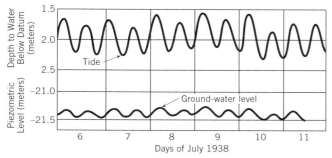

Figure 8.1 Tidal fluctuations and induced piezometric surface fluctuations observed in a well 30 m from shore at Mattawoman Creek, Maryland (from Meinzer, 1939).

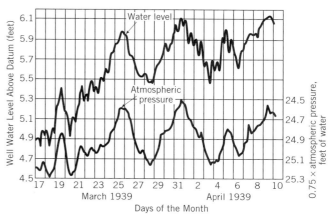

Figure 8.3 Water-level response to changes in atmospheric pressure (from Robinson, Trans. Amer. Geophys. Union, v. 20, p. 656–666, 1939. Copyright by Amer. Geophys. Union).

supported by both the grain structure and the fluid. As long as the atmospheric pressure is undergoing change, there must exist a pressure difference between the water in the well and that in the aquifer. When atmospheric pressure is increasing, the gradient is away from the well, which causes the water level to decline. When atmospheric pressure is decreasing, this decrease is fully removed from the water column in the well and only partially from the water in the confined aquifer. Hence, the gradient is toward the well, and the water-level rises.

A mathematical statement for barometric efficiency may be arrived at by reconsidering the statement of effective stress and our interpretation of the pore pressure coefficient. The effective stress concept may be restated as

$$1 = \frac{\partial P}{\partial \sigma} + \frac{\partial \bar{\sigma}}{\partial \sigma} \qquad (8.12)$$

where the pore pressure coefficient $\partial P/\partial \sigma$ represents the percentage of the load that is carried by the fluid, provided the fluid is not permitted to drain. Clearly, $\partial \bar{\sigma}/\partial \sigma$ must represent the percentage of the stress that is carried by the solid matrix, again under undrained conditions. This is the barometric efficiency, that is,

$$1 = \text{T.E.} + \text{B.E.} \qquad (8.13)$$

so that, from Eqs. 8.10 and 8.11

$$\text{B.E.} = 1 - \frac{\beta_p}{\beta_p + n\beta_w} = \frac{n\beta_w}{\beta_p + n\beta_w} \qquad (8.14)$$

Thus, as β_p approaches zero for a rigid aquifer, B.E. approaches one, and T.E. approaches zero. A barometric efficiency of one suggests that the well is a perfect barometer, which is possible only for a perfectly incompressible aquifer ($\beta_p = 0$). On the other hand, as β_p gets large, B.E. approaches zero and T. E. approaches one. The relationships among the tidal efficiency, barometric efficiency, and specific storage may be stated as

$$\text{T.E.} = \frac{\rho_w g \beta_p}{S_s} \qquad (8.15a)$$

$$\text{B.E.} = \frac{\rho_w g n \beta_w}{S_s} \qquad (8.15b)$$

As the sum of the barometric and tidal efficiencies is unity, the percentage of storage attributable to expansion of the water is B.E., and that attributed to compression of the matrix is T.E.

There are several examples of the application of the results just stated. Jacob (1941) was the first to employ these methods for a well in Long Island with a reported tidal efficiency of 0.42 and a porosity of 0.35. Given an estimated value of 0.58 for the barometric efficiency, the specific storage is readily calculated to be 8.6×10^{-7} ft^{-1}. Bredehoeft (1967) reported a barometric efficiency of 0.75 for a well in Iowa, along with an estimated porosity

of 0.178. Specific storage in this case is calculated to be 2.8×10^{-7} ft^{-1}. An interesting application for a composite rock section has been reported by Carr and van der Kamp (1969). These authors measured several values of both barometric and tidal efficiencies in sediments consisting of mostly sandstone and siltstone. The barometric efficiency averaged about 0.37 and the tidal efficiency averaged 0.57. The difference of 0.06 between the sum of these values is reported to be within experimental error. Forty rock samples provided a mean porosity of 0.177. The average specific storage was calculated to be about 6.3×10^{-7} ft^{-1}.

Barometric fluctuations are not commonly observed in wells tapping unconfined aquifers. The reason for this is that changes in atmospheric pressure are transmitted equally to the column of water in a well and to the water table through the unsaturated zone. Thus, there are no pressure gradients.

The last item to be concerned with is to demonstrate that the tidal and barometric efficiencies as described thus far—although defined as an undrained response—consist of a combination of drained and undrained parameters. It will be recalled that the pore compressibility β_p could be expressed in terms of a modulus of compression $1/H_p$ where the subscript p means we are dealing with pore compressibility only. It is convenient now to introduce other notations for some of the parameters. These are

$$\frac{1}{R_p} = \beta_p + n\beta_w \qquad (8.16a)$$

$$\frac{1}{Q_p} = n\beta_w \qquad (8.16b)$$

where the subscript p indicates compressibilities for the case of incompressible solid grains. The relationship between these constants is given as

$$\frac{1}{Q_p} = \frac{1}{R_p} - \frac{1}{H_p} \qquad (8.17)$$

With this new designation, it is clear that the tidal efficiency or pore pressure coefficient is equal to R_p/H_p and the barometric efficiency is equal to R_p/Q_p. In addition, Eq. 8.9c becomes

$$\left(\frac{\partial \varepsilon}{\partial \sigma}\right)_M = \frac{1}{Q_p} \frac{R_p}{H_p} = \frac{1}{H_p}\left[1 - \frac{R_p}{H_p}\right] = \frac{1}{H_p}\,[\text{B.E.}] \qquad (8.18)$$

As $\partial \varepsilon/\partial \sigma$ is a measurement at constant mass, it too is an undrained parameter and may be designated as $1/K_u$, where K_u is an undrained modulus of compression, still for the incompressible solid case. It thus follows that B.E. $= H_p/K_u$, which is a ratio of a drained to an undrained modulus, and T.E. $= 1 - H_p/K_u$. Further details on these relationships may be found in Domenico (1983).

The Drained Response of Water Levels to Natural Loading Events

There are numerous examples of the drained response of water levels to natural loading events, including the change in head in response to changes in river stage (Cooper and Rorabaugh, 1963), earthquakes (Cooper and others, 1965), and the inland propagation of sinusoidal fluctuations of ground-water levels in response to tidal fluctuations of a simple harmonic motion (Ferris, 1951; Werner and Noren, 1951). These problems are generally treated within the framework of the conventional one-dimensional diffusion equation given as Eq. 8.11 where the pressure-producing mechanisms are normally formulated within the boundary conditions. In other instances, it is frequently assumed that the stress causing the excess pressure is applied rapidly and then held constant so that the stress term of the one-dimensional consolidation equation (Eq. 8.7) is taken as zero. A typical example of this is the so-called Terzaghi consolidation equation (Terzaghi and Peck, 1948). In the application of this equation to consolidation problems, it is frequently assumed that the pore compressibility is considerably larger than the compressibility of the water so that the latter may be ignored, giving from Eq. 8.7

$$\frac{\partial^2 P_{ex}}{\partial z^2} = \frac{\rho_w g \beta_p}{K_z} \frac{\partial P_{ex}}{\partial t} = \left(\frac{1}{C_v}\right) \frac{\partial P_{ex}}{\partial t} \qquad (8.19)$$

where the hydraulic diffusivity C_v is termed the coefficient of consolidation and is given as

$$C_v = \frac{K_z(1 + e)}{a_v \rho_w g} = \frac{K_z}{S_s} \qquad (8.20)$$

In Eq. 8.20 the specific storage for the case of incompressible fluids is $a_v \rho_w g/(1 + e)$, where e is the void ratio and a_v is a coefficient of compressibility, defined as the rate of change of void ratio to the rate of change of effective stress causing the deformation.

Problems in basin loading or crustal deformation are normally described in terms of stress or strain rates applied over geologic intervals of time. For these purposes, Eqs. 8.7 and 8.8 may be restated here as

$$\frac{\partial P_{ex}}{\partial t} = \frac{K_z}{\rho_w g(1/R_p)} \frac{\partial^2 P_{ex}}{\partial z^2} + \left(\frac{R_p}{H_p}\right) \frac{\partial \sigma}{\partial t} \qquad (8.21)$$

and

$$\frac{\partial P_{ex}}{\partial t} = \frac{K_z}{\rho_w g(1/Q_p)} \frac{\partial^2 P_{ex}}{\partial z^2} - (Q_p) \frac{\partial \varepsilon}{\partial t} \qquad (8.22)$$

where R_p, H_p, and Q_p have been previously defined. It is noted that $1/R_p$ is incorporated within the specific storage for problems involving stress rates and $1/Q_p$ is incorporated within the specific storage for problems involving strain rates. Additionally, the pressure-producing stress term in Eq. 8.21 is multiplied by the pore pressure coefficient. Thus, the larger the pore pressure

coefficient, the greater the percentage of the stress change that is initially carried by the fluid and, consequently, the greater the initial fluid pressure production. Thus, for a pore pressure coefficient close to one, virtually all the stress change is initially carried by the fluid. As mentioned in Chapter 2, the pore pressure coefficient decreases with increasing rock rigidity and is therefore rock dependent.

Land Subsidence as a One-Dimensional Drained Response

Subsidence of the land surface in many areas of the world has been ascribed to several causes: tectonic movement; solution; compaction of sedimentary materials due to static loads, vibrations, or increased density brought about by water-table lowering; and changes in reservoir pressures with loss of fluids. Noteworthy cases of appreciable subsidence attributed to the last cause have been reported in Long Beach Harbor, California (Gilluly and Grant, 1949); the San Joaquin Valley, California (Poland and Davis, 1956; Poland, 1961); the upper Gulf coastal region, Texas (Winslow and Wood, 1959; Gabrysch and Bonnet, 1975); the Savannah area, Georgia (Davis and others, 1963); and Las Vegas Valley, Nevada (Domenico and others, 1966). Other localities include Mexico City (Cuevas, 1936) and London (Wilson and Grace, 1942). The geologic requisites and qualifying conditions for the occurrence of subsidence are so well adapted to alluvial basins that it is likely far more occurrences of this phenomenon are taking place than have been reported. The chief reason for this is the lack of close control of benchmarks necessary to detect small changes in land-surface altitude.

Serious problems can be caused by land-surface subsidence. The normal upward force of skin friction acting on piles or well casings may be reversed, which will subject these elements to "downdrag." This may result in failure or protrusion above land surface. Gradients of canals may be reduced, or even reversed, and so affect the normal flow of water. Cracking of concrete or brick structures is common in subsiding areas. Subsidence in coastal areas results in the conversion of pastures to tidelands, where areas previously elevated become subject to flooding from hurricanes (Winslow and Wood, 1959). Kreitler (1977) estimated that Hurricane Carla, which struck the Houston–Galveston region in 1961, would have flooded 146 square miles of land adjacent to Galveston Bay if it had struck in 1976. This is 25 square miles more than the area flooded in 1961.

Many of the observations of the rate of water-level decline and the rate of subsidence exhibit a fair degree of linearity. This linearity has been observed by Carrillo (1948) in Mexico City and by the Tokyo Institute of Civil Engineering (1975). Figure 8.4 demonstrates a reasonable linear trend between water-level decline and subsidence

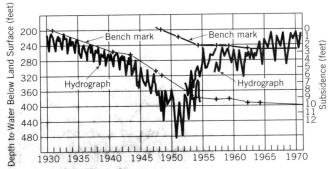

Figure 8.4 Subsidence and water-level decline in the San Joaquin Valley (from Poland and others, 1975).

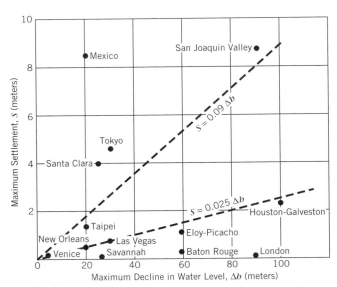

Figure 8.5 Relation between maximum settlement and maximum decline in water level.

in the San Joaquin Valley over the period 1930 to mid-1940s and then again to the late 1950s. From this diagram it is noted that subsidence actually stopped in response to the rise in water levels.

To investigate this linear trend further, consider Table 8.1 and Figure 8.5 constructed from it. It is important here to note the dates associated with the investigations reported in Table 8.1. As subsidence has continued beyond the period reported, it is not certain that the relationships demonstrated in Figure 8.5 still hold. At any rate, there appear to be two well-defined divisions, one where the subsidence–water-level decline ratio is larger than 0.09 (9 m of subsidence to 100 m of water-level decline) and the other where the ratio is equal to or less than 0.025 (2.5 m of subsidence to 100 m of water-level

decline). One possible reason for this may be related to a component of subsidence caused by tectonic factors that have not been filtered out of the calculations. In Las Vegas Valley, Nevada, for example, there is a component of subsidence due to the elastic yield of the bedrock in response to the load of Lake Mead (Figure 8.6). This regional deformation has affected 8000 square miles in southern Nevada and adjoining states. Yet another possi-

Table 8.1 **Summary of Cases of Subsidence**

Name of Place	Max. Settle. (m)	Area (sq. miles)	Head Decline (m)	Date of Investigation
United States				
San Joaquin Valley, CA	8.8	5212	90	1972
Santa Clara Valley, CA	4.0	251	25	1972
Houston–Galveston Area, TX	2.3	4710	100	1974
Eley–Picacho Area, AZ	1.1	—	30–60	1972
Las Vegas, NV	0.75	200	30	1972
Baton Rouge, LA	0.3	250	60	1970
New Orleans, LA	0.5	—	20	1968
Savannah, GA	0.1	19	27	1963
Japan				
Tokyo	4.6	77	30	1972
Nagoya	1.5	38	—	1976
Mexico				
Mexico City	8.5	58	20+	1964
Taiwan				
Taipei	1.35	Approx. 48	20	1969
United Kingdom				
London	0.1	—	90+	1942
Italy				
Venice	0.14	Approx. 3	5	1974

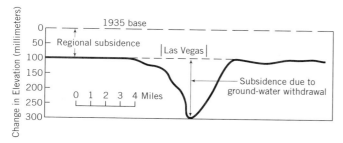

Figure 8.6 Subsidence profile, Las Vegas Valley, showing differential subsidence due to pumping superposed on regional subsidence (from Malmberg, 1960).

bility for the two divisions demonstrated in Figure 8.5 resides in the possibility of variations in the compressibility of the geologic units as well as in their thickness.

Mathematical Treatment of Land Subsidence

There are two related questions of general interest in the study of land subsidence. The first is concerned with the total amount of subsidence that might occur in a given hydrostratigraphic unit in response to a lowering of water levels. From what we have learned earlier, the amount can vary from point to point depending on the amount of water-level decline, the thickness of the more compressible units, and their compressibility. Thus, differential subsidence will be the general rule. The second question deals with the time rate of subsidence in response to the time rate of head change within the aquifer. This problem is not unlike the problem of disequilibrium compaction described in Chapter 2. At one extreme, the subsidence can occur concurrently with the head decline; at the other extreme, the rate of subsidence can lag considerably behind the rate of water-level decline so that the basin can be in a state of progressive subsidence for decades.

Vertical Compression We will focus on the first of these questions with an ideal depth–pressure diagram as given in Figure 8.7. Here an aquifer has an original head h that is lowered by the amount Δh, with water being removed from storage from the total system. Jacob (1940) was the first to attempt some quantification of this phenomenon when he postulated that the water so removed is derived from three sources: expansion of the confined water in the aquifer, compression of the aquifer, and compression of the adjacent and included clay beds. Jacob concluded that the third source is probably the chief one, a fact demonstrated by Poland in 1961. Jacob (1940) described the storativity of such systems as

$$S = \rho_w gm(\beta_p + n\beta_w + c\beta_p') \qquad (8.23)$$

where all terms should by now be familiar to the reader except for β_p', which is the compressibility for the clayey beds, and c, which is a dimensionless constant that de-

pends on the thickness, configuration, and distribution of the interbedded clay. For the assumption of Jacob (1940) that water is released from the clay beds instantaneously with no time lag, c is indeed a constant and will be equal to H/m where H is the thickness of the clayey beds and m is the thickness of the aquifers. Where this condition is not satisfied, c is no longer a constant but will approach a maximum value of H/m with time so that the storativity will increase with increasing time. These points will be cleared up later.

The first two components of storage in Eq. 8.23 are already familiar to the reader, at least in terms of what they mean with regard to the source of water in confined aquifers. Lohman (1961) was the first to quantify their meaning in terms of the elastic compression of aquifers. As the original head and the change in head in the aquifer of Figure 8.7 is invariant with respect to depth, the ratio of the change in thickness Δm to the original thickness m may be stated as

$$\frac{\Delta m}{m} = \beta_p \Delta \bar{\sigma} = \beta_p \Delta P \qquad (8.24)$$

where ΔP is the change in fluid pressure. The storativity for the elastic response may be written as

$$\frac{S}{\rho_w g} = \beta_p m + n\beta_w m \qquad (8.25)$$

Combining Eqs. 8.24 and 8.25 and solving for Δm gives

$$\Delta m = \Delta P \left(\frac{S}{\rho_w g} - \beta_w nm \right) \qquad (8.26)$$

This equation (or Eq. 8.24) gives the amount of vertical shortening of an aquifer of thickness m in response to

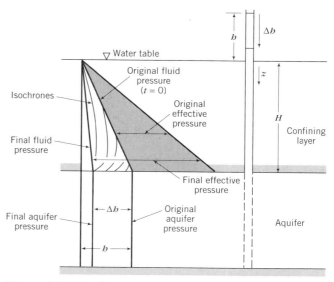

Figure 8.7 Depth–pressure diagram of the development of isochrones in a confining layer in response to lowered aquifer pressure.

a pressure change ΔP. As aquifers are elastic, or nearly so, this elastic component of compression is recoverable if the pressure is allowed to recover to its original (pre-pumping) value. Further, this compression takes place instantaneously with water-level decline.

Examination of the clay layer in Figure 8.7 shows that the original head distribution varies with depth z and is actually equal to the head within the aquifer at the contact between the two units. A rapid lowering of head in the aquifer is not accompanied by an immediate lowering of the head in the clay layer because of different hydraulic diffusivities. Thus, the head in the clay layer is converted to an "excess" head, at least with respect to the diminished head at its lower boundary. In response to this excess head, the clay layer will drain, with the head distribution at any time being shown by the so-called isochrones, which mean "same time." Eventually, the heads in the aquifer and in the clay layer are once more matched at the boundary between the two, and the drainage process comes to a halt. During this period of adjustment, the effective stress in the clay layer has increased in response to the drainage, resulting in a commensurate volume reduction in the layer itself. For the one-dimensional case we are considering where the solids and the fluids in the clay layer are assumed to be incompressible, the volume of fluid removed is equal to the volume of subsidence.

As the head change in the clay layer ranges from zero at the top to Δh at the bottom (Figure 8.7), the volume of water produced from each element of the clay layer will be different and can be expressed as

$$dq = S_s' h(z)\, dz \qquad (8.27)$$

where S_s' is the specific storage of the clay layer (ignoring fluid expansion) and $h(z)$ is the actual head change at some point z. The total volume of water extruded from a column of height H with a unit basal area is then

$$q = S_s' \int_0^H h(z)\, dz \qquad (8.28)$$

The solution of this equation is a general expression for the vertical shortening of a column of confining layer when steady flow conditions are reestablished within it. It is expressed as a volume of water passing through a unit surface area and has the units of length. As $h(z)$ varies linearly across the layer, $h(z) = \Delta h(z/H)$, where z varies from zero at the top to H at the bottom (Figure 8.7). Substituting this expression into Eq. 8.28 and integrating gives

$$q = S_s' \frac{\Delta h H}{2} \qquad (8.29)$$

Here we note that the quantity $\Delta h H/2$ represents the area of a triangle on the depth–pressure diagram of Figure 8.7 and is termed the "effective pressure area."

The result just given may be generalized for head changes in aquifers both above and below a compressible confining layer (Figure 8.8). The equation describing the head change in the confining layer is given as

$$h(z) = \Delta h_1 \left(\frac{z}{H}\right) + \Delta h_2 \left(\frac{H - z}{H}\right) \qquad (8.30)$$

Substitution of Eq. 8.30 into Eq. 8.28 and performing the integration gives

$$q' = S_s' H \left(\frac{\Delta h_1 + \Delta h_2}{2}\right) \qquad (8.31)$$

Examination of Eq. 8.31 and Figure 8.8 shows that when the transient flow process is terminated, the vertical shortening of the confining layer equals the product of the specific storage and the final area described by the increase in effective stress on a depth–pressure diagram. From Figure 8.8 and Eq. 8.31, this area is again a triangle for $\Delta h_2 = 0$, or a rectangle when $\Delta h_1 = \Delta h_2$, or a trapezoid for $\Delta h_1 \neq \Delta h_2$ (one-half the sum of the bases, $\Delta h_1 + \Delta h_2$, times the height H).

The concepts described pertain to the permanent compression of low-permeability layers in confined basins, the volume of water being a one-time reserve associated with permanent compression. This process does not affect the aquifer storage capacity in that the pore volume

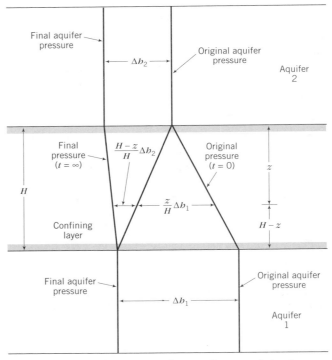

Figure 8.8 Depth–pressure diagram for a confining layer in response to lowered pressure in adjacent aquifers (from Domenico and Mifflin, Water Resources Res., v. 4, p. 563–576, 1965. Copyright by Amer. Geophys. Union).

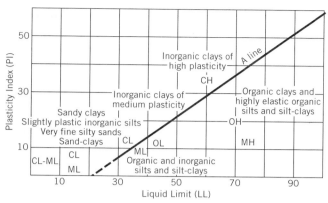

Figure 8.9 Plasticity chart for soil classification.

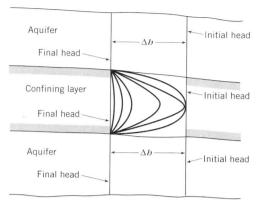

Figure 8.10 Schematic illustration of isochrone development in confining layer in response to instantaneous lowering of head at its upper and lower boundaries.

reduction takes place in the confining units. According to Poland (1961), the process of subsidence makes available a volume of water that would otherwise not be recoverable. Unlike elastic compressions, the lost pore volume is nonrecoverable.

Figure 8.9 in combination with Table 8.2 can be used to obtain approximate values of β_p, and the hydraulic diffusivity for various soil groups subject to consolidation. Figure 8.9 demonstrates the relationship between the Unified Soil Classification System and the plasticity chart. These same soil groups are represented in Table 8.2, which gives the results of over 1500 laboratory determinations by the Bureau of Reclamation (1960). The values for relative compression $\Delta H/H_0$ are from drained tests on samples that were laterally confined; that is, the results indicate vertical compression.

The Time Rate of Subsidence   The isochrone development on Figure 8.7 represents the degree of consolidation at various depths as a function of time. The terminal isochrone represents full consolidation for the given head decline in the aquifer. Hence, the rapidity of the drainage from the clay layer controls the rate of subsidence.

Figure 8.10 illustrates these points for a confining layer separating two aquifers. The schematic isochrone development for the confining layer is produced for a rapid (instantaneous) lowering of head in the dual aquifer system, here for simplicity taken to be Δh. At time equal to zero, the head distribution through the three-layered system is assumed to be invariant with depth. When the pumping starts, the heads in the aquifers are immediately

Table 8.2   **Laboratory Determination of Physical Properties of Sediments**

Soil Classification	Void Ratio, e	Hydraulic Conductivity (cm/s)	$\dfrac{\Delta H}{H_0}$ At Effective Stress of 2880 lb/ft²	$\dfrac{\Delta H}{H_0}$ At Effective Stress of 7200 lb/ft²
GW	—	$2.6 \times 10^{-2} \pm 1.25 \times 10^{-2}$	<0.014	—
GP	—	$6.2 \times 10^{-2} \pm 3.3 \times 10^{-2}$	<0.008	—
GM	—	—	<0.012	<0.03
GC	—	—	<0.012	<0.024
SW	0.37	—	0.014 ±	—
SP	0.5 ± 0.03	$>4 \times 10^{-6}$	0.008 ± 0.003	—
SM	0.48 ± 0.02	$7.2 \times 10^{-6} \pm 4.6 \times 10^{-6}$	0.012 ± 0.001	0.03 ± 0.004
SM-SC	0.41 ± 0.02	$7.7 \times 10^{-7} \pm 5.8 \times 10^{-7}$	0.014 ± 0.003	0.029 ± 0.01
SC	0.48 ± 0.01	$2.9 \times 10^{-7} \pm 1.9 \times 10^{-7}$	0.012 ± 0.002	0.024 ± 0.005
ML	0.63 ± 0.02	$5.7 \times 10^{-7} \pm 2.2 \times 10^{-7}$	0.015 ± 0.002	0.026 ± 0.003
ML-CL	0.54 ± 0.03	$1.26 \times 10^{-7} \pm 6.7 \times 10^{-8}$	0.01 ± 0.002	0.022 ± 0.0
CL	0.56 ± 0.01	$7.7 \times 10^{-8} \pm 2.9 \times 10^{-8}$	0.014 ± 0.002	0.026 ± 0.004
OL	—	—	—	—
MH	1.15 ± 0.12	$1.5 \times 10^{-7} \pm 9.6 \times 10^{-8}$	0.02 ± 0.012	0.038 ± 0.008
CH	0.8 ± 0.04	$4.8 \times 10^{-8} \pm 4.8 \times 10^{-8}$	0.026 ± 0.013	0.039 ± 0.015
OH	—	—	—	—

From Bureau of Reclamation (1960)

lowered an identical amount, converting the head in the confining layer to an "excess head" with respect to its boundaries. This starts the drainage process, both upward and downward, with the head (or pressure) decline in the confining layer described by the diffusion equation

$$\frac{\partial^2 P}{\partial z^2} = \frac{S_s'}{K_v}\frac{\partial P}{\partial t} \tag{8.32}$$

subject to the boundary and initial conditions

$$P(z, 0) = P_i$$

$$P(H, t) = 0$$

$$P(-H, t) = 0$$

The initial condition stipulates a constant value of pressure P_i with depth. The two boundary conditions correspond to decreased pressure (assumed 0, with respect to P_i) at both the top H and bottom $-H$ of the confining layer.

The solution to Eq. 8.32 for the conditions stated is well known in consolidation theory (Terzaghi and Peck, 1948) and in heat-flow theory (Carslaw and Jaeger, 1959). This solution provides for the pressure decline at different points in the confining layer, which for any time takes on maximum values near the two drainage boundaries and minimum values in the center. This is too much detail for this problem as we are here only interested in the average pressure decline (or consolidation) of the clay layer. This average is demonstrated in Figure 8.11. The ordinate represents the degree of consolidation, with values less than one indicating the presence of residual pressures. The abscissa is the familiar ratio of observation time and the time constant of the unit. As T^* gets small, corresponding to a small value for H and S_s in combination with a large value for K_v, the steady state is rapidly approached; that is, the unit approaches full consolidation so that the drainage process approaches termination shortly after the transient started. The quantity H in Figure 8.11 represents the distance to a drainage face and is taken as the thickness of the unit for single drainage (that is, the triangular effective pressure area of Figure 8.10) or $H/2$ for the double-drainage case given in Figure 8.10. Thus we notice that the time lag between the lowering of aquifer pressure and the consolidation of the interbedded clay layers contributing to subsidence is enhanced for thick clay layers of low permeability and high compressibility (storativity). Conversely, the time lag diminishes for thin units of moderate permeability and compressibility.

Last, we may now speculate that Jacob's c in Eq. 8.23 is of the form

$$c = \frac{t}{T^*}\frac{H}{m} \tag{8.33}$$

so that the storage component associated with the clay layer actually increases with time, reaching its maximum value when $t/T^* \cong 1$.

Example 8.1

Consider a confining layer between two aquifers. The specific storage of the confining layer is on the order of 1×10^{-3} ft^{-1}, and its thickness is 40 ft. The heads in the aquifers are lowered about 50 ft. Maximum vertical shortening of the confining layer is calculated

$$q = S_s'H\left(\frac{\Delta h_1 + \Delta h_2}{2}\right) = 1 \times 10^{-3} \times 40 \times \frac{100}{2} = 2 \text{ ft}$$

The value given (2 ft) is taken as the maximum subsidence in response to the 50-ft decline in water levels. The time to complete 50% of this subsidence is determined from Figure 8.11. Here, however, some additional information is required, namely, the vertical conductivity. Assume this value to be 1×10^{-3} ft/day. A consolidation of 50% (1 ft) is associated with a value of $t/T^* = 0.196$. Thus

$$\frac{t}{T^*} = \frac{t}{\dfrac{(H/2)^2 S_s'}{K_v}} = 0.196$$

$$t = \frac{0.196\,(40/2)^2(1 \times 10^{-3})}{1 \times 10^{-3}} = 78 \text{ days}$$

For 95% consolidation

$$t = \frac{1\,(40/2)^2\,(1 \times 10^{-3})}{1 \times 10^{-3}} = 400 \text{ days}$$

For a water-level decline of 5 ft/yr in each of the aquifers

$$q = S_s'H\left(\frac{\Delta h_1 + \Delta h_2}{2}\right) = 1 \times 10^{-3} \times 40 \times \frac{10}{2} = 0.2 \text{ ft}$$

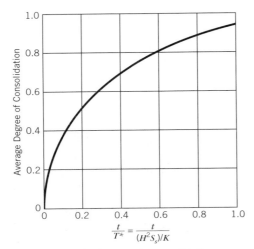

Figure 8.11 Average degree of consolidation versus the ratio of time to the basin time constant.

This represents an approximate measure of the subsidence to be expected with each year's pumping. Note now that it still requires 78 days to achieve 50% of the annual subsidence (0.1 ft) and 400 days to achieve 95% of the subsidence. If benchmarks are releveled every year, the results would give the appearance that there is no appreciable time lag between the lowering of the head and the attainment of full consolidation.

Simulation of Subsidence Frequently, subsidence in an area is sufficiently detrimental so as to warrant some detailed model study for prediction purposes. The first such study was conducted by Gambolati and Freeze (1973), who utilized a finite element model to obtain predicted drawdowns, and a finite difference model to determine the time rate of subsidence. The drawdowns obtained from the flow model were used as time-varying boundary conditions to solve a set of one-dimensional consolidation models. The model was applied to the subsidence problem in Venice, Italy, with the results reported by Gambolati and others (1974a, 1974b). Helm (1975) has also developed simulation models for application to land subsidence in California. Although the data demands for such models are large, the measured subsidence provides a means by which the models can be calibrated and improved over time. Such models are becoming more common where subsidence represents real as opposed to nuisance problems, for example, the Houston–Galveston area.

Three-Dimensional Consolidation

Elastic Properties in Deformational Problems

In our treatment of the consolidation equation, two assumptions stand out that may now be relaxed somewhat. First, it was assumed that the stress increment producing the abnormal fluid pressure was vertical only. Second, the individual solid components making up the rock matrix were incompressible. This last assumption means that the bulk volume changes are exactly equal to the pore volume changes. Or, stated another way, the volume of water squeezed out of a rock is equal exactly to the pore volume change of the rock. We have exploited this assumption, not only in our development of the consolidation equation, but in the development of the concept of specific storage.

The most complete model from which to obtain the important material properties is one that is fully compressible, that is, one in which all the components are compressible, including the solid grains β_s, the fluids β_w, the pores β_p, and the bulk volume β_b. When grain compressibilities are introduced into the material properties, the water volume squeezed out of a rock is invariably less than the bulk volume change. To incorporate such compressibilities in the material properties, it is necessary

to introduce a proportionality constant between the bulk and pore volume change. This may be accomplished with the following expression

$$\beta_b - \beta_s = \beta_p = \xi \beta_b \qquad (8.34)$$

where β_b is the bulk compressibility, β_s is the grain compressibility, and the difference $\beta_b - \beta_s$ is taken as the pore compressibility. In this expression, ξ is a constant of proportionality and may be defined as the ratio of the water volume squeezed out of a rock to the total volume change of the rock if the deformation takes place at constant fluid pressure, that is, the rock is free to drain. Thus, if ξ equals one, the bulk volume changes are equal to the pore volume changes, an assumption that has been inherent throughout our discussions of deformable media. From Eq. 8.34, the constant of proportionality may be described as

$$\xi = 1 - \left(\frac{K_b}{K_s}\right) = 1 - \left(\frac{\beta_s}{\beta_b}\right) \qquad (8.35)$$

where K_b is the bulk modulus of compression of the rock body and K_s is the bulk modulus of compression of the polycrystalline grains making up the porous rock. In this form, we note that if β_s is small with respect to β_b, ξ is approximately unity, and the volume of water removed from the rock is equal to the total volume change; that is, the incompressible grain case is recovered.

With the introduction of solid compressibilities, the elastic coefficients discussed in the previous section change considerably. These coefficients are given in Table 8.3. For the three-dimensional results, the stress term σ and the strain term ε refer to the mean confining stress and the volumetric dilation; that is,

$$\sigma_m = \frac{\sigma_{xx} + \sigma_{yy} + \sigma_{zz}}{3} \qquad (8.36a)$$

$$\frac{\Delta V}{V} = \varepsilon_d = \varepsilon_{xx} + \varepsilon_{yy} + \varepsilon_{zz} \qquad (8.36b)$$

where σ_m is the mean stress, V is volume, and ε_d is the volume strain or dilation. In addition, the following relationship now holds

$$\frac{1}{Q} = \frac{1}{R} - \frac{\xi}{H} \qquad (8.37)$$

which reduces to Eq. 8.17 for ξ equal to one. Indeed, all expressions for the three-dimensional coefficients given in Table 8.3 reduce to the one-dimensional coefficients already cited for the case where $\xi = 1$. This implies, in addition, that the testing for the one-dimensional condition is conducted under constrained conditions where the concern is with vertical deformations only.

Biot and Willis (1957) discuss the measurement techniques for the various coefficients described in Table 8.3, whereas a limited set of values have been reported

Table 8.3 Definitions of the Important Elastic Coefficients and Their Relation to the Coefficients Defined by Biot (1941)

$$\beta = \frac{1}{K} = -\left(\frac{1}{V}\right)\left(\frac{\partial V}{\partial \sigma}\right)_P$$

$$\beta_s = \frac{1}{K_s} = \left[\frac{(1-n)}{\rho_s}\right]\left(\frac{\partial \rho_s}{\partial \sigma}\right)_P$$

$$\beta_{s'} = \quad = -\left[\frac{(1-n)}{n\rho_s}\right]\left(\frac{\partial \rho_s}{\partial P}\right)_\sigma$$

$$\beta_n = \beta - \beta_s - n\beta_{s'} = \left(\frac{1}{V}\right)\left(\frac{\partial V_n}{\partial P}\right)_\sigma$$

$$\xi = \frac{K}{H} = \frac{(\beta - \beta_s)}{\beta}$$

$$\frac{1}{K_u} = \frac{1}{K} - \frac{R}{H^2} = -\left(\frac{1}{V}\right)\left(\frac{\partial V}{\partial \sigma}\right)_M$$

$$\frac{1}{H} = \beta - \beta_s = \left(\frac{1}{m_f}\right)\left(\frac{\partial m_f}{\partial \sigma}\right)_P$$

$$\frac{1}{R} = \beta_n + n\beta_f = \left(\frac{1}{m_f}\right)\left(\frac{\partial m_f}{\partial P}\right)_\sigma$$

$$\frac{1}{Q} = \frac{1}{R} - \frac{K}{H^2} = \left(\frac{1}{m_f}\right)\left(\frac{\partial m_f}{\partial P}\right)_\varepsilon$$

$$\frac{R}{H} = \quad = \left(\frac{\partial P}{\partial \sigma}\right)_M$$

$$\xi Q = \quad = \left(\frac{\partial P}{\partial \varepsilon}\right)_M$$

$$\frac{\xi Q}{(R/H)} = \quad = \left(\frac{\partial \sigma}{\partial \varepsilon}\right)_M$$

From Palciauskas and Domenico (1989).
Water Resources Res., v. 25, p. 203–213. Copyright by Amer. Geophys. Union.

by Palciauskas and Domenico (1982), Domenico (1983), Zimmerman and others (1986), and Green and Wang (1986). Table 8.4 presents an overview of these values. Laboratory data for both the Kayenta Sandstone and the Hanford basalts are reasonably good, whereas the clay and the limestone tabulations are "generic," that is, crude calculations based on tables of physical constants. The rocks are arranged from left to right in order of what is expected to be decreasing compressibility. Hence, ξ as a measure of the water volume squeezed out to the volume change of the material decreases from unity to 0.23 as rock rigidity increases. For clays, there is virtually a one-to-one relationship between volume changes and associated water volumes released. The coefficients $1/H$, $1/R$, and $1/Q$ are compressibilities of sorts and, in general, decrease in the direction of rock rigidity. Thus, $1/H$ as a measure of the pore compressibility decreases consistently in the direction of rock rigidity, as does $1/R$ as a

measure of the changes in fluid mass in response to changes in fluid pressure at constant stress. Except for a slight inconsistency with the generic limestone, this same statement holds for $1/Q$ as a measure in the change in fluid mass with respect to fluid pressure at constant strain. Also shown in the table are combinations of parameters, R/H and ξQ. The quantity R/H is already known as the change in fluid pressure with stress at constant fluid mass. As noted in the table, the more rigid the rock, the greater the proportion of the stress that is carried by the grain structure rather than the pore fluid. The combined parameter ξQ is defined as the change in fluid pressure with respect to strain at constant fluid mass. Here, this fluid pressure appears highest for rocks in the intermediate range and decreases toward both the soft and rigid-end members. The symbol K_u is used for an undrained modulus.

The coefficients described are for elastic, or reversible,

Table 8.4 **Typical Values for the Elastic Coefficients**

Parameter	Units	Clay	Mudstone	Sandstone	Limestone	Basalt
ξ		1	0.95	0.76	0.69	0.23
$\frac{1}{H}$	$(\text{Kbar})^{-1}$	1.6	0.045	0.0079	0.0024	0.00052
$\frac{1}{R}$	$(\text{Kbar})^{-1}$	1.62	0.054	0.012	0.0095	0.0044
$\frac{1}{Q}$	$(\text{Kbar})^{-1}$	0.02	0.011	0.0059	0.0078	0.0043
$\frac{R}{H}$		0.99	0.83	0.67	0.25	0.12
ξQ	(Kbar)	50	83	128	88	54
$\frac{1}{K}$	$(\text{Kbar})^{-1}$	1.6	0.047	0.011	0.003	0.0022
$\frac{1}{K_u}$	$(\text{Kbar})^{-1}$	0.02	0.01	0.0052	0.0024	0.0022

From Palciauskas and Domenico, 1989.
Water Resources Res., v. 25, p. 203–213. Copyright by Amer. Geophys. Union.

deformations. A reversible compression is one that disappears completely upon release of the stress that caused it. With geologic materials, many of the substances we treat within an elastic or reversible framework are not elastic at all. The point is that for some problems, the materials need not be elastic in terms of reversibility, but the strains have to be proportional to the stresses. That is, Hooke's law must hold so that a constant ratio exists between the stress and the strains. The degree of variation in this ratio dictates how far the real behavior departs from the ideal one, with clays and other plastic materials generally falling into this nonideal category. Reversibility also implies that any compressions occurring under load with the pore fluid free to drain are completely recoverable by injecting fluid back into the pores. Indeed, the "coefficient" describing the volume changes associated with both compression and expansion should, in theory, be the same. Again, this reversibility does not coincide with our observations for many materials.

Flow Equations for Deformable Media

The flow equations in a completely deformable media are of an identical structure of those presented for the one-dimensional consolidation equation, with the major differences residing in the material properties, or coefficients. These equations may be expressed as (Palciauskas and Domenico, 1989)

$$\frac{\partial P_{ex}}{\partial t} = \nabla \cdot \left[\frac{K}{\rho_w g (1/R)} \nabla P_{ex} \right] + \frac{R}{H} \frac{\partial \sigma}{\partial t} \quad (8.38)$$

and

$$\frac{\partial P_{ex}}{\partial t} = \nabla \cdot \left[\frac{K}{\rho_w g (1/Q)} \nabla P_{ex} \right] - (\xi Q) \frac{\partial \varepsilon}{\partial t} \quad (8.39)$$

where the coefficients R, Q, H, and ξ are described in Table 8.3. These equations may be compared with Eqs. 8.21 and 8.22.

As with the one-dimensional consolidation equation, the equations just given can be used to describe both the drained and undrained response of water levels to natural loads. Expressions for the undrained response to various loads were developed independently by van der Kamp and Gale (1983) and Domenico (1983). For the case of compressible solids, the tidal efficiency (pore pressure coefficient) and the barometric efficiency may be described as

$$\text{T.E.} = \left(\frac{R}{H} \right) = \frac{1}{\xi} \left(1 - \frac{R}{Q} \right) = \frac{1}{\xi} \left(1 - \frac{K_b}{K_u} \right) \quad (8.40)$$

and

$$\text{B.E.} = 1 - \frac{1}{\xi} \left(1 - \frac{R}{Q} \right) = 1 - \frac{1}{\xi} \left(1 - \frac{K_b}{K_u} \right) \quad (8.41)$$

respectively. Thus, these efficiencies are nothing more than partition coefficients that account for the distribution of incremental stress between the pore fluid and the solids under undrained conditions. For vertical strains only, the moduli K_b and K_u are described for the constrained condition. Similarly, Bredehoeft's (1967) fluid-pressure response to earth tides stated as $\partial P / \partial \varepsilon$ at constant fluid mass is given simply as ξQ, where Q is defined in Table 8.3.

8.2 Abnormal Fluid Pressures in Active Depositional Environments

Origin and Distribution

Sediment deposition and compaction occur concurrently in depositional environments, with a resultant porosity loss. Prior to porosity reduction, the added sediment load is carried at least in part by the water contained in the sediments, causing the fluid pressure to rise above its normal value. This has been schematically demonstrated with the spring analogy in Chapter 4 where, in the absence of fluid expulsion, the spring (rock matrix) cannot compress, and the added load is borne by the water. As demonstrated further with the example of water-level changes in response to passing trains, there is a diffusion of pore water to areas of lower pressure so that the sediment load is ultimately transferred to the matrix of skeletal grains. From the perspective of abnormally pressured environments, the main concern is with the length of time required for this stress transfer to take place, for this is the time period over which anomalously high fluid pressures may persist.

Bogomolov and others (1978) and Kissen (1978) have established the framework for categorizing abnormal pressure zones in depositionally active basins (Figure 8.12). The meteoric regime shown in the figure is the familiar topographic drive system where meteoric water enters the formations in the higher portions of the basins with discharge taking place at sea level. Hence, the shoreline is an active area of discharge, and this discharge area can recede or transgress depending on shifts in the shoreline. The compaction regime is characterized by upward migration of pore waters, as suggested in 1899 by King. The fluids so affected may have been buried with the sediment, and the abnormal fluid pressures are due in part to the added sediment load in parts of the basin undergoing active deposition.

To accumulate a thick sedimentary pile such as shown in Figure 8.12 the basement rock must be in some state of subsidence. Whatever the driving mechanism, such subsidence assures that the sediments and their con-

tained fluids become subjected to a thermal field. This is referred to as the thermobaric regime (Figure 8.12), where the fluid pressures may be caused by the thermal expansion of water in a low-permeability environment or by trapped waters of mineral dehydration. Phase transformations such as gypsum to anhydrite or smectite to illite are generally temperature dependent and will produce free water that can take on the pressure of the overburden provided the permeability is sufficiently low so as to inhibit movement to areas of low pressure. In addition, the sedimentary pile may undergo horizontal compression, as might occur because of uplift adjacent to the deposits undergoing burial. Such compression or tectonic strain is yet another pressure-producing mechanism.

The environments just discussed will persist as long as the basin continues to take on added sediment. Once the depositional period ends and tectonic forces contribute to uplift, the flow system will evolve eventually to a topographic drive system. Thus, in the evolution of hydrologic basins, it is useful to think of two end members (Coustau and others, 1975): the subsidence-induced abnormal pressure environment, as might be expected in young basins undergoing rapid deposition, and the more mature topographically driven systems that eventually evolve from them after uplift. Figure 8.13 represents these two end members. As noted in the abnormal pressure environment (Figure 8.13*a*), the lower water zone contains original formation waters, whereas the upper zone is continually invaded by downward-moving meteoric water. Following uplift and erosion and the evolution of a topographic drive system, there is one distinct hydrologic system that is characterized by recharge at the outcrop areas, cross-formational discharge, and a definite transition from meteoric water to highly saline water or brines with distance from the outcrop areas.

The relationships between normal fluid pressure and geostatic pressure are demonstrated in Figure 8.14. At any given depth z, the difference between the geostatic pressure (vertical stress) and the normal fluid pressure (neutral stress) has been defined as the effective stress in accordance with the effective stress law. In this figure,

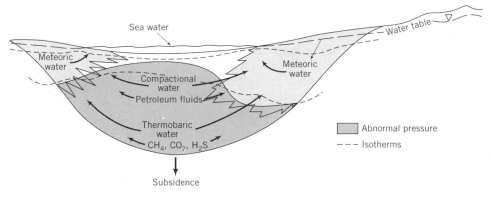

Figure 8.12 Conceptual diagram illustrating various hydrologic regimes in an active depositional environment (from Galloway and Hobday, 1983. Copyright © 1983 Springer-Verlag, New York. Reprinted with permission).

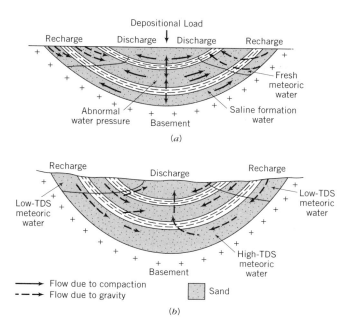

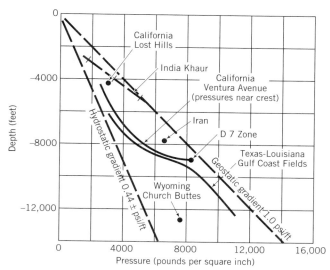

Figure 8.15 Relation of fluid pressure and depth in some excess-pressure oil pools (from Watts, 1948). Reprinted by permission of the Amer. Inst. of Mining, Engrs, Petroleum Div.

Figure 8.13 Hydrology of (a) compacting and (b) mature basins (modified from Coustau and others, 1975).

the fluid pressure P at any given depth is 0.43 times the vertical (geostatic) stress. In some geologic environments, fluid pressures have been noted to be as high as 90% of the geostatic pressure. In such cases, the effective stress is reduced considerably, which is generally reflected by higher than normal porosities and lower than normal bulk densities of the water-saturated rock. Figure 8.15 shows the relationship between pressure and depth in some abnormally pressured environments reported by Watts (1948). At relatively shallow depths, many of these areas are characterized by normal or near-normal fluid pressures. With increasing depth, fluid pressures are seen to be in excess of normal, in some cases approaching the total overburden pressure.

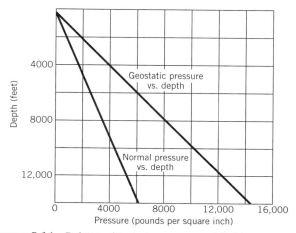

Figure 8.14 Relationship between normal fluid pressure and geostatic pressure.

According to Deju (1973), the occurrence of pressures in excess of normal is worldwide. In Europe, abnormal pressure has been noted in the Aquitaine Basin, France, in the Alps, Italy, offshore Norway (0.8 psi/ft, 1.8×10^4 Pa/m), Holland, northwest Germany (0.8 psi/ft, 1.8×10^4 Pa/m), the Carpathians (0.96 psi/ft, 2.17×10^4 Pa/m), and in the Soviet Union (0.8 psi/ft, 1.8×10^4 Pa/m). The sediment types include dolomite, evaporites, carbonates, marine shales, and sandstone. In Eastern Canada, an offshore zone was noted at about 3650 m with a pressure gradient of 0.6 psi/ft (1.36×10^4 Pa/m). The occurrences in the United States are well known and include California (both off- and onshore), the Green River Basin in Wyoming, the Uinta Basin in Utah (0.8 psi/ft, 1.8×10^4 Pa/m), and several locations in Texas and Louisiana. Some detailed studies are provided by Watts (1948) for the Ventura oil field in California and Dickinson (1953) for the Louisiana Gulf Coast. Other occurrences have been reported in and around Australia, Japan, West and East Africa, the Arabian Peninsula, South America, Southeast Asia, and the Indian subcontinent.

On the basis of this worldwide survey, Deju (1973) presented the following conclusions:

1. Two mechanisms appear to be responsible for most of the abnormal pressures observed: tectonic stress and rapid deposition of massive shales.

2. High fluid pressures of tectonic origin are usually associated with the major mountain systems of the world (Pyrenees, Alps, Himalayas, Andes).

3. High fluid pressures are often associated with mud volcanoes, diapirs, piercement folds, and dikes.

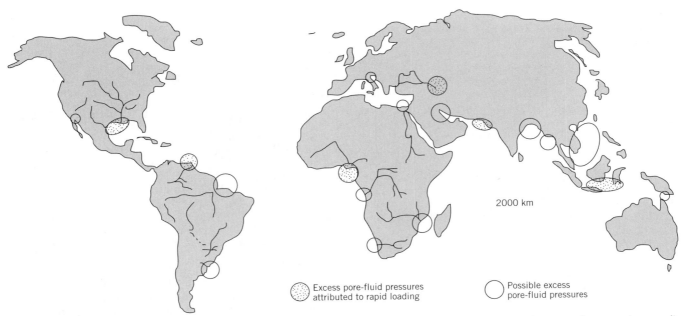

Excess pore-fluid pressures attributed to rapid loading

Possible excess pore-fluid pressures

Figure 8.16 Geographic areas where geopressures have been attributed to continuous loading and compaction, or where sediment deposition rates are such that geopressures would be predicted (from Sharp and Domenico, 1976, Geol. Soc. Amer. Publication, from Geol. Soc. Amer. Bull., v. 87, p. 390–400).

4. Rapidly deposited bentonitic shales and thick evaporite beds seem to form the best seal for fluid in the subsurface.

5. Most high fluid pressure occurrences are restricted to Tertiary sediments, or younger. Older sediments have apparently had sufficient time to dissipate the pressure.

Modern depositional environments associated with abnormal pressures (or the potential for abnormal pressures) are shown in Figure 8.16. It is obvious that such environments are rather common and were likely equally widespread throughout geologic time.

Mathematical Formulation of the Problem

From a physiomathematical viewpoint, the study of abnormal pressure development is a study of competing rates of pressure production and dissipation. The complexity of the problem is well illustrated in Figure 8.17. Gravitational loading and tectonic forces act to compress the pore fluid by compressing or otherwise reducing the pore itself, thus leading to the generation of abnormal fluid pressures. With continued burial in a thermal field, phase transformations produce free pore water that can take on the pressure of the overburden, and this magnitude of pressure can be continually regenerated in the "window" that defines the critical temperature range over which the transformation takes place. The pressure produced by either of these mechanisms can be aug-

mented by the thermal expansion of the fluid, which is most effective in low-permeability sediments under high confining pressures. With water flow to areas of lower pressure, some redistribution of the heat occurs by convective transport. Such convective transport can be of importance in the thermal evolution of the sedimentary basin. Given that the rate of pressure dissipation by fluid flow is slower than the rate of pressure generation, some sort of fracture is likely. This is demonstrated in Figure 8.17 with the coupling of pore pressure to the stress–strain characteristics of the rock. If the inelastic behavior is in the form of an extensive microfracture zone, the enhanced pore volume can accommodate the volumetric expansion of the heated water, providing pressure relief.

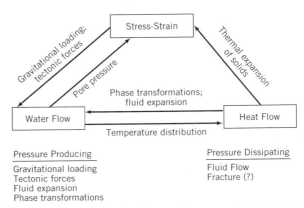

Figure 8.17 Coupling in abnormal pressure development.

If the microfracture zone coalesces to form macrofractures, the permeability may be enhanced and the pressures dissipate more readily. Thus, fracture generation may serve as a reserve mode of pressure dissipation that comes into play when the rate of pore water diffusion is inadequately slow compared to the rate of pressure generation.

The complexity and coupling of the processes described in Figure 8.17 are such that a complete physical description is not possible. First, the physics of many of the processes described are not sufficiently well known from a quantitative perspective. For example, if one wishes to describe the pressure production associated with phase transformations, the rate of transformation from layered or bound water at the low-temperature phase to free water at a higher-temperature phase must be known. If one wishes to examine fracturing as a pressure dissipative mechanism, some threshold pressure must first be identified and incorporated in such a way that describes the rate at which the fracture dissipates pressure. Given that these mechanisms and rates can be described, they can in general be incorporated into a flow equation of the form of the diffusion equation. However, at this stage of our knowledge, we must focus on these pressure-producing and -dissipating processes that can be well described.

For stress-related problems, the appropriate equation is (Palciauskas and Domenico, 1989)

$$\frac{\partial P_{ex}}{\partial t} = \nabla \cdot \left[\frac{K}{\rho_w g(1/R)} \nabla P_{ex} \right] + \frac{R}{H} \frac{\partial \sigma}{\partial t} + (\alpha_m R) \frac{\partial T}{\partial t}$$

(8.42)

where P_{ex} is the abnormal or excess fluid pressure, σ is the mean confining stress, and R and H are Biot's (1941) coefficients introduced in Table 8.3. A one-dimensional isothermal version of this equation was developed earlier. This three-dimensional version includes the fluid pressure changes due to the thermal expansion of water, where T is temperature and α_m is the change in fluid mass content at constant stress and pressure when the temperature varies.

When bulk volume strain ε is taken as the pressure-producing mechanism, the appropriate expression becomes

$$\frac{\partial P_{ex}}{\partial t} = \nabla \cdot \left[\frac{K}{\rho_w g(1/Q)} \nabla P_{ex} \right] - (\xi Q) \frac{\partial \varepsilon}{\partial t} + (\xi \alpha_b + \alpha_m) Q \frac{\partial T}{\partial t}$$

(8.43)

where Q and ξ are Biot's (1941) coefficients described in Table 8.3. The coefficients associated with the thermal effect include α_b, which is the thermal expansion of the rock body at constant stress and pressure. The combined term $\alpha_m R$ represents the change in pressure with temperature at constant mass and constant strain.

Isothermal Basin Loading and Tectonic Strain

In virtually all problems associated with basin loading, it is generally assumed that one of the principal stresses is oriented in a vertical direction, commonly expressed as

$$\sigma_v = \rho_b g z$$

(8.44)

where the subscript v designates that the stress is vertical, ρ_b is the density of the water-saturated rock, where the saturation is often assumed to persist from land surface downward, g is the acceleration of gravity, and z is depth. The product $\rho_b g$ is taken as the unit weight of the overburden, where the density can vary with lithology changes. The stress σ_v is the pressure exerted by the total weight of the overburden (solids plus water) and is commonly referred to as the geostatic pressure. The overburden pressure may be a major σ_{11} or a minor σ_{33} principal stress, depending on whether the rocks are in a state of compression or extension. The geostatic pressure gradient σ_v/z is normally taken as 1 psi/ft (2.26×10^4 Pa/m).

There are a few other designations for the state of stress in the Earth's crust, many of which are useful in the formulation of theoretical models. One assumption is that the stresses at depth are hydrostatic, that is, the principal components σ_{11}, σ_{22}, σ_{33} are all equal. This implies that principal stress differences do not increase with depth, which is generally not realistic from the field perspective. Such a state of stress may exist for very soft rocks such as shales or salts where plastic or viscoelastic deformation permits a partial or complete equalization of the principal stresses.

Yet another assumption pertains to a laterally constrained situation where horizontal displacements are not permitted. In the laterally constrained case, the effective horizontal stress is determined from the relationship

$$\bar{\sigma}_b = \bar{\sigma}_v \frac{\nu}{1 - \nu}$$

(8.45)

where ν is Poisson's ratio. For most rocks, Poisson's ratio ranges from 0.25 to 0.33. In accordance with Eq. 8.45, the ratio of horizontal to vertical stress should lie between 0.33 to 0.5. This is contrary to most in situ measurements, which lie between 0.5 and nearly one, with the lower-range characteristic of hard rocks and the upper-range more appropriate for soft rocks such as shale and salt. Thus the horizontally constrained model may not always describe the field situation, possibly because of plastic or viscoelastic deformation of soft rocks, or because tectonic forces can affect the horizontal stress but not the vertical stress so that the condition $\bar{\sigma}_b/\bar{\sigma}_v > 1$ can be obtained.

As both horizontal and vertical stress generally increase with increasing depth (McGarr and Gay, 1978), it is sometimes possible to follow the suggestion of Jaeger and

Cook (1969) and assume

$$\sigma_b = M\sigma_v \qquad (8.46)$$

Here, M is some constant that ranges from greater than zero to in excess of one. Thus, if M ranges between 0.3 and 0.5, we recover the laterally constrained case of Eq. 8.45. If M is taken as one, the hydrostatic case is recovered. On the other hand, if M is greater than one, the horizontal stress becomes the maximum principal stress, as might be the case in areas of tectonic compression. By tectonic compression, we mean the presence of stresses that are induced by other than the simple weight of the overburden. The relationship given as Eq. 8.46 is an oversimplification, but does permit an increase in the deviatoric stress ($\sigma_v - \sigma_b$) with depth, one of the few reliable facts known about principal stress differences (McGarr and Gay, 1978).

One-Dimensional Basin Loading

Consider the following assumptions applied to Eq. 8.42. Only fluid flow and gravitational loading will be considered in an isothermal environment. Both the flow and the compression take place in a vertical direction, and the fluids and the solids are incompressible so that $R/H = 1$. In addition, the stress rate $\partial\sigma/\partial t$ is assumed to be vertical and of constant value $\omega\gamma'$, where ω is a constant rate of sediment supply and γ' is the unit weight of the submerged sediment, that is, the difference between the bulk weight (sediments plus water) and the unit weight of water. Conceptually, the situation is depicted as a growing sedimentary pile in an oceanic environment (Figure 8.18). For these assumptions, Eq. 8.42 becomes

$$\frac{\partial P}{\partial t} = \left(\frac{K}{\rho_w g \beta_p}\right)\frac{\partial^2 P}{\partial z^2} + \gamma'\omega \qquad (8.47)$$

where $\gamma'\omega$ is the stress rate or pressure-generating term.

It is noted that the specific storage S_s equal to $\rho_w g \beta_p$ is now described in terms of the vertical pore compressibility only in that the fluid and the solids were assumed incompressible and the deformation vertical. In the absence of fluid flow, the rate of pressure development $\partial P/\partial t$ would equal $\gamma'\omega$, suggesting that the added load is

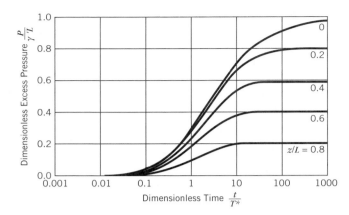

Figure 8.19 Graphic solution to the one-dimensional consolidation problem, where L is the thickness of the sedimentary pile, z/L is some dimensionless depth ranging from zero at the bottom of the pile to one at the top, t is time and T^* is the time constant for a basin, and $P/\gamma'L$ is the dimensionless excess pressure (modified from Bredehoeft and Hanshaw, 1968, Geol. Soc. Amer. Publication, from Geol. Soc. Amer. Bull., v. 79, p. 1097–1106).

transmitted totally and instantaneously to the water in the pores. With fluid flow, the pressure that develops at any time will depend on the rate at which fluid flow can dissipate this pressure.

Equation 8.47 was developed by Gibson (1958) and applied by Bredehoeft and Hanshaw (1968) for the conditions shown in Figure 8.18. As noted, fluid is allowed to escape from the top only, with the side and bottom boundaries being of the no-flow type. Further, as mentioned previously, the basin is laterally constrained so that no horizontal extensions are permitted. The solution to this problem is shown graphically in Figure 8.19.

Prior to discussing Figure 8.19 and the conclusions that might be reached from it, it is worthwhile to see if any useful information may be obtained from the governing differential equation. From Chapter 4, the inverse of the dimensionless group known as the Fourier number may be restated for this problem as

$$N_{FO}^{-1} = \frac{L^2/t}{K/S_s} = \frac{t}{K/S_s\omega^2} = \frac{t}{T^*} \qquad (8.48)$$

where T^* is the time constant for a basin and is equal to $K/S_s\omega^2$, and t is the depositional time to accumulate a given thickness of sedimentary pile. Note that the dimensionless group t/T^* is the variable making up the abscissa of the graphical solution given as Figure 8.19. Thus, we see that for times $t <<< T^*$ (e.g., typically less than 10^{-1}, Figure 8.19), there is no significant abnormal pressure production and the basin is in the normal pressure regime. The smaller the specific storage and the sedimentation rate, or the larger the hydraulic conductivity, the more likely the occurrence of normal pressure. For depositional times $t >>> T^*$ (e.g., typically greater than

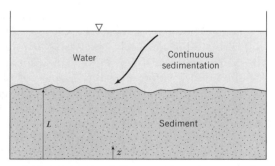

Figure 8.18 Conceptual diagram of continuous sedimentation in a fluid environment.

100, Figure 8.19), the pressures will be approaching the overburden pressure.

The conditions described here are reflected in Brede-hoeft and Hanshaw's (1968) analysis. These authors assumed a Gulf Coast rate of loading of 500 m/10^6 years for a period of 20×10^6 years, giving a sedimentary accumulation of 10,000 m. The results of this study are shown in Figure 8.20. For an upper limit of hydraulic diffusivity $K/S_s = 3.3 \times 10^{-2}$ cm^2/s, there is insignificant excess pressure production, and the fluid pressure remains near-normal throughout the sediment accumulation period (curve A). This corresponds to a time constant T^* equal to 4×10^8 years, whereas the depositional time t is on the order of 20×10^6, that is, $T^* >> t$. On the other hand, for a lower limit of hydraulic diffusivity of 3.3×10^{-5} cm^2/s, the fluid pressures approximate the overburden pressure (curve B). This corresponds to a time constant T^* equal to 4×10^5, that is, $t >> T^*$. Curve C lies intermediate between these extremes, with a hydraulic diffusivity of 3.3×10^{-4} and a time constant T^* equal to 4×10^6 years.

From the graphical solution of Figure 8.19, one can see that the limiting dimensionless numbers t/T^* of 10^{-2} and 10^2 apply to any case of loading irrespective of the hydraulic parameters used in the Bredehoeft–Hanshaw (1968) study. Values of t/T^* between 10 and 100 are associated with pressures approaching the overburden, whereas values on the order of 10^{-2} are associated with normal compaction.

To illustrate the relationships between the variables in the basin loading problem further, depositional time and the time constant have been plotted against various sedimentation rates for a wide range of hydraulic diffusivities (Figure 8.21). Also plotted is the time it takes to deposit

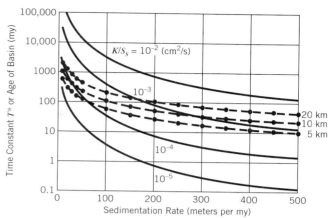

Figure 8.21 Time constant versus sedimentation rate. Solid lines represent the time constant as a function of sedimentation rate for various values of K/S_s. The dashed curves give the time required to accumulate 5, 10, and 20 km of sediment for the various sedimentation rates (from Palciauskas and Domenico, Water Resources Res., v. 25, p. 203–213, 1989. Copyright by Amer. Geophys. Union).

5, 10, and 20 km of sediments at the given sedimentation rate. Thus, we note that at a sedimentation rate of 500 m per million years (m/my) over a time period of 10 my, the sediment accumulation is 5 km. A depositional rate of 100 m/my over a time span of 100 my results in a 10 km deposit. It is noted that the time constant for sediments with a diffusivity of 10^{-2} cm^2/s is significantly larger than the time it takes to deposit 20 km of sediments, that is, the $K/S_s = 10^{-2}$ cm^2/s line lies above the 20 km line. As $T^* >> t$, these sediments will be normally pressured. For a hydraulic diffusivity of 10^{-3} cm^2/s, the time constant is greater than the depositional time only for sedimentation rates less than 100 m/my, whereas for larger sedimentation rates, t and T^* are approximately equal, suggesting modest abnormal pressure development. Hydraulic diffusivity values on the order of 10^{-4} and 10^{-5} cm^2/s are characterized by time constants that are considerably less than depositional times for most sedimentation rates and are expected to result in appreciably overpressured sediments.

Extensions of the One-Dimensional Loading Model

There are at least three extensions to the one-dimensional model presented that may provide further insight into the physics pertinent to abnormal pressure development. The first of these is for fluids and solids considered to be compressible. This seemingly insignificant addition has some important implications in areas of active deposition, especially with regard to the establishment of vertical gradients between units of differing lithology. The second of these extensions examines the basin loading problem within the framework of a mean stress problem.

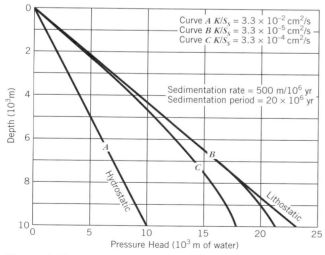

Figure 8.20 Pressure versus depth, Gulf Coast loading conditions (from Bredehoeft and Hanshaw, 1968, Geol. Soc. Amer. Publication, from Geol. Soc. Amer. Bull., v. 79, p. 1097–1106).

This implies the presence of deviatoric stresses and provides an opportunity to examine the possibility of rock fracture. The third extension is the one-dimensional constant strain problem, which is the other extreme of the one-dimensional constant stress problem.

Vertical Compression with Compressible Components

Given all the assumptions of the Bredehoeft–Hanshaw (1968) analysis with the exception that both the fluid and solid components are compressible ($R/H \neq 1$) gives the result stated already as Eq. 8.47:

$$\frac{\partial P}{\partial t} = \frac{K}{\rho_w g(1/R)} \frac{\partial^2 P}{\partial z^2} + \left(\frac{R}{H}\right) \gamma' \omega \qquad (8.49)$$

For clays, the pore pressure coefficient R/H will be sufficiently close to unity (Table 8.4) so that there is virtually no difference between this formulation and the previously described one. However, the coefficient R/H can range from 0.83 for mudstones to 0.25 for limestones. This suggests that any incremental load in a depositional environment containing various lithologies will not be totally borne by the fluid, but some part of it will be permanently carried by the solids, and the part carried by the solids will be different in different rocks. Depending on the magnitude of this parameter, it may be difficult to generate high fluid pressures in some rocks irrespective of the rate of loading and limitations on the rate of flow. For example, geologists have long noted that ancient limestones often show little sign of compaction during their diagenetic period. This may be the result of a pore pressure coefficient R/H considerably less than one if the lime mud crystallizes early in its depositional history so that most of the loading is carried by the grains. The increased stress on the grains can enhance pressure solution in such rocks.

The solution to Eq. 8.49 is likewise given by Figure 8.19, where the vertical axis is replaced by $P/\gamma' L$ (R/H). As R/H is less than one for rocks more rigid than clays, the initial fluid pressure will always be less than that predicted by the Bredehoeft–Hanshaw model. As the lithologies vary within the sedimentary pile, we anticipate large differences in vertical gradients as some units become more highly pressurized than others.

For compressible components, the inverse Fourier number is modified accordingly and can be expressed

$$N_{FO}^{-1} = \frac{t(R/H)}{T^*} \qquad (8.50)$$

which indicates that the pressure production is not nearly as pronounced for situations where R/H is less than one. It is emphasized, however, that the limits established for the Fourier number from the results of Figure 8.19 still hold. That is to say, it still requires a number greater than 10^{-2} to generate abnormal pressure due to gravitational loading; however, with the numerator decreased by sev-

eral percent for some rocks, it is just not as likely an event unless the rate of loading is very fast or the hydraulic diffusivity very small.

Horizontal Extension and Compression

The analog field condition of the one-dimensional loading problem is the basin that is constrained laterally. This means that the basin cannot relax tectonically in response to the sediment load, nor can any form of tectonic compression occur during the sedimentation period. Normally in a tectonically relaxed state, there is little reason to expect significant horizontal resistance at the basin borders so that the sedimentary accumulation may react passively to movements of the subadjacent basement, with normal faults forming along the margins. At a later stage, crustal stress can become compressional, or give way to those that are epidermal in origin. The latter can be the result of large differences in elevation between the bordering lands and the bottom of the subsiding basin. This may promote compressional settling or large-scale slumping from each side of the subsiding basin, leading to compressional folding.

The problem of tectonic extension or compression during loading may be examined within the framework of a mean stress problem (Palciauskas and Domenico, 1980). For these purposes, the stress rate $\partial\sigma/\partial t$ is assumed to be the result of a mean principal stress rate, not solely the vertical one. For the case where the horizontal stresses are equal and are designated σ_3, the mean principal stress expressed in Eq. 8.36a becomes

$$\sigma_m = \frac{\sigma_1}{3} + \frac{2\sigma_3}{3} \qquad (8.51)$$

The relationships between the horizontal and vertical stresses are given as Eq. 8.46

$$\sigma_3 = M\sigma_1 \qquad (8.52)$$

Here, as discussed previously, M can range from slightly greater than zero for horizontal extension to greater than one for crustal or epidermal compression. For these conditions, Eq. 8.47 becomes

$$\frac{\partial P}{\partial t} = \frac{K}{\rho_w g(1/R)} \frac{\partial^2 P}{\partial z^2} + C\gamma'\omega \left(\frac{R}{H}\right) \qquad (8.53)$$

where

$$C = \left(\frac{1}{3}\right) + \left(\frac{2M}{3}\right) \qquad (8.54)$$

It is seen that C acts as a multiplier for the effective rate of loading ($R/H)\gamma'\omega$, which is the effective rate of application of effective vertical stress. The larger C, the greater the abnormal pressure development. It is noted that the graphical solution given in Figure 8.19 applies here also where the ordinate is expressed as $P/\gamma' L(2M/$

$3 + 1/3)(R/H)$. Thus for $R/H = 1$ and $M = 0.6$ ($\sigma_3 = 0.6\sigma_1$), the mean stress problem yields 73% of the abnormal pressure that would result from the laterally constrained case. If $M = R/H = 1$, the laterally constrained case of Bredehoeft and Hanshaw (1968) is recovered.

An interesting feature of this development is that the smaller M, the larger the deviatoric stress, and the smaller the fluid pressure production. Later developments will show that failure can occur at low fluid pressures in association with large deviatoric stresses, or at higher fluid pressures in combination with smaller deviatoric stresses. Hence, some apparent conditions for failure would appear to be present for virtually all states of differential stress in compacting sediments.

Equation 8.53 is well suited for obtaining an approximate idea of the effects of horizontal compression and indicates that the pressure production can be quite large. If tectonic forces are involved, the multiplier C can exceed one. Thus, for $R/H = 1$ and $M = 1.5$ ($\sigma_3 = 1.5\sigma_1$), the pressure developed is 33% greater than one would expect from the laterally constrained case. Such pressures would never be achieved because of some pressure-induced inelastic rock failure during the compression.

Tectonic Strain as a Pressure-Producing Mechanism The other extreme of the one-dimensional constant stress problem is the problem of constant one-dimensional strain, visualized as occurring in a layer whose lateral extent is very large compared to its thickness. The layer is overlain and underlain by high-permeability material that is likewise undergoing deformation, but remains at normal or hydrostatic pressure due to large values of the hydraulic diffusivity. The main fluid migration is taken as vertical due to the horizontal compression of the bed. Clearly, such a one-dimensional strain problem is only an approximation of reality, but should have some insightful features as the other extreme of the one-dimensional stress problem. The governing equation in this case may be stated as

$$\frac{\partial P_{ex}}{\partial t} = \left[\frac{K}{\rho_w g (1/Q)}\right] \frac{\partial^2 P_{ex}}{\partial z^2} - \xi Q \frac{\partial \varepsilon}{\partial t} \quad (8.55)$$

Palciauskas and Domenico (1989) provide a solution to this problem for the boundary conditions of hydrostatic pressure at the upper and lower boundaries of the layer and an initial condition of $P_{ex} = 0$ at time $t = 0$. The details of this solution are not important if we are merely concerned with the magnitude of strain rates required to produce fluid pressures equivalent to those associated with basin loading as demonstrated in Figure 8.21. This may be done by equating the source terms of modified forms of Eqs. 8.55 and 8.49

$$\gamma' \omega = \xi R \frac{\partial \varepsilon}{\partial t} \quad (8.56)$$

Thus, for a sedimentation rate of 500 m/my and $1/R = 1.62$ Kbar^{-1} for clays (Table 8.4)

$$\frac{\partial \varepsilon}{\partial t} = \frac{\gamma' \omega}{\xi R} = \left(0.130 \frac{\text{Kbars}}{\text{km}}\right)\left(0.5 \frac{\text{km}}{\text{my}}\right)$$
$$\left(1.62 \frac{1}{\text{Kbars}}\right)\left(\frac{1 \text{ my}}{3 \times 10^{13} \text{ s}}\right) = 3 \times 10^{-15} \text{ s}^{-1}$$

Thus, given the information in Table 8.4 along with Figure 8.21, the relationship between the source terms given as Eq. 8.56 may be used to determine strain rates that develop excess pressures equivalent to those produced by basin loading. For the example above, strain rates on the order of 3×10^{-15} s^{-1} will produce significant abnormal pressure, provided the hydraulic diffusivity is on the order of 10^{-4} cm^2/s and smaller, and no abnormal pressure for diffusivities larger than 10^{-3} cm^2/s (Figure 8.21). For this type of material (clay) with hydraulic diffusivities approaching 10^{-5} cm^2/s, strain rates as low as 7×10^{-16} s^{-1} can produce abnormal pressures. For mudstones with small hydraulic diffusivities and $1/R = 0.05$ (Table 8.4), strain rates of less than 1×10^{-16} s^{-1} can produce abnormal pressure.

Thermal Expansion of Fluids

When rocks undergo burial in a depositional environment, they are subjected to increases in temperature that promote a thermal alteration of the component parts of the sediment, that is, the fluid, the solids, and the pores. If the heating takes place without expulsion of the pore water, the pressure changes with temperature increases are classified as constant mass phenomenon. This assumption is at the heart of much of the research in this area. The concept was introduced originally by Barker (1972) and advanced by Bradley (1975) and Magara (1975). A review of this work is given by Gretener (1981).

In Barker's (1972) theory, a geologic formation must first become "sealed" at some depth and thereafter buried in a thermal field with no expulsion of its pore water. Additionally, the assumptions of Barker (1972) require that the thermal expansion of water acts against a rigid, nonyielding matrix that remains at constant pore volume throughout the burial. The pressure change with temperature is expressed as

$$\left(\frac{\partial P}{\partial T}\right) = \frac{\alpha_f}{\beta_w} \quad (8.57)$$

where α_f is the thermal expansion coefficient of the water and β_w is the isothermal compressibility of water, both of which are temperature dependent (Figure 8.22). On the basis of these arguments, a temperature increase of 25°C over a change in depth of approximately 1 km (25°C/km) would cause the fluid pressure to increase by

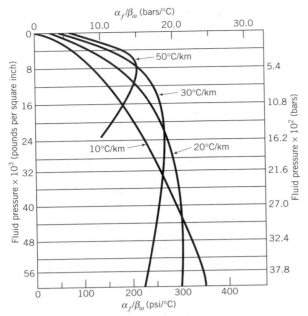

Figure 8.22 Values of α_f/β_w as a function of hydrostatic pressure for various geothermal gradients (modified from data of Knapp and Knight, J. Geophys. Res., v. 82, p. 2515–2522, 1977. Copyright by Amer. Geophys. Union).

300 bars (4400 psi). This corresponds to an incremental pressure gradient of 0.3 bars/m (1.34 psi/ft).

In examining Eq. 8.57, both Chapman (1980) and Daines (1982) suggested that the thermal expansion of water in depositional environments is probably not a very important pressure-producing mechanism. Their statements were not based on the physics of Eq. 8.57, but on heuristic arguments that discounted the possibility of maintaining constant fluid mass over long periods of geologic time. According to these authors, fluid flow would likely dissipate the pressures. Notwithstanding the importance of fluid flow, these arguments missed the main point of debate concerning the appropriateness of Eq. 8.57; that is, the equation depends solely on fluid properties and is medium independent. Thus, on the basis of Barker's (1972) result, one might conclude that

the fluid pressure response to a geothermal gradient in a granite under a large confining stress is no different from that in a shale in a tectonically relaxed state. If constant mass considerations are to have any realism as upper-bound calculations for pressure production, they must include the expansive properties of the medium responsible for accommodating some of the pressure, as well as the boundary conditions under which the heating takes place. Three possible boundary conditions present themselves: heating at constant stress, at constant strain, or under conditions that are neither constant stress nor constant strain.

When addressing the pressure changes at constant stress or constant strain, two avenues are open to us. We may derive the appropriate expressions directly, independent of fluid flow considerations, or they may be obtained by simply ignoring fluid flow in Eqs. 8.42 and 8.43 and treating these expressions within the framework of constant stress or constant strain. The end result is the same, giving the expressions

$$\left(\frac{\partial P}{\partial T}\right)_{M,\sigma} = R\alpha_m \tag{8.58}$$

and

$$\left(\frac{\partial P}{\partial T}\right)_{M,\varepsilon} = Q(\xi\alpha_b + \alpha_m) \tag{8.59}$$

respectively. Equation 8.58 gives the change in pressure with temperature increases at constant mass and constant stress, whereas Eq. 8.58 is the constant mass formulation for the boundary conditions of constant strain. The coefficients in these equations have been discussed earlier.

Table 8.5, taken from Hastings (1986), gives some representative values for various rock types. For each of the cases shown, the change in pressure with temperature (bars/°C) was multiplied by a geothermal gradient of 25°C/km to give a pressure change with depth. As expected, the constant pore volume case of Barker (1972) is rock independent, whereas the constant stress case yields the least fluid pressure production and the constant strain case the most. In the constant stress case, the rock

Table 8.5 **Pressure Changes with Depth for a Geothermal Gradient of 25°C/km for Various Rock Types and Boundary Conditions**

Rock Type	Constant Pore Volume	Constant Stress	Constant Strain
Shale	0.423 bars/m (1.87 psi/ft)	0.0045 bars/m (0.02 psi/ft)	0.475 bars/m (2.1 psi/ft)
Sandstone	0.423 bars/m (1.87 psi/ft)	0.249 bars/m (1.1 psi/ft)	0.479 bars/m (2.12 psi/ft)
Limestone	0.423 bars/m (1.87 psi/ft)	0.362 bars/m (1.6 psi/ft)	0.471 bars/m (2.08 psi/ft)

is free to expand elastically as well as thermally to partially accommodate the fluid pressure buildup. In the constant strain case, volumetric expansion of the porous medium is not permitted. Thus the importance of the boundary conditions.

Before dismissing the change in fluid pressure with changes in temperature at constant stress in soft rocks such as shale as insignificant, let us examine the nature of the coefficient (1/R). In elastic theory, 1/R is a compressibility that describes how the mass changes with pressure changes at constant stress (Table 8.3). In Eq. 8.58, 1/R represents an expansivity of the porous medium. Thus we are trapped in our elastic theory whereby reversibility is assumed a priori. Although this assumption of reversibility may be appropriate for most crystalline rocks and perhaps some sandstones and limestones, it hardly applies in the case of soft unconsolidated sediment. It is noted further that the constant stress case is part of the differential equation for basin loading (Eq. 8.42). Thus solutions to this flow equation or similar ones where temperature effects are included, such as those presented by Domenico and Palciauskas (1979), Sharp (1983), and Shi and Wang (1986), focus exclusively on the constant stress case and incorporate the assumption of reversibility.

The boundary conditions for constant stress or constant strain are not strictly valid in field situations where rocks at depth are being heated. Rather, the displacement field and the stress distribution should be derived from the stress equilibrium equation. This is generally a formidable task, but there is one situation that has been solved for the assumption that displacements far from a source of heat are zero. This permits the derivation of expressions that describe the change in fluid pressure P, the volume dilation ε, and the changes in hydrostatic stress σ purely as a function of temperature and the properties of the medium. Some typical results are presented in

Table 8.6 for a variety of rock types. The fluid properties in these calculations were taken as averages over an approximate 80°C rise in temperature. The elastic coefficients for these rocks are the same as those employed in Table 8.4. In all lithologies, the fluid pressure increases with temperature at a faster rate than do increases in stress caused mainly by the thermal expansion of the solids. The greater the rock rigidity, the greater the net decrease in effective stress. For the sandstone or basalt, the net decrease in effective stress is about 4 bars °C^{-1}, compared to 1.54 bars °C^{-1} for the mudstone and 0.09 bars °C^{-1} for the clay. Further, the elastic strain with temperature is obviously greater for the soft highly porous rocks, an indication of pressure accommodation by elastic volume increases. The small decrease in effective stress for soft rocks suggests that failure at some point may not be a likely event even under the maximum pressure-producing assumption of constant fluid mass. It is noted, however, that the concept of reversibility for the elastic coefficients is embedded in this analysis.

It is now useful to compare these results with other calculations for the Gulf Coast region. The calculations for the mudstone will suffice here. For a geothermal gradient of 30°C km^{-1}, the fluid pressure increase of 2.77 bars °C^{-1} translates into a thermal pressure production of 0.0831 bars m^{-1} (0.4 psi ft^{-1}). The fluid pressure production per meter of depth would obviously be much less for clays. Gretener (1981) notes that for the prevailing geothermal gradient, the thermal pressure development in the Gulf Coast can range between 0.255 bars m^{-1} (1.25 psi ft^{-1}) and 0.66 bars m^{-1} (3.2 psi ft^{-1}). Magara (1975) cites a value of 0.285 bars m^{-1} (1.4 psi ft^{-1}). These calculations indicate predictions of thermal pressure development that are between three and eight times greater than determined from Table 8.6 for mudstones. As noted earlier, the calculations of Gretener (1981) and Magara (1975) are based on the constant pore volume assump-

Table 8.6 **Pressure, Stress, and Strain with Temperature Changes for Various Rock Types**

	Clay	Mudstone	Sandstone	Limestone	Basalt
$\left(\dfrac{\partial P}{\partial T}\right)_{m,\sigma}$ (bar °C^{-1})	0.0956	1.73	5.9	9.13	8.72
$\left(\dfrac{\partial P}{\partial T}\right)_{m,\varepsilon}$ (bar °C^{-1})	11	11.23	15.5	12.58	10.13
$\dfrac{dP}{dT}$ (bar °C^{-1})	0.18	2.77	9.1	10.39	9.34
$\dfrac{d\sigma}{dT}$ (bar °C^{-1})	0.09	1.23	4.7	5.51	5.4
$\dfrac{d\varepsilon}{dT}$ (°C^{-1})	1.88×10^{-4}	9.47×10^{-5}	5×10^{-5}	2.34×10^{-5}	1.5×10^{-5}

From Domenico and Palciauskas (1988).
Geol. Soc. Amer. Pub. Geology of N. America, 0-2, p. 435-445.

tion. We conclude that thermal pressure development in very soft rocks is not very significant (unless they are heated at constant strain), but does become more significant with increasing rock rigidity.

The constant mass case gives an upper bound for the development of thermal pressure in depositional environments. When fluid flow occurs, we once more have competition between rates of pressure production and pressure dissipation. The reader is referred to Domenico and Palciauskas (1979), Sharp (1983), and Shi and Wang (1986) for details on this topic.

Fluid Pressures and Rock Fracture

The Coulomb theory of failure given in Eq. 5.17 describes the shear failure mode exclusively in terms of total and fluid pressures. This is not the only failure mode, nor are other failure theories restricted to a simple effective stress relationship. Irrespective of the failure mode, abnormally high fluid pressures can lead to some sort of rock fracture. On one scale is the inelastic behavior referred to as hydraulic fracture, the accepted condition for which is fluid pressures in excess of the least compressive stress in a regional stress field. The complete theory states that incipient fracture propagation will occur when the fluid pressure exceeds the least principal stress by an amount equal to the tensile strength of the rock. The direction of fracture propagation has a preferred orientation, normal or near normal to the least compressive stress. In the tectonically relaxed (extensional) state, the horizontal stresses are smaller than the vertical stresses, and the fracture growth is vertical; in the tectonically compressed state, they are horizontal.

At yet another scale are the pore volume increases that materialize in the form of microfractures, commonly at fluid pressures well below the least compressive stress (Handin and others, 1963). As with hydraulic fracture, this dilatancy has a preferred orientation, normal or near normal to the least compressive stress, and is a likely precursor in the formation of large throughgoing fractures. In this section we will examine a few of these failure theories.

The first and best understood is hydraulic fracture as described by Hubbert and Willis (1957). According to these authors, the pressure required to initiate fracture is

$$P^* = \sigma_v M - P(M - 1) \qquad (8.60)$$

where σ_v is the vertical stress, M is the ratio of horizontal to vertical stress and is some constant less than or equal to one, and P is the ambient (prefracture) fluid pressure. Dividing both sides by the depth z to obtain a critical pressure gradient gives

$$\frac{P^*}{z} = \frac{(\sigma_v M) - P(M - 1)}{z} \qquad (8.61)$$

Thus, if the horizontal stress is one-third of the vertical, Eq. 8.60 predicts a fracture pressure

$$P^* = \frac{\sigma_v + 2P}{3} \qquad (8.62)$$

This states that the fluid pressure required to initiate hydraulic fracture is one-third of the overburden pressure plus two-thirds of the ambient (prefracture) fluid pressure. As M approaches one, that is, as the deviatoric stress approaches zero, a greater fluid pressure is required to initiate fracture.

In addition to the tensional failure described, earth materials dilate in response to shearing strains, undergoing volume changes that may be positive or negative. This behavior has been observed in soils and both crystalline and water-saturated sedimentary rocks. For crystalline rocks, dilation begins at a deviatoric stress of one- to two-thirds of the stress difference ultimately required to fracture the rock (Brace and others, 1966). Fracture is accompanied by a porosity increase that ranges from 0.2 to 2 times the elastic volume change the material would have had were it perfectly elastic. This statement holds for both brittle and ductile rocks and for fracture at atmospheric pressure as well as under a large total confining pressure. For sedimentary rocks containing pore fluids, dilatancy has been observed when the ratio of the fluid to the confining pressure is equal to about 0.8 (Handin and others, 1963). This result is reported as being approximately correct for sandstone, limestone, siltstone, and probably shale.

Some of the features of these tests are shown in Figure 8.23. In order to interpret these results, the effective

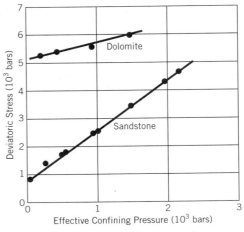

Figure 8.23 Relation among effective confining pressure, deviatoric stress, and failure. Dilatancy reportedly initiated at or previous to failure. Sandstone data from Handin and others (1963); dolomite data from Brace and others (1966) (from Palciauskas and Domenico, 1980. Reprinted by permission).

confining pressure is defined as

$$\sigma_e = \sigma_c - P \qquad (8.63)$$

where σ_e is the effective confining pressure, σ_c is the confining pressure in the triaxial apparatus, and P is the fluid pressure in the pores of the sample. As noted in the figure, when fluid pressure is low compared to the experimental confining pressure (that is, a high effective confining pressure), the onset of dilatancy requires a large deviatoric stress. As fluid pressure increases (that is, a lower effective confining pressure), the onset of dilatancy requires a smaller deviatoric stress. The important feature here is that failure is once more associated with high fluid pressures at small deviatoric stresses or low fluid pressures at large deviatoric stresses.

There are several other factors that dictate the onset of dilatancy, including the density of the sediments at the time of failure (Zoback and Byerlee, 1976) and the degree of packing as reflected by the porosity (Scott, 1963). Ideally, one might expect a physically based constitutive relation expressed in terms of a critical porosity for a given deviatoric and mean stress. Unfortunately, such a relationship is not available. There are, however, other approaches to the problem.

One approach is to consider the dilatant behavior in an empirical form suggested by Scott (1963). The basic assumption is that the inelastic volume change is a function only of the deviatoric stress. Assuming σ_{22} and σ_{33} are equal and designating the horizontal stress as σ_{33}, this may be expressed as

$$(dV/V)_{\text{inelastic}} = -Dd\tau^* \qquad (8.64)$$

where V is volume so that dV/V is dilatant strain, D is a dilation coefficient that can be positive or negative depending on the material, and τ^* is the deviatoric stress. The coefficient D is not likely to be a constant for a given material and will be somewhat strain dependent. When D is negative, the inelastic volume change acts to increase porosity with increasing deviatoric stress. However, to perform calculations, some threshold pressure has to first be identified and incorporated in such a way that describes the rate at which the fracture dissipates pressure.

A useful limiting expression for the initiation of fracture is the experimental observation of Handin and others (1963), where the onset of dilatancy occurs when the fluid pressure approaches 0.8 of the confining pressure, that is,

$$P_{\text{critical}} = 0.8\sigma_c \qquad (8.65)$$

where the critical designation implies a fluid pressure necessary for the onset of dilatancy. If the confining pressure is taken as equivalent to the horizontal stress, the fluid pressure required for the initiation of microfractures become

$$P_{\text{critical}} = 0.8[\sigma_v M - P(M - 1)] \qquad (8.66)$$

where the bracketed quantity is the Hubbert–Willis (1957) fracture criteria. It will be noted here that as M gets large (approaches one), the deviatoric stress decreases and the fluid pressure required for the onset of dilatancy increases. This is in accord with the experimental observations given in Figure 8.23. This feature is also incorporated in the mean stress formulation for abnormal pressure development (Eq. 8.53). Palciauskas and Domenico (1980) utilize this fracture criterion to examine the initiation of microfractures in oil-bearing source rocks at different stages of maturation. Last, Eq. 8.66 suggests that an upper limit of fluid pressure is about $0.8\,\sigma_v$. That is, microfracture generation at fluid pressures on the order of $0.8\,\sigma_v$ may represent an instability from which there may be no way to further increase fluid pressure.

Phase Transformations

Included among the main phase transformations that presumably lead to excess pressure development are the gypsum-anhydrite transformation (Heard and Ruby, 1966) and the montmorillonite-illite transformation (Powers, 1967; Burst, 1969; Prichett, 1980). Both transformations are thermally driven and result in the production of pore fluids. With montmorillonite, the expulsion of one interlayer of water decreases the layer spacing by about 2.5 Angstroms, resulting in a possible 20% decrease in the mineral grain volume per layer. Hence, the potential water volume release is considerable. In the Gulf Coast, the alteration of montmorillonite to a mixed layer clay begins at temperatures on the order of 60 to 70°C. Although it has not yet been possible to quantify the rate at which pressure is produced in this transformation, it can be speculated that the rate of transformation is proportional to the decrease in effective stress per unit time interval.

Figure 8.24 illustrates this last argument where the contained water goes from the interlayered state in the lattice to a state of free water. Bound water in a solid structure supports a finite effective stress that goes to zero in the free water state. Assuming the pore water produced is not free to drain, it should immediately take on the weight of the overburden. It appears further that this magnitude of pressure can be continually regenerated in the "window" that defines the critical pressure range in a subsiding basin. If it can be assumed that the transformation is described by first-order kinetics of the Arrhenius form, the size and depth of this window is a function of certain kinetic factors along with temperature and the rate of burial. In Figure 8.25, curve I indicates that the reaction occurs at near surface temperatures and the transformation is completed in the upper 1000 m of burial. In this development, N represents the number of expandable layers after time t, N_0 is the original amount, A is the frequency factor, E is the activation energy, and ω is the rate of burial. This rapid depletion of smectite

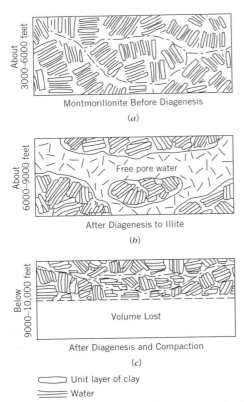

Figure 8.24 Schematic diagram of clay diagenesis showing (*a*) virtually all bound water with no porosity or permeability, (*b*) the generation of free water at the expense of the solids, (*c*) free water squeezed out upon completion of compaction (from Powers, 1967. Reprinted by permission).

pressible than the solids, we anticipate a net increase in pore space. This will act to lower fluid pressure. At the same time, the fluid will expand due to the reduced stress, but will also contract somewhat due to a lowered-temperature environment. Both these processes will act to lower the fluid pressure even more. The major question, as yet not satisfactorily answered, is whether the increased pore space is sufficient not only to reduce the pressures to their normal value, but to create a state of suction wherein the pressures are less than atmospheric. If such a state can indeed be produced, the driving forces that take over are dominantly those that we associate with the unsaturated zone. For such cases, we anticipate movement or drainage of the fluids downward due to gravity and some sort of redistribution within the unsaturated portions of the rock. Ultimately, with time, the upper parts of the rocks will be drained of their fluids (except for those held by capillary attraction) and—due to the accumulation of fluids at depth—the formation of a deep saturated zone with a water table significantly below land surface.

Neuzil and Pollock (1983) presented an analysis that demonstrates that subatmospheric pressures can develop due to the unloading of hydraulically tight rocks. It is noted that an initial condition for this analysis was hydrostatic conditions; that is, the process did not start from the abnormally pressured state, but from a normally pressured one. Senger and Fogg (1987) and Senger and others (1987) discuss the effects of uplift, deposition, and erosion on the development of underpressured basins.

is contrary to observations of smectite-illite occurrences in Gulf Coast sediments. Curve II, on the other hand, shows the reaction delayed to about 3000 m and then progressing to near completion over the next 1500 m of burial. Irrespective of the uncertainty regarding the exact location of the critical temperature range, Figure 8.25 suggests the presence of a rather narrow window associated with it.

Subnormal Pressure

If geologic loading can give rise to abnormally high fluid pressures, erosional unloading and eventual uplift may result in abnormally low pressures, referred to earlier as subnormal pressure (Russel, 1972). The mechanics can be explained by viewing the compaction process in reverse. With erosional unloading and uplift, the stress is removed from the porous matrix and from the solid grains. Hence, the pores may expand somewhat, as will the solid grains. As the horizontal stresses are generally increased relative to the vertical ones during erosional unloading, it is difficult to describe the details of this process. However, as the pores are generally more com-

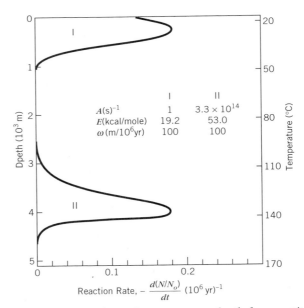

Figure 8.25 Transformation rate versus depth for smectite-illite (from Domenico and Palciauskas, 1988, Geol. Soc. Amer. Publication, from Geol. of N. Amer., 0–2, p. 435–445).

Irreversible Processes

Except for our section on rock fracture, the discussions thus far have been exclusively concerned with elastic deformations characterized by elastic material properties. It has been noted by Neuzil (1986) that different time scales of deformation are characterized by different moduli. Short-duration phenomena such as atmospheric pressure variations or tidal disturbances are described by elastic models that contain undrained moduli. Intermediate-duration phenomena are likewise described with elastic models, but as there is sufficient time for fluid flow to take place in response to the deformation, the appropriate moduli are the drained coefficients. The deformations associated with the drained parameters are significantly larger than in the undrained case. It is noted that these same elastic coefficients have been used to describe phenomena that operate over geologic time scales. However, we are aware that over time periods measured in terms of tens of thousands of years, irreversible nonelastic processes will occur at the same applied stress that is driving the elastic response. These irreversible processes will result in considerably greater deformations, for the same applied stress condition, than would be expected from their elastic counterparts. For the most part, it appears that we have ignored these greater deformations and have focused on the lesser ones, and some explanation is in order.

The reason for this apparent shortcoming is that it is most difficult to determine the microscopical laws governing irreversible deformations. On the other hand, Neuzil (1986) suggests that such new laws may not be necessary, and only different coefficients are required. These arguments suggest that both short- and long-term deformations can be described within elastic theory and the concept of effective stress, but that the constants applicable over long time frames will be different from those that apply to the short term. There still remains the problem of obtaining the constants for the long-term deformations. It is obvious that they cannot be measured in the laboratory but may be determined from field studies; for example, porosity versus depth determinations in thick sequences of rock. This has been the method of Shi and Wang (1986) with the use of Athy's (1930) relation for porosity versus depth in shales (Chapter 2). Yet another way is to devise a new microscopic law that "predicts" porosity variation with depth in response to some assumed process that is operating. This was the method of Palciauskas and Domenico (1989). Their calculations indicate that an irreversible "compressibility" coefficient for such deformation is almost two orders of magnitude greater than its elastic counterpart. This means that on a macroscopic scale, the irreversible deformational character of a rigid sandstone is similar to the elastic deformation of a rock whose stiffness lies intermediate between a soft clay and a mudstone.

8.3 Pore Fluids in Tectonic Processes

Walder and Nur (1984) posed three important questions regarding fluids in the earth's crust:

1. To what depths in the earth's crust is free water present?

2. What is the value of the fluid pressure at these depths?

3. What is the permeability of the rocks at these depths?

In that actual measurements of crustal properties exist only to depths of 2–3 km, any answers to these questions must be based on indirect evidence. Such evidence collected by these authors suggests that free water is present, at least episodically, at upper and midcrustal depths. The pressure of this fluid is thought by many to be abnormal, and by others to be at normal pressure. Brace (1980), for example, states that zones of permeability on the order of 10^{-15} m² must exist to depths of at least 10 km. He therefore concludes that this permeability is too high for the maintenance of abnormal fluid pressures. However, uniform values of permeability to midcrustal depths—high or low—are not to be expected, and not all rocks have direct permeable pathways to the Earth's surface. Further, from what has been learned in the section on active depositional environments, it is clear that if some process or processes can generate fluid pressure at a faster rate than fluid flow can dissipate it, the pressures will be maintained at abnormal proportions. We already discussed those processes that are likely responsible for such generation. Whether or not they or different ones operate at midcrustal depths is another question.

Fluid Pressures and Thrust Faulting

A thrust fault is a discontinuity in a rock mass characterized by a low dip and a net slip generally measured in miles, sometimes approaching 100 miles. The mechanism for overthrusting presents a problem in mechanics in that the force necessary for displacements of such magnitude far exceeds the strength of the rock itself. Thus, in the words of Smoluchowski (1909), "we may push the block with whatever force we like; we may eventually crush it, but we cannot succeed in moving it." With this constraint, two lines of reasoning were followed to explain such displacements; gravitational sliding down an inclined slope, and the lubrication of the thrust plane, usually by water. In other words, these propositions were concerned with reducing the sliding frictional resistance between two rock masses. As discussed by Terzaghi (1950) in our earlier treatment of landslides, water may not act as a lubricant but as an antilubricant.

Hubbert and Rubey (1959) invoked the Mohr–Coulomb theory and the same failure criteria as Terzaghi (1950) to present the case that high fluid pressures can sufficiently reduce the shearing resistance of a given rock mass so that large-scale displacements are possible. In their overthrust concept, fault blocks could be moved large distances by relatively small forces, provided the fluid pressures were sufficiently high to put the rocks in a state of floatation. The special geologic conditions required for the attainment and maintenance of such pressures were cited as (1) the presence of clay rocks; (2) interbedded sandstone; (3) large total thickness, and (4) rapid sedimentation. These conditions represent the essence of the depositional environments discussed in the previous section.

Since this original effort by Hubbert and Rubey, the embellishments added by others have taken two rather distinct lines. The first is a search for other causes of high fluid pressures needed in the overthrust concept within the same physical framework of the Mohr–Coulomb theory. Thus, Hanshaw and Zen (1965) propose that osmotically induced pressures across shale beds are an important cause of abnormal pressure where high- and low-pressure zones are induced on opposite sides of a shale bed. From a purely mechanistic viewpoint, it is not difficult to conceive of several other causes of abnormal pressure.

The second line of research is a search for mechanisms other than the fluid pressure hypothesis or, in some cases, a modification of the original concepts. In this regard, Chapman (1981) exhumes lubrication, and Guth and others (1982) argue that there are limits to the amount of fluid pressure development before some sort of fracture begins and releases the pressure. Gretener (1972) suggests that high fluid pressures could not be maintained under a thrust for long periods of time and argues that the thrust does not move all at once, but like a caterpillar, one segment at a time. The suggestion here is that abnormal pressure may be episodic.

It may be noted further that compressional forces associated with thrusting are sufficient to generate high fluid pressures that will inevitably fracture the rock, causing large-scale displacements. Thus, the origin of the fluid pressures may be tectonic compression as opposed to basin loading, which has the added attraction of containing the means by which the overthrust occurs. If the fracturing temporarily acts to relieve the pressure, continued compression may eventually build it up again, causing a repetition of the thrust and the caterpillar type of motion. In each event the pore volume that is generated at fracture may be just sufficient to equal the volume necessary to return the rock to its ambient pressure when the tensile strength of the rock was first exceeded. As long as the rocks remain in a state of compression, this process will continue, with the fractured state of the rock at any time representing an instability from which it is not possible to increase fluid pressures further. It is expected that with increased fracture, this instability will

be realized at progressively lower fluid pressures over geologic time. Hence, the process may be a decaying one that ultimately dies out when the compression finally takes place at or near constant fluid pressure. The analogy to this terminal situation is the normal compaction concept put forth earlier.

Whatever the final consensus—if any—there is little doubt that the Hubbert and Rubey hypothesis as developed from the original work of Terzaghi (1950) generated a flurry of interest in the mechanics of the overthrusting and the strength of rocks in general.

Seismicity Induced by Fluid Injection

Among the many thousands of fluid injection wells in the world, there are a few that have drawn attention as a result of the earthquakes that occur in conjunction with them. Three such wells or well sites in particular have been studied rather extensively: a deep waste injection well at the Rocky Mountain Arsenal near Denver, Colorado; experimental injection wells in the Chevron Oil Field near Rangeley, Colorado; and a high-pressure injection well used for hydraulic mining of salt in the Attica–Dale region of western New York.

In 1961, a deep disposal well (3638 m) was constructed to penetrate highly fractured Precambrian schist and granite gneiss bedrock underlying the Denver Basin. Injection was initiated in March 1962. During the first month of operation over 15,000 m^3 (4×10^6 gallons) were injected with well head pressure ranging from 0 to 72×10^5 Pa (72 bars, 10.4 psi). During the first month of operation, Denver experienced its first earthquake in 80 years (Simon, 1969). The injection period from 1962 to 1966 can be divided into four stages (Hsieh and Bredehoeft, 1981). From March 1962 to September 1963, the injection took place under pressure. Between October 1963 and September 1964, no injection took place. From October 1964 to March 1965, injection took place by gravity flow. Injection was stopped in April 1965, after D. Evans publicly suggested a relationship between fluid injection and over 700 earthquakes generated in the Denver area over this short three-year period (Figure 8.26).

Evans (1966) invoked the Mohr–Coulomb failure theory of the Terzaghi (1950) and Hubbert and Rubey (1959) hypothesis to explain the fault movements and induced seismicity along historically dormant faults. The effect of increased fluid pressure is to reduce the frictional resistance to fracture by decreasing the effective normal stress across the fracture (Byerlee, 1967). The continuous seismicity after injection stopped was explained by Healy and others (1968), who suggested that near vertical fractures, preearthquake in origin, have long been subjected to tectonic stress. This stored strain was released by the propagation of built-up fluid pressure throughout the reservoir after injections had ceased.

The association of the Denver earthquakes with fluid

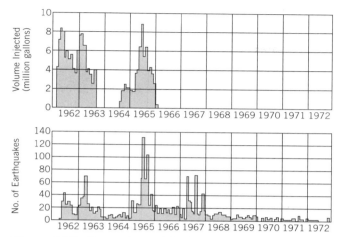

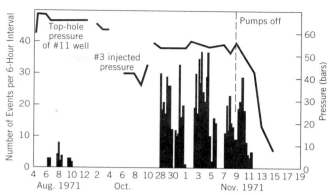

Figure 8.28   Number of earthquakes per 6-hr interval plotted with top hole pressure in injection well, western New York (from Fletcher and Sykes, J. Geophys. Res., v. 82, p. 3767–3780, 1977. Copyright by Amer. Geophys. Union).

Figure 8.26   Comparison of fluid injected and the frequency of earthquakes at the Rocky Mountain Arsenal. Upper graph shows monthly volume of fluid waste injected in the disposal well. Lower graph shows number of earthquakes per month (from Hsieh and Bredehoeft, J. Geophys. Res., v. 86, p. 903–920, 1981. Copyright by Amer. Geophys. Union).

injection prompted a renewed interest in the response of earth materials to abnormal pressures. An experiment was quickly established at the Rangeley Field where Chevron had been injecting fluids to enhance production. It was noted that a prevalent zone of earthquake activity persisted near the edge of the field. Four wells were chosen for the experiment, and testing started in 1969 and continued through 1973, in which the fluid pressures were alternately raised by injection and lowered by pumping (Figure 8.27). As noted, the frequency of earthquake activity was dramatically increased when the fluid pressure within the reservoir exceeded 275×10^5 Pa (275 bars). The experiments are described by

Raleigh and others (1972, 1976). Raleigh (1971) provides a good overview on this topic.

Fletcher and Sykes (1977) reported a rash of earthquakes of small magnitude in the vicinity of an injection well in western New York. This well is only 430 m deep. The relationship between the fluid pressures and the earthquakes are shown in Figure 8.28.

In examining the role of water in rock strength, Rojstaczer and Bredehoeft (1988) note that failure occurs at the three sites just cited when the fluid pressure is a mere one-third to three-fifths of the greatest principal stress. Such modest fracture pressures in rocks suggest that these rocks already possess a large deviatoric stress, perhaps to the extent that they were very close to failure at ambient fluid pressures. Hence, the perturbations necessary to induce failure were not large. On the other hand, with permeable materials, the pressure-producing mechanisms must operate rather quickly lest the fluid pressures drain off. Injection wells are perfectly suited for the establishment of a rapid fluid pressure response to increases in water volumes, especially in rocks of modest permeability and low compressibility.

Seismicity Induced in the Vicinity of Reservoirs

Since the fluid pressure mechanism of induced seismicity was first postulated in association with the Denver earthquakes, fluid pressure contributions to fault movements have been inferred at tens of sites throughout the world (Woodward and Clyde, 1979). The first such observations were at Lake Mead, where seismic activity started shortly after the filling period commenced. In 1937, with the lake 80% filled, 800 separate earthquakes were recorded over a two-month period, one of which had a magnitude greater than 5. As of 1975, 10,000 earthquakes have been recorded (Anderson and Laney, 1975). Induced seismicity of a magnitude of 5 or greater has the potential

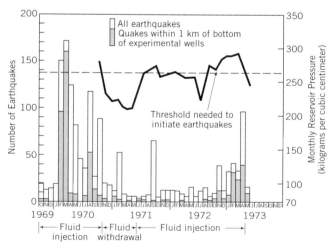

Figure 8.27   Earthquake frequency at Rangely Oil Field and its relation with induced reservoir pressures (from Wallace, 1974).

to cause failure of the dam itself. In 1967, the Konya Reservoir in India induced an earthquake of magnitude 6.5, which caused extensive damage and claimed 180 lives (Carder, 1970). Schleicher (1975) claims the Palisades Reservoir in Idaho also triggers earthquakes, with the number of occurrences increasing in the spring when the reservoir is at its highest level and in the fall when it is at its lowest level. Numerous other occurrences are cited by Carder (1970).

In examining earthquake activity in the vicinity of several dams, Roth (1969) drew the following conclusions:

1. Seismic activity seems to be more commonly associated with reservoirs deeper than 100 m, with the total water volume of less importance than the height.

2. Seismic activity generally reaches a maximum after the first filling of the reservoir and generally diminishes in the course of a few years.

3. In many cases the strongest shocks occur after many foreshocks, with the frequency of the foreshocks being greater in areas of least tectonic activity.

Seismic inducement at reservoirs has generally been associated with the following conditions:

1. The rocks at the site are under a large deviatoric stress and very near their failure strength or on the verge of sliding on preexisting fault planes.

2. The rocks may be associated with potentially active faults.

3. The triggering force is most likely an increase in fluid pressure.

Most authors again call on the Mohr–Coulomb theory to explain the phenomena. Some advanced reading on this topic is provided by Bell and Nur (1978) and Gupta and others (1972).

Seismicity and Pore Fluids at Midcrustal Depths

The information obtained from the "experiments" on induced seismicity suggests that failure along preexisting faults occurs at rather modest increases in fluid pressure. These "experiments" also suggest that in most if not all cases, the rocks are subjected to large deviatoric stresses, which accounts for the modest fluid pressure requirements. Given the same or similar states of deviatoric stress in the crust, fluid pressures that promote failure must be elevated at least to the levels as suggested by the induced seismicity studies. Given smaller deviatoric stresses, even greater fluid pressures are required. As little is known about the fluid pressure and stress conditions at crustal depths, little more can be stated about what is actually occurring there. Rojstaczer and Bredehoeft

(1988) summarize those factors upon which failure at such depths will depend: (1) the rate and duration of the pressurization mechanisms, (2) the permeability and compressibility of the rock in which it operates, (3) the degree to which the process is isolated from the surface of the earth, (4) the orientation of the fault planes relative to the principal stresses, and (5) the degree of difference between the greatest and least principal stresses. In this regard, items 1 and 2 dictate how close the pressure generation will take place to constant mass or constant pressure; item 3 refers to the ability of deep rocks to dissipate fluid pressures due to a hydraulic communication with the Earth's surface, item 4 is purely a geometric argument, and item 5 puts constraints on the pressures required to promote failure.

The Phreatic Seismograph: Earthquakes and Dilatancy Models

Earthquakes occur when elastic strain in a deforming medium is released, resulting in intense seismic vibrations. When such earthquakes occur, seismic waves are released, some of which (so-called P and Rayleigh waves) have compressional and dilational components, which causes both compression and dilation of the materials through which the waves pass. Such waves can travel great distances and can cause water-level fluctuations in wells located far from the source of the earthquake.

Blanchard and Byerly (1935) were the first to suggest that a float connected to a recording instrument would record the dilational components of passing seismic waves. These authors recognized that different wells would produce different responses to the same wave, depending on the character of the well itself, and the hydraulic properties at the point of recording. Their solution was to correct these responses empirically by comparing records from various well responses and seismographs. In this regard, Eaton and Takasaki (1959) established empirical relations between the maximum amplitude of a water-level fluctuation, and the type, wavelength, and amplitude of the waves producing the response. They noted that the response of the water level decreases rapidly as the wavelength of the passing seismic waves decreases.

Cooper and others (1965) were the first to address this problem quantitatively with an analytical solution. This solution describes the manner in which a water level will respond to a passing seismic wave, and this response is a function of the properties of the well, the hydraulic properties of the aquifer, and the type, period, and amplitude of the wave.

The work of Cooper and others (1965) was extended by Bodvarsson (1970), who attempted to examine confined fluids as strain meters. Bodvarsson noted, like Cooper and others (1956), that a well's seismograph cannot be compared to true seismographs unless the well-aquifer system is well known. A few years later, Nur (1972) suggested that water-level fluctuations to such waves may be helpful in predicting the occurrence of earthquakes. Nur based his theory on the observation by Sadovsky and others (1969) as well as others that prior to earthquakes, the ratio of the seismic velocity of the primary and secondary waves decreased dramatically and, following the earthquake, resumed a normal level. The velocity of the primary wave is greatly affected by the degree of water saturation, but the velocity of the secondary wave is not. Thus the intuitive conclusion that prior to an earthquake, some mechanism must be operative that causes a decrease in saturation of the rocks so affected. Preearthquake dilatancy is one mechanism that can account for this, where just prior to deformations that produce earthquakes, the rocks undergo inelastic volume increases produced by the formation of new fractures. As mentioned previously, dilation of crystalline rocks begins at a deviatoric stress of one- to two-thirds of the stress difference required to fracture the rock (Brace and others, 1966). This fracture is accompanied by a porosity increase that ranges from 0.2 to 2 times the elastic volume change the material would have had were it perfectly elastic. Hence, the flow of water from the preexisting pores or cracks into the newly formed pores or cracks can cause the rock to become temporarily undersaturated, which effects the velocity of the primary waves. Nur (1972) argued that as water moved into the dilatant region, the pore pressure would rise, lowering the effective confining pressure so that continued strain could cause the failure that triggers the earthquake. A further modification of this model is given by Scholz and others (1973).

It thus follows that if we had a water monitoring point in a rock mass during the sequence of events leading to an earthquake we might notice the following behavior. First, we assume the rocks are undergoing some tectonic strain and if the deformation is taking place at or near constant mass, or at least faster than the fluid can be expelled, the fluid pressure will rise accordingly. Thus we would notice a rise of the water level in the monitoring point. Now to go further we must speculate on the dilatancy model. We presume that the increased strain

is manifested by an increase in the mean confining stress acting on a saturated unit volume. There is little question that the stress contributing to strain is increasing at a faster rate than the fluid pressure within the unit volume. This is demonstrated from Eqs. 8.42 and 8.43 for constant fluid mass

$$\frac{\partial P}{\partial \varepsilon} = \xi Q \qquad \frac{\partial \sigma}{\partial \varepsilon} = \frac{\xi Q}{R/H} \qquad (8.67)$$

where R/H is generally less than one. This suggests that the effective stress is undergoing a net increase. However, if we speculate that the minimum principal stress is undergoing little change, the overall effect of this process is to increase the deviatoric stress progressively at a faster rate than the fluid pressure increase. When the critical deviatoric stress is achieved for the fluid pressure that is generated, the dilatant behavior is triggered. In this model, we have imposed the same conditions normally employed with the triaxial apparatus used in the laboratory. It is expected now that the water level in the well will fall as the fluid drains to fill the newly opened cracks. It is not possible to state the volume of new openings that are created during the dilatant period. One obvious estimate is that the pore volume that is generated is equal to the volume needed to return the fluid to its ambient pressure when the tensile strength of the rock was exceeded. With continued tectonic strain, the fluid pressure starts to rise again, but it can never rise to a level higher than it was when the rocks first became dilatant. That is, the fluid pressure that initiates dilatancy represents a theoretical upper limit to fluid pressure development lest the rocks become dilatant again. However, this is no longer the same rock with the same strength characteristics that it possessed previous to its dilatant behavior. For one thing, it appears likely that dilatancy represents a loss of cohesive strength, thereby reducing the resistance to shear. Hence, this second generation of fluid pressure rise may result in failure with consequent earthquake production.

Observations of the relationships between water well fluctuations and seismic events date back several years. Modern observations have been reported by Vorhis (1955), Gordan (1970), and Wesson (1981).

Problems

1. An elastic aquifer is overlain by 100 ft of clayey material. The aquifer has a storativity of 5×10^{-4} and a porosity of 0.3 and is 200 ft thick. The clayey material has a specific storage of 4×10^{-4} ft^{-1}.
a. Calculate the elastic compression of the aquifer in response to a 100-ft decline in head.
b. Calculate the inelastic compression of the clayey material for a 100-ft decline in head.

c. Ascertain the bulk modulus of compression of the aquifer.

2. Perform the calculation in Example 8.1 for a clay layer 60 ft thick. If the benchmarks are releveled every year, would the results give the appearance that there is no time lag between the lowering of the head and the attainment of full consolidation? Discuss your conclusions in terms of the time constant T^* and the observation time t.

3. The barometric efficiency of a well in an elastic aquifer is 0.4. If the porosity is 0.1 and the thickness is 91 m, estimate the storativity.

4. Reduce Eq. 8.38 to the more familiar form of the diffusion equation given as Eq. 4.31. State carefully all the relationships and assumptions required.

5. Derive Eq. 8.9 starting from Eqs. 8.7 and 8.8.

6. Considering sandstones, limestones, and basalts for the assumption of incompressible grains, prepare a chart that shows which of these rock types have the largest response to atmospheric pressure changes and ocean tides and which have the smallest. Considering the compressible grain case, which of these rocks have the largest response to earth tides and which have the smallest?

7. A sedimentary layer is 10,000 m thick and was deposited at a rate of 500 m/10^6 yr. The hydraulic diffusivity is 1 m²/yr. Calculate t/T^*. Determine the excess pressure as a percentage of the overburden pressure at 2000 m and 10,000 m below the surface of the sediment.

8. Answer the following from Figure 8.21.

a. For $K/S_s = 10^{-3}$ cm²/s for a 20-km deposit, what is the largest sedimentation rate required to develop excess pressure?

b. For $K/S_s = 10^{-3}$ cm²/s for a 10-km deposit, what is the lowest sedimentation rate required to develop excess pressure?

c. Comparing (a) and (b), which has the largest time constant T^*?

d. What is the value of T^*/t for $K/S_s = 10^{-2}$ cm²/s and a sedimentation rate of 500 m/my for a 5-km deposit?

e. For problem (d), will this situation be characterized by excess pressure? Why?

9. What assumptions are required to reduce Eqs. 8.42 and 8.43 to Eqs. 8.7 and 8.8?

10. Reproduce curves A, B, and C of Fig. 8.20. Take γ as 2.3 g/cm³. Then reproduce curve B if $R/H = 0.8$.

11. A repository for nuclear waste is under construction at the 3000-ft level in low-permeability basalts. The total vertical stress at this depth is 248 bars. For a water table about 200 ft below land surface, the neutral stress is about 83 bars at the 3000-ft level. What temperature increase due to radioactive decay will reduce the effective vertical stress to zero at the 3000-ft level? Assume a constant fluid mass situation. Everything you are required to know about this basalt is given in Table 8.4.

12. Prove the relations given in Eq. 8.67. Use Eqs. 8.42 and 8.43 as your starting point.

13. For basaltic rocks such as those in Table 8.4, calculate the change in total stress, fluid pressure, and effective stress with increased strain. Hint: See Eq. 8.67. Make similar calculations for the sandstone and mudstone of Table 8.4. Make some general statements regarding rock rigidity and the distribution of the total stress between the pore fluids and the matrix (effective stress).

Chapter 9

Heat Transport in Ground-Water Flow

9.1 *Conduction, Convection, and Equations of Heat Transport*
9.2 *Forced Convection*
9.3 *Free Convection*
9.4 *Energy Resources*
9.5 *Heat Transport and Geologic Repositories for Nuclear Waste Storage*

In the previous chapter, we discussed abnormal pressures that result from thermally expanded fluids and the role of temperature in phase transformations. For the most part, however, our interest has been in isothermal flow, simply because temperature effects were not important to the discussions. Our interest in this chapter shifts to hydrogeologic investigations where the nonisothermal aspects are the major ones that concern us.

Prior to getting to the main points of this chapter, a few well-known facts should be established. Lovering and Goode (1963) show that the effective perturbation depth of temperature fluctuations at the Earth's surface is on the order of 10 m. Thus, only very shallow ground water exhibits appreciable temperature changes in response to seasonal variations in the amount of solar energy reaching the Earth's surface. This is reflected by the fact that the temperature of ground water at depths ranging from 10 m to 20 m will generally be 1° to 2°C higher than the local mean annual temperature (Figure 9.1). Below this depth, the temperature increases more or less steadily. In general, the older and more compact the rock, the lower the geothermal gradient. The geothermal gradient in the Canadian Shield has been reported as 9.1°C/km, whereas gradients as high as 36°C/km have been reported for the Mississippi Embayment in Louisiana. In performing these measurements, it is normally

assumed that an equilibrium exists between the solid and liquid phases so that there is little or no temperature difference between them.

As the conductance of heat is directly proportional to the gradient in temperature, the higher the geothermal gradient, the higher the conductive heat flow at the surface. Heat flow on the shield areas has been determined to be on the order of 1 calorie per square centimeter per second (cal/cm²s), whereas at the midoceanic ridges, it is on the order of 8 cal/cm²s or greater. Goguel (1976) reports that the average heat flux by conduction at the Earth's surface is on the order of 1.3 cal/cm²s. Variations in surface heat flow and geothermal gradients have been attributed to differences in the thermal conductance of various media, geologically recent intrusions of magma, and the influence of ground-water flow on the thermal regime. Lachenbruch and Sass (1977) consider this last factor to be the most important in modifying the heat flow originating at great depths in the Earth.

9.1 *Conduction, Convection, and Equations of Heat Transport*

Heat can be transported from point to point in a porous medium by way of three processes: conduction, convec-

191

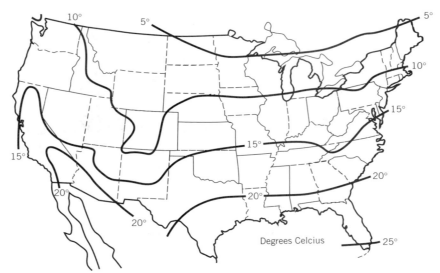

Figure 9.1 Temperature of ground water in the United States at depths of 10 to 20 meters (from Collins, 1925).

tion, and radiation. Conductive transport may be described by a linear law relating the heat flux to the temperature gradient. Convective heat transport is the movement of heat by a moving ground water. Radiation, better known as thermal electromagnetic radiation, is the radiation emitted because of the temperature of a body.

Fourier's Law

Fourier's law describes the conduction of heat from regions where the temperature is high to where it is low. This process is called heat conduction, or simply conduction, and is described for a solid or a liquid as

$$\mathbf{H} = -\kappa \text{ grad } T \qquad (9.1)$$

where $\mathbf{H}$ is the heat flux, T is temperature, κ is the proportionality constant relating the two, referred to as the thermal conductivity of the substance, and grad T is the temperature gradient. In the metric system, the unit of heat is the calorie, with temperature expressed in centigrades and distance in centimeters. The thermal conductivity has the units cal s^{-1} $cm^{-1}°C^{-1}$. In the English system, the unit of heat is the BTU (British thermal unit), and temperature is measured in degrees Fahrenheit. The thermal conductivity is expressed in (BTU) $hr^{-1}ft^{-1}°F^{-1}$. In SI units, the unit of heat is the joule with temperature expressed in degrees kelvin K so that thermal conductivity has the units joules $m^{-1}s^{-1}°K^{-1}$.

When dealing with fluids within solids, it is sometimes necessary to distinguish between energy transport in both the fluid and solid phase. We will assume that the solids and their contained fluids are at the same temperature so that we can define T in Eq. 9.1 as the average temperature of both phases. In addition, because both the fluids and the solids are conductors, it is necessary

to introduce an effective thermal conductivity so that Eq. 9.1 is expressed

$$\mathbf{H}_e = -\kappa_e \text{ grad } T \qquad (9.2)$$

where $\mathbf{H}_e$ is an effective energy flux vector, κ_e is an effective thermal conductivity, and T is understood to be the average temperature of the solid and fluid mass.

In two phase mixtures consisting of water and solids, both of which are conductors, an effective thermal conductivity can be described by considering the volume fractions and conductivities of the individual phases. For a parallel arrangement of fluids and solids, the accepted relationship is

$$\kappa_e = n\kappa_f + (1 - n)\kappa_s \qquad (9.3)$$

where κ_e is the effective thermal conductivity and the subscripts f and s refer to the fluids and the solids, respectively. In discussing this relationship, Slattery (1972) suggests that it must be amended to account for tortuosity (or a nonparallel arrangement of fluids and solids) in the porous medium. He distinguishes between two cases, one where the fluid is stationary and one where conduction is accompanied by fluid movement. For a nonmoving fluid, the effective thermal conductivity becomes

$$\kappa_e = n\kappa_f + (1 - n)\kappa_s - \kappa_s K^* \qquad (9.4)$$

where K^* accounts for the reduction in free transport because of the tortuosity of the porous medium. For a moving fluid, a convective effect referred to as thermal dispersion is noted. Thermal dispersion is a microscopic dispersal of heat due to the convective transport through the porous structure, and has the same effect, when viewed macroscopically, as an increase in the effective thermal conductivity. Dispersion will be taken up rather completely when we discuss mass transport. Here it is sufficient to say that this effect is normally incorporated in the transport parameter κ_e.

The special cases investigated by Slattery (1972) have been verified experimentally. Experimental data summarized by Green and Perry (1961) demonstrate that the value of the effective thermal conductivity is greater with a flowing fluid than with a stagnant one and, in general, increases with increasing velocity. Yagi and Kunii (1957) indicate that the effective thermal conductivity can be separated into two terms, one independent of fluid flow and the other dependent on fluid flow. In assessing the effect of several mechanisms on the thermal conductivity, Singer and Wilhelm (1950) include (1) molecular conduction through the fluid phase, (2) solid particle-to-particle molecular conduction, and (3) particle-to-particle radiation. None of these mechanisms is affected by fluid flow, and may be plotted as horizontal lines in a plot of thermal conductivity versus velocity (Figure 9.2). Two additional mechanisms control the dependence of the thermal conductivity on fluid velocity, (4) a series mechanism from fluid to solid to fluid, and so on, and (5) convection through the fluid phase. These become predominant at large velocities (Figure 9.2).

Some measured values for the effective conductivity in rocks are given in Table 9.1. In general the more porous the medium, the lower the effective thermal conductivity, mostly because water has a lower thermal conductivity than most solid minerals.

Because conductance can take place through both the solid and fluid phase, it is possible to propose upper and

Table 9.1 **Thermal Conductivities of Rocks**

Material	Thermal Conductivity (cal/m s °C)
Quartz	2
Sandstone	0.9
Limestone	0.5
Dolomite	0.4–1
Clay	0.2–0.3
Water	0.11
Air	0.006

lower bounds of the effective conductivity. For macroscopically homogeneous and isotropic materials, such limits may be determined by the method of Hashin and Shtrikman (1962). For the upper and lower limits, respectively,

$$\kappa_{eu} = \kappa_s + n \left[\frac{1}{(\kappa_f - \kappa_s)} + \frac{(1-n)}{3\kappa_s} \right] \qquad (9.5a)$$

and

$$\kappa_{e\ell} = \kappa_f + \frac{(1-n)}{1/(\kappa_s - \kappa_f) + n/3\kappa_f} \qquad (9.5b)$$

If porosity approaches zero, the upper and lower limits approach the conductivity of the solids. As porosity approaches one, both limits go to the conductivity of the fluid. Equations 9.5 hold equally well for the electrical conductivity as well as other two-phase conductive properties of porous media.

Convective Transport

In systems where the fluid is moving, there is a convective transport by the fluid motion. When the flow field is caused by external forces, the transport is said to occur by forced convection. Such is the case where ground water movement takes place in the absence of density gradients such that Darcy's law applies. A second type of transport, called free convection, occurs when the motion of the fluid is due exclusively to density variations caused by temperature gradients. Free convection is probably the dominant type of fluid motion in some hydrothermal systems where the bulk of liquid discharge is in the form of steam or hot water. A similar phenomenon can occur in mass transport, say, along fresh water–salt water interfaces in coastal areas where density differences reflect the various salinity differences. For these cases, there no longer exists a scalar potential so that Darcy's law does not describe the motion. It is emphasized that forced and free convection represent two limiting conditions. In the case of the former, buoyancy forces are assumed to be negligible. For the latter, fluid motion must be described entirely in terms of buoyancy.

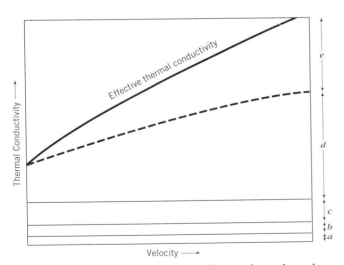

Figure 9.2 Relationship between effective thermal conductivity and the mechanisms that contribute to it. (*a*) Conduction by fluid, (*b*) conduction by solid particles, (*c*) particle-to-particle radiation, (*d*) conduction by a series mechanism from solid to fluid, and so on, (*e*) convection through fluid phase. (Modified from Schuler, Stallings, and Smith, Heat and mass transfer in fixed bed reactors, Chem. Eng. Prog. Symp. Ser., v. 48, p. 19–30, 1952). Reproduced by permission of the Amer. Inst. Chem. Engrs.

When buoyancy forces dominate, the velocity field and the energy field (temperature) are interdependent, and the equations must be solved iteratively.

The simplest system to demonstrate forced and free convection is shown in Figure 9.3, which is a hypothetical model for flow in hydrothermal areas. Note the transition from forced convection at shallow depths to a mixed convection pattern, and ultimately free convection. According to Bear (1972) the equation for ground water motion in such environments may be described in terms of two driving forces, fluid pressure changes and buoyancy, and is of the form

$$v = \frac{-k}{n\mu}\frac{\partial P}{\partial x} - \frac{kg\rho_0[1 - \alpha_f(T - T_0)]}{n\mu}\frac{\partial z}{\partial x} \quad (9.6)$$

where ρ_0 is some reference density, μ is the viscosity, and α_f is the coefficient of volume expansion for the water. If the pressure distribution is hydrostatic, $\partial P/\partial x = -\rho_0 g \partial z/\partial x$, and Eq. 9.6 becomes

$$v = \frac{kg\rho_0\alpha_f(T - T_0)}{n\mu}\frac{\partial z}{\partial x} \quad (9.7)$$

where, for vertical flow, $\partial z/\partial x = 1$. The quantity $\rho_0 g \alpha_f(T - T_0)$ is the driving force per unit volume of fluid.

The rate equation combining conduction with a Darcy type of fluid motion (forced convection) may be expressed

$$H_x = -\kappa_{e_x}\frac{\partial T}{\partial x} + n\rho_w c_w T v_x \quad (9.8a)$$

$$H_y = -\kappa_{e_y}\frac{\partial T}{\partial y} + n\rho_w c_w T v_y \quad (9.8b)$$

$$H_z = -\kappa_{e_z}\frac{\partial T}{\partial z} + n\rho_w c_w T v_z \quad (9.8c)$$

where v is the velocity and c_w is the specific heat of the fluid, defined as the heat necessary to raise the temperature of 1 g of fluid 1°C, with the units cal/g°C. The product $\rho_w c_w$ is the heat capacity per unit volume of fluid. For isotropic conditions

$$\mathbf{H} = -\kappa_e \text{ grad } T + n\rho_w c_w T\mathbf{v} \quad (9.9)$$

where $\mathbf{v}$ is a vector with components v_x, v_y, v_z.

Equations of Energy Transport

The starting place for development of conservation statements pertaining to energy transport is a word equation of the form

energy inflow rate − energy outflow rate =
change in energy storage with time   (9.10)

in units of energy per time. In general, this statement can be applied to a domain of any size. When applied to a representative unit volume with inflows through three sides, the left-hand side of this word equation is readily expressed as the divergence of the flux of interest (see Chapter 4):

$-\text{div } \mathbf{H}$ = net energy outflow rate per unit volume

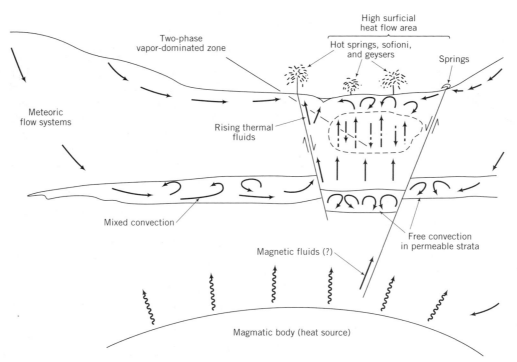

Figure 9.3   Forced, free, and mixed convection in a hydrothermal system (from Sharp and Kyle, 1988, Geol. Soc. Amer. Pub., from Geol. of N. America, 0–2, p. 461–483).

It remains to describe the nature of the flux, which could be conduction (Eq. 9.2) or conduction combined with convection (Eq. 9.8). The right-hand side of the word equation represents the gains or losses in energy per unit time for the unit volume. It will be recalled that for the conservation equation for fluid flow, it was assumed that the time rate of change of fluid volume per unit volume was proportional to the time rate of change of hydraulic head, thereby introducing a proportionality constant termed the *specific storage*. For the equations of energy transport we assume that the loss or gain of heat inside the unit volume is proportional to the temperature in the unit volume. Some proportionality constant is now needed to convert temperature changes to the gains or losses of heat. This proportionality constant is the heat capacity per unit volume, defined for water alone as the product of the fluid density ρ_w and the specific heat per unit volume of water c_w. The heat capacity per unit volume must express the quantity of heat gained or lost from the unit volume when the temperature changes by a unit amount (cal/L^3°C). For the fluid component in the unit volume, this becomes $n\rho_w c_w$. For the solid components, the expression becomes $(1 - n)\rho_s c_s$, where the subscript s stands for solids. Thus we define an effective heat capacity for the unit volume

$$\rho' c' = n\rho_w c_w + (1 - n)\rho_s c_s \qquad (9.11)$$

The conservative statement of Eq. 9.10 then becomes

$$-\text{div } \mathbf{H} = \rho' c' \frac{\partial T}{\partial t} \qquad (9.12)$$

where $\rho' c'$ is the heat removed or gained from the unit volume when the temperature changes a unit amount. Equation 9.12 states that the net energy outflow rate per unit volume equals the time rate of change of energy within the unit volume. Table 9.2 gives some values for the specific heat of various substances.

The Heat Conduction Equation

For a conductive system defined by Fourier's law of the form of Eq. 9.2

$$-\text{div } \mathbf{H} = \text{div}[\kappa_e \text{ grad } T] \qquad (9.13)$$

Here, κ_e has been defined as an effective thermal conductivity. The conservation statement then becomes

$$\text{div}[\kappa_e \text{ grad } T] = \rho' c' \frac{\partial T}{\partial t} \qquad (9.14)$$

By assuming that the effective thermal conductivity is constant

$$\frac{\partial^2 T}{\partial x^2} + \frac{\partial^2 T}{\partial y^2} + \frac{\partial^2 T}{\partial z^2} = \nabla^2 T = \frac{\rho' c'}{\kappa_e} \frac{\partial T}{\partial t} \qquad (9.15)$$

This equation is called the heat conduction (diffusion) equation and is analogous to the unsteady flow equation

Table 9.2 **Specific Heat of Substances**

Material	T(°C)	Specific Heat (kcal/kg°C)
Air	50	0.248
Water vapor	100	0.482
Methane	15	0.528
Benzene	20	0.406
Ethyl alcohol	25	0.581
Basalt (dry)	20–100	0.20
Chalk (dry)	20–100	0.214
Clay (dry)	20–100	0.22
Granite (dry)	12–100	0.192
Quartz (dry)	12–100	0.188

introduced in Chapter 4. The solution to this equation describes the value of temperature at any point in a three-dimensional field, or more precisely, how the temperature is changing with time.

If the net outflow is zero

$$\frac{\partial^2 T}{\partial x^2} + \frac{\partial^2 T}{\partial y^2} + \frac{\partial^2 T}{\partial z^2} = \nabla^2 T = 0 \qquad (9.16)$$

This equation is called Laplace's equation, as was the steady-state fluid flow equation introduced in Chapter 4. The solution of this equation describes the value of the temperature at any point in a three-dimensional field.

The Conductive-Convection Equation

For conduction and convection described by Eqs. 9.8

$$-\text{div } \mathbf{H} = -\text{div}[-\kappa_e \text{ grad } T + n\rho_w c_w T\mathbf{v}] = \rho' c' \frac{\partial T}{\partial t} \qquad (9.17)$$

or, with the usual assumptions of constant κ_e, n, ρ_w, c_w,

$$\kappa_e \nabla^2 T - n\rho_w c_w [\mathbf{v} \cdot \nabla T + T\nabla \cdot \mathbf{v}] = \rho' c' \frac{\partial T}{\partial t} \qquad (9.18)$$

For steady ground-water flow, $\nabla \cdot \mathbf{v} = 0$

$$\kappa_e \nabla^2 T - n\rho_w c_w \mathbf{v} \cdot \nabla T = \rho' c' \frac{\partial T}{\partial t} \qquad (9.19)$$

Note that when the velocity is zero we recover our transport equation for pure conduction (Eq. 9.15). When the temperature is steady

$$\kappa_e \nabla^2 T - n\rho_w c_w \mathbf{v} \cdot \nabla T = 0 \qquad (9.20)$$

The equations developed here are called the conduction-convection equations. They are conservation expressions that describe the manner in which energy is moved from one point to another by means of bulk fluid motion and by conduction. The one-dimensional form of Eq. 9.19 is written

$$\frac{\kappa_e}{\rho' c'} \frac{\partial^2 T}{\partial x^2} - \frac{n\rho_w c_w}{\rho' c'} v_x \frac{\partial T}{\partial x} = \frac{\partial T}{\partial t} \qquad (9.21)$$

Table 9.3 **Thermal Diffusivities of Some Common Rocks and Soils in cm²/s**

Soils and Unconsolidated Material	
Quartz sand, medium, dry	
Quartz sand, 8.3% moisture	0.0020
Sandy clay, 15% moisture	0.0033
Soil, very dry	0.0020–0.0030
Some wet soils	0.0040–0.0100
Wet mud	0.0022
Soil, Lexington, Kentucky	0.0021
Soil, Lexington, Kentucky (average 0–10 ft in place)	0.0072
Gravel	0.0057–0.0062
Rocks	
Shale	0.0040
Dolomite	0.0080
Limestone	0.0050–0.010
Sandstone	0.0113–0.0140
Granite	0.0060–0.0130

From Cartwright (1973).

where v_x is the mean ground-water velocity in the x direction. The first term on the left describes energy transport by conduction, and the second describes energy transport by convection. In all cases, it is assumed that the temperature of the fluid and the solids are equal. Slattery (1972) describes the case where distinctions are required between the fluid and solid phases.

Last, the combined term $\kappa_e/\rho'c'$ has the units of L^2/T and is referred to as the thermal dispersion λ. It is understood that the convective effect of thermodispersion is incorporated in the parameter κ_e. In the absence of fluid movement, this parameter (λ) becomes the thermal diffusivity. Table 9.3 gives some values for the thermal diffusivities of rocks.

Dimensionless Groups

The equations governing the transport of heat in groundwater flow have been called the conduction-convection equations. The solution to these equations for certain boundary and initial conditions provides us with the temperature distribution of a region. By applying the nondimensionalizing procedures introduced in Chapter 4, it is easily seen that the behavior of a thermal system is controlled by a few combinations of dimensionless variables. We have already seen this in our discussions of the Fourier number.

To the cited dimensionless quantities in Chapter 4 we add now a dimensionless temperature $T^+ = T/T_e$, where T_e is some characteristic temperature and the superscript + indicates a dimensionless quantity. The conduction-convection Eq. 9.19 becomes

$$\nabla^{2+}T^+ - \left[\frac{n\rho_w c_w vL}{\kappa_e}\right]\nabla^+T^+ = \left[\frac{L^2(\rho'c'/\kappa_e)}{t_e}\right]\frac{\partial T^+}{\partial t^+}$$

$$(9.22)$$

where the Peclet number for energy transport is

$$N_{PE} = \frac{n\rho_w c_w vL}{\kappa_e} = \frac{\rho_w c_w qL}{\kappa_e} \qquad (9.23)$$

where q is Darcy's specific discharge and L is some characteristic length. The Peclet number expresses the transport of energy by bulk fluid motion to the energy transport by conduction. In a practical sense this number reflects a competition between two rate processes, forced convection and conduction. Large numbers mean that convective transport dominates over conductive transport. The second dimensionless group on the right-hand side of Eq. 9.22 is recognized as the Fourier number for heat transport.

Suppose now that we assume a buoyancy-driven fluid system. A dimensional analysis of the steady-state transport Eq. 9.20 in combination with the buoyancy-driven velocity of Eq. 9.7 gives

$$\nabla^{2+}T^+ - \left[\frac{g\rho_0 c_w \rho_w Lk\alpha_f(T - T_0)}{\mu\kappa_e}\right]\nabla^+T^+ = 0 \qquad (9.24)$$

where the Rayleigh number for energy transport is

$$N_{RA} = \frac{g\rho_0 c_w \rho_w Lk\alpha_f \Delta T}{\mu\kappa_e} \qquad (9.25)$$

The Rayleigh number expresses the transport of energy by free convection to the transport by conduction and is generally used to establish the conditions for the onset of free convection. In most one-dimensional cases, the characteristic length is taken as equal to the thickness H of the formation over which the temperature difference ΔT is measured. In two-dimensional problems, it will be some length associated with fluid movement.

In problems of mixed convection, both the Peclet number and the Rayleigh number are important in the analysis. As we will see later, convective rolls at high Peclet numbers (substantial forced convection) take on a different geometry than those occurring at low Peclet numbers.

9.2 *Forced Convection*

Smith and Chapman (1983) recognize three main classes of forced convection problems of interest to hydrogeologists. The first of these is the vertical steady-state flow of ground water and its effect on a purely conductive vertical temperature distribution. If the resultant temperature profile can be measured, it can be matched against some curves representing the mathematical solution of the one-dimensional transport equation, and the Darcy velocity may be extracted. The second of the problems treats a two-dimensional vertical cross section of regional ground-water flow where the flow field alters the conductive temperature profile. Two versions are possible here. First, for reasonably shallow systems where it can be assumed that the fluid density and viscosity are not affected by temperature, the velocity field can be determined independently of the transport problem and subsequently used in the solution to the transport problem. If the density or viscosity is affected by temperature, the velocity field and the temperature field are "coupled," and the equations must be solved iteratively. The third class of problems is for the two-dimensional geometry, but the temperature field or the velocity field is assumed outright and is not part of the problem. This was the method of Kilty and Chapman (1980), who were interested in conceptual models to account for heat-flow variations in certain geologic settings. To this threefold classification, we may add yet a fourth, whereby the temperature field is affected by a three-dimensional flow field, and yet a fifth concerned with temperature distributions in evolving geologic basins.

Temperature Profiles and Ground-Water Velocity

Consider the three ideal cases shown in Figure 9.4, where a one-dimensional velocity field is imposed on a one-dimensional purely conductive thermal field. In Figures 9.4a and b, the direction of ground-water movement is taken as normal to the conductive isotherms. That is to say the hydraulic gradient and the conductive temperature gradient are collinear in the vertical direction so that the streamlines of fluid flow are normal to the isotherms of heat conduction. For these cases, a convection flux is established that will produce the greatest alteration of the conductive temperature distribution. In Figure 9.4a, the resultant temperature gradient will increase with in-

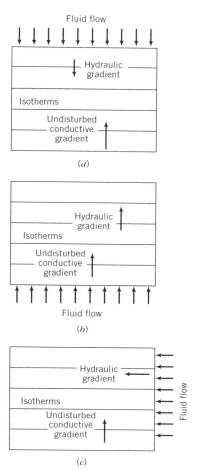

Figure 9.4 Three ideal cases where a fluid flow field is imposed on a conductive thermal gradient.

creasing depth, whereas in Figure 9.4b, it will decrease with increasing depth. In addition, the heat flow at the surface will be greater for Figure 9.4b. In Figure 9.4c, the hydraulic gradient and the temperature gradient are normal so that the streamlines of fluid flow are collinear with the isotherms of heat conduction. Here convective transport is eliminated as it is impossible to transport heat along an isotherm. Mathematically, the convective term in our transport equation becomes

$$\mathbf{v} \cdot \text{grad } T = 0$$

For this case there is no alteration of the conductive gradient. In between the extremes of streamlines normal to isotherms and streamlines collinear with isotherms, some alteration of the conductive gradient will occur, with the maximum alterations associated with the former situation. In addition, either of these extremes can be viewed from the perspective of a one-dimensional transport problem. This is the basis of the Bredehoeft–Papadopulos (1965) model for determining ground-water velocity from temperature profiles.

Figure 9.5 gives the conditions under which the model

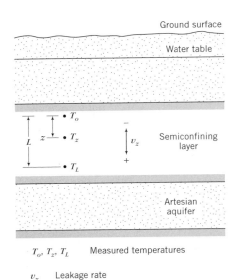

Figure 9.5 Diagrammatic sketch of typical leaky aquifer (from Bredehoeft and Papadopulos, Water Resources Res., v. 1, p. 325–328, 1965. Copyright by Amer. Geophys. Union).

applies. In this diagram, T_0 is an uppermost temperature measurement at $z = 0$, and T_L is a lower-most temperature measurement at $z = L$. These are the boundary conditions for the one-dimensional steady-state problem described by

$$\frac{\partial^2 T}{\partial z^2} - \frac{n \rho_w c_w v_z}{\kappa_e} \frac{\partial T}{\partial z} = 0 \qquad (9.26)$$

The solution to this problem is (Bredehoeft and Papadopulos, 1965)

$$T_z = T_0 + [T_L - T_0] \frac{\left[\exp\left(N_{PE} \frac{z}{L} \right) - 1 \right]}{[\exp(N_{PE}) - 1]} \qquad (9.27)$$

where T_z is the temperature at any depth over the thickness L (Figure 9.5) and

$$N_{PE} = \frac{\rho_w c_w q L}{\kappa_e} \qquad (9.28)$$

which is the familiar Peclet number for heat transport.

A graphical solution to this problem is given by the type curves of Figure 9.6 with the Peclet number taken as the solution parameter. The curves are convex upward or downward. A suggested method of analyses by Bredehoeft and Papadopulos (1965) requires a field plot of $(T_z - T_0)/(T_L - T_0)$ against the depth factor z/L at the same scale as the type curves and superimposing this plot on the type curve set. The value of N_{PE} determined from this match is then used to calculate the velocity q from Eq. 9.28. This assumes information on κ_e.

Sorey (1971) reports some data by Kunii and Smith (1961) that have been employed to verify the mathemati-

cal model just described. The experiment involved the measurement of the temperature distribution in a column of glass beads and sand through which fluid was flowing counter to an imposed temperature gradient. Sorey (1971) constructed the dimensionless plot shown as Figure 9.7, which matches the N_{PE} equals 3.3 curve of Figure 9.6. For the data reported by Kunii and Smith (1961) and the methods described, the mass flow rate was calculated to be $\rho_w q = 2.3 \times 10^{-4}$ g/cm²s, which compares favorably with the measured mass flow rate of 2.6×10^{-4} g/cm²s.

Field results have been reported by Cartwright (1970) and Sorey (1971). In Figure 9.8, Cartwright analyzes the temperature data from wells in the Illinois Basin and obtained a ground-water flow rate of 4.9×10^{-8} cm/s. The important limitations to this type of analysis reside

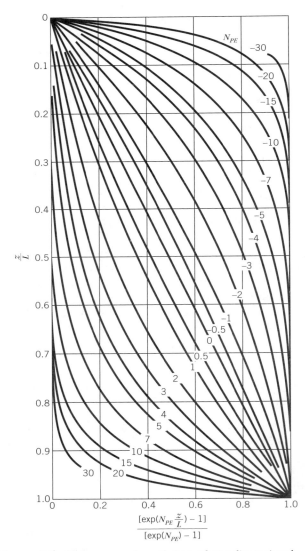

Figure 9.6 Type curves for solution of one-dimensional conductive advective equation (from Bredehoeft and Papadopulos, Water Resources Res., v. 1, p. 325–328, 1965. Copyright by Amer. Geophys. Union).

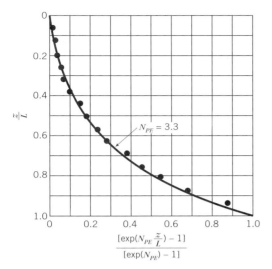

Figure 9.7 Type curve match for temperature profile in glass beads (from Sorey, Water Resources Res., v. 7, p. 963–970, 1971. Copyright by Amer. Geophys. Union).

in the one-dimensional assumptions for both the temperature field and the velocity field.

Heat Transport in Regional Ground-Water Flow

Let us now continue the "experiment" started in the previous section concerning the thermal outcome of various orientations of streamlines of fluid flow and isotherms of pure conduction. A conductive field is shown in the two-dimensional region of Figure 9.9a. A two-dimensional flow field as shown in Figure 9.9b is superimposed on this conductive field, with Figure 9.9c demonstrating the alteration of the temperature field. The greatest alteration occurs where the streamlines of flow are normal or nearly so to the conductive isotherms, that is, the recharge and discharge areas, and the least alteration takes place in the region of lateral flow. Because of this

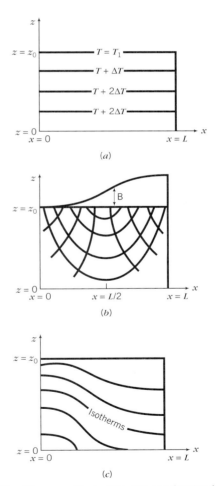

Figure 9.9 Diagrams illustrating (a) conductive heat transport in a two-dimensional region, (b) generalized flow system for the two-dimensional region, and (c) alteration of the isotherms (modified from Domenico and Palciauskas, 1973, Geol. Soc. Amer. Pub., from Geol. Soc. Bull., v. 84, p. 3803–3814).

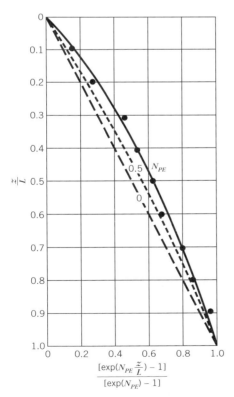

Figure 9.8 Type curve match for Illinois Basin (from Cartwright, Water Resources Res., v. 6, p. 912–918, 1970. Copyright by Amer. Geophys. Union).

convective alteration, the geothermal gradient will increase with increasing depth in the recharge area and decrease with depth in the discharge area, the latter associated with a greater amount of heat flow at the surface. This problem was solved by Domenico and Palciauskas (1973) for the following conditions:

1. The fluid properties are not affected by temperature and the medium is isotropic and homogeneous with respect to both fluid flow and heat conduction.

2. The two-dimensional flow problem is as described in Example 5.1, where the water table is unspecified but the lower and lateral boundaries are of the no-flow variety.

3. The temperature problem is described for the two-dimensional region of Figure 9.9 where the lateral boundaries are nonconductive ($\partial T/\partial x = 0$). The upper boundary is taken at some constant temperature T, and the lower boundary is described in terms of a constant temperature gradient T'_0.

Although the analytical solution to this problem was developed for any water table configuration, a special case was examined for the water table given in Figure 9.9b for which an analytical solution is presented in Example 5.1. For this special case, the solution to the two-dimensional transport equation is of the form

$$T(x, z) = [T_1 + T'_0(z - z_0)] - \frac{T'_0}{2} [N_{PE}] \left[f\left(\frac{z_0}{L}\right) \right]$$

(9.29)

where T'_0 is the temperature gradient across the bottom boundary, N_{PE} is the dimensionless Peclet number, and all other terms are described in Figure 9.9. In this form, the temperature distribution as described in Eq. 9.29 is controlled by three essential features. The first bracketed quantity describes the temperature distribution for pure conduction, that is, describes the conductive isotherms shown in Figure 9.9a. The other quantities obviously describe the perturbation of the purely conductive field and consist of a dimensionless Peclet number and a group of trigonometric and hyperbolic functions that deal exclusively with the ratio of basin depth z_0 to basin length $[f(z_0/L)]$. For this case the Peclet number is described

$$N_{PE} = \frac{KB\rho_w c_w}{\kappa_e}$$

(9.30)

where K is the hydraulic conductivity and B is the mean water table elevation as measured from the elevation of the discharge area (Figure 9.9). Thus if the hydraulic conductivity or the mean water table elevation B gets large with respect to the thermal conductivity (large Peclet number), forced convection dominates over conduction and the conductive field is altered significantly. Note that B is the characteristic length in this problem,

whereas for the one-dimensional problem of Bredehoeft and Papadopulos (1965), the characteristic length was determined to be the length of the vertical region over which perturbations were measured. The expression for $f(z_0/L)$ is cumbersome and will not be given here. However, as the basin length-to-depth ratio becomes large, the flow pattern is largely horizontal and $f(z_0/L)$ approaches zero so that there is a minimal interference with the conductive gradient, that is, $\mathbf{v} \cdot \operatorname{grad} T \cong 0$. Small length-to-depth ratios produce flow systems that contain vertical components of flow throughout and are associated with large perturbations.

For an approximate measure of the potential for convective alterations of the geothermal gradient, a modified Peclet number may be given as

$$N_{PE} = \frac{BK\rho_w c_w(z_0/L)}{\kappa_e}$$

(9.31)

Values less than one are associated with temperature distributions that are not far from the limiting case of pure conduction. Other geometric configurations of flow such as those describing local, intermediate, and regional flow (Figure 5.7) are also examined in terms of the interpretation of the characteristic length associated with Peclet numbers for two-dimensional flow systems (Domenico and Palciauskas, 1973). Van der Kamp and Bachu (1985) describe the use of dimensional analysis for studying thermal effects in various hydrogeologic regimes.

The study described was a theoretical one where the main interest was in determining the effect of fluid flow on temperatures within the framework of the classic regional flow model introduced in Chapter 5. Morgan and others (1981) have used this model to investigate the threshold hydraulic conductivity where temperature perturbations become significant in basins of fixed geometry within the Rio Grande Rift. Betcher (1977) solved the problem numerically with a finite element model and conducted a sensitivity analysis to ascertain the role of the various parameters in controlling the temperature distribution. Parsons (1970) used a numerical model to investigate the effect of hydraulic conductivity contrasts and included a case where the upper boundary was taken as a variable temperature boundary. Some results of his study are shown in Figure 9.10. Note in particular that the three-order-of-magnitude hydraulic conductivity contrast associated with the lower-permeability region more or less preserves the pure conduction isotherms, as does the rather large area of lateral flow where $\mathbf{v} \cdot \operatorname{grad} T \cong 0$.

Smith and Chapman (1983) include the effects of temperature on fluid density and viscosity in heterogeneous anisotropic media. Figure 9.11a gives some of their results for homogeneous material, showing the convective influence on heat flow at the surface, which is most pronounced for higher values of permeability. For homogeneous but anisotropic material, the heat flow at the

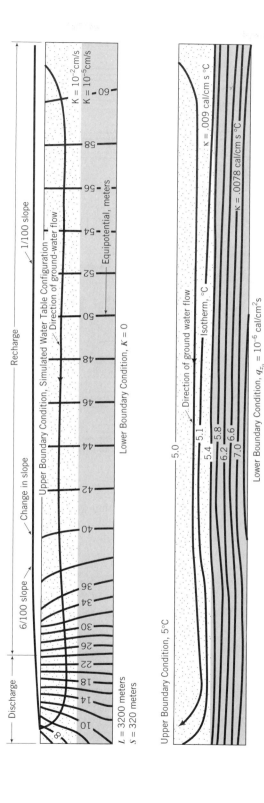

Figure 9.10 A simple hypothetical two-layered model of the ground water thermal regime. (*a*) Ground water basin. (*b*) Temperature distribution in the ground water basin (from Parsons, Water Resources Res., v. 6, p. 1701–1720, 1970. Copyright by Amer. Geophys. Union).

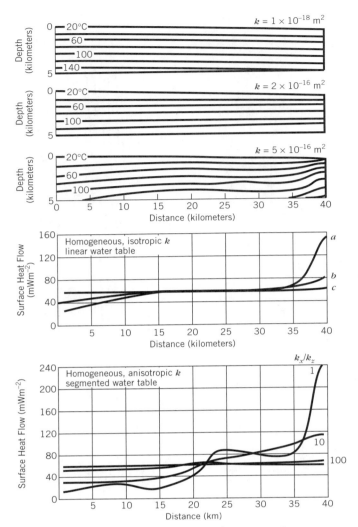

Figure 9.11 (*a*) Thermal effects of ground-water flow in a basin of homogeneous, isotropic permeability with a linear water table. Surface heat flow expressed in mWm⁻². (*b*) Thermal effects of ground-water flow in a basin of anisotropic homogeneous permeability. Horizontal permeability is 8×10^{-16} m², anisotropic ratio is 1, 10, and 100 for the three simulations shown. In all cases, ground-water flow is from left to right on the diagrams (from Smith and Chapman, J. Geophys. Res., v. 88, p. 593–608, 1983. Copyright by Amer. Geophys. Union).

surface increases with decreasing anisotropic ratios K_x/K_z (Figure 9.11*b*).

Woodbury and Smith (1985) present a solution for a three-dimensional flow field. Their analysis was concerned with the variations and areal distribution in heat flow at the surface as affected by relief on the water table and permeability variations.

The studies described are concerned with temperature distributions in modern environments. Temperature distributions have also been ascertained for paleoflow systems that may have been important in the formation of

ore deposits by topographically driven flow. Some results from Garven and Freeze (1984a, 1984b) are shown in Figure 9.12. Figures 9.12*a* and *b* depict temperature distributions at low and high Peclet numbers, respectively, due to differences in hydraulic conductivity. The advective perturbations are readily seen for the high conductivity condition. Figures 9.12*c* and *d* depict temperature distributions at high and low Peclet numbers, respectively, due to differences in the thermal conductivity. As noted, the higher the thermal conductivity the closer the temperature distribution approaches that of pure con-

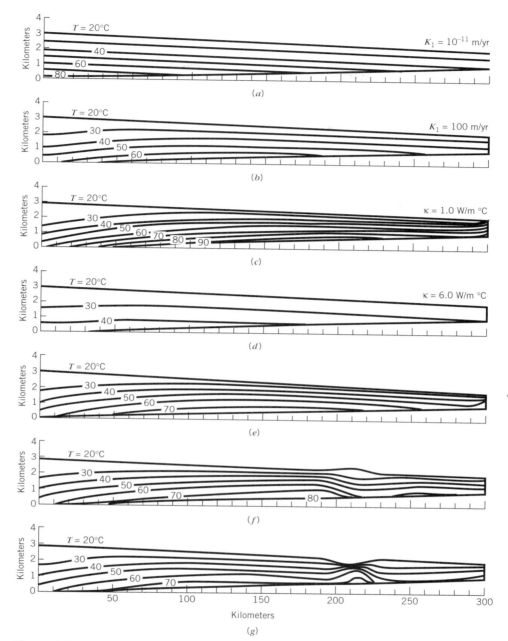

Figure 9.12 Temperature distributions at (*a*) low and (*b*) high Peclet numbers due to differences in hydraulic conductivity; (*c*) high and (*d*) low Peclet numbers due to differences in thermal conductivity and for different water table configurations (*e*, *f*, *g*). In all cases, ground-water flow is from left to right on the diagrams (from Garven and Freeze, 1984a, Amer. J. Sci., v. 284, p. 1085–1124. Reprinted by permission of the Amer. J. Sci.).

duction. Figures 9.12*e*, *f*, and *g* depict temperature distributions for different water table configurations, where the effects of recharge and discharge along the water table are reflected in the various distributions.

The field observations that support the general interference between isotherms and streamlines discussed in the theoretical work are provided in studies by Van Orstrand (1934) and Schneider (1964). Several case studies are presented in Geophysical Monograph 47 (Beck, Garren, and Stegena, 1989).

Heat Transport in Active Depositional Environments

The pressure evolution in active depositional environments was seen in Chapter 8 to be a transient phenomenon. As expected, the transient velocity field is responsible for the convective transport of heat. Examining this process from the perspective of the energy transport equations provides some information on the pressure-temperature history of a region. Although it is not possible to reconstruct all the factors that play a role in the pressure-temperature history of basins, mathematical models can be used to gain insights into certain processes and for establishing some limiting or threshold conditions.

In modeling studies, the conductive gradient is normally depicted as originating due to a heat flux across the lower boundary of an accumulating sedimentary pile (Figure 9.13). Most studies to date have assumed the condition of forced convection although some free convection cannot be ruled out. In addition, it is frequently assumed that the surface temperature remains constant during continued deposition of the sediment. Frequently, a moving coordinate system is used to study this problem. The results of such studies demonstrate that with uniform sedimentation—in the absence of fluid flow—the thermal gradient will constantly decrease, and for uniform erosion it will constantly increase. These facts are shown in Figure 9.14 by profiles 1 and 2 when the distance axis represents the depth from the sediment–water interface to the impermeable basement rock. Profiles 1 and 2 would also be expected from steady one-dimensional flow upward and downward, respectively, where the sedimentary layer remains of uniform thickness (Bredehoeft and Papadopulos, 1965). Profile 3 represents the combined effect of one-dimensional forced convection upward and the separation of the boundaries by the accumulation of sediment. In this case, a zero or near-zero velocity at the lower boundary and a maximum velocity at the growing sediment–water interface perturbs profile 1 into the reverse "S" pressure-depth profile 4.

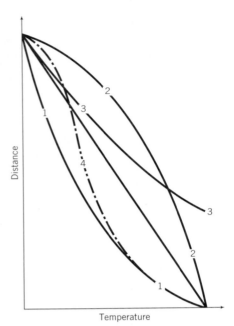

Figure 9.14 Temperature patterns with moving boundaries and convection.

Sharp and Domenico (1976) developed an energy transport model that incorporated one-dimensional fluid flow and one-dimensional heat transport. When applied to Gulf Coast conditions for the loading history of Figure 9.15, they concluded that abnormal pressures were low for much of the geologic record and were increased during the Eocene, the Oligocene, and most markedly during the Late Pliocene-Quaternary period in response to rapid sedimentation (Figure 9.16a). The period of low abnormal pressure was accompanied by pressure-depth profiles very close to the conductive geothermal gradient (assumed at 30°C/km). For the Late Pliocene-Quaternary period, the temperatures were uniformly lower than in the previous period below depths of 6000 m (Figure 9.16b). The reverse "S" profile dominates during this period of rapid sedimentation, mainly because of the moving boundary effect. They concluded that sediments at present in the near offshore are now at their maximum pore fluid pressure and minimum temperature at any depth. Chia (1979) extended this study by incorporating a three-dimensional loading or mean stress formulation within the confines of the vertical transport of fluid and heat. He also included the thermal expansion of fluids as a pressure-producing mechanism, and a fluid-pressure–fracture criterion to examine the conditions leading to fracture initiation.

Cathles and Smith (1983) developed a two-dimensional flow model in combination with energy transport to examine episodic basin dewatering and the genesis of mineral deposits. Bethke (1985) also has applied two-dimensional flow and energy transport models to accumulating sedimentary piles. His calculations indicate a tendency

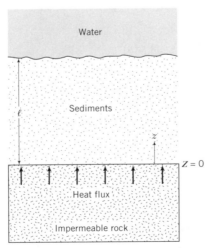

Figure 9.13 Schematic diagram of accumulating sediment with heat flux across the lower boundary.

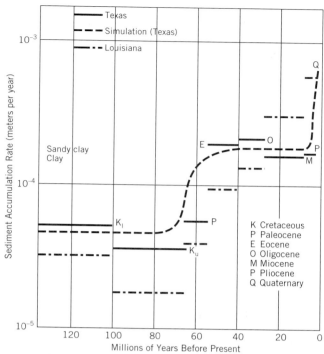

Figure 9.15 Estimated sediment accumulation rates for the offshore Texas and Louisiana coasts (from Sharp and Domenico, 1976, Geol. Soc. Amer. Pub., from Geol. Soc. Amer. Bull., v. 87, p. 390–400).

for fluids to migrate laterally toward the edge of the basins. Vertical flow occurs in the deeper parts of the basin where the ratio of vertical to lateral permeability exceeds the ratio of the lengths of vertical to lateral pathways to the surface. He demonstrated further that for small sedimentation rates the temperature gradient is not far removed from that expected from pure conduction, as would be expected for situations approximating normal compaction.

Heat Transport in Mountainous Terrain

The relief associated with mountainous terrain may promote an advective disturbance of the thermal regime that is somewhat more severe than discussed earlier in the chapter and may contribute to regional scale variations in surface heat flow. In the southern Great Basin, for example, Sass and others (1976) recognize a regional anomaly of low heat flux, termed the "Eureka low," which includes the Nevada Test Site and Yucca Mountain described earlier (Chapter 5). As the regional heat flow within the crust beneath the "Eureka low" is likely of the same intensity as occurs in the Great Basin in general, a hydrologic disturbance was postulated where "ground water is carrying much of the Earth's heat in the upper 3 km and is delivering it elsewhere" (Sass and Lachenbruch, 1982). Wherever that "elsewhere" is, there is likely to be yet another anomaly, this time a thermal high.

The nature of hydrologic disturbances in mountainous terrain has recently been investigated by Forster and Smith (1989). Their conceptual model is a two-dimensional region not unlike Figure 9.9, with the bedrock forming the upper surface and a conductive heat flux across the bottom boundary. As a continuation of their previous work described in Chapter 5 (Forster and Smith, 1988), the water table elevation is determined by using a free surface approach where an infiltration rate must be specified. Some of their results are shown in Figure 9.17*a*, *b*, and *c* for a fixed infiltration rate and a three-order-of-magnitude increase in permeability going from Figure 9.17*a* to 9.17*c*. In these figures, the isotherms are depicted with solid lines (°C), the heat lines are dashed and show the heat transport to the surface by both conduction and convection, and the dotted lines indicate

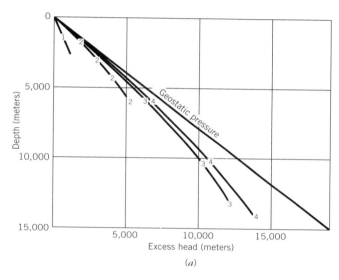

(*a*)

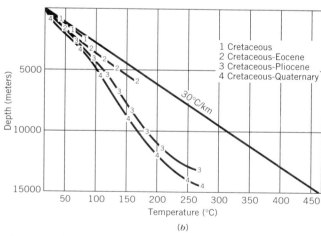

(*b*)

Figure 9.16 Calculated excess head and temperature distribution in Gulf of Mexico sediment at the end of each of the four specified time periods (from Sharp and Domenico, 1976, from Geol. Soc. Amer. Pub., from Geol. Soc. Amer. Bull., v. 87, p. 390–400).

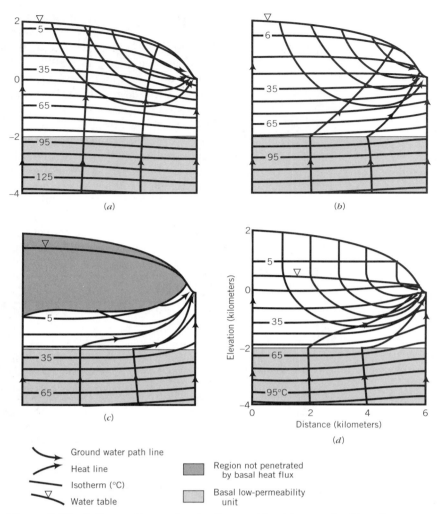

Ground water path line

Heat line

Isotherm (°C)

▽ Water table

Region not penetrated by basal heat flux

Basal low-permeability unit

Figure 9.17 Patterns of ground-water flow and heat transport with a three-order-of-magnitude increase in permeability when going from (*a*) to (*c*). Figure 9.17*d* has the infiltration rate decreased by a factor of 5 for the same permeability of Figure 9.17*c* (from Forster and Smith, J. Geophys. Res., v. 94, p. 9439–9451, 1989. Copyright by Amer. Geophys. Union).

pathlines for fluid flow. Figure 9.17*a* shows a conductive thermal regime, which becomes more disturbed as the permeability increases. In Figure 9.18*b* the disturbance is minor, whereas in Figure 9.17*c* several changes are noticeable. Such a permeable system is capable to transmit all the infiltration, and the water table falls below the bedrock surface. Further, almost all of the basal heat flux is transported to the valley. This means that for regions above the valley bottom, temperatures are governed more by thermal conditions at the bedrock surface than by heat flow from below.

The flow system in Figure 9.17*d* has the same permeability as the system in Figure 9.17*c*, but the infiltration rate has been reduced by a factor of 5, giving a fivefold decrease in the ground-water flux. The depth of the water table is increased markedly, and the reduced flux pro-

duces a rather weak thermal disturbance and a warmer thermal regime than for the higher-infiltration case of Figure 9.17*c*.

Figure 9.18 shows the effect of a high-permeability fracture on the flow system and thermal regime. A large percentage of the flow and basal heat are captured by the fracture zone, resulting in an upwelling of isotherms in the fracture zone. Such features may produce thermal springs at the fault outcrop. If the hydraulic conductivity of the rock matrix is increased by one order of magnitude, a greater flux is captured by the fracture, the 75°C isotherm of Figure 9.18 is reduced to 35°C, and the upwelling along the fault vanishes (Forster and Smith, 1989). In addition, the temperature in the upper regions approaches the temperature of the recharge because the upper unit is not penetrated by the basal heat flux.

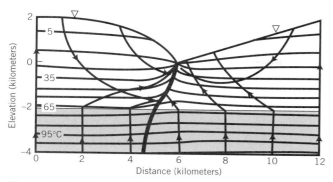

Figure 9.18 Patterns of ground-water flow and heat transport as influenced by a steeply dipping high-permeability fracture zone (from Forster and Smith, J. Geophys. Res., v. 94, p. 9439–9451, 1989. Copyright by Amer. Geophys. Union).

9.3 Free Convection

Free convection has been defined as a flow driven by density variations. As a buoyancy-driven flow, the visualization is of rising warm water, a cooling process, and the sinking of these same waters. Thus, with free convection we visualize ascending and descending currents that form cells of various geometries. The transport of heat is in the direction of the buoyant forces, and the isotherms depart markedly from their expected conductive distribution. Figure 9.19 is a typical example of a thermal profile established by free convection.

The Onset of Free Convection

Horton and Rogers (1945) were among the first to address the conditions for the onset of free convection in a porous medium. Their theoretical results were based on an imposed vertical temperature gradient across a horizontal layer, saturated with a nonmoving fluid, where the upper and lower boundaries were taken as perfect heat conductors. According to this work, convection currents may develop when the temperature gradient exceeds some critical value

$$\frac{\partial T}{\partial z} > \frac{4\pi^2 \lambda^2 \mu}{\kappa_e \rho_0 g \alpha_f H^2} \qquad (9.32)$$

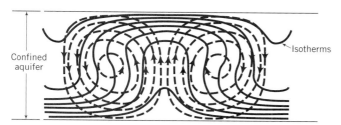

Figure 9.19 Free convection in a confined aquifer (from Donaldson, J., Geophys. Res., v. 67, p. 3449–3459, 1962. Copyright by Amer. Geophys. Union).

where λ has been introduced as the thermal diffusivity. When compared with results based on the Rayleigh criteria for pure fluids, this minimum temperature gradient exceeds the required temperature gradient for the onset of convection in fluids by a factor of $(16/27)H^2/\pi^2\kappa_e\rho_0$. Thus, temperatures that establish free convection in fluids are not sufficient to establish free convection in fluids contained within porous solids, largely because of the viscous drag of the solids on the fluids.

Lapwood (1948) demonstrated that the onset of free convection in porous media occurs at Rayleigh numbers on the order of $4\pi^2$; that is,

$$N_{RA} = \frac{g\rho_0(c_w\rho_w)Hk\alpha_f \Delta T}{\mu\kappa_e} \cong 40 \qquad (9.33)$$

This criterion is based on horizontal layers filled with a nonmoving fluid and upper and lower boundaries that are both impermeable to flow and isothermal. Other thermal and hydrologic boundary conditions for the horizontal layer case have been reported by Nield (1968).

Pratts (1966) investigated the effect of horizontal flow on the onset of free convection and found that the convection parameter $R_a \cong 40$ remains unchanged. The incorporation of horizontal flow, however, implies that both forced and free convection are occurring so it is expected that the nature of the convection rolls will be controlled in part by the magnitude of the horizontal flow. This is demonstrated in the review of Combarnous and Bories (1975), where for N_{PE} less than 0.75, the axes of convective cells appear to run perpendicular to the mean flow direction (Figure 9.20b). For higher values of

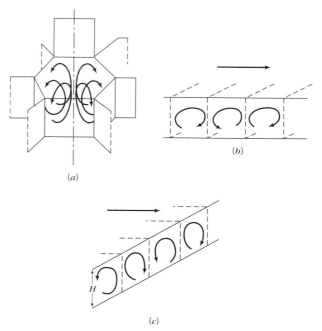

Figure 9.20 Forms of convective coils (from Combarnous and Bories, 1975. Reprinted with permission of Academic Press, from Advances in Hydroscience, v. 10, p. 231–307).

N_{PE}, helicoidal rolls form, with their axes parallel to the mean flow direction (Figure 9.20c). These can be compared with the polyhedral cells associated with free convection operating in the absence of a bulk fluid motion (Figure 9.20a).

Example 9.1

The following is an example of the Rayleigh criteria as demonstrated in Eq. 9.33. The parameters are as follows. The coefficient of thermal expansion for the fluid is $2 \times 10^{-4}\,°C^{-1}$, mass density is 1 g/cm³, acceleration of gravity is 980 cm/s², the permeability is 1 darcy or 10^{-8} cm², the sand thickness is 1000 ft or 3.048×10^4 cm, the temperature gradient is 0.023 °C/ft or 7.55×10^{-4} °C/cm so that the temperature change across 1000 ft of sand is 23°C, the heat capacity is 1 cal/cm³ °C, the fluid viscosity is 6×10^{-3} g/cm s, and the thermal conductivity is 4×10^{-3} cal/cm s °C.

$$N_{RA} = \frac{(980)(1)(1)(3.048 \times 10^4)(10^{-8})(2 \times 10^{-4})(23)}{(6 \times 10^{-3})(4 \times 10^{-3})}$$

$$\cong 57$$

which exceeds the critical value of 40.

Sloping Layers

Combarnous and Bories (1975) discuss two important aspects of free convection in sloping layers. For such layers the required Rayleigh number is higher than the critical value of 40 so that for isothermal bounding planes

$$N_{RA} \cos \psi > 4\pi^2 \qquad (9.34)$$

where ψ is the slope of the layer. For slopes of less than 15°, the convective movement takes the form of polyhedral cells shown in Figure 9.20. For greater slopes, the cells are in the form of longitudinal coils.

A second point made by these authors is that convection will occur for all conditions where the isotherms are not parallel to the geopotentials, that is, where the temperature gradient is not collinear with the body forces. It is thus recognized that there is always a convective movement in a sloping layer bounded by isothermal planes of different temperature. The geometry of the roll is unicellular and two dimensional (Figure 9.21). A hot current moves upslope along the bottom boundary and then turns and moves downslope along the upper cold boundary.

The three states of free convection just discussed may be viewed as transitional. For small Rayleigh numbers below the critical value, the unicellular convective flow occurs in sloping layers. At Rayleigh numbers above the critical value, the unicellular motion gives way to the polyhedral cells for slopes less than 15°. For steep slopes

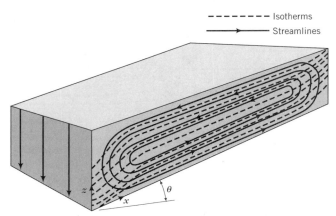

Figure 9.21 Geometry of unicellular flow in an inclined fluid layer (from Combarnous and Bories, 1975. Reprinted with permission of Academic Press, from Advances in Hydroscience, v. 10, p. 231–307).

and Rayleigh numbers above the critical value, the longitudinal coils form.

Geological Implications

In the study of sandstone diagenesis, it has become apparent that large amounts of mass must be transported to account for certain diagenetic changes. Free convection offers an attractive alternative as a mass transport mechanism as opposed to the "once-through" volumes of fluid associated with compaction or even the thousands of pore volumes of meteoric derived water through ancient uplifted rock. Horton and Rogers (1945) were probably the first to test this hypothesis for the Woodbine Sand in Texas to account for the NaCl distribution. Their method was to calculate the minimum value of permeability for which convection currents might occur (Eq. 9.33). Blanchard and Sharp (1985) took the same approach with the Rayleigh criterion (Eq. 9.33) in an attempt to account

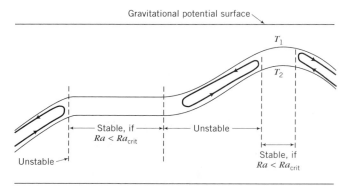

Figure 9.22 Schematic diagram illustrating regions of fluid stability in folded porous body. Boundaries of body assumed isothermal and impermeable. Reprinted with permission of Geochim. Cosmochim. Acta, v. 46, Wood and Hewett, Fluid convection and mass transfer in porous sandstone—a theoretical model. Copyright 1982. Pergamon Press PLC.

for the silica distribution in the Frio Sandstone in Texas. Wood and Hewett (1982) examined the stability analysis in sloping layers and made some general statements on the large-scale diagenetic effects that might be attributed to large-scale (kilometers) unicellular currents (Figure 9.22). They conclude that although the currents are slow (1 m per year), the mass flux associated with these currents operating over geologic time is sufficient to cause significant porosity changes. Cathles (1977) and Norton (1978) discuss convective fluid movements associated with intrusions and cooling plutons.

9.4 Energy Resources

Geothermal Energy

Three types of geothermal systems are generally recognized: (1) hydrothermal convection, (2) hot igneous rock, and (3) conductive dominated. The hydrothermal convective systems are either vapor dominated, which consist of dry or wet steam, or hot water dominated. Dry steam fields are normally of low permeability so that little recharge occurs. Temperatures on the order of 250°C are not uncommon. The wet steam field is characterized by a greater amount of recharge and consists of steam–water mixtures. The hot water systems may be classified as high ($> 150°C$), intermediate ($90-150°C$), and low ($< 90°C$) temperature systems. Both the vapor- and liquid-dominated systems require the following five characteristics: (1) a heat source, (2) sufficient permeability, (3) a source of replenishment, (4) a low-permeability bottom boundary, and (5) a low-permeability cap rock.

The temperature of the fluids in the foregoing classification dictates their use as an energy source (Nathenson and Muffler, 1975). The high-temperature systems are adequate for electrical generation. Nathenson and Muffler (1975) estimate the heat stored in these systems in the United States as 257×10^{18} calories. They suggest that this is five times the amount stored in undiscovered high-temperature reservoirs. To provide a perspective of this energy source, 10^{18} calories is equivalent to 690 million barrels of petroleum (White and Williams, 1975). The intermediate temperature systems are adequate for space and process heating. They have a stored heat content of about 345×10^{18} calories in the United States. For both the high- and intermediate-temperature systems, only about 25% of the stored energy is available at the surface. Low-temperature systems are useful only under locally favorable conditions, for example, the heating of homes or buildings by a local circulation system.

Hot igneous systems are of two types, molten magma and hot dry rock. The molten magma is currently not capable of exploitation due to the pressures and temperatures that would be encountered in drilling. Exploitation of the dry hot rock systems would require a circulation system for water to be injected and then recovered. The

temperatures are under 650°C, and the stored energy is in the several thousand times 10^{18} calories range. Renner and others (1975) describe the hydrothermal convection systems in the United States.

A conductive-dominated system generally relies on the thermal blanket effect where a low-conductivity cap rock acts as an insulator for heat storage below. Yet another type of conductive-dominated system is the abnormally pressured reservoirs in the Gulf Coast. Fluid pressures can reach as high as several thousand pounds per square inch, and temperatures can get above 300°C when one considers depths of 15–20 km. The potential thermal and mechanical energy is estimated to be between 46,000 and 190,000 megawatts for 20 years (Papadopulos and others, 1975).

It may be noticed that most of the references cited here are mid-1970s in vintage. This is no coincidence and reflects the effort to develop alternative energy sources in response to the oil embargo of the early 1970s. With a return to pre-1971 prices for oil and gas, the interest in geothermal resources has diminished considerably. Geothermal systems are discussed rather extensively in proceedings of a United Nations conference (1976) and in a bibliography by ERDA (1976).

The development of geothermal resources often encounters problems not unlike those associated with the use of ground water as a water supply. Water injection in hot aquifers is often considered as a viable management scheme to maintain the supply (Goguel, 1976). In other cases, the exploitation results in both land subsidence and induced tectonic activity (Herrin and Goforth, 1975). In response to a daily production of 130 million gallons of hot water from a reservoir located in Texas, four earthquakes of magnitudes 3.3 to 4.4 occurred within one month. Hundreds of microearthquakes were recorded but they gradually diminished.

The modeling of geothermal reservoirs must address the coupled movement of heat and ground water. Mercer and others (1975) have solved such a problem for the Wairakei Field, New Zealand. Reviews on the modeling methods are given by Pinder (1979) and Gary and Kassoy (1981). A numerical model for a hot-water-dominated system is given by Sorey (1978).

Energy Storage in Aquifers

Hot water storage in aquifers has been suggested as an attractive means for storing energy produced during periods of low consumption for use during peaking hours, or even seasons. According to Sauty and others (1982a), the important parameters in an energy storage problem include the recovery factor, which is the ratio of the quantity of energy recovered to that injected, and the temperature of the recovered water. These factors depend on the properties of the aquifer, the water contained within the aquifer, and the operating conditions such as injection rates and duration. This problem has been investigated numerically by Tsang and others (1981) and Sauty and others (1982a and b). Moltz and others (1981) have described the results of a long-term field test that attempts to assess the feasibility of an energy storage system.

Some of the physics involved in energy storage have been discussed by Sauty and others (1982a). The warmer body of water that develops during injection is normally restricted within a cylinder of radius r where

$$r = \frac{c_w}{c_s} \left(\frac{Qt}{\pi H} \right)^{1/2} \qquad (9.35)$$

where c_w is the specific heat of the fluid, c_s is the specific heat of the solids, Q is the injection rate, and H is the aquifer thickness. The lower the thermal conductivity, the smaller the heat losses, and the higher the efficiency of extraction. These features are demonstrated through the use of dimensionless parameters.

9.5 Heat Transport and Geologic Repositories for Nuclear Waste Storage

The Nuclear Waste Program

For the past few decades, one of the largest applied earth science programs ever attempted throughout much of the world has been the investigations for safe disposal of nuclear waste. Many countries have such a program, including Sweden, Belgium, France, Italy, Switzerland, Canada, and the United States. The purpose of most of these programs is the disposal of nuclear waste associated with commercial nuclear power plants. The wastes can be classified into three major categories. The first is the fission products resulting from the splitting of uranium and other fissile nuclei used as fuel. These radioactive elements emit both beta and gamma radiation and produce most of the heat in high-level waste. They are both short lived and long lived, the latter including ^{129}I and ^{99}Tc. The second category includes the transuranic elements such as plutonium and neptunium, many of which are associated with extremely long half-lives. Another category includes neutron activation products, for example, ^{63}Ni. These products are generated in the fuel claddings and structural materials in the core of the reactor. To provide some idea of the magnitude of the problem caused by these wastes, the Office of Technology Assessment (1982) has estimated that there will be something on the order of 72,000 metric tons of spent fuel by the year 2000. With reprocessing, these wastes would occupy a volume the size of a football field, one story high. Without reprocessing, the volume of an entire stadium could be filled.

The problem of nuclear waste disposal was first addressed in the United States in 1957 when the National Academy of Sciences released their report entitled "The disposal of radioactive wastes on land." Several options were considered, including disposal in space, in subseabeds, in ice sheets, in "very deep" holes, liquid injection in wells, rock melt disposal, transmutation, and a no-action alternative. The heat generation aspects associated with these alternatives did not go unnoticed. On the positive side the ice sheet disposal would utilize the heat for the waste to melt its way to the bottom. On the other hand, the seabed program ran into technical objections because of the uncertain thermal effects of the canisters on the sediments. With regard to the no-action alternative, a presidential message on February 2, 1980 stated that "civilian waste management problems shall not be deferred to future generations" (Department of Energy, 1980). The mined geologic repository has emerged as the most viable alternative.

The Rock Types

The scientists in the early phases of the disposal program were far more concerned with "geologic media" than with specific hydrogeologic environments. Among the first two rock types proposed were salt and granite, largely because the thermal conductivity of these substances is quite high. In fact, salt has one of the highest thermal conductivities of all natural rock types (Table 9.4). Thus, from the early beginnings there was concern with the long-term production of heat due to decaying radionuclides. Salt formations, both domed and layered, have been selected as possible candidate sites in Denmark, West Germany, the Netherlands, and Spain. Crystalline rocks have been selected in Canada, France, Japan, Switzerland, and the Scandinavian countries. On the other hand, Belgium and Italy are considering a clay-type rock. Some early potential candidates for the first repository for commercial wastes in the United States included the Michigan Basin (salt), the Gulf Coast (salt), the Permian basin in Texas (salt), and Vermont–New Hampshire (granite). In addition, an active program for the disposal of defense nuclear wastes is being pursued in bedded salt deposits near Carlsbad, New Mexico. This program is referred to as the Waste Isolation Pilot Plant (WIPP). Some 9 miles or so of tunnels are already in place, and testing has already begun.

In the United States, rock types and locations other than those cited earlier have been taken under consideration, including basalts at the Hanford Site in the state of Washington and volcanic tuff at the Nevada Test Site in southern Nevada. The Hanford Site has long been associated with subsurface storage and disposal of defense nuclear wastes, and the Nevada Test Site has been the location of subsurface nuclear detonations for decades. Hence, both are polluted already and both have the advantage of possessing vast tracts of land that have been removed from the public domain. Further, some

Table 9.4 **Thermal Conductivity of Potential Repository Rocks**

Material	Temperature (°C)	Thermal Conductivity (W/m°C)
Quartz monzonite	—	2.1–3.4
Salt (Louisiana)	300	2.2
Salt (Louisiana)	40	4.0
Tuff	—	2.4
Tuff	—	1.55
Salt	110	2.08–6.11
Granite	—	1.99–2.85
Shale	—	1.47–1.68
Tuff (welded)	—	1.2–1.9
Tuff (unwelded)	—	0.4–0.8
Basalt	—	1.16–1.56

preliminary planning is underway for a second repository sometime after the turn of the century. A "granite" program was formed, but now disbanded, and a sedimentary rock (shale) program is more or less inactive. Because of the political implications of nuclear waste disposal, whatever programs are active today may be abandoned in the future, or new ones added, but it is reasonably certain that the disposal problem will be with us for many decades.

Figure 9.23 shows the regions in the United States that were given consideration for the geologic disposal of nuclear waste. Nine sites were eventually selected for an environmental assessment, seven in salt, one in basalt, and one in tuff. The environmental assessments were published by the Department of Energy in 1986. From these assessments, three were picked for extensive site characterizations, and one of these three was to be se-

lected in the 1990s as the nation's first repository for commercial waste. The three candidates selected included a salt site in Texas and the two reservations just cited. In December 1987, Congress terminated site characterization plans at the Hanford site and the Texas site. At this time, only the Nevada Test Site remains as a candidate for the nation's first repository for commercial waste.

Figure 9.24 shows a cross section of the Nevada test site where the design repository is located in the unsaturated zone. Figure 9.25 shows the projected temperature within the repository rocks. The heat pulse produced will propagate through the host rock by conduction and convection. Other than establishing buoyancy-driven flows, the heat generated is expected to alter the preemplacement condition of the host rock, and this is one of the main concerns. These concerns have resulted in

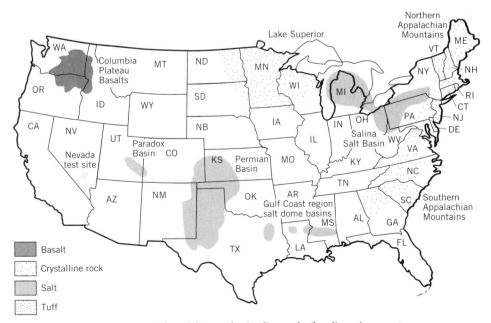

Figure 9.23 Regions considered for geologic disposal of radioactive waste.

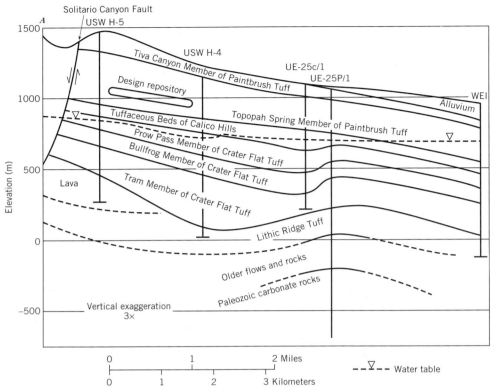

Figure 9.24 Cross section of the Nevada Test Site showing the location of proposed repository (from Department of Energy, 1988).

research in coupled phenomena, that is, the thermal-mechanical-hydrochemical processes associated with a nuclear repository (for example, Lawrence Berkeley Laboratory, 1984). This research has not been limited to the unsaturated zone at the Nevada Test Site but includes

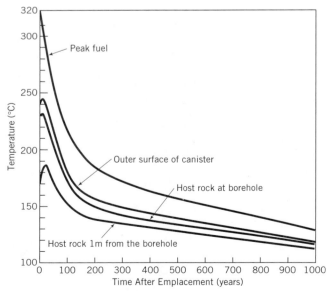

Figure 9.25 Temperature versus time after nuclear waste emplacement (from Department of Energy, 1988).

work on all potential repository rocks. Thus, for salt, we are not surprised to learn that the main thermal-mechanical coupling is the deformation or creep (flow) of the salt in response to heating. In addition, brine pockets have been known to migrate up the thermal gradient, that is, toward the heat source. For saturated rocks, one of the main concerns is the potential for fracturing by thermally induced fluid pressures. For unsaturated zones in general, the problems are complex and include the potential for upward gas transport of radionuclides due to the buoyant rise of hot air along with temperature-induced vapor phase transport. For all rocks in general, there is concern for the change in mechanical strength of the rock body and stress corrosion of the canisters.

A literature review of coupled thermal, hydrologic, and mechanical-chemical processes pertinent to the proposed nuclear waste repository at Yucca Mountain has been prepared by the Southwest Research Institute (1993).

Thermohydrochemical Effects

From the study of hot springs and ore deposits, it has long been recognized that fluids in hydrothermal systems react and subsequently alter the rocks involved in hydrothermal circulation (Browne, 1978). As the repository and adjoining rock will in effect be a "hydrothermal" system for a few to several hundred years after emplacement, there is concern over the nature of these alterations. Of special concern is the effect on the sorbing minerals or those that undergo exchange, which can provide a barrier to the migration of radionuclides. Many of these minerals are unstable at elevated temperatures and may be altered by reversible dehydration or mineral reactions. For example, certain zeolite minerals (hydrous aluminum silicates) are abundant in the volcanic tuffs at the Nevada Test Site. With few exceptions, natural zeolites are found in rocks that have formed at temperatures less than 200°C. According to Smyth (1982) reactions that may occur in these rocks below 200°C are of two types, simple reversible dehydration and mineralogical phase inversions. The first of these involves a loss of bound water with contraction of the unit cell by as much as 10% with the development and propagation of fractures and the loss of mechanical strength. Smyth (1982) concludes that this constraint may limit to 50°C the maximum temperature rise in zeolite horizons that occur 30 to 50 m from the emplacement horizons.

Another important mineral frequently associated with the retardation of radionuclide migration is montmorillonite. As noted in Chapter 8, the alteration of a montmorillonite to a mixed clay in the Gulf Coast begins at temperatures on the order of 60–70°C. Based on first-order kinetics of an Arrhenius form, Domenico and Palciauskas (1988) have examined the half-life (the time required for half of the expanded montmorillonite layer to transform to illite) as a function of temperature. At 40°C, the transformation requires 1.6×10^6 years; at 120°C, it requires 2400 years; and at 160°C, it requires 230 years. Thus, one might conclude that the process is too slow in comparison to the time scales that are involved in the thermal evolution of the host rock. However, the kinetics addressed the irreversible transformation only. Like zeolites, a reversible transformation can occur when the temperature reaches the boiling point of water. The energy required to break the bonds of the hydration envelope is significantly lower, so that the kinetics of the process involves time scales shorter than those of the

heating process. The potentially large volume of pore fluid released and enhanced permeability could allow migration of the pore fluid in response to the fluid pressures produced during the heating episode.

Thermomechanical Effects

If heating takes place in fractured rock that has an appreciable permeability, the thermal expansion of the fluid is easily accommodated by fluid movement, and the response may take place very close to the condition of constant fluid pressure (Chapter 8). However, these are not the types of rocks that are likely to be considered for a repository in that the permeability is already too high. A more likely selection might have discontinuous pores or fractures of various orientation saturated with water and of such a low permeability so that no fluid expulsion takes place during the heating episode. This is the extreme case of heating at constant fluid mass. This case was examined theoretically for rock properties at the Hanford Reservation (Palciauskas and Domenico, 1982). The sample calculations indicate that fracturing is inevitable.

Yet a third type of environment, and one that is presumably ideal from the perspective of nuclide migration, is the rock type that contains a few isolated pores or fractures and of such a low permeability that it contains essentially no water or at least is in a state of low fluid saturation. Heating in this case involves only the solid components and their thermal expansion. Significant porosity changes are anticipated only if the original porosity is very low, that is, on the order of 10^{-4} (Palciauskas and Domenico, 1982).

Other potential problems dealing with thermomechanical effects include the triggering of latent seismicity, spalling, and loss of mechanical strength. From the perspective of contamination, the safe disposal of nuclear waste is a very difficult problem; in the presence of heat generation from decaying nuclides, the problem is severely compounded.

Problems

1. Calculate upper and lower bounds for the thermal conductivity of saturated sandstone consisting largely of quartz with a porosity of 20% using Table 9.1. Repeat the calculations for a dry sandstone with the same porosity.

2. Figure 9.6 applies to both downward and upward moving ground water. Which half of the family of curves applies to upward moving ground water, and what is the significance of the curve where $N_{PE} = 0$?

3. From the type curve match of Figure 9.8, calculate the velocity q through a layer 10 m in thickness. Assume

a thermal conductivity of 5×10^{-3} cal s^{-1} cm^{-1} °C^{-1}.

4. From the data of Example 9.1

a. Calculate the minimum value of permeability to initiate free convection.

b. Repeat this calculation for a sand thickness of 500 ft.

c. For a Rayleigh number of 57, would free convection occur in a rock body with a slope of 30°?

5. Compare the Fourier number of Eq. 9.22 with the Fourier number for fluid flow (Eq. 4.41). State the analogous quantities and provide an interpretation of T^* for heat flow. For a repository rock, what are the advantages of having a high thermal conductivity and a low heat capacity?

6. Consider one-dimensional steady-state conduction across a layer of thickness L where the temperature at the bottom ($x = 0$) and at the top ($x = L$) is known. Solve this problem for the steady-state temperature as a function of x, L, T_0, and T_L (Hint: see Example 4.3).

7. A soluble toxic chemical is moving downward through a thick unsaturated zone in sands and gravels threatening an underlying water supply. An experimental program is being devised whereby refrigerants will be circulated through bore holes penetrating the unsatu- rated zone in an attempt to freeze the water and the contaminant in the unsaturated zone. Based on what you have learned in this chapter and in previous chapters, is it possible to make some approximate calculation of the effectiveness of one bore hole in contributing to this feat by the process of conduction alone? What equation or physical concepts may be modified to describe this problem, and what is the role of the time constant in considerations of the rate at which freezing can be accomplished? (Hint: If you are confused, consult Carslaw and Yeager, 1959, the chapter on the use of sources and sinks in cases of variable temperature, the section on the continuous line source.) Is this section (or equation) analogous to anything you have learned thus far?

Chapter 10

Solute Transport

Mass occurring in water as ions, molecules, or solid particles undergoes not only transport but also reactions. The reactions redistribute mass among ions or between liquid and solid phases. Figure 10.1 lists the important mass-transport and mass-transfer reactions in ground water. Over the next four chapters, we will study these processes, establishing the background necessary to understand the classical problems involved with ground-water geochemistry and contaminant hydrogeology. Chapter 10 begins with the mass transport processes, advection and dispersion, which spread dissolved mass in ground water. Advection is mass transport due simply to the flow of water in which the mass is dissolved. The direction and rate of transport coincide with the ground water. Dispersion is a process of fluid mixing that causes a zone of mixing to develop between a fluid of one composition that is adjacent to or being displaced by a fluid with a different composition.

10.1   Advection

The intimate relationship between ground-water flow and the process of advection means that knowledge about flow systems is directly transferable to understanding advection. For example, in the case of topographically driven flow systems (Chapter 5), the factors that influence ground-water flow patterns—water-table configuration, pattern of geologic layering, size of the ground-water basin, pumping or injection—control the direction and rate of mass transport. Thus, the background from the earlier treatment of flow allows us to consider advection in just a few pages.

The two flow nets in Figure 10.2 illustrate basic features about advection. First when only advection is operating, mass added to one or more stream tubes will remain in those stream tubes. Other processes (for example, dispersion) move mass between stream tubes. Second, the direction of mass spreading in steady-state systems is defined by pathlines even in relatively complex flow systems. Because advection is the dominant transport process, knowledge of ground-water flow patterns guides us in interpreting patterns of contaminant migration.

The discussion above assumes that mass transport does not influence ground-water flow. When this assumption fails, for example, with contaminated ground water having a density that is significantly different from the ambient ground water, the flow of water and mass may diverge. However, for most practical problems, ground water and the dissolved mass will move at the same rate (in the absence of other processes) and in the same

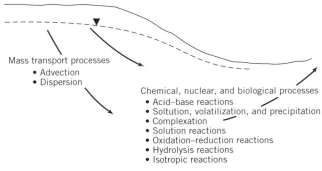

Figure 10.1 Conceptualization of mass transport in a ground-water flow system.

direction. Accordingly, the velocity of advective transport is described by the Darcy equation developed in Chapter 3

$$v = -\frac{K}{n_e}\frac{\partial b}{\partial l} \qquad (10.1)$$

where v is the linear ground-water velocity, K is the hydraulic conductivity, n_e is the effective porosity, and $\partial b/\partial l$ is the hydraulic gradient. As discussed in Chapter 3, the linear ground-water velocity and, therefore, the velocity of advective transport increases with decreasing effective porosity. This relationship is particularly im-

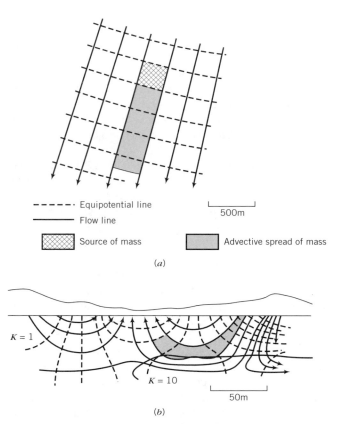

--- --- Equipotential line
——— Flow line

▨ Source of mass   ▨ Advective spread of mass

(a)

$K = 1$

$K = 10$

50m

(b)

Figure 10.2 Mass spreading by advection alone in (*a*) a shallow, unconfined aquifer and (*b*) a local flow system.

portant in fractured rocks, where the effective porosity can be much less than the total porosity, often as low as 1×10^{-4} or 1×10^{-5}.

There are situations where the mean ground-water velocity is different from the advective velocity of the mass. Studies by Corey and others (1963) and Krupp and others (1972) have shown that negatively charged ions can move faster than the water in which they are dissolved. Electrical charges due to the presence of clay minerals can force anions to remain in the center of pores, the location of the maximum microscopic velocity. The water itself may flow through the lower velocity regimes within a pore. Alternatively, a reduction in advective velocity is possible when the geologic medium takes on properties of a semipermeable membrane. Solutes avoid entering the membrane because of electrokinetic or in some cases size constraints. The process of membrane filtration has been implicated in the origin of brines in sedimentary basins.

There are other occasions where the linear velocity of advective transport is apparently less than that of the ground water. An example of retardation is provided by sorption, which will be taken up in Chapter 12.

10.2 Basic Concepts of Dispersion

Thinking again about flow tubes in a steady-state ground-water system is a good place to begin examining the concept of dispersion. Dispersion spreads mass beyond the region it normally would occupy due to advection alone. This idea can be illustrated with a simple column apparatus (Figure 10.3*a*). The apparatus is similar to a Darcy column with additional plumbing to permit the controlled and continuous addition of a tracer. Initially, a steady flow of water is set up through the column. The test begins with a dissolved tracer at a relative concentration $C/C_0 = 1$ added across the entire cross section of the column on a continuous basis. Monitoring the outflow from the column establishes the relative concentration of the tracer as a function of time.

The relative concentration of tracer varies in water going into and coming out of the column (Figure 10.3*a*). Such relative concentration-versus-time plots are known as the source loading curve and the breakthrough curve, respectively. The breakthrough curve does not have the same "step" shape as the source-loading function. Dispersion creates a zone of mixing between the displacing fluid and the fluid being displaced. Some of the mass leaves the column in advance of the advective front, which is defined as the product of the seepage velocity and time since displacement first started. The position of the advective front at breakthrough corresponds to a C/C_0 value of 0.5 (Figure 10.3*a*). In the absence of dispersion or other processes, the shapes of the loading function and the breakthrough curve are identical.

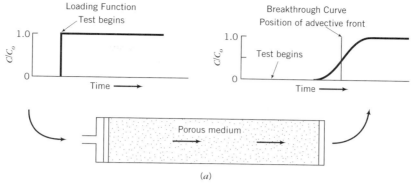

Figure 10.3(a) Experimental apparatus to illustrate dispersion in a column. The test begins with a continuous input of tracer $C/C_0 = 1$ at the inflow end. The relative concentration versus time function at the outflow characterizes dispersion in the column.

A zone of mixing gradually develops around the advective front (Figure 10.3b). Dispersion moves some tracer from behind to in front of the advective front. The size of the zone of mixing increases as the advective front moves further from the source.

This column experiment is an example of one-dimensional transport involving advection and dispersion. Similar mixing will occur in both two and three dimensions (Figure 10.4). Again dispersion spreads some of the mass beyond the region it would occupy due to advection alone. There is spreading both ahead of the advective front in the same flow tube and laterally into the adjacent flow tubes. This dispersion is referred to as longitudinal and transverse dispersion, respectively. Dispersion in three dimensions involves spreading in transverse and vertical directions as well as longitudinally.

There are a few other distinctions between Figures 10.3b and 10.4. At some distance behind the advective front for the case of one-dimensional dispersion, the original source concentration C_0 is encountered. The longer transport continues, the greater the length of column occupied by water with a tracer concentration C_0. In this zone, the concentration is at steady state. This result has been demonstrated hundreds of times in column experiments where transverse spreading is eliminated.

When transverse spreading is included with longitudinal dispersion (Figure 10.4), the concentration distribution changes significantly. There is a zone of dispersion beyond the advective front caused exclusively by longitudinal dispersion, and a resulting zone of mass depletion

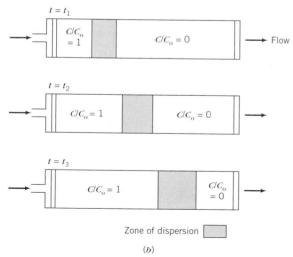

Figure 10.3(b) Schematic representation of dispersion within the porous medium at three different times. A progressively larger zone of mixing forms between the two fluids ($C/C_0 = 1$ and $C/C_0 = 0$) displacing one another.

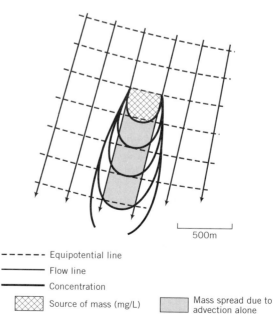

Figure 10.4 Comparison of mass distributions in a shallow unconfined aquifer due to advection alone and advection-dispersion.

behind the advective front. At some distance behind the advective front, depending on the extent of dispersion, the system is at steady state with respect to concentration. This part of the tracer plume is sufficiently far behind the zone of longitudinal dispersion that it does not contribute mass to the frontal portions of the plume. The steady-state concentrations are, however, less than the original concentration because mass also spreads transversely and occupies an area (taken in the vertical plane) that is greater than the area where the mass was injected. Transverse dispersion will reduce concentrations everywhere behind the advective front, while longitudinal dispersion, will do so only at the front of a plume.

Dispersion occurs in a porous medium because of two processes, diffusion and mixing due to velocity variations. This latter process is called mechanical dispersion. The following two sections provide an introduction to these very important processes.

Diffusion

In Chapter 3, Darcy's equation was introduced as a linear law relating fluid flow q to the gradient in hydraulic head. For mass transport, there is a similar law that describes how chemical mass flux is proportional to the gradient in concentration. This law is known as Fick's law and is expressed for a simple aqueous nonporous system as:

$$\mathbf{J} = -D_d \ \text{grad} \ (C) \tag{10.2}$$

In this formulation, $\mathbf{J}$ is a chemical mass flux with the negative sign indicating that the transport is in the direction of decreasing concentration. The porportionality constant D_d is termed the diffusion coefficient in a fluid environment. If the amount of diffusing material, J, and the concentration, C, in Eq. 10.2 are expressed in terms of the same quantity, the diffusion coefficient D_d has the units of L^2/T. In other words, if $\mathbf{J}$ is expressed in units of moles/L^2T, and the concentration is in moles/L^3, the proportionality constant has the units L^2/T in that the gradient operator grad( ) has the units L^{-1}.

Molecular diffusion is mixing caused by random molecular motions due to the thermal kinetic energy of the solute. Because of molecular spacing, the coefficient describing this scattering is larger in gases than in liquids and larger in liquids than in solids. The diffusion coefficient in a porous medium is smaller than in pure liquids primarily because collision with the solids of the medium hinders diffusion. Figure 10.5 provides a range in values for diffusion coefficients for various fluid environments.

In a porous medium, diffusion takes place in the liquid phase enclosed by a porous solid. Averaging techniques provide the following rigorous statement for Fick's law in the fluid phase of a porous sediment (Whitaker, 1967):

$$\mathbf{J} = -D_d \left[\text{grad} \ (Cn) + \frac{\tau}{V} \right] \tag{10.3}$$

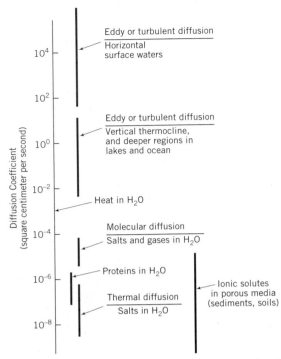

Figure 10.5 Diffusion coefficients characteristic of various environments. Reprinted with permission from Lerman, A., in Non-equilibrium Systems in Water Chemistry; Hem, J. D., ed.; ACS Advances in Chemistry Series 106: American Chemical Society: Washington DC, 1971, p. 32. Copyright 1971 American Chemical Society.

where V is the averaging volume, n is porosity, and τ is a tortuosity vector to account for the hindering of free diffusion by collision with the pore walls. The form of Eq. 10.3 has prompted the definition of a bulk diffusion coefficient D_d^* to account for the effects of tortuosity, giving the statement most commonly referred to as Fick's law for diffusion in sediments

$$\mathbf{J} = -D_d^* \ n \ \text{grad} \ (C) \tag{10.4}$$

In the absence of an evaluation for D_d^*, an effective diffusion coefficient $D_d' = D_d^* n$ is generally employed so that

$$\mathbf{J} = -D_d' \ \text{grad} \ (C) \tag{10.5}$$

Several different empirical approaches are useful in defining an effective diffusion coefficient (D_d'). The following equation relates D_d' for the fluid in a porous medium to the diffusion coefficient in a liquid D_d (Domenico, 1977):

$$D_d' = D^* D_d \tag{10.6}$$

where D^* is some constant less than one that accounts for the structure of the porous medium. In most cases D^* is a function of both porosity and tortuosity. According to Hellferich (1966), the effective diffusion coefficients in exchange columns fall in a range

$$D_d' = \frac{n}{2} D_d \quad \text{to} \quad D_d \left[\frac{n}{(2-n)} \right]^2 \quad (10.7)$$

where the factor D^* depends only upon porosity n. Other ways of expressing this relationship are (Greenkorn, 1983)

$$D_d' = u \, n D_d \quad (10.8)$$

where u is a number less than one, and (Greenkorn and Kessler, 1972)

$$D_d' = \frac{n}{\tau} D_d \quad (10.9)$$

where τ is the tortuosity. In this latter formulation, the tortuosity τ is defined as the ratio of the length of a flow channel for a fluid particle (L_e) to the length of a porous medium sample (L). The value of (L_e/L) is always greater than one except for perfect capillary type passages. Bear (1972) defines tortuosity as (L/L_e) resulting in values less than one, and ranging between 0.56 and 0.8 in granular media. He states further that the effective diffusion coefficient is the product of tortuosity and the diffusion coefficient in the bulk fluid. Whichever of these equations is employed, effective diffusion coefficients in porous media increase with increasing porosity and decrease with increasing path length to sample length ratios.

Table 10.1 cites diffusion coefficients in water for common cations and anions. With the exception of H^+ and OH^-, values of D_d range from 5×10^{-6} to 20×10^{-6}

Table 10.2 Geological Features Contributing to Nonidealities in a Porous Medium

A. Microscopic heterogeneity: pore to pore
  1. Pore size distribution
  2. Pore geometry
  3. Dead-end pore space
B. Macroscopic heterogeneity: well to well or intraformational
  1. Stratification characteristics
    a. Nonuniform stratification
    b. Stratification contrasts
    c. Stratification continuity
    d. Insulation to cross-flow
  2. Permeability characteristics
    a. Nonuniform permeability
    b. Permeability trends
    c. Directional permeability
C. Megascopic heterogeneity: formational (either fieldwide or regional)
  1. Reservoir geometry
    a. Overall structural framework: faults, dipping strata, etc.
    b. Overall stratigraphic framework: bar, blanket, channel fill, etc.
  2. Hyperpermeability-oriented natural fracture systems

From Alpay (1972), a practical approach to defining reservoir heterogeneity. Copyright 1972, Society of Petroleum Engineers Inc., JPT (July 1972).

cm^2/s with the smallest values associated with ions having the greatest charge.

Mechanical Dispersion

Mechanical dispersion is mixing caused by local variations in velocity around some mean velocity of flow. Thus, mechanical dispersion is an advective process and not a chemical one. With time, mass occupying some volume becomes gradually more dispersed as different fractions of mass are transported in these varying velocity regimes. Variability in the direction and rate of transport is caused by nonidealities in the porous medium. The most important variable in this respect is hydraulic conductivity. Table 10.2 lists some of the geologic features that produce nonidealities. Nonidealities exist at a variety of scales ranging from microscopic to megascopic, to an even larger scale (not included in the table) involving groups of formations.

Nonidealities at all of these scales are responsible for mechanical dispersion. For example, variability in velocity at the microscopic scale develops because of differing flow regimes across individual pore throats or variability in the tortuosity of the flow channels (Figure 10.6a). The heterogeneities shown in Figures 10.6b and 10.6c produce dispersion by creating variability in flow at a macroscopic scale. The first case (Figure 10.6b) is an

Table 10.1 Diffusion Coefficients in Water for Some Ions at 25°C

Cation	D_d (10^{-6} cm^2/s)	Anion	D_d (10^{-6} cm^2/s)
H^+	93.1	OH^-	52.7
Na^+	13.3	F^-	14.6
K^+	19.6	Cl^-	20.3
Rb^+	20.6	Br^-	20.1
Cs^+	20.7	HS^-	17.3
		HCO_3^-	11.8
Mg^{2+}	7.05		
Ca^{2+}	7.93	CO_3^{2-}	9.55
Sr^{2+}	7.94	SO_4^{2-}	10.7
Ba^{2+}	8.48		
Ra^{2+}	8.89		
Mn^{2+}	6.88		
Fe^{2+}	7.19		
Cr^{3+}	5.94		
Fe^{3+}	6.07		

From Li and Gregory (1974).
* Reprinted with permission from Geochim. Cosmochim. Acta, 38, Y.-H. Li and S. Gregory, Diffusion of ions in sea water and in deep-sea sediments, p. 703–714. Copyright 1974, with kind permission from Elsevier Science Ltd, The Boulevard, Langford Lane, Kidlington OX5 1GB, U.K.

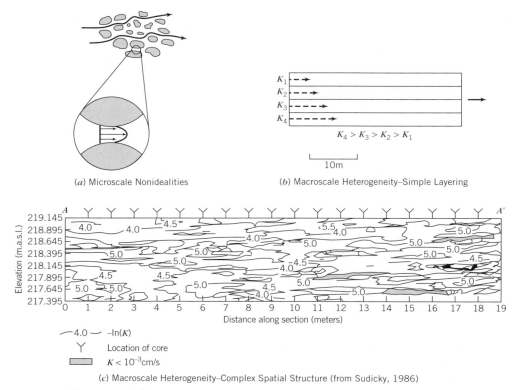

(a) Microscale Nonidealities

(b) Macroscale Heterogeneity–Simple Layering

$K_4 > K_3 > K_2 > K_1$

10m

(c) Macroscale Heterogeneity–Complex Spatial Structure (from Sudicky, 1986)

Figure 10.6 Examples of nonidealities at different scales giving rise to mechanical dispersion. Panel (c) is from Sudicky, E. A., Water Resources Res., v. 22, p. 2069–2082, 1986. Copyright by American Geophysical Union.

example of nonidealities created by layering. Contrasts in hydraulic conductivity among the various layers produce differing velocities. More complicated macroscopic heterogeneity is shown in Figure 10.6c. This example, adapted from Sudicky (1986), represents the plotted results of a large number of hydraulic conductivity measurements for the shallow unconfined aquifer at Canadian Forces Base Borden.

10.3 Character of the Dispersion Coefficient

Studies at the Microscopic Scale

The coefficient of hydrodynamic dispersion (D) is the sum of the coefficients of bulk diffusion (D_d^*) and mechanical dispersion (D'). Values of the bulk diffusion coefficient can be estimated reasonably well within an order of magnitude for granular media. However, questions remain as to what is a realistic range of values for the mechanical component D' and what is its relative contribution to hydrodynamic dispersion. Fortunately, results from laboratory column experiments can address these issues.

The fluid flow velocity and grain size are the main controls on the longitudinal dispersion in a column.

Pfannkuch (1962) explored these relationships by taking existing experimental data and casting them in the form of a series of dimensionless numbers D_L/D_d and vd_m/D_d, where v is the linear ground-water velocity, d_m is the mean grain size, D_L is the longitudinal dispersion coefficient, and D_d is the diffusion coefficient. The first of these numbers normalizes the observed dispersion in the column by dividing it by the coefficient of diffusion of the tracer in water. The second number is the Peclet number (N_{PE}), a ratio expressing advective to diffusive transport.

Figure 10.7 represents relationships derived by Pfannkuch (1962) with four classes of mixing related to the Peclet number. The first class (Class 1) is for small values of the Peclet number ($N_{PE} < 0.01$). For mixing in Class 1, D_L/D_d does not change as a function of N_{PE}, which indicates that diffusion is the main cause of mixing. The ratio D_L/D_d has a value of approximately 0.67 (Fried and Combarnous, 1971) because at this velocity D_L is equal to the bulk diffusion coefficient D_d^*, which is less than the diffusion coefficient (D_d) in water.

With increasing values of N_{PE} (Class 2, N_{PE} 0.1–4), mixing is influenced not only by diffusion but by mechanical mixing. D_L/D_d increases as a function of N_{PE} (Figure 10.7). With a further increase in velocity (Class 3, N_{PE} 4–10^4), mechanical dispersion dominates the mixing process except at the low end of the range of N_{PE}. Values

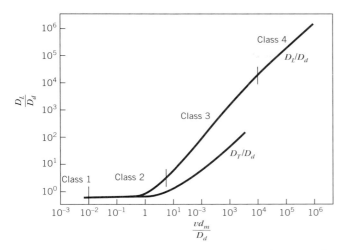

Figure 10.7 Behavior of D_L/D_d and D_T/D_d as a function of the Peclet number (N_{PE}). The classes identify regimes where mixing results from the same processes (after Pfannkuch, 1962 and Perkins and Johnston, 1963, A review of diffusion and dispersion in porous media). Copyright 1963, Society of Petroleum Engineers Inc., SPEJ (March 1963).

of D_L over this range are approximately proportional to $v^{1.2}$. Greenkorn (1983) explains this exponent of 1.2 as the outcome of several different mixing processes. With increasing N_{PE}, velocity profiles develop within the pores. This effect produces mixing analogous to dispersion in a capillary tube with $D_L \sim v^2$. Other forms of mixing produce a velocity dependence with $D_L \sim v$. A case in point is mixing due to the tortuosity of the flow paths due to the presence of grains. For values of N_{PE} in the range from 10^4–10^6 (Class 4), the effects of molecular diffusion are negligible. $D_L \sim v$ and the experimental data plot as a straight line with a slope of 45°.

One important result is the determination that longitudinal dispersion is proportional to velocity. This relationship has been generalized to describe dispersion at macroscopic and megascopic scales. Column experiments have also provided important information about transverse dispersion. Shown in Figure 10.7 is the ratio D_T/D_d plotted versus N_{PE} (from Perkins and Johnston, 1963). The two curves in Figure 10.7 are similar in shape, indicating that the process of transverse mixing is much the same as longitudinal mixing. However, longitudinal dispersion is greater than transverse dispersion at a given N_{PE}. For example, when $N_{PE} > 100$, values of D_L are approximately 10 times greater than D_T. This tendency for D_L to exceed D_T applies to dispersion not only at microscopic scales but at larger scales as well.

Dispersivity as a Medium Property

Column experiments have been useful in establishing relationships between the coefficients of mechanical dispersion and the mean ground-water velocity. The equa-

tions are

$$D_L = \alpha_L v \qquad \text{and} \qquad D_T = \alpha_T v \qquad (10.10)$$

where α_L and α_T are the longitudinal and transverse dispersivities of the medium. Dispersivities have units of length and, like hydraulic conductivity, are characteristic properties of a medium. In practice, they quantify mechanical dispersion in a system. There are actually two transverse dispersivities measured at 90° to each other. For example, with horizontal flow, one component of transverse spreading will occur in the horizontal plane, and the second will occur in the vertical plane.

Studies at Macroscopic and Larger Scales

Much of what is known about dispersion is based on early theoretical studies and column studies. However, important contributions have come from field studies at macroscopic and larger scales. The main emphasis in this research has been to estimate dispersivity values from field experiments. Gelhar and others (1992) undertook a critical review of field experiments at 59 sites around the world. Tests yielded some 106 values of longitudinal dispersivity ranging from 0.01 m to 5500 m at scales of 0.75 m to 100 km. It appears that longitudinal dispersivities increase indefinitely with scale. However, the detailed review of these experiments (Figure 10.8) determined that only 14 values were highly reliable. A further 31 values were of intermediate reliability. Looking at the most reliable results, however, the trend of increasing

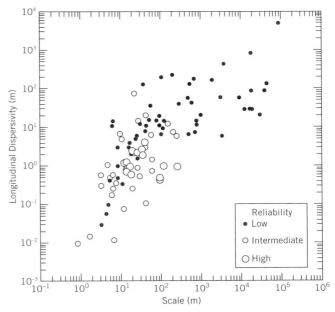

Figure 10.8 Longitudinal dispersivity versus scale with data classified by reliability (from Gelhar, and others, Water Resources Res., v. 28, p. 1955–1974, 1992. Copyright by Amer. Geophys. Union).

longitudinal dispersivity as a function of distance was more apparent than real (Gelhar and others, 1992). The most reliable dispersivity values are all at the low end of the range.

From such tests, a consistent view about macroscopic dispersion has emerged. Heterogeneity at the macroscopic scale contributes significantly to dispersion because it creates local-scale variability in velocity. Values of macroscopic dispersivity are in general two or more orders of magnitude larger than those from column experiments. A typical range of values in longitudinal dispersivity from column experiments is from 10^{-2} to 1 cm. In field experiments, values range from approximately 0.1 to 2 m over relatively short transport distances. At any given scale, longitudinal dispersivities can range over 2 to 3 orders of magnitude depending on the variability in hydraulic conductivity. Although no reliable, large-scale studies have been carried out, it is probable that longitudinal dispersivity values in excess of 10 m exist. Values, however, may not be as large as estimates determined using contaminant plumes and environmental isotopes, which are considered less reliable (Gelhar and others, 1992).

Many dispersion studies have suffered from errors related to the collection and interpretation of field data. These errors arise from (1) poor definition of the plume due to a small number of monitoring wells and nonpoint sampling, (2) failure to account for temporal variations in the advective flow regime, which can lead to lateral mass spreading, (3) incomplete information about the source loading, (4) inherent limitations in some test procedures, and (5) oversimplified techniques for interpreting the test results, for example, assuming that a source of finite size can be approximated as a point source. These problems typically force estimates of dispersivity values to be larger than actual values. Thus, many of the published estimates of dispersivity are calibration or curve-fitting parameters (Anderson, 1984).

Results from field experiments also support the general finding that horizontal transverse dispersivities are at least an order of magnitude smaller than longitudinal values (Gelhar and others, 1992). Further, vertical transverse dispersivities are typically some 1 to 2 orders of magnitude smaller than horizontal transverse dispersivities (Gelhar and others, 1992). In media with pronounced horizontal stratification, values of vertical dispersivity may be similar to the bulk diffusion coefficient (Sudicky, 1986).

Accompanying the increased understanding about large-scale dispersion is the realization that the fundamental concepts may not be as straightforward as they appear. One important question is whether constant values of macroscopic dispersivity exist. Several different field experiments have described dispersivities that increase as a function of the travel distance of the plume. In some cases, these variable dispersivities are simply an artifact

of the method of interpretation. For example, characterizing the two- or three-dimensional spread of a tracer with a one-dimensional model requires an apparent dispersivity that increases as a function of distance (Domenico and Robbins, 1984). However, the interpretation from a dimensionally correct model yields a constant dispersivity. There are results to suggest that scale-dependent dispersivities exist.

How can these nonconstant values be explained? One way is to consider them simply to be a result of incomplete spatial averaging. Consider a dispersion experiment run in a heterogeneous unit with relatively large lenses. The spread of tracer from a source to a nearby observation point in the same lens should yield a relatively small dispersivity that is comparable to a column value. The tracer must move a substantial distance before it is able to fully interact with the heterogeneity to the extent necessary to produce macroscale mixing. Thus, dispersivity values increase away from a source before becoming constant. Gelhar and others (1979) refers to this constant macroscale dispersivity as the asymptotic dispersivity. The approach to an asymptotic dispersivity might involve tracer spreading over 10s or 100s of meters.

Another way to explain spatially varying dispersivity values is in relation to some representative elemental volume (REV) for dispersivity. With the spreading of a tracer, eventually a transition will occur from the microscale to macroscale. The implicit assumption is that for each scale a REV exists (Figure 10.9).

In the zone of transition between various scales, it may be difficult to define a continuum value for dispersivity. Spreading may not yet have encompassed a volume that is equivalent to a REV. As a consequence, small changes in volume could result in significant variability in the spreading. Also, mass may not be normally distributed as classical theory would predict, and as a result, dispersivity values may not be definable. An example of this behavior is the concentration distribution from a tracer experi-

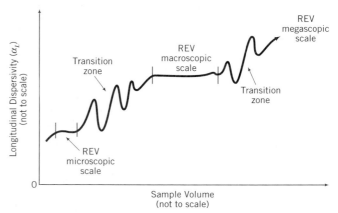

Figure 10.9 An example of the relationship between longitudinal dispersivity and sample volume with REVs defined at the microscopic and macroscopic scale.

ment shown in Figure 10.10 (Sudicky and others, 1983). Because the tracer was transported through different elements of the heterogeneous conductivity field in a sub-REV system, the tracer plume separated close to the source. Asymptotic values of dispersivity could conceivably exist for this system. However, in the early stages of spreading, defining a dispersivity value may be difficult.

The tendency for dispersivity values to be scale dependent has important implications for how macroscopic dispersivity values are estimated and used. For example, longitudinal dispersivity values from tracer tests run over relatively short distances may not be representative of asymptotic values. Further, it simply may not be practical to carry out dispersion experiments on a routine basis over distances large enough to yield an asymptotic value (Sudicky, 1986).

Problems associated with concepts of macroscale dispersivities have not invalidated their use. They remain, in spite of limitations, a useful conceptual model of a complicated process. However, one must be aware of limitations in extending classical concepts of dispersion from microscale to macroscale and to larger systems.

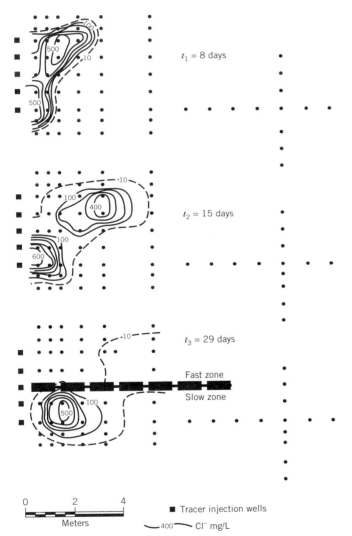

10.4 A Fickian Model of Dispersion

Dispersion or more formally hydrodynamic dispersion occurs as a consequence of two different processes: diffusion and mechanical dispersion. These two contributions to hydrodynamic dispersion are represented mathematically as

$$D = D' + D_d^* \qquad (10.11)$$

where D is the coefficient of hydrodynamic dispersion, D' is the coefficient of mechanical dispersion, and D_d^* is the bulk diffusion coefficient.

In most modeling studies, use of a coefficient of hydrodynamic dispersion implies that the behavior is Fickian. By Fickian, we mean the transport of a quantity in a direction down a concentration gradient at a rate that is proportional to the gradient. While this relationship describes molecular diffusion, it is not a mechanistically correct representation of mechanical dispersion. The justification of why the diffusional model is indeed used for mechanical dispersion is a practical one. The net effect of both processes operating together is to cause the concentration distribution to become normal within a short flow distance. In other words, although the microscopic mechanisms for diffusion and mechanical dispersion differ completely, the macroscopic outcome for both processes is similar, that is, a distribution that becomes normal. Given this outcome, we can represent dispersion within the framework of a diffusion model.

Consider the normal distribution of Figure 10.11 and

Figure 10.10 Measured Cl⁻ concentrations from a tracer experiment in the shallow aquifer at the Canadian Forces Base, Borden. After a short time, the Cl⁻ plume separated due to different rates of advection in a ''fast'' and ''slow'' zone (after Sudicky and others, 1983). Reprinted from J. Hydrol., v. 63, E. A. Sudicky, J. A. Cherry, and E. O. Frind, ''Migration of contaminants in groundwater at a landfill: A case study,'' p. 81–108, 1983 with kind permission from Elsevier Science, NL, Sara Burgerhartstraat 25, 1055 KV Amsterdam, The Netherlands.

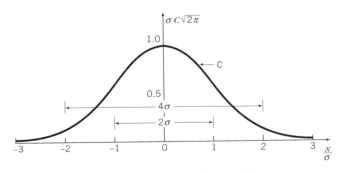

Figure 10.11 The normal or Gaussian distribution.

Table 10.3 Properties of the Normal Distribution

$\dfrac{x}{\sigma}$	Values from the Normal Distribution	Percentage of the Mass Contained (Calculated as Two Times the Value of the Normal Distribution)
0	0	0
0.1	0.0398	8
0.2	0.0793	15.8
0.3	0.1197	24
0.4	0.1554	31
0.5	0.1915	38
0.6	0.2257	45
0.7	0.2580	52
0.8	0.2881	57.6
0.9	0.3159	63.2
1.0	0.3413	68.3
1.2	0.3849	77
1.4	0.4192	83.8
1.6	0.4452	89
1.8	0.4641	92.8
2.0	0.4773	95.4
2.5	0.4938	98.7
3.0	0.4987	99.74
4.0	0.49996	99.99
infinity	0.5	100.00

the accompanying properties given in Table 10.3. The standard deviation σ or the variance σ^2 are measures of the spread in the data about the mean. Thus, from Table 10.3, a spread of 4σ ($x/\sigma = 2$) incorporates about 95.4% of the mass or area under the concentration distribution. One standard deviation contains about 68.3% of the total mass. Statistics of the Gaussian distribution are related to the dispersion coefficient as

$$\sigma = (2Dt)^{1/2} \qquad (10.12)$$

where t is time and D is the dispersion coefficient.

The relative concentration of a pulse or tracer added at a point changes as a function of time with transport in a constant-velocity flow system. After the distribution becomes Gaussian, the mean defines the position due to transport at the linear ground-water velocity and the variance describes the variation about the mean. Thus, for a one-dimensional column (Figure 10.12a),

$$D_L = \frac{\sigma_L^2}{2t} \qquad (10.13)$$

where D_L is the longitudinal dispersion coefficient. When the velocity is constant, Eq. 10.13 becomes

$$D_L = \frac{\sigma_L^2 v}{2x} \qquad (10.14)$$

where v is the linear ground-water velocity and x is the distance from the source. Both theoretical arguments and column experiments support these relationships.

When mass spreads in two or three dimensions, the distributions of mass sampled normal to the direction of flow are also normally distributed, with variances that increase in proportion to $2t$. Thus, coefficients of transverse dispersion are defined as

$$D_T = \frac{\sigma_T^2}{2t} \qquad (10.15)$$

The two-dimensional spread of a tracer in a unidirectional flow field results in an elliptically shaped concentration distribution (Figure 10.12b) that is normally distributed in both the longitudinal and transverse direc-

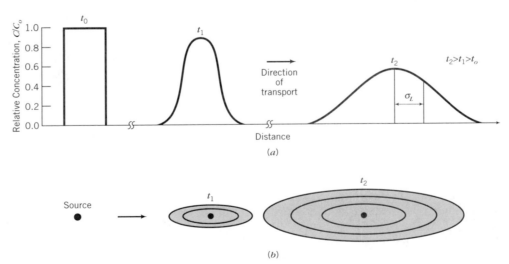

Figure 10.12 Variation in concentration of a tracer spreading in (a) one or (b) two dimensions in a constant-velocity flow system.

tions. Typically, longitudinal dispersion is greater than transverse dispersion (Figure 10.12*b*). Concentration distributions in three dimensions form ellipsoids of revolution (football shapes) when the two components of transverse spreading are the same and longitudinal dispersion is larger. When the vertical transverse dispersivity is small, as is often the case, plumes take on a surfboard shape.

The variances we have discussed differ in a subtle but important way. Those in Eqs. 10.13 and 10.15 are spatial distributions, while the one calculated from a breakthrough curve is the variance of a temporal distribution. The following equation (Robbins, 1983) interchanges space–time statistics relating to concentration distributions

$$\sigma_L^2 = v^2 \sigma_t^2 \qquad (10.16)$$

Thus, by recasting Eq. 10.3 in terms of σ_t

$$D_L = \frac{v^2 \sigma_t^2}{2t} \qquad (10.17)$$

the dispersion coefficient in the longitudinal direction can be calculated for a breakthrough curve.

For example, consider the two-dimensional plume shown in Figure 10.13*a* that is produced by a continuous point source. At any point along the midline of the plume (for example, *x*, 0) a breakthrough curve (Figure 10.13*b*) can be constructed by plotting the relative concentration as a function of time (Robbins, 1983). The relative concentration is C/C_{max}, where C_{max} is the highest concentration that will be observed at (*x*, 0). C_{max} is always less than the source concentration C_0.

This breakthrough curve is a cumulative normal distribution with a 2σ value that can be derived graphically from Figure 10.13*b* as

$$2\sigma_t = (t_{84} - t_{16}) \qquad (10.18)$$

where t_{84} and t_{16} are the breakthrough times corresponding to relative concentrations of 0.84, and 0.16, respectively, which accounts for about 68% of the mass (Table 10.3). With σ known, the dispersion coefficient is calculated with Eq. 10.17.

The same approach can establish values of dispersivity in the transverse direction. This time the relative concentration is plotted as a function of distance traveled transverse to the direction of flow at t_{50} (Robbins, 1983). The resulting concentration distribution is again normal (Figure 10.13*c*). Once the variance is estimated, the dis-

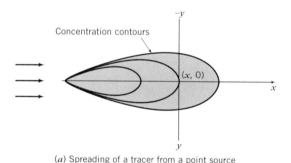

(a) Spreading of a tracer from a point source

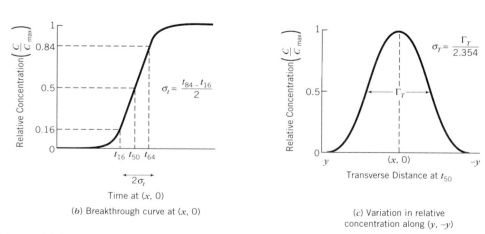

(b) Breakthrough curve at (x, 0)

(c) Variation in relative concentration along (y, −y)

Figure 10.13 Pattern of spreading of a tracer from a continuous point source in a two-dimensional flow field. Two graphical procedures for estimating the standard deviation of a normal distribution are shown in (*b*) and (*c*) (from Robbins, 1983).

persivity value is obtained from Eq. 10.14. The standard deviation in this case (Figure 10.13) is related to the half width of the distribution (Γ) at a relative concentration of $0.5C_{max}$ (Robbins, 1983)

$$\sigma_T = \frac{\Gamma}{2.354} \qquad (10.19)$$

This graphical approach can be used with any normal distribution. As before, space–time statistics can be interchanged with

$$\Gamma_y = \Gamma_t v \qquad (10.20)$$

10.5 Mixing in Fractured Media

Many of the concepts of mixing from the previous sections apply to fractured media. There is nevertheless sufficient added complexity to warrant treating these media separately. The discussion here will be focused on the only type of fractured medium that has been studied in detail. This is a case where fractures are the only permeable pathways through the rock. The hydraulic conductivity of the unfractured part of the medium is so small that fluid flow and hence advective transport are negligible. However, mass can diffuse into the unfractured rock matrix in response to concentration gradients.

Mixing in a fracture system is the result of both mechanical mixing and diffusion. Concentrations in the fractures will be affected by diffusion into the matrix, a chemical process (Figure 10.14a); variability within individual fractures caused by asperities, an advective dispersive process (Figure 10.14b); fluid mixing at the fracture intersections, a diluting or possibly diffusive process (Figure 10.14c); and variability in velocity caused by differing scales of fracturing (Figure 10.14d) or by variations in fracture density (Figure 10.14e). Some of these processes are coupled in a complex fashion and compete.

Diffusion into the matrix provides an important process to attenuate the transport of contaminants. This result contradicts experience with porous media, where the contribution of diffusion to dispersion is generally swamped by mechanical dispersion. Diffusion is more important in fractured media because localizing mass in fractures provides the opportunity for large concentration gradients to develop. Theoretical studies by Grisak and Pickens (1980) and Tang and others (1981) explored this process using numerical and analytical models of a single fracture bounded by an infinite porous matrix. Dispersion in a single fracture is caused by the variability in fracture aperture. This variability develops due to the roughness of the fracture walls and the precipitation of secondary minerals. At many locations, the fracture may

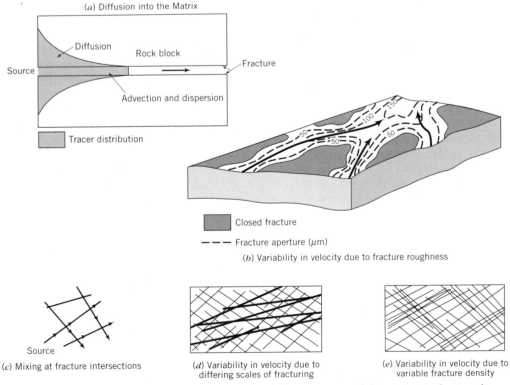

Figure 10.14 A series of sketches illustrating the different causes of dispersion in fractured rocks.

Table 10.4 **Tabulation of Dispersivity Values Based on Tracer Tests in Fractured Rocks**

Experiments	Migration Distance (m)	Dispersivity (m)	Fissure Width (mm)
Laboratory experiment with natural fissure (Neretnieks and others, 1982)	0.3	0.025	0.18
Laboratory experiment with natural fissure (Moreno and others, 1984)	0.19	0.005	0.15
	0.27	0.011	0.14
Stripa natural fissure, two different channels (Abelin and others, 1982)	4.5	2.0	0.11
	4.5	0.62	0.14
Studsvik site 2 (Landström and others, 1982)	14.6	—	—
Finnsjön site (Moreno and others, 1983)	30	0.35/5	0.47
			0.98
Studsvik site 1 (Landström and others, 1978)	22	6.1	—
	51	7.7	—
French site (Lallemand-Barrès, 1978)	11.8	0.8	—
U.S. site (Webster and others, 1970)	538	134	—

From Neretnieks (1985)

be closed to flow and transport. Such aperture topology (see Figure 10.14*b*) gives rise to channeling (Neretnieks, 1985) where mass moves predominantly along networks of irregularly shaped pathways in the plane of a fracture.

This channelization model has developed from tracer tests carried out in fractured rocks. For example, studies at the Stripa mine (Neretnieks, 1985) showed how the inflow of water to tunnels was extremely localized. Approximately one third of the flow entered from approximately 2% of the fractured rocks. These results are strongly indicative of localized channel flow within individual fractures.

Channels can be so poorly interconnected that they may not interact with one another over appreciable distances (Neretnieks, 1985). This behavior is termed *pure channeling*. From this discussion, it is easy to understand why the smooth, parallel plate model used by some investigators is really a simplified representation of a fracture.

Neretnieks (1985) has tabulated values of dispersivity from tracer experiments in fractured rocks (Table 10.4). Because of the relatively short transport distances, these values reflect mixing caused by diffusion into the matrix and mechanical dispersion within individual fractures. Thus, they are not true advective dispersivities. The values, in general, increase as a function of travel distance as in porous media. The larger dispersion lengths for the last four tests, however, may be artifacts of the interpretive model, which did not account for matrix diffusion (Neretnieks, 1985). This example emphasizes how published dispersivity numbers need to be examined in relation to the methods used for interpretation.

As is the case with porous media, there are conceptual problems that confound the application of classical theory. If pure channeling develops, the dispersivity will increase continuously as a function of travel distance and never reach an asymptotic value (Neretnieks, 1985). The lack of spatial averaging in channelized fracture systems may also inhibit the development of normal concentration distributions.

Dispersion at the next larger scale occurs when the geometry of the three-dimensional network begins to influence mass transport. A tracer moving along a fracture to an intersection (Figure 10.14*c*) partitions into two or more fractures. Because the water is also partitioned and because concentration is the mass per unit volume of solution, partitioning by itself does not affect the concentrations. However, concentrations will be affected if there is mixing with water that does not contain the tracer. This dilution depends directly on the type of mixing process at the intersection and quantity of water flowing along the fractures.

The dispersion in Figures 10.14*d* and *e* is analogous to that caused by heterogeneities in porous media. Differing scales of fracturing or spatial variability in fracture density create variability in local velocity. Even larger scale dispersion could develop if individual discontinuities exist on a regional scale or the variability in fracture density includes several units.

Many of the concerns about the applicability of classical concepts of dispersion apply to fractured rocks. For example, is it meaningful or even possible to describe asymptotic dispersivities? How does dispersivity change as a function of distance? Is it even possible in real networks to define a REV? Presently, these questions remain unresolved. Looking into the crystal ball, it is possible that many of the existing ideas about dispersion may have to be extended or reworked for fractured rocks.

10.6 A Geostatistical Model of Dispersion

Exciting new contributions to the study of dispersion are quantitative approaches that relate asymptotic dispersivities to geostatistical models of hydraulic conductivity. Eventually, these approaches will form the basis for a family of practical field techniques for characterizing dispersivity. To begin, a few basic ideas need to be developed concerning how a geostatistical model can represent heterogeneity in properties like hydraulic conductivity, porosity, or cation exchange capacity.

Medium properties in nature vary from place to place. Consider the equally spaced hydraulic conductivity measurements in Figure 10.15. Given such data, the goal of a geostatistical description is to represent the heterogeneity by a small number of statistical parameters. For the simplest case, three are sufficient, the mean, the variance, and the correlation length scales. This description implies that the medium is stationary. In other words, there is no trend in the statistical character of the medium across the field.

For a series of data, the mean is the level around which the parameter values fluctuate, while the variance describes the spread about this level (Box and Jenkins, 1976). When data vary over several orders of magnitude, they are often best described by a log-normal distribution (Freeze and others, 1989). In the case of hydraulic conductivity, the following equation describes the distribution:

$$Y_i = \ln K_i \qquad (10.21)$$

where Y_i is the log hydraulic conductivity. The mean and variance for the population are μ_Y and σ_Y^2, respectively.

The correlation length is a measure of the spatial persistence of zones of similar properties. For example, hydraulic conductivity usually is correlated such that values immediately adjacent to one another are similar and those much further away are dissimilar. Because of the way

Figure 10.15 A series of equally spaced ln K values (from Freeze and others, 1989). Reprinted by permission of Ground Water. Copyright © 1989. All rights reserved.

sediments are deposited, the correlation is usually anisotropic. For example, spatial persistence in hydraulic conductivity is greater in the direction of bedding than across the bedding. Chapter 3 introduced the concept of statistical homogeneity or heterogeneity as descriptors that describe this characteristic of a medium.

Mean and Variance

The population statistics for a heterogeneous field are usually estimated from a finite number of samples. In terms of notation, the statistics of the actual population are different than the sample estimates (Table 10.5). Thus, for a series of log hydraulic conductivity values, the mean estimated from the samples is

$$\overline{Y} = \frac{1}{n} \sum Y_i \qquad i = 1 \ldots n \qquad (10.22)$$

where $\overline{Y}$ is the mean hydraulic conductivity and Y_i are n individual log hydraulic conductivity values. The sample variance is

$$S_Y^2 = \frac{1}{n} \sum (Y_i - \overline{Y})^2 \qquad i = 1 \ldots n \qquad (10.23)$$

where S_Y^2 is the variance in hydraulic conductivity.

Table 10.5 Summary of Notation for the Geostatistical Description of a Medium

Parameter	Population Statistic	Population Estimate	Sample Estimate
Mean	μ_Y	$\hat{\mu}_Y$	$\overline{Y}$
Variance	σ_Y^2	$\hat{\sigma}_Y^2$	S_Y^2
Autocovariance	τ_{Y_k} or $\tau_Y(b)$	$\hat{\tau}_{Y_k}$ or $\hat{\tau}_Y(b)$	C_{Y_k} or $C_Y(b)$
Autocorrelation	ρ_{Y_k} or $\rho_Y(b)$	$\hat{\rho}_{Y_k}$ or $\hat{\rho}_Y(b)$	r_{Y_k} or $r_Y(b)$
Correlation length	λ_Y	$\hat{\lambda}_Y$	

From Freeze and others (1989). Reprinted by permission of Ground Water. Copyright © 1989. All rights reserved.

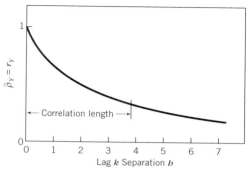

Figure 10.16 Example of an autocorrelation function with the correlation length defined as separation at which $\rho = 0.37$ (modified from Freeze and others, 1989). Reprinted by permission of Ground Water. Copyright © 1989. All rights reserved.

Autocovariance and Autocorrelation Functions

A population can be described in terms of autocovariance (τ) or autocorrelation (ρ) functions. These functions are related mathematically and describe how the correlation between any two hydraulic conductivity values decays with the separation or lag. A lag is simply some constant separation interval. For example, with lag one, values being compared are a distance h apart, and with lag two, $2h$, etc. Although there is flexibility in defining an autocorrelation function, an exponential model is used commonly (Freeze and others, 1989)

$$\rho(h) = \exp(-|h|/\lambda) \qquad (10.24)$$

where $\rho(h)$ is the autocorrelation of the population at a separation h. The correlation length scale (λ) (or λ_x, λ_y, λ_z for statistically anisotropic media) is the separation in the given direction at which ρ takes on a value of e^{-1} or

0.37 (Figure 10.16) or τ declines to 0.37 of the covariance at lag zero.

For example, with a finite number of vertical values of log hydraulic conductivity, sample autocovariance (C_Y) and autocorrelation (r_Y) functions can be calculated from pairs of equally spaced data. The autocovariance function at a separation h is

$$C_Y(h) = \frac{1}{m} \sum (Y(y_i) - \overline{Y})(Y(y_i + h) - \overline{Y}) \quad i = 1 \ldots m$$

$$(10.25)$$

where m is the number of pairs of sample points at a given separation, y_i is the spatial coordinate of the measurement $Y(y_i)$, and $Y(y_i + h)$ is the measurement h units away from $Y(y_i)$. The autocorrelation function $r_Y(h)$ is formed by taking $C_Y(h)$ values and dividing by $C_Y(0)$ or the variance S_Y^2 of the sample. For unevenly spaced data, values are grouped into a series of class intervals with the separation distance given by the midpoint of each interval. The following example illustrates the calculation of the mean, variance, and correlation length for a stationary medium.

Example 10.1

Listed on Table 10.6 are values of hydraulic conductivity Y (where $Y = \ln K$) collected at 5-cm intervals along a core. Calculate the mean, variance, and correlation length scale.

The solution is outlined on Table 10.6. The mean and variance are 1.0 and 0.182, respectively. The autocovariance coefficients at separations of 5 and 10 cm are 0.149 and 0.07, respectively, which when divided by the variance give autocorrelation coefficients 0.81 and 0.38, respectively. Because the autocorrelation coefficient at 10 cm is nearly 0.37, λ is about 10 cm.

Table 10.6 **Work Sheet for Example 10.1**

i	Y_i	$Y_i - \overline{Y}$	$(Y_i - \overline{Y})^2$	$(Y_{i+1} - \overline{Y})$	$(Y_{i+2} - \overline{Y})$	$(Y_i - \overline{Y})(Y_{i+1} - \overline{Y})$	$(Y_i - \overline{Y})(Y_{i+2} - \overline{Y})$
1	1.5	0.50	0.25	0.70	0.50	0.35	0.25
2	1.7	0.70	0.49	0.50	0.10	0.35	0.07
3	1.5	0.50	0.25	0.10	-0.2	0.05	-0.10
4	1.1	0.10	0.01	-0.20	-0.60	-0.02	-0.06
5	0.8	-0.20	0.04	-0.60	-0.40	0.12	0.08
6	0.4	-0.60	0.36	-0.40	-0.50	0.24	0.30
7	0.6	-0.40	0.16	-0.5	-0.10	0.20	0.04
8	0.5	-0.50	0.25	-0.1	0.0	0.05	0.0
9	0.9	-0.10	0.01	0.0		0.0	
10	1.0	0.0	0.0				
Σ	10		1.82			1.34	0.58
$\frac{1}{n}\Sigma$	1		0.182			0.149	0.07

Generation of Correlated Random Fields

Some applications (for example, optimization of a pump and treat system) depend on being able to generate synthetic hydraulic conductivity or permeability fields with specified geostatistical properties (μ_Y, λ_x, λ_y, λ_z). Various mathematical approaches can be used for this purpose such as nearest-neighbor models (Smith and Schwartz, 1981) or turning bands methods (Delhomme, 1979; Mantoglou and Wilson, 1982). A new Fourier transform approach is capable of generating three-dimensional, cross-correlated fields. The term *cross-correlated* means that in addition to generating, say, a hydraulic conductivity field, the algorithm also generates a second field like porosity, where porosity values are correlated to the hydraulic conductivity values in some specified manner (Robin and others, 1993).

The algorithm of Robin and others (1993) has created the example of a two-dimensional, correlated permeability field in Figure 10.17. The field has a mean permeability of 5.8×10^{-11} m², a variance in ln k of 0.27, and correlation length scales $\lambda_x = 0.1$ m, and $\lambda_z = 0.02$ m. The field is statistically anisotropic because the correlation length in the x direction (or horizontal direction) differs from the correlation length in the z direction (or vertical direction).

Estimation of Dispersivity

Gelhar and Axness (1983) provide the following theoretical equation for predicting longitudinal dispersivity from a geostatistical description of the hydraulic conductivity field:

$$A_L = \frac{\sigma_Y^2 \lambda}{\gamma^2} \qquad (10.26)$$

where A_L is the asymptotic longitudinal dispersivity, σ_Y^2 is the variance of the log-transformed hydraulic conductivity [i.e., $Y = \ln K$], λ is the correlation length in the mean direction of flow, and γ is a flow factor, which Dagan (1982) considers equal to one. This equation assumes (1) unidirectional mean flow and (2) an exponential covariance. The longitudinal asymptotic macrodispersivity (A_L^*) includes contributions from (1) the heterogeneous structure of the medium (A_L); (2) a local or pore-scale component (α_L); and (3) diffusion

$$A_L^* = A_L + \alpha_L + D_d^*/v \qquad (10.27)$$

where D_d^* is the bulk diffusion coefficient and v is the linear ground-water velocity. The transverse asymptotic macrodispersivity is

$$A_T^* = \alpha_T + D_d^*/v \qquad (10.28)$$

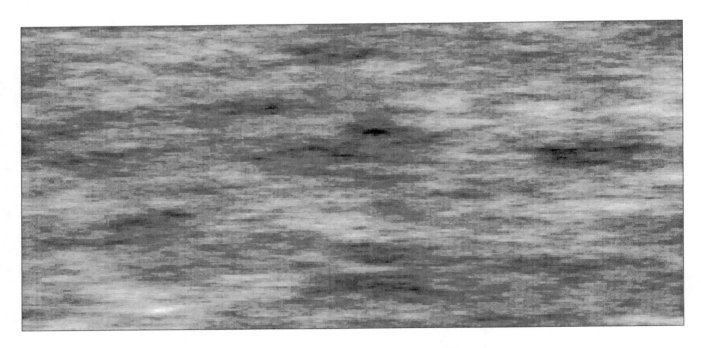

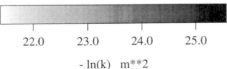

- ln(k) m**2

Figure 10.17 Example of a correlated permeability field with dimensions 1.06 × 0.5 m having a mean k of 5.8×10^{-11} m², σ^2 of 0.27, λ_x of 0.1 m and λ_z of 0.02 m. Provided by R.A. Schincariol.

The transverse asymptotic dispersivity (A_T) is zero, implying that the heterogeneous structure of the medium does not create transverse dispersion.

Given the variance in ln K and the correlation length, one can estimate the overall contribution to dispersion that is related to the macroscale heterogeneity. The other two components, α_L and D_d^*/v, are relatively small and can be estimated or in many cases neglected. The sum gives us A_L^*. Unfortunately, the characterization of the statistical properties of a medium requires considerable hydraulic conductivity data, which practically are not available. Borehole flow meters or geophysical approaches are being developed to provide ways of characterizing the statistical structure.

An example in the estimation of longitudinal asymptotic dispersivity is provided with data collected by Sudicky (1986). Shown in Figure 10.18 is the contoured hydraulic conductivity field based on 1279 measurements along a single cross section. This section is oriented parallel to the direction of tracer migration at Canadian Forces Borden (Freyberg, 1986; and Mackay and others, 1986). The distribution in hydraulic conductivity when expressed as ln K is normally distributed with $\sigma_Y^2 = 0.29$, with horizontal and vertical correlation length scales (λ_x and λ_z) of 2.8 and 0.12 m, respectively (Sudicky, 1986). These values confirm what is apparent from looking at the contoured field, namely that the correlation structure is anisotropic.

The contribution of heterogeneity to the asymptotic longitudinal dispersivity can be calculated from Eq. 10.23 as

$$A_L = \frac{\sigma_Y^2 \lambda_x}{\gamma^2} = \frac{0.29 \times 2.8}{1^2} = 0.81 \text{ m}$$

where the flow factor γ is taken to be one. With the contribution from diffusion essentially negligible and the local-scale dispersivity (α_L) estimated to be 0.05 m, most of the longitudinal dispersion arises due to heterogeneity in the hydraulic conductivity field.

An attractive feature of this method is that it does not require that dissolved mass pass through the test volume. Thus, it has the potential to be applied to large-scale systems—too big to be tested using more conventional

methods. One limitation is the relatively large number of hydraulic conductivity values that are required to define the spatial structure. The only practical way to address this limitation is through indirect determination of hydraulic conductivity using various kinds of borehole logs or a borehole flowmeter.

10.7 *Tracers and Tracer Tests*

Tracers and tracer tests are useful in understanding dispersion and advection. The most important tracers include (1) ions that occur naturally in a ground-water system such as Br^- or Cl^-, (2) environmental isotopes such as ^2H, ^3H, or ^{18}O, (3) contaminants of all kinds that enter a flow system, and (4) chemicals added to a flow system as part of an experiment. This latter group could include radioisotopes such as ^3H, ^{131}I, ^{82}Br; ions such as Cl^-, Br^-, I^-; and organic compounds such as rhodamine WT, fluorescein, and uranine. Many of these ions or compounds do not react to any appreciable extent with other ions in solution and the porous medium. They are considered as "ideal tracers." Reactive tracers are used more specifically to define the nature of reactions.

Given this variety of tracers and testing strategies, confusion may arise as to which to use in site studies. In most cases, the choice is linked inexorably to the scale of the study or the presence of a contaminant plume. For example, tracing flow and dispersion in a unit of regional extent (10s to 100s of kilometers) will involve either naturally occurring ions or environmental isotopes. Conducting a tracer experiment on such a large scale is simply not feasible because of the long times required for tracers to spread regionally. On a more localized or site-specific basis (several kilometers), the presence of a contaminant plume that has spread over a long time automatically makes it the tracer of choice. Again, time is insufficient (except perhaps in karst) to run an experiment at the scale of interest. Only for small systems (for example, some fraction of a kilometer) is there a possibility of running a field tracer experiment. It is for these experiments when a selection must be made among the various tracers.

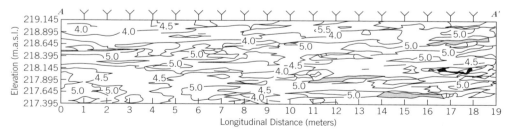

Figure 10.18 Variation in hydraulic conductivity (as $-\ln K$) along the cross section *A-A'* at Canadian Forces Base Borden (modified from Sudicky, Water Resources Res., v. 22, p. 2069–2082, 1986. Copyright by American Geophysical Union).

Field Tracer Experiments

Tracers are transported in two kinds of flow environments. In a natural gradient system, tracers move due to the natural flow of ground water. In a system stressed by injection and/or pumping, transport occurs in response to hydraulic gradients typically much larger than those in natural systems. A small-scale tracer experiment can involve either of these flow conditions.

Success in these tests depends on adequately characterizing concentration distributions in space and time. Typically, a large three-dimensional network of monitoring points is necessary to define the plume accurately. Point sampling is essential to avoid concentration averaging within the well bore. For example, estimated values of dispersivity tend to be larger when the number of individual sampling points is small or samples are collected over relatively large vertical intervals. This excess dispersion represents error due to inadequacies in the sampling network.

In theory, mass-transport parameters can be estimated from almost any test where a tracer is added in a controlled way. However, a few more or less standard tests are preferred because there are simple procedures for interpreting the results. Unfortunately, many of these tests are subject to error because when run in their basic configuration there is often an insufficient number of monitoring points.

Natural Gradient Test

The natural gradient test involves monitoring a small volume of tracer as it moves down the flow system.

Keeping the quantity of tracer small minimizes the disturbance of flow conditions. The resulting concentration distributions provide the data necessary to determine advective velocities, dispersivities, and occasionally equilibrium and kinetic parameters.

Such an experiment typically requires a dense network of sampling points. This chapter concludes with a brief description of some of the important tracer tests of this type run over the last decade.

Single Well Pulse Test

The single well pulse test (Figure 10.19a) involves first injecting a tracer followed by water into an aquifer at some constant rate. After an appropriate period of injection, the aquifer is pumped at the same rate. The concentration of tracer in water being withdrawn is monitored as a function of time or total volume of water pumped. These concentration/time data provide a basis for estimating the longitudinal dispersivity and chemical parameters like K_d within a few meters of the well (for example, Pickens et al., 1981). Fried (1975) describes a more refined version of this test where concentration/time data are collected at different positions in the well with a down-hole probe.

While historically of some interest, this test has limited applicability in estimating values of dispersivity. It is generally not possible to scale up dispersivity measurements made at such a small scale to the larger scales of interest in most problems. In addition, the lack of detailed observations on how the tracer is spreading in the vicinity of the well makes any dispersivity estimate quite crude. This test holds more promise for evaluating geochemical

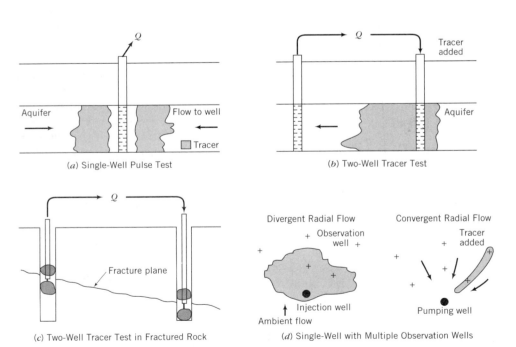

Figure 10.19 Examples of field tracer tests.

processes. Because reactions operate on a local scale, there is no real scaling problem.

Two-Well Tracer Test

In the two-well test, water is pumped from one well and injected into another at the same rate to create a steady-state flow regime (Figure 10.19b). The tracer is added continuously at a constant concentration at the injection well and monitored in the withdrawal well. The simplest way of running this test is by recirculating the pumped water back to the injection well, as is shown on the figure. However, the tracer concentration will begin to increase at the injection well once breakthrough occurs at the pumped well. The test can also be run without recirculation by providing the water for the injection well from an alternative source. The resulting concentration versus time data are interpreted in terms of processes and parameters. These tests can be conducted over several hundred meters in highly permeable systems. However, with only a single monitoring point, the test provides at best only a crude estimate of dispersivity. Estimates can be improved by adding more observation wells between the pumping–injection doublet.

Modified versions of a two-well test have been conducted in fractured rocks (Raven and Novakowski, 1984). Packing off a small section of the borehole (Figure 10.19c) isolates a single fracture plane between pumping and injection wells. This test is technically more demanding because it requires working with packers. Nevertheless, it involves the same interpretive techniques and is subject to the same limitations that we just discussed.

Interpretive techniques for two-well tracer tests are quite sophisticated. Grove and Beetem (1971) present a graphical, curve-matching procedure that is appropriate for idealized flow conditions that develop in homogeneous and isotropic units. A computer code for this purpose is provided by Grove (1971). Güven and others (1986) have developed an alternative approach to evaluating two-well tracer tests. Their procedure interprets variation in breakthrough at the pumped well due to variability in the velocity field caused by geologic layering. Essentially this analysis assumes that mixing due to local hydrodynamic dispersion is small relative to that caused by the heterogeneity of the system. The evaluation involves varying porosity and the hydraulic conductivity of the different units to match the simulated and observed breakthrough curves. Other approaches can account for the effects of dispersion (for example, Huyakorn and others, 1986).

Single Well Injection or Withdrawal with Multiple Observation Wells

These tests create a transient radial flow field by injection or withdrawal. The radially divergent test (Fig. 10.19d) involves monitoring the tracer as it moves away from the well. The radially convergent test involves adding the tracer at one of the observation wells and monitoring as it moves toward the pumped well. Parameters can be estimated at scales of practical interest with accuracy, given a reasonable number of observation wells. Between these two tests, the divergent flow test is preferable (Gelhar and others, 1985). The converging flow field (Figure 10.19d) counteracts spreading due to dispersion and is thought to be less useful.

Estimates from Contaminant Plumes and Environmental Tracers

Estimates of dispersivities at scales greater than several hundreds of meters are based on mixing patterns caused by the presence of a dissolved contaminant plume or temporal variations in the chemistry of natural recharge. These latter changes can be caused by the atmospheric testing of nuclear weapons, which impacts tritium and chlorine-36 levels in rainfall, or by longer term changes in climate, which impact oxygen-18 and deuterium (for example, Pleistocene to present). The detailed examination and interpretation of data from contaminant plumes and environmental isotopes will be reserved for several later chapters.

Estimates of advection and dispersion based on data from plumes or environmental measurements are less reliable than field tracer tests. There is uncertainty in defining the loading function for the constituent of interest and, often, there is an inadequate number of sampling points (Gelhar and others, 1992).

Massively Instrumented Field Tracer Tests

Over the past decade, there have been several large-scale natural gradient tracer tests (Table 10.7) run in field settings. Three tests, Borden, Cape Cod, and Columbus, are unique because between five to ten thousand monitoring points provided precise three-dimensional pictures of the evolving tracer plumes over travel times of approximately two to three years. These tests were run to learn more about the manifestations of physical and chemical transport processes at large scale in shallow unconfined aquifers, and to validate the geostatistical concepts of dispersion. The study of chemical processes involved the use of reactive tracers in addition to conservative tracers. The test at Borden used chloride and bromide as the conservative (that, is nonreactive) tracers and five organic compounds as reactive tracers (Table 10.7). The test at Cape Cod used bromide as the nonreactive tracer, and lithium (Li^+) and molybdate (MoO_4^{2-}) as reactive tracers. The test at Columbus used only conservative tracers. The following two sections summarize some results from the Borden test and examine how well theories work that predict longitudinal dispersivity from statistical properties of the hydraulic conductivity field.

Table 10.7 Information on Massively Instrumented Tracer Tests

Site	Aquifer Material	Test Scale (m)	References
Canadian Forces Base Borden, Ontario	Glaciofluvial sand	90	Mackay and others, 1986 Freyberg, 1986 Sudicky, 1986
Cape Cod, Massachusetts	Sand and gravel	250	LeBlanc and others, 1991 Garabedian and LeBlanc, 1991 Hess and others, 1992
Columbus, Mississippi	Sandy gravel/gravelly sand	280	Boggs and others, 1992 Adams and Gelhar, 1992 Rehfeldt and others, 1992

Borden Tracer Experiment

The natural gradient tracer test at Canadian Forces Base Borden illustrated how ground-water velocities and the extent of dispersion could be interpreted from "snapshots" of the conservative tracer distributions. The Cl^- and Br^- concentrations were monitored on 14 occasions over 1038 days at approximately 5000 observation points. Here, we examine Cl^- data. The addition of 10.7 kg of Cl^- provided a mean concentration of 892 mg/l (Freyberg, 1986). Figure 10.20 illustrates the vertically averaged Cl^- distribution at four times. Advection moved the tracer plume along the shallow flow system, and longitudinal dispersion elongated the plume in the direc-

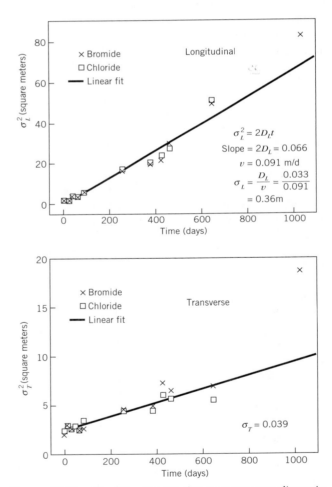

Figure 10.20 Vertically averaged concentration distribution of Cl^- at various times after injection (from Mackay and others, Water Resources Res., v. 22, p. 2017–2029, 1986. Copyright by American Geophysical Union).

Figure 10.21 Fit of the Borden data to a constant dispersivity model (from Freyberg, Water Resources Res., v. 22, p. 2031–2046, 1986. Copyright by American Geophysical Union).

Table 10.8 **Estimates of Mass in Solution, Location of the Center of Mass, and Spatial Covariance for the Chloride Plume**

Date	Elapsed Time (days)	Mass in Solution (kg)	Center of Mass[1] x_c (m)	y_c (m)	z_c (m)	Spatial Covariance[2] $\sigma_{x'x'}$ (m²)	$\sigma_{y'y'}$ (m²)	$\sigma_{x'y'}$ (m²)
		Chloride						
Aug. 24, 1982	1	6.7	0.2	0.1	2.78	2.1	2.4	0.5
Sept. 1, 1982	9	9.2	0.7	0.4	3.02	1.7	2.4	0.7
Sept. 8, 1982	16	9.2	1.6	0.7	3.06	2.3	2.8	0.8
Sept. 21, 1982	29	11.5	2.9	0.9	3.27	2.5	2.6	0.9
Oct. 5, 1982	43	11.3	4.1	1.6	3.34	4.4	2.7	1.2
Oct. 25, 1982	63	9.0	5.7	2.0	3.50	4.4	2.4	1.1
Nov. 16, 1982	85	11.2	7.7	3.2	3.75	5.7	3.3	0.8
May 9, 1983	259	11.5	22.7	11.6	4.52	17.8	4.4	3.7
Sept. 8, 1983	381	9.6	32.3	15.3	5.18	20.6	4.4	3.9
Oct. 26, 1983	429	9.2	35.9	17.2	5.25	24.3	6.0	3.2
Nov. 28, 1983	462	8.2	38.2	17.4	5.33	27.8	5.5	2.1
May 31, 1984	647	9.1	53.1	23.9	5.55	51.5	5.5	3.0

[1] Given in the field coordinate system.

[2] Given in rotated coordinates: x' is parallel to linear horizontal trajectory, y' is perpendicular to linear horizontal trajectory. From Freyberg, Water Resource Res., v. 22, p. 2031–2046, 1986. Copyright by American Geophysical Union.

tion of flow. Horizontal-transverse spreading was, however, less marked (Figure 10.20).

Freyberg (1986) calculated the center of mass of the Cl⁻ plume in three dimensions (Table 10.8) as well as the variance in concentration distributions in the longitudinal and horizontal-transverse directions. Transverse-vertical spreading was negligible.

The data in Table 10.8 indicate that the center of mass traveled 58.2 m in 647 days, which equates to a linear ground-water velocity of 0.09 m/day. The simplest calculation of the dispersivity is based on the relationships between variance, time, and dispersivity (Eq. 10.12), assuming that dispersivities are constants. Fitting a line

Table 10.9 **Geostatistical Description of Aquifers at the Three Sites and Comparison of Dispersivity Values Estimated from Field Tracer Tests and Stochastic Theory**

Site	Aquifer Properties λ_b	λ_v	S_y^2	Longitudinal Dispersivity (m) Theory[1]	Test
Borden	2.8	0.12	0.29	0.45–0.6	0.36
Cape Cod	5.1	0.26	0.26	0.35–0.78	0.96
Columbus	5.3	0.7	2.8	1.5	5–10

[1] These calculations involve a variety of stochastic approaches and the values are not the same as those which would come from the simple equation we presented.

λ_b = correlation length in the horizontal direction

λ_v = correlation length in the vertical direction

S_y^2 = sample estimate of the variance in ln K

through σ_L^2 and σ_T^2 versus time data (Figure 10.21) provided an estimate of D_L and D_T. From these values, Freyberg (1986) calculates α_L and α_T to be 0.36 and 0.039 m, respectively. At later times, the straight-line approximation does not fit the data too well (Figure 10.21) indicating that dispersivities may not be constant with time or travel distance.

Validation of the Stochastic Model of Dispersion

Section 10.6 introduced the idea of how asymptotic dispersivities could be estimated from parameters of the heterogeneous hydraulic conductivity field. The theory is based on stochastic continuum theories developed by Dagan (1982, 1984) and Gelhar and Axness (1983). This theoretical approach was tested at each of the three sites by comparing the longitudinal dispersivity value obtained in the tracer test with that estimated from geostatistical information on the hydraulic conductivity fields. These statistical data came from a large numer of measurements made using permeameters and borehole flow meters.

The summary statistics for these sites (Table 10.9) show that the aquifers at both Borden and Cape Cod are relatively homogeneous compared to that at Columbia. These differences are reflected in variances in the natural logarithm of hydraulic conductivity, which for Borden and Cape Cod are 0.29 and 0.26, respectively, versus 2.7 for Columbus (Rehfeldt and others, 1992). The correlation length scales in the horizontal and vertical directions are all quite similar (Table 10.9). The greater variance in ln K at Columbus results in a larger estimated asymptotic dispersivity at this site. This result generally is confirmed by the results from the actual tracer tests, although at

Columbus, the dispersivity estimated from the tracer test is somewhat larger than would be predicted by stochastic theory (Adams and Gelhar, 1992). These results suggest that it will be feasible to estimate dispersivity values from statistical information about the hydraulic conductivity field. Further, borehole flow meters provide a practical approach to collecting the large number of hydraulic conductivity measurements required for such an approach.

Problems

1. Consider two different media, one fractured and the other unfractured. The hydraulic conductivity is the same for each (1×10^{-8} m/s) but the effective porosities are different (3×10^{-4} and 0.30, respectively). Under an imposed hydraulic gradient of 0.01, determine how much time is required for a tracer to be advected 30 m in both the fractured and porous media.

2. The hydrogeologic cross section on the figure illustrates the pattern of groundwater flow along a local flow system. Assume that a source of contamination develops with advection as the only operative transport process. Describe the pathway for contaminant migration and estimate at what time in the future the plume will reach the stream.

3. The approximate coefficient of diffusion for Cl^- in water is 2×10^{-5} cm²/s. Estimate its value in a saturated porous medium having a porosity of 0.25 and a tortuosity of 1.25.

4. A useful rule of thumb is that mass has to spread several tens of correlation lengths before the asymptotic dispersivity is reached. Explain why.

5. Ground water flows through the left face of a cube of sandstone (1 m on a side) and out of the right face with a linear ground-water velocity of 10^{-5} m/s. The porosity of the sandstone is 0.10, and the effective diffusion coefficient is 10^{-10} m²/s. Assume that a tracer has concentrations of 120, 100, and 80 mg/l (1.2, 1.0, and 0.8×10^5 mg/m³) at the inflow face, the middle of the block, and the outflow face, respectively. Calculate the mass flux through the central plane due to advection and due to diffusion.

6. The essence of a column experiment for measuring dispersivity is to obtain a breakthrough curve, which can be evaluated to provide an estimate of the variance in concentration as a function of time (σ_t^2).

a. Modify Eq. 10.17 to a form in which the longitudinal dispersivity α_L is expressed as a function of the length of the column (L), time (t), and σ_t^2.

b. A breakthrough curve for a column experiment provides a σ_t^2 value of 5.5 min² at a time of 82 min (i.e., time at which the advective front arrives). Knowing that the column is 56 cm long, calculate the longitudinal dispersivity of the medium (α_L).

7. A contaminant is added as a point source to ground water flowing with a constant velocity of 4×10^{-6} m/s. Assuming longitudinal and two transverse dispersivities (y and z directions) of 1.0, 0.1, and 0.01 m, determine the spatial standard deviations ($\sigma_{x,y,z}$) in the plume size after 400 m of transport.

8. Following is a series of 15 hydraulic conductivity measurements (K) measured on cores collected at a spacing of 0.2 m. Estimate the sample mean, variance, and vertical correlation length (using two lags). Assume an exponential autocorrelation function.

Hydraulic Conductivity × 10^{-4} cm/s

1	. . . 4	8	. . . 36
2	. . . 13	9	. . . 12
3	. . . 48	10	. . . 13
4	. . . 64	11	. . . 18
5	. . . 62	12	. . . 36
6	. . . 58	13	. . . 48
7	. . . 38	14	. . . 52
		15	. . . 64

9. A series of hydraulic conductivity measurements for an unconfined aquifer provide a mean hydraulic conduc-

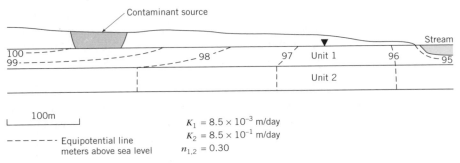

Problem 2

Contaminant source

Stream

100
99
98
97   Unit 1   96
95

Unit 2

100m

$K_1 = 8.5 \times 10^{-3}$ m/day
$K_2 = 8.5 \times 10^{-1}$ m/day
$n_{1,2} = 0.30$

- - - - - Equipotential line
meters above sea level

tivity (Y) of 0.004 m/s, where $Y = \ln K$, a variance in the log-transformed hydraulic conductivity (σ_Y^2) of 1.0, and a correlation length in the direction of the mean flow (λ_L) of 10.0 m. Estimate the asymptotic macroscale dispersivity for the aquifer.

10. Following is an enlargement of the Cl$^-$ plume measured after 462 days during the tracer experiment at Canadian Forces Base Borden and in Figure 10.20.

a. Determine σ_L^2 from the field data using the graphical procedure and a conventional statistical calculation.

b. Assuming that the longitudinal dispersivity and velocity are constant and that the plume originates as a point source at the coordinates (0.0, 0.0) in Figure 10.20, estimate α_L.

c. Explain in this case why estimates of σ_T will not yield estimates of the transverse dispersivity.

Hint: Examine Figure 10.20

d. Examine the character of transverse dispersion in Figure 10.20. What qualitative assessment can you make about the importance of transverse dispersion?

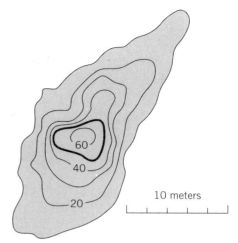

Problem 10

Chapter 11

Principles of Aqueous Geochemistry

Chapters 11 and 12 discuss the basic principles of aqueous geochemistry, and the theory and processes of mass transfer. This information is a necessary prerequisite to discussions of the chemistry of natural ground water and contaminants that come later. More specifically, Chapter 11 examines fundamental ideas about the properties of water, the way mass occurs in water, and equilibrium and kinetic concepts in aqueous systems. It also discusses the chemistry of organic compounds, the common constituents in natural ground water, and the unique graphical and statistical techniques developed to describe their occurrence.

11.1 Introduction to Aqueous Systems

Mass can exist in the subsurface as (1) separate gas or solid phases, for example CO_2 gas in the soil zone or minerals that form the porous medium; (2) a separate component of the liquid phase, for example, crude oil and liquid organic contaminants; or (3) mass dissolved in the water itself (that is, solutes), for example, Na^+ or Cl^-. While all of these occurrences are important in a geologic context, this chapter will focus on (3). The nonaqueous phase liquids, gases, and solids will also be discussed because they are often the sources of solutes in ground water or potential contaminants.

Ions, molecules, or solid particles in water undergo not only transport (Figure 11.1) but also reactions, which redistribute mass among various ion species, or between the liquid and solid phases. Our list of reactions is not exhaustive but it includes the most important ones affecting the chemistry of ground water. These chemical processes operate in a system consisting of one or more solid phases, an aqueous solution phase, and a gas phase (Figure 11.1) and redistribute mass within a phase or between phases.

Most inorganic substances are electrolytes, dissolving in water to form ions. Positively charged species, such as Ca^{2+} or K^+, are called cations. Negatively charged species, such as HCO_3^- or Cl^-, are called anions. Complex ions form by combining simpler cations and anions. Organic substances can also dissolve to form organic cations or anions. However, organic liquids are usually nonelectrolytes and dissolve in water as nonionic molecules. In the aqueous phase, the distribution of species is described in terms of a molar concentration or an activity, which is an effective concentration accounting for chemical nonidealities in reactions. The notation adopted throughout the book is to represent molar concentrations with round brackets, for example, (Ca^{2+}), and activities with square brackets, $[Ca^{2+}]$. Gases are represented in terms of partial pressures of constituent gases (for example, P_{CO_2}) or in nonideal systems by fugacities f_{CO_2}. Solid

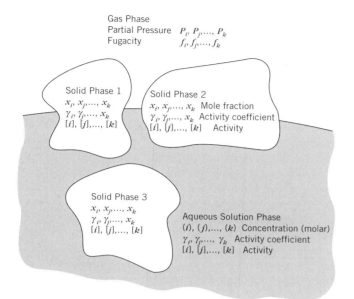

Gas Phase
Partial Pressure $P_i, P_j, ..., P_k$
Fugacity $f_i, f_j, ..., f_k$

Solid Phase 1
$x_i, x_j, ..., x_k$
$\gamma_i, \gamma_j, ..., x_k$
$[i], [j], ..., [k]$

Solid Phase 2
$x_i, x_j, ..., x_k$ Mole fraction
$\gamma_i, \gamma_j, ..., x_k$ Activity coefficient
$[i], [j], ..., [k]$ Activity

Solid Phase 3
$x_i, x_j, ..., x_k$
$\gamma_i, \gamma_j, ..., x_k$
$[i], [j], ..., [k]$

Aqueous Solution Phase
$(i), (j), ..., (k)$ Concentration (molar)
$\gamma_i, \gamma_j, ..., \gamma_k$ Activity coefficient
$[i], [j], ..., [k]$ Activity

Figure 11.1 Generalized model of a chemical system for ground water at some temperature and pressure (modified from Stumm and Morgan, 1981, Aquatic Chemistry). Copyright © 1981 by John Wiley & Sons, Inc. Reprinted by permission of John Wiley & Sons, Inc.

phase compositions (Figure 11.1) are represented as mole fractions and activities.

Concentration Scales

Molar concentration (M) defines the number of moles of a species per liter of solution (mol/L). A mole is the formula weight of a substance expressed in grams. For example, a one liter solution containing 1.42 g of Na_2SO_4 has a (Na_2SO_4) molarity of $1.42/(2 \times 22.99 + 32.06 + 4 \times 16.00)$ or 0.010 M. Because Na_2SO_4 dissociates completely in water according to the following reaction

$$Na_2SO_4 = 2Na^+ + SO_4^{2-}$$

the molar concentrations of Na^+ and SO_4^{2-} are 0.02 and 0.01 M, respectively. This reaction states that one mole of Na_2SO_4 dissolves to produce two moles of Na^+ and one mole of SO_4^{2-}. Because substances react in molar proportions, thermodynamic calculations use concentrations expressed in moles.

Molal concentration (m) defines the number of moles of a species per kilogram of solvent (mol/kg). This scale for concentrations in dilute solutions is almost the same as molar concentrations because a one-liter solution has a mass of approximately 1 kg. For more concentrated solutions, the two scales become increasingly different.

Equivalent charge concentration is the number of equivalent charges of an ion per liter of solution with units such as eq/L or meq/L. The equivalent charge for

an ion is equal to the number of moles of an ion multiplied by the absolute value of the charge. For example, with a singly charged species such as Na^+, 1 M Na^+ = 1 eq/L, and with a doubly charged species such as Ca^{2+}, 1 M Ca^{2+} = 2 eq/L. Equivalent concentrations can also be represented as equivalent charges per unit mass of solution with units such as eq/kg or equivalents per million (epm).

Mass per unit mass concentrations define a scale in terms of the mass of a species or element per total mass of the system. Many older analyses have been reported using this scale with concentrations in parts per million (ppm) or parts per billion (ppb). More recently, these units of concentration have given way to corresponding concentrations in mg/kg or μg/kg.

Mass per unit volume concentration is the most common scale for concentration. It defines the mass of a solute dissolved in a unit volume of solution. Concentrations are reported in units such as mg/L or μg/L. Again, there is a close correspondence between these last two scales of concentration. For dilute solutions, 1 ppm = 1 mg/kg = 1 mg/L. As the total salinity of a solution begins to exceed approximately 10^4 mg/L, the conversion equation

$$mg/kg = \frac{mg/L}{\text{solution density}} \tag{11.1}$$

is necessary to move between these concentration scales.

Concentration conversions involve a few simple equations. The most common scale change takes the reported values of chemical analyses in mg/L (also mg/kg or ppm) and converts them to molar concentrations:

$$\text{molarity} = \frac{mg/L \times 10^{-3}}{\text{formula weight}} \tag{11.2}$$

Conversion from mg/L to meq/L is sometimes necessary to present chemical data graphically or to check chemical analyses. This conversion equation is

$$meq/L = \frac{mg/L}{\text{formula wt/charge}} \tag{11.3}$$

To understand the conversion of scales, consider the following example.

Example 11.1

The concentration of SO_4^{2-} in water is 85.0 mg/L. Express this concentration as molarity, and meq/L.

$$mol/L = \frac{85 \times 10^{-3}}{32.06 + 4 \times 16.0} = 0.89 \times 10^{-3}$$

$$meq/L = \frac{85}{(32.06 + 4 \times 16.0)/2} = 1.77$$

Gas and Solid Phases

In thermodynamic calculations, concentrations of gases are given as partial pressures. In a mixture of gases at a fixed temperature and in equilibrium with the aqueous phase, the partial pressure of a gas is the pressure the gas would exert if it occupied the whole volume of the mixture by itself. For example, CO_2 gas in the atmosphere has a P_{CO_2} of about $10^{-3.5}$ atmospheres (atm). It is also common to report gas concentrations using a scale such as ppm or mg/L.

Concentrations for solids are represented as a mole fraction, which is the ratio of the number of moles of the phase of interest to the total number of moles in the system. However, thermodynamic calculations often do not require the solid phase concentrations because by convention the solid phase concentration or activity for a pure solid is one.

11.2 Structure of Water and the Occurrence of Mass in Water

The structure of water influences the solubility of solids and organic liquids. The water molecule is comprised of two hydrogen and one oxygen atom. Shared electrons form covalent bonds between atoms. Hydrogen atoms are located asymmetrically with respect to the oxygen atom (Figure 11.2a). Because of this structure and the covalent bonding, the electrical center of the negative charges (electrons) has a different location than that of the positive charges. This separation makes the molecule polar in terms of electrical charge. When a molecule is perfectly symmetrical (for example, carbon tetrachloride, benzene, and other hydrocarbons), there is no charge separation and the molecule is nonpolar.

The water molecules are joined by a series of hydrogen bonds (Figure 11.2b). These bonds form from electro-

static interaction between a hydrogen and an oxygen atom. These interactions produce an irregular tetrahedral network (Figure 11.2c). This molecular structure explains some of the unique characteristics of water such as its freezing and boiling points.

The polar nature of water also contributes to the relative solubility of electrolytes and the insolubility of nonelectrolytes in aqueous solution. Charged species (for example, ions or molecules) bind with the polar molecules in the water structure, which is an important factor contributing to solubility. In water, a charged species forces a local rearrangement in the structure (Figure 11.2d). Water molecules immediately adjacent bind to the charged species. Those nearby reorient themselves because of the charge. The zone of influence immediately adjacent to the ion is known as the hydration sheath. A so-called random region separates the hydration sheath from the bulk solution. Within the random region, the water molecules are less oriented than either in the bulk solution or with the hydration sheath.

In the case of an uncharged species, such as many nonpolar organic molecules, there is much less interaction with the water molecule, and the nonpolar molecule in effect has to find a space for itself in the water structure. These effects contribute to the relative insolubility of nonpolar molecules in water. Moreover, many of the thermodynamic relationships governing the dissolution process have molecule size as a variable. Solubility turns out to be inversely proportional to molecule size (see Section 12.2).

11.3 Equilibrium Versus Kinetic Descriptions of Reactions

The state of chemical equilibrium for a closed system describes a position of maximum thermodynamic stability. At equilibrium, there is no chemical energy available to alter the relative distribution of mass between the reactants and products in a reaction. Away from equilibrium, energy is available to spontaneously drive a system toward equilibrium by allowing the reaction to progress.

Theoretical approaches are available for modeling the chemical composition of a solution at equilibrium. However, these calculations provide no information about the time required to reach equilibrium or the reaction pathways involved. A different approach, based on kinetic models of reaction, is required.

From a chemical perspective, ground water is best thought of as a partial equilibrium system. Some reactions are at equilibrium and are describable in terms of equilibrium concepts. However, other reactions not at equilibrium are best represented by kinetic concepts. How a reaction is finally modeled depends on its rate. An equilibrium reaction is "fast" in relation to the mass transport processes that are working to change concentrations. A

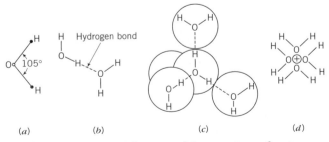

Figure 11.2 Essential features of the structure of water and adjustments due to the presence of an ion (adapted from Pytkowicz, 1983): (a) the position of the hydrogen and oxygen atoms in the polar water molecule; (b) hydrogen bonding between water molecules, giving rise to (c) the tetrahedral structure of water; (d) water molecules rearrange themselves around a cation due to the polar nature of the water.

kinetic reaction is "slow" in relation to the transport processes. Thus, in applying an equilibrium model to a reaction, we assume mass is transferred instantaneously between reactants and products to move the system to equilibrium. If the system transfers mass in a reaction at a rate that is slower than physical transport, it requires a kinetic description.

The degree of competition between the rates of reaction and the mass transport processes influences how one might model a system. For example, it may be appropriate to model ion exchange in a sluggish regional flow system as an equilibrium process. However, the same exchange process operating in an aquifer with high flow velocities could require a kinetic description.

Not all of the reactions are of the equilibrium type. Irreversible reactions proceed in the forward direction until the reactants are used up. Examples include radioactive decay, some redox reactions, and some organic reactions. Such reactions require a kinetic description.

The kinetic approach to modeling reactions is much more general than the equilibrium one. Nevertheless, most geochemical work involves models based on equilibrium concepts. The greatest limitation in the application of kinetic models is the lack of basic data concerning the order and rate of the reactions. The strength of the equilibrium approaches is the ability to apply concepts of chemical thermodynamics to estimate equilibrium and other constants as a function of temperature (Nordstrom and Munoz, 1986).

Reaction Rates

There are sufficient data on reaction rates to provide guidance in deciding whether to use an equilibrium or a kinetic approach. Figure 11.3 compares the half-time or half-life of many of the reactions on our list in Figure 10.1. As a generalization, the fastest reactions are the solute–solute or solute–water type with half-times of

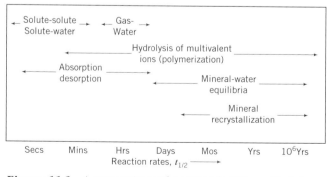

Figure 11.3 A comparison of reaction half-times ($t_{1/2}$) for many of the common reactions in aqueous systems (from Langmuir and Mahoney, 1984). Reprinted by permission First Canadian/American Conference on Hydrogeology. Copyright © 1984. All rights reserved.

fractions of seconds to at most a few minutes. These reactions are examples of homogeneous reactions, reactions that occur in a single phase. Of the list in Figure 10.1, the acid–base, and complexation reactions are homogeneous. The dissolution–precipitation reactions are examples of heterogeneous reactions or reactions involving more than one phase. These reactions have half-times varying from days to 10^6 years (Figure 11.3). Heterogeneous solution or exsolution reactions involving a gas phase have relatively larger half-times than liquid–solid reactions. Reactions on surfaces, which Langmuir and Mahoney (1984) term adsorption–desorption reactions, have half-times ranging from seconds to days. The larger half-times describe surface reactions in the porous matrix of rock fragments. Radioactive decay and isotopic fractionation reactions are not included on the figure because the range in half-times is extremely variable.

Redox reactions are often relatively slow because they are mediated by microorganisms (Morel and Hering, 1993). Typical half-times for these reactions range from hours to a few years based on data for marine systems (Berner, 1980). Transformation reactions involving organic contaminants in ground water depend on the physical and chemical properties of the particular compounds (for example, hydrolysis), and/or the abundance of microbes (for example, biodegradation).

This summary of reaction rates illustrates why the equilibrium approaches have proven so successful in modeling the chemistry of natural ground-water systems. Almost all of the important reactions have rates that are faster than the typical rates of ground-water flow.

11.4 Equilibrium Models of Reaction

Equilibrium concepts are useful in modeling many homogeneous and heterogeneous reactions. Consider the following reaction, where ions C and D react to produce ions Y and Z

$$cC + dD = yY + zZ \qquad (11.4)$$

with c, d, y, and z representing the number of moles of these constituents. For a dilute solution, the law of mass action describes the equilibrium distribution of mass between reactants and products as

$$K = \frac{(Y)^y (Z)^z}{(C)^c (D)^d} \qquad (11.5)$$

where K is the equilibrium constant and $(Y),(Z),(C)$, and (D) are the molal (or molar) concentrations for reactants and products. Depending on the reaction, we refer to this equilibrium constant as (1) an acidity or dissociation constant in acid–base reactions, (2) a complexation constant in complexation reactions, (3) a solubility constant in solid dissolution reactions or a Henry's law constant in gas dissolution, or (4) an adsorption constant for surface reactions.

Standard geochemical texts, for example, Morel and Hering (1993) tabulate values for these constants. Equilibrium constants expressed as a function of temperature are also included in the data bases of computer codes for modeling aqueous systems such as SOLMNEQ (Kharaka and Barnes, 1973), EQ3/6 (Wolery, 1979), and PHREEQE (Parkhurst and others, 1980).

Activity Models

Activity models for dissolved species represent the nonideal behavior of components in nondilute solutions. Nonidealities result from electrostatic interactions among ions in solution. Mass action equations account for nonidealities by replacing molar concentrations by activities or thermodynamically effective concentrations. For the case of species Y, the following equation relates activity to molar concentration:

$$[Y] = \gamma_Y (Y) \tag{11.6}$$

where γ_Y is the activity coefficient. Typically, γ is close to one in dilute solutions and decreases as salinity increases. Activities of species with more than one charge usually are smaller than those with a single charge. At relatively high salinities, the activity coefficients in many cases begin to increase and even exceed one.

We can calculate γ for ions from one of several different activity models. The simplest is the Debye–Hückel equation

$$\log \gamma_i = -Az_i^2(I)^{0.5} \tag{11.7}$$

where A (Table 11.1) is a constant that is a function of temperature, z_i is the ion charge, and I is the ionic strength of the solution. Mathematically, the ionic strength in mol/L is

$$I = 0.5 \sum M_i z_i^2 \tag{11.8}$$

with (M_i) as the molar concentration of species i having a charge z. Ionic strength is a measure of the total concentration of ions that emphasizes the increased contribution of species with charges greater than one to solution nonideality.

Use of the Debye–Hückel equation is limited to solutions with ionic strengths less than 0.005 M, which is fresh, potable ground water. It is inappropriate for use with more saline water.

The extended Debye–Hückel equation provides estimates of γ_i to a maximum ionic strength of about 0.1 M or a total dissolved solids content of approximately 5000 mg/L (Langmuir and Mahoney, 1984). This equation is

$$\log \gamma_i = \frac{-Az_i^2(I)^{0.5}}{1 + Ba_i(I)^{0.5}} \tag{11.9}$$

where A and B are temperature-dependent constants and a_i is the radius of the hydrated ion in centimeters (Table 11.1).

Geochemical studies of basinal brines require other activity models to deal with the higher salinity water. The simplest model takes an equation like Eq. 11.9 and adds correction or curve-fitting parameters. The Davies equation is an example. Establishing activities in even

Table 11.1 **Parameters Used in the Extended Debye–Hückel Equation at 1 Atmosphere Pressure**

Temperature (C°)	A	B (× 10⁸)	$a_i \times 10^{-8}$ cm	Ion
0	0.4883	0.3241	2.5	Rb^+, Cs^+, NH_4^+, Ag^+,
5	0.4921	0.3249	3.0	K^+, Cl^-, Br^-, I^-, NO_3^-
10	0.4960	0.3258	3.5	OH^-, F^-, HS^-, BrO_3^-
15	0.5000	0.3262	4.0–4.5	Na^+, HCO_3^-, $H_2PO_4^-$,
20	0.5042	0.3273		HSO_3^-, Hg_2^{2+}, SO_4^{2-},
25	0.5085	0.3281		HPO_4^{2-}, PO_4^{3-}
30	0.5130	0.3290	4.5	Pb^{2+}, CO_3^{2-}, MoO_4^{2-},
35	0.5175	0.3297	5.0	Sr^{2+}, Ba^{2+}, Ra^{2+}, Cd^{2+},
40	0.5221	0.3305		Hg^{2+}, S^{2-}, WO_4^{2-}
50	0.5319	0.3321	6	Li^+, Ca^{2+}, Cu^{2+}, Zn^{2+},
60	0.5425	0.3338		Sn^{2+}, Mn^{2+}, Fe^{2+}, Ni^{2+}, Co^{2+}
			8	Mg^{2+}, Be^{2+}
			9	H^+, Al^{3+}, Cr^{3+}, trivalent rare earths
			11	Th^{4+}, Zr^{4+}, Ce^{4+}, Sn^{4+}

Reprinted with permission from Manov, G. G. and others, J. Am. Chem. Soc., v. 65, 1943, p. 1765–1767. Copyright © 1943 American Chemical Society

From Klotz, 1950 with permission.

more saline water requires the use of ion interaction models of a type proposed by Pitzer and Kim (1974). This sophisticated activity model is more difficult to use than the Debye–Hückel or Davies-type equations, but provides accurate estimates of activity up to 20 M (Langmuir and Mahoney, 1984).

On an activity versus ionic strength plot for Cl^- and Ca^{2+}, note the typical curve shapes for monovalent and divalent ions (Figure 11.4). We have shown activities calculated using the Pitzer model, but they are not strictly comparable to the other three. At low ionic strengths, all four methods give approximately the same activity coefficients. However, as the ionic strength increases (Figure 11.4), estimates of activity from the Debye–Hückel, extended Debye–Hückel, and Davies equations differ noticeably.

Because the extended Debye–Hückel equation or the Davies equation are applicable to charged ionic species, they cannot determine γ's for neutral species that exist as ion pairs or other complexes. For solutions with ionic strengths less than about 0.1 M, activity coefficients are assigned a value of one. In solutions with higher ionic strengths, activity estimates are difficult. Possible approaches include making activities a proportion of the ionic strength or equivalent to the activity of CO_2.

Typically for ground waters, activity coefficients are calculated using the extended Debye–Hückel equation (Eq. 11.9) or the Davies equation. The following example illustrates the use of the extended Debye–Hückel equation.

Example 11.2

Water contains ions in the following molar concentrations: Ca^{2+} 3.25×10^{-3}, Na^+ 0.96×10^{-3}, HCO_3^- 5.75×10^{-3}, and SO_4^{2-} 0.89×10^{-3}. Calculate activity coefficients for Na^+ and Ca^{2+} at 25°C using Eq. 11.9.

First, calculate the ionic strength from Eq. 11.8 or

$$I = 0.5 \times (3.25 \times 2^2 + 0.96 \times 1^2 + 5.75 \times 1^2 + 0.89 \times 2^2) \times 10^{-3} = 0.012\,M$$

This value of I is at the upper end of the range of applicability of the extended Debye–Hückel equation. Now, determine the γ values

$$\log \gamma_{Na^+} = \frac{-0.508 \times 1^2 (.012)^{0.5}}{1 + .328 \times 10^{-8} \times 4.0 \times 10^8 (.012)^{0.5}}$$

$$= -0.049$$

$$\gamma_{Na^+} = 10^{-0.49} = 0.89$$

$$\log \gamma_{Ca^{2+}} = \frac{-0.508 \times 2^2 (.012)^{0.5}}{1 + .328 \times 10^{-8} \times 6.0 \times 10^8 (.012)^{0.5}}$$

$$= -0.183$$

$$\gamma_{Ca^{2+}} = 10^{-1.83} = 0.66$$

Nonidealites relate to both gas and solids. However, for most problems, gas phase nonidealities can be neglected. Nonideal behavior in minerals is due to elemental substitution in a crystal lattice. Some applications may require treatment of nonidealities. However, all of the problems in this book assume that pure solids have an activity of one. Readers interested in a discussion of solid activities should refer to Stumm and Morgan (1981).

11.5 Deviations from Equilibrium

Viewing ground water as a partial equilibrium system implies that some reactions may not be at equilibrium (for example, dissolution or precipitation reactions). The departure of a reaction from equilibrium is determined as the ratio of the ion activity product to the equilibrium constant (IAP/K). The ion activity product is calculated by substituting sample activity values in the mass law expression for a reaction. For example, given a ground water with known activities of $[C]$, $[D]$, $[Y]$, and $[Z]$, the ion activity product (IAP) for the reaction in Eq.

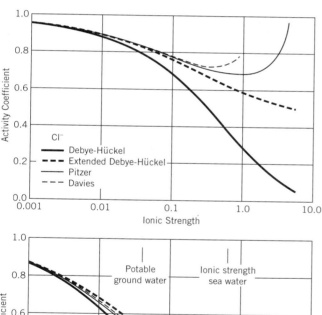

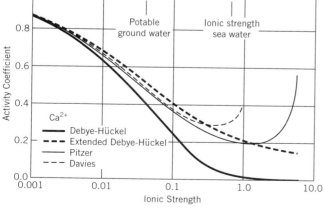

Figure 11.4 Comparison of activities of Cl^- and Ca^{2+} as a function of ionic strength, calculated using the Debye–Hückel, extended Debye–Hückel, Davies, and Pitzer equations (unpublished data from A. S. Crowe).

11.4 is

$$IAP = \frac{[Y]^y[Z]^z}{[C]^c[D]^d} \qquad (11.10)$$

If the $IAP > K$, the reaction is progressing from right to left, reducing $[Y]$ and $[Z]$ and increasing $[C]$ and $[D]$. If $IAP < K$, the reaction is proceeding from left to right. The IAP at equilibrium is equal to the equilibrium constant.

This theory provides the saturation state of a ground water with respect to one or more mineral phases. When $IAP/K < 1$, the ground water is undersaturated with respect to the given mineral. When $IAP/K = 1$, the ground water is in chemical equilibrium with the mineral, and when $IAP/K > 1$, the ground water is supersaturated. Undersaturation with respect to a mineral results in the net dissolution provided the mineral is present. Supersaturation results in the net precipitation of the mineral should suitable nuclei be present.

The saturation state is often expressed in terms of a saturation index (SI), defined as $\log(IAP/K)$. When a mineral is in equilibrium with respect to a solution, the SI is zero. Undersaturation is indicated by a negative SI and supersaturation by a positive SI.

Example 11.3 illustrates how to calculate the state of saturation with respect to a mineral.

Example 11.3

Given a ground water with the molar composition shown below, calculate the saturation state with respect to calcite and dolomite. The activity coefficients for Ca^{2+}, Mg^{2+}, and CO_3^{2-} are 0.57, 0.59, and 0.56, respectively. The equilibrium constants defining the solubility of calcite and dolomite are 4.9×10^{-9} and 2.7×10^{-17}, respectively.

$(Ca^{2+}) = 3.74 \times 10^{-4}$	$(Mg^{2+}) = 4.11 \times 10^{-6}$
$(Na^+) = 2.02 \times 10^{-2}$	$(H^+) = 10^{-7.9}$
$(K^+) = 6.14 \times 10^{-5}$	
$(HCO_3^-) = 1.83 \times 10^{-2}$	$(CO_3^{2-}) = 5.50 \times 10^{-5}$
$(SO_4^{2-}) = 1.24 \times 10^{-3}$	$(Cl^-) = 1.19 \times 10^{-3}$

$$IAP_{cal} = [Ca^{2+}][CO_3^{2-}]$$

$$= 0.57 \times 3.74 \times 10^{-4} \times 0.56 \times 5.50 \times 10^{-5}$$

$$= 6.56 \times 10^{-9}$$

$$\{IAP/K\}_{cal} = \frac{6.56 \times 10^{-9}}{4.90 \times 10^{-9}} = 1.34$$

The sample is slightly oversaturated with respect to calcite.

$$IAP_{dol} = [Ca^{2+}][Mg^{2+}][CO_3^{2-}]^2 = 4.89 \times 10^{-19}$$

$$\{IAP/K\}_{dol} = \frac{4.89 \times 10^{-19}}{2.7 \times 10^{-17}} = 0.018$$

The sample is strongly undersaturated with respect to dolomite.

Calculations are easily carried out using one of several different computer codes such as EQ3 (Wolery, 1979) or WATEQF (Plummer and others, 1976). The codes correct equilibrium constants for temperature and account for the reduced activities of free ions due to the formation of complexes.

11.6 Kinetic Reactions

Chemical kinetics provide a useful framework for studying reactions in relation to time and reaction pathways. A kinetic description is applicable to any reaction, but is necessary for irreversible reactions or reversible reactions that are slow in relation to mass transport. Consider this reaction

$$aA + bB + \cdots \text{(other reactants)} \underset{k_1}{\overset{k_2}{=}} rR + sS \qquad (11.11)$$
$$+ \cdots \text{(other products)}$$

where k_1 and k_2 are the rate constants for the forward and reverse reactions, respectively. Each constituent in Eq. 11.11 has a reaction rate (for example, r_A) that describes the rate of change of concentration as a function of time $\{d(A)/dt\}$. Because of the stoichiometry, these rate expressions are related

$$-\frac{r_A}{a} = -\frac{r_B}{b} = \cdots = \frac{r_R}{r} = \frac{r_S}{s} = \cdots \qquad (11.12)$$

where a, b, r, and s are the stoichiometric coefficients.

Reaction rate laws describe the time rate of change of concentration as a function of rate constants and concentration. The rate law for a component (say A) in Eq. 11.11 has this form:

$$r_A = -k_1(A)^{n1}(B)^{n2} \cdots \text{(other reactants)}$$
$$+ k_2(R)^{m1}(S)^{m2} \cdots \text{(other products)} \qquad (11.13)$$

where $n1$, $n2$, . . . and $m1$, $m2$, . . . are empirical or stoichiometric coefficients. This equation expresses the rate of change in (A) as the difference between the rate at which the component is being used in the forward reaction and generated in the reverse reaction.

Although reactions are written as a single, overall expression, they usually occur as a series of elementary reactions. In other words, a reaction like Eq. 11.11 represents what is really a set of reactions that involves intermediate species not included in the overall reaction. Elementary reactions are defined as reactions that take place in a single step, for example (Langmuir and Mahoney, 1984),

$$H^+ + OH^- = H_2O$$

$$CO_2(aq) + OH^- = HCO_3^-$$

$$H_4SiO_4 = SiO_2 \text{ (quartz)} + 2H_2O$$

For an elementary reaction, the coefficients in Eq. 11.13 are the stoichiometric coefficients. Thus, rate laws for elementary reactions are easy to establish. Unfortunately, most reactions are not of this type and the rate laws contain empirical coefficients. Their values are established through laboratory experiments.

Many of the rate laws involve simple forms of Eq. 11.13. For example, (1) the irreversible decay of ^{14}C

$$^{14}C \rightarrow {}^{14}N + e \qquad (11.14)$$

has this rate law

$$\frac{d(^{14}C)}{dt} = -k_1(^{14}C) \qquad (11.15)$$

and (2) the oxidation of ferrous iron between pH 2.2 and 3.5

$$Fe^{2+} + \tfrac{1}{4}O_2 + \tfrac{1}{2}H_2O \rightarrow FeOH^{2+} \qquad (11.16)$$

has this rate law in Fe^{2+} (Langmuir and Mahoney, 1984)

$$\frac{d(Fe^{2+})}{dt} = -k_1(Fe^{2+})(Po_2) \qquad (11.17)$$

Both Eq. 11.14 and Eq. 11.16 are examples of irreversible reactions, which progress until the reactants are consumed by the reaction. In this case, the rate-law expressions for ^{14}C and Fe^{2+} in Eqs. 11.15 and 11.17 describe these components as a consequence of the overall reaction.

Kinetic reactions are described according to the order of the reaction. For the reaction,

$$2A + B \rightarrow C \qquad (11.18)$$

the rate law in terms of (A) is

$$\frac{d(A)}{dt} = \frac{2d(B)}{dt} = \frac{-2d(C)}{dt} = -k_1(A)^{n1}(B)^{n2} \quad (11.19)$$

The coefficients $n1$ and $n2$ define the order of the reaction in A and B. The order of the overall reaction is the sum of all the n's. Thus, Eq. 11.19 is of order $n1 + n2$. Applying this scheme to the previous examples, Eq. 11.14 is a first-order reaction, and Eq. 11.16 is second order.

As Eq. 11.13 illustrates, the rate law for an equilibrium reaction depends not only on how fast a constituent is being consumed in the forward reaction but how fast it is being created in the reverse reaction. For this reaction (Langmuir and Mahoney, 1984)

$$Fe^{3+} + SO_4^{2-} \underset{k_1}{\overset{k_2}{\rightleftharpoons}} FeSO_4^+ \qquad (11.20)$$

rate laws for Fe^{3+} and SO_4^{2-} in the forward direction are of second order and written

$$\frac{d(Fe^{3+})}{dt} = \frac{d(SO_4^{2-})}{dt} = -k_1(Fe^{3+})(SO_4^{2-}) \quad (11.21)$$

The reverse reaction has the following rate law:

$$\frac{d(FeSO_4^+)}{dt} = k_2(FeSO_4^+) \qquad (11.22)$$

The change in (Fe^{3+}) with time for the overall reaction is

$$\frac{d(Fe^{3+})}{dt} = -k_1(Fe^{3+})(SO_4^{2-}) + k_2(FeSO_4^+) \qquad (11.23)$$

Rate laws are even more complex for reactions in parallel or series and which cannot be written as a single overall reaction.

Solutions to rate equations like Eq. 11.23 describe how the concentrations of various components change from some initial condition. This exercise is useful mainly for characterizing reactions in laboratory experiments, but not in modeling chemical changes in ground-water systems. In practice, kinetic equations like Eq. 11.23 must be integrated into a complete equation for mass transport, which includes all of the other processes.

11.7 Organic Compounds

Organic compounds are classified in terms of functional classes. The classes are distinguished by particular functional groups and the unique structural relationships between the functional group and other atoms in the compound. A functional group (for example, hydroxyl group, OH; carbonyl group, $C=O$) is a simple combination of two or more of the following atoms: C, H, O, S, N, and P, that gives the chemical compound unique chemical and physical properties.

The large number of different organic compounds, the complexity of their structure, and the specific requirements of particular studies have prompted the use of a variety of classification schemes. Our version (Figure 11.5) is a condensed and slightly reordered version of the Garrison and others (1977) classification. It consists of 16 major classes with all subdivisions except the first defined on the basis of functional classes.

The actual steps involved in classifying a compound are complicated because most can be included in more than one class. Garrison and others (1977) proposed a simple set of rules to guide classification. A hierarchical search of classes begins with Class 1 and proceeds through Class 16. For example, any hydrocarbon, defined as a compound containing only carbon and hydrogen, containing a halogen atom (chlorine, fluorine, bromine, or iodine) is placed in Class 2 except the few that fall in Class 1. Any compound containing phosphorous is placed in Class 4 and so on. Readers can refer to Garrison and others (1977) for a more complete discussion of rules for classification and a summary of subclasses.

The brief description of the various major classes includes a sketch of the various functional classes where

1. Miscellaneous Nonvolatile Compounds

2. Halogenated Hydrocarbons

Aliphatic

Trichloroethylene

Aromatic

Cl

Chlorobenzene

3. Amino Acids

Basic structure

NH₂
|
C

Aspartic acid

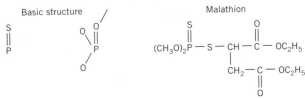

4. Phosphorous Compounds

Basic structure

Malathion

$(CH_3O)_2P$—S—CH—C—OC_2H_5

5. Organometallic Compounds

Tetraethyllead

C_2H_5
|
C_2H_5—Pb—C_2H_5
|
C_2H_5

6. Carboxylic Acid

Basic structure

O
||
R—C—OH

Acetic acid

O
||
CH_3—C—OH

7. Phenols

Basic structure

OH

Cresol

CH_3
OH

8. Amines

Basic structure

Aliphatic   Aromatic

N—   NH₂
|
CH_3

Dimethylamine

CH_3
|
CH_3—N—H

9. Ketones

Basic structure

O
||
R—C—R'

Acetone

O
||
CH_3—C—CH_3

Figure 11.5 Classification of organic compounds (*continued on next column*).

10. Aldehydes

Basic structure

O
||
R—C—H

Formaldehyde

O
||
H—C—H

11. Alcohols

Basic structure

R—OH

Methanol

CH_3—OH

12. Esters

Basic structure

O
||
R—C—OR'

Vinyl acetate

13. Ethers

Basic structure

C—O—C

1,4-Dioxane

O

O

14. Polynuclear Aromatic Hydrocarbons

Phenanthrene

15. Aromatic Hydrocarbons

Basic structure

Toluene

CH_3

16. Alkane, Alkene, and Alkyne Hydrocarbons

Ethane

H H
| |
H—C—C—H
| |
H H

Ethene

H   H
\   /
C==C
/   \
H   H

Ethyne

H—C≡C—H

Figure 11.5 (*Continued.*)

appropriate, a brief discussion of typical compounds in each class and their origin, and a sketch illustrating how the functional class is represented in the structure of various organic compounds (Figure 11.5).

1. **Miscellaneous Nonvolatile Compounds** are a diverse class of compounds that are difficult to classify in terms of functional classes. Naturally occurring compounds include fulvic acid, humic acid, chlorophyll, xanthophylls, and enzymes. Industrially produced compounds include tannic acid, most dyes, and optical brighteners. Fulvic and humic acids are major components of a complex set of compounds called aquatic humic substances. According to Thurman (1985), 40 to 60% of the dissolved organic carbon occurring naturally in ground water is comprised

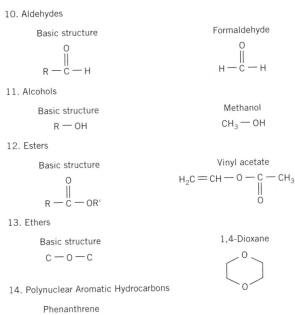

of aquatic humic substances. The natural plant pigments, chlorophylls and xanthophylls, generally have a low solubility. They are minor constituents in ground water, as are enzymes. Tannic acids, dyes and optical brighteners occur in ground water mainly as contaminants.

2. *Halogenated Hydrocarbons* are one of the largest and most important groups of contaminants found in ground water. This class, divided into the aliphatic and aromatic subclasses, is characterized by the presence of one or more halogen atoms (Figure 11.5). The term *aliphatic* describes a structure where carbon atoms are joined in open chains. An aromatic hydrocarbon has a molecular structure based on benzene, C_6H_6, characterized by six carbon atoms arranged in a ring and radially bonded hydrogen atoms. Three of the carbons are doubly bonded and three are singly bonded. Included in the aliphatic subclass are solvents such as methylene chloride, chloroform, carbon tetrachloride, 1,1,1-trichloroethane, and trichloroethylene; pesticides such as aldrin and dieldrin; and industrial chemicals such as vinyl chloride, methyl chloride, and methyl iodide. The aromatic subclass contains common pesticides, such as DDD and DDE, and industrial chemicals such as 1,2,4-dichlorobenzene and the various PCBs.

3. *Amino Acids* are characterized structurally by an amino group, NH_2 on the carbon atom adjacent to the carbonyl carbon, COOH (Figure 11.5). These compounds represent a relatively small proportion of the naturally occurring organic matter in water. According to Thurman (1985), the dominant amino acids in ground water (for example, glycine, serine, alanine, and aspartic acid) are present at concentrations of a few hundred μg/L.

4. *Phosphorous Compounds* are distinguished by the presence of phosphorous in phosphorous-sulfur, phosphate, or phosphite bonds (Figure 11.5). The most important group of compounds within this class is a large group of pesticides. The unhalogenated compounds include diazanon, malathion, parathion, and tributyl phosphate and halogenated compounds include clorthion, dicapthon, and dursban.

5. *Organometallic Compounds* are a highly toxic group of organic contaminants. They have a broadly diverse structure, distinguished by the presence of a metal atom. Examples of compounds in this class include tetraethyllead, diethylmercury, and copper phthalocyanine.

6. *Carboxylic Acids* are an important class of naturally occurring compounds and contaminants. These organic acids are part of a family of functional classes that includes keytones, aldehydes, and esters, which are characterized by the presence of a carbonyl COO functional group. Carboxylic acids are distinguished by a hydroxyl group bonded to the carbonyl group (Figure 11.5). In natural systems, carboxylic acids ionize and, therefore, are quite soluble, contributing 5 to 8% of the organic carbon in surface water (Thurman, 1985). Common compounds in this class include acetic acid, benzoic acid, butyric acid, formic acid, palmitic acid, propionic acid, stearic acid, and valeric acid. Carothers and Kharaka (1978) report concentrations of carboxylic acid in excess of 5000 mg/L in formation waters. Carboxylic acids are also important components of landfill leachates due to anaerobic fermentation reactions. Many of these acids are produced for industrial uses. The most important products are phenoxy acetic pesticides such as 2,4-D, silvex, and 2,4,5-T. These compounds are also widely used in the food industry.

7. *Phenols* are a small but important class of naturally occurring and manufactured compounds. They are characterized by an aromatic ring with an attached hydroxyl group. In uncontaminated ground water, Thurman (1985) cites natural abundances of 1 μg/L or less with compounds such as phenol, cresol, and syringic, vanillic, and *p*-hydroxybenzoic acids. While phenols are commonly associated with petroliferous rocks, they originate in ground water mostly as contaminants from industrial wastes or biocides. The wastes include cresol, phenol, pyrocatechol, and napthol, while pesticides include dinitrocresol and 2,4-dinitrophenol. Chlorinated phenolic pesticides, included in Class 2, are important contaminants near wood-preserving facilities.

8. *Amines* considered here are mainly a group of industrial contaminants. In this system of classification some of the naturally occurring amino acids contain the amine functional class. The aliphatic functional class can be considered as derivatives of ammonia (NH_3) with one, two, or three of the hydrogens replaced by alkyl groups (CH_3) (Figure 11.5). In the aromatic amines, a hydrogen is replaced by an aromatic ring (Figure 11.5). Sources of potential contamination in ground water are wastes from herbicide and synthetic-rubber production. Some common contaminants include diethylamine, dimethylamine, ethylamine, methylamine, aniline, benzidine, napthylamine, and pyridine.

9. *Keytones* are also distinguished by the presence of the carbonyl group (Figure 11.5). In the keytones, the carbonyl group is bonded to two alkyl groups, which may be alike or different. These compounds occur naturally but are mainly contaminants in industrial effluent and municipal sewage. Acetone, fluorenone, fenchone, 2-butane, butyl propyl ketone, and methyl propenyl ketone are examples of these compounds.

10. *Aldehydes* have a structure similar to the keytones except that the carbonyl group is bonded to one alkyl group and one or two hydrogens (Figure 11.5). They occur naturally at relatively low concentrations and are occasionally contaminants in ground water. Formaldehyde and iso-butyraldehyde are examples of these compounds.

11. *Alcohols* are some of the most soluble organic compounds in water. They can be considered as derivatives of a hydrocarbon in which a hydroxyl group replaces a hydrogen (Figure 11.5). In natural systems, alcohols are present only in trace quantities because they are nontoxic and broken down easily in the food chain. They occasionally are found as contaminants in ground water. A few of the alcohols are methanol, glycerol, terpinol, ethyleneglycol, and butanol.

12. *Esters* like the carboxylic acids, keytones, and aldehydes contain the carbonyl functional group and have the basic structure shown in Figure 11.5. Esters are usually found in ground water as the result of contamination. Some commonly occurring compounds include pesticides such as butyl mesityl oxide, dimethrin, dimethy carbate, dinobuton, omite, tabutrex, vinyl acetate, and warfarine plus a variety of industrial chemicals.

13. *Ethers* occur in water both naturally and as contaminants. Structurally, the ether functional group consists of an oxygen bonded between two carbons as shown in Figure 11.5. Examples of ethers found as contaminants are tetrahydrofuran and 1,4-dioxane.

14. *Polynuclear Aromatic Hydrocarbons (PAH)* form from a series of benzene rings (Figure 11.5). These compounds are important components of ancient sediments and crude oils. They occur in potable water in trace concentrations and probably originate from coal tar or creosote contamination. Thurman (1985) lists phenanthrene, fluoranthrene, anthracene, and benzopyrene as examples of PAHs found in rivers and rainwater.

15. *Aromatic Hydrocarbons* are compounds with a molecular structure based on that of benzene, C_6H_6 (Figure 11.5). These hydrocarbons are a major constituent of petroleum and in ground water are usually indicative of contamination from a spill of crude oil or a petroleum distillate. Some typical benzoid hydrocarbons are benzene, ethylbenzene, toluene, methylbenzene, and *o*-xylene.

16. *Alkane, Alkene, and Alkyne Hydrocarbons* are important components of oil and natural gas. Alkanes are characterized by the CH_3 functional class and conform to the general formula C_nH_{2n+2} (where *n* is the number of carbon atoms). Common alkanes are methane, ethane, propane, butane, and the various structural isomers (same chemical formula but different structure). The cycloalkanes (C_nH_{2n}) are characterized by a ring structure that contains single carbon–carbon bonds. Cyclopropane and cyclohexane are examples.

The alkenes are aliphatic hydrocarbons with a carbon–carbon double bond (Figure 11.5). Alkenes have a general formula C_nH_{2n} with members of the series including ethene, propene, named similarly to the alkane series.

The alkynes are characterized by a carbon triple bond and the general formula C_nH_{2n-2} (Figure 11.5). The alkyne series is named like the alkanes giving, for example, ethyne, propyne, butyne, etc. These hydrocarbons are usually found in ground water as a result of crude oil or refined petroleum products.

11.8 Ground-Water Composition

Inorganic and organic solids, organic liquids, and gases found in the subsurface dissolve in ground water to some extent. Thus, the variety of solutes in ground water (Table 11.2) is not surprising. Inorganic constituents are classified as major constituents with concentrations greater than 5 mg/L (Table 11.2), minor constituents with concentrations ranging from 0.01 to 10 mg/L, and trace elements with concentrations less than 0.01 mg/L (Davis and DeWiest, 1966). Naturally occurring, dissolved organic compounds in ground water could number in the hundreds. They are typically present in minor or trace quantities. By far the most abundant organic compounds in shallow ground water are the humic and fulvic acids. However, other organic compounds are also present (Table 11.2). Deep ground water or formation water can contain concentrations of organic matter up to 2000 mg/L (Hull and others, 1984). Examples of organic acids are acetate and propionate, anions that contribute to titratable alkalinity. The important ground water gases include oxygen, carbon dioxide, hydrogen sulfide, and methane.

The Routine Water Analysis

It is neither feasible nor necessary to measure the concentration of all constituents that conceivably might occur in water. A "routine" analysis involves measuring the concentration of a standard set of constituents. Such tests form the basis for assessing the suitability of water for human consumption or various industrial and agricultural uses. The routine analysis typically includes the major constituents with the exception of silicon and carbonic acid, and the minor constituents with the exception of boron and strontium. Laboratory results are reported as concentrations in mg/kg or mg/L. The reported concen-

Table 11.2 **The Dissolved Constituents in Potable Ground Water Classified According to Relative Abundance**

Major constituents (greater than 5 mg/L)

Bicarbonate	Silicon
Calcium	Sodium
Chloride	Sulfate
Magnesium	Carbonic acid
Nitrogen	

Minor constituents (0.01–10.0 mg/L)

Boron	Nitrate
Carbonate	Potassium
Fluoride	Strontium
Iron	Bromide
Oxygen	Carbon dioxide

Trace constituents (less than 0.1 mg/L)

Aluminum	Nickel
Antimony	Niobium
Arsenic	Phosphate
Barium	Platinum
Beryllium	Radium
Bismuth	Rubidium
Cadmium	Ruthenium
Cerium	Scandium
Cesium	Selenium
Chromium	Silver
Cobalt	Thallium
Copper	Thorium
Gallium	Tin
Germanium	Titanium
Gold	Tungsten
Indium	Uranium
Iodide	Vanadium
Lanthanum	Ytterbium
Lead	Yttrium
Lithium	Zinc
Manganese	Zirconium
Molybdenum	

Organic compounds (shallow)

Humic acid	Tannins
Fulvic acid	Lignins
Carbohydrates	Hydrocarbons
Amino acids	

Organic compounds (deep)

Acetate
Propionate

Modified from Davis, S. N., and R. J. M. DeWiest, 1966, Hydrogeology. Reprinted by permission.

trations for metals (for example, Ca, Mg, Na, and so on) is the total concentration of metals irrespective of whether they are complexed or not.

A routine analysis often includes a few other items in addition to concentrations (Table 11.3). Of note are pH, total dissolved solids (TDS) reported in mg/L, and specific conductance reported in microsiemens per centimeter. The TDS content is the total quantity of solids when a water sample is evaporated to dryness. Specific conductance is a measure of the ability of the sample to conduct electricity and provides a proxy measure of the total quantity of ions in solution. The measurement is approximate because the specific conductance of a fluid with a given TDS content varies depending on the ions present.

The routine analysis identifies nearly all of the mass dissolved in a sample. Unanalyzed ions and organic compounds are usually a negligible proportion. One simple check on the quality of a routine analysis is to compare the sum of concentrations of cations and anions in milliequivalent per liter. Given that water is electrically neutral, the ratio of the sums should be one. Table 11.4 illustrates this calculation. While not exactly one, the value is within the $\pm$ 0.05 range of acceptability used by most laboratories. The cation/anion ratio provides one check that concentration determinations are not grossly in error.

Large numbers of routine analyses collected over time provide basic data for research studies in many areas. These data must be used with care. Errors can occur because of a failure to measure rapidly changing parameters in the field, to preserve the samples against deterioration due to long storage, and to assure the quality of the laboratory determinations. These issues are discussed in more detail in Chapter 17.

Specialized Analyses

Laboratories can carry out a routine analysis in a rapid and cost-effective manner. Occasionally, there is a need for more specialized analyses, for example, trace metals, radioisotopes, organic compounds, various nitrogen-containing species, environmental isotopes or gases. Such work is often related to problems of ground-water contamination or research and regulatory needs. Com-

Table 11.3 **Example of a Routine Water Analysis**

Parameter	mg/L	Parameter	mg/L
pH	7.7	Conductivity	2300
Calcium	1[1]	Magnesium	1
Sodium	550	Potassium	3.5
Iron	8.7	NO_2, NO_3[2]	0.1[1]
Nitrite	0.1[1]	Chloride	45
Sulfate	59	Fluoride	0.25
Bicarbonate	1315	Hardness, *T*	8
Alkalinity, *T*	1078		
TDS[3]	1321		
Balance 1.01			

[1]Indicates concentration "less than." Conductivity reported in μS, pH in pH units. All metal parameters expressed as totals. Alkalinity and hardness expressed as calcium carbonate. Nitrate, nitrite, and ammonia expressed as N.
[2]NO_2 = nitrite, NO_3 = nitrate.
[3]Total dissolved solids.

Table 11.4 **Evaluating the Electroneutrality of the Example Routine Analysis**

	Cation Concentration			Anion Concentration	
	mg/L	meq/L		mg/L	meq/L
Ca^{2+}	1.0	0.05	HCO_3^-	1315	21.6
Mg^{2+}	1.0	0.08	SO_4^{2-}	59	1.22
Na^+	550	23.9	Cl^-	45	1.27
K^+	3.5	0.09	F^-	0.25	0.01
Fe	8.7	0.31	Total		24.1
	Total	24.4	cation/anion ratio = 1.01		

mercial laboratories should be able to analyze for all but the most exotic constituents. However, given the cost of this more specialized work, serious attention has to be paid to quality-assurance issues.

11.9 Describing Chemical Data

Describing the concentration or relative abundance of major and minor constituents and the pattern of variabil-ity is part of many ground-water investigations. Over time, different graphical and statistical techniques have been developed to assist with this task. Each technique has particular advantages and disadvantages in represent-ing features of the data. Thus, in working with a set of analytical results, one should examine alternative ways of presenting results to select the most appropriate.

The methods divide into two major groups. First is a group of graphical approaches for describing abundance or relative abundance. Second are approaches that pre-

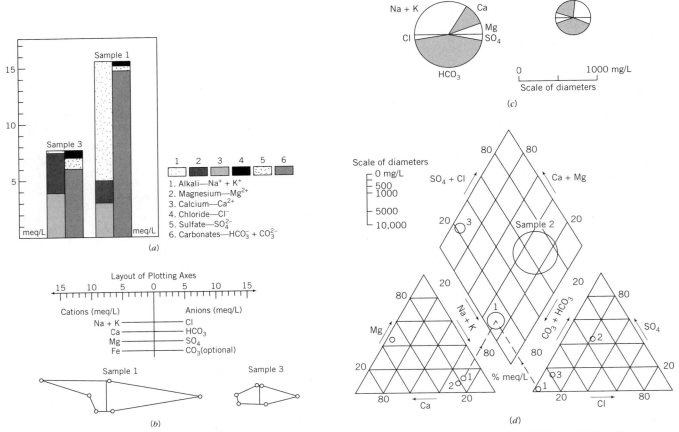

Figure 11.6 Four different ways of plotting major ion data (modified from Zaporozec, 1972). Panel (*a*) is a Collins diagram, Panel (*b*) is a Stiff diagram, Panel (*c*) is a pie diagram, and Panel (*d*) is a Piper diagram. Reprinted by permission of Ground Wa-ter. Copyright © 1972. All rights reserved.

Table 11.5 **Sample Chemical Data Used to Demonstrate Various Techniques for Plotting Chemical Data**

Chemical Analyses	Sample 1: Tertiary, Czechoslovakia			Sample 2: Upper Cretaceous, Czechoslovakia			Sample 3: Upper Cambrian, Wisconsin		
	mg/L	meq/L	meq (%)	mg/L	meq/L	meq (%)	mg/L	meq/L	meq (%)
Cations									
$Na^+ + K^+$	266.2	10.68	68.24	1913.7	81.54	70.24	7.9	0.34	4.4
Mg^{2+}	21.9	1.80	11.4	132.8	10.95	9.41	43.0	3.54	45.6
Ca^{2+}	61.7	3.08	18.5	468.5	23.38	20.50	78.0	3.89	50.1
Mn^{2+}	Traces	—	—	0.67	0.02	0.01	0.14	0.004	0.05
Fe	2.3	0.08	—	0.15	0.005	0.004	0.11	—	—
Sum	—	15.64	100.0	—	115.89	100.0	—	7.77	100.0
Anions									
Cl^-	11.3	0.32	2.05	850.0	23.98	20.63	17.0	0.48	6.4
NO_3^-	0.0	—	—	0.0	—	—	0.7	0.0	0.1
HCO_3^-	906.1	14.85	95.1	2568.5	42.10	36.23	364.0	5.96	79.5
SO_4^{2-}	21.2	0.44	2.82	2406.5	50.10	43.11	50.0	1.04	13.9
Sum	—	15.61	100.0	—	116.18	100.0	—	7.50	100.0

From Zapovozec (1972).

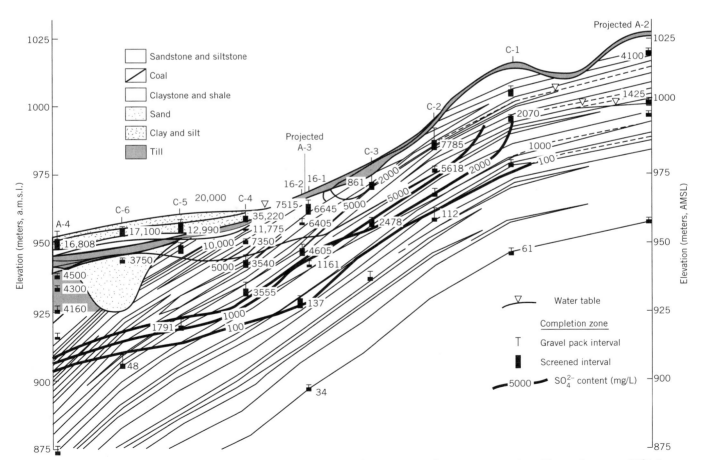

Figure 11.7 Example of how concentration data for an ion can be represented on a cross section. Shown here are SO_4^{2-} data from Blackspring Ridge, Alberta, Canada (from Stein and Schwartz, 1990). Reprinted from J. Hydrol. v. 117 by R. Stein and F. W. Schwartz, On the origin of saline soils at Blackspring Ridge, Alberta, Canada, p. 99–131, 1990, with kind permission from Elsevier Science, NL, Sara Burgerhartstraat 25, 1055 KV Amsterdam, The Netherlands.

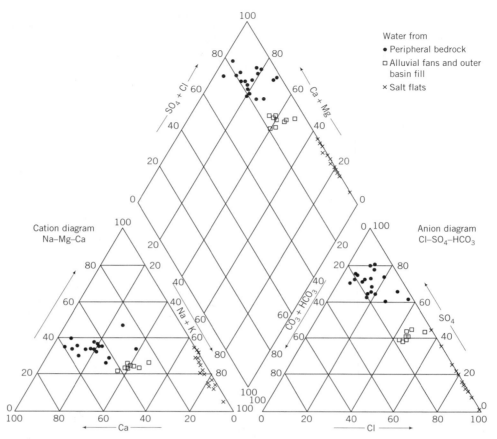

Figure 11.8 Example of the use of a Piper diagram for classifying ground water and defining a pathway of chemical evolution. The diagram shows how samples from the Salt Basin in west Texas can be classified according to setting for example, limestone (dots), alluvial fans, and basin fills (squares), and the salt flats (crosses). The evolutionary pathway is defined by ground water moving from the limestone, through the alluvium to the salt flats (from Boyd and Kreitler, 1986).

sent patterns of variability in addition to abundance. This group subdivides further into graphical/illustrative and statistical approaches.

Abundance or Relative Abundance

Several different graphical approaches can depict the abundance or relative abundance of ions in individual water samples. The most common are (1) the Collins (1923) bar diagram, (2) the Stiff (1951) pattern diagram, (3) the pie diagram, and (4) the Piper (1944) diagram. Data in Table 11.5 will be used to illustrate these approaches. For plotting chemical data, concentrations need to be expressed as meq/L or %meq/L. The Collins and Stiff diagrams require absolute concentrations (meq/L), while the pie and Piper diagrams require relative concentrations (%meq/L).

Figures 11.6*a*, *b*, *c*, and *d* illustrate the sample data plotted in different ways. The Collins, Stiff, and pie diagrams are relatively simple to construct. They require

only that concentrations be plotted as a bar segment, a point on a line, or a percentage of the pie. The appropriate fields are shaded and possibly labeled in the case of the Collins and Piper diagram (Figure 11.6). The Stiff diagrams can be plotted with or without the labeled axes.

Plotting data for a Piper diagram is complicated because there are three separate diagrams (Figure 11.6*d*). The relative abundance of cations with the %meq/L of $Na^+ + K^+$, Ca^{2+}, and Mg^{2+} assumed to equal 100% is first plotted on the cation triangle. Similarly the anion triangle displays the relative abundance of Cl^-, SO_4^{2-}, and $HCO_3^- + CO_3^{2-}$. Straight lines projected from the two triangles into the quadrilateral field define a point on the third field (Figure 11.6*d*). To provide some indication of the absolute quantity of dissolved mass in the sample, the size of the data point is sometimes related to the salinity (TDS). One advantage of all four techniques is that they present the major ion data for a sample on one figure. However, with exception of the Piper diagram, these approaches are useful only in displaying the results

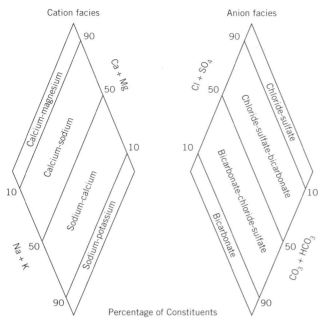

Figure 11.9 Templates for classifying waters into facies for cations and anions (from Back, 1961).

for a few analyses, which are often "type" waters from an area. Presenting a large number of these diagrams together is confusing and not much more helpful than a table of concentration values.

Abundance and Patterns of Change

Graphical/illustrative type diagrams or statistics can define the pattern of spatial change among different geologic units, along a line of section, or along a pathline. One simple way of representing spatial change in a single geologic unit is to take the single sample diagrams (for example, pie or Stiff) and place them on a map. Such maps can convey a sense of how the pattern of ion abundances change within a unit (for example, Swenson, 1968). Including all of the constituents on a single map can be advantageous. However, making sense out of a collection of geometric shapes is difficult.

When chemical data vary systematically in space, it is often best to plot and contour concentrations (or other data) on maps or cross sections (Figure 11.7). This presentation makes it obvious how individual parameters vary. Any measured or calculated chemical parameter can be

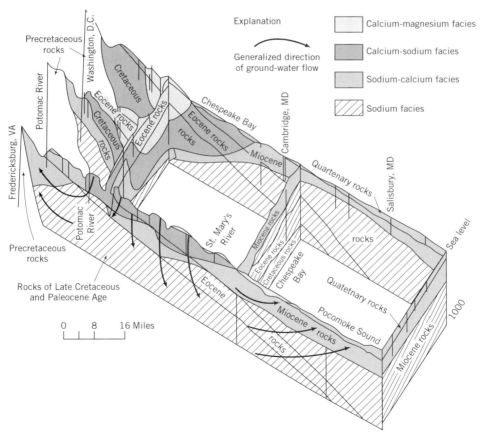

Figure 11.10 Example of hydrochemical facies mapping. Shown on the fence diagram are cation data from part of the northern Atlantic Coastal Plain (modified from Back, 1961).

represented in this way. One problem is the large number of figures that could be required to describe the chemistry of an area completely. However, given the usefulness of these diagrams, this limitation is not serious.

Another problem arises with "noisy" data. Simply contouring a set of such data can produce a complex and cluttered figure. A Piper diagram is preferable to concentration maps when data are noisy. By classifying samples on the Piper diagram, one can identify geologic units with chemically similar water, and define the evolution in water chemistry along a flow system (Figure 11.8).

Noisy data can also be smoothed before plotting on a map or cross section. The facies mapping approach (Back, 1961) provides one way of smoothing chemical data. Samples are classified according to facies with two templates for the Piper diagram (Figure 11.9), one for the cations and the other for the anions. The limited number of possibilities for classifying the chemical data effectively eliminates local variability yet preserves broad trends. The example (Figure 11.10) shows the progres-

sive change in cation chemistry from calcium-magnesium water in the upland part of the Atlantic Coastal Plain to sodium water in deeper units located at the downstream end of the flow system.

In closing, we mention two common statistical approaches. Summary statistics (for example, mean and standard deviation) for various subsets of a chemical data set provide a simple way of describing patterns of variability. A useful classification technique is cluster analysis (Davis, 1986). This procedure involves sorting a set of individual samples into smaller groups. Presenting the results on a two-dimensional diagram called a dendogram enables one to represent the pattern of clustering graphically down to individual pairs of samples and the natural breaks between major groups. One advantage with this approach is that water can be classified on the basis of any desired combination of chemical data. Unlike the Piper diagram, one is not limited to the same group of major ions. For a complete discussion of the method, readers can refer to Davis (1986).

Problems

1. The routine analysis of a water sample provides the following concentrations (as mg/L): $Ca^{2+} = 93.9$; $Mg^{2+} = 22.9$; $Na^+ = 19.1$; $HCO_3^- = 334.$; $SO_4^{2-} = 85.0$; $Cl^- = 9.0$ and a pH of 7.20.

a. Calculate the concentrations of Ca^{2+}, Na^+, HCO_3^-, and Cl^- in terms of molarity and milliequivalents per liter (meq/L).

b. What is the ionic strength of the sample?

c. Calculate the activity coefficients for Ca^{2+} and HCO_3^- using the extended Debye–Hückel equation.

2. Write mass law expressions for the following equilibrium reactions.

a. $CaCO_3 = Ca^{2+} + CO_3^{2-}$

b. $CO_2(g) = CO_2(aq)$

c. $Mn^{2+} + Cl^- = MnCl^+$

3. For the water sample in question (1), determine the saturation index for calcite at 25°C given that $[CO_3^{2-}] = 0.34 \times 10^{-5}$ and that the equilibrium constant for calcite

dissolution is 4.9×10^{-9}. What does the saturation index indicate about the state of saturation with respect to calcite?

4. Assess in a preliminary way the quality of the analytical results in question (1) by determining the cation/anion balance.

5. Determine the order of the following kinetic expressions.

a. degradation of organic matter (org): $d(org)/dt = -k_1(org)$

b. oxidation of pyrite: $d(Fe^{3+})/dt = -k_1\Sigma(Fe^{3+})$, where Σ is surface area of reacting mineral

c. general reaction: $d(A)/dt = -k_1$

d. general reaction: $d(C)/dt = -k_1(A)(B)^2$

6. Represent the analytical results in question (1) graphically using (a) a Collins diagram and (b) a Piper diagram. (c) Classify the water using the concept of hydrochemical facies.

Chapter 12

Chemical Reactions

A variety of physical, chemical, and biological processes operate in the subsurface to transfer mass among the fluids, gases, and solids found there. In this chapter we classify and describe them, breaking down this relatively large and complex topic into manageable chunks. Further, this chapter will develop a sense of why reactions are important in problems of contamination and the geochemistry of natural waters.

12.1   *Acid–Base Reactions*

Acid-base reactions are important in ground water because of their influence on pH and the ion chemistry. pH is a master variable controlling chemical systems. It is defined as the negative logarithm of the hydrogen ion activity and describes whether a solution is acidic (pH < 7), neutral (pH $= 7$), or basic (pH > 7). Hydrogen ion activity is represented here as $[H^+]$ so that pH $= -\log[H^+]$. In solution, hydrogen ion exists as a proton associated with a molecule of water (H_3O^+). Thus, we can express hydrogen ion activity as $[H_3O^+]$. Writing the hydrogen activity as H_3O^+ differentiates between a proton (a hydrogen atom that has lost an electron: H^+) and a hydrogen ion in solution.

An acid is a substance with a tendency to lose a proton.

A base is a substance with a tendency to gain a proton. Acids react with bases in what are called acid-base reactions. In general, because no free protons result from an acid-base reaction, there must be two acid-base systems involved:

$$\text{Acid}_1 + \text{Base}_2 = \text{Acid}_2 + \text{Base}_1 \qquad (12.1)$$

In the forward reaction, the proton lost by Acid_1 is gained by Base_2, and in the reverse reaction the proton lost by Acid_2 is gained by Base_1. The following equilibrium reaction illustrates this point:

$$HCO_3^- + H_2O = H_3O^+ + CO_3^{2-} \qquad (12.2)$$

In Eq. 12.2 protons are transferred from HCO_3^- to H_2O, which in this reaction functions as a base, and from the acid H_3O^+ to the base CO_3^{2-}.

Applying the law of mass action to Eq. 12.2 yields

$$K = \frac{[H_3O^+][CO_3^{2-}]}{[HCO_3^-][H_2O]} \qquad (12.3)$$

This equation simplifies by assuming that $[H_2O] = 1$ and $[H_3O^+] = [H^+]$, which gives

$$K = \frac{[H^+][CO_3^{2-}]}{[HCO_3^-]} \qquad (12.4)$$

By making this substitution, the acid–base reaction involving H_2O–H_3O^+ is neglected. This simplification is usually made in writing acid–base reactions in water. Nevertheless, thinking of acid–base reactions in terms of reaction pairs helps to understand these reactions.

In the previous example, the solvent water functions as a base. However, bases such as ammonia (NH_3) also ionize in water

$$NH_3 \; + \; H_2O \; = \; NH_4^+ \; + \; OH^- \qquad (12.5)$$
$$\text{Base}_1 \quad \text{Acid}_2 \quad \text{Acid}_1 \quad \text{Base}_2$$

which implies that water also functions as an acid. The concept of an acid or base, thus, can be quite abstract.

The strength of an acid or a base refers to the extent to which protons are lost or gained, respectively. Strength is often reported in qualitative terms of strong or weak. Consider the following generalized ionization reaction for an acid HA in water (Glasstone and Lewis, 1960):

$$HA \; + \; H_2O \; = \; H_3O^+ \; + \; A^- \qquad (12.6)$$

where A^- is an anion like Cl^-. The strength of an acid or base depends on whether equilibrium in a reaction like Eq. 12.6 is established to the right or left side. For example, a strong acid will ionize and establish an equilibrium to the right because of the significant proton transfer to H_2O in spite of the fact that H_2O is a weak base. When the acid (HA) is strong, its conjugate base A^- will be weak. If HA is a weak acid, then its conjugate base will be strong. This relationship with just a few exceptions applies to acids and bases of all types (Glasstone and Lewis, 1960).

Natural Weak Acid–Base Systems

The few weak acid–base reactions that are important in natural ground waters are listed in Table 12.1. The first reaction describes the dissociation of water into hydrogen and hydroxyl ions, where the constant K_w is the ion product for water. This reaction is a simplified form of the acid–base reaction

$$H_2O \; + \; H_2O \; = \; H_3O^+ \; + \; OH^- \qquad (12.13)$$

Writing the reaction this way implies that protons will be transferred between water molecules and between H_3O^+ and OH^- even in pure water. The dissociation reaction for water is fundamental in establishing the relationship between (H^+) and (OH^-) according to the law of mass action.

Next in Table 12.1 are weak acid–base reactions that originate mainly from adding CO_2 gas to water. The first reaction describes how CO_2 gas dissolves to produce carbonic acid $H_2CO_3^*$. By convention, the concentrations of $CO_2(aq)$ and H_2CO_3 are represented by the single species $H_2CO_3^*$ or

$$(H_2CO_3^*) = (CO_2 \cdot aq) + (H_2CO_3) \qquad (12.14)$$

The next pair of reactions in Table 12.1 describes the two-stage ionization of carbonic acid, which is controlled by the first and second dissociation constants K_1 and K_2, respectively. The last group of reactions describes the dissociation of silicic acid in water. This acid originates mainly from the dissolution of silicate minerals.

CO_2–Water System

The reactions involving CO_2 in Table 12.1 show that CO_2 dissolved in water partitions among $H_2CO_3^*$, HCO_3^- and CO_3^{2-}. When the pH of a solution is fixed, the mass law equations in Table 12.1 let us determine the concentrations of the species. Following is an example illustrating this calculation.

Table 12.1 **Important Weak Acid–Base Reactions in Natural Water Systems**

Reaction	Mass Law Equation	Eq.	$-\log$ $K(25°C)$
$H_2O = H^+ + OH^-$	$K_w = (H^+)(OH^-)$	12.7	14.0
$CO_2(g) + H_2O = H_2CO_3^*$	$K_{CO_2} = \dfrac{(H_2CO_3^*)}{P_{CO_2}(H_2O)}$	12.8	1.46
$H_2CO_3^* = HCO_3^- + H^+$	$K_1 = \dfrac{(HCO_3^-)(H^+)}{(H_2CO_3^*)}$	12.9	6.35
$HCO_3^- = CO_3^{2-} + H^+$	$K_2 = \dfrac{(CO_3^{2-})(H^+)}{(HCO_3^-)}$	12.10	10.33
$H_2SiO_3 = HSiO_3^- + H^+$	$K = \dfrac{(HSiO_3^-)(H^+)}{(H_2SiO_3)}$	12.11	9.86
$HSiO_3^- = SiO_3^{2-} + H^+$	$K = \dfrac{(SiO_3^{2-})(H^+)}{(HSiO_3^-)}$	12.12	13.1

Example 12.1

Assume that CO_2 is dissolved in water so that $(CO_2)_T = 10^{-3}$ *M* and

$$10^{-3} = (H_2CO_3^*) + (HCO_3^-) + (CO_3^{2-}) \quad (12.15)$$

What is the concentration of the carbonate species at a pH of 6.35? We chose this particular pH to simplify the calculation.

Substitution of the known pH into Eq. 12.9 gives

$$10^{-6.35} = \frac{(HCO_3^-)(10^{-6.35})}{(H_2CO_3^*)} \quad (12.16)$$

Simplification gives $(HCO_3^-) = (H_2CO_3^*)$. Substitution into Eq. 12.10 provides

$$10^{-10.33} = \frac{(CO_3^{2-})(10^{-6.35})}{(HCO_3^-)} \quad (12.17)$$

and shows that $(HCO_3^-) \gg (CO_3^{2-})$. Assuming for the moment that (CO_3^{2-}) is negligible, a solution to Eq. 12.15 gives $(HCO_3^-) = (H_2CO_3^*) = 10^{-3.31}$ *M*. Substitution into Eq. 12.10 gives $(CO_3^{2-}) = 10^{-7.29}$ *M*.

Similar calculations over a broad pH range would show how the concentrations of the carbonate species change. A useful way of illustrating this relationship is with a Bjerrum plot. It is a plot of the logarithm of activities (or concentrations) of various species versus pH. A log C–pH diagram for the carbonate system is shown in Figure 12.1 for $(CO_2)_T = 10^{-4}$. The arrows on the figure depict the crossover points, where at pH 6.35 $(H_2CO_3^*) = (HCO_3^-)$ and at 10.33 $(HCO_3^-) = (CO_3^{2-})$. These are known as the pK_a values for aqueous CO_2 (i.e., pK_1 = 6.35 and pK_2 = 10.33).

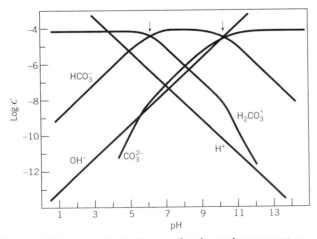

Figure 12.1 Log C–pH diagram for the carbonate system with $(CO_2)_T = 10^{-4}$ *M*. From F. M. M. Morel and J. G. Hering, 1993. Principles and Applications of Aquatic Chemistry. Copyright © 1993 by John Wiley & Sons, Inc. Reprinted by permission of John Wiley & Sons, Inc.

A Bjerrum diagram illustrates the dominance of particular species over some pH range. For example, below pH 5, $H_2CO_3^*$ is the dominant carbonate species (Figure 12.1). In the range from approximately pH 7 to 9, HCO_3^- is the most abundant species. The carbonate ion (CO_3^{2-}) is dominant when pH is above 10.

Alkalinity

So far, we have discussed only cases where the pH determines the distribution of carbonate species. In natural water systems, the situation is more complex. The pH and carbonate speciation are interdependent, a function of not only the ionization equilibria for the carbonate species and water but also strong bases added through the dissolution of carbonate and silicate minerals.

Formulating these processes mathematically requires an understanding of the concept of alkalinity. For our purposes, alkalinity is defined as the net concentration of strong base in excess of strong acid with a pure CO_2–water system as the point of reference. When CO_2 is dissolved in water at a fixed P_{CO_2}, the charge balance is

$$(H^+) = (OH^-) + (HCO_3^-) + 2(CO_3^{2-}) \quad (12.18)$$

This charge balance will change, for example, by adding a strong acid such as HCl and a strong base such as NaOH to the solution. Assuming that NaOH and HCl ionize fully, the charge balance is

$$(Na^+) + (H^+) = (OH^-) + (HCO_3^-) + 2(CO_3^{2-}) + (Cl^-) \quad (12.19)$$

Rearranging Eq. 12.19 to group together ions contributed from the strong acids and bases yields

$$(Na^+) - (Cl^-) = -(H^+) + (OH^-) + (HCO_3^-) + 2(CO_3^{2-}) \quad (12.20)$$

The left side of Eq. 12.20 is the net concentration of strong base in excess of strong acid and defines alkalinity when $Na^+ > Cl^-$. A more general form of this equation is

$$Alk = \sum (i^+)_{sb} - \sum (i^-)_{sa}$$
$$= -(H^+) + (OH^-) + (HCO_3^-) + 2(CO_3^{2-}) \quad (12.21)$$

where (i^+) and (i^-) are the positively and negatively charged species from the strong acids and bases. When the alkalinity is zero, Eq. 12.21 becomes the charge balance equation for a pure CO_2–water system.

In natural ground water, the generation of net positive charges through the dissolution of carbonate and silicate minerals is always greater than the contribution of net negative charges from the ionization of strong acids. Thus, as a general rule, ground waters are alkaline. Strong acids are rare in natural waters and when they do occur are the result of contamination. In these cases, the charge contribution from the strong acid may exceed that due to the strong base and create mineral acidity.

So far, the pure CO_2–water system with zero alkalinity is the point of reference in defining alkalinity. In practice, it may be more convenient to use other reference points and to define carbonate or caustic alkalinity (Morel and Hering, 1993). Further, establishing the relative contribution of noncarbonate alkalinity may be useful.

As Eq. 12.21 shows, increasing alkalinity increases the net positive charge on the left side of the equation. This increase is not simply balanced by an increase in one of the negative ions on the right side, because equilibrium relationships in the solution must be maintained. Part of the increase in the concentration of negatively charged species comes from the ionization of $H_2CO_3^*$ to HCO_3^- to CO_3^{2-} and an increase in pH. Thus, increasing the alkalinity with a strong base ultimately leads to an increase in pH. This behavior is commonly observed as ground waters evolve by dissolving minerals along a flow path.

12.2 Solution, Exsolution, Volatilization, and Precipitation

Because water is an excellent solvent, it dissolves many of the gases, liquids, and solids in the subsurface. Dissolution more than any other process is responsible for the large solute loading to ground water. Other processes like gas exsolution, volatilization, and mineral precipitation remove mass from water.

Gas Solution and Exsolution

Gas solution and exsolution can transfer significant quantities of mass between soil gases and ground water. Commonly, we model these processes using equilibrium concepts based on Henry's law. This mass-law equation relates the concentration of dissolved gas in solution to the partial pressure of the same gas in an atmosphere in contact with the solution. Henry's law strictly speaking does not apply to gases such as CO_2 or NH_3 that react in solution. However, in the case of CO_2, so little of the $CO_2(aq)$ reacts that Henry's law approximates the distribution of gas between the two phases. Thus, the concentration of $H_2CO_3^*$ is almost all $CO_2(aq)$ with a negligibly small concentration of H_2CO_3.

The Henry's law equation for CO_2 is

$$K_H = \frac{P_{CO_2}}{(CO_2 \cdot aq)} \qquad (12.22)$$

where K_H is the Henry's Law constant with units like atm-L/mol or atm-m^3/mol, P_{CO_2} is the partial pressure of CO_2 (atm) in the gas phase and $(CO_2 \cdot aq)$ is the molar concentration of CO_2 gas in the solution (mol/L or mol/m^3). In applications, the concentration of $CO_2(aq)$ in Eq. 12.22 is the same as $(H_2CO_3^*)$.

Example 12.2 illustrates how Henry's law determines the concentration of CO_2 in a solution in equilibrium with an atmosphere of a given composition.

Example 12.2

What is the concentration of $H_2CO_3^*$ dissolved in a small drop of rain at 25°C falling through the atmosphere? Assume that the CO_2 in the droplet is in equilibrium with the atmosphere with a P_{CO_2} of $10^{-3.5}$ atm. The Henry's law constant is $10^{1.47}$ atm-L/mol.

$$(H_2CO_3^*) = (CO_2 \cdot aq) = \frac{P_{CO_2}}{K_H}$$

$$(H_2CO_3^*) = \frac{10^{-3.5}}{10^{1.47}}$$

$$= 10^{-4.97} \, M$$

This example shows that CO_2 gas is related to the carbonate system through $H_2CO_3^*$. Thus, a change in the partial pressure of CO_2 gas not only changes $(H_2CO_3^*)$ according to Henry's law but also (HCO_3^-), (CO_3^{2-}), (H^+), and (OH^-), which are related through the carbonate equilibria. The concentrations of some cations may also change because of equilibria with solid phases.

The addition or removal of gases from solution, thus, can play a major role in controlling the chemistry of ground water. For example, a significant step in the chemical evolution of natural ground water involves the solution of CO_2 gas in the soil zone. This process not only increases the concentration of HCO_3^- but also enhances the aggressiveness of the water in dissolving solids.

Solution of Organic Solutes in Water

Liquids other than water in the subsurface include oil or organic contaminants. These nonaqueous components can migrate as a separate liquid phase or be dissolved in ground water. This latter process is of interest here.

Organic compounds differ widely in their overall solubilities (Mackay and others, 1985). Some solutes like methanol are extremely soluble, while others such as PCBs or DDT are sparingly soluble or hydrophobic. As a general rule, the most soluble organic compounds are charged species or those containing oxygen or nitrogen. Examples of this latter group are alcohols or carboxylic acids, which hydrogen bond with the water and are easily accomodated in solution. Thus, one important factor influencing the solubility of organic compounds is the functional groups that are part of the molecule. Even minor changes in the structure of an organic compound such as the position of the Cl atom on a benzene ring can have a noticeable influence on solubility. Further, solubility of alcohols decreases as the size of the molecule increases.

Without hydrogen bonding, solubility is much reduced. The energetics of dissolution, related to the thermodynamics accompanying the organic–water interac-

Table 12.2 A Comparison of Molecular Mass Versus Solubility for Various Aromatics

Compound	Molecular Mass (g/mol)	Solubility (g/m³)
Benzene (C_6H_6)	78.0	1780
Toluene (C_7H_8)	92.0	515
o-Xylene (C_8H_{10})	106.0	175
Cumene (C_9H_{12})	120.0	50
Naphthalene ($C_{10}H_8$)	128.0	33
Biphenyl ($C_{12}H_{10}$)	154.0	7.48

Reprinted with permission from Mackay, D., and Leinonen, P.J., Environ. Sci. Technol., v. 9, 1975, p. 1179. Copyright © 1975 American Chemical Society.

tions and the formation of the space in the water for the molecule, do not favor dissolution. Many of the terms describing this dissolution process are size dependent. The larger the organic molecule is, the larger the space that is required in the aqueous solution, and the less soluble the compound is (Schwarzenbach and others, 1993). For example, with aromatic compounds (Table 12.2), solubility is inversely proportional to the molecular mass or size of the molecules. Temperature and the presence of other organic liquids are factors that also influence solubility.

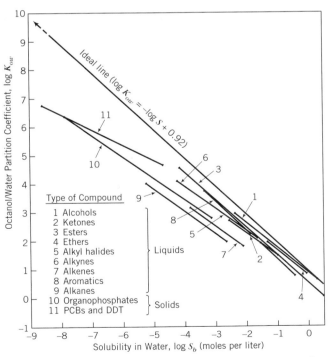

Figure 12.2 Relationship between solubility in water and octanol–water partition coefficients for various organic compounds. Reprinted with permission from C. T. Chiou, D. W. Schmedding, and M. Manes, Environ. Sci. Technol., 16, 1982, p. 6. Copyright © 1982 American Chemical Society.

Verschueren (1983) tabulated solubility data for many organic compounds. When data are unavailable or unreliable, measured octanol/water partition coefficients provide a basis for estimating solubility. The octanol/water partition coefficient (K_{ow}) is a dimensionless equilibrium constant that characterizes partitioning of an organic solute (org) between octanol (an organic liquid) and water. The mass law expression for this reaction is

$$K_{ow} = \frac{(org \cdot oct)}{(org \cdot aq)} \qquad (12.23)$$

where (org · oct) is the concentration of the solute in octanol and (org · aq) is the concentration in water. Values of K_{ow} for many organic compounds have been tabulated by Hansch and Leo (1979). Because the partitioning of an organic solute between water and octanol is conceptually not much different from the partitioning of an organic compound between itself and water, a correlation exists between K_{ow} and solubility. This relationship is illustrated for various classes of liquid and solid organic compounds in Figure 12.2. The regression equation for all compounds is

$$\log K_{ow} = 7.30 - 0.747 \log Sb \qquad (12.24)$$

where Sb is the solubility (*M*) of the compound. Chiou and others (1982) developed equations of this form for all the classes of compounds shown in Figure 12.2.

Volatilization

Volatilization is a process of liquid or solid phase evaporation that occurs when contaminants present either as nonaqueous phase liquids or dissolved in water contacting a gas phase. This situation commonly arises with organic contaminants in the saturated and unsaturated zones (Figure 12.3) or during the sampling and analysis of volatile compounds. The process itself is controlled by the vapor pressure of the organic solute or solvent. The vapor pressure of a liquid or solid is the pressure of the gas in equilibrium with respect to the liquid or solid

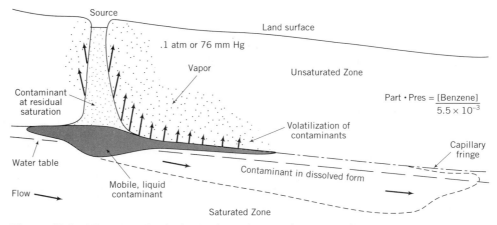

Figure 12.3 Migration of volatiles in the soil atmosphere away from an organic contaminant spill (modified from Schwille, 1985). Reprinted by permission Second Canadian/American Conference on Hydrogeology. Copyright © 1985. All rights reserved.

at a given temperature. Vapor pressure represents a compound's tendency to evaporate and is essentially the solubility of an organic solvent in a gas. Raoult's law describes the equilibrium partial pressure of a volatile organic in the atmosphere above an ideal solvent like benzene

$$P_{org} = x_{org}P^o_{org} \qquad (12.25)$$

where P_{org} is the partial pressure of the vapor in the gas phase, x_{org} is the mole fraction of the organic solvent, and P^o_{org} is the vapor pressure of the pure organic solvent.

Volatilization of dissolved organic solutes from water is described by Henry's law. Commonly, such Henry's law equations are written as

$$P_{org} = K_H (\text{organic} \cdot \text{aq}) \qquad (12.26)$$

where P_{org} is the partial pressure of the organic compound in the vapor phase (typically, atm), and (organic · aq) is the concentration in the aqueous phase (typically, mol/m^3 or mol/L). With the equation written in this manner, K_H has units atm-m^3/mol or atm-L/mol. Thus, the Henry's law constants for reactions with volatile organic compounds are the same as those for gases in Eq. 12.22.

The Henry's law constant, expressed in units of atm-L/mol, is approximately equal to P^o in atmospheres divided by the solubility of the compound in water (mol/L). Thus, the most common convention treats volatilization of pure solvents (such as benzene) using Raoult's law, and the volatilization of solutes (such as benzene dissolved in water) using Henry's law.

An alternative form of Eq. 12.26 can be written as

$$(\text{organic} \cdot \text{vapor}) = K'_H (\text{organic} \cdot \text{aq}) \qquad (12.27)$$

where the concentration of the organic compound in the vapor phase, (organic · vapor), is expressed in units of mol/L and the concentration in the aqueous phase is

also mol/L to make K'_H a dimensionless Henry's law constant. The two different Henry's law constants are related to each other by the following equation:

$$K'_H = K_H/RT \qquad (12.28)$$

where R is the gas constant 0.0821 L-atm/mol-°K, and T is absolute temperature in °K. Zero on the Celsius scale is equal to 273.15°K.

On occasion, one may be interested in representing the partial pressure of an organic compound in the vapor phase as a concentration. The following expression from Johnson and others (1990), which is based on the ideal gas law, serves this purpose:

$$C_{org} = \frac{x_{org}P^o_{org}fw \cdot 1000}{RT} \qquad (12.29)$$

where C_{org} is the approximate concentration of an organic compound in mg/L, x_{org} is mole fraction of the compound in the liquid phase, P^o_{org} is the pure component vapor pressure in atmospheres, and fw is the formula weight of the compound in g/mol.

Example 12.3

Toluene with a formula weight of 92.1 g/mol is found as a pure liquid phase in the subsurface. Given that the vapor pressure of pure toluene is 0.029 atm, calculate its partial pressure and approximate concentration in the vapor phase at 20°C or 293.15°K.

The partial pressure is determined by substituting known information into Eq. 12.25:

$$P_{org} = 1.0 \times 0.029$$

$$= 0.029 \text{ atm}$$

The approximate concentration is determined by substituting known information into Eq. 12.29 or

$$C_{org} = \frac{1.0 \times 0.029 \times 92.1 \times 1000}{0.0821 \times 293.15}$$

$$= 111 \text{ mg/L}$$

Table 12.3, from Jackson and others (1985), lists vapor pressures for a selected group of organic contaminants. Even a limited set of data shows vapor pressures ranging over 9 or 10 orders of magnitude from the most volatile

compounds such as dichloromethane or 1,1-dichloroethane to nonvolatile compounds such as pentachlorophenol or DDT.

Volatilization is important for problems of organic contamination within or close to the unsaturated zone, and for issues related to sample collection and analysis. Soil gases containing volatiles, when allowed to accumulate in the basements of structures, can cause fires, explosions, or health problems. Contamination can be so serious that venting of the soil may be required. Alternatively, the tendency for some organic contaminants to volatilize forms the basis of a useful technique for detecting and

Table 12.3 **Vapor Pressure Data for a Selected Group of Organic Contaminants**

Compound	Formula	V.P.[a]	Henry's Law[b] Constant
Halogenated hydrocarbons			
Dichloromethane	CH_2Cl_2	349	3×10^{-3}
Trichloromethane	$CHCl_3$	160	4.8×10^{-3}
Tetrachloromethane	CCl_4	90	2.3×10^{-2}
Bromoform	$CHBr_3$	5.6 (25°C)	5.8×10^{-4} c
1,1-Dichloroethane	$CHCl_2CH_3$	180	4.3×10^{-3} c
1,2-Dichloroethane	CH_2ClCH_2Cl	61	9.1×10^{-4} c
1,1-Dichloroethene	$H_2C{=}CCl_2$	500	—
Trans-1,2-dichloroethene	$CHCl{=}CHCl$	200 (14°C)	4.2×10^{-2} c
1,1,1-Trichloroethane	CCl_3CH_3	100	1.8×10^{-2}
Trichloroethene	$Cl_2C{=}CHCl$	60	1×10^{-2}
1,1,2-Trichloroethane	$CH_2ClCHCl_2$	19	7.4×10^{-4} c
Tetrachloroethane	$Cl_2C{=}CCl_2$	14	8.3×10^{-3}
Aromatic hydrocarbons			
Benzene	C_6H_6	76	5.5×10^{-3}
Phenol	C_6H_5OH	0.2	3.0×10^{-7} c
Chlorobenzene	C_6H_5Cl	8.8	2.6×10^{-3} c
Ethylenebenzene	$C_6H_5C_2H_5$	7	8.7×10^{-3}
Toluene	$C_6H_5CH_3$	22	5.7×10^{-3}
o-Xylene	$C_6H_4(CH_3)_2$	5	5.3×10^{-3}
Other organic solvents			
Acetone	$CH_3{-}CO{-}CH_3$	89 (5°C)	—
Diethyl ether	$C_2H_5OC_2H_5$	442	5.1×10^{-4} c
Tetrahydrofuran	C_4H_8O		—
1,4-Dioxane	$O(CH_2{-}CH_2)_2O$	30	—
Biocides			
Pentachlorophenol	C_6Cl_5OH	1.1×10^{-4}	3.4×10^{-6}
DDT	$(ClC_6H_4)_2CHCCl_3$	1×10^{-7}	3.8×10^{-5}
Lindane	$C_6H_6Cl_6$	9.4×10^{-6}	4.8×10^{-7}

[a] Vapor pressure in mm Hg at 20°C; 1 atm = 760 mm Hg.
[b] atm-m³/mole; c indicates values calculated from vapor pressure and solubility data.
Modified from Jackson and others, 1985, Contaminant Hydrogeology of Toxic Organic Chemicals at a Disposal Site, Gloucester, Ontario. 1. Chemical Concepts and Site Assessment, IWD Scientific Series 141, Environment Canada, 114 p. Reproduced with the permission of Environment Canada.

monitoring contamination in the subsurface and for in situ approaches for recovering contaminants like gasoline from the subsurface.

Volatilization also causes problems in sampling and analysis. When samples have access to the atmosphere, volatilization reduces the concentration of the contaminant exponentially. If exposed to the atmosphere long enough, the concentration of the contaminant can fall below detection limits.

Dissolution and Precipitation of Solids

Of all the processes that influence solute transport, the dissolution and precipitation of solids are two of the most important in terms of their control on ground-water chemistry. Extremely large quantities of mass can be transferred under some conditions between the water and solid mineral phases. For example, recharging ground water derives almost its entire solute load through the dissolution of minerals along flow paths. Mineral precipitation removes much of the metal present in a low

pH contaminant plume as dispersion and other processes increase the pH of the ground water. These examples show how ground water proceeds toward chemical equilibrium with respect to various minerals from undersaturation in the case of evolving natural ground water and from supersaturation in the case of metal transport.

Solid Solubility

The solubility of a solid reflects the extent to which the reactant (solid) or products (ions and/or secondary minerals) are favored in a dissolution–precipitation reaction. In many reactions where the activity of the reacting solid is equal to one, a comparison of the relative size of the equilibrium constant provides an indication of the solid solubility in pure water. For example, Table 12.4 shows the chloride and sulfate salts to be the most soluble phase and the sulfide and hydroxide groups to be the least soluble. Minerals in the carbonate and in the silicate and aluminum silicate groups have a small but significant solubility.

Table 12.4 **Some Common Mineral Dissolution Reactions and Associated Equilibrium Constants**

Mineral or Solid	Reaction	$\log K(25°C)$	Source
Chlorides and sulfates			
Halite	$NaCl = Na^+ + Cl^-$	1.54	1
Sylvite	$KCl = K^+ + Cl^-$	0.98	1
Gypsum	$CaSO_4 \cdot 2H_2O = Ca^{2+} + SO_4^{2-} + 2H_2O$	−4.62	1
Carbonates			
Magnesite	$MgCO_3 = Mg^{2+} + CO_3^{2-}$	−7.46	1
Aragonite	$CaCO_3 = Ca^{2+} + CO_3^{2-}$	−8.22	1
Calcite	$CaCO_3 = Ca^{2+} + CO_3^{2-}$	−8.35	1
Siderite	$FeCO_3 = Fe^{2+} + CO_3^{2-}$	−10.7	1
Dolomite	$CaMg(CO_3)_2 = Ca^{2+} + Mg^{2+} + 2CO_3^{2-}$	−16.7	2
Hydroxides			
Brucite	$Mg(OH)_2 = Mg^{2+} + 2OH^-$	−11.1	1
Ferrous hydroxide	$Fe(OH)_2 = Fe^{2+} + 2OH^-$	−15.1	1
Gibbsite	$Al(OH)_3 = Al^{3+} + 3OH^-$	−33.5	1
Sulfides			
Pyrrhotite	$FeS = Fe^{2+} + S^{2-}$	−18.1	1
Sphalerite	$ZnS = Zn^{2+} + S^{2-}$	−23.9	3
Galena	$PbS = Pb^{2+} + S^{2-}$	−27.5	1
Silicates and aluminum silicates			
Quartz	$SiO_2 + H_2O = H_2SiO_3$	−4.00	1
Na-Montmorillonite	$3\text{Na-Mont} + 11\frac{1}{2} H_2O =$		
	$3\frac{1}{2} \text{Kaol} + 4H_4SiO_4 + Na^+$	−9.1	2
Kaolinite	$\text{Kaol} + 5H_2O = 2Al(OH)_3 + 2H_4SiO_4$	−9.4	2

[1] Morel and Hering (1993).
[2] Stumm and Morgan (1981).
[3] Matthess (1982).

When other ions are present in a solution, the solubility of a solid can be different from its value in pure water. Solubility increases due to solution nonidealities and decreases due to the common-ion effect. The previous chapter explains this issue of nonidealities in relation to activity models. Generally, the solubility of a solid increases with increasing ionic strength, because other ions in solution reduce the activity of the ions involved in the dissolution–precipitation reaction.

The common-ion effect occurs when a solution already contains the same ions that will be released when the solid dissolves. The presence of the common ion means that less solid is able to dissolve before the solution reaches saturation with respect to that mineral. Thus, the solubility of a solid is less in a solution containing a common ion than it would be in water alone.

12.3 Complexation Reactions

With the explosion of knowledge in contaminant hydrogeology, the prevailing view about the importance of complexation reactions has changed. For example, complexation facilitates the transport of potentially toxic metals such as cadmium, chromium, copper, lead, uranium, or plutonium. Such reactions also influence some types of surface reactions. Thus, a process that was mainly of academic importance now has taken on practical significance in contamination problems.

A complex is an ion that forms by combining simpler cations, anions, and sometimes molecules. The cation or central atom is typically one of the large number of metals making up the periodic table. The anions, often called ligands, include many of the common inorganic species found in ground water such as Cl^-, F^-, Br^-, SO_4^{2-}, PO_4^{3-}, and CO_3^{2-}. The ligand might also comprise various organic molecules such as amino acids.

The simplest complexation reaction involves the combination of a metal and ligand such as Mn^{2+} and Cl^- as follows:

$$Mn^{2+} + Cl^- = MnCl^+ \qquad (12.30)$$

A more complicated manifestation of complexation is the reaction series that forms when complexes themselves combine with ligands. An example is the hydrolysis reaction of Cr^{3+}

$$Cr^{3+} + OH^- = Cr(OH)^{2+}$$
$$Cr(OH)^{2+} + OH^- = Cr(OH)_2^+ \qquad (12.31)$$
$$Cr(OH)_2^+ + OH^- = Cr(OH)_3^0$$

and so on. The metal is distributed among at least three complexes. Such series involve not only hydrolysis reactions but other ligands such as Cl^-, F^-, and Br^-.

Stability of Complexes and Speciation Modeling

Most inorganic reactions involving complexes are kinetically fast. Thus, they can be examined quantitatively using equilibrium concepts. For example, the mass-law equations for Eq. 12.30 and Eq. 12.31 are

$$K_{MnCl^+} = \frac{[MnCl^+]}{[Mn^{2+}][Cl^-]}$$

and

$$K_{Cr(OH)^{2+}} = \frac{[Cr(OH)^{2+}]}{[Cr^{3+}][OH^-]}$$

$$K_{Cr(OH)_2^+} = \frac{[Cr(OH)_2^+]}{[Cr(OH)^{2+}][OH^-]}$$

and so on.

Calculation of the distribution of metals among various complexes involves the solution of a series of mass-law equations. We write the reactions in terms of the formation of the complex from some combination of the metal and an appropriate number of ligands. For example, writing Eq. 12.31 as a series of association reactions gives the series of reactions and mass-law equations in Table 12.5. These so-called mononuclear complexation reactions have this general form

$$M + lL + bH = ML_lH_b: \quad \beta_i = \frac{[ML_lH_b]}{[M][L]^l[H]^b} \quad (12.32)$$

where M is a metal, L is a ligand, and H is a hydrogen ion.

The stability constants, β_i, for the reaction determine the strength of the complex. Large values of β_i are associated with stronger or more stable complexes. Stability constants are related to the equilibrium constants for the series of reactions. Morel and Hering (1993, p. 332) have tabulated stability constants for a host of metal–ligand reactions.

Table 12.5 Association Reactions and Mass-Law Expressions Describing the Hydrolysis of Cr

Equation	Mass-Law Expressions
$Cr^{3+} + OH^- = Cr(OH)^{2+}$	$\beta_1 = \dfrac{(Cr(OH)^{2+})}{(Cr^{3+})(OH^-)}$
$Cr^{3+} + 2OH^- = Cr(OH)_2^+$	$\beta_2 = \dfrac{(Cr(OH)_2^+)}{(Cr^{3+})(OH^-)^2}$
$Cr^{3+} + 3OH^- = Cr(OH)_3^0$	$\beta_3 = \dfrac{(Cr(OH)_3^0)}{(Cr^{3+})(OH^-)^3}$

In some cases, polynuclear complexes form. This complex is characterized by more than one metal (for example, $Cr_3(OH)_4^{5+}$ or $Cu_2(OH)_2^{2+}$). However, because these complexes are usually rare and difficult to include in speciation models, they are not often treated.

The stability constants for a series of metal ligand reactions provide the basic information necessary to determine how the total concentration of a metal in solution $(M)_T$ is distributed as a metal ion and various complexes. The following example illustrates the speciation calculation when Cr hydrolyzes.

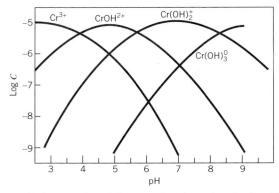

Figure 12.4 Log C-pH diagram for chromium hydroxide complexes.

Example 12.4

A solution with a pH of 5.0 contains a trace quantity of Cr $\{(Cr)_T = 10^{-5}\ M\}$. Determine the speciation for Cr among the various hydroxy complexes and Cr^{3+}. Stability constants (as base 10 logarithms) are 10.0, 18.3, and 24.0 for $Cr(OH)^{2+}$, $Cr(OH)_2^+$, and $Cr(OH)_3^0$, respectively. Assume that the pH does not change with the addition of the metal, no solids form, and the solution behaves ideally.

The mole balance equation for $(Cr)_T$ is

$$(Cr)_T = (Cr^{3+}) + (Cr(OH)^{2+}) + (Cr(OH)_2^+) + (Cr(OH)_3^0)$$

Substitution of the appropriate mass-law equations for the association reactions into this equation gives

$$(Cr)_T = (Cr^{3+}) + \beta_1(Cr^{3+})(OH^-) + \beta_2(Cr^{3+})(OH^-)^2 + \beta_3(Cr^{3+})(OH^-)^3$$

$$(Cr)_T = (Cr)^{3+}\{1 + \beta_1(OH^-) + \beta_2(OH^-)^2 + \beta_3(OH^-)^3\}$$

Because pH is fixed, $(OH^-) = 10^{-9}\ M$ and all the terms in the brackets are known. Substitution of the known values solving for (Cr^{3+}) gives

$$(Cr^{3+}) = \frac{10^{-5}}{1 + 10^{10}(10^{-9}) + 10^{18.3}(10^{-18}) + 10^{24}(10^{-27})}$$

$$= \frac{10^{-5}}{1 + 10 + 10^{0.3} + 10^{-3}}$$

$$= 10^{-6.12}\ M$$

The concentration of the complexes is calculated by substituting known values of (Cr^{3+}) and (OH^-) in the mass-law equations

$$(Cr(OH)^{2+}) = \beta_1(Cr^{3+})(OH^-)$$

$$= 10^{10.0}(10^{-6.12})(10^{-9})$$

$$= 10^{-5.12}\ M$$

$$(Cr(OH)_2^+) = 10^{-5.82}\ M$$

$$(Cr(OH)_3^0) = 10^{-9.12}\ M$$

Thus, most of the Cr occurs as $Cr(OH)^{2+}$ and $Cr(OH)_2^+$. Only about 7.5% of $(Cr)_T$ is the free ion Cr^{3+}.

In Example 12.4, the speciation of Cr is controlled by solution pH. The log C-pH plot for $(Cr)_T = 10^{-5}$ (Figure 12.4) shows that at a pH below 4, Cr^{3+} is the dominant Cr species. As the pH increases, the various hydroxy complexes dominate.

This use of stability constants to determine the speciation of metals can be extended to a mixed group of mononuclear complexes. For example, if trace quantities of Pb are present in a ground water, a mixed group of chloride, hydroxy, and carbonate complexes can form. Speciation is determined by taking the appropriate mole-balance equation

$$(Pb)_T = (Pb^{2+}) + (PbCl_2^0) + (PbCl_3^-) + (PbOH^+) + (PbCO_3^0) \qquad (12.33)$$

and substituting the appropriate mass-law equations. This approach requires that metal concentration in the ground water be sufficiently small so as not to influence the major ion chemistry or pH of the solution. At some point, however, adding large quantities of metal affects the pH and carbonate chemistry. The speciation becomes much more difficult to determine in this case because the system of mole-balance and mass-law equations is larger and more complex. An example of this more complex situation is the case where the chemistry of a ground water changes when the most abundant ions, Ca^{2+}, Mg^{2+}, Na^+, K^+, and H^+, complex with the common ligands SO_4^{2-}, Cl^-, HCO_3^-, and OH^-.

Major Ion Complexation and Equilibrium Calculations

So far, in calculating mineral saturations, we have not considered that the free ion or ligand concentration is less than the total concentration when complexes form. The calculated *IAP/K* ratios for some minerals can decrease depending on the extent of complex formation.

Table 12.6 Calculated Molar Concentrations of Complexes Formed from Major Ion Species (Based on Data from Example 11.2)

Metal	Ligand		
	HCO_3^-	CO_3^{2-}	SO_4^{2-}
Ca^{2+}	5.46×10^{-5}	8.03×10^{-6}	2.16×10^{-5}
Mg^{2+}	4.98×10^{-7}	1.46×10^{-7}	2.78×10^{-7}
Na^+	1.54×10^{-4}	1.13×10^{-5}	6.66×10^{-5}

Let us reexamine Example 11.2 to illustrate this point. Assume now that the molar concentrations for the ions in Example 11.2 are total concentrations. Table 12.6 presents the concentrations of the major complexes. Even though the concentrations of the complexes are relatively small, the reduction in $[Ca^{2+}]$ and $[CO_3^{2-}]$ due to complexation reduces the *IAP/K* ratio for calcite from 1.34 to 1.03. For dolomite the ratio decreases from 0.018 to 0.011. This example shows that even in relatively fresh ground water (ionic strength = 0.023 *M*), the error in determining mineral saturation can be substantial when complexation is not considered. This error can become even larger as the water becomes more saline. In seawater, for example, only approximately 40% of $(SO_4)_T$ exists as SO_4^{2-}, while the remainder exists as $NaSO_4^-$ (37%) and $MgSO_4^0$ (19%) (Morel and Hering, 1993).

Enhancing the Mobility of Metals

In general, metals in ground water are most mobile in water with a low pH where most of the mass occurs as a charged metal ion. Ignoring for a moment the effects of sorption, mobility begins to decline once pH increases to the point where equilibrium is reached with respect to a solid phase, usually a metal–hydroxide, metal–carbonate, or metal–sulfide. At this point, equilibrium with the solid determines that most of the metal will be associated with the solid phase.

Over the pH range common to most natural ground waters, the concentration of most metals is small, reflecting this control. Complexes in these circumstances can enhance the solubility of metals. The transport of uranium in ground water is a good example. Figure 12.5 shows the relative abundance of various uranyl complexes in ground water with a composition typical of the Wind River Formation in Wyoming (Langmuir, 1978). Across the entire range of pH represented on the figure, a significant proportion of the total uranium is complexed. Below a pH of about 4, a uranyl–fluoride complex is the most dominant species (Figure 12.5). In the range of typical ground water for this formation (pH 6.6–8.3),

both phosphate and carbonate complexes are important. This extensive complexing, especially over the neutral to alkaline pH range, increases the solubility of some uranium minerals by several orders of magnitude. Such increases in solubility apparently are necessary to form uranium ore deposits.

Organic Complexation

Complexation can also involve organic ligands present naturally in ground water or added as contaminants. For example, at a site near Ottawa, Canada, Killey and others (1984) showed how up to 80% of the ^{60}Co present as a contaminant in ground water occurred as weakly anionic complexes. These complexes formed with naturally occurring organic compounds. This and similar work emphasizes the need to describe organic systems quantitatively. The task is difficult due to the diversity of organic materials existing in water and problems in identifying them in laboratory analyses. However, it may be essential. Killey and others (1984) indicate that concentrations of organic matter as low as 2 mg/L can increase contaminant mobility.

We will restrict the overview of organic complexation of metals to specific ligand groups including humic substances, artificial complexing agents such as nitrilotriacetic acid (NTA) or ethylenediaminetetraacetic acid (EDTA), and specific compounds such as amino or carboxylic acids. The term *humic substance* (Section 11.7) refers to a group of organic acids, which constitute the

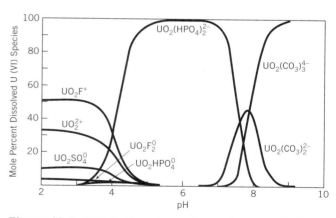

Figure 12.5 Distribution of uranyl complexes vs. pH for some typical ligand concentrations in ground waters of the Wind River Formation at 25°C. $P_{CO_2} = 10^{-2.5}$ atm, $\Sigma F = 0.3$ ppm, $\Sigma Cl = 10$ ppm, $\Sigma SO_4 = 100$ ppm, $\Sigma PO_4 = 0.1$ ppm, $\Sigma SiO_2 = 30$ ppm (from Langmuir, 1978). Reprinted from D. Langmuir, Uranium solution-mineral equilibria at low temperatures with applications to sedimentary ore deposits, Copyright 1978, p. 547–569, with kind permission from Elsevier Science Ltd., The Boulevard, Langford Lane, Kidlington OX5 1GB, UK.

most abundant fraction of naturally occurring organic matter dissolved in water. Complexation of humic substances is usually modeled as a simple mixture of one or two "type" ligands. These could include well-characterized components such as fulvic acid and humic acid or ideal classes of ligands within the humic acid group.

NTA and EDTA are used in detergents and cleaning agents. By virtue of their strong complexing capability, these two organic compounds are useful in studying metal–organic complexation. It is unlikely that these ligands would be encountered in ground water, except perhaps in problems related to waste-water disposal into the subsurface. However, as Morel and Hering (1993) point out, these ligands are important end-members approximating the behavior of the most powerful ligands one would likely encounter in nature.

A variety of other naturally occurring organic compounds such as amino acids and carboxylic acids have significant complexing capabilities. However, the most common amino acids, serine, glycine, and aspartic acid, are not present in sufficient concentrations to influence metal transport in ground waters. In deep ground-water systems, the concentration of organic compounds can be relatively large. For example, carboxylic acids, forming as organic matter undergo diagenesis in a geologic basin, can reach concentrations as high as 10,000 ppm (Surdam and others, 1984). The presence of these acids could be important in enhancing the solubility of aluminum in formation waters.

12.4   Reactions on Surfaces

Reactions between solutes and the surfaces of solids play an important role in controlling the chemistry of ground water. In natural systems, these reactions can completely change the cation chemistry, particularly in the case of the water-softening reactions. When contaminants are being transported, they can retard the spread of some constituents or virtually immobilize them.

Sorption Isotherms

When water containing a trace constituent with a concentration C_i is mixed with granular solids and allowed to equilibrate, mass often partitions between the solution and the solid. The following equation represents this process:

$$S = \frac{(C_i - C)(\text{solution volume})}{sm} \tag{12.34}$$

where the equilibrium concentration, C, and the starting concentration, C_i, have units such as mg/L or μg/L; sm is the sediment mass (for example, g); and S is the quantity of mass sorbed on the surface (for example, mg/g or μg/g). Such an experiment provides the single point a

on the S versus C plot (Figure 12.6). By repeating the procedure at the same temperature (hence, isotherm) with different values of C_i, the family of points forms a sorption isotherm (Figure 12.6). This experiment is known as a batch test.

Real isotherms have no prescribed shape. They can be linear, concave, convex, or a complex combination of all these shapes. Sorption is modeled by fitting an experimentally derived isotherm to theoretical equations. Two of the most common relationships are the Freundlich isotherm:

$$S = KC^n \tag{12.35}$$

and the Langmuir isotherm:

$$S = \frac{Q^0 KC}{1 + KC} \tag{12.36}$$

where K is a partition coefficient reflecting the extent of sorption, n is another constant usually ranging between 0.7 and 1.2, and Q^0 is the maximum sorptive capacity for the surface.

Figure 12.7 depicts Freundlich isotherms calculated with a K value of 1.5 and values of n of 0.5, 1.0, and 1.5, and Langmuir isotherms with values of K of 0.5 and 1.5 and a Q^0 value of 30 mg/g. These examples illustrate the range in curve shapes that can be fitted with these two equations. If the fit is not satisfactory, there are other equations.

A Freundlich isotherm with $n = 1$ is a special case because this linear isotherm is easy to incorporate into mass-transport models. The following equation relates S to C:

$$S = K_d C \tag{12.37}$$

where K_d with units like cm³/g is the distribution coefficient (i.e., the slope of the linear sorption isotherm). Increasing values of K_d are indicative of a greater tendency for sorption.

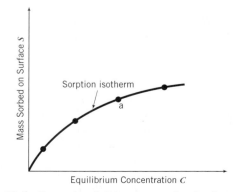

Figure 12.6   Example of a simple sorption isotherm constructed using data points from batch tests.

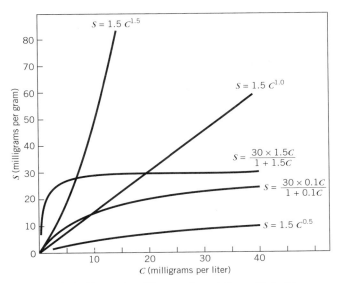

Figure 12.7 Example of Langmuir and Freundlich isotherms.

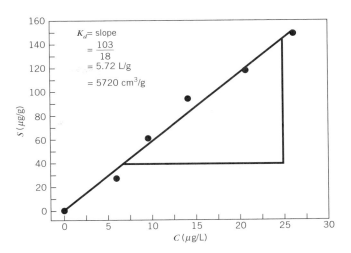

S is plotted versus C, as shown in the figure to establish the isotherm. K_d is the slope of the isotherm. The calculation illustrates a strong affinity for the solid by pyrene.

Hydrophobic Sorption of Organic Compounds

The K_d-based approach for modeling sorption is most commonly applied to hydrophobic (or water-hating) organic molecules as they sorb onto solid particles. These mainly nonpolar molecules tend to partition preferentially into nonpolar environments provided by small quantities of solid organic matter (for example, humic substances and kerogen). Organic matter can occur as discrete solids, as films on individual grains, or as stringers of organic material in grains. Overall, the more hydrophobic a compound is, the greater is its tendency to partition into a solid phase. The following example illustrates the results of batch experiments with pyrene.

Example 12.5

Table E12.5 summarizes data collected from batch sorption experiments with pyrene using 25 mg of porous media in 50 mL of solution. Construct a sorption isotherm and calculate the distribution coefficient in cm³/g.

$$S = \frac{(C_i - C) \times 0.050}{0.025}, \qquad \text{units} \rightarrow \frac{\mu g/L \cdot L}{g}$$

Table E12.5

Test No.	C_i (μg/L)	C (μg/L)	S (μg/g)
1	19.4	5.9	27.0
2	39.7	9.5	60.4
3	60.9	14.1	93.6
4	79.1	20.7	117.8
5	100.3	26.1	148.2

Several studies (for example, Karickhoff and others, 1979; Schwarzenbach and Westall, 1981) have shown how the distribution coefficient can be expressed as the product of constants describing the contaminant and the porous medium, or

$$K_d = K_{oc} f_{oc} \qquad (12.38)$$

where K_{oc} is the partition coefficient of a compound between organic carbon and water with typical units such as cm³/g, and f_{oc} is the weight fraction of organic carbon (dimensionless), defined as g_{oc}/g_s or grams of solid organic carbon to grams of total aquifer solids.

When values of f_{oc} and K_{oc} are known, the K_d value determined from Eq. 12.38 can be used to predict the extent of partitioning. Organic carbon contents can be measured in the laboratory on porous-medium samples. However, this parameter is not well characterized. The range of values reported for geologic materials is extremely variable (Table 12.7). In estimating K_{oc}, it is fortunate that a good correlation exists between log K_{oc} and log K_{ow}, the octanol–water partition coefficient. This correlation is expected because the partitioning of an organic compound between water and organic carbon is not much different than between water and octanol.

Regression equations in practice describe the relationship between K_{oc} and K_{ow} (Griffin and Roy, 1985).

Karickhoff and others (1979): log K_{oc}

$$= -0.21 + \log K_{ow} \qquad (12.39)$$

Schwarzenbach and Westall (1981): log K_{oc}

$$= 0.49 + 0.72 \log K_{ow} \qquad (12.40)$$

Table 12.7 **A Synthesis of Some Data on the Organic Carbon Content of Sediments**

Site Name	Type of Deposit	Texture	Organic Carbon Content
Borden, Ontario[1]	Glaciofluvial	Fine-medium sand	0.0002
Gloucester, Ontario[2]	Glaciofluvial	Sands and gravels	0.0006
North Bay, Ontario[3]	Glaciofluvial	Medium sand	0.00017
Woolwich, Ontario[3]	Glaciofluvial	Fine-medium sand	0.00023
Chalk River, Ontario[3]	Glaciofluvial	Fine sand	0.00026
Cambridge, Ontario[3]	Glaciofluvial	Medium sand	0.00065
Rodney, Ontario[3]	Glaciofluvial	Fine sand	0.00102
Wildwood, Ontario[3]	Lacustrine	Silt	0.00108
Palo Alto, Baylands[4]	?	Silty sand	0.01
River Glatt, Switzerland[5]	Glaciofluvial	Sand, gravel	<0.000–0.01
Oconee River, Georgia[6]	River sediment	Sand,	0.0057
		coarse silt,	0.029
		medium silt,	0.02
		fine silt	0.0226

[1] Mackay and others (1986).
[2] Jackson (personal communication, 1989).
[3] J. Baker, University of Waterloo (personal communication, 1987).
[4] Mackay and Vogel (1985).
[5] Schwarzenbach and Giger (1985).
[6] Karickhoff (1981).

Hassett and others (1983):

$$\log K_{oc} = 0.088 + 0.909 \log K_{ow} \quad (12.41)$$

where K_{ow} is a tabulated constant (dimensionless), and K_{oc} has units of cm³/g. The similarity in these equations is evident in Figure 12.8. Overall, they provide a preliminary estimate of K_{oc} when other information is lacking. In field studies, K_{oc} values estimated from these equations need to be refined with field or laboratory experiments.

The following example illustrates the application of the empirical equations in estimating a distribution coefficient.

Example 12.6

An aquifer has an f_{oc} of 0.01. Estimate the K_d value characterizing the sorption of 1,2-dichloroethane having a log $K_{ow} = 1.48$.

Starting with the basic equation $K_d = f_{oc}K_{oc}$ and taking logs of both sides gives

$$\log K_d = \log f_{oc} + \log K_{oc}$$

Substitution for $\log K_{oc}$ with the Schwarzenbach and Westall (1981) equation yields

$$\log K_d = \log 0.01 + 0.49 + 0.72(1.48) = -0.445$$

The K_d value is $10^{-0.445}$ or 0.36 cm³/g.

It is tempting to extend this model of organic sorption to more-polar and less-hydrophobic (hydrophilic) com-

pounds (for example, methanol). Such compounds have large aqueous solubilities and relatively low values of log K_{ow} (for example, −0.66 for methanol). In contrast to the nonpolar compounds, these are preferentially partitioned to the aqueous phase with much less affinity for organic carbon. Karickhoff (1984) recommends that the model

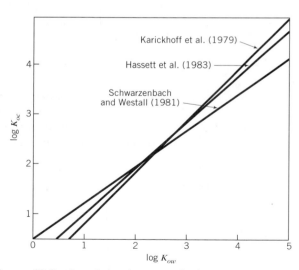

Figure 12.8 Correlation between the log octanol–water partition coefficient (K_{ow}) and the log organic carbon/water partition coefficient as determined by three different studies (modified from Griffin and Roy, 1985). Reprinted by permission of the Environmental Instit. for Waste Management Studies, The Univ. of Alabama.

of hydrophobic sorption should only be used for compounds with solubilities less than 10^{-3} *M*.

When organic molecules carry a charge, hydrophobic sorption represents only part of the total sorption. Electrostatic forces caused by interaction with charged surfaces are also important. Organic acids and bases, for example, ionize to some extent depending on the pH of the ground water. Above a specified pH, a significant proportion of the compound will occur in ionic form. The pH range over which ionization is significant depends on the equilibrium constant for the reaction (pK_a or pK_b). For example, with the following acid and base dissociation reactions

$$HA = H^+ + A^- \qquad BOH = B^+ + OH^-$$

the organic acid dissociates to a negatively charged compound (A^-), and the base to a positively charged compound (B^+). The extent of acid or base dissociation is defined in terms of the ionization fractions (α and β) where

$$\alpha = \frac{[A^-]}{[A^-] + [HA]} \quad \text{and} \quad \beta = \frac{[B^+]}{[B^+] + [BOH]} \quad (12.42)$$

By substitution of the mass action equations

$$K_a = \frac{[H^+][A^-]}{[HA]} \quad \text{and} \quad K_b = \frac{[B^+][OH^-]}{[BOH]}$$

Eq. 12.42 becomes

$$\alpha = \frac{K_a}{K_a + [H^+]} \quad \text{and} \quad \beta = \frac{K_b}{K_b + [OH^-]}$$

These expressions provide a convenient way to determine the ionization fractions for the organic acid or base. For example, in the case of organic acids, HA is sorbed but A^- moves through the system without retardation. Thus, a large ionization fraction (say > 0.5) implies a deterioration in the ability for hydrophobic sorption to retard contaminant spread. In the case of organic bases, BOH is sorbed and B^+ can undergo cation exchange. Thus, the effect of ionization on transport is more complicated for a base than for an acid.

Another instance where the hydrophobic model of sorption begins to break down is with low f_{oc} values. Organic compounds can sorb to a small but significant extent on inorganic surfaces. This sorption is negligible in media with a large f_{oc}. At the critical level of organic matter (McCarty and others, 1981), the sorption due to organic and inorganic solids is equal. The critical level of organic matter is not constant but a function of K_{ow} and the surface area of inorganic solids. Thus, for organic compounds with a high affinity for organic carbon (high K_{ow}), the critical level is smaller than it would be for less readily sorbed compounds. Similarly, the critical level will be greater for a medium with a larger surface area (i.e., a large clay content). McCarty and others (1981)

have developed an empirical equation to estimate the critical organic fraction with K_{ow} and surface area as controlling parameters.

There are other limitations in the model of hydrophobic sorption. A linear isotherm does not describe all sorption especially at large concentrations. Commonly, nonlinear forms of the Freundlich or other isotherms best describe experimental data, especially when the concentration of the organic compounds is relatively high. Another problem relates to the question of desorption, the partitioning of organic compounds back into the aqueous phase. Several studies show that isotherms for desorption may be quite different from the comparable ones for sorption. This failure for reversible processes to behave in the same way is termed *hysteresis*. Often, the hysteresis results in an isotherm that during desorption exhibits a greater affinity for partitioning to the solid than was the case during sorption. Other limitations of the model include problems of competition between compounds and kinetic effects in sorption–desorption reactions.

K_d-based Approaches for Modeling the Sorption of Metals

K_d-based models for sorption have also been applied to characterize the sorption of metals on mineral surfaces. One of the most concerted efforts in this respect has been in the evaluation of various sites as potential host rocks for the disposal of nuclear wastes. The extent to which sorption retards the spread of radionuclides has an extremely important bearing on the suitability of a given site.

Moody (1982) discusses the myriad difficulties in obtaining consistent K_d values from experiments. The problem is that the metal-sorption processes are far too complex to be accurately represented by a simple one-parameter model. Thus, in spite of appealing and apparent simplicity, one should be cautious in using the simple K_d approach in describing the sorption of metals. These limitations have promoted more sophisticated approaches.

Multiparameter Equilibrium Models

A more rigorous framework for surface reactions exists that can account for properties of the solution and the solid surfaces. To illustrate this approach, we will examine classical cation exchange reactions on clay minerals and trace metal sorption on variably charged surfaces.

Cation exchange is the most well known of the surface reactions. Driving the reaction is the electrostatic attraction between charged cations in solution and the surface charge on clay minerals or other oxide surfaces. Clay minerals in particular have a significant fixed negative surface charge because of substitutions of cations of a lower valence in the lattice, and to a lesser extent because of broken bonds at the edges of the mineral (Table 12.8). Kaolinite differs from the rest of the clay minerals. It has a variable charge as hydroxylated edges change their charge as a function of pH.

Cations binding to the exchange sites balance the negative surface charges of the mineral. These cations can exchange with other cations present in solution. The term *cation exchange capacity* describes the quantity of exchangeable cations sorbed on the surface or indirectly the negative charge of the soil (Yong, 1985). Cation exchange capacity (CEC) has units of milliequivalents per 100 g of sample, and for the fixed-charge clay minerals is an important descriptive parameter. Values of CEC vary considerably depending on the type of clay mineral (Table 12.8). Illite and chlorite fall at the lower end of the range of observed values while the vermiculite and smectite groups are at the upper end of the range. Much of the variability in CEC is a function of the surface area of the exchanger (Table 12.8).

Clay minerals exhibit a preference as to which ion occupies an exchange site. Attempts have been made over the years to establish a selectivity sequence, where equivalent amounts of cations are arranged according to their relative affinity for an exchange site. Many variations of this sequence have been published, such as the following one proposed by Yong (1985):

$$Li^+ < Na^+ < H^+ < K^+ < NH_4^+ < Mg^{2+} < Ca^{2+} < Al^{3+}$$

In general, the greater the charge on a cation, the greater the affinity for an exchange site. Although the idea of an affinity sequence provides a useful concept for addressing exchange preference, it is too simplified to be applied rigorously to evaluate exchange processes (Sposito, 1984).

The general form of a cation exchange reaction is

$$nMX + mN^{n+} = nM^{m+} + mNX \qquad (12.43)$$

where M and N are metal cations with charges m^+ and n^+, respectively, and MX and NX represent the metals sorbed on the solid phase. The mass-law expression for this reaction is

$$K = \frac{[M^{m+}]^n [NX]^m}{[N^{n+}]^m [MX]^n} \qquad (12.44)$$

where the bracketed quantities are the activities of species in solution and on the exchanger. As with ions in solution, activity models are required for ions on the exchanger. The most common model assumes that the activities of adsorbed species are equal to their mole fraction, for example $[NX]^m = x_{NX}^m$, and $[MX]^n = x_{MX}^n$, where the mole fraction of NX is

$$x_{NX} = \frac{(N^*)}{(N^*) + (M^*)}$$

with (N^*) and (M^*) representing the molar concentration of species sorbed on the solid. With an appropriate activity model, the mass-law expression is

$$K_s = \frac{[M^{m+}]^n x_{NX}^m}{[N^{n+}]^m x_{MX}^n} \qquad (12.45)$$

where K_s is now known as a selectivity coefficient. Once an activity model is assumed for the adsorbed species, the equilibrium constant and the selectivity coefficient are no longer equal.

As an example of how this theory is applied, consider the following exchange reaction between an ion in solution P^+ and an ion Q^+ sorbed on the surface as Q-clay:

$$P^+ + Q\text{-clay} = Q^+ + P\text{-clay}$$

Table 12.8 **Properties of the Common Clay Minerals**

Clay Mineral	Lattice Description	CEC (meq/100 g)	Surface Area (m²/g)	Source of Charge	Charge Characteristics
Kaolinites	1:1, strong H bonds	5–15	15	Edges, broken bonds (hydroxylated edges)	Variable charge
Illites	1:2, strong K bonds	25	80	Isomorphous substitution, some broken bonds at edges	Mostly fixed charge
Chlorites	2:2, strong bonds	10–40	80	Isomorphous substitution	Fixed charge
Vermiculites	1:2, weak Mg bonds	100–150	n.d.	Isomorphous substitution	Fixed charge
Montmorillonites	1:2, very weak bonds	80–100	800	Isomorphous substitution, some broken bonds at edges	Mostly fixed charge

n.d.—not determined.

The mass law expression assuming that concentration is equal to activity is

$$K_s = \frac{(Q^+)\, x_{\text{P-clay}}}{(P^+)\, x_{\text{Q-clay}}} \qquad (12.46)$$

where P^+ and Q^+ are molar concentrations and $x_{\text{P-clay}}$ and $x_{\text{Q-clay}}$ are mole fractions.

We can manipulate Eq. 12.46 further to learn more about the distribution coefficient, K_d. Assume that $(P^+) \ll (Q^+)$, and that no other ions are present in solution. Because both ions are singly charged, the molar concentrations and equivalent concentrations of the sorbed species are the same so that CEC = $(P^*) + (Q^*)$ and because (P^+) is small CEC $\sim (Q^*)$ and $\tau = (P^+) + (Q^+) \sim (Q^+)$. If P^+ and Q^+ are the only ions in solution, $x_{\text{P-clay}} = (P^*)/(P^* + Q^*)$, and $x_{\text{Q-clay}} = (Q^*)/(P^* + Q^*)$. Substitution of the approximations into Eq. 12.46 gives

$$K_s = \frac{\tau(P^*)}{(P^+)\text{CEC}} \qquad (12.47)$$

Given that $K_d = (P^*)/(P^+)$, Eq. 12.47 becomes

$$K_d = \frac{K_s \text{CEC}}{\tau} \qquad (12.48)$$

Thus, K_d is a function of both the properties of the exchanger and the solution. Values are not unique constants and must be determined for particular ions on the exchanger as a function of differing solution concentrations. This is one of the main weaknesses in the K_d approach for modeling sorption.

One important example of cation exchange is the natural water-softening reactions. Ground water with relatively high concentrations of Ca^{2+} and Mg^{2+} (hard water) and other minor species such as Fe^{2+} changes its chemical composition as it moves into marine clay or shale with Na^+ on the exchange sites. The following reaction describes this change

$$\begin{matrix} Ca^{2+} & & Ca \\ Mg^{2+} + 2Na\text{-clay} = 2Na^+ + & Mg\text{-clay} \\ Fe^{2+} & & Fe \end{matrix}$$

Selectivity coefficients for these reactions are such that in clay-rich sediments, the ground water is softened with nearly all of the Ca^{2+} and Mg^{2+} replaced by Na^+. Iron, which is often associated with problems of taste and staining, is also removed. Chapter 15 will present more specific examples of how this process affects the composition of natural ground waters.

Another important type of sorption reaction involves solids whose surface charges change as a function of ground-water composition. Examples of such solids include kaolinite, metal oxides (for example, SiO_2, Al_2O_3), and metal oxyhydroxides [for example, $Al(OH)_3$, $Si(OH)_4$].

Typically, the surface on one of these solids is posi-

tively charged at low pHs. However, at higher pHs, the surface becomes negatively charged and functions as a cation exchanger. There is a pH in between, known as the zero point of charge (pH_{zpc}), where the surface has zero net charge. The term *isoelectric point* defines the pH_{zpc} when the binding and dissociation of protons (H^+) are the only reactions affecting surface charge. These surface–charge relationships, unique for given solids and solutions, usually must be determined experimentally. Tabulated in Table 12.9 are estimates of the isoelectric points for various solids.

The variable surface charge is explained by the presence on the surface of species such as XOH, XO^-, and XOH_2^+, where XOH represents the hydroxylated surface and XO^- the ionized surface site (Morel and Hering, 1993). The following set of equilibrium reactions describes the concentrations of these surface species:

$$XOH = H^+ + XO^- \qquad XOH + H^+ = XOH_2^+$$

These equations illustrate how the species composing the surface depend on the pH of the ground water. At low pH, XOH_2^+ is the dominant surface species, while at high pH XO^- is dominant. The surface species are free to react with metals or metal complexes. For example, the sorption of a metal ion (M^{2+}) is described by an equilibrium reaction similar to the preceding one or

$$M^{2+} + XO^- = XOM^+$$

By solving the resulting system of equilibrium reactions, one can calculate the concentration of the metal on the surface (XOM^+). As Morel and Hering (1993) show, this or other surface species bind the metal or metal complexes. Choosing the appropriate species to include usually involves matching experimental data to theoretical models. This model for exchange is sufficiently general to account for the sorption of metal complexes.

Stollenwerk (1991) studied the sorption of molybdate (MoO_4^{2-}) on ferrihydrite, which coats quartz grains of an outwash aquifer. A variety of reactions involving the

Table 12.9 **Examples of the Isoelectric Points (pH) for Various Solids**

Solid	pH	Source
Quartz α-SiO_2	2–3.5	1
Albite $NaAlSi_3O_8$	2.0	1
Kaolinite $Al(Si_4O_{10})(OH)_8$	<2–4.6	1
Montmorillonite	≤2.5	1
Hematite Fe_2O_3	5–9	2
Magnetite Fe_3O_4	6.5	2
Goethite FeOOH	6–7	2
Corundum Al_2O_3	9.1	2
Gibbsite $Al(OH)_3$	~9	2

[1] Murray and Parks (1980).
[2] Drever (1988).

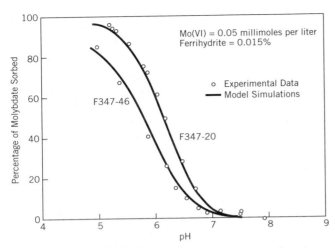

Figure 12.9 Variation in molybdate sorption as a function of pH. The circles present data from actual measurements. The line presents the results of model simulations. The difference in the results is due to the increased competition caused by elevated SO_4^{2-} and PO_4^{3-} present in the water from the deeper well F347-46 (from Stollenwerk, 1991).

$\{S\}FeOH^0$ complexation site causes the sorptive properties of the aquifer to change as a function of pH. At pHs below about 7, the ferrihydrite coatings are capable of sorbing molybdate (Figure 12.9). In this pH range, below the pH_{zpc} for ferrihydrite, the surface has a net positive charge. At very lower pHs, almost 100% of the total molybdate ends up sorbed on the surface. Once the pH climbs above the pH_{zpc}, the surface has a net negative charge and loses its capability of sorbing molybdate.

Stollenwerk's experiments also showed how other anions like phosphate (PO_4^{3-}) and sulfate (SO_4^{2-}) compete

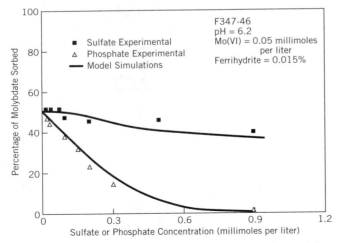

Figure 12.10 Measured and simulated variations in molybdate sorption as a function of competition from SO_4^{2-} and PO_4^{3-} (from Stollenwerk, 1991).

with MoO_4^{2-} in sorption reactions. Figure 12.10 shows how the increasing concentration of phosphate significantly reduces the percentage of molybdate sorbed on the surface. Sulfate, however, affects molybdate sorption to a much smaller extent. As implied in the figure, the sorptive behavior of the surface can be modeled well with a series of surface complexation reactions.

It is feasible to model surface reactions in a realistic way. However, the mathematical approaches are complex and require considerable data. Such models are just now being incorporated into state-of-the-art contaminant transport models and, as Stollenwerk's study shows, are being evaluated in relation to actual field problems.

12.5 Oxidation–Reduction Reactions

Oxidation–reduction or redox reactions are unlike any of the other reactions so far because they are mediated by microorganisms. The microorganisms act as catalysts speeding up what otherwise are extremely sluggish reactions. Microorganisms, mainly bacteria, occur ubiquitously in the subsurface. While they can be found in ground water, they usually form colonies attached as films on the porous medium or fracture surfaces. Microorganisms use redox reactions as a source of energy. In spite of the help from microorganisms, redox reactions are slow in relation to other reactions and typically are treated from a kinetic viewpoint.

Oxidation Numbers, Half-Reactions, Electron Activity, and Redox Potential

Our overview begins with an explanation of the oxidation number. To take into account bond polarity, charges defining the oxidation number are assigned to a compound. For example, in the case of CO_3^{2-}, the oxidation number for carbon is (+IV) and for oxygen (−II), giving the molecule a net negative charge of −2. By convention, oxidation numbers are written as roman numerals.

The term *oxidation* refers to the removal of electrons from an atom, forcing a change in the oxidation number of an element. For example, the following reaction

$$Fe^{2+} = Fe^{3+} + e^-$$

describes the oxidation of Fe^{2+}, where e^- is an electron. In this reaction, the oxidation number for iron changes from (+II) to (+III).

Reduction refers to the addition of an electron to lower the oxidation number:

$$Fe^{3+} + e^- = Fe^{2+}$$

All redox reactions transfer electrons and involve elements with more than one oxidation number. Listed in Table 12.10 are some important elements, their typical oxidation states, and a few of the ions and solids that

Table 12.10 Some Elements Found with More Than One Oxidation State and Examples of Ions or Solids Formed from Those Elements

Element and Oxidate State in Brackets	Example
C(+IV)	HCO_3^-, CO_3^{2-}
C(0)	CH_2O, C
C(−IV)	CH_4
Cr(+VI)	CrO_4^{2-}, $Cr_2O_7^{2-}$
Cr(+III)	Cr^{3+}, $Cr(OH)_3$
Fe(+III)	Fe^{3+}, $Fe(OH)_3$
Fe(+II)	Fe^{2+}
N(+V)	NO_3^-
N(+III)	NO_2^-
N(0)	N_2
N(−III)	NH_4^+, NH_3
S(+VI)	SO_4^{2-}
S(+V)	$S_2O_6^{2-}$
S(+II)	$S_2O_3^{2-}$
S(−II)	H_2S, HS^-

From Morel, F. M. M., and J. G. Hering, 1993. *Principles and Applications of Aquatic Chemistry.* Copyright © 1993 by John Wiley & Sons, Inc. Reprinted by permission of John Wiley & Sons, Inc.

form. A host of trace metals not included on the table also have variable oxidation numbers.

There is a direct analogy between oxidation–reduction reactions and the acid–base reactions studied in Section 12.1. Instead of a transfer of protons from an acid to a base, there is a transfer of electrons from a reductant (electron donor) to an oxidant (electron acceptor). No free electrons result from a redox reaction because each complete reaction involves a pair of redox reactions

$$Ox_1 + Red_2 = Red_1 + Ox_2 \qquad (12.49)$$

where Ox and Red refer to oxidants and reductants, respectively. This equation is analogous to Eq. 12.1, which applies to acid–base reactions.

Following is an example of a complete redox reaction:

$$O_2 + 4Fe^{2+} + 4H^+ = 2H_2O + 4Fe^{3+}$$

O(0) Fe(+II)	O(−II) Fe(+III)	(12.50)
Ox_1 Red_2	Red_1 Ox_2	

consisting of two separate half-redox reactions Ox_1-Red_1 and Red_2-Ox_2:

$$O_2 + 4H^+ + 4e^- = 2H_2O$$

$$4Fe^{2+} = 4Fe^{3+} + 4e^- \qquad (12.51)$$

Adding these two reactions together gives the original reaction Eq. 12.50. Half-reactions are unique in that electrons are reactants or products. Otherwise, these reactions are similar to other equilibrium reactions.

Mass-law equations can be written in terms of the concentration or activities of reactants and products (including the electrons) and an appropriate equilibrium constant. For example, the mass-law expression for the half-reaction

$$Ox + ne^- = Red$$

is

$$K = \frac{[Red]}{[Ox][e^-]^n} \qquad (12.52)$$

Rearrangement of Eq. 12.52 gives the electron activity $[e^-]$ for a half-reaction:

$$[e^-] = \left\{ \frac{[Red]}{[Ox]K} \right\}^{1/n} \qquad (12.53)$$

In the same way that pH defines $[H^+]$, pe defines electron activity, $[e^-]$. Rewriting Eq. 12.53 by taking the logarithm of both sides gives

$$pe = -\log[e^-] = \left(\frac{1}{n}\right)\left\{ \log K - \log \frac{[Red]}{[Ox]} \right\}$$

When a half-reaction is written in terms of a single electron or $n = 1$, the log K term is written as pe^o so that

$$pe = pe^o - \log \frac{[Red]}{[Ox]} \qquad (12.54)$$

Morel and Hering (1993, p. 430) and Pytkowicz (1983, p. 249) tabulate values of pe^o or log K. Some of the important half-reactions together with values of pe^o are presented in Table 12.11.

Another way of characterizing redox conditions is in terms of E_H, the redox potential. Values of E_H have units of volts, acknowledging that a redox reaction involves the transfer of electrons. E_H is related to pe by the following equation

$$E_H = \frac{2.3RT}{F} pe \qquad (12.55)$$

where F is the Faraday constant defined as the electrical charge of one mole of electrons (96,500 Coulombs) with $2.3RT/F$ equal to 0.059 V at 25°C. In practice, pe is taken to be the calculated value of electron activity and E_H the measured electrode potential for an electrochemical cell. In other words, pe is a calculated quantity and E_H is a measured one. The following example illustrates how pe of a ground water is calculated assuming redox equilibrium.

Table 12.11 **Some Important Half-Redox Reactions**

	$pe^o = \log K$
Hydrogen	
$H^+ + e^- = \frac{1}{2}H_2(g)$	0.0
Oxygen	
$\frac{1}{2}O_3(g) + H^+ + e^- = \frac{1}{2}O_2(g) + \frac{1}{2}H_2O$	+35.1
$\frac{1}{4}O_2(g) + H^+ + e^- = \frac{1}{2}H_2O$	+20.75
$\frac{1}{2}H_2O_2 + H^+ + e^- = H_2O$	+30.0

(Note also $HO_2^- + H^+ = H_2O$; $\log K = 11.6$)

Nitrogen	
$NO_3^- + 2H^+ + e^- = \frac{1}{2}N_2O_4(g) + H_2O$	+13.6

(Note also $N_2O_4(g) = 2NO_2(g)$; $\log K = -0.47$)

$\frac{1}{2}NO_3^- + H^+ + e^- = \frac{1}{2}NO_2^- + \frac{1}{2}H_2O$	+14.15

(Note also $NO_2^- + H^+ = HNO_2$; $\log K = 3.35$)

$\frac{1}{3}NO_3^- + \frac{4}{3}H^+ + e^- = \frac{1}{3}NO(g) + \frac{2}{3}H_2O$	+16.15
$\frac{1}{4}NO_3^- + \frac{5}{4}H^+ + e^- = \frac{1}{8}N_2O(g) + \frac{5}{8}H_2O$	+18.9
$\frac{1}{5}NO_3^- + \frac{6}{5}H^+ + e^- = \frac{1}{10}N_2(g) + \frac{3}{5}H_2O$	+21.05
$\frac{1}{8}NO_3^- + \frac{5}{4}H^+ + e^- = \frac{1}{8}NH_4^+ + \frac{3}{8}H_2O$	+14.9
Sulfur	
$\frac{1}{2}SO_4^{2-} + H^+ + e^- = \frac{1}{2}SO_3^{2-} + \frac{1}{2}H_2O$	-1.65

(Note also $SO_3^{2-} + H^+ = HSO_3^-$; $\log K \simeq 7$)

$\frac{1}{4}SO_4^{2-} + \frac{5}{4}H^+ + e^- = \frac{1}{8}S_2O_3^{2-} + \frac{5}{8}H_2O$	+4.85
$\frac{1}{6}SO_4^{2-} + \frac{4}{3}H^+ + e^- = \frac{1}{48}S_8^0(s.\ ort.) + \frac{2}{3}H_2O$	+6.03

(Note also S_8^0 (s. ort.) $= S_8^0$ (s. col.); $\log K = -0.6$)

$\frac{3}{19}SO_4^{2-} + \frac{24}{19}H^+ + e^- = \frac{1}{38}S_6^2 + \frac{12}{19}H_2O$	+5.40
$\frac{5}{32}SO_4^{2-} + \frac{5}{4}H^+ + e^- = \frac{1}{32}S_5^2 + \frac{5}{8}H_2O$	+5.88

(Note also $S_5^{2-} + H^+ = HS_5^-$; $\log K = 6.1$ and $HS_5^- + H^+ = H_2S_5$; $\log K = 3.5$)

$\frac{2}{13}SO_4^{2-} + \frac{16}{13}H^+ + e^- = \frac{1}{26}S_4^{2-} + \frac{8}{13}H_2O$	+5.12

(Note also $S_4^{2-} + H^+ = HS_4^-$; $\log K = 7.0$ and $HS_4^- + H^+ = H_2S_4$; $\log K = 3.8$)

$\frac{1}{8}SO_4^{2-} + \frac{5}{4}H^+ + e^- = \frac{1}{8}H_2S(aq) + \frac{1}{2}H_2O$	+5.13

(Note also $H_2S(g) = H_2S(aq)$; $\log K_H = -1.0$, and other acid–base, coordination, and precipitation reactions)

Trace Metals	
Cr	
$\frac{1}{3}HCrO_4^- + \frac{7}{3}H^+ + e^- = \frac{1}{3}Cr^{3+} + \frac{4}{3}H_2O$	+20.2

(Note also $HCrO_4^- = \frac{1}{2}Cr_2O_7^{2-} + \frac{1}{2}H_2O$, $\log K = -1.5$; $HCrO_4^- = H^+ + CrO_4^{2-}$, $\log K = -6.5$; and various Cr(III) precipitation and coordination reactions.)

Mn	
$\frac{1}{5}MnO_4^- + \frac{8}{5}H^+ + e^- = \frac{1}{5}Mn^{2+} + \frac{4}{5}H_2O$	+25.5
$\frac{1}{2}MnO_2(s) + 2H^+ + e^- = \frac{1}{2}Mn^{2+} + H_2O$	+20.8
Fe	
$Fe^{3+} + e^- = Fe^{2+}$	+13.0
$\frac{1}{2}Fe^{2+} + e^- = \frac{1}{2}Fe(s)$	-7.5
$\frac{1}{2}Fe_3O_4(s) + 4H^+ + e^- = \frac{3}{2}Fe^{2+} + 2H_2O$	+16.6
Co	
$Co(OH)_3(s) + 3H^+ + e^- = Co^{2+} + 3H_2O$	+29.5
$\frac{1}{2}Co_3O_4(s) + 4H^+ + e^- = \frac{3}{2}Co^{2+} + 2H_2O$	+31.4
Cu	
$Cu^{2+} + e^- = Cu^+$	+2.6
$\frac{1}{2}Cu^{2+} + e^- = \frac{1}{2}Cu(s)$	+5.7

Table 12.11 *(Continued)*

	pe° = log K
Trace Metals	

Se

$\frac{1}{2}SeO_4^{2-} + 2H^+ + e^- = \frac{1}{2}H_2SeO_3 + \frac{1}{2}H_2O$ +19.4

$\frac{1}{4}H_2SeO_3 + H^+ + e^- = \frac{1}{4}Se(s) + \frac{3}{4}H_2O$ +12.5

$\frac{1}{2}Se(s) + H^+ + e^- = \frac{1}{2}H_2Se$ −6.7

    (Note also $H_2Se = H^+ + HSe^-$, log $K = -3.9$;

    $H_2SeO_3 = H^+ + HSeO_3^-$, log $K = -2.4$;

    $HSeO_3^- = H^+ + SeO_3^{2-}$, log $K = -7.9$;

    $SeO_4^{2-} + H^+ = HSeO_4^-$, log $K = +1.7$)

Ag

$AgCl(s) + e^- = Ag(s) + Cl^-$ +3.76

$Ag^+ + e^- = Ag(s)$ +13.5

Hg

$\frac{1}{2}Hg^{2+} + e^- = \frac{1}{2}Hg(l)$ +14.4

$Hg^{2+} + e^- = \frac{1}{2}Hg_2^{2+}$ +15.4

Pb

$\frac{1}{2}PbO_2 + 2H^+ + e^- = \frac{1}{2}Pb^{2+} + H_2O$ +24.6

    (Note many other reactions for Mn, Fe, Co, Cu, Se, Ag, Hg, and Pb)

* From Morel, F. M. M. and J. G. Hering, 1993. *Principles and Applications of Aquatic Chemistry.* Copyright © 1993 by John Wiley & Sons, Inc. Reprinted by permission of John Wiley & Sons, Inc.

Example 12.7

A ground water has $(Fe^{2+}) = 10^{-3.3}$ M and $(Fe^{3+}) = 10^{-5.9}$ M. Calculate the pe at 25°C assuming that the activities of Fe-species are equal to their concentrations. What should be the measured E_H of this solution?

From Table 12.11, the half-reaction for the reduction of Fe^{3+} to Fe^{2+} is

$$Fe^{3+} + e^- = Fe^{2+} \quad \text{with} \quad pe° = 13.0$$

Substitution into Eq. 12.54 gives

$$pe = pe° - \log \frac{(Fe^{2+})}{(Fe^{3+})}$$

$$pe = 13.0 - \log \frac{10^{-3.3}}{10^{-5.9}}$$

$$pe = 13.0 - 2.6 = 10.4$$

$$E_H = \frac{2.3RT}{F} = 0.059 \times 10.4 = 0.61 \text{ V}$$

Kinetics and Dominant Couples

So far, redox phenomena have been modeled as equilibrium processes. The implied assumption is that some of the half-redox reactions are in equilibrium. However, redox reactions often are slow relative to physical processes because the number of active microorganisms is small or some of the reactants are not metabolized easily by microorganisms. Another feature of many redox reactions is that with extremely large equilibrium constants they are essentially irreversible. Take for example a reaction where an excess of dissolved oxygen oxidizes organic matter (as CH_2O):

$$CH_2O + O_2 \rightarrow CO_2 + H_2O$$

This reaction progresses to the right until all of the organic matter disappears. This example raises the question of how redox phenomena apply to nonequilibrium systems. A constant value of pe or E_H exists when the concentration of one of the couples, O_2/H_2O in this case, is much greater than the other. Because utilization of all of the organic matter will require a negligibly small proportion of the oxygen, the redox potential defined by the O_2/H_2O couple does not change as a function of the reaction progress. Thus, the measured E_H may be dominated by one couple, although mixed potentials are the rule (Stumm and Morgan, 1981).

The concept of a dominant couple is useful in understanding the redox chemistry of natural and contaminated systems. However, research has shown that redox equilibrium among all of the couples does not occur (Lindberg and Runnells, 1984). Effectively, pe values computed from each couple might span a broad range of values. Thus, while the concept of master pe is alluring, the practice is not recommended (Lindberg and Runnells, 1984). Calculations of pe provide indications of the direction in which a system is evolving (Stumm and Morgan, 1981). This general absence of internal equilibrium implies that care must be taken in assessing pe values. Further, the difficulty in interpreting E_H values measured with electrodes means that a quantitative description of redox reactions remains elusive for ground water.

Control on the Mobility of Metals

Redox reactions influence the mobility of metal ions in solution. One example is the behavior of metals in sulfur systems involving the sulfate/sulfide couple. Assume that the total sulfur in a system (S_T) exists as four possible species, SO_4^{2-}, H_2S, HS^-, and S^{2-}, where in the first sulfur occurs as $S(+VI)$ and in the last three as $S(-II)$. The relationship between H_2S, HS^-, and S^{2-} is fixed by equilibrium relationships, much like the carbonate system. In oxic environments, SO_4^{2-} is the dominant species of the couple with essentially none of the $S(-II)$ species present. Sulfate ion is reduced in an anoxic environment and the other three species dominate.

As the sulfur species change with changing redox conditions, so too do the solids that precipitate. When SO_4^{2-} dominates, metal concentrations can be relatively high because there are no solubility constraints to force the removal of solids from solution. Once the system becomes reducing, almost all of the metal and sulfur are removed from the system as metal-sulfides, which are usually relatively insoluble. Thus, in one case, the total concentration of the metal is found as a metal ion, and in the other as a solid.

In a more complex system where pH can change, other ions are present, and one of several solids could form, it would be useful to have a way of presenting the essential features of the redox and solid phase chemistry. E_H–pH or pe–pH diagrams serve this purpose nicely. Shown in Figure 12.11 is an E_H–pH diagram for iron. The upper and lower lines on the field represent the oxidation of water to O_2 and the reduction of water to H_2. These two half-reactions define the upper and lower limits for oxidation or reduction. The other lines are boundaries for what are known as stability fields. These fields are labeled with the one dominant ion or solid for the specified E_H and pH. When an ion dominates (for example, Fe^{2+}), the element is mobile because a relatively small proportion of the metal occurs as a solid. In the stability fields for solid phases (for example, FeS_2 and Fe_2O_3), nearly all of the iron will exist as that phase, making the element essentially immobile.

Biotransformation of Organic Compounds

A large number of half-reactions involve the oxidation of organic compounds into simpler inorganic forms such as CO_2 and H_2O. These reactions are referred to as biodegradation or biotransformation reactions because they are

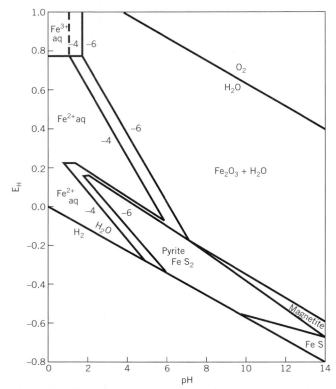

Figure 12.11 Example of an E_H–pH diagram showing the stability relationships of iron oxides and sulfides (from Macumber, 1984). Reprinted by permission First Canadian/American Conference on Hydrogeology. Copyright © 1984. All rights reserved.

microbially catalyzed. In some cases, oxygen acts as the electron acceptor:

$$\tfrac{1}{4}CH_2O + \tfrac{1}{4}O_2(g) = \tfrac{1}{4}CO_2(g) + \tfrac{1}{4}H_2O$$

However, when O_2 is unavailable, other species, for example NO_3^-, $Fe(+III)$, $Mn(+IV)$, SO_4^{2-}, and CO_2, accept electrons. A few of the most important of these biotransformation reactions include

Fe(III) reduction: $\quad \tfrac{1}{4}CH_2O + Fe(OH)_3 + 2H^+$
$$= \tfrac{1}{4}CO_2(g) + Fe^{2+} + \tfrac{11}{4}H_2O$$

denitrification: $\quad CH_2O + \tfrac{4}{5}NO_3^- + \tfrac{4}{5}H^+$
$$= CO_2(g) + \tfrac{2}{5}N_2 + \tfrac{7}{5}H_2O$$

sulfate reduction: $\quad CH_2O + \tfrac{1}{2}SO_4^{2-} + \tfrac{1}{2}H^+$
$$= \tfrac{1}{2}HS^- + H_2O + CO_2(g)$$

methane formation: $\quad CH_2O + \tfrac{1}{2}CO_2(g)$
$$= \tfrac{1}{2}CH_4 + CO_2(g)$$

where CH_2O is a "type" organic compound.

Biotransformation reactions are important when the organic compound is a ground-water contaminant. If biotransformation occurs rapidly, the reaction can attenuate contaminant concentrations and perhaps alter the major ion chemistry. The following chapter examines the important features of microbial systems and how they control the kinetics of redox reactions.

12.6 Hydrolysis

Hydrolysis is another kind of transformation reaction that operates on organic compounds. Unlike biodegradation, hydrolysis is not catalyzed by microorganisms. Hydrolysis involves the reaction of an organic molecule with water or a component ion of water, for example,

$$R-X \xrightarrow{H_2O} R-OH + X^- + H^+$$

where R—X is an organic molecule with X representing an attached halogen, sulfur, phosphorous or nitrogen. According to Neeley (1985), the introduction of a hydroxyl group into the parent molecule makes the product more susceptible to biodegradation, and more soluble.

Not all organic compounds will undergo hydrolysis. Table 12.12 from Harris (1982) (reprinted by Neely, 1985) lists the susceptibility of various functional groups to hydrolysis. First-order rate laws generally describe hydrolysis with the rate of disappearance of RX proportional

Table 12.12 **Summary of the Susceptibility of Various Functional Groups to Hydrolysis**

Resistant to Hydrolysis[1]	
Alkanes	Aromatic nitro compounds
Alkenes	Aromatic amines
Alkynes	Alcohols
Benzene	Phenols
Biphenyl	Glycols
Polycyclic aromatic hydrocarbons	Ethers
	Aldehydes
Halogenated aromatics	Ketones
Halogenated hydrocarbon pesticides (DDT, etc.)	Nitriles
	Carboxylic acid
	Sulfonic acid

Susceptible to Hydrolysis[2]	
Alkyl halides	Epoxides and lactones
Amides[a]	Phosphoric acid esters
Carbamates[b]	Sulfonic acid esters
Carboxylic acid esters	

[a] Most amides have a half-life greater than a year at pH 7 and 25°C. Structures such as $Cl_2CHCONH_2$ have a half-life less than a year.
[b] All carbamates having only alkyl or aromatic or N and O are persistent (i.e., half-life > 1 year).
[1] Reprinted from Harris, J. C., Handbook of Chemical Property Estimation Methods, W. J. Lyman, W. F. Reehl, and D. H. Rosenblatt, eds. Originally McGraw-Hill, 1982, subsequently published by American Chemical Society, 1990. Copyright © 1990 American Chemical Society.
[2] Reprinted with permission from Neely, W. B. 1985, Hydrolysis. In Environment Exposure from Chemicals, eds. W. B. Neely and G. E. Blau, CRC Press, p. 157–173. Copyright CRC Press, Inc. Boca Raton, FL.

to the concentration of RX. As is the case with many kinetic descriptions of reactions, this kind of simple rate law oversimplifies what is often a much more complicated relationship (Neeley, 1985).

Both biotic and abiotic transformation reactions can operate on organic compounds. Examples of these systems are presented in the following chapter.

12.7 Isotopic Processes

Isotopes are atoms of the same element that differ in terms of their mass. For example, hydrogen with an atomic number of 1 has three isotopes, ^1_1H, ^2_1H, and ^3_1H, with mass numbers (superscripts) of 1, 2, and 3, respectively. The first of these isotopes is stable, while the last one ^3_1H (usually written ^3H) decays radioactively to ^3_2He. Thus, radioactive decay is one of the important reactions that involves isotopes. In radioactive decay, atoms of a particular isotope change spontaneously to a new, more stable isotope. Isotopic concentrations change also due to processes like evaporation, condensation, or water–rock interactions. Typically, these processes favor one of the isotopes of a given element over others, producing fractionation.

Radioactive Decay

Radioactive decay occurs mainly by the emission of an α particle (^4_2He) or a β particle (electron $_{-1}^0e$). Often accompanying the emission of these particles is γ radiation, which is electromagnetic energy of short wavelength. This radiation forms when nuclides produced in an excited state (noted by *) revert back to a so-called groundstate. As the following examples illustrate, α-decay changes both the mass number and the atomic number, β-decay changes the atomic number only, and γ-emission changes neither:

α-decay: $^{232}_{90}Th \rightarrow ^{228}_{88}Ra + ^4_2He$

β-decay: $^{228}_{88}Ra \rightarrow ^{228}_{89}Ac + ^0_{-1}e$

γ-emission: $^{236}_{92}U^* \rightarrow ^{236}_{92}U + \gamma$

The arrows in these equations indicate that radioactive decay is irreversible. The quantity of the reacting or parent isotope continually decreases, while the product or daughter isotope increases. Reactions become more complex when the daughter itself decays through a series of other products until a stable form is finally created. We refer to these reactions as *decay chains* or *disintegration series* (for example, $^{232}_{90}Th$, $^{235}_{92}U$, and $^{238}_{92}U$ series).

The decay of radioactive isotopes is independent of temperature and follows the first-order rate law:

$$\frac{dN}{dt} = -kN \tag{12.57}$$

where N is the number of atoms of radioactive material, t is time, and k is the rate constant for radioactive decay. Integration of Eq. 12.57 yields an expression for the quantity of parent remaining at any time following its formation:

$$N(t) = N_0 e^{-kt} \qquad (12.58)$$

where N_0 is the initial number of atoms of parent isotope.

The decrease in the activity of a radioactive substance is commonly expressed in terms of a radioactive half-life ($t_{1/2}$), which is the time required to reduce the number of parent atoms by one-half. The half-life relates to the rate constant for decay as follows:

$$t_{1/2} = \frac{0.693}{k} \qquad (12.59)$$

Listed in Table 12.13 are radioactive half-lives for many radionuclides of interest in ground-water investigations.

Radioactive decay is important for two reasons. When radioactive isotopes occur as contaminants, decay through several half-lives can reduce the hazard when the residence time in a flow system is much greater than the half-life for decay. This capacity to attenuate radioactivity provides the rationale to support the subsurface disposal of some radioactive contaminants. Radioactive decay also forms the basis for various techniques used to age-date ground water. The most important radioactive isotopes used for this purpose are ^{3}H (tritium), ^{14}C, ^{32}Si, and ^{36}Cl.

Table 12.13 Radioactive Half-Lives for Elements of Interest in Age Dating and Contaminant Studies

Nuclide	$t_{1/2}$ (yr)	Nuclide	$t_{1/2}$ (yr)
^{3}H	12.3	^{231}Pa	3×10^4
^{14}C	5×10^3	^{241}Am	432
^{36}Cl	3.1×10^5	^{243}Am	7×10^3
^{63}Ni	100	^{79}Se	6.5×10^4
^{90}Sr	29	^{93}Mo	3.5×10^3
^{93}Zr	1.5×10^6	^{99}Tc	2×10^5
^{94}Nb	2×10^4	^{99}Tc	2×10^5
^{107}Pd	7×10^6	^{126}Sn	1×10^5
^{129}I	2×10^7	^{151}Sm	90
^{135}Cs	3×10^6	^{147}Sm	1.3×10^{11}
^{137}Cs	30	^{106}Ru	1.0
^{154}Eu	8.2	^{235}U	7×10^8
^{210}Pb	22	^{238}U	4.5×10^9
^{226}Ra	1.6×10^3	^{237}Np	2×10^6
^{227}Ac	22		
^{230}Th	8×10^4		
^{232}Th	1.4×10^{10}		

Modified from Moody (1982).

Table 12.14 Isotopes of Environmental Significance and Their Relative Abundances

Element	Isotopes		Average Abundance % of Stable Isotopes
Hydrogen	$^1_1$H		99.984
	$^2_1$H[a]		0.015
	$^3_1$H[b]	radioactive	10^{-14} to 10^{-16}
Oxygen	$^{16}_8$O		99.76
	$^{17}_8$O		0.037
	$^{18}_8$O		0.10
Carbon	$^{12}_6$C		98.89
	$^{13}_6$C		1.11
	$^{14}_6$C	radioactive	$\sim 10^{-10}$
Sulfur	$^{32}_{16}$S		95.02
	$^{33}_{16}$S		0.75
	$^{34}_{16}$S		4.21
	$^{36}_{16}$S		0.02

[a] Deuterium, often referred to by D.
[b] Tritium, often referred to by T.
Modified from Fritz and Fontes (1980). Reprinted from Handbook of Environmental Isotope Geochemistry, v. 1, by P. Fritz and J. C. Fontes, "Introduction," p. 1–19, 1980 with kind permission from Elsevier Science, NL, Sara Burgerhartstraat 25, 1055 KV Amsterdam, The Netherlands.

Isotopic Reactions

The isotopes of hydrogen, oxygen, carbon, and sulfur are useful in studying chemical processes. Related to these elements are seven isotopes of environmental significance (Table 12.14). Commonly, ground-water studies involve ^{2}H or D, ^{18}O, ^{13}C, and ^{34}S.

Because one isotope dominates the rest in terms of relative abundances (Table 12.14) and the changes due to fractionation are too small to measure accurately, isotopic abundances are reported as positive or negative deviations of isotope ratios away from a standard. This convention is represented in the following general equation (Fritz and Fontes, 1980):

$$\delta = \frac{R_{sample} - R_{standard}}{R_{standard}} \times 1000 \qquad (12.60)$$

where δ, reported as permil (‰), represents the deviation from the standard, and R is the particular isotopic ratio (for example, ^{18}O/^{16}O) for the sample and the standard. For example, we would express sulfur-isotope ratios as

$$\delta^{34}S_{sample} = \frac{(^{34}S/^{32}S)_{sample} - (^{34}S/^{32}S)_{standard}}{(^{34}S/^{32}S)_{standard}} \times 1000$$

A δ^{34}S value of $-20‰$ means that the sample is depleted in ^{34}S by 20‰ or 2% relative to the standard.

This way of expressing isotopic compositions takes advantage of the ability of mass spectrometers to measure isotopic ratios accurately. The errors involved in determining δ values is typically a small proportion of the possible range of values.

Deuterium and Oxygen-18

This section discusses the occurrence of D and ^{18}O in water, gases (for example, CH_4), and ions (for example, HCO_3^- and SO_4^{2-}). Deuterium and oxygen-18 compositions of water are usually measured with respect to the SMOW (Standard Mean Ocean Water) standard (Fritz and Fontes, 1980). This choice of standard is particularly appropriate because precipitation that recharges ground water originated from the evaporation of ocean water. However, the isotopic composition of rain or snow in most areas is not the same as ocean water. Evaporation and subsequent cycles of condensation significantly change the isotopic composition of water vapor in the atmosphere.

Water vapor in equilibrium with water is depleted in δD and $\delta^{18}O$ by 80‰ and 10‰, respectively (Wallick and others, 1984). Thus, for ocean water (δD and $\delta^{18}O \sim 0$‰), a vapor in equilibrium should have $\delta D = -80$‰ and $\delta^{18}O = -10$‰. The first few drops of rain falling from this vapor should theoretically have an isotopic composition that is the same as the starting water. This change in isotopic ratios (for example, $^{18}O/^{16}O$) because of a chemical reaction (for example, water = water vapor) is termed *fractionation*. This process is described mathematically by a fractionation factor (α) (Fritz and Fontes, 1980), for example

$$\alpha = \frac{(^{18}O)/^{16}O)_{water}}{(^{18}O/^{16}O)_{vapor}}$$

or in terms of del notation

$$\alpha = \frac{1000 + \delta^{18}O_{water}}{1000 + \delta^{18}O_{vapor}} \qquad (12.61)$$

The isotopic ratios of water vapor in air masses associated with oceans, are however, not in equilibrium with the water. The vapor is more depleted than expected. The extent of depletion is correlated mainly with temperature (Gat, 1980). Rainfall sampled in coastal areas near the equator has an isotopic composition most like seawater, but it is still slightly depleted. Moving away from the equator to higher latitudes, the rainfall becomes more and more depleted relative to seawater. For example, along the western coast of the United States and Canada, values of δD range from approximately -22 to -101‰ and ^{18}O from -4 to -14‰ (Gat, 1980). This so-called latitude effect is related to progressive temperature-controlled depletion of heavier isotopes in the vapor as they are preferentially removed during precipitation events. This continuous removal of a condensate enriched in heavier isotopes is known as a Rayleigh process.

Because fractionation effects at lower temperatures are increasingly more pronounced, the depletion becomes more marked as air masses begin to move across cooler, continental land masses, or are forced to higher and colder elevations over mountain ranges. The temperature control on fractionation also explains why snow is more depleted in D and ^{18}O.

If we assume that differences in the isotopic composition of rainwater reflects the increasing effects of "rainout" from the vapor, then a relationship could exist among samples. When the δD and $\delta^{18}O$ content of rainwater from sampling sites around the world are plotted together, they lie along a straight line (Figure 12.12), known as the meteoric water line (Craig, 1961). The equation for this line is approximately $\delta D = 8\delta^{18}O + 10$‰.

The meteoric water line provides an important key to the interpretation of deuterium and oxygen-18 data. Water with an isotopic composition falling on the meteoric water line is assumed to have originated from the atmosphere and to be unaffected by other isotopic processes. Deviations from the meteoric water line result from other isotopic processes. In most cases, these processes affect the relationship between δD and $\delta^{18}O$ in such a unique way that the position of the data points can help to identify a process. Figure 12.12 illustrates the direction away from the meteoric water line in which various processes push the composition of water.

Two of the more commonly observed processes, evaporation from open water and exchange with rock minerals (Figure 12.12), exhibit this deviation (Fontes, 1980). When water in a pond or lake evaporates, there is an enrichment in the heavier isotopes. However, because of the dynamics of the process, the isotopic composition of the pond follows an "evaporation line" with a slope

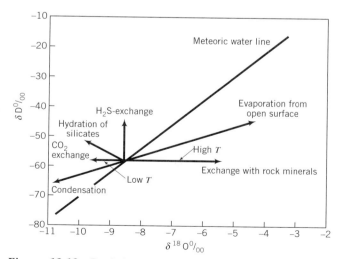

Figure 12.12 Deviations in isotopic compositions away from the meteoric water line as a consequence of various processes (from IAEA Report No. 228, 1983).

ranging from one to five depending on the local rates of evaporation. The point of intersection of the evaporation line with the meteoric water line usually is taken as the isotopic composition of unaltered precipitation. In general, the farther along the evaporation line the data points lie, the greater the evaporation.

Isotopic exchange between minerals and ground water is important in deep, basinal flow systems or in geothermal systems. The relatively high temperatures enable oxygen and hydrogen to exchange between phases and to achieve an equilibrium distribution. This equilibrium is described by the fractionation factor for the mineral. Savin (1980) has summarized fractionation factors for a variety of different minerals and described how they change as a function of temperature.

Exchange between meteoric water and minerals containing oxygen results in the kind of deviation depicted in Figure 12.12. Because only ^{18}O is involved in exchange, the samples typically fall along a horizontal line, reflecting an "oxygen shift" away from the meteoric water line. Truesdell and Hulston (1980) indicate that the size of the oxygen shift is proportional to the difference in original $\delta^{18}O$ between the water and rock, the temperature, and the time of contact, and inversely proportional to the water/rock ratio. Yellowstone Park, Steamboat Springs, Wairaki, and Salton Sea are well-known examples. Oxygen and hydrogen isotopes are extremely valuable in hydrogeologic studies. The most important applications include flow system tracing and indirect age determinations, which are discussed later.

Problems

1. Often (CO_3^{2-}) is not reported in water-quality analyses because the concentration is too low to measure directly. Develop an expression from the appropriate mass laws to calculate the concentration of (CO_3^{2-}) given pH and (HCO_3^-).

2. A water sample has the following ionic composition at 25°C:

Ionic Concentration (M) × 10^{-3}

Ca	Mg	Na	HCO₃	SO₄	Cl	pH
4.17	1.32	11.7	12.0	5.49	0.08	7.60

Given that the activity coefficients for Ca^{2+}, Mg^{2+}, and HCO_3^- are 0.52, 0.55, and 0.85, respectively, answer the following questions.

a. What is the partial pressure of CO_2 in equilibrium with the sample?

b. Assuming no complexation, what is the saturation state of the sample with respect to calcite?

c. Assume that the concentration reported for calcite represents $(Ca)_T$ and that three significant ion pairs form $CaCO_3^0$, $CaHCO_3^+$, and $CaSO_4^0$. Develop an equation to calculate the concentration of free Ca^{2+} in the sample, assuming the solution is ideal.

3a. Given that the concentration of benzene (molecular weight 78.1) dissolved in ground water at the water table is 280 $\mu g/L$, calculate the equilibrium partial pressure in the adjacent soil gas. The Henry's Law constant for the partitioning reaction is 5.5×10^{-3} atm-m³/mol.

b. If benzene was present as the pure liquid having a vapor pressure of 0.1 atm, what would be the partial pressure?

4. Following are the results of a batch sorption experiment in which 2 g of porous media were mixed with 10 mL of a solution containing various concentrations of Cd.

Initial (Cd) Concentration (mg/mL)	Equilibrium (Cd) Concentration (mg/mL) × 10^{-4}
0.0005	0.040
0.001	0.093
0.010	1.77
0.050	39.3
0.096	205.2

a. Find the equation for the Freundlich isotherm that best fits these data.

Hint: Transform the basic equation and data in terms of the logarithm of concentration.

b. Assuming that only the lowest concentrations are of interest, fit the same set of data with a linear Freundlich isotherm (i.e., $n = 1$) to obtain a distribution coefficient (K_d). Be sure to indicate the units for K_d.

5. At a site, the following contaminants are found (i) parathion (log K_{ow} 3.80), (ii) chlorobenzene (log K_{ow} 2.71), and (iii) DDT (log K_{ow} 6.19).

a. Using the equation of Hassett and others (1983) in Eq. 12.41, estimate the partition coefficient between solid organic carbon and water (K_{oc}).

b. With an f_{oc} of 0.01, estimate the distribution coefficients (K_d) for these compounds.

c. On the basis of sorption alone, which of these compounds is the most mobile and which is the least mobile?

6. Compare the range in variability of the distribution coefficients for carbon tetrachloride (log K_{ow} 2.83) estimated using the three different regression equations (Eqs. 12.39, 12.40, and 12.41) and an f_{oc} of 0.01.

7. The properties of six organic compounds are listed below.

Compound	log K_{ow}	Henry's Law Constant (atm-m³/mol)
1	2.04	5×10^{-3}
2	0.89	5×10^{-4}
3	0.46	7×10^{-4}
4	−.27	3×10^{-4}
5	2.83	5×10^{-3}
6	1.48	4×10^{-3}

a. List the organics in order of decreasing solubility.
b. List the organics in order of decreasing mobility.

8. Consider the following half-redox reaction involving NO_3^- with a concentration of 10^{-5} M and ammonium with a concentration of 10^{-3} in ground water with a pH of 8.

$$\tfrac{1}{8}NO_3^- + \tfrac{5}{4}H^+ + e^- = \tfrac{1}{8}NH_4^+ + \tfrac{3}{8}H_2O$$

What are the pe and E_H of the system at 25°C assuming redox equilibrium?

9. The couple O_2/H_2O is found to be the dominant couple, producing a pe of 12.0. By assuming redox equilibrium, determine how $(Fe)_T = 10^{-5}$ M would be distributed as Fe^{2+} and Fe^{3+}.

10. An organic compound has a tendency to hydrolyze with $t_{1/2} = 3.5$ yr. What would an initial concentration of 100 $\mu g/L$ be reduced to after 10 years?

11. Explain what is meant by the SMOW standard and the convention for reporting isotopic results.

12. These isotopic measurements were made on several prairie lakes and potholes.

Sample	$\delta^{18}O$ (‰)	δD (‰) (SMOW)
1	−6.5	−98.0
2	−8.5	−102.0
3	−2.6	−83.0
4	−4.0	−86.0

What is the best explanation for the deviation in these results from the meteoric water line? What is the estimated concentration of unaltered meteoric water in this area?

Chapter 13

Colloids and Microorganisms

This chapter is concerned with the occurrence and transport of colloids and microorganisms. A colloid is a particle ranging in size from 1 nm to 1000 nm. Particles can include bacteria, viruses, large macromolecules of dissolved organic carbon (for example, humic substances), small droplets of nonaqueous phase liquids, inorganic rock, or mineral fragments (for example, clay minerals, ferric oxyhydroxides). Bacteria are small single-celled organisms that occur ubiquitously in ground-water and surface-water systems. A virus is considered to be a genome (that is, genetic information stored on DNA or RNA) within some kind of protein coating (Chapelle, 1993).

Particle transport is important in ground water for at least two reasons. First, the migration of particles provides another way for mass to spread in the subsurface. This process may play a role in the genesis of soils or ore deposits (McDowell-Boyer and others, 1986), but is of particular interest in relation to contaminant migration. The particles themselves may be contaminants, as is the case with bacteria and viruses, or the contaminants can be trace metals, radionuclides, or organic compounds sorbed onto clay or organic particles. Because of their small size, particles have relatively large surface areas, which makes them efficient contaminant collectors. In cases where some contaminant has a low mobility in solution, facilitated transport on particles may create unexpectedly high contaminant fluxes.

Second, particle transport is important because of the related surface processes. For example, the original colonization of some porous media by bacteria may have depended on their transport from above or laterally. The biofilm created by the attached bacteria and other organisms can have an enormous influence on the transport of nutrients and organic compounds dissolved in water and often on the chemistry of ground water.

13.1 A Conceptual Model of Colloidal Transport

Figure 13.1 is a conceptual model for colloid transport developed by McCarthy and Zachara (1989). The transport of colloids requires (1) a source for colloids in the medium, (2) conditions that promote the suspension of particles or stability of colloids, and (3) advective transport without significant size filtration or surface interaction.

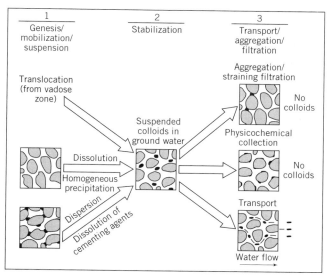

Figure 13.1 A conceptual model for the origin and transport of colloids in ground water (from McCarthy and Zachara, 1989). Reprinted with permission from J. F. McCarthy and J. M. Zachara, Environ. Sci. Technol., v. 23, no. 5, p. 496–502. Copyright © 1989 American Chemical Society.

Occurrence of Colloidal Material

Colloidal transport begins with a source of particles of the correct size. One possibility is the translocation of colloids from other locations. Examples could include humic substances flushed from the vadose zone, viruses and bacteria present in waste waters from septic tanks, or organic macromolecules generated in landfill leachates (Gounaris and others, 1993). Colloids also can form from precipitation of solids from solution. The cause is supersaturation of a mineral phase stemming from abrupt changes in the water chemistry due to varying pH, ion concentrations, or redox potential (McCarthy and Zachara, 1989). For example, Gschwend and Reynolds (1987) describe the formation of iron phosphate colloids in the sewage plume at the Otis Air Force Base due to the reaction of phosphate, present as a dissolved contaminant in solution, and ferrous iron released from aquifer solids. Other colloids formed by precipitation are small metal solids or radionuclides that can form in hydrolysis reactions. Figure 13.1 also illustrates schematically how colloids can form when small particles attached to the surface of solids are remobilized.

There are other possibilities for generating colloids besides those shown in Figure 13.1. For example, it is well known in the petroleum industry that injection of fresh water into saline formations can lead to the electrostatic mobilization of clay coatings on grains. The solubilization of organic matter such as kerogen or bitumen, or the breakdown of biofilms, can also form colloids (McCarthy and Zachara, 1989).

Stabilization

The simple presence of colloids in ground water does not assure that they are transported. As Figure 13.1 implies, transport requires that suspended colloids be stable. In a suspension of colloids, stability is determined as the outcome of competition between van der Waals forces that cause similar particles to aggregate and electrostatic forces that cause particles to separate. Van der Waals forces are weak electrical forces that cause attraction between atoms and molecules. These forces are created by oscillations in the charge distribution involved with a positively charged nucleus and negatively charged electron swarm. Electrostatic forces are produced by the strong electrical attraction or repulsion between point charges of differing or similar charges, respectively. McCarthy and Zachara (1989) discuss the complexities of surface chemistry, particle mineralogy, and other factors that determine stability. While these processes are complex, certain generalizations are possible. For example, stability is promoted by a decreased ion concentration in the water, a decreased abundance of divalent ions (for example, Ca^{2+}) relative to monovalent ions (for example, Na^+), and a pH that is different from the pH of the zero point of charge (pH_{zpc}) for the colloid surface.

Puls and Powell (1992) present experimental results that illustrate the concept of stabilization. The colloids they studied were 150-nm Fe_2O_3 particles. As we explained in Chapter 12, the surface charges of an iron oxide changes as a function of pH. At the pH_{zpc}, the net surface charge is zero. Figure 13.2 is a plot of colloid diameter versus pH. Over the pH ranges of 2.0 to 6.5 and 9.7 to 11.0, the measured colloid diameters were about 150 nm—the diameter of the starting colloids. In the pH range from 6.5 to 7.6 (close to the measured pH_{zpc} of 7.0), the diameter of the colloids increased, indicating

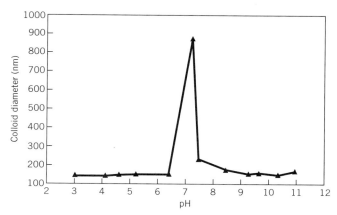

Figure 13.2 Stability of 150-nm Fe_2O_3 particles as a function of pH in 0.005 M NaClO$_4$ (from Puls and Powell, 1992). Reprinted with permission from R. W. Puls and R. M. Powell, Environ. Sci. Technol., v. 26, no. 3, p. 614–621. Copyright © 1992 American Chemical Society.

aggregation and suspension instability. Near the pH$_{zpc}$ the particles are uncharged and van der Waals forces cause them to aggregate.

Transport and Filtration

Stable colloids in solution are potentially transportable as long as they do not interact substantially with the porous medium. One possible mode of interaction is through physical straining (also known as straining filtration), where the particles are too large to migrate through small pore throats. However, colloids and most microorganisms are commonly much smaller than the grain size (and pores) of gravels, sands, and silts (Figure 13.3). Physico-chemical collection on solid surfaces (McCarthy and Zachara, 1989) is a more important process of filtration. The term *physico-chemical collection* applies to a variety of different surface-active attractive forces. One example is electrostatic interaction, where a positive particle is attracted to a negative surface. In general, negatively charged colloids would be most mobile, given the tendency for most aquifer solids to be negatively charged (McCarthy and Zachara, 1989).

Puls and Powell (1992) provide an interesting example of how Fe$_2$O$_3$ colloids interact with a porous medium in column experiments. The medium was a quartz-rich material with a relatively low pH$_{zpc}$ because of the abundance of quartz (pH$_{zpc} \simeq 2$), and manganese oxides and layered silicates (pH$_{zpc} < 4$). Thus, when the pH is less than about 3, surfaces of the porous medium will be positively charged. Above pH 3, surfaces exhibit a net negative charge. Recall that the surfaces of the Fe$_2$O$_3$ colloids would be positively charged below pH 7 and

negatively charged above pH 7. As expected, in the range of pH 3 to 7, experiments showed there to be no colloid migration because the surfaces are positively charged and the colloids are negatively charged. At pHs above about 7.6, the colloids are mobile because the colloid and porous-medium surfaces are both negatively charged.

13.2 Colloidal Transport in Ground Water

Colloids can be mobile in ground water. One common health issue is related to the transport of viruses in ground water. A dominant source of viral contamination is raw sewage disposed in the subsurface through septic tanks. Other public health problems are related to waste-water renovation schemes that involve the recharge of secondary sewage effluents. Keswick and Gerba (1980) summarize case studies that indicate the possibility for viruses to be transported several hundred meters in ground water. In exceptional cases, travel distances of close to one kilometer are possible. Other literature has examined the potential for radionuclide transport in colloidal form. For example, at the Nevada Test Site, Buddemeier and Hunt (1988) describe the possibility of facilitated transport of radionuclides away from a nuclear detonation cavity. At Chalk River in Canada, significant radionuclide transport was found associated with particulates (Champ and others, 1984).

Sampling and Measuring

The colloid characterization process is complicated because most of the steps involved in sampling and analysis—drilling a well, pumping out a sample, handling a sample at the ground surface or in the laboratory—can create particles not found in the original sample. For example, McCarthy and Zachara (1989) explain how drilling can add small particles when drilling mud is used. Pumping often creates unnaturally high flow velocities that remobilize surface-bound colloids. Oxygenation of the samples at the surface or during preparations for laboratory analyses could change the redox conditions, leading to the precipitation of Fe(III) oxides.

Good sampling practice involves pumping wells for several hours to eliminate fines near the well screen. Pumping rates are normally kept low, on the order of 100 mL/min, to avoid detaching colloids (Gounaris and others, 1993). A gas-driven, positive displacement pump avoids contact of the sample with oxygen. At the surface, samples are typically collected in nitrogen-flushed containers or filtered for analysis in field glove boxes containing nitrogen.

Several different approaches are available to establish the concentration and size distribution of colloids in samples. Ultrafiltration systems process the samples through

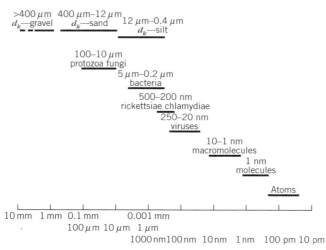

Figure 13.3 Comparison of the sizes of common microorganisms with the grain sizes of sediments, molecules, and atoms (modified from Matthess and Pekdeger, 1985, Groundwater Pollution Microbiology, eds. C. H. Ward, W. Giger, and P. L. McCarty). Copyright © 1985 by John Wiley & Sons, Inc. Reprinted by permission of John Wiley & Sons, Inc.

a succession of smaller and smaller nanometer-sized filters. Knowing the volume of water passed through the system and the mass of colloid trapped on each of the filters, one can determine colloid concentrations (Gounaris and others, 1993). The mineralogy of the particles trapped on the various filters can be determined using energy-dispersive scanning-electron microscopes and x-ray diffraction. Gschwend and Reynolds (1987) used a laser light-scattering approach, which can be interpreted to provide information on colloid size distributions and colloid concentrations. A variety of other light-scattering approaches can be used for colloid determinations as well as centrifugation.

Studies at Cape Cod

Throughout this book, we make use of field studies that illustrate both theoretical and practical concepts. One of these important sites is the Cape Cod Toxic-Substances Hydrology Research site, located near Falmouth Massachusetts (Figure 13.4). The U.S. Geological Survey (USGS) has operated this site since the early 1980s. It has been the focal point for studies concerned with the fate and transport of contaminants and the chemical and biological behavior of dissolved and colloidal material in ground water. Garabedian and LeBlanc (1991) provide a detailed overview of studies in relation to a large sewage plume from the Otis Air Base and smaller scale process studies conducted at an abandoned gravel pit, which is shown on Figure 13.4 as the tracer-test site. Our interest here is with the behavior of bacteria and other colloids. Later, we will examine other interesting features related to contamination at the site.

At Otis Air Base, approximately 1740 m³/day of treated sewage has been disposed through infiltration beds into the subsurface since 1936. The disposal unit is an unconfined aquifer comprised of sand and gravel outwash approximately 35 m thick. The sewage plume is about 915 m wide, 23 m thick, and 3.35 km long (Figure 13.4). The treated sewage contains above-background concentrations of sodium, chloride, ammonium, nitrate, phosphate, detergents, and several different volatile organic compounds.

Gschwend and Reynolds (1987) looked at features of colloid transport. They found that samples of sewage and ground water up to approximately 360 m down-gradient from the infiltration beds contained colloids. There are two distinctly different types of colloids present in the ground water. Colloids in the actual sewage are organic macromolecules with a diameter of 6 nm. However, they are not found in the nearest down-gradient well, approximately 60 m away. Thus, the abundant colloids in the sewage are not mobile (Gschwend and Reynolds, 1987). Colloids in the ground water are approximately 100 nm in diameter and have a vivianite-type mineralogy {Fe₃ (PO₄)₂·8H₂O}. Gschwend and Reynolds (1987) suggest

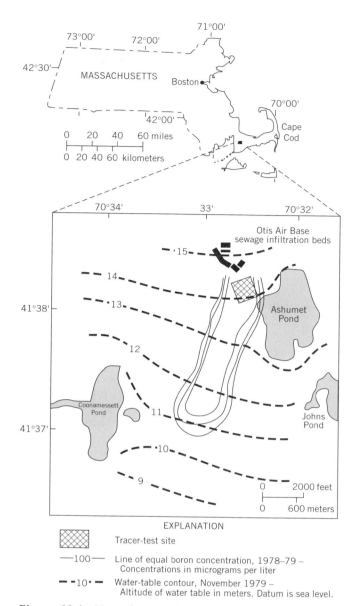

Figure 13.4 Maps showing the location of the Cape Cod Toxic-Substances Hydrology Research Site, the tracer-test site, and the extent of the sewage plume from Otis Air Base (from Garabedian and LeBlanc, 1991).

that these colloids formed by precipitation from aqueous solution. Reducing conditions in the plume promoted the dissolution of iron coatings from grains, while the required phosphate is a contaminant in the plume. These colloids are observed up to 300 m from the infiltration beds.

13.3 Microbiological Systems

Microorganisms occur ubiquitously in the subsurface. Bacteria predominate, but protozoa and fungi are also common. Bacteria are small single-celled organisms. To live, they metabolize dissolved organic matter present naturally or due to contamination. Section 12.5 provided examples of redox reactions that are bacterially catalyzed. Protozoa are somewhat larger, single-celled organisms that can feed on bacteria. Fungi live by degrading other organic matter and effectively recycle plant and animal debris (Chapelle, 1993).

Bacteria occur in the subsurface in different ways. Some are adapted to move freely in the water due the presence of flagella that can propel them. The direction of motion can be either random or directed in response to substrate concentrations (chemotactic). This property of motility can be important for the local redistribution of bacteria, and possibly influences larger scale transport. Most commonly, bacteria are immotile, bound to the surfaces of the aquifer solids. The attached population forms what is called a biofilm, which consists of bacteria held together and to the particles by extracellular polymers (McCarty and others, 1984).

In pristine ground-water systems, the bulk of the bacterial population is attached to the particle surface (Ghiorse and Wilson, 1988). Chapelle (1993) draws on theoretical work by Kelly and others (1988) to suggest that when substrate concentrations are low, immotile bacteria have a competitive advantage over motile forms. There is some indication that the proportion of motile bacteria may be larger in contaminated systems containing metabolizable organic compounds (Ghiorse and Wilson, 1988). With increasing substrate concentrations, motile bacteria may become more competitive (Chapelle, 1993).

Studies by Harvey and his co-workers on Cape Cod

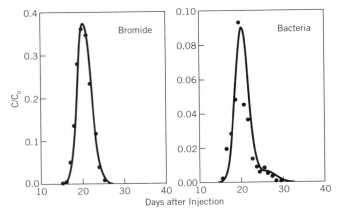

Figure 13.6 Breakthrough curves for Br$^-$ and stained bacteria in a sampling device located 6.8 m down-gradient from the injection well. The solid line illustrates the fit of a theoretical model to the observed data (from Harvey and Garabedian, 1991). Reprinted with permission from R. W. Harvey and S. P. Garabedian, Environ. Sci. Technol., v. 25, no. 1, p. 178–185. Copyright © 1991 American Chemical Society.

have produced systematic observations on the transport behavior of unattached bacteria at several scales. Harvey and Garabedian (1991) describe the results of a natural gradient tracer test conducted using Br$^-$ and bacteria (0.2–1.4 μm long). The bacteria were stained for identification and injected slowly into the aquifer (Figure 13.5). Samples were collected with time at a multilevel sampling device located approximately 6.8 m down-gradient. The breakthrough curves for Br$^-$ and the stained bacteria (expressed as dimensionless concentrations) are shown in Figure 13.6 at the sampling device. The timing of the breakthrough was similar. However, the peak abundance was much less for the bacteria than for Br$^-$ and the curve exhibited tailing. The reduced peak abundance was attributed to the loss of bacteria by sorption on the aquifer over the 6.8-m-long travel path. However, these bacteria behaved much like colloids because motility was apparently not a factor in the transport.

The simplest conclusion from these experiments is that unattached bacteria will not be particularly mobile in the outwash aquifer over much more that several tens of meters. However, other field observations point to the transport of unattached bacteria over a substantial length of the contaminant plume (Harvey and George, 1986; Harvey and Garabedian, 1991). The main determinant for transport appears to be whether the bacteria come from an indigenous population that is actually living in the aquifer or a nonindigenous population that is not living in the aquifer but is added with the sewage. Indigenous bacteria are potentially much more transportable because bacteria being lost by predation or due to sorption to the surfaces of grains in the aquifer are being replaced by rapidly growing bacterial populations (Harvey and others, 1989). Organic contaminants within the

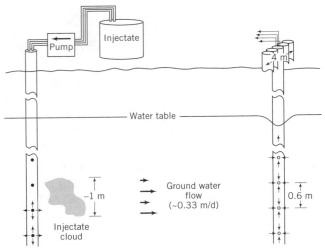

Figure 13.5 Schematic figure showing the design of the natural-gradient tracer experiment with Br$^-$ and stained bacteria (from Harvey and Garabedian, 1991). Reprinted with permission from R. W. Harvey and S. P. Garabedian, Environ. Sci. Technol., v. 25, no. 1, p. 178–185. Copyright © 1991 American Chemical Society.

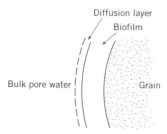

Figure 13.7   Conceptualization of a biofilm at a pore scale.

plume provide a source of organic matter necessary for the growth of the bacteria.

Although indigenous bacteria were used for the small-scale tracer test, growth rates were negligible during transport. Thus, the reduction in bacterial abundances were dramatic. Harvey and Garabedian (1991) suggest that staining the bacteria to observe them may have inhibited their growth, or that conditions for growth were not appropriate in the part of the aquifer where the test was conducted. In summary, bacterial transport over more than 1000 meters is possible in plumes containing organic contaminants (Harvey and others, 1989).

Biofilms

Biofilms are the site for transformation reactions that are important in reducing the concentrations of some organic contaminants. They are commonly conceptualized as a simple layered structure wherein molecular diffusion operates to bring metabolizable organic compounds from the bulk pore fluid to clusters of bacteria cells in the biofilm (Figure 13.7). Real biofilms are more complex. For example, in cases where organic substrate concentrations are relatively low, patchy or discontinuous biofilms form (Rittmann, 1993). Detailed microscopic studies of model biofilms demonstrated fluid flow within biofilm channels (Stoodley and others, 1994). These channels provide a delivery system for nutrients into the biofilm that is more efficient than diffusive transport.

Sampling and Enumerating Microbial Populations

A variety of techniques have been developed to enumerate microorganisms in aquatic systems. The first critical step is collecting the sample. Depending on the purpose of the study, one could consider either water samples or samples of soil material. Water samples can be collected from conventional monitoring wells. For example, at Cape Cod, Harvey and others (1984) collected water samples from conventional monitoring wells using a submersible pump connected to the surface by Teflon tubing. After purging three to five casing volumes, samples were collected in sterile sample bottles. NcNabb and

Mallard (1984) discuss these sampling techniques in detail and provide useful guidance on how to avoid sample contamination.

Detailed microbiological studies require the collection of soil cores because most of the microbial population is surface attached. Beyond the difficulties in actually collecting the core is the problem of how to avoid microbial contamination during the drilling and sampling process. Boreholes should be drilled using techniques (for example, hollow-stem augering) that do not require drilling fluids (McNabb and Mallard, 1984). Conventional coring equipment—solid tube samplers, split-tube samplers, Shelby tube samples, and so on—can be sterilized and used to collect the sample.

Because of the near impossibility in avoiding microbial contamination on the outside surface of a core, its surface is pared away as it is extruded. Figure 13.8a illustrates an anaerobic glove box that can be used in the field. The glove box is flushed with nitrogen for about 30 minutes at a rate of about 2500 L/hr (Armstrong and others, 1988). The sample tube, sealed temporarily after collection, is

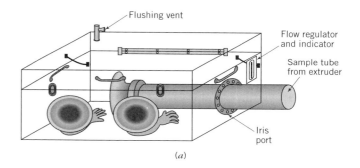

(a)

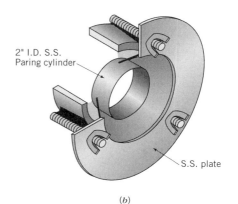

(b)

Figure 13.8   Panel (a) A field-sampling glove box for processing a sediment core for miocrobial analysis. (b) The sampling tube is inserted into the field glove box and the core paring tool is attached to the end of the tube. The cores extruded from the sample tube are collected in sterile containers (from Armstrong and others, 1988). Reprinted with permission of NWWA/API Conference Petroleum Hydrocarbons and Organic Chemicals in Ground Water: Prevention, Detection and Restoration: National Ground Water Assoc., Copyright © 1988. All rights reserved.

inserted into the box through an iris port and a sterile paring tool (Figure 13.8b) is screwed on to the sample tube. The core is extruded using a rig-mounted extruder and appropriate subsamples are placed in sterile containers (Armstrong and others, 1988).

Once the sample of water or solid is collected, techniques are available to characterize the microbial population. In the case of a soil sample, however, an additional step is required to remove bacteria from the solid materials. The most common procedure involves shaking a sediment-distilled water slurry. Both chemical additives (Chapelle, 1993) and glass beads (Kölbel-Boelke and others, 1988) are sometimes added to assist.

Plate Counts

A plate count involves smearing a sample (water or sediment–water slurry) containing the bacteria on a growth medium, incubating the plate for some time, and counting the bacterial colonies. The growth medium can vary but typically contains nutrients and a carbon "food source" for the bacteria, which are solidified by agar to form a gel (NRC, 1993). The approach requires serial dilutions because often a sample will create so many colonies that they cannot be counted. By diluting the starting water sample or sediment slurry more and more with sterile water, the number of bacteria will also be reduced until the number of colonies can be counted. Accounting for the dilutions, the number of bacteria can be expressed, for example, as colony-forming units per milliliter (CFU/mL) in the case of water samples, or CFU/(g dry weight of sediment).

This procedure has limitations because not all bacteria in a sample grow in the environmental conditions afforded by the agar medium. One accommodation in the testing procedure involves using sterilized soil extract that is stabilized by agar as a component in the growth medium. This organic-poor mixture can increase the number of colony-forming units because growing conditions more closely resemble the natural environment. Similarly, when enumerating anaerobic populations, anaerobic conditions must be maintained with appropriate electron acceptors such as nitrate or sulfate.

We noted that not all bacteria present in a sample grow. Further, as the samples are slurried, a significant proportion of the bacteria will remain on the sediment surface and not be counted (Chapelle, 1993). Thus, plate counts provide a minimum estimate of bacterial abundance.

In site cleanups, a concern is not the overall size of the bacterial population, but the size of the population of microorganisms capable of biotransforming a specific organic contaminant. Plate counts can provide this information. The growth medium is modified to utilize the organic compound of interest as the only carbon source. Thus, by reporting the abundance of contaminant-degrading organisms, one can examine the biodegradation ca-

pacity of the system and the potential for contaminant-specific toxicity.

Plate counts provide information on the number of colony-forming units, but not the actual number of cells. A procedure that provides this information is the most-probable-number (MPN) approach. The approach creates many incubations that result from plating a large number of successively diluted samples and appropriate replicates of each. The basic idea is to find out how much a sediment slurry or water sample can be diluted before no growth takes place. The dilution information provides a basis for estimating the number of cells per volume (or mass) of the original sample. While this procedure is somewhat different from a standard plate count, it can be adapted as the standard plate count method to consider specific contaminants. It has most of the same limitations as well (NRC, 1993).

Direct Counting Procedures

Direct counting procedures use a microscope to count the bacteria. Stains like acridine orange are used in conjunction with epifluorescence microscopy to facilitate counting (NRC, 1993). Following this procedure, cells are separated from the solids as before in a sediment–water slurry. The fluid fraction containing the cells is separated, stained with the fluorescent dye, and filtered so that bacterial cells are collected on a membrane filter (Chapelle, 1993). Under ultraviolet light provided with the microscope, the fluorescing cells can be counted. This procedure provides an estimate of the number of cells per gram of material.

Biochemical Techniques

Several tools are able to identify bacteria (NRC, 1993). DNA probes identify genetically similar materials in a given cell. The degree of match helps to identify the cell. One, however, needs relatively detailed information on the genetic sequences that are likely present in the unknown cell to create a useful probe. Another useful tool is fatty acid analysis. This procedure identifies bacteria on the basis of the unique fatty acids found in the cell membranes.

Rates of Microbial Reactions

Beyond knowing the numbers of bacteria present or those able to degrade specific organic compounds, one often may be interested in the rates of biotransformation reactions. This information is particularly useful in the design and evaluation of bioremediation systems. Microcosms are one tool to examine the rates of microbial processes. The microcosm provides a laboratory setting for isolating and studying subsurface environments (Wilson and Noonan, 1984).

In its simplest form, a microcosm consists of a porous medium, water, and a functioning population of microorganisms isolated in a container. The particular design is a trade-off between the convenience of operation in maintaining a given environment, cost, and the faithfulness with which given processes are simulated (Wilson and Noonan, 1984). With the flow-through systems (Figure 13.9*a*), fluids are moved continuously through the microcosm in a column-style experiment.

Figure 13.9*b* illustrates a static microcosm. These experiments typically involve a set of identical microcosms, including three replicates, that contain the sample of aquifer material spiked with a known concentration of the organic compound or contaminant of interest. These are incubated in the dark at a fixed temperature. At specific time intervals, three replicated microcosms in the set are sacrificed to provide a measurement of the concentration of the organic compound. A set of sterilized microcosms is also monitored as a control to establish that no other processes or sample-handling problems cause a loss of the organic tracer. These concentrations, plotted as a function of time, describe the rate of loss of the organic compound from the microcosm, which usually can be represented by a kinetic model.

Figure 13.9*c* illustrates a laboratory batch microcosm. This microcosm is constructed from a large glass bottle.

The mixture of aquifer solids and ground water is spiked with a mixture of contaminants and incubated in the dark. (Nielsen and Christensen, 1994a). Periodically, samples are collected for analysis. The large volume of sample in the container assures that the effect of removing the samples is minimal.

Examples of degradation curves for four aromatic hydrocarbons, which were developed with the laboratory batch microcosm, are presented in Figure 13.10. The experiment was performed with sediment and aerobic ground water from a shallow glaciofluvial aquifer in Denmark (Nielsen and Christensen, 1994a). The microcosms were incubated in the dark for 149 days. From Figure 13.10, considerable variability in the rate of degradation is apparent. The rate for biphenyl is approximately 50 $\mu g \ L^{-1} \ day^{-1}$ while that for *p*-dichlorobenzene is 1.6 $\mu g \ L^{-1} \ day^{-1}$.

A much more difficult procedure for estimating biodegradation rates involves tracer tests run with biodegradable organic compounds. Such a test was conducted by Barker and others (1987) at Canadian Forces Base Borden. About 1800 L of ground water were collected from the shallow aquifer and spiked with 2360 $\mu g/L$ benzene, 1750 $\mu g/L$ toluene, 1080 $\mu g/L$ *p*-xylene, 1090 $\mu g/L$ *m*-

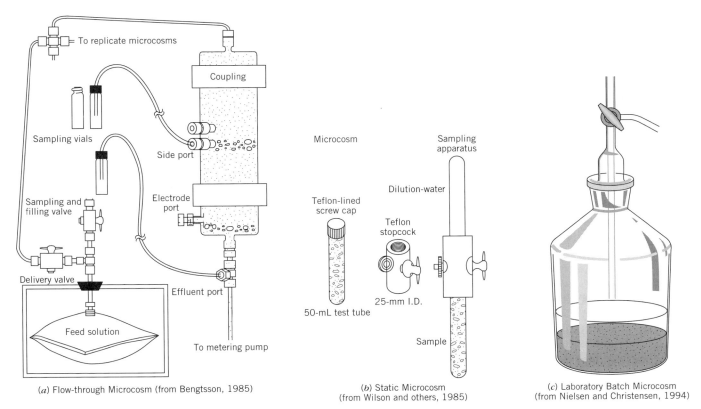

(*a*) Flow-through Microcosm (from Bengtsson, 1985)

(*b*) Static Microcosm
(from Wilson and others, 1985)

(*c*) Laboratory Batch Microcosm
(from Nielsen and Christensen, 1994)

Figure 13.9 Examples of (*a*) flow-through (*b*) static microcosms, in Ground Water Quality, eds. C. H. Ward, W. Giger, and P. L. McCarty. Copyright © 1985 by John Wiley & Sons, Inc. Reprinted by permission of John Wiley & Sons, Inc., and (*c*) a laboratory batch microcosm reprinted from J. Contam. Hydrol., v. 15, by P. H. Nielsen and T. H. Christensen, Variability of biological degradation of aromatic hydrocarbons in an aerobic aquifer determined by laboratory batch experiments, p. 305–320, 1994 with kind permission from Elsevier Science, NL, Sara Burgerhartstraat 25, 1055 KV Amsterdam, The Netherlands.

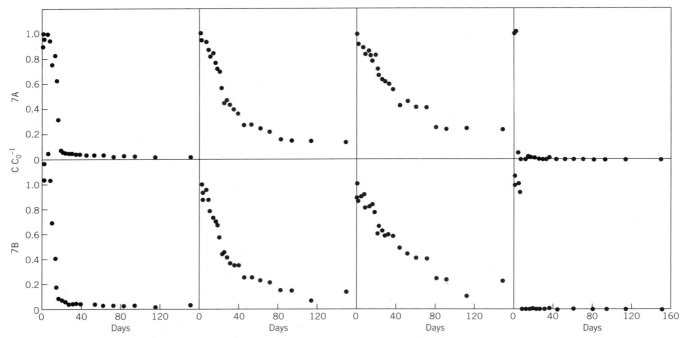

Figure 13.10 Normalized concentrations of benzene, *o*-xylene, *p*-dichlorobenzene, and biphenyl as a function of time in laboratory batch microcosms for two replicate sediment samples (from Nielsen and Christensen, 1994). Reprinted from J. Contam. Hydrol., v. 15, by P. H. Nielsen and T. H. Christensen, "Variability of biological degradation of aromatic hydrocarbons in an aerobic aquifer determined by laboratory batch experiments, p. 305–320, 1994 with kind permission from Elsevier Science, NL, Sara Burgerhartstraat 25, 1055 KV Amsterdam, The Netherlands.

xylene, 1290 μg/L *o*-xylene, and 1280 mg/L Cl⁻. This water was injected into the aquifer just below the water table. The spreading of the various organic compounds with time was monitored using a dense network of multilevel piezometers. Figure 13.11 compares the vertically integrated concentrations at three different times. Both benzene and toluene are biodegrading. The zero-order biodegradation rates were -30 mg day^{-1} for benzene, -37 mg day^{-1} for toluene, and -33, -47, and -55 mg day^{-1} for *o*-, *m*-, and *p*-xylene, respectively. Again, rates are quite variable depending on the organic compound (Barker and others, 1987).

Microbial Ecology of the Subsurface

The subsurface is home to a diverse set of microorganisms. In pristine ground-water systems, population densities range between 10^5 and 10^7 cells per gram dry weight (gdw) (Ghiorse and Wilson, 1988). Generally, these population densities are small relative to those found in the soil zone and the unsaturated zone. This difference in the size of microbial populations is caused by a decrease in the concentration of organic matter as recharging ground water moves through the soil.

The distribution of microorganisms depends in a complex way on the residence time of water in the system and the geochemistry of organic matter transported along the flow system. As Ghiorse and Wilson (1988) suggest, the quantity of organic matter and nutrients passed along

a pristine ground-water system is relatively meager. Thus, microbial populations appear adapted for growth and survival in nutrient-poor (oligotrophic) conditions (Balkwill and others, 1989). Theoretically, in a system closed to organic carbon, the concentration of organic matter should decline until it is no longer possible to support the microbial population (Ghiorse and Wilson, 1988). However, this end point has not been discovered in large flow systems.

Results from several studies show that even down to several hundred meters, microbial abundance does not diminish with depth. However, there can be considerable spatial variability both vertically and laterally. For example, in holes drilled as part of the Deep Probe Project of the U.S. Department of Energy, Levine and Ghiorse (1990) found a highly significant correlation between bacterial abundance and hydraulic conductivity. Bacteria were more abundant in sandy sediments and much less abundant in clayey sediments. Similar distributions are observed in shallow environments as well. This pattern may reflect the much greater difficulty in originally colonizing the fine-grained sediments because of straining filtration of migrating organisms. In addition, the larger fluxes of water through the more permeable units would likely make significantly more dissolved organic matter available. Fredrickson and others (1990) found significant diversity in the ability of bacteria to utilize various types of organic compounds even when nearby samples were compared. Their data indicate a great diversity in bacteria

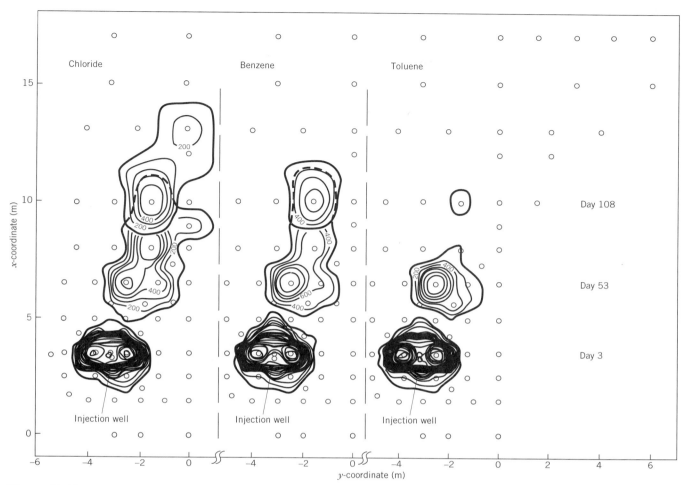

Figure 13.11 Vertically integrated concentrations of chloride, benzene, and toluene. Concentration units are mg · m/L for chloride and μg · m/L for benzene and toluene (from Barker and others, 1987). Reprinted with permission of Ground Water Monitoring Review, p. 64–71. Copyright © 1987. All rights reserved.

at the population and organism levels even within the same geologic unit.

13.4 Microbial Processes

There are two different modes in which organic substrates participate in biodegradation reactions. With primary utilization, a dissolved organic compound provides the main source of energy and carbon for the microorganisms. In effect, the microorganisms produce energy by the electron transfer involved with redox reactions. In other cases, an organic compound or substrate of interest is cometabolized along with some other primary substrate. The secondary substrate effectively is not capable

of providing sufficient energy to maintain the bacterial population.

A variety of organic compounds can serve as the primary substrate for microbial metabolism (NRC, 1993). Table 13.1 lists examples of primary substrates. Generic examples of these primary utilization reactions were presented in Section 12.5 wherein the organic compound (the electron donor) reacts with various possible electron acceptors (for example, oxygen, nitrogen, sulfate). Following are examples of overall reactions involving these common organic compounds:

benzene: $C_6H_6 + 7.5O_2 \rightarrow 6CO_2 + 3H_2O$

hexane: $C_6H_{14} + 9.5O_2 \rightarrow 6CO_2 + 7H_2O$

Table 13.1 **Examples of Primary Substrates**

Aerobic and anaerobic	Glucose, acetone, isopropanol, acetate, benzoate, phenol
Aerobic primarily	Alkanes, benzene, toluene, xylene, vinyl chloride, methane, propane

Writing the reactions in this way, however, hides the detailed biochemistry of the reaction. For example, benzene degradation involves several intermediate compounds that are related to various enzyme systems. Enzymes react with the organic substrates to make compounds that are able to react at rates appropriate to maintain life (Chapelle, 1993).

Some chlorinated solvents, like trichloroethene (TCE), dichloroethene (DCE), dichloroethane (DCA), and vinyl chloride (VC), are relatively oxidized compounds and resist biotransformation because they do not provide much energy. They, however, can be cometabolized with other organic compounds. For example, Roberts and others (1989) demonstrate how methanogenic bacteria are stimulated through the addition of methane and oxygen. The oxidation of methane provides the energy and carbon necessary for bacterial growth and leads to the production of enzymes required to degrade the secondary substrates like TCE and VC. Chemically, both TCE and VC end up being mineralized to CO_2. This possibility for cometabolizing chlorinated solvents holds promise as a remedial approach for aquifers contaminated by TCE.

Other organic solvents like 1,1,1-trichloroethane (TCA), tetrachloroethene (PCE), TCE, DCE, DCA, and CT undergo secondary utilization in reduction reactions. The organic compound in this case is reduced and another compound (for example, a transition metal) is oxidized. However, unlike the oxidation-type reactions that lead to the mineralization of the compound, reduction reactions typically produce a less chlorinated compound.

Issues in Biodegradation

Before leaving this discussion on microbial reactions, we will mention briefly some of the factors that come into play in determining whether and how organic compounds biodegrade in ground-water settings. Table 13.2 summarizes several of the most important factors in this respect.

Biofilm Kinetics

Biotransformation reactions take place within the biofilm. There are two coupled processes of interest, the transport of substrates or organic compounds from the bulk ground water into the biofilm, and the growth (or decline) in the mass of the biofilm. The processes are coupled because transport of the substrate depends in part on how much biofilm there is and the mass of the biofilm depends in part on the availability of the substrate as a source of energy (McCarty and others, 1984).

Rittman and McCarty (1980a) developed a model to simulate biofilm processes where the concentration of a single substrate is sufficiently small to be rate limiting. Later, we will examine a case where the availability of the electron acceptor (for example, O_2) limits the reaction rate.

The organic compound moves by diffusion through the diffusion layer and into the biofilm and ultimately is lost by bacterial utilization. The following empirical kinetic law describes the rate of loss (Bouwer and McCarty, 1984):

$$\frac{dS_f}{dt} = \frac{kX_fS_f}{K_s + S_f} \tag{13.1}$$

where S_f is the concentration of the rate limiting substrate, X_f is the active cell density in the biofilm (mass of bacteria L^{-3}), k is the maximum specific rate of substrate utilization (mass of substrate/mass of bacteria per unit time), and K_s is the Monod half-maximum-rate concentration (mass of substrate L^{-3}). K_s is the concentration at which the specific rate of utilization is one-half of k.

Table 13.2 **Summary of Complicating Factors Affecting the Microbial Degradation of Organic Compounds**

Factor	Explanation
Toxicity	Relatively large concentrations of organic substrates can prevent or slow metabolic reactions (NRC, 1993). How this factor comes into play depends on the particular substrate and its concentration, as well as features of the microbial population.
Acclimation	Often a time lag of days to months occurs before the microbial population begins to degrade the organic compound. Explanations for the lag include time necessary to expand an initially small population, adaptation to a change in electron acceptor, and possibly time for genetic changes to occur (Chapelle, 1993).
Low concentrations	Microorganisms often cannot reduce substrate concentrations below some minimum level. Rittman and McCarty (1980a, 1980b) define the minimum substrate concentration as S_{min}. During primary utilization, substrate concentrations below S_{min} cannot support the growth of the biofilm.
Sequestration	Although organic contaminants may be abundant in the subsurface, they may be unavailable to microorganisms. This situation can develop when contaminants occur as nonaqueous phase liquids or are sorbed to aquifer solids. Biodegradation often requires that substrates enter the cell, a difficult process in these cases (Alexander, 1994). Enzymes and/or inorganic nutrients are subject to sorption, thereby interfering with the biodegradation process.
Mixtures	Often with a mixture of organic contaminants, some compounds are selectively utilized, while others are ignored until the first compound is completely utilized.

Substrate utilization is a function of two constants (k and K_s), the substrate concentration, and the active cell density X_f. McCarty and others (1984) present expressions for bacterial growth, in essence describing how X_f changes with time as the difference between growth rates (that is, substrate conversion) and loss rates. By solving the coupled mathematical problem of substrate transport and bacterial growth, it is possible to model a steady-state biofilm system.

For a secondary substrate with S_f much less than K_s, Eq. 13.1 simplifies to the following first-order rate law (Bouwer and McCarty, 1984):

$$\frac{dS_f}{dt} = \frac{kX_f S_f}{K_s} \qquad (13.2)$$

Equation 13.2 states that substrate utilization within the biofilm is proportional to the concentration of bacteria and the ratio k/K_s. Under some conditions, bacterial utilization, as described by Eq. 13.2, can be the rate limiting process controlling transport of the secondary substrate within the biofilm. In other words, diffusion is relatively fast through the biofilm and boundary layer relative to this loss rate. In this situation, we can model the biofilm at a macroscopic scale by a simple kinetic expression (that is, Eq. 13.2). The reaction is characterized by a half-life of substrate decomposition (Bouwer and McCarty, 1984):

$$t_{1/2} = \frac{\ln 2}{kX_f/K_s} \qquad (13.3)$$

where Eq. 13.3 comes from the integration of Eq. 13.2. This model is appropriate when the concentration of active organisms is small and ground-water movement is slow (Bouwer and McCarty, 1984).

Conditions exist when the biodegradation of a substrate is limited by the availability of an electron acceptor. For example, the aerobic biodegradation of an organic liquid in water can utilize the small quantities of oxygen naturally dissolved in the water. In spite of an abundance of organic substrate, the lack of sufficient oxygen, as dictated by the stoichiometry of the redox reaction, limits biodegradation. The reaction rate increases when additional oxygen is provided.

The mathematical description of material transport to the biofilm in this case would involve two transport equations, one for the substrate and one for the electron acceptor. Again, both reactants move by diffusion in the diffusion layer and film and disappear through utilization by bacteria. The two governing transport equations are coupled through the reaction term because the rate of substrate utilization (following Bailey and Ollis, 1986) is a function of both the substrate and electron acceptor. This rate expression is a modified version of Eq. 13.1. The kinetic equation for the utilization of the electron acceptor depends on both concentrations as well.

More sophisticated biofilm models can account for contributions of both the substrate and electron acceptors to cell growth. Borden and Bedient (1986) and Molz and others (1986) have embedded biofilm processes in a contaminant transport model where both substrate and oxygen transport are modeled together with changing cell mass.

13.5 *Biotransformation of Common Contaminants*

Chapter 17 examines families of organic compounds that are prevalent in problems of ground-water contamination (Table 13.3). Here, we will briefly survey biotransformation processes in relation to these families. Commonly, biotransformation processes are examined in relation to two prototypical settings. In aerobic settings, oxygen is the major electron acceptor for the various biotransformation reactions. In anaerobic settings, oxygen is absent, and other compounds, like nitrate, sulfate, and carbon dioxide, function as electron acceptors.

Hydrocarbons and Derivatives

Hydrocarbons and related compounds constitute one of the most important contaminant families in ground wa-

Table 13.3   **Important Families of Organic Contaminants Found in Ground Water**

Chemical Family	Examples of Compounds
Hydrocarbons and derivatives	
fuels	benzene, toluene, *o*-xylene, butane, phenol
PAHs	anthracene, phenanthrene
alcohols	methanol, glycerol
creosote	*m*-cresol, *o*-cresol
ketones	acetone
Halogenated aliphatics	tetrachloroethene, trichloroethene, dichloromethane
Halogenated aromatics	chlorobenzene, dichlorobenzene
Polychlorinated biphenyls	2,4'-PCB, 4,4'-PCB

ter. In aerobic ground-water systems, aromatic hydrocarbons with up to two benzene rings are mineralized relatively rapidly with minimal lag times. For example, the microcosm studies of Nielsen and Christensen (1994a) found that benzene, toluene, and o-xylene degraded with short lag times (about 4 days). Initial concentrations of approximately 150 μg/L of benzene, and toluene decreased to <2 μg/L in less than one month. The degradation of o-xylene required about three months and was substantially less complete. These laboratory findings of relatively rapid degradation rates are also corroborated by field tests. For example, the rapid biodegradation of BTX compounds in field tests at Canadian Forces Base Borden was discussed previously.

Hydrocarbons degrade in anaerobic systems as well. Convincing field data come from a study of an oil spill at Bimidji, Minnesota (Cozzarelli and others, 1988). Detailed chemical sampling indicated that under anaerobic conditions concentrations of benzene, toluene, and o-, m-, and p-xylene decreased in ground water over a distance of approximately 10 m. Ethylbenzene, however, was resistant to biodegradation. Lyngkilde and Christensen (1992) followed the transport of a variety of organic compounds within a landfill-leachate plume. Along the first 100 m of the plume, a redox sequence from methanogenic, sulfidogenic, to ferrogenic conditions prevailed. Over the first 20 m (methanogenic–sulfidogenic environment), there was some indication that various isomers of benzene as well as toluene and xylenes were able to degrade. Within the ferrogenic zone, there was obvious degradation of these compounds as well as ethylbenzene.

Other components of fuels (or coal-tar derivatives) are polyaromatic hydrocarbons (PAHs) such as naphthalene, anthracene, or pyrene. These compounds are comprised of a series of benzene rings. In general, PAHs are biodegradable in aerobic settings, with the rate decreasing as the molecular weight of the compound increases. Durant and others (1995) found through microcosm studies that bacteria were capable of aerobically mineralizing naphthalene and phenanthrene at the site of a former manufactured gas plant. In anaerobic systems, PAHs may be somewhat more resistant to degradation. However, Lyngkilde and Christensen (1992) report the reduction of naphthalene concentrations under ferrogenic conditions.

Phenolic compounds are also found as components in fuels or as industrial contaminants. As discussed in Section 11.7, phenols are characterized by an aromatic ring with an attached hydroxyl group. Examples of these compounds include phenol, nitrobenze, o-cresol, and o-nitrophenol. Microcosm studies by Nielsen and Christensen (1994b) show that these compounds are mineralized in aerobic systems. Indications are that they biodegrade relatively rapidly with broad variation in the rates. At a former wood treatment facility in Pensacola, Florida, Bekins and others (1993) reported the degradation of phenol and 2-, 3-, and 4-methylphenol under meth-

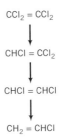

Figure 13.12 Reductive dehalogenation from PCE to TCE to DCE and VC.

anogenic conditions. Smolenski and Sulflita (1987) found that p-, m-, and o-cresol also degraded under anaerobic conditions. The remaining hydrocarbons in Table 13.3, alcohols and ketones, are readily degradable in groundwater.

Halogenated Aliphatic Compounds

Halogenated aliphatic compounds include the common industrial solvents like carbon tetrachloride (CT), tetrachloroethylene (PCE), trichloroethylene (TCE), trichloroethane (TCA), vinyl chloride (VC), and dichloroethylene (DCE). These compounds are formed from carbon atoms joined together in chains with attached atoms. In aerobic systems, they resist degradation and are extremely persistent. There are no known bacteria that can oxidize these compounds as a primary substrate because the reaction yields so little energy (Chapelle, 1993). Thus, a major activity for contaminant hydrogeologists is in cleaning up solvent spills.

An important exception to this general result is when the aerobic system contains methane either naturally or applied in a bioremediation scheme. Compounds like TCE may be metabolized as a secondary substrate when methane is present as a primary substrate. Under anaerobic conditions (methanogenic), halogenated compounds are commonly biotransformed. The reactions involve a sequential loss of chloride atoms in a processes termed *reductive dehalogenation*. Figure 13.12 illustrates reductive dechlorination reactions beginning with PCE. The last step from vinyl chloride to CO_2 is quite slow relative to the other dechlorination reactions, so that vinyl chloride often tends to accumulate.

Halogenated Aromatic Compounds

Halogenated aromatic compounds have a structure characterized by a benzene ring and attached halogens. Examples of these compounds include chlorobenzene, dichlorobenzene, and trichlorobenzene. In general, these compounds are degradable under aerobic conditions (Chapelle, 1993). However, the ability to biodegrade is reduced as the molecule becomes more highly chlorinated (NRC, 1993). In anaerobic systems, these compounds degrade via reductive dechlorination reactions, although there often may be a time lag (Chapelle, 1993). Nielsen and others (1995) examined the fate of *o-* and *p*-dichlorobenzene in the leachate plume of the Vejen landfill. Under anaerobic conditions ranging from methanogenic to iron(III) reducing, these compounds were not transformed.

Polychlorinated Biphenyls (PCBs)

PCBs were used historically as a dielectric compound in transformers. Although the manufacture and use of these compounds has ceased, they are still occasionally discovered at sites. The biphenyl structure involves two linked benzene rings with between one and 10 attached chlorine atoms per molecule. Generally, only less chlorinated PCBs oxidize under aerobic conditions. The more chlorinated PCBs are generally very resistant to oxidation under natural conditions. Under anaerobic conditions, highly chlorinated PCBs transform via reductive dehalogenation reactions.

Complex Transformation Pathways

Both biotic and abiotic transformation reactions can operate on organic compounds in a complicated way. Consider the common industrial solvent 1,1,1-trichloroethane (TCA), which is often a contaminant in ground water (Vogel and McCarty, 1987). Under abiotic conditions, TCA transforms to 1,1-dichloroethylene (1,1-DCE) along one reaction pathway and to acetic acid along another (Figure 13.13). The pseudo-first-order rate constants for the formation of 1,1-DCE and acetic acid from TCA are about $0.04 \ yr^{-1}$ and $0.20 \ yr^{-1}$, respectively (Vogel and McCarty, 1987). These intermediates will transform biologically under methanogenic conditions to vinyl chloride (VC) and CO_2, respectively (Figure 13.13). In the presence of a highly active population of methanogenic bacteria, most of the TCA would biotransform along the

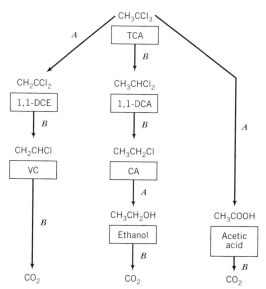

Figure 13.13 Probable fate of TCA under methanogenic conditions: biotic transformation pathways are denoted by lines marked *B* and abiotic transformation pathways by lines marked *A*. Reprinted with permission from Vogel, T. M., and McCarty, P. L., Environ. Sci. Technol., v. 21, 1987, p. 1209. Copyright © 1987 American Chemical Society.

middle pathway (Figure 13.13) by reductive dehalogenation to 1,1-dichloroethane (1,1-DCA) and chloroethane (CA). The next reaction along this pathway, the hydrolysis of CA to ethanol, has a pseudo-first-order rate constant of about $0.37 \ yr^{-1}$, giving CA a half-life of about 1.9 yr (Vogel and McCarty, 1987). In biologically active samples, ethanol would be transformed to CO_2.

Exactly which of these pathways will be dominant and which intermediates of TCA are found at a site depend on the potential for methanogenic biotransformation. Irrespective of the pathway, however, the intermediates should be present because of the rate limitations imposed by the various reactions.

Problems

1. For an aerobic reaction, k/K_s is estimated to be 5.0 L/mg cells per day. Given a concentration of microorganisms of 0.01 mg cells/L or 10^4 bacteria/mL, estimate the half-life of substrate utilization.

2. In a laboratory experiment, the concentration of an organic compound is found to decline over a period of 15 days.

Elapsed Time (days)	Concentration (g/L)
0	123.0
1	111.7
2	102.2
4	84.5
8	58.9
15	30.5

Estimate the decay rate constant and the half-life for this particular compound.

3. Assume TCA occurs in an aquifer with minimal biological activity at a concentration of 100 μg/L. (a) After five years, what other compounds besides TCA would be present? (b) Use the transformation rate data provided in this chapter to estimate which of the various compounds would be most abundant.

Chapter 14

The Equations of Mass Transport

14.1   *Mass Transport Equations*
14.2   *Mass Transport with Reaction*
14.3   *Boundary and Initial Conditions*

Hydrogeologists traditionally have worked to extend their understanding of ground-water flow through the precise language of mathematics. Not surprisingly, this same approach has been applied to problems of mass transport. The equations developed in this chapter provide a comprehensive framework for quantitatively describing mass transport with and without accompanying chemical reactions. These equations are at the heart of approaches used to model the chemical evolution of natural ground water and contaminant migration.

14.1   Mass Transport Equations

As with flow, the starting point for developing the mass transport equations is a conservation equation. This equation in words is

mass inflow rate − mass outflow rate =
change in mass storage with time

in units of mass per unit time. This statement applies to a domain of any size. For a representative volume with inflow through three sides, the left-hand side of this word equation is the divergence of the flux of interest (see Chapter 4), that is,

− div **J** = net mass outflow rate per unit volume

It remains to describe the nature of the flux J, which could be diffusive transport or advective transport combined with diffusion.

The right-hand side of the conservation statement assumes that the gains or losses in mass inside the volume are proportional to changes in concentration in mass inside the unit volume. Thus, a proportionality constant is required that expresses the mass lost or gained when the concentration changes a unit amount. In this case, the proportionality constant is unity because mass occupies the pores, where concentration is mass per unit volume of fluid (m/V_w) and porosity n for a fully saturated medium is the fluid volume per unit total volume (V_w/V_T). Thus, the product Cn is mass per unit total volume and the conservation statement becomes

$$-\operatorname{div}\mathbf{J} = \frac{\partial (Cn)}{\partial t} \qquad (14.1)$$

This equation states that the net mass outflow rate per unit volume equals the time rate of change of mass within the unit volume.

The Diffusion Equation

For a diffusive system defined by Fick's law of the form in Eq. 10.9, Eq. 14.1 becomes

$$\text{div } [D_a^* n \text{ grad } C] = \frac{\partial (Cn)}{\partial t} \qquad (14.2)$$

With the further assumption that porosity does not vary spatially or temporally, and that the diffusion coefficient is constant spatially, we can write

$$D_a^* \text{ div } [\text{grad } C] = \frac{\partial C}{\partial t}$$

or

$$\frac{\partial^2 C}{\partial x^2} + \frac{\partial^2 C}{\partial y^2} + \frac{\partial^2 C}{\partial z^2} = \nabla^2 C = \left(\frac{1}{D_a^*}\right)\frac{\partial C}{\partial t} \qquad (14.3)$$

This diffusion equation is analogous to the unsteady flow equation introduced in Chapter 4 and the unsteady conductive heat transport equation in Chapter 9. The solution to this equation determines how the concentration at any point in a three-dimensional field is changing with time. If the net outflow is zero so that the concentration is not changing in time

$$\frac{\partial^2 C}{\partial x^2} + \frac{\partial^2 C}{\partial y^2} + \frac{\partial^2 C}{\partial z^2} = \nabla^2 C = 0 \qquad (14.4)$$

Equation 14.4 is Laplace's equation, as were the steady-state fluid flow equation and the steady-state heat conduction equation. The solution of this equation describes the value of the concentration at any point in a three-dimensional field.

The Advection–Diffusion Equation

For systems with fluid motion, mass transport can be due to both advection and diffusion. This combined flux can be described mathematically by combining the advective flux in Eq. 10.1 and the diffusive flux in Eq. 10.4, or

$$J_x = -nD_a^* \frac{\partial C}{\partial x} + v_x Cn \qquad (14.5a)$$

$$J_y = -nD_a^* \frac{\partial C}{\partial y} + v_y Cn \qquad (14.5b)$$

$$J_z = -nD_a^* \frac{\partial C}{\partial z} + v_z Cn \qquad (14.5c)$$

where $v_i Cn$ is the advective flux.

For isotropic conditions,

$$\mathbf{J} = -nD_a^* \text{ grad } C + \mathbf{v}Cn \qquad (14.6)$$

where $\mathbf{v}$, the linear ground-water velocity, is a vector with components v_x, v_y, v_z. Substitution of the rate expression (Eq. 14.6) into Eq. 14.2 and carrying out the necessary operations gives

$$\nabla \cdot (nD_a^*\nabla C) - \nabla \cdot (\mathbf{v}Cn) = \frac{\partial(nC)}{\partial t} \qquad (14.7)$$

Expanding the advection terms and assuming constant D_a^* and n give

$$D_a^*\nabla^2 C - \mathbf{v} \cdot \nabla C + C\nabla \cdot \mathbf{v} = \frac{\partial C}{\partial t} \qquad (14.8)$$

For steady ground-water flow, $\nabla \cdot \mathbf{v} = 0$ and

$$D_a^*\nabla^2 C - \mathbf{v} \cdot \nabla C = \frac{\partial C}{\partial t} \qquad (14.9)$$

When the velocity goes to zero, that is, no advection, we recover the diffusion equation for a stationary fluid, Eq. 14.4. When the concentration profile is steady, Eq. 14.9 becomes

$$D_a^*\nabla^2 C - \mathbf{v} \cdot \nabla C = 0 \qquad (14.10)$$

Equations 14.9–14.10 are called the advection-diffusion equations. They describe the manner in which mass is moved from one point to another by advection as modified by diffusion. The one-dimensional form of Eq. 14.9 is

$$D_a^* \frac{\partial^2 C}{\partial x^2} - v_x \frac{\partial C}{\partial x} = \frac{\partial C}{\partial t} \qquad (14.11)$$

The first term on the left describes mass transport by diffusion, and the second describes mass transport by advection.

Before leaving the advection–diffusion equation, we would like to consider the concept of dimensional analysis as applied to mass transport. In Chapter 4, the method of dimensional analysis was introduced and applied to the unsteady flow equation. The result was a dimensionless group known as the Fourier number. Applying these same ideas to the advection–diffusion equation lets us define a dimensionless concentration $C^+ = C/C_e$, where C_e is some characteristic concentration and the superscript $+$ indicates a dimensionless quantity. The diffusion–advection equation in dimensionless form is

$$\nabla^{+2}C^+ - \frac{vL}{D_a^*}\nabla^+ C^+ = \frac{L^2}{t_e D_a^*}\frac{\partial C^+}{\partial t^+} \qquad (14.12)$$

where the dimensionless quantity vL/D_a^* is the Peclet number for mass transport (N_{PE}) with L as some characteristic length. The Peclet number expresses the transport by bulk fluid motion or advection to the mass transport by diffusion. The dimensionless quantity $L^2/t_e D_a^*$ is the Fourier number for mass transport.

The Advection–Dispersion Equation

Diffusion alone does not fully account for mixing during mass transport. Scale-dependent mechanical dispersion in many systems is usually more important. Logically, if dispersion is to be incorporated in the advection–diffusion equation, it should be reflected in the velocity term. Unfortunately, it has not been possible to do this in a simple way. Instead, the coefficient of hydrodynamic dispersion incorporates the combined effects of diffusion and mechanical dispersion. If the dispersion coefficient is constant, this equation is

$$D\nabla^2 C - \mathbf{v} \cdot \nabla C + C \nabla \cdot \mathbf{v} = \frac{\partial C}{\partial t} \qquad (14.13)$$

where $D = D' + D_d^*$. The justification for treating dispersion in this manner is purely a practical one and stems from the fact that the macroscopic outcome is the same for both diffusion and mechanical dispersion. The actual physical processes, however, are entirely different. By replacing the diffusion coefficient by a coefficient of hydrodynamic dispersion, the mixing process is Fickian. Thus, in cases where mechanical dispersion is small compared to diffusion, Eq. 14.13 reverts to the advection–diffusion equation. In the absence of ground-water flow, Eq. 14.13 reduces to the diffusion equation.

The advection–dispersion equation is the workhorse of modeling studies in ground-water contamination. We will have much to say about its solution and application later in the book.

14.2 Mass Transport with Reaction

The equations presented so far describe the processes of advection, mechanical dispersion, and diffusion. With reactions, the statement of mass conservation expands to incorporate sources of material within the porous volume. The equation is modified by source or sink terms depending on whether a constituent is being added or removed by chemical processes. The appropriate statement of mass conservation when reactions are considered is

mass inflow rate − mass outflow rate ±

mass production rate = change in mass storage with time

The plus or minus term designates either a source or a sink and takes on different forms for different reactions. At this point, it is convenient to represent such sources or sinks in a symbolic form. For example, in the case of one-dimensional transport, the word equation above becomes

$$D_x \frac{\partial^2 C}{\partial x^2} - v_x \frac{\partial C}{\partial x} \pm \frac{r}{n} = \frac{\partial C}{\partial t} \qquad (14.14)$$

where r is taken symbolically as the mass produced or consumed per unit volume per unit time, moles/L^3T. This equation applies only to a single constituent. Involving several dissolved species in the transport requires a system of such equations, one for each constituent.

Source terms are usually specified in terms of a rate

law. This law, for example, can be expressed in terms of the rate of decrease of a reactant, or the rate of increase of a product, depending on which constituent is being described by the transport equation. It may be necessary to track both a reactant and a product using two coupled transport equations. One example is particular kinds of biotransformation reactions where both the reactant and the products are contaminants.

Prior to examining some rate laws, let us consider the concept of dimensional analysis as applied to transport with reaction. In order to do so we replace the quantity r/n in Eq. 14.14 by the product of a reaction rate coefficient k and concentration C, that is, kC where k is the volume reacted per unit volume per unit time. Applying the dimensionalizing procedures to the steady-state form of Eq. 14.14, we arrive at two dimensionless groups

$$\frac{kL}{v} \text{ (Damköhler number I)}$$

$$\frac{kL^2}{D} \text{ (Damköhler number II)}$$

where the Damköhler numbers may be taken as a measure for the tendency for reaction to the tendency for transport. Damköhler number I is important at high Peclet numbers, and Damköhler number II is important at low Peclet numbers. Note the ratio of number II to number I reduces to the Peclet number.

Let us now examine some rate laws and show how they are included in mass transport equations. Before proceeding, however, we will review the terminology used to organize the discussion. Reactions are classified as homogeneous, operating within a single phase, or heterogeneous, operating between two phases. In addition, reactions can be described from an equilibrium or kinetic viewpoint depending on the rate of the reaction relative to the mass transport process.

First-Order Kinetic Reactions

One example of a simple kinetic reaction is the first-order decay of a constituent due to radioactive decay, biodegradation, or hydrolysis.

$$r = \frac{d(nC)}{dt} = -\lambda nC \qquad (14.15)$$

where λ is the decay constant for radioactive decay or some reaction rate coefficient for biodegradation or hydrolysis. In all cases, λ has units of time^{-1} and is the volume reacted (or disintegrated) per unit volume per unit time. With this formulation, the one-dimensional transport equation becomes

$$D_x \frac{\partial^2 C}{\partial x^2} - v_x \frac{\partial C}{\partial x} - \lambda C = \frac{\partial C}{\partial t} \qquad (14.16)$$

Equilibrium Sorption Reactions

An example of a heterogeneous equilibrium reaction is the sorption of mass from solution. The rate law is

$$r = \frac{\partial C^*}{\partial t} \qquad (14.17)$$

where C^* is the concentration of the solute on the solid phase. The one-dimensional transport equation that incorporates this reaction is

$$D_x \frac{\partial^2 C}{\partial x^2} - v_x \frac{\partial C}{\partial x} - \frac{1}{n} \frac{\partial C^*}{\partial t} = \frac{\partial C}{\partial t} \qquad (14.18)$$

The rate law incorporated in Eq. 14.18 is a general one, appropriate for both equilibrum and kinetic sorption reactions. A solution to this equation requires that a specific rate law be incorporated. For kinetic nonequilibrium sorption reactions,

$$\frac{\partial C^*}{\partial t} = f(C, C^*) \qquad (14.19)$$

or, in words, the rate of sorption is a function of both the concentration of mass in solution and the mass sorbed on the solid. For equilibrium sorption reactions,

$$C^* = f(C) \qquad (14.20)$$

Because the concentration of sorbed mass is a function of mass in solution, the $\partial C^*/\partial t$ term in Eq. 14.19 can be expressed in a more tractable $\partial C/\partial t$ term. This substitution produces a single differential equation containing one dependent variable, which is solvable by analytical methods.

The following derivation explains how the transport equation for the simple case of linear sorption is obtained. The rate of mass sorption per unit volume of porous medium is

$$\frac{\partial C^*}{\partial t} = \rho_b \frac{\partial S}{\partial t} \qquad (14.21)$$

where S is the quantity of mass sorbed on the surface and ρ_b is the bulk density. The bulk density can also be defined as

$$\rho_b = \rho_s(1 - n) \qquad (14.22)$$

where ρ_s is the mass density of the minerals making up the rock or soil, normally 2.65 g/cm^3 for most sandy soils. The quantity $\rho_s(1 - n)$ is the total mass of solids per unit volume of porous medium.

What now is required is an expression for an equilibrium sorption isotherm. Although several are available, we will use the simplest (and likely most useful) one, the linear Freundlich isotherm. By taking this isotherm and differentiating with respect to time, we obtain

$$\frac{\partial S}{\partial t} = K_d \frac{\partial C}{\partial t} \qquad (14.23)$$

where K_d is the distribution coefficient. Combining Eqs. 14.21, 14.22, and 14.23 gives

$$\frac{\partial C^*}{\partial t} = (1 - n)\rho_s K_d \frac{\partial C}{\partial t} \qquad (14.24)$$

which is a function of the fluid concentration alone. Substitution of Eq. 14.24 into the general transport equation and rearranging terms give

$$D_x \frac{\partial^2 C}{\partial x^2} - v_x \frac{\partial C}{\partial x} = \frac{\partial C}{\partial t}\left[1 + \frac{(1 - n)}{n}\rho_s K_d\right] \qquad (14.25)$$

The bracketed quantity on the right-hand side is a constant termed the *retardation factor, R_f*, or

$$R_f = \left[1 + \frac{(1 - n)}{n}\rho_s K_d\right] \qquad (14.26)$$

Equation 14.25 becomes

$$\frac{D_x}{R_f}\frac{\partial^2 C}{\partial x^2} - \frac{v_x}{R_f}\frac{\partial C}{\partial x} = \frac{\partial C}{\partial t} \qquad (14.27)$$

Thus, the retardation factor merely serves to decrease the values of the transport parameters D and v. As local equilibrium implies that the net rate of reaction is zero, we anticipate similar effects for other equilibrium-controlled isotherms of the form $f(C)$ although the mathematics do not work out so simply. Aris and Amundson (1973) cite several other forms of sorption isotherms, which all have this property. Equilibrium-controlled reactions form a basis for much of the modeling conducted in contaminant hydrogeology. Similar assumptions have also been employed by geochemists in models describing the diagenesis of marine sediments (Berner, 1974), by chemical engineers interested in the effects of diffusion and longitudinal dispersion in ion exchange and chromatographic columns (Lapidus and Amundson, 1952; Kasten and others, 1952), and by soil scientists interested in the effects of dispersion and ion exchange in porous media (Lai and Jurinak, 1971; Rubin and James, 1973).

Heterogeneous Kinetic Reactions

The general case of a reaction between a solid and a solution involves several steps in series, with the slowest being rate controlling. The rate-determining step can be transport controlled or surface controlled, depending on the relative magnitude between the rate at which the mass moves to or reacts on the surface.

The starting point for discussing transport-controlled reactions is a model of the reaction process. Assume that a thin stationary layer occurs at the interface between a solid and the bulk fluid and that diffusion is slow relative to actual reactions at the surface so that the coming and going of atoms is rate controlling. If the process of transport across the thin layer is diffusion, there must be a concentration gradient. This gradient can be assumed linear and represented as $(C_{eq}^* - C)/\sigma$ where C_{eq}^* is the saturation concentration at the solid surface, C is the concentration in the bulk fluid, and σ is the thickness of the stationary layer separating the two. The flux J across the surface in moles vol^{-1} time^{-1} is

$$J = \frac{D_d A(C_{eq}^* - C)}{\sigma} \qquad (14.28)$$

where A is the surface area of the material per unit volume of water and D_d is the diffusion coefficient. The ratio D_d/σ is the mass transfer coefficient k_m and has the units of velocity. In this form, the rate of dissolution of the solid is the product of a driving force, which is the departure from saturation $(C_{eq}^* - C)$, a mass transfer coefficient, and a surface area of the material per unit volume. Diffusion-controlled reactions of the form represented in Eq. 14.28 can be expressed as

$$\frac{1}{n}\frac{\partial C^*}{\partial t} = \frac{D_d S^*}{\sigma}(C_{eq}^* - C) \qquad (14.29)$$

where S^* is the specific surface of the porous medium (Domenico, 1977). It is possible to include Eq. 14.29 in a mass transport model.

Now let us consider the surface-controlled reactions. The best examples of these reactions are the sorption and desorption reactions. According to Klotz (1946), one of the earliest formulations assumes that the irreversible local rate of sorption is

$$\frac{1}{n}\frac{\partial C^*}{\partial t} = k_1 C(C_{eq}^* - C^*) \qquad (14.30)$$

where k_1 is a constant and C_{eq}^* is the saturation concentration on the solid. Thomas (1944) considered the form

$$\frac{1}{n}\frac{\partial C^*}{\partial t} = k_1 C(C_{eq}^* - C^*) - k_2 C^* \qquad (14.31)$$

where k_2 is a constant. For a porous medium, the following equation describes a reversible sorption process:

$$\frac{1}{n}\frac{\partial C^*}{\partial t} = S^*(k_1 C - k_2 C^*) \qquad (14.32)$$

where k_1 is a forward (sorption) rate constant and k_2 is a backward (desorption) rate constant with units of velocity. This equation states that sorption and desorption occur simultaneously and at different rates. This equation can be rewritten as

$$\frac{1}{n}\frac{\partial C^*}{\partial t} = S^* k_1 \left(C - \frac{k_2}{k_1} C^* \right) = S^* k_1 (C - C_{eq}^*) \qquad (14.33)$$

for the case where $(k_1/k_2)C^* = C_{eq}^*$. This equation is of the same form as the diffusion-controlled reaction discussed earlier. Here the rate constant k_1 has the units LT^{-1} and the specific surface has the units L^{-1} so that the combined term has the units of T^{-1}. With diffusion-controlled reactions, the combined parameter D_d/σ is referred to as the mass transfer coefficient LT^{-1} and is equivalent to k_1 in Eq. 14.33. The combined parameter DS^*/σ has the units T^{-1}.

The use of the reversible equation for sorption is widespread in transport problems of an academic nature. Transport equations incorporating this term have been solved first for the condition of advection only (Amundson, 1950), and later for advection and dispersion (Lapidus and Amundson, 1952). Deju (1971), supported by earlier work (Deju and Bhappu, 1970), used this form of adsorption to describe the chemical weathering of sediments. More recent theoretical studies dealing with sorption in relation to contamination problems includes the work of Relyea (1982), Melnyk and others (1983), Valocchi (1985) and Bahr and Rubin (1987). However, kinetic models for sorption are rare in field applications because of the difficulty in determining the rate expressions.

14.3 Boundary and Initial Conditions

Solving time-dependent differential equations requires boundary and initial conditions. The task of the boundary conditions is to account for the effects of the system outside of the region of interest on the system being modeled. Once the boundary conditions are specified, the outside effects are accounted for and need not be directly modeled. The boundary conditions discussed earlier for flow have their counterparts in relation to mass transport. The first boundary condition is a specified concentration along a given boundary. An important special case for this boundary condition is where the specified concentration is constant, termed a *constant concentration boundary*. A second type of boundary is where the normal component of the concentration gradient is specified. Given a normal component of the concentration gradient along a boundary, there must be a diffusive flux across the boundary. Thus, we use the term *mass flux boundary* where mass enters or leaves a region across a boundary. One special type of mass flux boundary is a no-flux boundary, that is, a region across which no mass is permitted to leave. Expressed mathematically

$$\frac{\partial C}{\partial N} = 0 \qquad (14.34)$$

where N is the normal to the boundary. This means there is no concentration gradient normal to the boundaries so mass cannot leave the region. A second special type of mass flux boundary is a variable-flux boundary.

The one-dimensional problem in Figure 14.1a, involving contaminant migration in an aquifer away from a long narrow ditch, provides an example of boundary conditions. The contaminants enter the system at an upper boundary at $x = 0$. A lower boundary not shown in

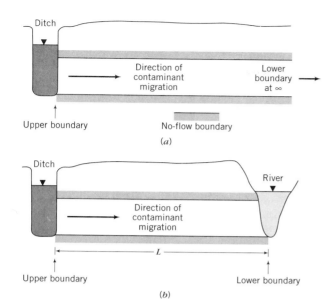

Figure 14.1 Contaminant migration in a one-dimensional system that is (*a*) semi-infinite or (*b*) finite in length.

the figure is located far down the system at $x = \infty$. In this figure, the source (ditch) can be depicted as a constant concentration boundary and the lower impermeable bottom as a no-flux boundary. A right-hand boundary need not be specified as the boundary is too far removed from the interior to have any effect. The location of this boundary makes the system semi-infinite. These same conditions for the ditch and impermeable bottom boundary apply in Figure 14.1*b*, but now something must be said about the right-hand side boundary at the river. A con-

stant concentration or constant mass flux boundary overspecifies what the final result will look like so that a variable-flux boundary is called for and is of the form

$$vC - \frac{D\,\partial C}{\partial x} \qquad (14.35)$$

This equation states that at the exit the rate at which some contaminant leaves the system must be equal to the combined rate of advection and dispersion by which the species is transported through the system.

In many mass transport problems (e.g., Figures 14.1*a*, *b*), contaminants enter the system through one or several boundaries. This loading can be specified using either a fixed concentration or a variable-flux boundary condition. Equation 14.35 is an example of a variable-flux condition. A constant-concentration condition for a one-dimensional case is

$$C(0, t) = g(t) \qquad (14.36)$$

This equation indicates that at the boundary ($x = 0$) the concentration is given for all time as a function of time.

Examples of some commonly used boundary conditions for one-dimensional systems are listed on Table 14.1. The second condition shown on the table is an example of pulse loading. The boundary condition changes after some time (t_0) has elapsed. Thus, it is necessary to provide two boundary conditions and the time for which each applies.

An initial condition describes the distribution of a contaminant within the domain when the simulation begins (time = 0). For a one-dimensional problem, a general form of an initial condition is written

$$C(x, 0) = f(x) \qquad (t = 0) \qquad (14.37)$$

Table 14.1 Examples of Some Boundary Conditions That Are Used to Add Contaminants to a One-Dimensional System

Name	Type	Form
Constant concentration	Fixed concentration	$C(0, t) = C_o$
Pulse-type loading with constant concentrations	Fixed concentration	$C(0, t) = \begin{cases} C_o, 0 < t \le t_o \\ 0, t > t_o \end{cases}$
Exponential decay with source concentration $\to 0$	Fixed concentration	$C(0, t) = C_o e^{-\alpha t}$
Exponential decay with source concentration $\to C_a$	Fixed concentration	$C(0, t) = C_a + C_b e^{-\alpha t}$
Constant flux with constant input concentration	Variable flux	$\left(-D\dfrac{dc}{dx} + vC\right)\Big|_{x=0} = vC_o$
Pulse-type loading with constant input fluxes	Variable flux	$\left(-D\dfrac{dc}{dx} + vC\right)\Big|_{x=0} = vC_o, \ 0 < t \le t_o$
		$\left(-D\dfrac{dc}{dx} + vC\right)\Big|_{x=0} = 0, \ t > t_o$

C_o, C_a, C_b = various constant concentrations.
$\quad \alpha$ = decay constant.
$\quad t_o$ = time at which concentration changes due to pulse loading.

where $f(x)$ is a function describing the one-dimensional variation in concentration in the x direction. Some common initial conditions are $C(0, x) = 0$, $C(0, x) = C_i$, which provide for a constant concentration along the system.

Formulating the transport problem in terms of equations and boundary conditions is the first step in modeling mass transport. The step of solving the resulting equation using various modeling approaches will be taken up in Chapter 18.

Problems

1. Consider one-dimensional steady-state diffusion across a layer of thickness L where the concentration C at the bottom ($x = 0$) and at the top ($x = L$) is known. Solve this problem for the steady-state concentration as a function of x, L, C_0, and C_L. (*Hint:* See Example 4.3.)

2. Provide a dimensional analysis of the steady-state version of Eq. 14.14 where r/n is replaced by kC, and obtain two Damköhler numbers.

3. The diffusion coefficient has units of L^2/T. Can you name two other parameters or groups of parameters that have these same units, one treating fluid flow and the other for the transport of heat by conduction?

4. Compare the Fourier number of Eq. 14.12 with the Fourier number for fluid flow (Eq. 4.83) and the Fourier number for heat conduction (Eq. 9.22). State the analogous quantities and provide an interpretation of T^* for mass transport by diffusion.

Chapter 15

Mass Transport in Natural Ground-Water Systems

The mass transport processes and chemical concepts presented in the last four chapters provide a framework for studying the chemically related problems in hydrogeology. This general framework applies equally well to a variety of core hydrogeological problems like the chemical evolution of ground water along flow systems or ground-water contamination. Moreover, this process approach lets us address other important problems like the diagenesis of carbonate rocks, karst formation, the origin of some kinds of ore deposits, and the secondary migration of petroleum, which historically have resided in other areas of geology.

This chapter examines the processes that control the chemical evolution of ground water and some manifestations of these processes in relation to the geologic "work" of ground water. This chapter also examines approaches used in the dating of ground water.

15.1 Mixing as an Agent for Chemical Change

Solutes and stable isotopes are affected by advection and dispersion. These processes become apparent along a flow system when a water of one chemical composition displaces or simply ends up next to water with a different composition. Mechanical dispersion and diffusion create a zone of mixing between the two waters. Accordingly, phenomena thought of most often in relation to laboratory columns operate on a much larger scale and strongly influence the chemistry of natural ground waters.

The Mixing of Meteoric and Original Formation Waters

There are several reasons why ground-water mixing can occur. A common situation is related to large-scale structural changes in a sedimentary basin when, following uplift, meteoric water has the opportunity to flush units that previously contained connate water or formation water. Meinzer (1923) defined connate water as water that is trapped in the pores of a sediment during deposition. For marine deposits, this suggests formation water of the composition of seawater. Whether or not this composition is retained up to the beginning of burial depends strongly on the depositional environment as well as numerous other geologic factors. For example, Land and Prezbindowski (1981) point out that the Edwards Formation was massively altered by meteoric diagenesis prior to burial, where the original connate water was first mixed with several pore volumes of meteoric water.

Whatever its chemical composition at the time of burial, the displacement of formation water will generally begin when the following two conditions are satisfied: (1) the permeable unit is uplifted so that its outcrop occupies a position that permits recharge by meteoric water, and (2) the down-dip portions of the unit have outlets through which the formation water can be displaced. When permeable beds pinch out with depth, definite outlets are not readily available and the water must escape through overlying less permeable formations.

The typical pattern is for fresh, unaltered meteoric water to be found in and near regions of outcrop with progressive increases in salinity down the flow system. Fresh water or "noses" of relatively fresh water grading into saline water have been noted quite frequently. Some examples include the Madison Formation west and south of the Williston Basin (Downey, 1984), the Inyan Kara Formation in southwestern North Dakota (Butler, 1984), and the Dakota–Newcastle Formation of the Dakota Sandstone in western South Dakota (Peter, 1984). These formations have limited recharge areas that coincide with the uplifted portions of the aquifers.

Two field examples of the development of freshwater noses are shown in Figures 15.1 and 15.2. Figure 15.1 from Downey (1984) demonstrates how freshwater is beginning to invade regions around areas of limited outcrop in western United States. Figure 15.2 from Mitsdarffer (1985) shows a well-developed freshwater nose in the Mission Canyon Formation in the Williston basin with the source of recharge probably the Black Hills of South Dakota. An interesting question is whether the entire freshwater portions of the Dakota Sandstone represent an immense freshwater nose surrounded by more saline waters in the neighboring basins.

Another type of mixing occurs whenever brines are

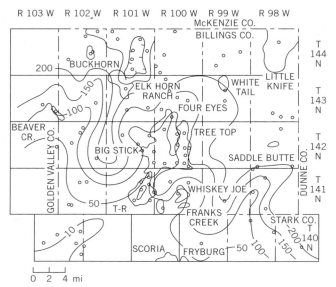

Figure 15.2 Dissolved solids concentration for the Mission Canyon Formation, in parts per thousand (after Mitsdarffer, 1985).

forced upward in a compaction-driven flow system. An example of this mixing occurs in the Gulf of Mexico basin (Bethke and others, 1988), where abnormal fluid pressure is driving pore fluids out of the basin. In the last two million years, when the effects of overpressuring have been most pronounced, brine migration has apparently pushed back an apparently much more extensive freshwater incursion into the basin (Bethke and others, 1988).

The implication of this process in terms of groundwater chemistry is that brine can move into parts of a basin where it otherwise may not have formed. An example is the Na-Ca-Cl brine found in Lower Cretaceous shelf carbonates of the Edwards Group in south-central Texas (Land and Prezbindowski, 1981). This brine may have formed in the overpressured portion of the Gulf of Mexico basin from evaporites known to occur there, and subsequently migrated into shallower, up-dip parts of the Edwards Group.

Not all mixing in ground water is related to the large-scale tectonic deformation of sedimentary basins. In units having a low hydraulic conductivity, the porewater can often be relatively old and chemically different from modern-day recharge. Examples of such materials are the clay-rich tills found in the northern part of the United States and in Canada (Desaulniers and others, 1981; Bradbury, 1984). These deposits often have hydraulic conductivities of 10^{-8} cm/s and lower. Because of the relatively short time since deposition, they can contain water incorporated when the till was deposited. Even where this is not the case, the low permeability of till means that any long-term changes in the chemistry of recharge will be

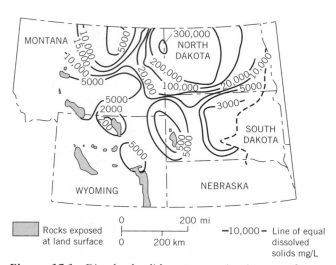

Figure 15.1 Dissolved solids concentration in water from the Cambrian–Ordovician aquifer (after Downey, 1984).

reflected in the chemical patterns along the flow system because of the age differences of the water. Section 15.5 describes a site in Wisconsin where changing values of δD and $\delta^{18}O$ document significant differences in the isotopic composition of modern-day recharge and original pore water, possibly present when the glacial till was deposited about 10,000 years ago (Bradbury, 1984).

Diffusion in Deep Sedimentary Environments

Diffusion has been described as the transport of mass in response to concentration gradients. As the diffusive flux is proportional to the concentration gradient, this process can cause significant mass redistribution whenever large concentration gradients can occur and can be maintained over geologic time. Thus, this process should be most effective where zones of hypersalinity underlie brackish water. Diffusion acts to eliminate the concentration gradient and, thus, becomes less significant as this equalization occurs.

The occurrence of brines in sedimentary basins has been attributed to three causes: dissolution of original evaporites, retention of original formation waters buried with the sediment, or membrane filtration. Whatever the mechanism, the resulting concentration gradients can give rise to significant mass redistribution over geologic

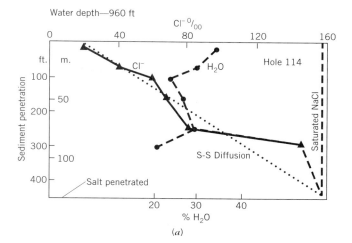

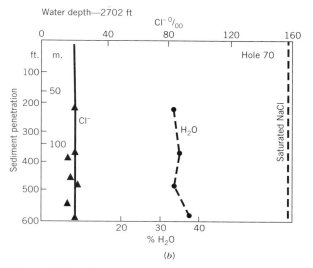

Figure 15.4 Plots of depth versus chlorinity and water content in shallow sediments of the Gulf of Mexico. (*a*) Rock salt is encountered at 433 feet. The dashed line represents the distribution of Cl⁻ at steady state due to diffusion between two boundaries of constant Cl⁻. (*b*) Away from diapiric structures there is no significant variation in Cl⁻ concentration with depth (from Manheim and Bischoff, 1969). Reprinted from Chem. Geol., v. 4, by F. T. Manheim and J. F. Bischoff, Geochemistry of pore waters from Shell Oil Company drill holes on the continental slope of the northern Gulf of Mexico, p. 63–82, 1969, with kind permission from Elsevier Science, NL, Sara Burgerhartstraat 25, 1055 KV Amsterdam, The Netherlands.

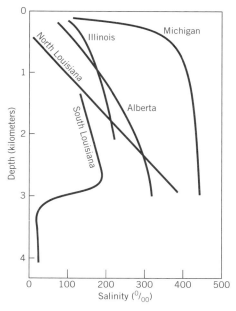

Figure 15.3 Variation in maximum porewater salinity as a function of depth in the Alberta, Illinois, and Michigan basins as well as the Louisiana Gulf Coast (from Ranganathan and Hanor, 1987). Reprinted from J. Hydrol., v. 92, by V. Ranganathan and J. S. Hanor, A numerical model for the formation of saline waters due to diffusion of dissolved NaCl in subsiding sedimentary basins with evaporites, p. 97–120, 1987, with kind permission from Elsevier Science, NL, Sara Burgerhartstraat 25, 1055 KV Amsterdam, The Netherlands.

time. The common pattern that emerges is a consistent increase in porewater salinity (mainly Na^+ and Cl^-) with depth (Figure 15.3). The data from north Louisiana, in particular, show a linear increase in NaCl over 2700 m to a depth at which salt is known to occur.

One interpretation as to how these chemical patterns have evolved comes from theoretical studies by Ranganathan and Hanor (1987) and field studies by Manheim and Bischoff (1969) and Manheim and Paull (1981). Ranganathan and Hanor (1987) solved one-dimensional forms of the diffusion and diffusion–advection equation with realistic sets of transport parameters and concluded that salinity–depth relationships develop over several tens of millions of years that are similar to those observed in a variety of sedimentary basins. Manheim and Bischoff (1969) examined the porewater composition from six holes on the continental slope of the Gulf of Mexico and showed how the presence or absence of salt diapirs affected porewater salinity. Samples from holes drilled near salt plugs showed systematic increases in salinity with depth that was interpreted as being caused by salt diffusion (Figure 15.4*a*). The development of the diffusion profile in Figure 15.4*a* required about 400,000 years. Samples from holes drilled away from diapiric structures showed little change in chemistry as compared to seawater except for minor diagenetic changes and a loss of SO_4^{2-} (Figure 15.4*b*). Similar studies along the southeastern Atlantic coast of the United States showed the same result over a greater range in depth (Manheim and Paull, 1981).

15.2   Chemical Reactions in the Unsaturated Zone

In the unsaturated zone, it is mainly the mass transfer processes that ultimately exert the greatest control over the concentrations of major and minor ions and organic compounds in ground water. Almost all the reactions that we discussed in Chapter 12 operate to some extent. Following is a summary of the important chemical and biological processes affecting the inorganic and organic constituents.

(1)   gas dissolution and redistribution

$$CO_2(g) + H_2O = H_2CO_3^*$$

$$H_2CO_3^* = HCO_3^- + H^+$$

$$HCO_3^- = CO_3^{2-} + H^+$$

(2)   weak acid–strong base reactions, for example,

calcite: $CaCO_3(s) + H^+ = Ca^{2+} + HCO_3^-$

anorthite: $CaAl_2Si_3O_8(s) + 2H^+ + H_2O = $ kaolinite $+ Ca^{2+}$

albite: $2NaAlSi_3O_8(s) + 2H^+ + 5H_2O = $ kaolinite $+ 4H_2SiO_3 + 2Na^+$

enstatite: $MgSiO_3(s) + 2H^+ = Mg^{2+} + H_2SiO_3$

(3)   sulfide mineral oxidation, for example,

$$4FeS_2(s) + 15O_2 + 14H_2O = 4Fe(OH)_3 + 16H^+ + 8SO_4^{2-}$$

(4)   precipitation/dissolution of gypsum

$$CaSO_4 \cdot 2H_2O(s) = Ca^{2+} + SO_4^{2-} + 2H_2O$$

(5)   cation exchange, for example,

$$Ca^{2+} + 2Na\text{-}X = 2Na^+ + Ca\text{-}X$$

$$Ca^{2+} + Mg\text{-}X = Mg^{2+} + Ca\text{-}X$$

$$Mg^{2+} + 2Na\text{-}X = 2Na^+ + Mg\text{-}X$$

where Na-X is Na adsorbed onto a clay mineral.

(6)   organic reactions

This list is generally the conceptual model for unsaturated zone processes proposed by Moran and others (1978a). While their model was not intended to be general, combinations of these processes explain the chemistry of water in different settings.

(1) Gas Dissolution and Redistribution

The dissolution and redistribution of $CO_2(g)$ are important soil-zone processes. Rainwater or melted snow contains relatively small quantities of mass, is somewhat acidic, and has a P_{CO_2} of about $10^{-3.5}$ atm. As this water moves downward, it rapidly dissolves CO_2; this process occurs in soil at partial pressures larger than the atmospheric value. Elevated CO_2 pressures are due primarily to root and microbial respiration and to a lesser extent the oxidation of organic matter (Palmer and Cherry, 1984). Values for soils range generally from $10^{-3.5}$ atm to more than 500 times larger (Palmer and Cherry, 1984). CO_2 dissolved in water is further redistributed among the weak acids of the carbonate system. For a pH range of 4.5 to 5.5, $H_2CO_3^*$ is the dominant carbonate species with HCO_3^- and H^+ the most dominant anion and cation, respectively.

One direct result of dissolving $CO_2(g)$ in water is a rapid increase in the total carbonate content of the water and a decrease in pH. Calculations show the extent to which the P_{CO_2} in the unsaturated zone influences the

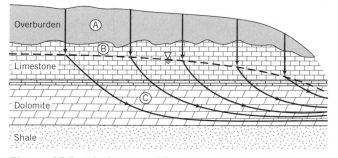

Figure 15.5   Idealization of flow through a hypothetical carbonate sequence. The circled letters are reference points for calculations discussed in the text (modified from Palmer and Cherry, 1984). Reprinted from J. Hydrol., v. 75, by C. D. Palmer and J. A. Cherry, Geochemical evolution of groundwater in sequences of sedimentary rocks, p. 27–65, 1984, with kind permission from Elsevier Science, NL, Sara Burgerhartstraat 25, 1055 KV Amsterdam, The Netherlands.

Table 15.1 **Calculated pH and Total Carbonate Concentration for Water in Equilibrium with Specified P_{CO_2} Values**

Partial Pressure CO_2 (atm)	pH	Total Carbonate[a] (mM)
$10^{-3.5}$	5.62	0.019
$10^{-2.5}$	5.12	0.178
$10^{-1.5}$	4.62	1.72

[a] $(H_2CO_3^*)$ + (HCO_3^-) + (CO_3^{2-}) in millimoles per liter; temperature = 10°C.

Modified from Palmer and Cherry (1984). Reprinted from J. Hydrol., v. 75, by C. D. Palmer and J. A. Cherry, "Geochemical evolution of groundwater in sequences of sedimentary rocks," p. 27–65, 1984 with kind permission from Elsevier Science, NL, Sara Burgerhartstraat 25, 1055 KV Amsterdam, The Netherlands.

evolving porewater chemistry (Palmer and Cherry, 1984). Consider the hypothetical system in Figure 15.5. In moving to position *A* (Figure 15.5), the only process is the open system dissolution of CO_2 (P_{CO_2} remains fixed). As expected, increasing the P_{CO_2} from $10^{-3.5}$, to $10^{-2.5}$ to $10^{-1.5}$ atm increases the total amount of carbonate in solution by about two orders of magnitude and lowers the pH from 5.62 to 4.62 across the range of partial pressures (Table 15.1).

Another important soil zone–atmospheric process is the dissolution of $O_2(g)$. The resulting levels of dissolved oxygen are sufficiently large at least initially to control the redox chemistry in shallow ground water.

(2) Weak Acid–Strong Base Reactions

CO_2-charged ground water is effective in dissolving minerals. The most common reactions involve the weak acids of the carbonate and silicate systems and strong bases from the dissolution of carbonate, silicate, and aluminosilicate minerals. As we showed in Section 12.2, this process causes the weak acids to dissociate. In the carbonate system, the relative abundance of HCO_3^- and

CO_3^{2-} increases at the expense of $H_2CO_3^*$. Overall both the alkalinity and cation concentrations increase.

Calculations again illustrate these effects (Palmer and Cherry, 1984). Referring to Figure 15.5, the point of interest is now *B*, where water encounters limestone bedrock. The system is assumed open (i.e., the P_{CO_2} remains fixed) with sufficient residence time to be at saturation with respect to calcite. P_{CO_2} again is a key variable in determining the dissolved carbonate content, the total Ca content (mainly Ca^{2+}), and the total quantity of calcite dissolved to reach equilibrium. As the P_{CO_2} is increased from $10^{-3.5}$ to $10^{-2.5}$ to $10^{-1.5}$, pH decreases from 8.30 to 7.64 to 6.99 (Table 15.2). The dissolved carbonate content (mainly HCO_3^-) and total Ca increase. The tendency for weak acids to ionize is reflected by the dominance of HCO_3^-.

This simple system involves the dissolution of only one mineral. Langmuir (1971) calculated how HCO_3^- and pH change as the water approaches equilibrium with respect to a second mineral (dolomite). For a system that is open with respect to CO_2, water evolves to equilibrium with calcite along lines of constant P_{CO_2} (Figure 15.6). For example, moving along the P_{CO_2} line of $10^{-2.5}$ atm, the pH of a solution in equilibrium with calcite is 7.65. The second solubility curve labeled "dolomite" illustrates the pH and HCO_3^- concentration for soil water in equilibrium with dolomite alone or calcite and dolomite together. At the temperature of calculation, the solubility curves for dolomite and dolomite plus calcite are the same (Langmuir, 1971).

Another feature of this diagram is the series of paths labeled "no CO_2 added." These lines describe how the concentration of HCO_3^- and pH change under closed system conditions. In calculating these pathways, the P_{CO_2} decreases as CO_2 is depleted by the dissolution of carbonates. Whether the system is open or closed with respect to CO_2 can have an important bearing on the porewater chemistry. For example, an open system (P_{CO_2} = 10^{-2} atm) has a HCO_3^- concentration of about 260 ppm and pH of about 7.34 at calcite saturation (Figure 15.6).

Table 15.2 **Calculated pH, Total Calcium Concentration, Total Carbonate Concentration and Amount of Calcite Reacted in Water at Equilibrium with Respect to Calcite and with Specified P_{CO_2} Values**

Partial Pressure CO_2 (atm)	pH	Total Ca (mM)	Total CO_3 (mM)	Calcite[b] Reacted (mM)
$10^{-3.5}$	8.30	0.628	1.26	0.628
$10^{-2.5}$	7.64	1.40	2.96	1.40
$10^{-1.5}$	6.99	3.22	8.13	3.22

[a] $(H_2CO_3^*)$ + (HCO_3^-) + (CO_3^{2-}) in millimoles per liter.
[b] Amount of calcite dissolved in the unsaturated zone; temperature = 10°C.

Modified from Palmer and Cherry (1984). Reprinted from J. Hydrol., v. 75, by C. D. Palmer and J. A. Cherry, "Geochemical evolution of groundwater in sequences of sedimentary rocks," p. 27–65, 1984 with kind permission from Elsevier Science, NL, Sara Burgerhartstraat 25, 1055 KV Amsterdam, The Netherlands.

Under closed system assumptions, these values are 170 ppm and 7.70, respectively.

A model involving just the two processes considered so far, the dissolution of CO_2 gas accompanied by calcite dissolution, describes the chemistry of water in the unsaturated zone in carbonate terranes. Shown in Table 15.3 is a chemical analysis for water from a near-surface spring in a limestone in central Pennsylvania (Langmuir, 1971). This analysis probably approximates the composition of water from the unsaturated zone. The most abundant ions are Ca^{2+} and HCO_3^-, with relatively small concentrations of Na^+ and Cl^-. The calculated Pco_2 for this sample is $10^{-2.08}$ atm and like nearly all of the spring water is undersaturated with respect to calcite and dolomite. Presumably, transport through fractures has been so rapid that calcite equilibrium was not achieved. Farther along the flow system, calcite equilibrium finally constrains increases in Ca^{2+} and HCO_3^- concentrations.

As our sample list of weak acid–strong base reactions shows, silicate and alumino-silicate minerals will also react to some extent when they are present. The relatively low solubility of these minerals, however, means that their contribution to the mass dissolved in porewater will be relatively small when soluble minerals like calcite are present. In systems where there are no carbonates present, these reactions control the ion chemistry.

Garrels and MacKenzie (1967) looked at a case where waters rich in CO_2 reacted with a suite of minerals found in igneous rock (Table 15.3). Again by using water from ephemeral springs as a surrogate for soil water, we can observe the behavior of a system with no carbonate min-

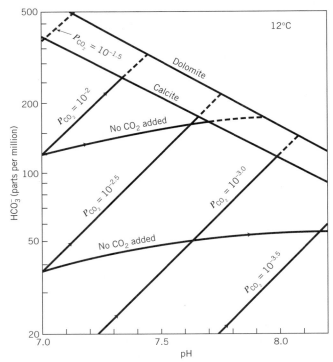

Figure 15.6 Possible approaches to equilibrium of ground water in contact with calcite and (or) dolomite at 12°C. The solubility curves of dolomite and calcite plus dolomite are the same at this temperature. Reprinted from D. Langmuir, The Geochemistry of some carbonate ground waters in central Pennsylvania, Copyright © 1971, p. 1023–1045, with kind permission from Elsevier Science Ltd, The Boulevard, Langford Lane, Kidlington 0X5 1GB, UK.

Table 15.3 Comparison of the Chemistry of Waters from Four Different Recharge Environments and the Unsaturated Zone Processes Giving Rise to These Waters

Study	Ca^{2+}	Mg^{2+}	Na^+	K^+	HCO_3^-	SO_4^{2-}	Cl^-	pH	log Pco_2	SI_c	SI_d
1. Langmuir (1971)	41.	4.4	6.5	2.1	134	—	12	7.13	−2.08	−0.78	−1.16
2. Garrels and MacKenzie (1967)[a]	3.1	0.7	3.0	1.1	20	1.0	0.5	6.2	−1.8	—	—
3. Kimball (1984)	57	21	26	0.4	210	57	2.6	7.9	−2.57	0.26	0.14
4. Moran et al. (1978b)[a]	62	25	469	7.9	748	577	4.7	8.0	—	—	—

1. Processes	2. Processes	3. Processes	4. Processes
• CO_2 dissolution and redistribution	• CO_2 dissolution and redistribution	• CO_2 dissolution and redistribution	• CO_2 dissolution and redistribution
• weak acid–strong base calcite	• weak acid–strong base plagioclase biotite K-feldspar	• O_2 dissolution	• O_2 dissolution
		• weak acid–strong base calcite dolomite albite	• weak acid–strong base calcite
		• pyrite oxidation	• pyrite oxidation
			• precipitation–dissolution of gypsum
			• cation exchange

Notes: Concentration in milligrams per liter.
Partial pressures in atmospheres.
SI, saturation indices for calcite and dolomite.
[a] Mean values for several samples.

erals (Table 15.3). The concentration of all ions is low because only plagioclase feldspar, biotite, and K-feldspar dissolve. In fact the concentration of most ions is so low that the initial composition of precipitation had to be considered when Garrels and MacKenzie worked with the data. Dissolution apparently takes place under closed-system conditions with about half of the CO_2 consumed.

(3) Sulfide Oxidation

Sulfide oxidation is one of the important redox reactions within the unsaturated zone. Minerals like pyrite or marcasite are oxidized to produce $Fe(OH)_3(s)$, SO_4^{2-}, and H^+. In fact, pyrite oxidation is one of the most important acid-producing reactions in geologic systems (Moran and others, 1978a). In coal mining areas like the Appalachians, this reaction can be the cause of serious acid-mine drainage problems (Moran and others, 1978a).

An example of how various chemical reactions, including pyrite oxidation, control recharge chemistry is a study by Kimball (1984) in the Piceance Creek basin of northwestern Colorado. The initial acquisition of solutes in the recharge to the Uinta Formation is controlled by three processes considered so far, namely: open-system dissolution of CO_2 and O_2, weak acid–strong base reactions, and sulfide mineral oxidation (Table 15.3). The main difference as compared to the essentially monomineralic system described by Langmuir (1971) is that the dissolution of dolomite and albite, and the oxidation of pyrite, add Mg^{2+}, Na^+, and SO_4^{2-} to the water (Table 15.3).

(4) Gypsum Precipitation and Dissolution

Conditions exist when the solute load in soil waters can be much higher than we have considered so far. The cyclical precipitation and dissolution of gypsum is probably the most important process in this respect. Over much of the Plains region of the United States and Canada, annual potential evaporation exceeds annual precipitation by a considerable amount. Thus, water that infiltrates in normal precipitation years evaporates and deposits a small quantity of gypsum. With repeated rain or snowmelt, gypsum accumulates in the upper part of the soil horizon (Moran and others, 1978a). Exceptional recharge can dissolve some of this soluble material and move it down and into the ground-water system (Figure 15.7). In some arid areas, recharge water could have SO_4^{2-} concentrations in excess of 5000 mg/L (Hendry and others, 1986).

(5) Cation Exchange

The precipitation and dissolution of gypsum in the Great Plains region is also accompanied by cation exchange (Moran and others, 1978a). The most important ex-

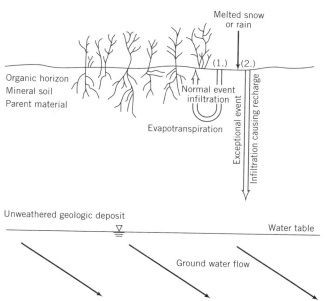

Figure 15.7 A conceptual model of subsurface flow in the Plains region of the United States and Canada. (1) Annual potential evapotranspiration greatly exceeds annual precipitation, and most infiltration is lost by evapotranspiration. (2) Exceptional precipitation events produce recharge (from Moran and others, 1978a).

change reactions are the water-softening reactions, where Ca^{2+} and Mg^{2+} in the water exchange with sorbed Na^+ as ground water moves through clayey material. Shallow ground water from a study area in North Dakota (Moran and others, 1978a) provides an example of how all of the processes we have considered so far work together. The chemical data on Table 15.3 represent mean values for 39 water samples collected from various near-surface drift units. The abundance of Na^+ and SO_4^{2-} in the shallow ground water is attributed to gypsum dissolution, sulfide oxidation, and cation exchange. The cation exchange process is important here because it has effectively increased Na^+ concentrations at the expense of Ca^{2+} and Mg^{2+}.

(6) Organic Reactions

Work by Wallis and others (1981), Thurman (1985), and Hendry and others (1986) has identified a variety of important organic reactions such as dissolution of organic litter at the ground surface, complexation of Fe and Al, sorption of organic compounds, and oxidation of organic compounds.

The dissolution of organic litter at or close to the ground surface is the major source of dissolved organic carbon (DOC) in soil water and shallow ground water. DOC concentrations typically fall in a range from 10 to 50 mg/L in the upper soil horizons and less than 5 mg/

L deeper in the unsaturated zone (Thurman, 1985). Concentrations are highest at their source and decline with depth through sorption and oxidation. The most important fraction of the DOC is a group of humic substances, consisting mainly of humic and fulvic acids. Tannins and lignins, amino acids, and phenolic compounds are often present in smaller concentrations (Wallis and others, 1981).

The complexation of Fe and Al with organic matter is an important process facilitating the transport of these poorly soluble metals from the *A* horizon of the soil to the *B* horizon (Thurman, 1985). This is one key feature of the soil-forming process called podzolization (Thurman, 1983). In terms of ground-water systems, this reaction is not particularly important because most of these complexed metals sorb in the *B* horizon of the soil.

Dissolved organic compounds originating in the upper part of the soil horizon are not particularly mobile due to sorption. Many of the sorption models we discussed in Chapter 12 operate within the soil zone. Hydrophobic sorption occurs because of the relatively large quantities of solid organic matter present in the upper part of many soil horizons (Thurman, 1985). Similarly, the abundance of metal oxides, hydroxides, and clay minerals leads to surface complexation reactions and electrostatic interactions. Again in terms of the chemistry of shallow ground water, this is one of the processes that keeps the quantity of DOC in recharge at relatively low concentrations.

Oxidation reactions involving organic matter can influence the chemistry of shallow ground water. For example, the oxidation of dissolved organic matter (represented as CH_2O) provides a source of CO_2 gas within the unsaturated zone, which is readily dissolved in soil water. A second reaction, involving the oxidation of a sulfur-containing compound (represented by the amino acid cysteine), is thought to play a major role in the accumulation of gypsum in shallow soils (Hendry and others, 1986). This reaction is the organic counterpart to the pyrite oxidation reaction. In arid areas, this process can also produce recharge with large SO_4^{2-} concentrations (Hendry and others, 1986).

15.3 Chemical Reactions in the Saturated Zone

The chemistry of ground water depends on not only the chemistry of the recharge but also the reactions operating along the flow system. Processes in the saturated zone are more complex than in the unsaturated zone because the mass transport processes play a more important role, and because geologic, hydrogeologic, and geochemical settings are much more diverse. However, most of the same processes affecting ion concentrations in the unsaturated zone are also operative in the saturated zone including:

(1) weak acid–strong base reactions, for example,

carbonate minerals + H^+ = cations + HCO_3^-

silicate minerals + H^+ = cations + H_2SiO_3

alumino-silicate minerals + H^+ = cations + H_2SiO_3 + secondary minerals (for example, clay minerals)

(2) dissolution of soluble salts, for example,

halite: $NaCl(s) = Na^+ + Cl^-$

anhydrite: $CaSO_4(s) = Ca^{2+} + SO_4^{2-}$

gypsum: $CaSO_4 \cdot 2H_2O(s) = Ca^{2+} + SO_4^{2-} + 2H_2O$

carnalite: $KCl \cdot MgCl_2 \cdot 6H_2O(s) = K^+ + Mg^{2+} + 3Cl^- + 6H_2O$

kieserite: $MgSO_4 \cdot H_2O(s) = Mg^{2+} + SO_4^{2-} + H_2O$

sylvite: $KCl(s) = K^+ + Cl^-$

(3) redox reactions, for example,

$$\tfrac{1}{4}O_2(g) + H^+ + e^- = \tfrac{1}{2}H_2O$$

$$\tfrac{1}{2}Fe_2O_3(s) + 3H^+ + e^- = Fe^{2+} + \tfrac{3}{2}H_2O$$

$$\tfrac{1}{2}MnO_2(s) + 2H^+ + e^- = \tfrac{1}{2}Mn^{2+} + H_2O$$

$$\tfrac{1}{8}SO_4^{2-} + \tfrac{9}{8}H^+ + e^- = \tfrac{1}{8}HS^- + \tfrac{1}{2}H_2O$$

$$\tfrac{1}{8}CO_2(g) + H^+ + e^- = \tfrac{1}{8}CH_4(g) + \tfrac{1}{4}H_2O$$

$$\tfrac{1}{4}CO_2(g) + H^+ + e^- = \tfrac{1}{4}CH_2O + \tfrac{1}{4}H_2O$$

(4) cation exchange, for example,

$$Ca^{2+} \qquad\qquad\qquad Ca$$
$$Mg^{2+} + 2Na\text{-clay} = 2Na^+ + Mg\text{-clay}$$
$$Fe^{2+} \qquad\qquad\qquad Fe$$

(1) Weak Acid–Strong Base Reactions

If the ground water is not yet in equilibrium with carbonate, silicate, and alumino-silicate minerals, they will continue to dissolve in the saturated zone. Because of their relative abundance, reasonably fast reaction rate with H^+, and reasonable solubilities, these reactions increase the cation concentrations, alkalinity, and pH.

The examples presented in the previous section are again useful in illustrating these effects. Consider the flow system in Figure 15.8, where water flows through a sequence of overburden, limestone, and then dolomite (Palmer and Cherry, 1984). Recall that as the water moved to reference point *A*, it dissolved CO_2 at a fixed partial pressure, and at point *B* it had become saturated with respect to calcite. In the saturated zone at point *C*, ground water reaches equilibrium with respect to both calcite and dolomite. In moving to *C*, the pH, the total Mg concentration, and the total CO_3 concentrations increase (Figure 15.8). The P_{CO_2} declines from $10^{-2.5}$ atm to $10^{-2.79}$, because once the system closes below the water table,

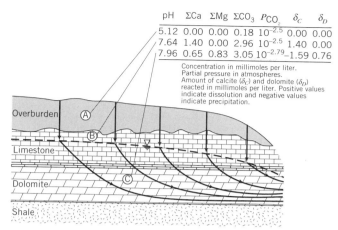

Figure 15.8 Pattern of mass transfer as water proceeds through an idealized overburden-carbonate rock sequence (modified from Palmer and Cherry, 1984). Reprinted from J. Hydrol., v. 75, by C. D. Palmer and J. A. Cherry, Geochemical evolution of groundwater in sequences of sedimentary rocks, p. 27–65, 1984, with kind permission from Elsevier Science, NL, Sara Burgerhartstraat 25, 1055 KV Amsterdam, The Netherlands.

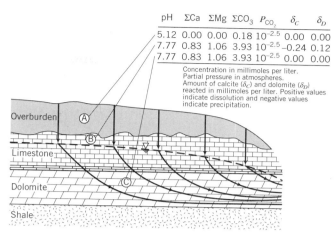

Figure 15.9 Pattern of mass transfer as water proceeds through an idealized overburden-carbonate rock sequence (modified from Palmer and Cherry, 1984). Reprinted from J. Hydrol., v. 75, by C. D. Palmer and J. A. Cherry, Geochemical evolution of groundwater in sequences of sedimentary rocks, p. 27–65, 1984, with kind permission from Elsevier Science, NL, Sara Burgerhartstraat 25, 1055 KV Amsterdam, The Netherlands.

CO_2 depletes by the dissolution of dolomite. The total Ca concentration decreases because, in dissolving dolomite, the ground water became supersaturated with respect to calcite. In moving from B to C, 1.59 mmol of calcite precipitates and 0.76 mmol of dolomite dissolves (Figure 15.8).

The chemistry of the ground water at point C depends on not only the processes involved but also the order of encounter (Freeze and Cherry, 1979). A slight change in geology, placing the dolomite above the limestone, has a significant impact (compare the data in Figure 15.8 with those in Figure 15.9). The pH in the second case is lower, and Σ Ca, Σ Mg, Σ CO_3, and P_{CO_2} are all higher. Holding the system open while dolomite dissolves to equilibrium enables more mass to dissolve. Further, in moving from point B to point C, the chemical composition of the ground water does not change (Figure 15.9).

The example of evolving composition in a predominantly carbonate terrane (Langmuir, 1971) illustrates how the composition simply proceeds toward equilibrium with respect to those minerals available for dissolution (for example, calcite and dolomite). Once in the zone of saturation, the ground water approaches saturation with respect to both calcite and dolomite because of the longer residence time (Figure 15.10; Table 15.4). Also, water from a predominantly limestone source rock more closely approaches saturation with respect to calcite than dolomite. This fact is illustrated in Figure 15.10 by the abundance of data points for limestone source rocks (triangles) lying above the dashed line ($SI_c = SI_d$). Even in this simple carbonate system, there are differences in the pathways of evolution apparently

depending on what carbonate minerals were encountered and in what order.

Looking further at the example presented by Garrels and MacKenzie (1967), evolution through acid–base reactions increases the quantity of mass dissolved in the

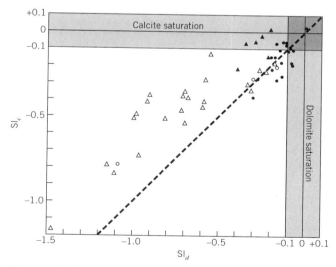

Figure 15.10 Saturation index for calcite versus the saturation index for dolomite. Spring waters are designated by open symbols; well waters by solid symbols. Triangles denote a limestone source rock; circles denote a dolomitic source rock. Cross-hatched area shows the limits of uncertainty in SI_c and SI_d. Reprinted from D. Langmuir, The Geochemistry of some carbonate ground waters in central Pennsylvania, Copyright © 1971, p. 1023–1045, with kind permission from Elsevier Science Ltd, The Boulevard, Langford Lane, Kidlington OX5 1GB, UK.

Table 15.4 Averages of Various Chemical Parameters for Spring and Well Waters from a Carbonate Terrane

Parameter	Average Value	
	Spring Water	Well Water
Specific conductance[a]	347	499
Ca^{2+} (ppm)	47.5	55.0
Mg^{2+} (ppm)	13.9	29.6
HCO_3^- (ppm)	183	265
pH	7.37	7.46
SI_c[b]	−0.41	−0.15
SI_d	−0.63	−0.18

[a] Millimhos per centimeter.
[b] Saturation index.
Modified from Langmuir (1971). Reprinted from D. Langmuir, The Geochemistry of some carbonate ground waters in central Pennsylvania, Copyright © 1971, p. 1023–1045, with kind permission from Elsevier Science Ltd, The Boulevard, Langford Lane, Kidlington 0X5 1GB, UK.

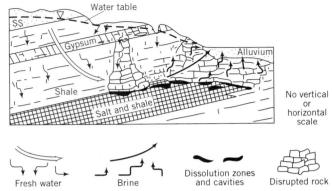

Figure 15.11 Schematic representation of the circulation of fresh water and brine in areas of salt dissolution in western Oklahoma. There is no scale, but the length of the section could range from 1 to 15 km and the thickness from 30 to 300 m (modified from Johnson, 1981). Reprinted from J. Hydrol. Special Issue: Symposium on Geochemistry of Groundwater, v. 54, by K. S. Johnson, Dissolution of salt on the east flank of the Permian Basin in the southwestern United States; p. 75–94, 1981 with kind permission from Elsevier Science, NL, Sara Burgerhartstraat 25, 1055 KV Amsterdam, The Netherlands.

ground water. The effects of increased residence time can be evaluated by comparing the chemistry of ephemeral and perennial springs (Table 15.5). The system behaves as expected. The concentrations of cations increase due to the continued hydrolysis of biotite and plagioclase as does the alkalinity (reflected in the HCO_3^- concentration and pH). An important source of Ca^{2+} is the dissolution of small quantities of carbonate minerals. In the deeper parts of the system, montmorillonite occurs as a weathering product of plagioclase in addition to kaolinite.

(2) Dissolution of Soluble Salts

Section 15.1 explained how the presence of evaporites in a sequence could affect the water chemistry through diffusion. There are other situations (for example, Figure 15.11) where active ground-water flow directly encounters evaporites. Mineral salts are extremely soluble and dissolve to produce a brine whose composition depends on the particular minerals present (for example, halite, anhydrite, gypsum, carnalite, kieserite, and sylvite).

This process (Figure 15.11) leads to the formation of saline brines in the shallow ground water of western Oklahoma and the southeastern part of the Texas Panhandle (Johnson, 1981). Freshwater recharged through permeable units moves downward until it encounters salts at depths ranging from 10 to 250 m. The dissolving salt produces cavities at the up-dip limit or the top of the salt (Johnson, 1981). Periodically, the rocks overlying the cavities collapse. The process is apparently self-perpetuating because the collapse and fracturing of overlying units provide improved access to the salt for fresh water.

The evaporites, mainly halite and gypsum/anhydrite, are interbedded with a thick sequence of red-beds. Given the particular salts involved, high concentrations of Na^+ and Cl^- are not surprising. Clearly when ground water encounters large quantities of soluble salts in the subsurface, the impact on the chemistry is considerable.

Table 15.5 Chemical Composition of Ephemeral and Perennial Springs of the Sierra Nevada

Sample Source	Ca^{2+}	Mg^{2+}	Na^+	K^+	HCO_3^-	SO_4^{2-}	Cl^-	SiO_2	pH
Ephemeral springs	3.11	0.70	3.03	1.09	20.0	1.00	0.50	16.4	6.2
Perennial springs	10.4	1.70	5.95	1.57	54.6	2.38	1.06	24.6	6.8

[a] Concentrations in milligrams per liter.
Reprinted with permission from R. M. Garrels, and F. T. MacKenzie, in Equilibrium Concepts in Natural Water Systems; Gould, R. F. ed., American Chemical Society: Washington, D.C., 1967, p. 224–225. Copyright 1967 American Chemical Society.

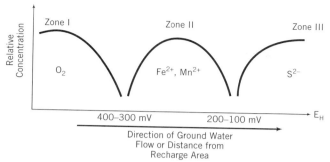

Figure 15.12 Different redox zones existing along a confined aquifer. Reprinted from Jackson and Inch (1980). Hydrogeochemical processes affecting the migration of radionuclides in a fluvial sand aquifer at Chalk River Nuclear Laboratories, NHRI Scientific Series, Paper 7, Environment Canada, 58 p. Reproduced with the permission of Environment Canada, 1997.

(3) Redox Reactions

The redox conditions encountered along a flow system are important in controlling the chemistry of metal ions and solids (for example, Fe^{2+}, Mn^{2+}, and Fe_2O_3), species or solids containing sulfur (for example, SO_4^{2-}, H_2S, and FeS_2), and dissolved gases containing carbon (for example, CO_2, CH_4). It is possible in some flow systems to define redox zones. These zones are parts of an aquifer in which pe is controlled by a dominant redox couple. Field studies (for example, Champ and others, 1979; Jackson and Patterson, 1982) have shown that oxygen, iron-manganese, and sulfides zones will often be present. The probable half-reactions controlling pe in these zones are the reduction of oxygen to water, the reduction of iron or manganese oxides, and sulfate reduction to HS^- or H_2S. In a few cases, a methane zone can form from the reduction of CO_2.

Reduced sulfide and methane zones tend to develop in confined flow systems containing excess oxidizable DOC and a lack of recharge containing oxygen downgradient in the system. Thus, redox zones will be observed in an extensive artesian aquifer that receives recharge from a limited area of outcrop or in an aquifer with confining units.

Figure 15.12 summarizes the common changes in redox chemistry from zone to zone (Jackson and Inch, 1980). In zone I, oxygen initially present in recharge will decline through reduction by organic carbon

$$CH_2O + O_2 = CO_2 + H_2O \qquad (15.1)$$

The concentrations of Fe^{2+} and Mn^{2+} in zone II increase because the Fe(III) and Mn(IV) minerals, which are oxidized solids, are not stable in the more reducing environment. These reactions are

$$CH_2O + 8H^+ + 4Fe(OH)_3(s) = 4Fe^{2+} + 11H_2O + CO_2 \qquad (15.2)$$

$$CH_2O + 4H^+ + 2MnO_2(s) = 2Mn^{2+} + 3H_2O + CO_2 \qquad (15.3)$$

Once the pe is sufficiently reduced, sulfide species appear from the reduction of SO_4^{2-} or

$$2CH_2O + SO_4^{2-} + H^+ = HS^- + 2H_2O + 2CO_2 \qquad (15.4)$$

When the rate of sulfide production exceeds the dissolution rate of iron and manganese oxides, Fe^{2+} and Mn^{2+} concentrations fall (Figure 15.12) as metal sulfides precipitate as the stable phase.

A study at Chalk River Nuclear Laboratories near Ottawa, Canada, demonstrates how redox conditions can change along a flow system. By careful measurements of pH, E_H, dissolved oxygen (DO), and the total concentration of sulfides (S_T^{2-}) as well as other key chemical parameters, Jackson and Patterson (1982) defined three redox zones. The main geologic units were two fluvial sand aquifers separated by a thin layer of interbedded clay about 1 m thick (Figure 15.13). Ground water flows southward from an upland area toward Perch Lake. Inflow to the aquifers is near the area labeled "disposal area" (Figure 15.13) with none further down-gradient.

The lower aquifer in the recharge area contains dissolved oxygen and has an E_H of about 0.55 V (Figure 15.14), which is quite close to the theoretical range of 0.71 to 0.83 V. Both Fe^{2+} and Mn^{2+} concentrations are low, and sulfide is undectable. The declining E_H along the deep flow system coincides with the reduction of oxygen. Increasing concentrations of iron and manganese are evident once the oxygen is depleted (Figure 15.14). Zone II is probably not extensive because sulfide is becoming relatively abundant at E_H values below 0.2 V. The downstream end of the flow system is a sulfide zone with E_H values approaching the theoretical range for the sulfate–sulfide couple (i.e., -0.21 to -0.32 V). Concentrations of Fe^{2+}, Mn^{2+}, and SO_4^{2-} decline in zone III from the precipitation of ferrous sulfides (Jackson and Patterson, 1982).

The reactions that oxidize organic matter (Section 12.5) generate CO_2 as a product. This CO_2 is redistributed among $H_2CO_3^*$, HCO_3^-, and CO_3^{2-}. In aquifers where Ca^{2+} and Mg^{2+} are exchanged onto clay minerals for Na^+, the possibility exists for carbonate dissolution and even higher HCO_3^- concentrations. With CO_2 generated by redox reactions, ion exchange, and carbonate dissolution, the water will evolve chemically to a sodium-bicarbonate type like that from the Atlantic coastal plain (Foster, 1950), the Eocene aquifers of the east Texas basin (Fogg and Kreitler, 1982), and the Milk River aquifer (Hendry and Schwartz, 1990).

Chapelle and others (1987) investigated the source of CO_2 in ground water from the Atlantic Coastal Plain in Maryland. The units of interest included the Magothy–Upper Patapsco aquifer and the Lower Patapsco aquifer. These aquifers crop out between Washington and Balti-

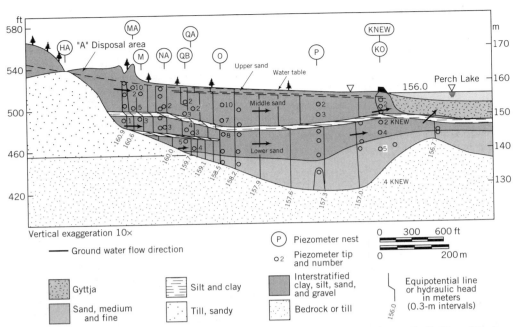

Figure 15.13 Cross section through the lower Perch Lake basin aquifer at Chalk River Nuclear Laboratories, Canada, showing equipotential lines, piezometer tips, and the position of the water table. Hydraulic head measurements averaged over the period of 1973–1975 (from Jackson and Patterson). Water Resources Res., v. 18, p. 1255–1268, 1982. Copyright © by American Geophysical Union.

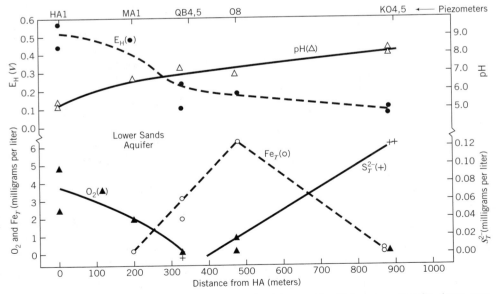

Figure 15.14 The pH, E_H, O_2, and total dissolved iron and sulfide values for the deep, confined Lower Sands aquifer. The piezometer numbers are shown on the top abscissa (from Jackson and Patterson). Water Resources Res., v. 18, p. 1255–1268, 1982. Copyright © by American Geophysical Union.

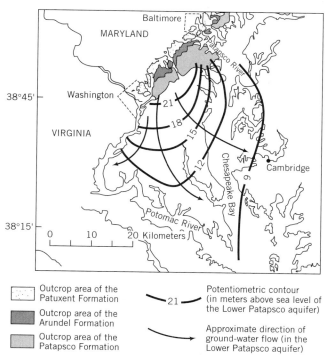

Figure 15.15 Map showing the regional outcrop area for the Patuxent, Arundel, and Patapsco Formations on the Atlantic coastal plain in Maryland. Ground-water flow in the Lower Patapsco Formation is away from the outcrop area (from Chapelle and others). Water Resources Res., v. 23, p. 1625–1632, 1987. Copyright © by American Geophysical Union.

(4) Cation Exchange

The most important exchange reactions are the natural water softening reactions, which take Ca^{2+} and Mg^{2+} out of water and replace them with Na^+. The main requirement for this process is a large reservoir of exchangeable Na^+, which is most often provided by clay minerals deposited in a marine environment. One does not have to go far in the United States or Canada to find clays or shales capable of ion exchange. A case in point is bedrock in the Wabamun Lake area of central Alberta, Canada (Schwartz and Gallup, 1978). Shale, sandstone, and coal of the Edmonton Group are overlain by two younger bedrock units that include the Pembina coals and sandstone of the Paskapoo Formation. Bedrock in turn is mantled by sand and gravel. The shales of the Edmonton Group and those interbedded with the Pembina coals were deposited in a marine environment. Figure 15.16 is a plot of Ca^{2+} versus Na^+ concentrations for samples from the drift, Pembina coal, and Edmonton Group. It shows how the pattern of cation dominance shifts from Ca^{2+} (and Mg^{2+}, not shown) to Na^+ as water moves deeper into the marine shale.

When ion exchange takes place, its effects on the cation chemistry of water should be unmistakable. However, as the following case study shows, an equivalent increase in Na^+ concentration that is matched by an increasing Cl^- concentration may mean that other processes are at work.

more (Figure 15.15) and dip toward the east. Recharge occurs along the outcrop areas with flow toward the Potomac River and under Chesapeake Bay (Figure 15.15). Although these units contain no calcareous minerals, HCO_3^- concentrations increase significantly in the direction of flow. Where the Patapsco aquifer crops out, HCO_3^- concentrations range from 0 to 50 mg/L. Approximately 30 km down-gradient, they have increased to about 150 to 200 mg/L, and finally near Cambridge, Maryland (Figure 15.15), HCO_3^- concentrations range from 400 to 500 mg/L (Chapelle and others, 1987). This pattern is similar to that observed by Foster (1950) farther south in Virginia.

The potential sources of CO_2 are the bacterially mediated oxidation of solid organic matter or the abiotic decarboxylation of these materials. The presence of both sulfate-reducing and methanogenic bacteria in cores, however, points to the generation of CO_2 in redox reactions. Although sulfate-reducing bacteria are present, the relatively small quantities of SO_4^{2-} present make the contribution of CO_2 from sulfate reduction small relative to that from methanogenesis. The bacteria facilitating these reactions occur together because the theoretical pe of a system is nearly the same with either reaction.

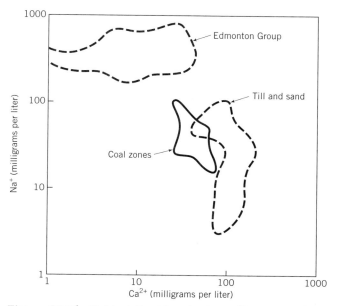

Figure 15.16 Fields defined by plotting Ca^{2+} versus Na^+ data for three different geologic units in the Wabamun Lake area of central Alberta, Canada (from Schwartz and Gallup, 1978).

15.4 Case Study of the Milk River Aquifer

We have singled out the Milk River aquifer for detailed discussion because the interpretation of chemical patterns requires the integration of concepts of mass transport and mass transfer. Several studies have documented the geologic and hydrogeologic setting including Meyboom (1960), Schwartz and Muehlenbachs (1979), and Hendry and Schwartz (1988, 1990). The aquifer is part of a thick sequence of Cretaceous rocks (Figure 15.17). Units of interest from oldest to youngest are the shale of the Colorado Group, the Milk River Formation whose lower member is the aquifer, and shale of the Pakowki Formation (Figure 15.17). The aquifer pinches out to the north and east as the result of a facies change.

Ground-water recharge occurs where the aquifer crops out in northern Montana and along an area of outcrop–subcrop in southern Alberta. Flow is mainly northward or down-dip in the aquifer. Leakage from the aquifer moves upward through the Pakowki Formation and overlying units. Before extensive development of the ground-water resource, flowing wells were common.

An exciting feature of this aquifer system is the well-defined patterns of variability in the major and minor ions and the stable or radiogenic isotopes. Cl^-, $\delta^{18}O$, and δD change markedly along the flow system (Figure 15.18) and appear to act as conservative tracers (Hendry and Schwartz, 1988). Cl^- concentrations increase from less than 1 mM to more than 100 mM at the downstream end of the flow system. $\delta^{18}O$ and δD values increase from about $-20°/_{oo}$ and $-150°/_{oo}$, respectively, in the recharge areas to about $-8.5°/_{oo}$ and $-85°/_{oo}$, respectively, at the northern end of the aquifer.

Several different studies have been concerned with explaining these patterns, such as Schwartz and Muehlenbachs (1979), Domenico and Robbins (1985a), and Phillips and others (1986). A reexamination of these ideas prompted the development of one that involves dispersion and advection in the aquifer and diffusion in the aquitard (Hendry and Schwartz, 1988). This model takes into account the effects of both mass transport processes and geologic change in controlling ground-water chemistry.

Tóth and Corbet (1986) showed how the configuration of the land surface changed with time. As recently as five

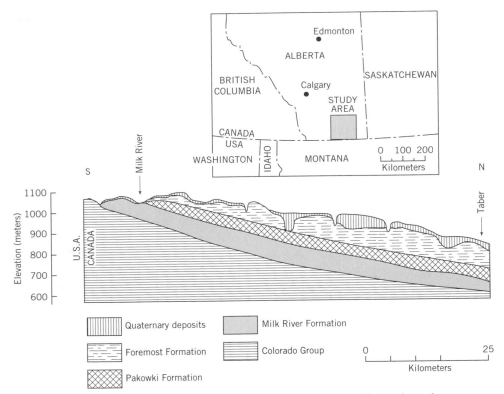

Figure 15.17 Study area for the Milk River aquifer investigation. The geological cross section illustrates the most important drift and bedrock units in southern Alberta, Canada (from Hendry and Schwartz, 1990. Reprinted by permission of Ground Water. Copyright © 1990. All rights reserved.

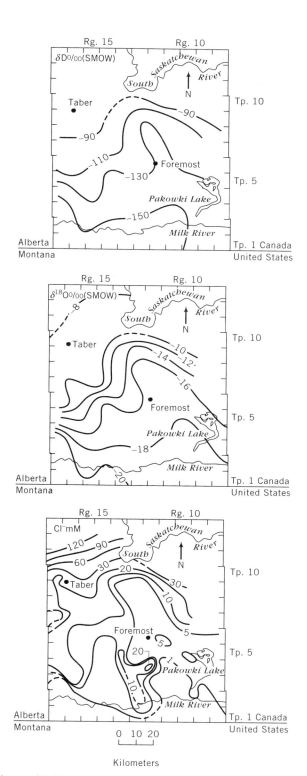

Figure 15.18 Areal variation in Cl⁻, δ¹⁸O, and δD in water from the Milk River aquifer (from Hendry and Schwartz). Water Resources Res., v. 24, p. 1747–1764, 1988. Copyright © by American Geophysical Union.

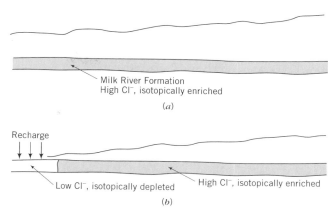

Figure 15.19 The effect of geologic changes on the chemistry of ground water in the Milk River aquifer. In (a) the aquifer is deeply buried. With time, erosion uncovers upstream end of the aquifer, causing inflow of meteoric water (modified from Hendry and Schwartz, 1990). Reprinted by permission of Ground Water. Copyright © 1990. All rights reserved.

million years ago, the land surface was probably 700 m higher than now. Even in the early Pleistocene, the land surface was about 200 m higher than present (Tóth and Corbet, 1986). When the aquifer was deeply buried (Figure 15.19), ground water probably was enriched isotopically relative to present-day meteoric water ($\delta^{18}O \sim -20^{\circ}/_{\circ\circ}$ and $\delta D \sim -150^{\circ}/_{\circ\circ}$) and had a mean Cl⁻ concentration of about 85 mM (Hendry and Schwartz, 1988). Continuing erosion eventually exposed a relatively large recharge area that provided a new source of water to the aquifer. Proximity to the surface would result in recharge with a low Cl⁻ concentration (Figure 15.19b). This change in the flow dynamics of the aquifer probably occurred about 1×10^6 years ago. Somewhat later, the stable isotope composition of the precipitation became similar to present-day meteoric water due to a changing climate.

This chemically different water with time has displaced the original formation water. However, in so doing, it has created diffusion gradients between the isotopically depleted, low chlorinity water in the aquifer and the original water remaining in the shale of the Colorado Group. Thus water moving down the aquifer experienced an increase in Cl⁻ concentration and enrichment in $\delta^{18}O$ and δD due to aquitard diffusion. This simple model of advection and dispersion in the aquifer and diffusion in the aquitard was tested mathematically by Hendry and Schwartz (1988).

Na⁺ and HCO₃⁻ concentrations increase in the direction of flow (Figure 15.20). SO₄²⁻ behaves oppositely with the highest concentrations in the recharge area and a systematic decrease down-gradient in the confined part of the aquifer (Figure 15.20). Not shown are maps for Ca²⁺ and Mg²⁺. Except for a few areas where the Milk River crops

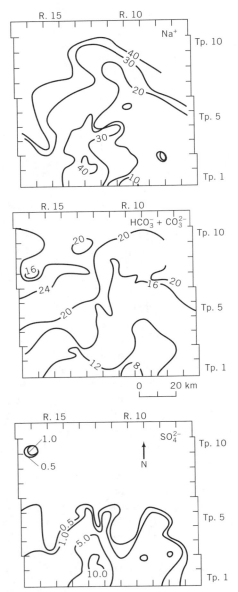

Figure 15.20 Concentration distribution (m*M*) of Na^+, $HCO_3^- + CO_3^{2-}$, and SO_4^{2-} in waters of the Milk River aquifer (modified from Hendry and Schwartz, 1990). Reprinted by permission of Ground Water.

out, concentrations of these ions are less than a few milligrams per liter.

The classical interpretation of these data would point to sulfate reduction and ion exchange as the key processes responsible for generating these patterns. However, other processes are probably more important. The aquitard diffusion model, explaining the distribution of Cl^-, also constrains the behavior of the cations. Diffusion of only Cl^- would produce a charge imbalance. Thus, counter ions must diffuse from the shale to maintain electroneutrality. Na^+ would be the likely species given

its abundance in formation waters in the basin (Hendry and Schwartz, 1990). If the Na^+ distribution is in fact related to Cl^-, a relationship should exist between Na^+ and Cl^- concentrations. The lines in Figure 15.21 are hypothetical pathways of chemical evolution generated by adding equivalent amounts of Na^+ and Cl^- to water from the recharge area. The lines fit the observed trends in the data extremely well, suggesting that the process controlling the Na^+ ion concentration is also the one affecting Cl^-.

Sulfate reduction cannot explain the increase in HCO_3^- and decline in SO_4^{2-}. There is a lack of measurable sulfide in either ionic or gaseous forms. The most reasonable explanation of the SO_4^{2-} distribution is a geologic one. Approximately 35,000 years ago, glaciation continued to open up the recharge area of the aquifer and deposited till. The sulfate chemistry of the aquifer in areas of subcrop simply reflects unique processes that operate within the till to produce high SO_4^{2-} (Hendry and others, 1986). Thus, high SO_4^{2-} water is limited to the vicinity of the subcrop because time has not been sufficient for this chemically distinctive water to move further. On average, the transport of a tracer through the confined part of the aquifer requires about 3×10^5 yr (Hendry and Schwartz, 1988).

Mass transfer calculations down-gradient of the SO_4^{2-} bulge suggest that minor cation exchange and sulfate reduction is occurring but not nearly to the extent suggested initially by the major ion distributions. The increasing HCO_3^- concentrations and large concentrations of methane gas in the down-dip end of the aquifer come from the reduction of organic matter. It is not certain whether the reaction is occurring in the aquifer or whether both CH_4 and HCO_3^- are diffusing from the Colorado shale. Given that a major gas field occurs at the down-dip end of the aquifer, it is most likely that the shale unit is the source rock for the gas.

More recently, the Milk River aquifer was the focus for a large multidisciplinary study designed to test isotopic and geochemical techniques for testing old ground

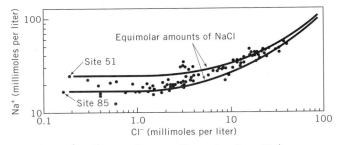

Figure 15.21 Scatter diagram illustrating how Na^+ concentration varies in relation to Cl^-. The lines depict how water from sites 51 and 85 would evolve by adding equivalent amounts of Na^+ and Cl^- (from Hendry and Schwartz, 1990). Reprinted by permission of Ground Water.

water (Ivanovich and others, 1991). Beyond standard major ion and environmental isotope analysis (that is tritium, D, ^{18}O), samples were measured for ^{14}C, ^{36}Cl, various noble gases (He, Ne, Ar, Kr, Xe, and ^{222}Ra), Ar isotopes (^{37}Ar, ^{39}Ar), uranium and thorium isotopes, and various isotopes of krypton.

An important result of this work is yet another model to explain the origin of salinity patterns within the Milk River aquifer. Based on the geochemistry of halides (Cl, Br, and I) and their radioactive isotopes ^{36}Cl and ^{129}I, Fabryka-Martin and others (1991) suggest that the major influence on the major ion chemistry is the diffusion of modified seawater from low-permeability lenses or units within the Milk River aquifer. Thus, while diffusion remains a dominant process, this model looks to sources within the aquifer to provide dissolved mass as opposed to sources outside the aquifer (Hendry and Schwartz, 1988).

15.5 Age Dating of Ground Water

There are two different approaches for dating ground water. The direct approaches involve interpreting the concentration distribution of radioactive elements that occur as environmental tracers. The most common environmental tracers are tritium, carbon-14, and chlorine-36. The indirect dating methods rely on knowledge of how particular isotopes or organic chemicals change as a function of time in recharge. Examples are δ^{18}O, which has changed on a continent-wide basis as a consequence of deglaciation, and short-range tracers, the chlorofluorocarbons, which have been increasing in the atmosphere since they began to be manufactured in the 1930s.

Direct Methods

The direct dating techniques interpret the distribution of a radioactive species in terms of a first-order kinetic rate law for decay. The residence time of mass in the system or the ground-water age (t) is described mathematically as

$$t = \frac{t_{1/2}}{\ln 2} \ln \frac{A_0}{A_{obs}} \qquad (15.5)$$

where $t_{1/2}$ is the half-life for decay, A_0 is the activity assuming no decay occurs, and A_{obs} is the observed or measured activity of the sample.

Tritium (^{3}H, $t_{1/2}$ = 12.35 yr) and carbon-14 (^{14}C, $t_{1/2}$ = 5730 yr) are commonly used for age dating. However, both suffer from limitations. The relatively short half-life for ^{3}H makes it is useful for dating water only less than about 40 years old. In addition, the cessation of nuclear testing in the atmosphere has eliminated the global source of new tritium. Within a few more decades the tritium levels in precipitation will decline to an extent that tritium will be less useful as a tracer. ^{14}C with a much longer half-life has the potential to date water up to about 40,000 years old. However, interpreting ^{14}C data is difficult because of the need to account for other processes besides radioactive decay that influence the measured ^{14}C activity.

Phillips and others (1986) have demonstrated the potential of using chlorine-36 (^{36}Cl) in dating water. Particu-larly attractive with this radionuclide is the long half-life ($t_{1/2}$ = 3.01 $\times$ 10^5 years) and the smaller number of potential reactions (as compared to ^{14}C) that need to be accounted for in calculating an age date (Bentley and others, 1986). One limitation with this radioisotope is the need for tandem accelerator mass spectrometry for measurements, which at present limits the availability of the approach.

Tritium

Tritium concentrations are reported in terms of tritium units (TU) with 1 TU corresponding to one atom of ^{3}H in 10^{18} atoms of ^{1}H (Fontes, 1980). Tritium occurs naturally in the atmosphere, with concentrations in precipitation usually less than 20 TU. However, tritium generated by thermonuclear testing in the atmosphere between about 1952 and 1963 has swamped the natural production of tritium. The long-term record of ^{3}H in precipitation at Ottawa, Canada, shows that during the period of nuclear testing concentrations were often greater than 1000 TU (Figure 15.22). Tritium levels declined once weapons testing stopped in 1963 but present-day levels remain above the natural background. Details on the seasonal variation and latitude dependance of ^{3}H levels in precipitation are presented by Gat (1980) and Fontes (1980).

Ideally, by knowing the concentration of ^{3}H in precipi-

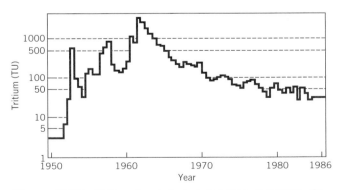

Figure 15.22 Tritium levels in precipitation at Ottawa, Ontario (modified from Robertson and Cherry). Water Resources Res., v. 25, p. 1097–1109, 1989. Copyright © by American Geophysical Union.

tation (the source) and its distribution in ground water, one should be able to date the water. In most cases, however, tritium cannot be used in such a quantitative way. The main problems stem from the uncertainty and complexity of atmospheric loading. For example, at most places, there are usually insufficient data to establish the historical pattern of ^3H loading. Sometimes this limitation can be overcome by correlating partial local records to stations like Ottawa, Canada, with long-term records. The complex loading function also provides a problem in interpreting ^3H data. Without a great deal of information about tritium distributions in the ground water, it is difficult to determine whether a sample with 30 TU is a late 1950s water that has decayed through three half-lives or a 1970s water originally with 75 TU that has decayed through one half-life.

The main application of tritium is to differentiate pre-1952 water from younger water. Tritium, then, is often used as an indirect method that relies mainly on knowledge of how it is loaded to ground-water systems. The logic is that assuming pre-1952 water to have had an original ^3H concentration of 5 TU, the concentration in 1988 would be at maximum 0.6 TU, which is close to the detection limit even using enrichment techniques of analysis. Thus, any detectable tritium in a sample implies that the water contains some component of more recent or post-1952 water (Fontes, 1980).

Some of the limitations of tritium dating can be overcome by adding measurements of helium-3 (^3He), the stable daughter of tritium decay. The isotopic decay reaction is written as

$$^3H \rightarrow {}^3He^* + \beta^-$$

where the * indicates a tritiogenic source of ^3He (Solomon and others, 1995). Other sources of ^3He exist that must be appropriately accounted for. When both ^3H and ^3He measurements are available, summing ^3H (TU) and ^3He (as TU) eliminates the decay of ^3H (Solomon and others, 1995). Thus, one could determine, for example, the peak concentration coinciding with 1960's bomb testing. In addition, one could determine the $^3H/^3He$ age of the water. Solomon and others (1995) demonstrated this dating technique at a site on Cape Cod.

Carbon-14

Measurements of ^{14}C are reported as percent modern ^{14}C (pmc) determined as the ratio of the sample activity to that of the international standard expressed as a percentage. ^{14}C originates naturally in the upper atmosphere through a reaction involving nitrogen and neutrons. Like ^3H, weapons testing in the atmosphere has affected its concentration in recent years. However, except for young waters, this increase does not affect the interpretation.

^{14}C in ground water comes from the solution of $CO_2(g)$ in the soil zone. The activity of ^{14}C in CO_2 gas is approxi-

mately 100 pmc (Fritz and Fontes, 1980) and slightly higher in the ions coming from the dissolution of CO_2. The dating method works because once carbonate species move below the water table ^{14}C begins to decay and there are no additional sources.

The main problem in applying this method is that some reactive minerals contain carbon, and carbon transfers in and out of the ground water. These interactions can reduce the ^{14}C activity in the water and, thus, need to be accounted for in estimating the age. In terms of Eq. 15.5, the value of A_0 (the ^{14}C activity assuming no decay) would be lower than 100 pmc, reflecting the fact that other processes besides radioactive decay influence the ^{14}C activity of the sample. Any age calculation is meaningful as long as A_0 and A_{obs} differ only due the effects of radioactive decay.

Mook (1980), Reardon and Fritz (1978), and Wigley and others (1978) list the following processes that can alter the ^{14}C activity of ground water:

1. The congruent dissolution of carbonate minerals, which adds "dead carbon" or carbon without ^{14}C activity to the ground water. Overall, this process lowers the ^{14}C activity measured for the sample.

2. The incongruent dissolution of carbonate or other Ca-containing minerals, accompanied by the precipitation of calcite. This process will remove ^{14}C as calcite precipitates, and if dolomite is the mineral dissolving, additional dead carbon is added through (1) above. This process could occur in the zone of saturation following the rapid solution of calcite to equilibrium with subsequent precipitation as dolomite slowly dissolves.

3. The addition of dead carbon from other sources such as the oxidation of old organic matter, sulfate reduction, and methanogenesis. These again reduce the ^{14}C activity of the sample.

4. Possible isotopic exchange involving CO_3^{2-} and carbonate minerals, which could lower the ^{14}C activity. This process is generally considered to have a negligible effect at normal ground-water temperatures.

Two approaches are available for estimating a ground-water age (Kimball, 1984). One way is to interpret ages on the basis of the ion and isotopic data for a single sample without information from other samples. The second approach involves using the ion or isotopic data from many samples in an integrated way and mass balance modeling to sort out all the major inputs and outputs of carbon.

The simplest way to establish A_0 is to account only for the most important process affecting ^{14}C activity, which is the congruent dissolution of calcite. The reaction between water containing CO_2 and calcite is

$$CO_2 + H_2O + CaCO_3(s) = Ca^{2+} + 2HCO_3^- \quad (15.6)$$

At equilibrium according to this reaction, half the bicarbonate would be generated from a source containing ^{14}C (CO_2) and the other half would be generated from a dead source (calcite). Assuming that the activity of the CO_2 is 100 pmc and the calcite is 0 pmc, A_0 would be 50 pmc, reflecting the equal contribution of carbon from both sources.

In many cases, it is unlikely that the reaction would be at equilibrium due to the lack of carbonate minerals or kinetic effects (Pearson and Hanshaw, 1970). The following form of the reaction describes this more realistic situation (Mook, 1980)

$$(a + 0.5b)\ CO_2 + 0.5b\ CaCO_3 + H_2O$$
$$= 0.5b\ Ca^{2+} + b\ HCO_3^- + a\ CO_2 \quad (15.7)$$

Excess CO_2 in this reaction results in a ^{14}C activity greater than 50 pmc and perhaps close to 85 pmc.

Two of the simpler techniques for correcting A_0 values are based on these ideas. For example, one empirical approach simply assumes A_0 has a value of 85 ± 5 pmc. Measurements of dissolved carbon in ground water of northwest Europe suggested that this value was in fact representative of soil water and shallow ground water in temperate climates (Vogel, 1967, 1970). Another less empirical approach is based on Eq. 15.7 and provides an A_0 value that is the weighted contribution of CO_2 and calcite. Mathematically, this correction is a simple mixing equation or

$$(a + b)A_0 = (a + 0.5b)A_{CO_2} + 0.5b\ A_c \quad (15.8)$$

where A_{CO_2} is the estimated activity of CO_2 in the soil zone (usually 100 pmc), A_c is the estimated activity of calcite (usually 0 pmc), and $(a + b)$ is the total moles of carbon in the water (C_T). Rearranging Eq. 15.8 and assuming A_c is 0 provides the desired correction equation for A_0

$$A_0 = (a + 0.5b)\frac{A_{CO_2}}{(a + b)} \quad (15.9)$$

or written in terms of C_T

$$A_0 = (C_T - 0.5b)\frac{A_{CO_2}}{C_T} \quad (15.10)$$

Equation 15.10 is known as the Tamers equation (Tamers, 1967, 1975). The first term on the right side of Eq. 15.10 assumes that for water with a normal pH the total carbon is found as CO_2 (or $H_2CO_3^*$ depending on conventions) and HCO_3^-. C_T is determined using the major ion data together with a speciation program like WATEQF. The molar concentration of HCO_3^- (i.e., b) is known from the water analysis, and A_{CO_2} is taken as 100 pmc.

Another way of establishing how much dead carbon has been contributed from carbonate dissolution is to use $\delta^{13}C$ as a measure of the extent of carbonate reactions (Ingerson and Pearson, 1964). Again, a carbon-mixing equation can be written (Mook, 1980)

$$(a + b)\delta^{13}C_T = (a + 0.5b)\delta^{13}C_{CO_2} + 0.5b\delta^{13}C_c \quad (15.11)$$

where $\delta^{13}C_T$, $\delta^{13}C_{CO_2}$, and $\delta^{13}C_c$ are the measured values of $\delta^{13}C$ in the ground water (as total carbon), soil gas, and calcite, respectively, and the other terms are molar concentrations. With some algebraic manipulation, we can express Eq. 15.11 in terms of $(a + 0.5b)/(a + b)$ or

$$\frac{(a + 0.5b)}{(a + b)} = \frac{(\delta^{13}C_T - \delta^{13}C_c)}{(\delta^{13}C_{CO_2} - \delta^{13}C_c)}$$

where the left side of the equation is the unknown factor in Eq. 15.9 correcting the A_0 value (Mook, 1980). The $\delta^{13}C$ content of CO_2 gas in the soil zone is known reasonably well (i.e., $-27 \pm 5^\circ/_{oo}$ PDB in temperate regions and $-13 \pm 4^\circ/_{oo}$ in tropical regions; see Mook, 1980) as is the value for marine carbonate (0 to $+2^\circ/_{oo}$).

All of these approaches are useful to some extent. However, they do not account for many of the processes affecting the ^{14}C content of the water (Pearson and Hanshaw, 1970). More comprehensive schemes exist that we will briefly mention. Fontes and Garnier (1979), for example, account for mineral dissolution and isotopic exchange reactions. The most comprehensive model to date is that of Wigley and others (1978). Their scheme, which is complicated enough to require a computer code, takes into account an arbitrary number of carbon sources (for example, dissolution of carbonate minerals and oxidation of organic matter) and sinks (for example, mineral precipitation, CO_2 degassing, and methane production), and the equilibrium fractionation between phases.

One way in which the Wigley and others (1978) model can be used is in conjunction with the code BALANCE (Kimball, 1984; Plummer, 1984). As a first step, BALANCE estimates the transfer of carbon into and out of solution. These estimates provide input data for the model of Wigley and others (1978) to calculate fractionation factors, the predicted $\delta^{13}C$ and the corrected value of A_0.

Kimball (1984) examined several methods for interpreting ^{14}C ages of ground water in the Piceance Creek basin in Colorado. He compared four different techniques: the mass balance model coupled with the procedure of Wigley and others (1978), the Tamers equation, the Ingerson and Pearson equation, and the Fontes and Garnier equation. The results in particular for waters from the lower aquifer illustrate some important features about ^{14}C dating. First, there is a clear requirement to account for the chemical processes in producing a date. In all cases, the corrected ages are less than the analytical ages and in a few cases more than an order of magnitude less. Second, there is a marked variability in the dates depending on which correction technique is used. For example, the dates derived from mass balance modeling sometimes were an order of magnitude less than those obtained from the other methods (i.e., 1600 versus 18,000 yr). It appears from Kimball's work that the Tamers equation, which accounts only for mineral dissolution, gives the oldest dates. The other approaches yield younger dates. Beyond this, there is little consistency among the results.

These results raise the question as to what interpretive method yields the best date. Can we say that because the mass balance procedure accounts for more processes the date is better? Probably not. Maybe the reaction model could be wrong, or perhaps important processes were left out. In spite of a long history of development, the ^{14}C method is at best a semiquantitative tool. Confident predictions can be made only when the processes affecting the carbon chemistry are absolutely defined— the exception rather than the rule. The hope of collecting a single sample of water and extracting a date seems to have faded in light of the effort required for process identification.

Chlorine-36

There is strong evidence (Bentley and others, 1986; Phillips and others, 1986) that ^{36}Cl will emerge as a useful radioisotope for dating water up to two million years old. Again, the most important source of this isotope is fallout from the atmosphere, although small quantities are produced in the subsurface. The results of analyses are reported as the ratio of ^{36}Cl/Cl, with a typical value for meteoric water lying in a range from 100 to 500 $\times 10^{-15}$. Because there appear to be comparatively few processes that affect the ^{36}Cl/Cl ratio, the dating equations are straightforward modifications of Eq. 15.11. The simple geochemistry of Cl in ground water avoids the complexity inherent with ^{14}C. About the only processes contributing dead Cl to ground water are the dissolution or mixing of older water containing Cl$^-$. Thus far, ^{36}Cl has been used to date ground water in the Great Artesian basin of Australia (Bentley and others, 1986) and the Milk River aquifer in Alberta, Canada (Phillips and others, 1986).

Like tritium, levels of ^{36}Cl have been elevated up to two or three orders of magnitude as a consequence of global fallout from high-yield nuclear weapons tests in the 1950s. The presence of so-called bomb-pulse ^{36}Cl in a water sample provides a clear indication of the relatively young age of the sample. To date, it has been useful in sorting out the patterns of complex unsaturated zone recharge (Fabryka-Martin and others, 1993).

Indirect Methods

Indirect chemical methods depend on interpreting systematic changes in the chemical composition of indicator species or isotopes along ground-water flow paths. Unlike the radiogenic isotopes, some chemical pattern is interpreted in light of events or other independent data

that provide a basis for establishing times. This section will review the use of two promising approaches that involve environmental isotopes (δ^{18}O and δD) and anthropogenic contaminants with a known history of use (chlorofluorocarbons).

$\delta^{18}O$ and δD

The record of changing δ^{18}O and δD composition of precipitation recharging an aquifer can be preserved in ground water from areally extensive aquifer systems or from units having an extremely low hydraulic conductivity. In either case, water recharging the aquifer may be several thousand or tens of thousands of years younger than the ground water.

Desaulniers and others (1981) showed the shift in the isotopic composition as a function of depth in ground water collected from glacial till in southwestern Ontario, Canada. The δ^{18}O content of shallow ground water falls in a range of -9 to $-10^{\circ}/_{\circ\circ}$ (SMOW), which is similar to present-day precipitation. Water samples collected deeper in the various till units are isotopically lighter (that is, -14 to $-17^{\circ}/_{\circ\circ}$). Corrected ^{14}C ages for these water samples are in excess of 8000 yr B.P. This age indicates that the deeper water was either the original pore fluid in the till or was recharged soon thereafter. The so-called isotopic shift that is recorded with depth thus represents the gradual climatic warming following deglaciation.

Bradbury (1984) found similar results in studies of tills along the south shore of Lake Superior. At a site near Duluth, he documented a $10^{\circ}/_{\circ\circ}$ shift in δ^{18}O over approximately 50 m of low-permeability till. Again the shallow samples were isotopically enriched ($-10^{\circ}/_{\circ\circ}$, δ^{18}O) relative to the deepest samples ($-17^{\circ}/_{\circ\circ}$, δ^{18}O).

By carefully documenting the isotopic character of samples from low-permeability sites in formerly glaciated areas, it may be possible to date ground water indirectly. Water that is isotopically lighter than present-day meteoric water is likely relatively old, originating during colder glacial climates. This approach to indirect dating appears to have promise in selecting sites for the long-term containment of hazardous wastes (Desaulniers and others, 1981).

Chlorofluorocarbons

Chlorofluorocarbons (CFCs or Freons) are a family of organic compounds used widely as propellants in aerosol cans and refrigerants. In recent years, their use is being phased out because of concerns about the destruction of ozone in the Earth's atmosphere. Two of the (CFCs) in particular, dichlorodifluoromethane (CCl_2F_2 or CFC-12) and trichlorofluoromethane (CCl_3F or CFC-11) account for the greatest use commercially (Busenberg and Plummer, 1992). CFC-12 and CFC-11 have been manufactured since the 1930s and 1940s, respectively.

The potential of environmental CFCs in dating ground

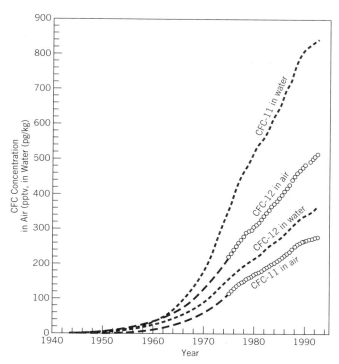

Figure 15.23 Reconstruction of atmospheric concentrations of CFC-11 and CFC-12 since 1940 in parts per trillion volume per volume air (after Busenberg and others, 1993; Elkins and others, 1993) and calculated corresponding solubilities of CFC-11 and CFC-12 in water at 9°C in picograms per kilogram water. Reprinted from Dunkle and others, Water Resources Res., v. 29, p. 3837–3860, 1993. Copyright © by American Geophysical Union.

pected CFC-12 and CFC-11 concentrations in the atmosphere. Figure 15.23, taken from Dunkle and others (1993), shows how the atmospheric concentration of these compounds has changed as a function of time. Busenberg and Plummer (1992) discuss in detail how these atmospheric concentration values were reconstructed.

The dating method assumes that the concentration of CFCs in precipitation recharging the ground water is in equilibrium with the atmospheric gas-phase concentrations according to Henry's law. The equilibrium constant describing the partitioning of gases into water is temperature dependent. Thus, the actual concentrations of CFC-11 and CFC-12 found in the ground water depends on the ground-water temperature. Figure 15.23 depicts the expected concentration of the two main CFCs in ground water assuming equilibrium with the atmospheric concentrations at 9°C.

To employ CFC-11 and CFC-12 in dating, one would collect a sample of ground water, determine the concentrations of CFCs, and use the observed concentrations with Figure 15.23 to obtain a date of recharge. All of these steps are complicated. To assure that samples are completely isolated from the air, Busenberg and Plummer (1992) sealed samples in glass ampules by heat fusing immediately following their collection. The laboratory measurement of CFCs is also difficult because it is imperative to avoid contamination of samples (especially old samples) with modern air, which contains high CFC concentrations. Readers interested in a detailed discussion of the analytical techniques can refer to Busenberg and Plummer (1992).

An example in the application of CFC dating is the study of the shallow ground water of the Delmarva Peninsula, which is located along the eastern side of Chesapeake Bay (Dunkle and others, 1993). Table 15.6 summarizes data for a nest of wells (Figure 15.24) along a cross section (F1–F1') completed in a shallow surficial aquifer near the town of Fairmount. The concentrations of CFC-11 and CFC-12 (pg/kg) are used with the relationship presented in Figure 15.23 to provide the estimated age in years. Usually, the oldest of the pair of dates (CFC-11

water has been recognized since the mid-1970s (Thompson, 1976; Thompson and Hayes, 1979). These compounds have also been used in tracer experiments (Randall and Schultz, 1976). Interest in CFC dating has been renewed due mainly to efforts of scientists at the U.S. Geological Survey (Busenberg and Plummer, 1992; Dunkle and others, 1993).

Using the history of production of these compounds and direct atmospheric measurements since 1977, it is possible to develop relatively accurate estimates of ex-

Table 15.6 Sample Data for Ground Water from the Delmarva Peninsula

Well	Depth to Screen (Top) (m)	CFC-11		CFC-12	
		Concentration (pg/kg)	Age (Year)	Concentration (pg/kg)	Age (Year)
57	6.1	661.3	1985.0	281.9	1985.8
58	12.2	582.5	1982.3	260.4	1984.0
59	18.3	409.1	1976.0	216.0	1980.0
60	24.4	232.3	1971.5	152.4	1974.8
61	29.6	207.1	1970.5	95.3	1970.3

From Dunkle and others (1993).

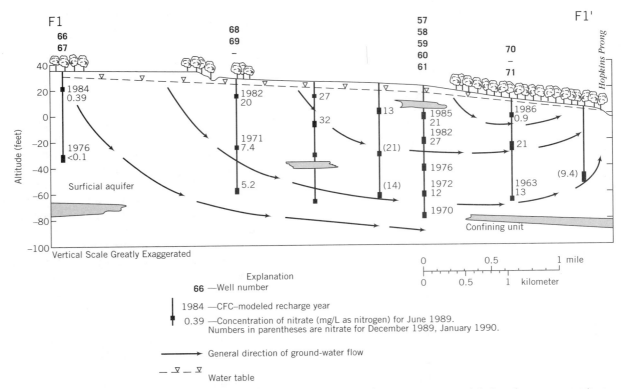

Figure 15.24 Cross section of the Fairmount network along F1–F1' showing CFC-modeled recharge years, nitrate concentrations (in milligrams per liter as N), and generalized ground-water flow paths. Reprinted from Dunkle and others, Water Resources Res., v. 29, p. 3837–3860, 1993. Copyright © by American Geophysical Union.

and CFC-12) is selected as the best estimate of the age. The distribution of ages (Figure 15.24) agrees with relative ages based on hydrologic arguments (Dunkle and others, 1993). In other words, samples furthest along the flow system are oldest.

As with many other of the age dating techniques, there are limitations. For example, CFC concentrations in anaerobic environments may be reduced through microbial degradation. Processes like dispersion and sorption can also influence CFC concentrations (Dunkle and others, 1993). In spite of limitations, the approach has potential for dating relatively young waters.

Problems

1. Explain why the isotopic composition of ground water found in glacial till in Canada and parts of the United States may be depleted in $\delta^{18}O$ and δD relative to precipitation sampled at the same locations.

2. Over the Plains region of the northern United States and Canada, carbonate-rich till overlies marine shale and sandstone. In nonarid areas, recharge from snowmelt might take on the following chemical composition as it moves downward through till and into shale bedrock (see Table 15.7).

Interpret the chemical evolution of the water in terms of the most likely mass transport processes.

3. Explain why the order in which ground water encounters minerals can be important in determining how the major ion chemistry evolves.

4. In carbonate rocks subject to recharge, the most dramatic changes in major ion chemistry occur over a relatively short distance as infiltration first enters the unit. Later changes are often almost insignificant by comparison. Explain why we use arguments related to kinetics and mineral equilibrium.

Table 15.7 Chemical Composition of Shallow Ground Water

Unit	Ca²⁺	Mg²⁺	Na⁺	HCO₃⁻	SO₄²⁻	Cl⁻	pH
			Concentration (milligrams per liter)				
till	79.0	50.0	210.0	436.0	61.0	14.0	7.80
shale	5.0	0.5	450.0	1044.0	6.0	53.0	8.10

5. Using the concept of redox zones, explain why H_2S gas is rarely found in ground water close to the water table in recharge areas.

6. Explain how tritium can be used in hydrogeological investigations.

7. Isotopic measurements on a sample yield a measured carbon-14 activity of 6.9% modern carbon, and a measured $\delta^{13}C_T$ as total carbon of 4.6‰. Assuming a $\delta^{13}C$ of 8‰ for calcite and a $\delta^{13}C$ of -27‰ for soil gas, estimate the age of the ground water.

8. Use the Tamers equation to determine the age of a water sample having a total inorganic carbon content (C_T) of 16.7 mM and a HCO_3^- concentration of 997 mg/L. The measured carbon-14 activity of the sample is 2.7% modern carbon.

9. The concentration of CFC-11 in four ground-water samples was measured as 1.9, 76.9, 176.1, and 430.4 pg/kg (data from Dunkle and others, 1993). Estimate the year in which recharge occurred, assuming a recharge temperature of 9°C.

Chapter 16

Mass Transport in Ground-Water Flow: Geologic Systems

The ability of ground water to dissolve rocks and minerals and to redistribute large quantities of dissolved mass has broad geologic implications. The so-called geologic work of ground water that is related to mass transport thus includes classical problems such as chemical diagenesis, the formation of some types of ore deposits, soil salinity, and evaporite formation. Our objective in treating these topics here is to develop a sense of how mass transport processes in ground water work at various scales and to describe specific physical and chemical factors that relate to these problems.

16.1 Mass Transport in Carbonate Rocks

Hydrogeologists for years have had a fascination with carbonate rocks. Not only are they often productive aquifers, but with karst give rise to spectacular surface landforms and cave systems. However, karst is but one manifestation of chemical diagenesis. The term *chemical diagenesis* refers to any chemical change that occurs in a sediment following deposition. Now, let us examine

in more detail some diagenetic changes that affect carbonate rocks. The process responsible for most of these changes is mixing—the displacement of connate water by recharging meteoric water, the same process we discussed in Chapter 15.

The environment of diagenesis for emerging carbonates is demonstrated in Figure 16.1. This idea was advanced by Vernon (1969) and modified by Hanshaw and others (1971). This type of environment develops everywhere on an emerging carbonate platform where fresh water drives out the marine fluids to some depth below sea level. Three diagenetic settings are evident: (1) a fresh-water zone, replenished by meteoric water; (2) a mixing zone where fresh water mixes with seawater; and (3) a deep marine zone, dominated by seawater. Environments such as those depicted in Figure 16.1 do not represent small-scale phenomena as this type of environment must, at some point in time, exist everywhere on an emerging carbonate bank. The slower the rate of emergence, the longer each part of the bank becomes subjected to meteoric water. Diagenesis progressively advances in the direction of the retreat of the sea from the area.

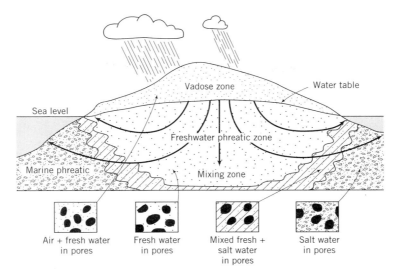

Figure 16.1 Diagrammatic cross section showing a mixing zone in an emerging carbonate rock (after Longman, 1982, as modified from Vernon, 1969). Reprinted by permission.

The Approach Toward Chemical Equilibrium in Carbonate Sediments

Modern carbonate sediment in marine environments consists of the metastable phase aragonite ($CaCO_3$), high-magnesium calcite, and a limited amount of low-magnesium calcite. Ancient carbonate rocks, on the other hand, are virtually void of aragonite and high-magnesium calcites and consist chiefly of low-magnesium calcite and dolomite. Hence, sometime after emergence from the marine environment, some major transformations take place, some accompanied by a gain in porosity and others by a loss in pore space. Virtually hundreds of references on this topic are available, with the broad review in Bathurst (1971) being the most extensive.

Inversion is the replacement of a mineral by its polymorph, for example, aragonite by calcite. Inversion occurs by either a solid-state transformation, which makes it very slow, or by a solution-reprecipitation process, which is likely faster (Carlson, 1983). Land and others (1967) view the process as occurring in five progressive stages, each of which is associated with a loss in porosity (Figure 16.2): stage I is unconsolidated sediment consisting of aragonite and high-magnesium calcite; stage V is a stabilized limestone consisting of mostly calcite.

If the transformation from aragonite to calcite is one of solution and reprecipitation, it may be viewed within a simple thermodynamic framework. Comparing the phases calcite and aragonite, calcite is the less soluble. However, we cannot stipulate a solubility for calcite without further information for some of the environmental parameters. For example, calcite has a solubility of about 100 mg/L if the partial pressure of CO_2 is about 10^{-3} bars

and 500 mg/L if the partial pressure of CO_2 is 10^{-1} bars, again at a pH of 7. Whatever the solubility of calcite, the solubility of aragonite is 1.38 times higher (D. Langmuir, 1987, personal communication). Thus, fresh water entering the emerged aragonite is undersaturated with respect to both calcite and aragonite and starts to dissolve the aragonite. When a sufficient amount of $CaCO_3$ is in solution, the solubility of calcite is achieved so that the solution becomes saturated with respect to calcite but remains undersaturated with respect to aragonite. With continued dissolution of aragonite, the waters become

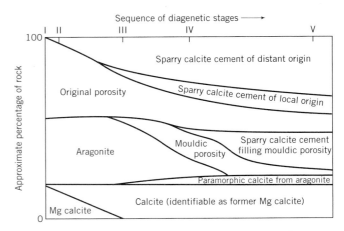

Figure 16.2 Sequential stages in the calcitization of a marine carbonate sand consisting of aragonite and Mg calcite (from Geol. Soc. Amer. Bull., Land and others, 1967. Reproduced with permission of the publisher, The Geological Society of America, Boulder, Colorado USA. Copyright © 1967, The Geological Society of America, Inc.).

supersaturated with respect to calcite, initiating its precipitation. With only aragonite as our initial starting material, and with calcite having a lower solubility than aragonite, it is patently impossible to achieve saturation with respect to the high-solubility phase, and aragonite is always in a state of dissolution. Hence, in the absence of an infinite supply, it disappears from the rocks, with all the CaCO₃ going to calcite. With high-magnesium calcites and aragonite in the starting material, the dissolution–reprecipitation process continues down the solubility gradient until the transformation to the lowest-solubility phase (calcite) is complete. Thus, with both aragonite and high-magnesium calcites in the initial starting material, the waters can be supersaturated with respect to the two lowest-solubility phases but never supersaturated with respect to all three. When the sediment becomes depleted of the high-solubility phase, the saturation concentration shifts downward, becoming lower than the saturation concentration of the remaining high-solubility phase. Thus the ultimate transformation to calcite.

The foregoing arguments are based purely on thermodynamic grounds. What cannot be ascertained from thermodynamic reasoning is the distance over which the flow system must persist until saturation with respect to the lowest-solubility phase (calcite) is achieved. If this state of saturation and, ultimately, supersaturation cannot be achieved, the whole aragonite deposit will be completely dissolved and removed from the geologic record with no coprecipitation of calcite. Because calcitic limestone is ubiquitous in the geologic record and absent in modern sediments, supersaturation with respect to calcite is undoubtedly achieved over distances that are short relative to the size of the emerging carbonate body. The determination of "how short" is a problem for transport theory.

Consider Figure 16.3a, which shows the spatial distribution of CaCO₃ in solution for the problem being addressed. Meteoric water enters the emerged aragonite deposit undersaturated with respect to aragonite and starts to dissolve the CaCO₃ deposit. The concentration eventually builds up to the saturation concentration of calcite. Continued dissolution of aragonite gives rise to a state of supersaturation with respect to calcite, and calcite starts to precipitate. As the concentration continually rises, aragonite is dissolving faster than calcite is precipitating. Finally, where the concentration no longer changes with distance in the system, the net dissolution is zero, that is, as much calcite precipitating as aragonite is dissolving. Note that this occurs at a solubility less than the solubility of aragonite. Eventually, aragonite will be totally removed from the system with precipitation of calcite. The distance to saturation with respect to calcite shown on Figure 16.3a is described as (Palciauskas and Domenico, 1976)

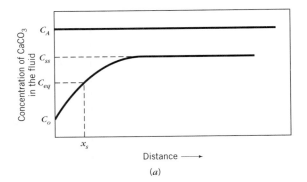

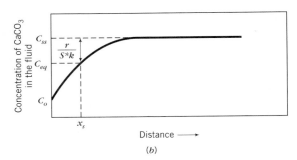

Figure 16.3 Spatial variation of CaCO₃ in solution where the entering fluid has a concentration of C_0, the equilibrium concentration for calcite is C_{eq}, the concentration C_{ss} denotes the position where the net dissolution is zero, and C_A indicates the saturation concentration of aragonite (from Geol. Soc. Amer. Bull., Palciauskas and Domenico, 1976. Reproduced with permission of the publisher, The Geological Society of America, Boulder, Colorado USA. Copyright © 1976, The Geological Society of America, Inc.).

$$x_s = \frac{-2\alpha_x}{1 - \left(1 + \dfrac{4S^*k\alpha_x}{v}\right)^{1/2}} \log \frac{C_{ss} - C_0}{C_{ss} - C_{eq}} \quad (16.1)$$

where x_s is the saturation distance; α_x is the longitudinal dispersivity; v is the ground-water velocity; S^* is the specific surface of the original material, defined as the surface area of the pores per unit bulk volume of sediment; k is a reaction rate coefficient, defined as the volume of the original deposit dissolved per unit surface area of rock per unit time; and the concentrations C are as defined in Figure 16.3

The numerator on the right-hand side of Eq. 16.1 suggests that the saturation distance increases with increasing values of longitudinal dispersivity. Within the denominator, the important dimensionless quantity is the Damköhler number

$$\frac{S^*k}{v}\alpha_x \quad (16.2)$$

The larger this quantity, the smaller the saturation distance. Hence, this interesting interplay of variables suggests that if the dissolution rate of the original material is fast with respect to the velocity at which the dissolution products can be carried away by moving ground water, the saturation distance is rapidly achieved, and the transformation to calcite takes place within a relatively short distance. If the rate of dissolution is slow compared to the velocity at which the products can be carried away, more of the sediment is dissolved with no coprecipitation of calcite; that is, the solution remains undersaturated with respect to calcite. If the reaction rate coefficient is close to zero, the saturation distance occurs at infinity; that is, the fluid remains undersaturated throughout the flow domain. Thus, a highly dispersive system thins the mass in solution, whereas a strongly advective one reduces the contact time between the dissolving agent (water) and the solid surfaces making up the porous medium, and both intrude upon the kinetics.

Equation 16.1 also applies to the approach to calcite equilibrium in terrestrial limestones where meteoric water enters the formations in an undersaturated state. In limestones, supersaturation may be achieved because of the dissolution of Ca- and CO_3-bearing minerals other than calcite (for example, gypsum), which are more soluble than calcite. The solution to this problem is given in graphical form in Figure 16.3b. As noted, the concentration once more increases exponentially with distance from C_0 to C_{ss}, passing through the saturation concentration at some intermediate point. In this diagram r is the production rate of the high-solubility phase, which produces ions in solution contributing to the supersaturation, and S^*k is the dissolution rate constant for the low-solubility phase. The quantity r has the units M/L^{3}T, or mass produced per unit volume per unit time, and S^*k has the units T^{-1}, or volume dissolved per unit volume per unit time. As the production rate r becomes small with respect to S^*k, C_{ss} approaches C_{eq}, and the saturation distance x_s becomes larger. Conversely, if r is large relative to S^*k, the saturation distance decreases. In a monomineralic terrain, that is $r = 0$, the terminal concentration is the saturation concentration and supersaturation is not possible.

The competition between kinetic and physical processes has been observed at the laboratory scale for dissolution–precipitation reactions at large Peclet numbers (advective systems). This is demonstrated in Figure 16.4, which shows the departure from saturation in two calcite packs that differ only in particle diameter. In these experiments, water that is undersaturated with respect to calcite is sent through a calcite pack at different velocities, and the concentration at the end of the pack is determined for each test. Given the exit concentration, the degree of saturation with respect to calcite is then determined. The departure from saturation is plotted

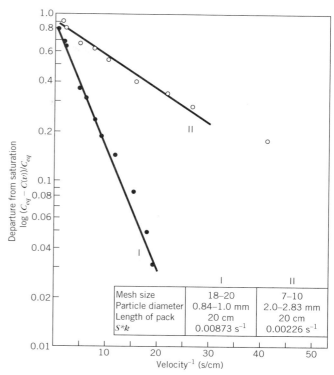

Figure 16.4 Departure from saturation versus velocity^{-1} from calcite-pack experiments (from data presented by Weyl, 1958, pp. 173–174) (from Geol. Soc. Amer. Bull., Palciauskas and Domenico, 1976. Reproduced with permission of the publisher, The Geological Society of America, Boulder, Colorado USA. Copyright © 1976, The Geological Society of America, Inc.).

against $1/v$ so that the slope of the line equals S^*kL. As shown in the figure, the departure from saturation increases (approaches one) with increasing velocity. For a given velocity, the degree of saturation increases with increasing values of S^*k. The dimensionless Damköhler group S^*kL/v for experiment I is 0.713 and for the second experiment is only 0.206. The most likely reason for this is that experiment I had finer particles and, consequently, a higher specific surface.

The Problem of Undersaturation

Up to now our main interest has been in the rate at which states of saturation and supersaturation are achieved in carbonate terrain. However, we recognize that the existence of karst is proof of the ability of water to remain in some undersaturated state over relatively large flow distances. The mechanism for this undersaturation was understood only when a number of dissolution experiments on limestone were carried out (Berner and Morse, 1974; Plummer and Wigley, 1976). It was found that dissolution rates decrease sharply as saturation is approached. For instance, Plummer and Wigley (1976) found dissolution rates could be approximated by a second-order equation at low solute concentrations

$$F = K(C_s - C)^2 \qquad (16.3)$$

where F is the dissolution rate, K is a constant that was found to vary by two orders of magnitude between specimens, C is the concentration of dissolved calcite, and the subscript s denotes a saturation concentration. However, closer to equilibrium, the reaction was found to vary between a fourth-order process and an approximate eighth-order process. Two specimens of Iceland spar were used in these experiments, the second of which was not as clear as the first. Dissolution rates of the second specimen were markedly slower than the first, and the authors suspected the calcite to have trace inhibitors. Trace concentrations of lead, copper, and phosphate in particular have been found to significantly reduce dissolution rates close to saturation (Terjesen and others, 1961; Berner and Morse, 1974). Thus it is the asymptotic approach to saturation that allows slightly undersaturated water to penetrate for considerable distance in carbonate rocks.

Dolomitization

The mineral dolomite, $CaMg(CO_3)_2$, frequently forms by replacement of calcium carbonate, $CaCO_3$. This obviously requires some input of Mg, normally considered to be provided by a moving fluid. The transformation ultimately results in a 13% reduction of the space occupied by the minerals. Land (1973) performed some calculations that focus on the number of pore volumes of fluid required to dolomitize completely 1 m^3 of sediment with 40% porosity without any loss of porosity. His calculations are as follows:

44 pore volumes of hypersaline brine that has precipitated gypsum

807 pore volumes of normal seawater

8070 pore volumes of normal seawater diluted by a factor of 10 by fresh water

From these calculations, dolomites may form in a variety of environments where the conditions just described hold. Folk and Land (1975) focus on these environments with a diagram that plots salinity against Mg/Ca ratios of the aqueous environment in which the transformation takes place (Figure 16.5). Thus, for dolomites to form in high-salinity *sabkhas* (the Arab word for salt flats), the Mg/Ca ratio must be on the order of 5:1 to 10:1, whereas for low-salinity lakes, the ratio can be less than 1:1. With freshwater aquifers with mixing zones similar to that shown in Figure 16.1, the ratio must be on the order of 1:1 or 2:1, depending on the salinity.

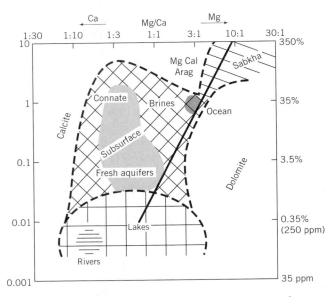

Figure 16.5 Fields of occurrence of common natural waters plotted on a graph of salinity versus Mg/Ca ratios. As salinity rises, higher Mg/Ca ratios are required to promote dolomitization (from Folk and Land, 1975). Reprinted by permission.

From the preceding discussion, the lower the salinity, the lower the required Mg/Ca ratio required for dolomitization. Three models have been proposed where ground water is the carrier fluid for the magnesium. Figure 16.6*a* is the reflux model proposed by Zenger (1972), where hypersaline waters in the *sabkha* environment have a density greater than that of the lagoonal seawater, so that waters with a high Mg/Ca ratio seep downward, dolomitizing the carbonates beneath the lagoon. A variation of this model, also proposed by Zenger (1972), is the evaporative pumping model, where marine waters are pulled landward toward the *sabkha* due to evaporation on the *sabkha* during dry spells (Figure 16.6*b*). Yet a third conceptual model has been proposed for the mixing zone concept of Figure 16.1 by Hanshaw and others (1971). In these environments, dolomite is able to form at low Mg/Ca ratios at progressively reduced salinities. At salinities on the order of 90 to 95% fresh water and 10 to 5% seawater, dolomite forms at a Mg/Ca ratio of about 1:1 (Scoffin, 1987). Figure 16.6*c* is taken from Land (1973) to account for the dolomitization of Pleistocene limestones of Jamaica.

16.2 Economic Mineralization

White (1968) has identified four factors responsible for the formation of ore deposits by ground water:

1. A source for the mineral constituents

2. Dissolution of the minerals in water

3. Migration of the fluids

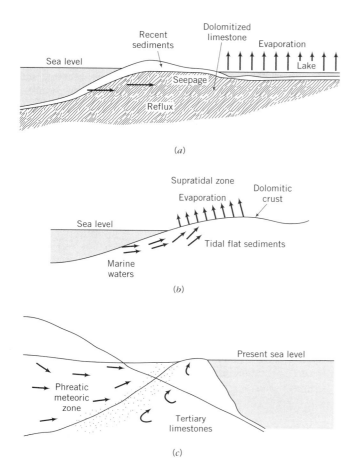

(a)

(b)

(c)

Figure 16.6 (*a*) Seepage reflux and (*b*) evaporative pumping models for dolomitization (after Zenger, 1972). Figure (*c*) shows mixing zone dolomite on the north coast of Jamaica (from Land, 1973). Panels (*a*) and (*b*) reproduced with permission of J. of Geological Education. Panel (*c*) reprinted from Sedimentology v. 20, by L. S. Land, Holocene meteoric dolitimization of Pleistocene limestone in North Jamaica, p. 411–422, 1973, with kind permission from Elsevier Science, NL, Sara Burgerhartstraat 25, 1055 KV Amsterdam, Netherlands.

4. Precipitation of the minerals in response to physical and chemical changes in the fluid and/or the porous medium.

With some rewording, a similar set of conditions can be applied to the migration and entrapment of hydrocarbons.

Origin of Ore Deposits

Mass precipitating from ground water or left as a product of weathering can in some cases form valuable mineral deposits. Examples of deposits formed in this way are listed in Table 16.1. These deposits form in a variety of different ground-water settings that range from local water table conditions in the case of nickel laterites or super-

gene sulfides, to regional scale convection deep in the crust in the case of porphyry copper or lode gold deposits. Space is not sufficient here to examine all these deposits in detail. However, a glance at the factors contributing to precipitation clearly demonstrates the role of the mixing zone in the precipitation process. The dominant contributing factors include abrupt chemical changes (changing E_H–pH conditions, decomplexation), mixing of oxidizing and reducing waters with precipitation at the redox front, the mixing of meteoric and magmatic fluids, and declining temperatures.

Roll-Front Uranium Deposits

A classic study on the relationship between the occurrence of uranium and ground-water flow is the work of Galloway (1978) on the uranium deposits of the Catahoula Formation of the Texas Coastal Plain. This Oligocene to Miocene–aged formation ranges in thickness from 60 to 300 m and was deposited in a complex fluvial environment. The uranium deposits in cross section are commonly crescent shaped and are formed by mineral precipitation at a redox boundary. An important aspect of this study involved mapping the distribution of fine- and coarse-grained facies because the distribution of uranium coincided with the distribution and orientation of permeable sand belts.

Galloway's conceptual model (Figure 16.7) proposed that leaching of volcanic ash layers provided the source of uranium. This process began shortly after the fluvial sediments were deposited. Flow through semiconfined sands transported uranium through the most permeable parts of the flow system. The flow pattern was controlled by the geometry of the channel sands and the presence of down-dip faults (Figure 16.7). The transport continued until uranium and other trace metals like iron precipitated somewhere close to the sulfate–sulfide redox boundary. The redox front itself would have continued to move down-gradient with existing minerals continually solubilized and precipitated. Effectively, the redox front provides a place for initially dispersed trace metals to be concentrated. The migration of the front must eventually stop because of postdepositional changes to the flow system that reduce the overall permeability and the flux of water. These changes could have included (1) the compaction and sealing of boundary aquitards, (2) the continued displacements along faults, and (3) the diagenetic modification of porous, permeable units (Galloway, 1978).

In oxidizing environments, the uranyl (+VI) species (for example, UO_2^{2+}, $UO_2CO_3^0$, $UO_2SO_4^0$, and UO_2OH^+) are mobile. Various uranyl complexes contribute to this mobility in an important way (Langmuir, 1978). Thus, uranium would be transported at early stages when the system is oxidizing. Mobility would be maintained as E_H declined through oxygen and iron manganese redox zones. Galloway (1978) suggested that reductants might

Table 16.1 **Some of the Different Kinds of Ore Deposits Whose Origin Depends in Part on Flowing Ground Water**

Type of Deposit	Example	Type of Flow System	Factors Contributing to Precipitation
Nickel laterite	New Caledonia	Shallow, water table	Weathering and changing E_H–pH at the water table
Laterite bauxite	Jamaica	Shallow, water-table drainage helped by karst	Accumulation as residual deposit accompanying weathering
Supergene sulfide	Chuquicamate, Chile	Shallow, water table	Weathering and changing E_H–pH at the water table
Calcrete uranium	Yeelirrie, Australia	Discharge end of shallow ground-water flow system	Dissolution from source rock, transport, and precipitation due to evaporation and decomplexation
Roll-front uranium	Texas Coastal Plain	Shallow ground water	Leaching of ash, transport, and precipitation at redox front
Unconformity-related uranium	Athabasca district Saskatchewan, Canada	Deep ground-water flow related to faulting	Mixing of oxidizing uraniferous and reducing waters
Mississippi Valley type lead zinc deposits	Pine Point, Northwest Territories, Canada	Gravity or compaction flow of brines from deep sedimentary basins	Leaching from sedimentary source rocks, transport, and deposition due to declining temperatures and possibly changing E_H–pH
Porphyry copper	San Manuel, Kalamazoo, Arizona	Convection in response to intrusion of a stock or dike	Mixing of meteoric and magmatic fluids and cooling
Lode gold deposits	Carlin, Nevada	Fluid convection of meteoric water deep in the crust	Leaching of source rocks, transport and deposition in fractured rocks due to declining temperature

include organic debris and sulfides in the sandstone or gases migrating vertically. The highly insoluble solids such as coffinite ($USiO_4$) or uraninite (UO_2) would be the stable phase as redox fell to a point close to where the sulfate–sulfide or a similar couple became controlling. The chemical behavior of uranium is shown graphically in Figure 16.8. The stability field for amorphous and crystalline forms of uraninite (not shown) lies in a similar position on the E_H diagram as coffinite (Galloway and Hobday, 1983). Details concerning what minerals precipitate and the origin of the strongly reducing conditions in the sandstones are discussed by Galloway (1978) and Galloway and Hobday (1983).

The distribution of other metals besides uranium is controlled by these same redox processes. For example, paralleling the occurrence of uranium in the roll front is a zone of iron disulfide mineralization. The E_H–pH diagram for iron shows that the stability field for pyrite lies in a similar position to that of coffinite or uraninite.

Mississippi Valley–Type Lead–Zinc Deposits

Carbonate-hosted lead–zinc deposits are found in the midcontinent region of the United States and in Canada and are often referred to as Mississippi Valley–type deposits. Ohle (1959), and later Anderson and Macqueen (1982), summarized the similarities between these depos-

its. All occur primarily, though not exclusively, in preferred horizons in carbonate rocks and are sufficiently conformable to the bedding so as to be described as stratiform or stratabound. All contain fluid inclusions filled with brines that indicate a temperature of formation of 100°C to 150°C. Most have similar structural associations, being located in regional highs such as the Ozarks, Cincinnati Arch, and so on, where they are found near the top or along the flanks of the domal structures (Figures 16.9 and 16.10).

A postulated ground water origin for these deposits was stated as early as 1854 by Whitney. By the late 1800s, most investigators agreed on the meteoric origin. Their differences, however, had to do with the source of the meteoric water. One group favored a downward percolation theory from overlying source beds whereas the other group favored lateral- or upward-moving waters. Some difficulties with both these theories came about with the later introduction of fluid inclusion data, which suggested a formation temperature of 100°C to 150°C. With the current geothermal gradient, the lowest temperature cited requires a depth in excess of 5000 ft whereas the deposits known to date are above 2000 ft in depth.

Water from Compaction
Stoiber (1941) was one of the first to propose that the fluids that formed the lead–zinc deposits in the Tristate region were closely related

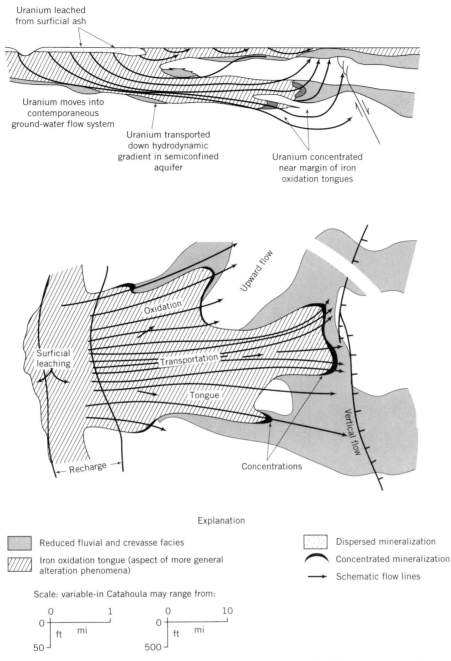

Uranium leached
from surficial ash

Uranium moves into
contemporaneous
ground-water flow system

Uranium transported
down hydrodynamic
gradient in semiconfined
aquifer

Uranium concentrated
near margin of iron
oxidation tongues

Upward flow

Oxidation

Surficial
leaching

Transportation

Tongue

Vertical flow

Recharge

Concentrations

Explanation

Reduced fluvial and crevasse facies

Iron oxidation tongue (aspect of more general
alteration phenomena)

Dispersed mineralization

Concentrated mineralization

Schematic flow lines

Scale: variable-in Catahoula may range from:

0   1   0   10
0   ft   mi   0   ft   mi
50   500

Figure 16.7 Diagrammatic representation of the origin of roll-front-type uranium
deposits in the Catahoula fluvial systems of Texas (modified from Galloway, 1978).
Reproduced from Econ. Geol., 1978, v. 73, p. 1656–1676.

to oil field brines and were derived from compaction. Nobel (1963) advanced this idea in general to account for all Mississippi Valley–type deposits. This idea was accepted and amplified by Jackson and Beals (1967), Dunham (1970), and Dozy (1970). Dozy (1970) also noted that both oil and mineral deposits were common around the Ozark uplift area, indicating that the processes that control the migration of oil and ore should be in

agreement. The conceptual model of Nobel (1963) is shown in Figure 16.11. As noted, the ore-forming fluids move out as hot brine through permeable zones to precipitate eventually on basin margins that currently act as ground-water recharge areas.

The hypothesis of Nobel (1963) and others has been examined from the perspective of mathematical (numerical) models by Sharp (1978) and Cathels and Smith

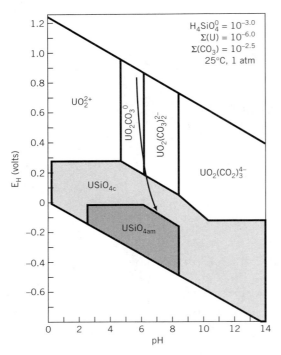

Figure 16.8 E_H–pH diagram showing the stability fields of crystalline and amorphous coffinite ($USiO_4$) in carbonate-rich ground water containing approximately 60 mg/L of H_2SiO_3, 200 mg/L total carbonate species, and 240 μg/L uranium. The arrow depicts the possible variation in E_H–pH along the flow system (modified from Galloway and Kaiser, 1980).

(1983). Sharp (1978) used a one-dimensional compactional model coupled to heat transport with provisions for faulting for the overpressured fluids of the Ouachita basin to account for lead–zinc deposits near the Ozark Dome. Faulting was presumed to rupture the overpressured basin during the Late Pennsylvanian–Permian period, providing hot pore fluids to the northern Arkansas–southeastern Missouri area where lead–zinc deposits are known to occur. The main emphasis of Sharp's (1978) work was to determine if critical temperatures could be maintained along with peak flow rates required in the mineralization. Cathels and Smith (1983) suggested a pulselike release of compactional fluids based on the modeling of a basin similar to the Illinois Basin. The consensus of these studies indicates that the postulated compactional origin of Mississippi Valley-type lead–zinc deposits is possible.

Gravity Flow Origins The Pine Points district in the Northwest Territories of Canada contains carbonate-hosted lead–zinc deposits similar in many respects to those of the midcontinental United States. In this region, Jackson and Beals (1967) postulated that the metals were derived from an argillaceous sediment or from metal sulfides in associated black shales. A compaction model was proposed to account for the Pine Point ore deposit. Garven and Freeze (1984a,b) examined a gravity-driven flow system as a potential origin for these stratabound ore deposits. Their conceptual model is shown in Figure 16.12, where fluid flow is directed across a thick shale unit and focused into a basal carbonate unit. The ore-forming constituents are leached from the shales and are deposited at the discharge end of the basin. The

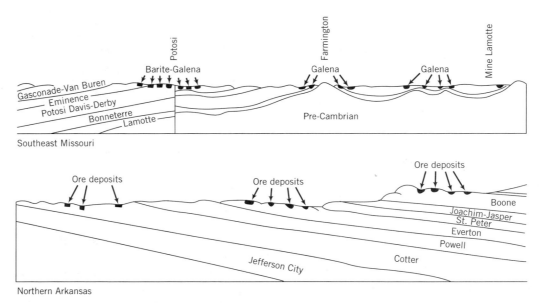

Figure 16.9 Generalized sections of two lead–zinc mining districts showing structural and stratigraphic relations of the ore deposits (from Geol. Soc. Amer. Special Paper, v. 24, Bastin, 1939. Reproduced with permission of the publisher, The Geological Society of America, Boulder, Colorado USA. Copyright © 1939, The Geological Society of America, Inc.).

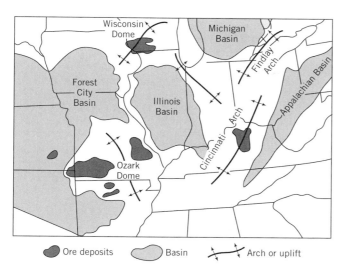

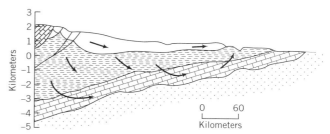

Figure 16.12 Conceptual model of fluid flow in a carbonate unit that could be the locus of a stratabound orebody (from Amer. J. Sci., Garven and Freeze, 1984a. Reprinted by permission of Amer. J. Sci.).

Figure 16.10 Map showing the relationship of the ore deposits to sedimentary basins and arches and uplifts (modified from Heyl and others, 1970).

noted in the figure, the fluid in the margin sediments may be forced into the adjacent continent, that is, according to Oliver (1986), the thrust sheet acts like a squeegee driving fluids ahead of it.

As noted in Figure 16.13, metamorphism of coal decreases with distance from the orogenic belt and gas fields are closer to the orogenic belt than oil fields, both presumably due to higher temperatures. The lead–zinc deposits of the type we have been discussing are farthest from the belt and presumably were carried by the same fluids.

It has been mentioned that Dozy (1970) noticed the close association between lead–zinc deposits of the Mississippi Valley–type and liquid hydrocarbons. This association is shown in Figure 16.14. The hypothesis of Oliver (1986) suggests that both oil and minerals are transported by the tectonic brines from the main orogenic belts on the continent. Such a system may be seen to be compatible with the Garven–Freeze (1984a, b) gravity flow model. Here the tectonic brines originating from the precollision marginal sediments may mix with meteoric water driven by the hydraulic head established by tectonically created topographic relief (Figure 16.15).

In recent years we have seen parts of the hypothesis by Oliver (1986) embraced by members of the modeling community—at least those parts that can be readily programmed for a digital computer. Thus, in 1988, Garven proposed that oil migration in the Alberta Basin throughout Tertiary time was the result of a foreland basin uplift. The similarity between the Appalachian orogen and the Cordilleran orogen in Alberta has been reasonably well established. Garven (1988) states that the transport of oil waned in the Late Tertiary as the regional flow dissipated due to further erosion of the landscape. Koons (1988) discusses the fluid flow regime due to a continental collision in New Zealand and the occurrence of active zones of mineralization. Bethke (1988) and Bethke and Marshak (1990) conclude on the basis of model studies that deep basin brines have migrated for hundreds of kilometers through the interior basin of North America. The brines redistributed hydrocarbons in their present reservoirs and formed Mississippi Valley–type deposits.

mathematical models included not only fluid and energy transport, but mass transport as well. Their simulation results indicate that gravity-driven flow can provide favorable flow rates, temperatures, and metal concentrations for the formation of an ore deposit in a relatively short geologic time. Bethke (1986) also considered a topographic drive system to account for Mississippi Valley–type deposits from Illinois basin brines.

The Expulsion of Fluids from Orogenic Belts and Continental Collisions

In 1986, Oliver proposed a mineralization hypothesis of continental proportions, the essence of which calls for the expulsion of fluids from buried and deformed continental margin sediments to foreland basins and the continental interior during overthrusting associated with plate collisions. A simplified history of the Appalachian orogen in response to the closing of the ancestral Atlantic Ocean and collision of North America and Africa is shown on Figure 16.13. As

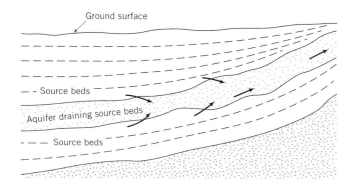

Figure 16.11 Idealized section showing aquifer transmitting water of compaction from source beds (from Noble, 1963). Reproduced from Econ. Geol., 1963, v. 58, p. 1145–1156.

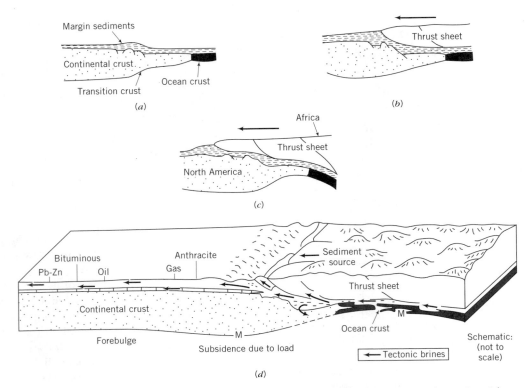

Figure 16.13 Simplified history of Appalachian orogen. (*a*) Passive continental margin with margin sediments. (*b*) Thrust sheet overrides margin sediments; load causes subsidence. (*c*) Continent–continent collision advances thrust sheet and terminates orogeny. (*d*) Block diagram of orogen at time corresponding to panel (*b*). Heavy arrows schematically illustrate flow of the brines expelled to buried sediments. Gas and anthracite deposits are closer to orogen than oil and bituminous coal, respectively. Continental crust is 35 km thick; horizontal dimension in diagram is 500 km. (from Geology, Oliver, 1986. Reproduced with permission of the publisher, The Geological Society of America, Boulder, Colorado USA. Copyright © 1986, The Geological Society of America, Inc.).

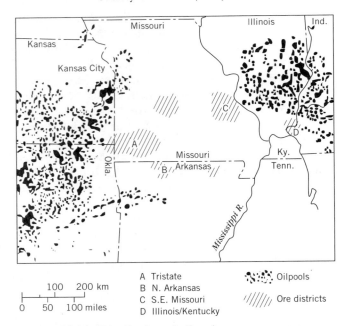

Figure 16.14 Distribution of oil and ore occurrences around Ozark uplift (from Dozy, 1970). Reprinted by permission of the Institution of Mining and Metallurgy.

They mention further that the migration events coincide both temporally and spatially with episodes of orogenic deformation along the margins of the craton, which provided the topographic relief to establish the paleo flow systems. Garven (1995) presents a good overview of continental scale ground-water flow and geologic processes.

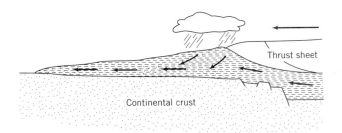

Figure 16.15 Simplified illustration showing injection of tectonic brines into hydrologic flow induced by tectonically created topography (from Geology, Oliver, 1986. Reproduced with permission of the publisher, The Geological Society of America, Boulder, Colorado USA. Copyright © 1986, The Geological Society of America, Inc.).

Noncommercial Mineralization: Saline Soils and Evaporites

Any time that mineralized ground water gets within a few meters of the ground surface in an arid climate, the potential exists for saline soils or thick evaporite deposits to form. Evaporation concentrates whatever salts were initially present in the water. The quantity of salts generated at or near the ground surface is a function of the mass flux and the time over which the flux is operative.

Minerals precipitate in the inverse order of solubility as the solution reaches saturation with respect to the various solid phases. The particular mineral assemblage that forms depends very much on the composition of the starting water. Figure 16.16, which is a slightly modified version of one presented by Hardie and Eugster (1970), describes the possible pathways along which the major ion chemistry of an evaporating water could proceed. As is apparent on the figure, there are critical points at which the solution will evolve in one of two possible ways. The direction of change depends on the relative abundance of key cations and anions. Drever (1982) discusses this model in detail along with a few more recent refinements.

Here we will consider two examples. The first is a case of saline soil development in east-central Saskatchewan, Canada (Meyboom, 1966b), and the second is an evaporite deposit in the northern Salt Basin in west Texas and New Mexico (Boyd and Kreitler, 1986). Shown in Figure 16.17 are features of "the Great Salt Plain" that occupies a broad depression between hills on the east and west sides of the study area in Saskatchewan. The shading outlines an area with large patches of salt efflorescence (mainly Na_2SO_4) on which only a few salt-tolerant plants are able to survive. The area of salt accumulation coin-cides almost exactly with an area of ground-water discharge. We can see the evidence in Figure 16.17 by observing that salt accumulates in places where the head gradient between the shallow ground water and a deeper artesian system is at a maximum.

The Texas example illustrates many of the same features, only on a slightly larger scale. Recharge originating in rocks and alluvial fans peripheral to the salt flats is eventually evaporated in areas where the water table is less than 1 m from the ground surface (Boyd and Kreitler, 1986). Gypsum is the most commonly found mineral, occurring in the form of gypsum mud and gypsum sand. However, dolomite, calcite, magnesite, halite, and native sulfur are also found.

Boyd and Kreitler (1986) have documented the pattern of major ion evolution along the flow system. Ground-water flow from the Permian limestone bedrock is initially quite fresh. Salinity increases progressively down the flow system with a shift toward Na^+ and Cl^- dominance as the effects of evaporation become evident. Ultimately, the ground water evolves to a Na-Mg-SO$_4$-Cl–type brine with TDS values between 50,000 and 300,000 mg/L (path IV in the Hardie-Eugster model) in the most evaporated parts of the salt flats. Halite, in fact, occurs locally in areas where the brine concentration is close to the maximum observed.

Hardie (1991) presents a good overview on the formation of evaporites.

16.3 Migration and Entrapment of Hydrocarbons

Whenever two immiscible fluids like oil and water occupy the same porous medium, there will be a simultaneous flow of each fluid, with each propelled by its own driving force. The treatment of this topic is complex so we will use here the abrupt interface approximation. This interface is assumed to segregate the fluids rigidly but will be subject to adjustments with the flow of one or both of the fluids. As with the problem discussed for the intrusion of salt water, this approximation will generally be a good one if the mixing zone between the fluids is relatively narrow compared to the region occupied by the fluids.

Displacement and Entrapment

When two homogeneous fluids occupy adjacent regions in space (Figure 16.18), each is characterized by its own hydraulic head (Hubbert, 1953)

$$b_1 = z + \frac{P}{\rho_1 g} \qquad (16.4a)$$

$$b_2 = z + \frac{P}{\rho_2 g} \qquad (16.4b)$$

where ρ_2 is greater than ρ_1. Because the pressure is continuous across the interface, the interface is shared by both

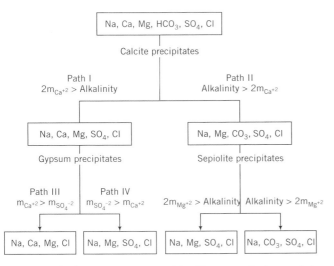

Figure 16.16 Conceptual model of brine evolution and the sequence of mineral precipitation (from Boyd and Kreitler, 1986; modified from Hardie and Eugster, 1970). Mineral. Soc. Am. Spec. Pub., v. 3, p. 273–290. Reprinted with permission.

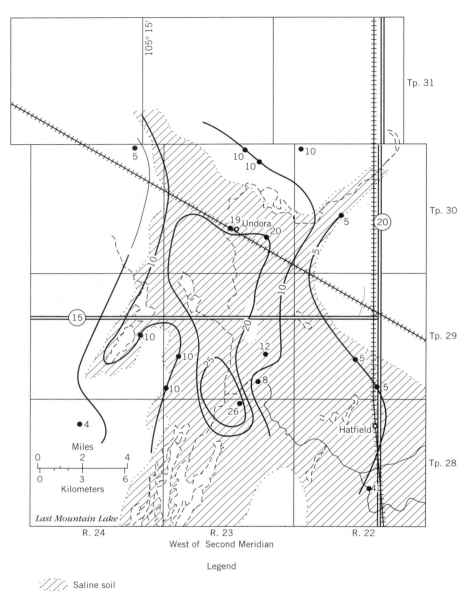

Legend

///// Saline soil

●12   Head difference, in feet, between piezometric surface of artesian aquifer and water table

—25— Approximate line of equal head difference, in feet, between piezometric surface of artesian
       aquifer and water table

Figure 16.17 Saline soils on the Great Salt Plain, Saskatchewan, and their relationship to the hydrogeology. (Reprinted from Meyboom, P., 1966b. Groundwater Studies
in the Assiniboine River Drainage Basin, Part 1: The Evaluation of a Flow System in
South-central Saskatchewan. Geological Survey of Canada, Bull. v. 139, 65 p.).

fluids. This makes it possible to solve for the pressure in
either of Eqs. 16.4 and substitute the result in the other
equation. Solving for the less dense fluid

$$b_1 = \left(\frac{\rho_2}{\rho_1}\right) b_2 - \left[\frac{\rho_2 - \rho_1}{\rho_1}\right] z \qquad (16.5)$$

which is an expression for the head of a lighter fluid b_1
in terms of the head of the more dense fluid. Solving this
expression for z gives

$$z = \left(\frac{\rho_2}{\rho_2 - \rho_1}\right) b_2 - \left(\frac{\rho_1}{\rho_2 - \rho_1}\right) b_1 \qquad (16.6)$$

where z is the elevation of the interface given in Figure 16.18.

A question now arises as to the slope of the interface
if one or both of the fluids are in movement. The slope
of the interface is given as the change in its elevation z
along the interface s

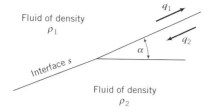

Figure 16.18 Slope of the interface between two immiscible fluids (from Hubbert, 1940). Reprinted by permission of the J. of Geol. Univ. of Chicago Press. Copyright 1940.

$$\sin \alpha = \frac{\partial z}{\partial s} = \left(\frac{\rho_2}{\rho_2 - \rho_1}\right)\frac{\partial h_2}{\partial s} - \left(\frac{\rho_1}{\rho_2 - \rho_1}\right)\frac{\partial h_1}{\partial s} \quad (16.7)$$

Substituting Darcy's law, where the gradient is taken as $\partial h/\partial s$, gives

$$\sin \alpha = \frac{\partial z}{\partial s} = -\left(\frac{\rho_2}{\rho_2 - \rho_1}\right)\frac{q_2}{K_2} + \left(\frac{\rho_1}{\rho_2 - \rho_1}\right)\frac{q_1}{K_1} \quad (16.8)$$

If the dense fluid q_2 is not in motion

$$\sin \alpha = \frac{\partial z}{\partial s} = \left(\frac{\rho_1}{\rho_2 - \rho_1}\right)\frac{q_1}{K_1} \quad (16.9)$$

As $\rho_2 > \rho_1$, $\sin \alpha > 0$ and the slope of the interface will increase upward in the direction of flow. The greater the flow rate q_1 of the lighter fluid, the greater the slope. It follows that a light fluid will always displace a denser fluid if the latter is not in motion, provided the displacement is done immiscibly. Consider, for example, that the heavy fluid is taken as a liquid hydrocarbon immobilized within an anticlinal structure and the light fluid is taken as a gas. If the anticlinal structure is filled with the liquid hydrocarbon and a later generated gas phase enters the reservoir, the liquid hydrocarbon can be totally displaced by the gas. A similar type of displacement is shown in Figure 16.19 for salt water and fresh water in a synclinal structure where the salt water is not in motion and the displacement is assumed to be immiscible. Normally, highly saline water would tend to lie at the bottom of such synclines and less dense moving fluids would tend to rise and pass over or around the more dense mass, giving the pattern shown in Figure 16.19.

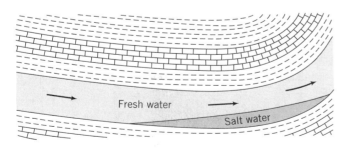

Figure 16.19 Position of salt water in syncline under conditions of freshwater movement (from Hubbert, 1953). Reprinted by permission.

If the denser fluid is in motion and the light fluid static, $\sin \alpha$ is negative, and the slope of the interface will increase downward in the direction of flow. The greater the flow rate of the denser fluid q_2, the greater the negative tilt. This is demonstrated in Figure 16.20 for groundwater movement in an anticlinal structure containing a nonmoving liquid hydrocarbon. The relationship between the slope of the oil–water interface and the hydraulic gradient for the fluid can be expressed (Hubbert, 1953)

$$\tan \alpha = \frac{dz}{dx} = \left(\frac{\rho_w}{\rho_w - \rho_o}\right)\frac{dh}{dx} \quad (16.10)$$

where the subscript w is for water and o is for oil. As the hydraulic gradient decreases and approaches zero, the slope of the oil–water interface approaches zero (horizontal). According to Hubbert (1953), structures with leeward-closing dips less than the tilt of the oil–water interface cannot hold oil. Hence, in the presence of low-velocity ground water (flat hydraulic gradients), even minor structures become more efficient in their capability to hold the hydrocarbon. On the other hand, for steep hydraulic gradients, the hydrocarbon may be flushed out of the trap.

Basin Migration Models

Tóth (1988) has summarized the various forces that give rise to hydrocarbon migration. These include sediment

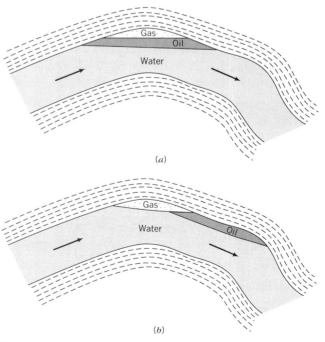

Figure 16.20 Relationship between the tilt of the oil–water interface and the intensity of ground-water flow (from Hubbert, 1953). Reprinted by permission.

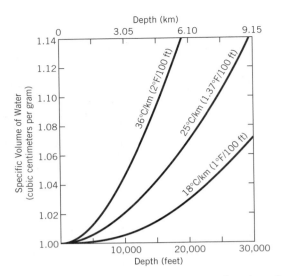

Figure 16.21 Specific volume of water as a function of depth for specified geothermal gradients (from Magara, 1978). Reprinted from Compaction and Fluid Migration: Practical Petroleum Geology, by K. Magara, 1978, with kind permission from Elsevier Science, NL, Sara Burgerhartstraat 25, 1055 KV Amsterdam, Netherlands.

compaction and rebound, buoyancy, gravity flow, confined flow, gas expansion, thermal expansion of liquids, molecular diffusion, and osmosis. Most of these driving forces have been previously discussed. Tóth (1988) then arranges basin migration models into four classes according to the dominant force driving the fluids. These include (1) compaction, (2) compaction-heat, (3) compaction-gravitational, and (4) gravitational.

The compaction models selected for review are those of Jacquin and Poulet (1970, 1973), Bonham (1980), and Bethke (1985). Jacquin and Poulet (1970, 1973) set out to determine the flux, the pressure and the flow direction

as a function of time in a hypothetical basin. They conclude that the potential for migration is greatest in reservoir rocks during the active sedimentation period. Bonham (1980) considers progressive compaction wherein during the early years of deposition, water is expelled upward from the older beds to the younger ones. With increased time, the older sediments achieve their minimum porosity with no further expulsion of the pore waters. In the later stages of compaction, no additional water for migration is provided from the already compacted units. Thus the driving force diminishes. Bethke (1985) examines these conditions with a numerical model (see Chapter 9).

The compaction-heat drive model merely recognizes that the specific volume of water increases with increasing depth (temperature) (Figure 16.21) and the direction of movement of this water coincides with the direction of movement caused by isothermal compaction.

The compaction-gravity drive system has been discussed by Coustau and others (1975) along with several other investigators and is based essentially on the evolution of abnormal pressure basins to mature gravity-flow systems as demonstrated in Chapter 8. The compaction system is thought to be the cause of hydrocarbon accumulation in traps. The gravity flow superimposed on this system can flush these hydrocarbons, except in the centrally located parts of the basin. Tóth (1988) presents an interesting modification of this system based on the hydrogeologic regimes shown in Figure 8.4. Given a suitable cap rock, no vertical or lateral escape of hydrocarbons is possible in the vicinity of converging flow fields. This type of flow is exemplified by the convergence of continental gravity flow and marine compaction flow.

Figure 16.22 demonstrates the various hydraulic, hydrodynamic, and structural traps that can develop within the gravity-driven flow systems. Structural traps occur

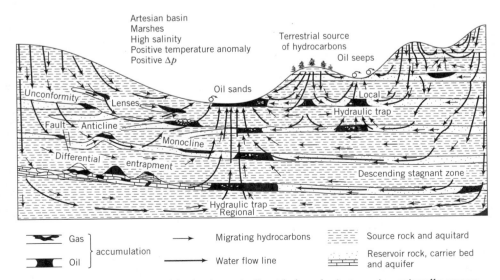

Figure 16.22 Hydraulic and hydrodynamically aided geologic traps in regionally unconfined fields of gravity flow of formation water (from Tóth, 1980). Reprinted by permission.

where faulting result in offset of permeable horizons, causing a potential reservoir rock to be bounded by an impermeable one. Hydrodynamic traps can occur in anticlinal structures whereas hydraulic traps occur where two flow systems converge.

16.4 Self-Organization in Hydrogeologic Systems

As illustrated in the previous sections on mixing zone phenomena and economic mineralization, the interactions between fluid flow and chemical reactions in disequilibrium can result in changes in the types, amounts, and locations of minerals and solutes within a flow system. These changes fall into the general category of self-organization, defined as the patterning of one or more descriptive variables that result from reaction-transport feedbacks (Ortoleva and others, 1987a,b). Self-organization in ground-water flow systems may commonly involve patterning associated with dissolution, with precipitation, and with interactions between dissolution and precipitation and other processes.

Patterning Associated with Dissolution

Dissolution-induced patterning is the simplest of the self-organizational mechanisms noted and, as such, has been most widely modeled, both numerically and in the laboratory. Consider the most basic case, that of a medium containing a single reactive component into which un-

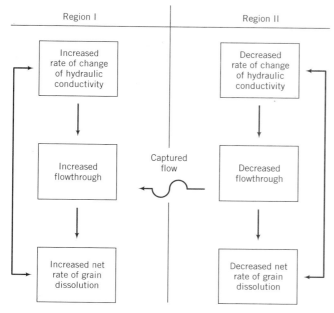

Figure 16.23 Dissolution-transport feedback loop for a region (I) of initially higher porosity and hydraulic conductivity adjacent to a region (II) (from Amer. J. Sci., Ortoleva and others, 1987a. Reprinted by permission of Amer. J. Sci.).

dersaturated ground water is flowing. Zones of initially higher porosity and hydraulic conductivity become preferentially more porous and permeable as a result of dissolution. In turn, flow is increasingly directed toward those zones, and dissolution can be enhanced there (Figure 16.23). The feedback loop thus established is integral to self-organization. Numerical modeling of a generic, single-component system by Steefel and Lasaga (1988) indicates that the propagation of preferential channels in zones of dissolution, or patterning of porosity and hydraulic conductivity, is a function of the rate of reaction relative to the rate of fluid flow. When the reaction is sufficiently rapid relative to flow, the length of such channels is proportional to the ratio of advective to dispersive transport, which is embodied in the Peclet number. That is, when there is no competition between flow and reaction, the Damköhler number plays no role in describing the process. Where the reaction is slow relative to flow, channel length directly depends on the ratio of the reaction rate constant to the flow rate, embodied in the Damköhler number. Not surprisingly, in both situations, the relative spatial extent of patterning is dependent on the scale of transport being considered.

An obvious natural analog to the single-component system is a carbonate medium. (It should be remembered that dissolution of the single-phase aragonite creates a multicomponent problem involving precipitation; this more complicated case will be touched on later.) For a calcitic limestone without a relatively inert matrix (as would be provided by quartz grains), so that dispersion is minimal, significant channel lengths should be possible. The formation of karst conduits is a logical consequence. In a related vein, petroleum engineers have conducted experimental studies of flow enhancement by acid injection into carbonate reservoir rocks. Hoefner and Fogler (1988) introduced dilute hydrochloric acid into limestone and dolomite cores, then, after evacuating the cores, injected an alloy into each to make a casting of the channel network resulting from dissolution. Each core was subsequently dissolved in concentrated acid, leaving only the casting. The relatively rapid dissolution of calcite at moderately acidic pH was reflected in the limestone castings by the persistence of a single, long channel. For dolomite, whose dissolution was significantly slower, pronounced channelization was evident only at low flow rates. These results, while forced by lower values of pH than those typical of natural systems, confirm the relations of patterning to the Peclet and Damköhler numbers obtained by Steefel and Lasaga (1988).

Patterning Associated with Precipitation and Mixed Phenomena

It is conceivable that repetitive patterns of precipitates and, in turn, of reduced porosity and hydraulic conductivity could result from ground-water mixing in a porous medium without concurrent dissolution. This situation might exist, for example, where different flow systems meet at a common discharge zone. However, given the prevalence of multimineralic media, it is possible that precipitation-induced patterning linked to local dissolution is more widespread. The systematic inversion of aragonite to calcite is one such water–rock interaction. The supersaturation–nucleation–depletion cycle (Ortoleva and others, 1987a) represents a theoretical mechanism for repetitive banding of precipitates. Consider water containing an aqueous species A as infiltrating a reactive medium, dissolving a species B from disseminated grains of a mineral X, so that supersaturation results with respect to a solid AB. Precipitation of AB at *a* site within the medium can repress nucleation of the solid at another site down-gradient in favor of continued growth of the solid at the first site. Eventually, though, up-gradient diffusion of B will become negligible, permitting nucleation and growth at a site even farther down-gradient, and so forth.

The localization of precipitation by up-gradient diffusion may yet be overwhelmed by considerations of hydraulic efficiency. If precipitation of a relatively insoluble solid significantly reduces the available pore space, initial dissolution channels can be abandoned in favor of forming new, adjacent channels. Rege and Fogler (1989) have demonstrated this process by injecting an acidic solution of ferric chloride into limestone cores, inducing calcite dissolution and ferric hydroxide precipitation. Rege and Fogler (1989) observed fluctuations in hydraulic conductivity during runs by monitoring changes in the pressure gradient along each core. After each run, following Hoefner and Fogler (1988), a casting of the reaction-induced channel network was created. At low initial flow rates (high Damköhler numbers), precipitation caused eventual refocusing of flow into a few channels in which flow rates increased sufficiently to prevent further plugging. For high initial flow rates (low Damköhler num-

bers), plugging and diversion of initial channels were minimized.

Beyond the simple cases of reaction-transport feedbacks discussed thus far, self-organization phenomena may extend to multiple, interactive precipitate bands and dissolution channels in "impure" media. Mechanical feedbacks can additionally come into play, as dissolution may induce compaction, thereby potentially reducing porosity. Other phenomena operative with fluid flow may include buoyancy-driven convective cells resulting from density differences associated with dissolution and precipitation. Dewers and Ortoleva (1988) have considered the role of mechanical feedbacks and convective cells in the generation of relatively impermeable seals in sedimentary basins.

Self-organization phenomena may also extend to reactions that are strongly dependent on temperature. For heterogeneous chemical processes, it can be argued that it is not possible for the overall rate of reaction to increase without limit if heat is added. If surface reactions are normally small compared to diffusion rates at a given temperature, an increase in temperature can cause an exponential rise in the rate of surface reaction, but will have a small effect on the mass transport controlled by diffusion. Surface kinetics will then speed up and eventually overhaul and become fast with respect to transport, thereby forcing the reaction process from the kinetic to the diffusion-controlled regime (Frank-Kamenetskii, 1969). Homogeneous liquid–liquid reactions, on the other hand, may increase without limit in high-temperature environments. These same ideas apply to exothermic reactions taking place in the absence of outside energy. At least in theory, chemically controlled exothermic reactions can pass from the kinetic to the diffusion regime, whereas endothermic reactions, which are self-regulating in a thermal sense, cannot. An overall view of self-regulatory or cybernetic chemical systems is given by Frank-Kamenetskii (1969).

Problems

1. Consider the following reactions:

$$A \text{ (solid)} \underset{r_2}{\overset{kC_b}{\rightleftharpoons}} B \text{ (solution)} \overset{r_1}{\longleftarrow} D \text{ (solid)}$$

a. In the absence of the solid D, write an equation for the equilibrium concentration C_{eq} if there is equilibrium between A and B.

b. Provide an expression for the concentration when the net dissolution rate is zero.

2. For oil and gas in an anticline consider the following:

$$\frac{\rho_w}{\rho_w - \rho_o} = 10 \qquad \frac{\rho_w}{\rho_w - \rho_g} = 2$$

The hydraulic gradient is 10 ft/mile and the leeward

closing-dip of the anticline is 80 ft/mile. Can the structure hold oil? Can the structure hold gas?

3. From the following figure, b_w may be taken as a point on a piezometric surface and z as a point on a structure contour map. After multiplying these by appropriate density amplification factors, the difference gives the head at this point in an oil body (b_o). Explain (hint: see Eq. 16.5).

Problem 3

4. Develop a hydrodynamic approach for petroleum exploration that would take advantage of the typical kinds of reservoir data that become available once a sedimentary basin becomes extensively drilled.

5. Assume an interface between two fluids of different density in a homogeneous, isotropic medium where $\rho_2 > \rho_1$. Comment on whether or not the following statements are true. Include the reasoning you apply in arriving at the conclusion.

a. A sloping interface always indicates movement of one or both of the fluids.

b. A horizontal interface is only possible if $q_1 > q_2$.

c. If one of the fluids is a liquid hydrocarbon and the other is a dense salt water in motion, $\sin \alpha < 0$ when the hydrocarbon is trapped in an anticlinal structure and the interface slopes downward in the direction of flow. The slope will be steeper with increasing q_2.

d. If one of the fluids is a gaseous hydrocarbon and the other is a moving salt water, $\sin \alpha < 0$ when the hydrocarbon is trapped in an anticlinal structure and the interface slopes downward in the direction of flow. The slope, however, will be less steep than in (c).

e. A light fluid will always displace a heavier fluid provided the heavy fluid is not in motion and the displacement is done immiscibly.

Chapter 17

Introduction to Contaminant Hydrogeology

The next three chapters examine theoretical and practical concepts related to the occurrence of contaminants in the subsurface. The term *contaminant* is used generally to refer to dissolved constituents or nonaqueous phase liquids (NAPLs) (gasoline, oil, or industrial solvents) added to the water as a consequence of man's activities.

To an important extent, the complexity of a contamination problem is determined by whether or not NAPLs are present. Figure 17.1*a* is an example of the simplest case of ground-water contamination where there is only a plume of dissolved contaminants and no NAPLs. Contamination by NAPLs raises the level of complexity because contaminants can migrate as a separate liquid phase, a dissolved phase, and a vapor phase.

NAPLs are classified according to whether they are more or less dense than water. Dense nonaqueous phase liquids (DNAPLs) have a specific gravity greater than one and sink through water. Light nonaqueous phase liquids (LNAPLs) have a specific gravity less than one and float on water. Figure 17.1*b* illustrates contamination developed due to a DNAPL spill. DNAPL, present as a residual fluid within the pores and/or "ponded" on low-permeability layers, dissolves into the flowing ground water to create a large plume. Within the unsaturated zone, volatilization promotes the spreading of DNAPL as a vapor phase.

The behavior of NAPLs in saturated and unsaturated fluid systems is sufficiently complicated that we have reserved Chapter 19 for a more comprehensive discussion. Chapter 17, then, focuses mainly on plumes of contaminants dissolved in water and techniques available for assessing the spatial distribution of such contaminants. Chapter 18 presents examples of analytical and numerical procedures available to model the spreading of dissolved contaminant plumes.

17.1 Sources of Ground-Water Contamination

Three important attributes distinguish sources of ground-water contamination: (1) their degree of localization, (2) their loading history, and (3) the kinds of contaminants emanating from them. Given the large number of ways of contaminating ground water, there is a spectrum of source sizes ranging from an individual well to areas of 100s km^2 or more. In practice, the terms *point* and *nonpoint* describe the degree of localization of the source. A point source is characterized by the presence of an identifiable, small-scale source, such as a leaking storage tank, one or more disposal ponds, or a sanitary landfill. Usually, this source produces a reasonably well-

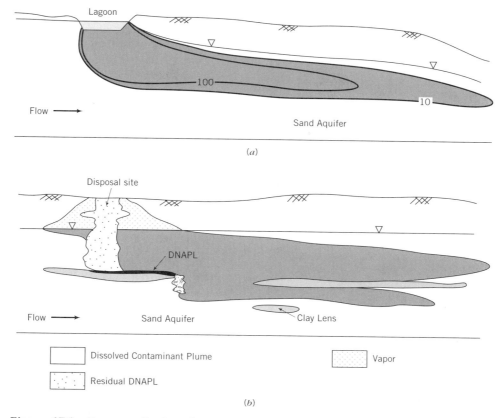

Figure 17.1 Conceptualization of contamination problems of varying complexity. In Panel (*a*), waste dissolved in water leaking from a lagoon creates a large plume of dissolved contaminants. In Panel (*b*), the presence of DNAPLs results in a much more complex problem, where the contamination occurs as a pure organic liquid, a vapor in the soil zone, and a dissolved phase in the ground water.

defined plume. A nonpoint problem refers to larger-scale, relatively diffuse contamination originating from many smaller sources, whose locations are often poorly defined. Examples of nonpoint contaminants could include herbicides or pesticides that are used in farming, nitrates that originate as effluents from household disposal systems, salt derived from highways in winter, and acid rain. Typically, there are no well-defined plumes in these cases but a large enclave of contamination with extremely variable concentrations.

The loading history describes how the concentration of a contaminant or its rate of production varies as a function of time at the source. A spill is an example of pulse loading, where the source produces contaminants at a fixed concentration for a relatively short time (Figure 17.2*a*). This loading could occur from a one-time release of contaminants from a storage tank or storage pond. Long-term leakage from a source is termed *continuous source loading*. Figure 17.2*b* is one type of continuous loading where the concentration remains constant with time. This loading might occur, for example, when small quantities of contaminants are leached from a volumetrically large source over a long time. LNAPLs or DNAPLs

present at a source can dissolve at a slow rate over many decades to provide another type of continuous source.

Most sources of long-term leakage cannot be described in terms of a constant loading function. For example, the concentration of chemical wastes added to a storage pond at an industrial site can vary with time due to changes in a manufacturing process, seasonal or economic factors, or the addition of other reactive wastes (for example, Figure 17.2*c*). Leaching rates for solid wastes at a sanitary landfill site could be controlled by seasonal factors related to recharge, or by a decline in source strength as components of the waste (for example, organics) biodegrade. This latter source behavior could result in the loading history shown on Figure 17.2*d*. Typically, increasing complexity in the loading function translates directly into increasingly variable concentration distributions.

An amazingly large list of potential activities can result in ground-water contamination (Table 17.1). As a consequence of these many different industrial, agricultural, and domestic activities, the list of potential contaminants can number in the thousands or tens of thousands of compounds. The question of how to organize this list

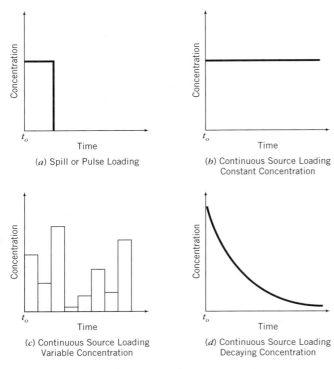

Figure 17.2 Examples of functions used to characterize contaminant loading from (*a*) a spill or (*b*, *c*, and *d*) long-term leakage.

for study is a difficult one. One approach has been to concentrate on a subset of this list and to pay special attention to contaminants that commonly occur in effluents and drinking water, that produce adverse health effects, or that persist within the food chain. An example is the U.S. Environmental Protection Agency's list of 129 priority pollutants, containing 114 organic compounds and 15 inorganic species, mainly trace metals (Table 17.2).

The organic priority pollutants are subdivided into four groups: volatiles, base-neutral extractables, acid extractables, and pesticides (Table 17.2). The method of analysis is different for each group but all compounds are finally examined using combined gas chromatography–mass spectrometry (GC-MS). The inorganic group is analyzed with other techniques. One advantage of working with a standard list (for example, Table 17.2) is that highly automated procedures screen water samples for subsets of priority pollutants at a reasonable cost. Custom analyses of other species are often more costly.

The way that we organize the contaminants is by grouping according to reaction type and mode of occurrence. Thus, dissolved compounds that are affected similarly by chemical, nuclear, or biological processes are grouped together. Contaminants that occur as an NAPL are a separate subgroup. The major groups of contaminants include (1) radionuclides, (2) trace elements, (3) nutrients, (4) other inorganic species, (5) organic contaminants, and (6) microbial contaminants. The NAPLs

that are included in (5) naturally subdivide into LNAPLs and DNAPLs.

All of these contaminants have the potential to produce health problems. In moving through the groups, we will point out the most serious ones. As a broad generalization, too much of anything in water can produce health problems in humans. For some contaminants, particularly the radionuclides, increasing exposure results in increasing health consequences. Thus, any exposure above background levels can be of concern. For other contaminants, such as the major ions Na^+ or Cl^-, there is often a threshold below which no serious health effects will occur.

Radioactive Contaminants

The nuclear industry is the main generator of radioactive contaminants. Potential sources occur throughout the nuclear fuel cycle, which involves the mining and milling of uranium, uranium enrichment and fuel fabrication, power plant operation, fuel reprocessing, and waste disposal. The kinds of contaminants depend on the type of reactor and the extent of spent-fuel reprocessing. For example, in the United States, Japan, France, and Germany, there are large numbers of light-water reactors (LWRs) that utilize enriched uranium (^{235}U) as the predominant fuel source, and possibly ^{239}Pu and ^{238}U. In Canada, heavy-water reactors (HWRs) use natural or enriched uranium as a fuel, and heavy water (D_2O) as a coolant and moderator.

During mining when the raw ore is processed ^{238}U, ^{230}Th, ^{226}Ra, and ^{222}Rn(gas) are potential contaminants along with nonradioactive contaminants that include trace constituents, and major ions such as SO_4^{2-} or Cl^-. Areas where this contamination occurs are Colorado, New Mexico, Texas, Utah, Wyoming, northern Saskatchewan, and Ontario. The enrichment and fuel fabrication step treats raw uranium concentrate to increase the concentration of ^{235}U in the fuel relative to the more abundant ^{238}U, and actually produces UO_2, which is the actual fuel. The most common contaminants are ^{238}U, ^{235}U, ^{90}Sr, and ^{137}Cs.

Many radionuclides are generated as fission products from the decay of ^{235}U or ^{239}Pu and from neutron activation of stable elements in the coolant or metallic components of the reactor. Fission is the power-generating part of a nuclear reaction whereby the heavy nucleus is split into nuclei of lighter elements and neutrons. Neutron activation is a process wherein neutrons are added to the nucleus of a stable isotope to produce a radioactive one. These processes produce radionuclides like ^{137}Cs, ^{134}Cs, ^{58}Co, ^{51}Cr, ^{54}Mn, ^{55}Fe, ^{3}H, and ^{131}I. Overall in excess of 75 radionuclides could potentially be produced. Fortunately, most of these radioisotopes, except for ^{3}H, remain in the spent fuel or the reactor and are not a serious source of contamination at this stage in the fuel cycle.

Table 17.1   **Sources of Ground-Water Contamination**

CATEGORY I—SOURCES DESIGNED TO DISCHARGE SUBSTANCES
Subsurface percolation (e.g., septic tanks and cesspools)
Injection wells
  Hazardous waste
  Nonhazardous waste (e.g., brine disposal and drainage)
  Nonwaste (e.g., enhanced recovery, artificial recharge, solution mining, and in situ mining)
Land application
  Wastewater (e.g., spray irrigation)
  Wastewater by-products (e.g., sludge)
  Hazardous waste
  Nonhazardous waste

CATEGORY II—SOURCES DESIGNED TO STORE, TREAT, AND/OR DISPOSE OF SUBSTANCES; DISCHARGE THROUGH UNPLANNED RELEASE
Landfills
  Industrial hazardous waste
  Industrial nonhazardous waste
  Municipal sanitary
Open dumps, including illegal dumping (waste)
Residential (or local) disposal (waste)
Surface impoundments
  Hazardous waste
  Nonhazardous waste
Waste tailings
Waste piles
  Hazardous waste
  Nonhazardous waste
Materials stockpiles (nonwaste)
Graveyards
Animal burial
Above-ground storage tanks
  Hazardous waste
  Nonhazardous waste
  Nonwaste
Underground storage tanks
  Hazardous waste
  Nonhazardous waste
  Nonwaste
Containers
  Hazardous waste
  Nonhazardous waste
  Nonwaste

Open burning and detonation sites
Radioactive disposal sites

CATEGORY III—SOURCES DESIGNED TO RETAIN SUBSTANCES DURING TRANSPORT OR TRANSMISSION
Pipelines
  Hazardous waste
  Nonhazardous waste
  Nonwaste
Materials transport and transfer operations
  Hazardous waste
  Nonhazardous waste
  Nonwaste

CATEGORY IV—SOURCES DISCHARGING SUBSTANCES AS CONSEQUENCE OF OTHER PLANNED ACTIVITIES
Irrigation practices (e.g., return flow)
Pesticide applications
Fertilizer applications
Animal feeding operations
De-icing salts applications
Urban runoff
Percolation of atmospheric pollutants
Mining and mine drainage
  Surface mine-related
  Underground mine-related

CATEGORY V—SOURCES PROVIDING CONDUIT OR INDUCING DISCHARGE THROUGH ALTERED FLOW PATTERNS
Production wells
  Oil (and gas) wells
  Geothermal and heat-recovery wells
  Water-supply wells
Other wells (nonwaste)
  Monitoring wells
  Exploration wells
Construction excavation

CATEGORY VI—NATURALLY OCCURRING SOURCES WHOSE DISCHARGE IS CREATED AND/OR EXACERBATED BY HUMAN ACTIVITY
Ground water–surface water interactions
Natural leaching
Saltwater intrusion/brackish water upconing (or intrusion of other poor-quality natural water)

From Office of Technology Assessment (1984)

Reprocessing treats spent fuel to remove ^{235}U and ^{239}Pu. However, the other radionuclides remain to produce the high-level waste problems common to many countries of the world.

The health hazards associated with ionizing radiation are well known. An exposed individual can be affected by cancer or by genetic defects that could affect his or her offspring. These risks are difficult to assess at low levels of exposure. However, most risk assessments consider that the probability of inducing cancer or genetic

damage increases as a function of dose without a threshold. Other health problems associated with exposure to radiation include, for example, cataracts, nonmalignant skin damage, depletion of bone marrow, and infertility.

Trace Metals

The next major group of contaminants is the trace metals (Table 17.3). As a group, trace metals contain the largest proportion of elements found on the periodic table. The

Table 17.2 **Environmental Protection Agency List of Priority Pollutants. Organic Compounds Are Subdivided into Four Categories According to the Method of Analysis**

BASE-NEUTRAL EXTRACTABLES

Acenaphthene	Diethyl phthalate
Acenaphthylene	Dimethyl phthalate
Anthracene	2,4-Dinitrotoluene
Benzidine	2,6-Dinitrotoluene
Benzo[*a*]anthracene	Di-*n*-octyl phthalate
Benzo[*b*]fluoranthene	1,2-Diphenylhydrazine
Benzo[*k*]fluoranthene	Fluoranthene
Benzo[*ghi*]perylene	Fluorene
Benzo[*a*]pyrene	Hexachlorobenzene
Bis(2-chloroethoxy)methane	Hexachlorobutadiene
Bis(2-chloroethyl) ether	Hexachlorocyclopentadiene
Bis(2-chloroisopropyl) ether	Hexachloroethane
Bis(2-ethylhexyl) phthalate	Indeno[1,2,3-*cd*] pyrene
4-Bromophenyl phenyl ether	Isophorone
Butyl benzyl phthalate	Naphthalene
2-Chloronaphthalene	Nitrobenzene
4-Chlorophenyl phenyl ether	*N*-Nitrosodimethylamine
Chrysene	*N*-Nitrosodiphenylamine
Dibenzo[*a,h*]anthracene	*N*-Nitrosodi-*n*-propylamine
Di-*n*-butyl phthalate	Phenanthrene
1,2-Dichlorobenzene	Pyrene
1,3-Dichlorobenzene	2,3,7,8-Tetrachlorodibenzo-*p*-dioxin
1,4-Dichlorobenzene	1,2,4-Trichlorobenzene
3,3'-Dichlorobenzidine	

ACID EXTRACTABLES

p-Chloro-*m*-cresol	2-Nitrophenol
2-Chlorophenol	4-Nitrophenol
2,4-Dichlorophenol	Pentachlorophenol
2,4-Dimethylphenol	Phenol
4,6-Dinitro-*o*-cresol	2,4,6-Trichlorophenol
2,4-Dinitrophenol	Total phenols

VOLATILES

Acrolein	1,2-Dichloroethane
Acrylonitrile	1,1-Dichloroethylene
Benzene	*trans*-1,2-Dichloroethylene
Bis(chloromethyl) ether	1,2-Dichloropropane
Bromodichloromethane	*cis*-1,3-Dichloropropene
Bromoform	*trans*-1,3-Dichloropropene
Bromomethane	Ethylbenzene
Carbon tetrachloride	Methylene chloride
Chlorobenzene	1,1,2,2-Tetrachloroethane
Chloroethane	Tetrachloroethene
2-Chloroethyl vinyl ether	Toluene
Chloroform	1,1,1-Trichloroethane
Chloromethane	1,1,2-Trichloroethane
Dibromochloromethane	Trichloroethylene
Dichlorodifluoromethane	Trichlorofluoromethane
1,1-Dichloroethane	Vinyl chloride

PESTICIDES

Aldrin	Dieldrin	PCB-1016[a]
α-BHC	α-Endosulfan	PCB-1221[a]
β-BHC	β-Endosulfan	PCB-1232[a]
γ-BHC	Endosulfan sulfate	PCB-1242[a]
δ-BHC	Endrin	PCB-1248[a]
Chlordane	Endrin aldehyde	PCB-1254[a]
4,4'-DDD	Heptachlor	PCB-1260[a]
4,4'-DDE	Heptachlor epoxide	Toxaphene
4,4'-DDT		

[a]*not pesticides*

INORGANICS

Antimony	Chromium	Nickel
Arsenic	Copper	Selenium
Asbestos	Cyanide	Silver
Beryllium	Lead	Thallium
Cadmium	Mercury	Zinc

most common sources of contamination include (1) effluents from mining, (2) industrial waste water, (3) runoff, solid wastes, or waste water contributed from urban areas, (4) agricultural wastes and fertilizers, and (5) fossil fuels. Readers can refer to an excellent survey of sources and background concentrations in a book by Forstner and Wittman (1981).

Trace metals can be toxic and even lethal to humans even at relatively low concentrations because of their tendency to accumulate in the body. Some studies have found positive correlations between the concentration of trace metals in water (for example, Be, Cd, Pb, and Ni) and death rates from some cancers. Bioaccumulation of trace metals in the food chain has produced the most well-known cases of metal poisoning (for example, in Minamata, Japan). Organisms higher up the food chain progressively accumulate metals. Eventually, humans at the top of the chain can experience severe health problems.

Table 17.3 **Examples of Trace Metals Occurring in Ground Water**

Aluminum	Gold	Silver
Antimony	Iron	Strontium
Arsenic	Lead	Thallium
Barium	Lithium	Tin
Beryllium		
Boron	Manganese	Titanium
Cadmium	Mercury	Uranium
	Molybdenum	Vanadium
Chromium	Nickel	Zinc
Cobalt	Selenium	
Copper		

Nutrients

This group of potential contaminants includes those ions or organic compounds containing nitrogen or phosphorus. By far, the dominant nitrogen species in ground water is nitrate (NO_3^-), then to a lesser extent ammonium ion (NH_4^+). Agricultural practices including the use of fertilizers containing nitrogen, cattle feeding operations, and the cultivation of virgin soils (leading to the oxidation of large quantities of nitrogen existing in organic matter in the soil) are important sources of contamination.

Other sources are sewage that could enter ground water from septic tank systems, or irrigated waste water. According to Bouwer (1985), effluent from sewage treatment plants in the United States contains 30 mg/L of nitrogen, mainly as NH_4^+, as well as organic nitrogen in the form of nitrobenzenes or nitrotoluenes.

The main health effects related to contamination by nitrogen compounds are (1) methemoglobinemia, a type of blood disorder in which oxygen transport in young babies or unborn fetuses is impaired or (2) the possibility of forming cancer-causing compounds (for example, nitrosamines) after drinking contaminated water. Typically, phosphorus contamination is considered together with nitrogen. However, it is less important because of the low solubility of phosphorus compounds in ground water, the limited mobility of phosphorus due to its tendency to sorb on solids, and the lack of proven health problems. The major sources of phosphorus are again soil-applied fertilizers and waste water.

Other Inorganic Species

This miscellaneous group includes metals present in nontrace quantities such as Ca, Mg, and Na plus nonmetals such as ions containing carbon and sulfur (for example, HCO_3^-, HS^-, CO_3^{2-}, SO_4^{2-}, and H_2CO_3) or other species such as Cl^- and F^-. Many of these ions are major contributors to the overall salinity of ground water. Extremely high concentrations of these species make water unfit for human consumption and for many industrial uses. The health-related problems are not as serious as those caused by the other contaminant groups. However, high concentrations of even relatively nontoxic salts, particularly Na^+, can disrupt cell or blood chemistry with serious consequences. At lower concentrations, an excessive intake of Na^+ may cause less serious health effects such as hypertension.

The potential sources of major ion salinity include (1) saline brine that is produced with oil, (2) leachate from mine tailings, mine spoil, or sanitary landfills, and (3) industrial waste water that often has large concentrations of common ions in addition to heavy metals or organic compounds. Fluoride is probably the best example of a trace nonmetal occurring as a contaminant. Like the metals, trace quantities of these contaminants can produce health problems at relatively low concentrations. With fluoride, an increase in concentration to as little as 7 or 8 times the levels for combating tooth decay can cause skeletal fluorosis.

Organic Contaminants

Contamination of ground water by organic compounds is a logical consequence of the large quantities of unrefined petroleum products and man-made organic compounds used today. Of the list of sources we considered previously, almost every one either is known to contribute or has the potential to contribute organic contaminants to ground water.

Table 17.4 lists the most important families of organic contaminants. The following sections examine these constituents in detail.

Petroleum Hydrocarbons and Derivatives

Petroleum hydrocarbons and derivatives are made up of carbon and hydrogen that are derived from crude oil, natural gas, and coal. The organic compounds in crude oil can be divided into three main groups (Zemo and others, 1993). The alkanes or paraffins are present in most petroleum products. Common alkanes (*n*-alkanes) have the general formula C_nH_{2n+2} and include *n*-butane, *n*-pentane, *n*-hexane, and so on. The cycloalkanes have carbon atoms arranged in a circle containing either five or six carbon atoms. Examples include cyclopentane and cyclohexane. The second major group, the alkenes or olefins, are not constituents of crude oil but formed during the refining process. These molecules have the general formula C_nH_{2n} and include compounds like ethene and propene (see examples in Section 11.7).

The third major group of compounds in crude oil is the aromatic hydrocarbons. These compounds contain at least one benzene ring (that is, C_6H_6). Examples of these compounds include benzene (C_6H_6), toluene ($C_6H_5CH_3$), ethylbenzene (C_8H_{10}), and xylene ($C_6H_4(CH_3)_2$). These so-called BTEX compounds are both extremely soluble in water and toxic.

The polynuclear aromatic hydrocarbons (PAHs) are also components of concern in petroleum hydrocarbons. As indicated in Section 11.7, these compounds form from a series of benzene rings. Examples include anthracene and phenanthrene. The physical and chemical properties for the most important organic compounds in fuels are summarized in Table 17.5.

The fuels are produced from crude oil through a refining process. Distillation separates crude oil into different fractions according to the temperature of boiling. Zemo and others (1993) provide the following general descriptions of the chemistry of these fractions:

- Gasolines (low boiling fractions) C_4 to C_{12} alkanes, C_4 to C_7 alkenes, aromatic compounds including the BTEX compounds, C_3 benzenes, and C_4 benzenes

Table 17.4 **Important Families of Organic Contaminants Found in Ground Water**

Chemical Family	Examples of Compounds
Hydrocarbons and derivatives	
fuels	benzene, toluene, *o*-xylene, butane, phenol
PAHs	anthracene, phenanthrene
alcohols	methanol, glycerol
creosote	*m*-cresol, *o*-cresol
ketones	acetone
Halogenated aliphatics	tetrachloroethene, trichloroethene, dichloromethane
Halogenated aromatics	chlorobenzene, dichlorobenzene
Polychlorinated biphenyls	2,4'-PCB, 4,4'-PCB

- Middle distillates (kerosene, diesel, home heating oil, jet fuel) C_{10} to C_{24} alkanes, slightly soluble aromatic compounds including C_3 to C_5 benzenes, C_0 to C_8 naphthalenes, and C_0 to C_5 anthracenes, and

- Residual products (diesel Nos. 4 and 6, Bunker C, motor oil), C_{20} to C_{78} alkanes, nonsoluble aromatics (predominantly PAHs)

Figure 17.3*a* shows most of the major components of crude oil as they would appear on a gas chromatograph (Senn and Johnson, 1985). The bars indicate the fractions represented in the various petroleum products. The examples of chromatograms in Figures 17.3*b* and *c* illustrate that products like gasoline or jet fuel contain a very different set of compounds. Similarly, the subtle differences among the chromatograms for fuels provide a basis for "geochemical fingerprinting" for source identification (Senn and Johnson, 1985).

Overall, the compositional differences among the fuels have important implications for monitoring and cleanup. For example, the low-boiling fractions like gasoline contain many highly volatile components. The oils are much less volatile. The chemistry of fuels is also complicated by the various additives designed to improve combustion and to remove combustion deposits from engines.

Halogenated Aliphatic Compounds
These compounds are formed from chains of carbon and hydrogen atoms where certain hydrogens may be replaced by chlorine, fluorine, or bromine atoms. Examples of these compounds include tetrachloroethene (PCE), trichloroethene (TCE), and carbon tetrachloride (CT). Historically, the use of these solvents or their improper disposal has given rise to many of the most serious problems of contamination encountered in hydrogeologic practice. As Table 17.5 indicates, most of these compounds have specific gravities greater than 1, and thus can be found as DNAPLs.

Halogenated Aromatic Compounds
These compounds are formed from benzene rings with substituted halogens. Examples include chlorobenzene and dichlorobenzene (DCB), which are used in various industrial and agricultural applications. Again, by virtue of their relatively large specific gravity, these compounds occur as DNAPLs in ground water.

Polychlorinated Biphenyls
Polychlorinated biphenyls were widely used in the 1960s and 1970s in transformers and capacitors. Their environmental persistence and toxicity have made them an important contaminant even though their production has been curtailed. Chemically they consist of chlorine-substituted benzene rings joined together.

Health Effects
There are important health effects related to drinking water contaminated by organic compounds. However, as Craun (1984) points out, it is difficult to establish which compounds are most toxic because not all have been tested, and health risks are inferred from studies of laboratory animals, poisonings or accidental ingestion, and occupational exposures. Furthermore, there is a serious lack of information on the health effects related to the combined effect of several compounds and on the epidemiology of populations consuming contaminated water. Organic contamination may cause cancer in humans and animals, as well as a host of other problems including liver damage, impairment of cardiovascular function, depression of the nervous system, brain disorders, and various kinds of lesions. More detailed information on health effects related to organic materials can be obtained from Craun (1984).

Biological Contaminants
The important biological contaminants include pathogenic bacteria, viruses, or parasites. It does not take a

Table 17.5 **Properties for Selected Organic Compounds**

Compound/Family	Formula	Specific Gravity	Solubility (mg/L)	K_{ow}	Vapor Pressure (mm Hg)
Fuels and derivatives					
Benzene	C_6H_6	0.879	1750	130	60
Ethylbenzene	C_8H_{10}	0.867	152	1400	7
Phenol	C_6H_6O	1.071	93,000	29	0.2
Toluene	$C_6H_5CH_3$	0.866	535	130	22
o-xylene	$C_6H_4(CH_3)_2$	0.880	175	890	5
PAHs					
Acenaphthene	$C_{12}H_{10}$	1.069	3.42	10,000	0.01
Benzopyrene	$C_{20}H_{12}$	1.35	0.0012	1.15×10^6	—
Benzoperylene	$C_{22}H_{12}$	—	0.0007	3.24×10^6	—
Naphthalene	$C_{10}H_8$	1.145	32	2800	0.23
Methyl naphthalene	$C_{10}H_7CH_3$	1.025	25.4	13,000	—
Ketones					
Acetone	CH_3COCH_3	0.791	inf	0.6	89
Methyl ethyl ketone	$CH_3COCH_2CH_3$	0.805	2.68×10^5	1.8	77.5
Halogenated aliphatics					
Bromodichloromethane	$CHBrCl_2$	2.006	4400	76	50
Bromoform	$CHBr_3$	2.903	3010	250	4
Carbon tetrachloride	CCl_4	1.594	757	440	90
Chloroform	$CHCl_3$	1.49	8200	93	160
Chloroethane	CH_3CH_2Cl	0.903	5740	35	1000
1,1-Dichloroethane	$C_2H_4Cl_2$	1.176	5500	62	180
1,2-Dichloroethane	$C_2H_4Cl_2$	1.253	8520	30	61
1,1-Dichloroethene	$C_2H_2Cl_2$	1.250	2250	69	495
cis-1,2-Dichloroethene	$C_2H_2Cl_2$	1.27	3500	5	206
trans-1,2,-Dichloroethene	$C_2H_2Cl_2$	1.27	6300	3	265
Hexachloroethane	C_2Cl_6	2.09	50	39,800	0.4
Methylene chloride	CH_2Cl_2	1.366	20,000	19	362
1,1,2,2-Tetrachloroethane	$CHCl_2CHCl_2$	1.600	2900	250	5
Tetrachloroethene	CCl_2CCl_2	1.631	150	390	14
1,1,1-Trichloroethane	CCl_3CH_3	1.346	1500	320	100
1,1,2-Trichloroethane	$CH_2ClCHCl_2$	1.441	4500	290	19
Trichloroethene	C_2HCl_3	1.466	1100	240	60
Vinyl chloride	CH_2CHCl	0.908	2670	24	266
Halogenated aromatics					
Chlorobenzene	C_6H_5Cl	1.106	466	690	9
2-Chlorophenol	C_6H_4ClOH	1.241	29,000	15	1.42
p-Dichlorobenzene (1,4)	$C_6H_4Cl_2$	1.458	79	3900	0.6
Hexachlorobenzene	C_6Cl_6	2.044	0.006	1.7×10^5	1×10^{-5}
Pentachlorophenol	C_6OHCl_5	1.978	14	1.0×10^5	1×10^{-4}
1,2,4-Trichlorobenzene	$C_6H_3Cl_3$	1.446	30	20,000	0.42
2,4,6-Trichlorophenol	$C_6H_2Cl_3OH$	1.490	800	74	0.012
PCBs					
Aroclor 1254		1.5	0.012	1.07×10^6	7.7×10^{-5}
Other					
2,6-Dinitrotoluene	$C_6H_3(NO_2)_2CH_3$	1.283	1320	100	—
1,4-Dioxane	$C_4H_8O_2$	1.034	4.31×10^5	1.02	30
Nitrobenzene	$C_6H_5NO_2$	1.203	1900	71	0.15
Tetrahydrofuran	C_4H_8O	0.888	0.3	6.6	131

Specific gravity at various temperatures; refer to Nyer and others (1991) for details.

inf is infinite solubility.

Vapor pressure about 20°C; 1 atm = 760 mm Hg.

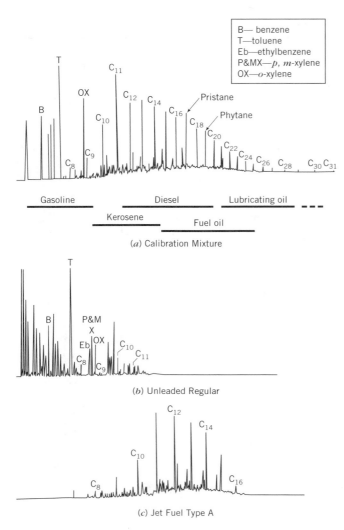

B— benzene
T—toluene
Eb—ethylbenzene
P&MX—*p, m*-xylene
OX—*o*-xylene

(a) Calibration Mixture

(b) Unleaded Regular

(c) Jet Fuel Type A

Figure 17.3 Chromatograms for various organic compounds. Panel (*a*) is a calibration mixture spanning the range of compounds found in gasoline. The various ranges indicate how boiling selects the various compounds in fuels. Panels (*b*) and (*c*) are examples of chromatograms for unleaded regular gasoline and jet fuel (modified from Senn and Johnson, 1985). Reprinted by permission of Conference on Petroleum Hydrocarbons and Organic Chemicals in Ground Water—Prevention, Detection and Restoration. Copyright © 1985. All rights reserved.

degree in medicine to be aware of the serious health problems from typhoid fever, cholera, polio, and hepatitis. Other less serious abdominal disorders are often too well known by travelers to countries with poor sanitation. As a group, these health effects are some of the most significant related to the contamination of ground water.

The main source of biological contamination is from human and animal sewage, or waste water. Ground waters become contaminated due to (1) land-disposal of sewage from centralized treatment facilities or septic tank systems, (2) leachates from sanitary landfills, and (3) vari-

ous agricultural practices such as the improper disposal of wastes from feedlots.

Particulate contaminants are less mobile than those dissolved in water. For this reason, most reported problems are of a local nature, related for example to (1) poor well construction, which enables surface runoff or sewage to enter the well, and (2) sewer line breaks or septic tank fields located close to a well.

17.2 Solute Plumes as a Manifestation of Processes

The important processes and parameters that influence the transport of mass should be familiar. Now, we will examine how these processes work together to control the spread of contaminants. The natural starting place is with the mass transport processes because they determine the maximum extent of plume spread and the geometric character of the concentration distribution. The chemical, nuclear, and biological processes mainly attenuate the spread of contaminants, reducing the size of the contaminated region to a fraction of that attributable to mass transport alone. Advection is by far the most dominant mass transport process in shaping the plume. Hydrodynamic dispersion is usually a second-order process, except in some cases involving fractured rocks.

The magnitude and direction of advective transport are controlled by (1) the hydraulic conductivity distribution within the flow field, (2) the configuration of the water table or potentiometric surface, (3) the presence of sources or sinks (for example, wells), and (4) the shape of the flow domain. All of these parameters are important in controlling the ground-water velocity, which drives advective transport. We can illustrate this concept with the simulation results in Figure 17.4. In these examples, the water-table configuration and the other boundary conditions for the steady-state flow system remain fixed. Contaminants are added at the same place in the recharge area by fixing concentration along part of the inflow boundary.

Because there is no dispersion or reactions, the plumes have a uniform concentration equal to the source concentration. Their shape depends on the hydraulic conductivity distribution. A reduction in hydraulic conductivity reduces the extent of plume spread by simply lowering the ground-water velocity (compare Figures 17.4*a* and *b*). Adding two or three layers also influences plume shape (Figures 17.4*c, d,* and *e*) because in these cases both the magnitude and direction of ground-water flow change. Again, there is a link between advective transport and the pattern of flow.

We will not take time to examine the other three parameters in the list that influence advection. Earlier parts of the book, hopefully, have developed a sense of how patterns of flow respond to water-table configura-

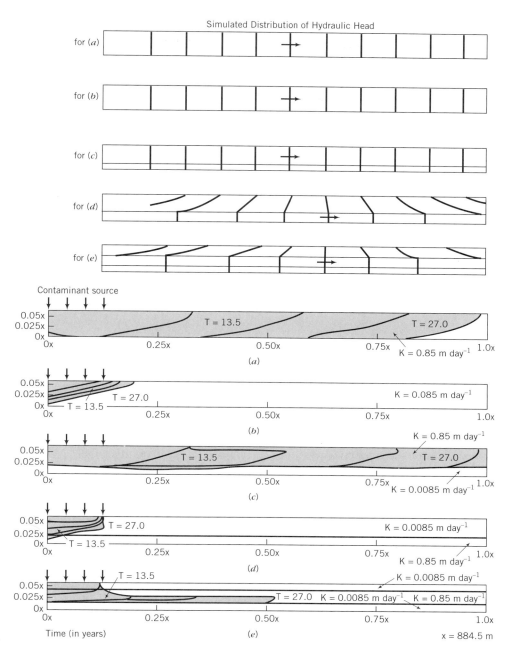

Figure 17.4 Changing plume shapes with time (*T*) in years as a function of the different patterns of layering. In all cases, advection is the only process operating (modified from Schwartz, 1975). Reprinted from J. Hydrol., v. 27, by F. W. Schwartz, On radioactive waste management: An analysis of the parameters controlling subsurface contaminant transfer, p. 51–71, 1975 with kind permission from Elsevier Science, NL, Sara Burgerhartstraat 25, 1055 KV Amsterdam, The Netherlands.

tion, pumping, and region shape. The extension of these basic concepts to advective transport simply involves visualizing the movement of a contaminant along stream tubes, at least for steady-state flow.

Adding dispersion to advective transport can cause important changes in the shape of a plume. Consider the nonlayered system shown in Figure 17.5. All of the

transport parameters except the longitudinal (α_L) and transverse (α_T) dispersivities are held constant. As the magnitude of the dispersivities (α_L, α_T) increases from (0.0, 0.0) m to (0.03, 0.006) m to (0.3, 0.06) m, and finally (3.0, 0.6) m (Figures 17.5*a*, *b*, *c*, and *d*), the size of the plume increases markedly. However, as the plume size increases, the maximum concentration decreases.

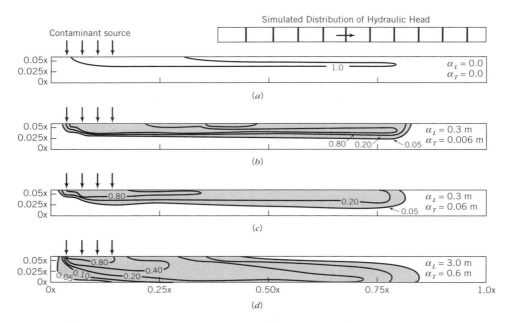

Figure 17.5 Changing plume shapes as a function of the longitudinal and transverse dispersivities: (a) $\alpha_L = 0.0$; $\alpha_T = 0.0$, (b) $\alpha_L = 0.03$ m; $\alpha_T = 0.006$ m, (c) $\alpha_L = 0.3$ m; $\alpha_T = 0.06$ m, and (d) $\alpha_L = 3.0$ m; $\alpha_T = 0.6$ m (modified from Schwartz, 1975). Reprinted from J. Hydrol., v. 27, by F. W. Schwartz, On radioactive waste management: An analysis of the parameters controlling subsurface contaminant transfer, p. 51–71, 1975 with kind permission from Elsevier Science, NL, Sara Burgerhartstraat 25, 1055 KV Amsterdam, The Netherlands.

Increased dispersion mixes the contaminant with an increasing proportion of the uncontaminated water.

Because both α_L and α_T increase proportionately, we cannot separate their overall effects on concentration distributions. A series of model trials published by Frind and Germain (1986) shows how the shape of a continuous plume changes only as a function of the transverse dispersivity. Flow is downward and laterally away from a source at the left end of the system (Figure 17.6). The transport simulations in Figure 17.6 document the change in plume shape as values of α_T are increased from 0.001 m, to 0.01 m, to 0.1 m. For comparative purposes, the position of the advective front is shown by the dashed lines. As transverse dispersion increases, the plume changes from one that is compact and not much larger than the stream tube initially containing the contaminant (Figure 17.6a) to one that has spread laterally into adjacent stream tubes (Figure 17.6c). Accompanying this transverse spreading is a reduction in the overall length of the plume, exemplified by the position of the 0.5 contour line relative to the position of the advective front.

Another important group of processes includes (1) a simple, first-order kinetic reaction that could account for radioactive decay, biodegradation, or hydrolysis, and (2) an equilibrium, binary exchange reaction. Let us compare a series of plumes in which the half-life of the kinetic reaction decreases from 30 yr, to 3.0 yr, to 0.33 yr (Figures 17.7a, b, and c). In general, the more rapidly a reaction

removes a contaminant the smaller the plume will be at a given time. Thus, the smallest plume is associated with the smallest half-life (Figure 17.7c).

Ion exchange can also produce the same dramatic attenuation in concentration. The changing plume shapes in Figure 17.8 are due to a changing selectivity coefficient (K_s). When $K_s = 0$, there is no exchange, and the contaminant moves due to mass transport alone. However, as K_s increases over four orders of magnitude, the plume becomes smaller as exchange retards spreading (compare Figures 17.8a, b, c, and d). Other parameters that control ion exchange reactions (for example, cation exchange capacity or ion charge) also influence the distribution of contaminants. Irrespective of the model describing sorption, the process is of paramount importance in controlling contaminant transport.

We will not present examples of all reactions that might occur in a contaminant plume because the results are generally similar. The greater the tendency for any reaction to remove a contaminant from solution, the smaller the plume will be relative to its unaltered size. In some instances, the geochemical processes have the capability of immobilizing particular contaminants at the source. In terms of the transport processes as a whole, the reactions are as important as advection in finally determining the disposition of contaminants.

The loading function also influences plume shape. One modeler humorously pointed out that given a proper loading function he could generate a plume that "looked

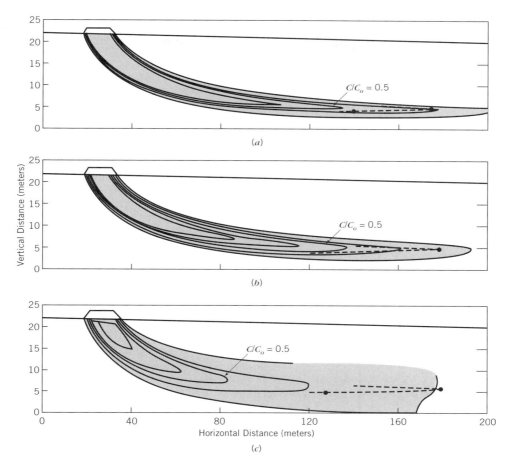

Figure 17.6   Changing plume shapes as a function of the transverse dispersivity: (*a*) α_T = 0.001 m; (*b*) α_T = 0.01 m; and (*c*) α_T = 0.1 m. Dashed curves mark the position of the front due to advection alone (modified from Frind and Germain, Water Resources Res., v. 22, p. 1857–1873, 1986. Copyright by American Geophysical Union).

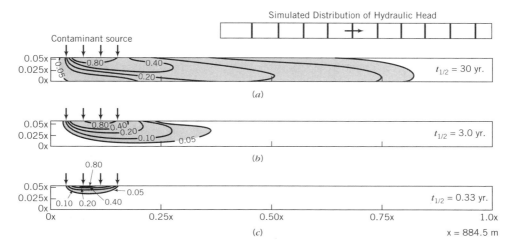

Figure 17.7   Changing plume shapes as a function of the half-life of a kinetic reaction: (*a*) $t_{1/2}$ = 30 yr, (*b*) $t_{1/2}$ = 3.0 yr, (*c*) $t_{1/2}$ = 0.33 yr (modified from Schwartz, 1975). Reprinted from J. Hydrol., v. 27, by F. W. Schwartz, On radioactive waste management: An analysis of the parameters controlling subsurface contaminant transfer, p. 51–71, 1975 with kind permission from Elsevier Science, NL, Sara Burgerhartstraat 25, 1055 KV Amsterdam, The Netherlands.

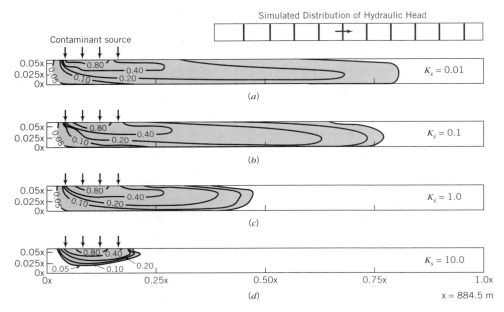

Figure 17.8 Changing plume shapes as a function of the selectivity coefficient for a binary exchange where in (*a*) $K_s = 0.01$, (*b*) $K_s = 0.1$, (*c*) $K_s = 1.0$, and (*d*) $K_s = 10.0$ (modified from Schwartz, 1975). Reprinted from J. Hydrol., v. 27, by F. W. Schwartz, On radioactive waste management: An analysis of the parameters controlling subsurface contaminant transfer, p. 51–71, 1975 with kind permission from Elsevier Science, NL, Sara Burgerhartstraat 25, 1055 KV Amsterdam, The Netherlands.

like an elephant." Let us look at some plume shapes in relation to nonconstant loading functions. Adding the same quantity of mass to a flow system over increasingly longer times or, in effect, reducing the rate of pulse loading changes the center of mass of the plume and the internal concentration distribution. Figure 17.9*a* illustrates a case where a contaminant is added for 5.1 years at a relative concentration of 1.0 and after this time at a relative concentration of 0.0. This plume is compact and located toward the discharge end of the system. As the initial period of loading is lengthened while at the same time reducing the concentration (for example, $C_0 = 0.5$

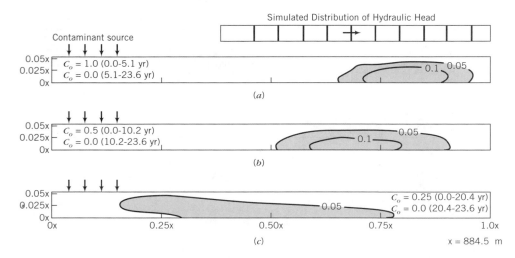

Figure 17.9 Changing plume shapes as a function of the nonconstant loading rates indicated on each cross section (modified from Schwartz, 1975). Reprinted from J. Hydrol., v. 27, by F. W. Schwartz, On radioactive waste management: An analysis of the parameters controlling subsurface contaminant transfer, p. 51–71, 1975 with kind permission from Elsevier Science, NL, Sara Burgerhartstraat 25, 1055 KV Amsterdam, The Netherlands.

for 10.2 yr in Figure 17.9*b*; and $C_0 = 0.25$ for 20.4 yr in Figure 17.9*c*), the plume is larger in size but has a smaller maximum internal concentration, and is located closer to the source.

With mathematical simulations like these, there was no real problem in controlling the transport with an appropriate choice of model parameters. In real problems where the parameters are not known at all, it may not be easy to interpret contaminant distributions in terms of individual processes. For example, a broad zone of dispersion at the front of a plume might be explained in terms of a large longitudinal dispersivity and a constant loading function, or of a small longitudinal dispersivity and a source concentration that increased with time, at least initially. To explain the geometry of a plume fully requires a detailed characterization of the various processes and the loading function.

Fractured and Karst Systems

Life would be simpler if all plumes were as "ideal" as those just discussed. Often, plumes are irregularly shaped with nonsystematic concentration distributions. These plumes form mainly because of extreme variability in the pattern of flow. Although such variability can occur in

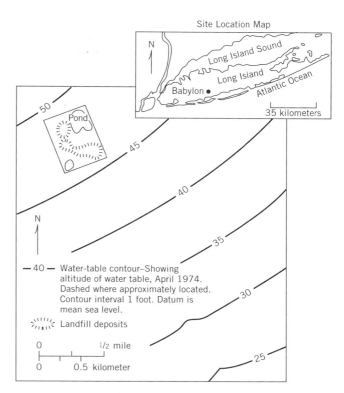

Figure 17.11 Maps showing the location of the Babylon landfill and the water-table contours in April 1974 (modified from Kimmel and Braids, 1980).

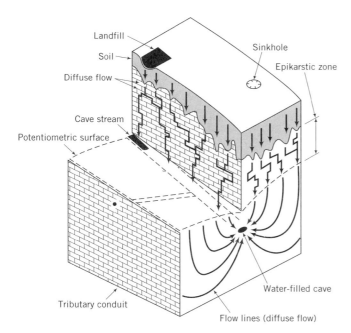

Figure 17.10 The complex pattern of flow and contaminant migration in a maturely karsted terrane. Flow converges to a cave stream, which is the ultimate pathway for contaminant migration. Diffuse flow in the ground-water system is occurring along bedding planes and other joints (from Quinlan and Ewers, 1985). Reprinted by permission of Fifth National Symposium and Exposition on Aquifer Restoration and Ground Water Monitoring. Copyright © 1985. All rights reserved.

porous media, karst and some types of fractured rocks provide a setting where changes in hydraulic conductivity of many orders of magnitude are possible over very small vertical and horizontal distances.

The complexity of mass spreading in such a system is exemplified by karst (Quinlan and Ewers, 1985). Ground-water flow in a mature karst aquifer is typically convergent toward major conduits. Figure 17.10 illustrates how diffuse recharge moving through the unsaturated zone gradually converges to a major conduit draining most of the subbasin. Thus unlike the simulated examples, a plume will not gradually spread as it moves down the flow system but converge toward a cave stream and move rapidly to the spring or springs at the downstream end of the basin (Quinlan and Ewers, 1985). The geometry of the network of small fractures and conduits determines the plume geometry. This problem is in many respects more analogous to those commonly encountered in surface water rather than ground water, with rapid turbulent flow in a network of channels and tributaries.

Babylon, New York, Case Study

Now let us consider some real examples of how concentration distributions depend on processes. At Babylon,

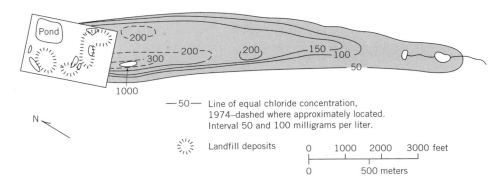

—50— Line of equal chloride concentration,
1974–dashed where approximately located.
Interval 50 and 100 milligrams per liter.

Landfill deposits

```
0      1000   2000   3000 feet
├───┼───┼───┼
0            500 meters
```

Figure 17.12 Map of the Cl⁻ plume in 1974 at depths below the water table ranging between 9.1 and 12.1 m (modified from Kimmel and Braids, 1980).

New York (see insert Figure 17.11), Cl⁻, nitrogen compounds, trace metals, and various organic compounds originating from a landfill have contaminated shallow ground water (Kimmel and Braids, 1980). Landfilling at the site began in 1947 with urban refuse, incinerated garbage, cesspool waste, and industrial refuse. The refuse is 18 to 24 m thick and often placed below the water table. The cesspool wastes were treated to some extent before being discharged into lagoons on the north and west parts of the property.

The surficial sand aquifer is approximately 27.5 m thick and has a hydraulic conductivity of 1.7×10^{-3} m/s. Flow in the aquifer is generally to the south-southeast (Figure 17.11). Contaminants include major ions Ca^{2+}, Mg^{2+}, Na^+, K^+, HCO_3^-, SO_4^{2-}, and Cl^-; nitrogen species such as NH_4^+ and NO_3^-; heavy metals, particularly iron and manganese; and organic compounds.

Shown in Figure 17.12 is the Cl⁻ plume between 9.1 to 12.1 m below the water table. Because Cl⁻ does not react, the plume is a manifestation of only the mass transport processes. Of these, the most important is advection. Longitudinal dispersion is relatively significant but transverse dispersion is negligible. A three-dimensional analysis of the main portion of the Cl⁻ plume by Kelly (1985) determined an advective velocity of 2.9×10^{-4} cm/s,

which is similar to the velocity reported by Kimmel and Braids (1980). Values of dispersivity in the longitudinal and two transverse directions (y and z) are estimated to be 18.6 m, 3.1 m, and 0.6 m, respectively. The tendency for α_x to be about six times larger than α_y and for α_z to be much smaller than either of the others is in line with expectations from carefully run tracer tests. The relatively smooth decline in Cl⁻ concentration away from the source suggests a relatively constant rate of loading. However, there is no way to establish with the given information whether the estimate of α_x contains a component due to variable source loading that might inflate the dispersivity value.

The ratio of NO_3^-–N to total N (Figure 17.13) indicates that at the source most of the N is present as NH_4^+, which is indicative of reducing conditions near the landfill. The source of NH_4^+ is the microbial decomposition of organic wastes containing nitrogen. The increase in the proportion of NO_3^-–N as a function of the travel distance away from the source is probably due to the oxidation of NH_4^+ to NO_3^- as mixing brings oxygen into the plume. Mapping the distribution of various nitrogen species often can be helpful in assessing redox conditions.

This reduced zone also explains why metals, particularly iron and manganese, are so mobile. The E_H–pH dia-

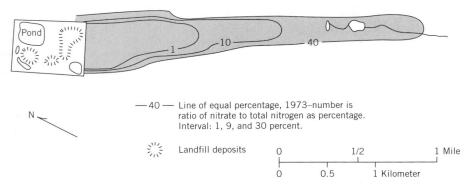

—40— Line of equal percentage, 1973–number is
ratio of nitrate to total nitrogen as percentage.
Interval: 1, 9, and 30 percent.

Landfill deposits

```
0              1/2            1 Mile
├───┼───┼
0      0.5    1 Kilometer
```

Figure 17.13 Map showing the ratio of NO_3^-–N to total N expressed as a percentage. Low values close to the source reflect the presence of NH_4^+–N as the dominant N species (modified from Kimmel and Braids, 1980).

gram for Fe (Figure 12.11) illustrates that with moderate reducing conditions in a pH range of 6.0 to 6.5 (as observed near the source), Fe^{2+} is the stable form of Fe. Similarly, Mn^{2+} is the stable form of manganese. The gradual transition to more oxidizing conditions down the plume creates a situation where solids (for example, $Fe(OH)_3$ and $MnO(OH)_2$ are the most stable form. The mobility of the metal ions is dramatically reduced.

Alkali Lake, Oregon, Case Study

The second case examines contamination of a shallow ground-water system at Alkali Lake, Oregon (Figure 17.14) by chlorophenolic compounds (Pankow and others, 1984; Johnson and others, 1985). It illustrates important features of transport involving dissolved organic compounds subject to sorption. In November 1976, 25,000 barrels of wastes from the manufacture of 2,4-D (2,4-dichlorophenoxyacetic acid) and MCPA (4-methyl-2-chlorophenoxyacetic acid) were crushed and disposed of in 12 shallow, unlined trenches. The eight different chlorophenolic compounds listed in Table 17.6 were of particular interest.

Ground water flows westward in fractured, fine-grained, eolian, or lacustrine deposits with a maximum thickness of approximately 30 m (Figure 17.14). These deposits contain $2.4 \pm 0.3\%$ solid organic matter. Hydraulic conductivities fall in a range from 0.01 to 0.10 cm/s, while effective porosities are from 1 to 5%. These porosity values are much less than the measured total porosities of 0.6 to 0.7. With the given hydraulic gradient, the mean ground-water velocity ranges from 3.9 to 1040 cm/day.

All eight contaminants have spread to some extent. The plumes for 2,4-DCP (Figure 17.15a), 2,6-DCP, and 2,4,6-TCP are the largest. Spreading is less with TeCP (Figure 17.15b) and PCP. The CPP plumes are the smallest, especially CL4D2 shown in Figure 17.15c. Contaminants are moving approximately in the direction of the hydraulic gradient. However, because of anisotropism, the direction of flow and the gradient do not exactly coincide. The plume is elongated in the direction of the hydraulic gradient, implying that advection is the dominant mechanism of spreading. The 2,4-DCP plume is unaf-

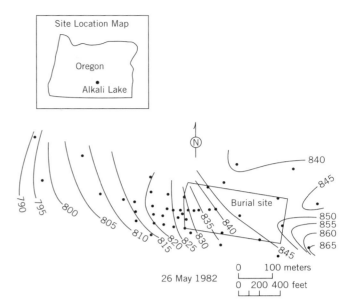

Figure 17.14 Maps showing the site location (inset) and the configuration of the water table in May 1982 (modified from Pankow and others, 1984). Reprinted by permission of Ground Water. Copyright © 1984. All rights reserved.

fected by sorption or other retardation processes and thus is useful for evaluating the mass transport processes. Qualitatively, the shape of the plume indicates that longitudinal dispersion is more significant than transverse dispersion. An unpublished analysis of this plume has determined a mean flow velocity of 1.25×10^{-4} cm/s and values of α_x, α_y, and α_z of approximately 700 cm, 70 cm, and 7 cm, respectively.

Of the eight contaminants studied at the site, 2,4-DCP, 2,6-DCP, and 2,4,6-DCP were shown through batch experiments to have a K_d of zero. There is thus no tendency for sorption on solid organic carbon in the aquifer. In ground water of pH 10 at this site, these compounds ionize to negatively charged species. They are, therefore, much more mobile than their published K_{ow} values would suggest. This example points out the need to understand fully the limitations in predicting K_{oc} values from regression relationships. The other contaminants TeCP, PCP, and the three CPPs are also largely ionized in this ground water. However, these compounds do sorb to some extent on organic carbon. Even so, the retardation in these

Table 17.6 **List of Contaminants Investigated by Johnson and Others (1985) at Alkalai Lake, Oregon**

2,4-DCP	2,4-Dichlorophenol	CPPs including CL2D2, CL3D3, CL4D2
2,6-DCP	2,6-Dichlorophenol	chlorophenoxyphenol dimers with two,
2,4,6-TCP	2,4,6-Trichlorophenol	three, and four chlorines
TeCP	2,3,4,6-Tetrachlorophenol	
PCP	Pentachlorophenol	

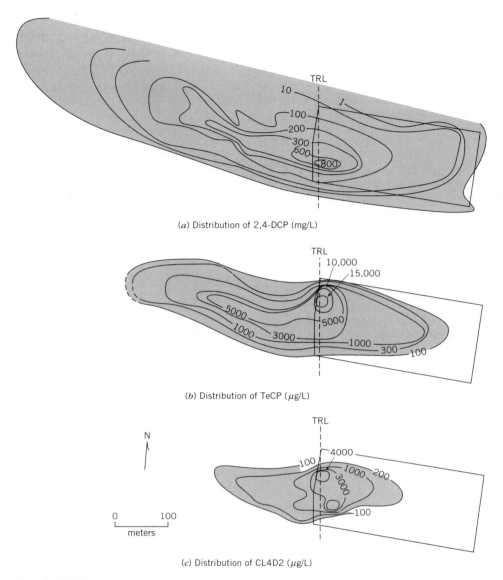

(a) Distribution of 2,4-DCP (mg/L)

(b) Distribution of TeCP (μg/L)

(c) Distribution of CL4D2 (μg/L)

Figure 17.15 Maps illustrating the pattern of spreading for (a) 2,4-DCP, (b) TeCP, and (c) CL4D2 at Alkali Lake (modified from Johnson and others, 1985). Reprinted by permission of Ground Water. Copyright © 1985. All rights reserved.

plumes is much less than predicted from K_d values determined from batch experiments. This behavior may be due in part to fractures in sediments at the site. The data from the batch experiments do predict correctly the relative order of retardation for all eight compounds. This study is instructive because it is one of the first to document compound-dependent retardation for organic compounds.

17.3 Design and Quality Assurance Issues in Solute Sampling

The problem of how to characterize the distribution of contaminants is of great practical concern because plume definition is such an integral part of any field study. A lesson too often learned the "hard way" is that all the basic theory and background knowledge in the world cannot correct problems with poor measurements. This section deals with network design and issues of quality assurance in sampling water.

Design of Sampling Networks

A major concern in sampling is the design of the network. Important considerations in design include the need for close interval point sampling and sample locations that take into account the character and complexity of flow. With piezometers used for measuring water levels, often little attention is paid to screen length. Consider the aquifer in Figure 17.16. As far as hydraulic head is concerned, there is little difference in whether we measure head in a small volume of the aquifer (point sample) or in a larger volume (nonpoint sample). The hydraulic head is about the same in either case (Figure 17.16*a*) because in a permeable unit the vertical gradients in hydraulic head are usually small. Using the piezometers for chemical sampling, however, can produce dramatically different results (Figure 17.16*b*). For this example, only the point samples provide concentrations suitable for interpreting the contaminant distribution. Because of mixing within the sampling device, estimates of the plume geometry based on the nonpoint samples would be in error. Thus, all measurements not concerned simply with presence/absence indications should use point sampling.

Because point sampling usually involves removing a small volume of water, systems should be designed with a minimum internal volume. This consideration requires a small sampling chamber or small-diameter casing (tubing) connecting the intake system to the surface. Keeping the internal volume small avoids the possibility of mixing a sample with a large preexisting volume of water of different composition. Such mixing in large-diameter monitoring wells could be a major source of error. Unfortunately, a system that is ideal for chemical sampling is usually not very good for measuring hydraulic heads with an electric tape.

The location of sampling points should account for the character and complexity of flow. In many respects, defining the vertical extent of the plume is most difficult because often plumes are not very thick. For example, Smith and others (1987) compare the vertical variation in specific conductance and nitrate concentration as deduced from four standpipe piezometers with results from 15 multiport samplers installed at the same site over an

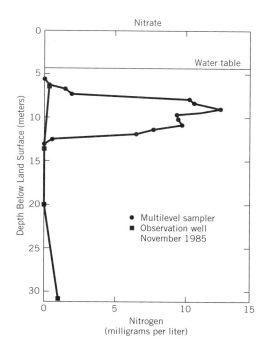

Figure 17.17 Demonstration of the need for close-interval vertical sampling. Data for nitrate (modified from Smith and others, 1987).

interval of about 25 m (Figure 17.17). The loss in resolution in defining the plume with a small number of vertical samples is obvious. With NO_3^-, four piezometers failed even to detect the plume (Figure 17.17).

For determining plume boundaries, wells need to be installed within and closely adjacent to the plume. One line of samplers installed along the midline of the plume will establish its overall length and thickness. Lines of samplers installed normal to the midline serve to define the plume's width and thickness. This simple design is not universally applicable. Care must be taken to design a system that is appropriate for the overall complexity of flow conditions. A good illustration comes from Quinlan and Ewers (1985), who looked at the problem of how to define plumes in karst terranes. As we showed with Figure 17.13, contaminants from a surface source could migrate through a succession of larger and larger conduits to one large conduit possibly 3 to 10 m wide. With a simple network design such as we just described, there is almost no chance of intersecting this single conduit and collecting samples along the critical pathway for contaminant migration. Quinlan and Ewers (1985) proposed a site-specific monitoring scheme that requires the definition of major elements of the karst system. Spring mapping and dye tracing are some of the tools that can establish connections among the disposal site, springs, and the conduit systems. Ultimately, a few critical springs and wells on the major conduits provide the proper monitoring sites.

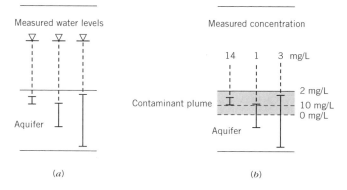

Figure 17.16 (*a*) In many cases there may not be a substantial difference in hydraulic head values measured in an aquifer using point and nonpoint sampling procedures. (*b*) Concentrations of contaminants can change markedly over relatively small vertical distances. In this example, nonpoint sampling will yield concentrations that are much less than the true concentration.

Assuring the Quality of Chemical Data

Another concern in sampling is with the overall quality of the chemical data. Reliable chemical data are not the result of chance but the care and precautions taken in study programs. There are four potential problems that can have a serious impact on chemical data, including (1) contamination of samples with fluids that were used to drill the hole, (2) changes in water quality caused by the presence of the well, (3) sample deterioration, and (4) sloppy field and laboratory practices.

The first problem has to do with the question of exactly what fluid is being sampled. Occasionally, samples will contain some of the gas, water, or oil in combination with foams, emulsions, and muds that were required to drill the hole (Hull and others, 1984). Once these exogenous fluids are present in the subsurface, it is often difficult to remove them completely. One way to avoid sampling drilling water is to develop the well. A tracer, added to the drilling fluid, can be monitored as the well is developed to check that development is complete.

Changes in the chemistry of the ground water can also be related to the presence of the well. The most serious potential problem arises from the use of cement (or other soluble materials) as a seal above the intake of a well or piezometer. High sample pHs (for example, above 9) mostly reflect a continuing interaction between the cement seal and the water being sampled. This problem can persist for a long time and is difficult to fix. Samples can also be affected when water inside or outside of the well casing dissolves small quantities of the casing or glues used at joints. A possible solution to these problems is to remove several standpipe volumes of water before sampling. However, a proper selection of materials for the sampling device is the only sure fix for these problems.

Another problem is sample deterioration. Once the temperature, pressure, and gas content of a water sample change following collection, the chemistry can change as well. For example, the addition of oxygen to an anoxic sample results in a rapid change in E_H that can cause dissolved trace metals to precipitate as solids. Loss of CO_2 can raise the pH and decrease the HCO_3^- concentration and in some cases cause $CaCO_3$ to precipitate. The extent of the change in composition depends in many respects on the chemistry of the particular sample. These problems have been addressed by specialized sampling equipment for high-pressure–high-temperature environments (Hull et al., 1984), the use of flow cells at the well head (Jackson and Inch, 1980), and methods of sample treatment to avoid deterioration. In designing a sampling program, it is a good idea to refer to information on standard practice, for example, Rainwater and Thatcher (1960), Ball and others (1976), Wood (1976), Gibb and

others (1981), Scalf and others (1981), Claasen (1982), and Lico and others (1982). Methods for preserving samples containing organic compounds are described by Goerlitz and Brown (1972), Wershaw and others (1973), and Minear and Keith (1984).

Problems can also be related to the handling of samples in the field or laboratory analyses. In many studies, these problems go unrecognized and have an important impact on the quality of the chemical data. Sample contamination caused by improper bottle washing, filtering, or the use of impure preservatives is the main concern in handling samples in the field. It is prudent to check for this problem on an ongoing basis by running blanks of ultrapure distilled water through the various field treatments and submitting these samples to a laboratory. In the case of organic compounds, adsorption onto containers or filtering devices, and loss of volatiles from the sample, can also pose problems (Gibb and Barcelona, 1984; Jackson et al., 1985).

Likewise, the quality of the laboratory analyses should be checked on an ongoing basis. Some common approaches are (1) submitting "spiked" samples with known concentrations, (2) submitting duplicate samples to different laboratories, and (3) submitting replicate samples. The first two of these checks generally assess the analytical accuracy of the laboratory, while the third tests analytical repeatability. Guidance for developing quality assurance programs is provided by Wilson (1952), the American Chemical Society (1980), and Friedman and Erdman (1982).

17.4 Sampling Methods

This section examines the various approaches available to hydrogeologists in collecting ground-water samples for chemical analysis. The range of possible approaches is surprisingly broad, from simply pumping water from a well to new and technologically innovative approaches.

Conventional Wells or Piezometers

Commonly, water samples are pumped or bailed from wells or piezometers. In most site investigations, a cluster of wells or piezometers—referred to as a nest—is installed so as to obtain vertically distributed water samples (Figure 17.18a). This vertical coverage is critical in defining contaminant distributions.

The most important feature with this approach is that each well or piezometer is completed in its own borehole. In spite of technologies that facilitate completion of multiple sampling wells in a single piezometer, there is a compelling simplicity with a permanent single-well completion. Seals are easier to place as compared to multilevel, and water levels can be measured in a straight-forward manner. These installations are also extremely durable, with little possibility of failure once the stand-pipes have been emplaced. The main disadvantage of piezometers for sampling is the higher cost of drilling additional boreholes versus the installation of many multi-level samplers in a single borehole.

Figure 17.18*b* illustrates a conventional standpipe piezometer. The design with filter sand placed around the intake is most appropriate for shallower depths. When these samplers are installed in relatively deep boreholes, the seal is set above a metal-petal basket (or similar device) rather than a sand pack. In deep boreholes, placing sand around the intake is much more difficult than at shallow depths. Standpipes placed in deposits that cave are often installed without a filter sand or seal (Figure 17.18*c*). They are usually emplaced using hollow stem augers, which hold the borehole open until the casing is run in to the proper depth. A variation on this design involves placing a bentonite seal above the intake by passing material through the hollow-stem auger. While a good idea, this installation procedure is tedious and may not work.

Water samples can be collected from these wells most simply using bailers. To avoid the possibility of cross-

contaminating wells, companies are marketing lines of disposable bailers. Bailing, however, may pose problems in many applications due to loss of gas or volatile organic compounds. In most cases, water is pumped from wells using permanently installed gas drive or submersible electric pumps. A detailed survey of all of the common sampling devices, and extensive discussion of the appropriateness in sampling for various types of chemical constituents, is provided by U.S. EPA (1993).

Multilevel Samplers

Multilevel sampling involves placing several samplers at various depths in a single borehole (Cherry, 1983). The advantage of this method is the relatively large number of discrete sampling points that can be installed at a relatively low cost. The method is most economical for near-surface investigations in noncohesive sand or silts, where filter sands and seals are not required. For cohesive deposits, it is necessary to place seals between intakes. Whether or not filter sand is required depends on the type of seal. Extra difficulties related to correctly placing the seals and filter sands can affect the cost and reliability of these installations. The small size of casing typically used in these samplers makes it difficult to develop the screen and sandpack and may promote plugging. Further, the design of an appropriate sounder for measuring water levels in these samplers can challenge one's ingenuity. However, the small internal volume of these devices is a significant benefit. Because small quantities of water are required for flushing the system and the actual sample, sampling will probably not perturb the contaminant plume.

A variety of multilevel devices (from Cherry, 1983) is shown in Figures 17.19*a*, *b*, *c*, *d*, and *e*. These can be installed in cohesive or noncohesive units, but the examples show either one or the other type. The first system (Figure 17.19*a*) consists of a large number of small-diameter standpipes placed in a single borehole. The main requirement for installing this system is patience in carefully placing the filter sand and seals.

Boreholes that cave usually require a hollow stem auger to install the multilevel device. To facilitate the installation, the standpipes (sometimes polyethylene tubing) are bundled around a more rigid casing (Figure 17.19*b*). Another variation on this design places the tubing inside a solid casing to protect the tubing against damage (Figure 17.19*c*). When there is some possibility of units not caving to prevent vertical communication in the borehole, seals need to be provided between the intakes. Cherry (1983) discusses ways of installing these seals.

Another innovative procedure involves sampling through specially designed valved couplings (Figure 17.19*d*) that join sections of PVC or stainless steel casing.

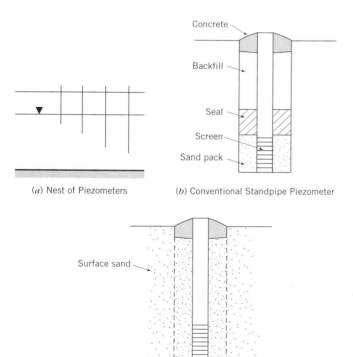

(a) Nest of Piezometers (b) Conventional Standpipe Piezometer

(c) Piezometer Installed in Caving Materials

Figure 17.18 Basic designs for standpipe piezometers. Panel (*a*) illustrates the concept of how a nest of piezometers provides spatially distributed concentration. Panels (*b*) and (*c*) are examples of typical piezometers for noncaving and caving materials (modified from Cherry, 1983).

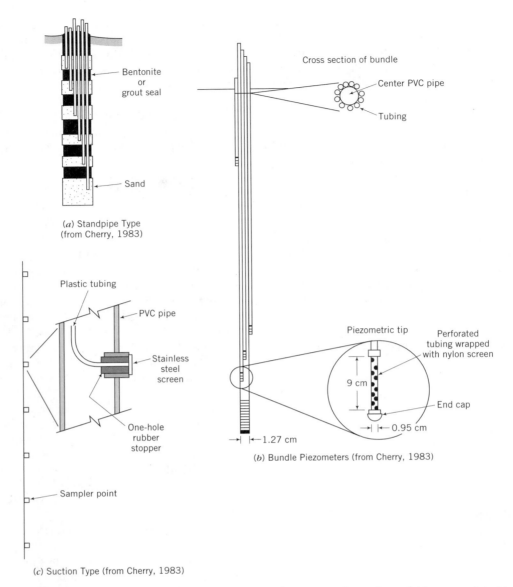

(a) Standpipe Type
(from Cherry, 1983)

(c) Suction Type (from Cherry, 1983)

(b) Bundle Piezometers (from Cherry, 1983)

Figure 17.19   Examples of multilevel sampling devices. Portions of panel (e) are reprinted by permission of Ground Water Monitoring Review. Copyright © 1981. All rights reserved.

The sampling device passes through a valve or port to sample fluids outside of the casing. Another tool containing a down-hole transducer measures hydraulic head through these same ports. Packers prevent communication in the borehole between the sampling ports. To reduce the complexity of the installation, a special tool inflates the packers through the casing. A stainless steel version of this sampling system is particularly well suited for deep boreholes in bedrock.

The final type of multilevel device is what Cherry (1983) refers to as gas-drive samplers. They consist of a sample chamber with narrow-diameter tubes connecting the device to the surface (Figure 17.19e). Compressed gas forced down one of the tubes will empty the sample

chamber for flushing or sampling. The figure illustrates the completion details for these devices.

Solid and Fluid Sampling

Sometimes it is worthwhile to collect samples of both the porous medium and the fluids by drilling and coring. Fluids squeezed out of the sample can be analyzed in the normal manner. Another approach is to determine contaminant concentrations in the dried solid and report contaminant concentrations as mass of contaminant per unit mass of solid. This latter type of analysis is used most commonly in sampling for LNAPLs and DNAPLs.

The main advantages of this sampling technique are,

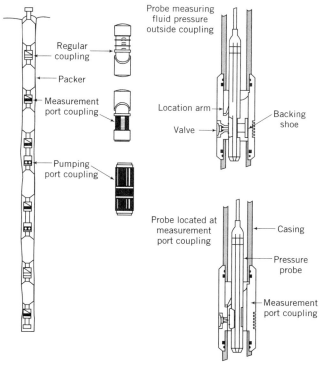

(*d*) Westbay-Type Samplers (from Cherry, 1983)

Figure 17.19d

first, the spatial control in being able to relate specific concentrations of contaminant to specific lithologies, and, second, the ability to minimize contamination due to the sampling method. In all of the direct sampling approaches we discussed earlier, the possibility exists for a ''memory effect.'' Some of the contaminants brought into the sampling device at an earlier time may remain sorbed on the internal surfaces. These compounds can desorb during later sampling and result in errors.

The approach of solid and fluid sampling can eliminate this possible problem. There are limitations. First, it is often quite difficult to retain typical aquifer materials using split-spoon sampling or thin-wall tube samplers. A variety of designs to overcome this limitation is presented in U.S. EPA (1993). A second limitation is the cost of bringing a rig back onto the site if more than one round of sampling is required.

Cone Penetrometry

Cone penetrometers are used in standard geotechnical engineering practice. In its most basic form, the cone penetration test involves pushing a cone-shaped instrument into the soil and measuring the tip resistance to penetration and friction (Chiang and others, 1989). Figure 17.20 illustrates how a hydraulic ram is used to force the cone penetrometer or other tools into the ground. Typically, the information is used for assessments of soil strengths and stratigraphic interpretations of site conditions.

Refinements to this technology have provided the capability to make water-pressure measurements (yielding hydraulic heads) as a function of depth, and to estimate hydraulic conductivity from measurements of the rate of pore pressure dissipation (Chiang and others, 1989). Other tools provide a capability of collecting water samples with depth.

Shown in Figures 17.21*a*, *b* are two samplers that are used to replace the standard cone penetrometer. With the BAT® system, the sampling probe is pushed to the depth of interest. A sliding sleeve protects the filter tip (that is, intake) during penetration. At the desired depth, the filter is exposed and the water sample enters the sampler up to the septum at the top filter. The sample is collected by lowering a sampling probe consisting of an evacuated sampling tube and the double-ended

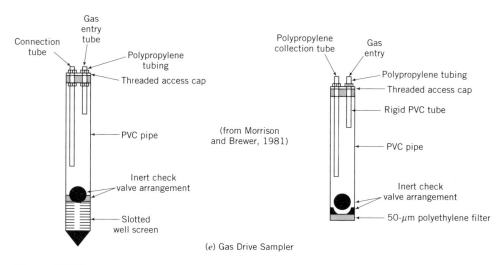

(*e*) Gas Drive Sampler

Figure 17.19e

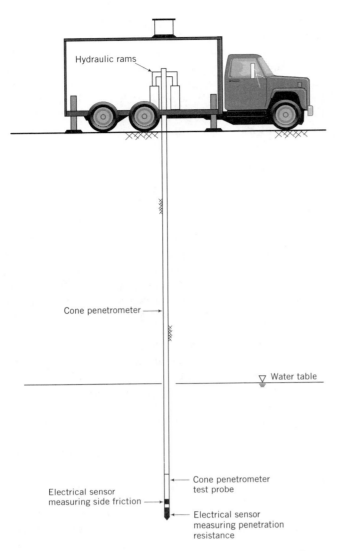

Figure 17.20 Schematic of a cone penetrometer rig (from Smolley and Kappmeyer, 1991). Reprinted by permission of Ground Water Monitoring Review. Copyright © 1991. All rights reserved.

provides a rapid and cost-effective approach for collecting ground-water samples. It can be used in its own right for site investigation or in a reconnaissance mode to assist in siting conventional water-sampling wells. The BAT® sampler can be installed permanently for future sampling. Cone penetrometry methods are limited to unlithified materials and are difficult to use in units containing gravel (U.S. EPA, 1993).

New sampling approaches based on this technology are under active development. Probes have been developed for soil sampling, electrical resistivity and ground-penetrating radar measurements, and the direct detection of NAPLs.

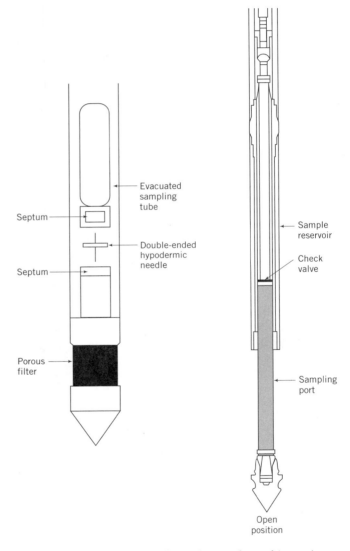

Figure 17.21 Schematic of sampling tools used in conjunction with a cone penetrometer rig. Panel (*a*) illustrates the BAT® tool (from Mines and others, 1993). Panel (*b*) illustrates the HydroPunch® (from Smolley and Kappmeyer, 1991). Reprinted by permission of Ground Water Monitoring and Remediation. Copyright © 1993 and 1991. All rights reserved.

hypodermic needle shown in Figure 17.21*a*. Once the sampling probe is connected to the filter tip, the needles penetrate the septa, and the sample is drawn into the sample vial. The sampling probe may then be withdrawn. The method appears to minimize the loss of volatiles (Torstensson and Petsonk, 1988; Mines and others, 1993).

The HydroPunch® is a stainless steel and Teflon® sampling tool that also can be used with a cone penetrometer rig to collect water from unconsolidated sediments (Edge and Cordry, 1989; Smolley and Kappmeyer, 1991). As before, the sampling tool is pushed closed to the desired sampling depth. Retraction of the sample chamber exposes the intake tube and water enters the sample chamber through a check valve (Figure 17.21*b*). Once the sample is collected, the tool is retrieved. This type of sampling

Other Sampling Methods

A variation on the cone penetrometer approach involves the use of a portable hammer to drive a small wellpoint attached to a metal casing to various depths. Water samples can be collected from this well at desired depths, and the installation can be left in place permanently.

An extremely versatile vehicle-mounted sampling system based on this sampling method, called the Geoprobe®, was developed in 1988 by Kejr Engineering, Inc. This system uses a vehicle-mounted, hydraulically powered percussion probing hammer to advance the tool string to depths of 30 m or more. Specialty tools provide a capability for soil sampling and the collection of water samples through a stainless steel screen. Water is collected using a bailer or with plastic tubing and a peristaltic pump.

The screened auger approach involves the use of a slotted hollow-stem auger to collect samples periodically while drilling is in progress (Taylor and Serafini, 1988). Having reached a desired depth, the screened portion of the auger is developed by air-lift pumping. The actual sample is collected by using a submersible pump inside the hollow-stem augers. Case studies (Taylor and Serafini, 1988) indicate that this sampling approach is useful for reconnaissance-type investigations of ground-water problems.

Dissolved Contaminants in the Unsaturated Zone

Nearly all sampling techniques in the unsaturated zone involve pulling the fluid under vacuum through a porous ceramic cup into a container (Figure 17.22). Once a sufficient volume of fluid has been collected, the sample is removed by suction or by gas displacement depending on how deep the sampler is installed. Signor (1985) describes varieties of these samplers.

These devices function in both unsaturated and saturated zones. This characteristic makes them useful for sampling when the state of saturation changes frequently. An example might be a facility for artificial recharge that is operated with only periodic flooding of the infiltration basins. The main problem with this method is the possibility for the fluid composition to change from reaction with the ceramic cup, or from the loss of gases as the sample is collected under vacuum. These effects can be examined in the laboratory and hopefully accounted for in the interpretation of results.

17.5 Indirect Methods for Detecting Contamination

The indirect methods for plume delineation can provide a rapid and inexpensive alternative to conventional sam-

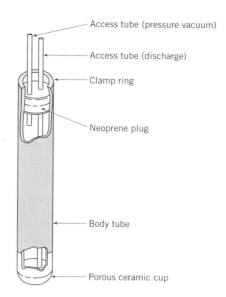

Figure 17.22 A typical device for collecting soil-water samples from the unsaturated zone (from Soil-moisture Equipment Corp). Reprinted with permission.

pling. These approaches are successful when contamination is manifested in a secondary way, for example, by the presence of volatile organic compounds in the soil gas found above a spill or the change in electrical properties of a unit caused by the presence of the plume. A variety of published case histories points to the usefulness of these approaches in site evaluation. However, like many of the indirect approaches, they do not always provide an unambiguous interpretation.

Soil-Gas Characterization

Characterizing the composition of soil gases has emerged as an industry-standard technique for tracing volatile organic compounds in ground water. The approach involves defining zones of ground-water contamination based on the presence of volatile components in the soil gas. With time, volatiles at the capillary fringe partition into the soil gas and gradually diffuse upward to the ground surface. The presence of volatiles is established commonly by collecting soil gas from some fixed depth and analyzing the sample with a gas chromatograph. However, this approach does not detect all organic contaminants because not all are volatile. In addition, the organic compound should not be too soluble. Highly soluble volatiles moving through the unsaturated zone will dissolve into any water present. Reisinger and others (1987) illustrate the typical components of gasoline that might be successfully detected in a soil-gas survey (Figure 17.23).

A metal probe with a perforated tip is driven from 2 to 4 m into the unsaturated zone, and a small sample of soil gas is extracted by pumping. A variety of commercially available equipment is available to assist with placing of sampling probes. Gas samples are usually analyzed on site with a gas chromatograph (GC). The GC measurement provides a quantitative estimate of the mass of a particular volatile compound per volume of soil gas (for example, μg/L). When plotted on a map, these data can be used to infer zones of contamination in the unsaturated and saturated zones. By relating contaminant concentrations in the soil gas to measured concentrations in the ground water, the soil-gas data can be transformed to provide a quantitative estimate of concentrations in ground water.

A variation on this approach involves sampling soil gases by using a grid of static collectors (Malley and others, 1985). The sampling device, which contains activated charcoal, is buried and left for several weeks, giving time for the collector to equilibrate with the soil atmosphere. A grid of samplers provides an estimate of the spatial variability. Once the collectors are returned to the laboratory, the volatiles are desorbed and analyzed using mass spectrometry. This analytical procedure provides for the detection and identification of compounds up to mass 240 (Malley and others, 1985). Maps of mass fluxes (or relative ion counts) provide an estimate of the plume position.

Soil-gas sampling provides a rapid and economical way of surveying large sites for contamination. Thus, it is attractive for reconnaissance studies aimed at discovering what volatile contaminants are present and where they are located. This information often assists in designing

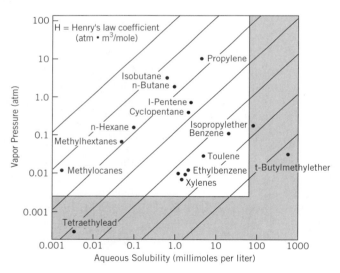

Figure 17.23 The unshaded area shows the range in vapor pressure and aqueous solubility for the common constituents in gasoline for which soil-gas surveying will be successful (from Reisinger and others, 1987). Reprinted by permission of First Outdoor Action Conference on Aquifer Restoration, Ground Water Monitoring and Geophysical Methods. Copyright © 1987. All rights reserved.

a conventional sampling program. In some studies, the information provided by a soil-gas survey may be all that is required to establish whether or not a contamination problem exists, or to identify a source.

There are some circumstances when these approaches may not work well. A low-permeability layer in the unsaturated zone can inhibit the upward diffusion of vapor and promote extensive horizontal spreading. This situation will produce an estimated distribution of contaminants that is much larger than the actual one. Other problems relate to how the contaminant occurs. With LNAPLs, vapor concentrations in the unsaturated zone will be larger and easier to detect than if the contaminant is dissolved in the ground water (Reisinger et al., 1987). Furthermore when contaminants are present in localized fracture zones (for example, in sandstone or limestone), the rates of diffusion away from the fractures may be so slow that vapor-phase transport is limited in extent (Reisinger et al., 1987). Thus, care must be exercised in interpreting the results, and where necessary, conclusions should be confirmed using an independent approach.

Whittmann (1985) describes a case study that demonstrates the potential of soil-gas sampling for delineating contaminated ground water. The case considered the origin of contamination in the Verona Well Field at Battle Creek, Michigan. Conventional ground-water monitoring pointed to the presence of an additional but unidentified source. Soil-gas surveying was used to explore a large railyard. The sampling approach involved driving a metal probe approximately 1 m into the soil and extracting the soil gas with a hand pump. Sample analyses were performed on-site using a gas chromatograph. Figure 17.24 shows PCE results for 43 samples. The survey detected three areas of elevated soil gas in the railyard. The most concentrated of the three was a small solvent-disposal area that had gone undetected in previous investigations. Subsequently, two wells confirmed the PCE source. Overall, the approach worked well, except that it did not reveal the full extent of the PCE plume. Migrating away from the source, the plume moved deeper and lost touch with the water table. Under such conditions, volatiles were unable to partition into the soil gas.

Geophysical Methods

Exploration geophysics is the science (or art) of inferring the distribution of physical properties beneath the surface, given the measurements of associated fields over a surface above—or along—a borehole within the medium. Such methods can ascertain the presence of some well-defined target (Table 17.7). Targets can range from characteristics of the geologic environment to properties of the fluid(s) within pores.

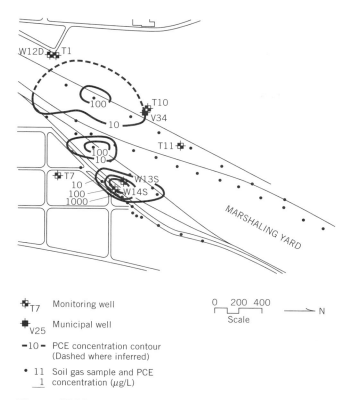

Geophysical methods have been used for many years as a standard tool in the petroleum and environmental industries. These approaches use a variety of instruments to measure changes in the physical and chemical property of materials (geologic or otherwise) and pore fluids. In the next few pages, we will survey the geophysical approaches that are commonly used in contamination studies, but only mention those that are used for geologic investigations. We will leave it to other texts to discuss borehole logging techniques.

Table 17.7 **Relationship Between Geophysical Properties and Fields**

Physical Properties	Fields
Density	Gravity
Elasticity: Bulk modulus	Seismic wave
Shear modulus	
Magnetic permeability	Magnetic
Electrical conductivity	Electromagnetic
(low and high frequency)	
Dielectric permittivity	

Electrical Methods

Electrical methods are commonly applied to studies of ground-water contamination. The various electrical methods usually measure either conductivity or resistivity (the reciprocal of conductivity). Traditionally, resistivity is measured from the surface by adding electrical current to the ground with a pair of electrodes and measuring the resulting drop in voltage with a second set of electrodes. Apparent resistivity values are either read directly from the instrument or calculated. In lateral profiling, the survey is run by maintaining a fixed electrode spacing (producing a constant depth of current penetration) and moving from station to station following a grid. Anomalies on the apparent resistivity map may coincide with contaminant distributions.

Early applications (for example, Cartwright and McComas, 1968) used electrical resistivity surveys to detect changes caused by the presence of ground-water plumes with elevated salinity. In recent years, these types of surveys have fallen out of favor because terrain conductivity methods (that is, electromagnetic methods) are much more rapid and offer possibilities for higher resolution.

Electromagnetic methods induce a current in the ground with an alternating-current transmitting coil. The magnetic field around the coil induces an electrical field in the earth to depths that are largely controlled by the background properties of the medium, the moisture content, and the relative difference in the conducting properties of the medium and the target. In the Geonics EM31, the two coils are mounted 3.7 m apart on rigid boom. The equipment is highly portable, and an operator can collect a large number of closely spaced electrical conductivity measurements in a short time. The depth of signal penetration is about 0.75 to 1.5 times the coil separations (Zalasiewicz and others, 1985). Other instruments provide larger coil separations and greater signal penetration.

Terrain conductivity methods have been frequently applied to subsurface mapping and particularly to the investigation of contaminated sites. Stewart and Gay (1986) have evaluated electromagnetic soundings for deep detection of conducting fluids, and Greenhouse and Slaine (1983) have applied the technique to mapping contaminant migration.

The most important application of electromagnetic methods is to detect buried objects or other features at waste-disposal sites. Examples of buried objects are filled waste trenches or lagoons, buried steel drums, and lost underground piping or storage tanks. A case study of the application of very-high-resolution electromagnetic surveying is provided by Jordan and others (1991). The method was used to examine a vacant lot that was previously used for industrial hazardous-waste processing. During its operation, several unlined lagoons may have existed. The survey was designed to identify the filled lagoons and possibly buried metals. Glacial till covered the site.

The conductivity survey collected 4823 EM data points along lines 3.8 m apart in areas suspected of contamination. Stations were about 0.6 to 0.9 m apart. The resulting conductivity map (Figure 17.25) indicates a broad range of conductivity. Suspected lagoon locations are indicated by conductivity values greater than 50 mS/m (that is, milliSiemens per meter), while near-zero conductivities are thought to represent areas of buried metallic debris. The locations of the lagoons correspond with locations determined from old aerial photographs.

Ground-Penetrating Radar

Ground-penetrating radar (GPR) is a rapidly evolving technology for subsurface investigations (Daniels, 1989). The method is used to map the distribution of buried objects (drums, pipelines, etc.), to define the configuration of the water table and stratigraphic boundaries, and to establish the distribution of LNAPLs. Thus, GPR is well suited for surveying abandoned waste-disposal sites.

A transmitting antenna at the surface radiates short pulses of radio waves into the ground. Reflected energy from the subsurface is recovered by an antenna that is

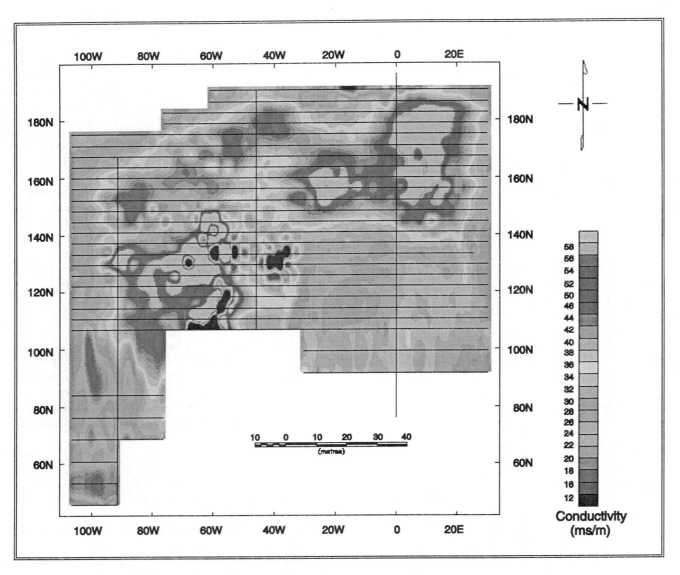

Figure 17.25 Results of a very high resolution electromagnetic survey. Data were collected along lines approximately 3.8 m apart (from Jordan and others, 1991). Reprinted by permission of Fifth National Outdoor Action Conference on Aquifer Restoration, Ground Water Monitoring and Geophysical Methods. Copyright © 1991. All rights reserved. Electronic data file kindly provided by Geosoft Inc., Toronto.

moved across the surface. In principle, then, GPR works like reflection seismic methods except that electromagnetic reflections, rather than acoustic returns, are recorded at the surface. Radar energy is reflected due to changes in dielectric constants and electrical conductivity. Such changes are usually related to variation in properties of the minerals present, the degree of saturation, and material density. By moving the GPR along at a slow speed, it is possible to obtain an almost continuous profile of the subsurface. Under ideal conditions (that is, dry sandy units or bedrock), GPR provides a highly resolved picture of subsurface-site conditions. It is less useful in clayey, moist lake clay or glacial till.

The state-of-practice in radar technology is moving toward highly resolved three-dimensional studies involving a large number of parallel survey lines on the ground surface. With the help of appropriate visualization software, one can interpret and display the three-dimensional relationships among various radar reflectors on a site (Thompson and others, 1995).

Magnetometry

Magnetic surveys measure variations in the Earth's total magnetic field. NAPLs, unlike ferromagnetic materials, do not affect the magnetic properties of the geologic medium. Consequently, direct magnetometry is of no use in detecting the various types of ground-water contamina-tion. However, magnetometry is widely used to locate containers and pipelines that may have contained contaminants. The response of the magnetometer is directly proportional to the mass of the target and inversely proportional to the cube of the distance to the target. Ferromagnetic substances can be detected at depths approaching 20 m. Examples in the application of magnetometer surveys at various waste-disposal sites include Gilkeson and others (1986) and Cochran and Dalton (1995).

Seismic Methods

Seismic techniques use both reflection and refraction to measure the travel time of seismic waves that are generated from some surface source (sledgehammer, explosives). Reflection seismic approaches tend to be used most frequently in environmental applications. Refinements in seismic methods have permitted highly refined investigations at relatively shallow depths (Hunter and others, 1984, 1988; Kaida and others, 1995). Reflection seismic methods are often most useful in reconnaissance-type studies for determining the top of the bedrock surface, the topology of structural features, and the pattern of stratigraphic layering. An example of the application of seismic methods to determine the continuity of a clay layer underlying a proposed hazardous-waste facility is provided by Slaine and others (1990).

Problems

1. Shown on the attached figure is a series of plumes from a sanitary landfill. Examine these plumes in detail, and answer the following questions.

a. Qualitatively evaluate the extent to which advection and dispersion are important in controlling contaminant spread at the site.

b. Given the type of source and the resulting plume shapes, what can you say about the type of source loading?

c. Suggest what processes could be operating to cause the increasing pH away from the source.

d. Metals, for example, Fe, tend to be relatively abundant in landfill leachates. However, Fe^{2+} tends to be strongly attenuated relative to mobile species like Cl^-. Explain why iron behaves in this way.

2. Assume that some very preliminary studies identified that the contamination problem in question 1 had developed in a surficial sand aquifer about 15 m thick. Develop a plan for a detailed site investigation that involves both direct and indirect methods of detecting contamination. The plan should provide specific details about the types of instrumentation/surveys required and plans for water-quality analyses.

Cl⁻ (mg/L)

pH

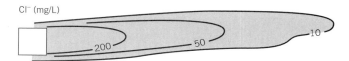

Dissolved Oxygen (mg/L)

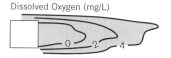

Fe²⁺ (mg/L)

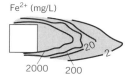

Problem 1.

Chapter 18

Modeling the Transport of Dissolved Contaminants

Chapter 17 explained the nature of contaminant plumes as a manifestation of physical and chemical processes. The approach was a conceptual one, simply illustrating what effect processes could have on contaminant transport. This chapter develops approaches for modeling the spread of contaminants dissolved in water. The starting point in modeling is with the various mass transport equations that were developed in Chapter 14. The end point will be a set of powerful mathematical tools for analyzing problems of contaminant transport.

18.1   Analytical Approaches

Analytical models have proven to be powerful tools for studying transport processes. Because analytical solutions for contaminant transport abound in the literature, the discussion here will focus on the more practical ones that can be used in the field. Normally, this would preclude all one-dimensional solutions. However, in some cases, the pertinent features of a one-dimensional solution may be embedded in a more complex three-dimensional equation. Take for example three one-dimensional solutions of the form

$$\frac{C(x, t)}{C_0} = F_1(\alpha_x, x, t);$$

$$\frac{C(y)}{C_0} = F_2(\alpha_y, y);$$

$$\frac{C(z)}{C_0} = F_3(\alpha_z, z)$$

where C is the concentration; C_0 is the source concentration; α_x, α_y, and α_z are the dispersivities; x, y, z are spatial coordinates; and t is time. If the three one-dimensional solutions can be found, an approximate solution to a three-dimensional problem in some cases will be

$$\frac{C(x, y, z, t)}{C_0} = F_1(\alpha_x, x, t) F_2(\alpha_y, y) F_3(\alpha_z, z)$$

In words, the three-dimensional solution is the product of the one-dimensional solutions. Thus, understanding the three-dimensional result follows from knowledge about the one-dimensional results. This is exactly the approach we will follow in this section.

Advection and Longitudinal Dispersion

Advective transport is frequently demonstrated with a plug-flow model that neglects all longitudinal and lateral mixing. The contaminant moves with the velocity defined by Darcy's equation corrected for flow through the pores. In Figure 18.1a, water containing a tracer with concentration C_0 enters the flow tube and displaces the original fluid. The advective front is located at the position $x = vt$. The concentrations are at steady state and everywhere equal to the source concentration C_0. Such behavior is unrealistic in the field or the laboratory, but the concept is useful in describing other processes.

Figure 18.1b shows advection with longitudinal dispersion. In the absence of lateral or transverse dispersion, the displacing fluid mixes with and displaces the original fluid in the x direction strictly within the flow tube. The concentration at the advective front $x = vt$ is now less than the original concentration C_0. Longitudinal dispersion causes some of the dissolved mass to move ahead of the advective front at the expense of material behind the advective front. However, at some point $x \ll vt$, the concentration is steady and is the same as the source C_0. This steady-state part of the plume possesses all the properties of the plug-flow model of Figure 18.1a.

The situation described above can be reproduced in a laboratory column (Figure 10.5). The resulting breakthrough curve (Figure 18.2a) can be thought of as a window fixed at some point x along the column. For some time after the experiment is begun, the concentration ratio at this point (that is, C/C_0) will be zero, as only water initially present in the column passes by. Prior to the breakthrough of the advective front, the concentration ratio C/C_0 becomes finite and equals 0.5 at the instant the advective front breaks through. Eventually, with increasing time, $C/C_0 = 1$, indicating that the original fluid has been totally displaced and only the tracer of concentration C_0 is passing through the column.

An alternative way to demonstrate the column experiment is shown in Figure 18.2b. This concentration distribution is the spatial distribution of the ratio C/C_0 throughout the column at one point in time. Here, we see that $C/C_0 = 1$, 0.5, and 0 at some $x \ll vt$, $x = vt$, and $x \gg vt$, respectively.

The concentration distribution in a column is completely described by the Ogata–Banks equation (1961),

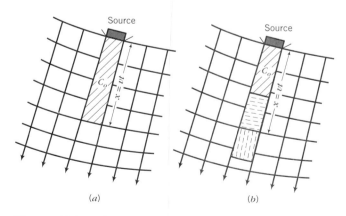

Figure 18.1 Schematic diagram showing the essential features of (a) plug flow, and (b) plug flow accompanied by longitudinal dispersion.

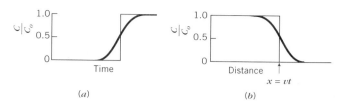

Figure 18.2 (a) Breakthrough curve at the end of a column describing the variation in relative concentration as a function of time and (b) the spatial variation in relative concentration along the column at one time.

which is a solution to the one-dimensional dispersion advection equation introduced in Chapter 14

$$D_x \frac{\partial^2 C}{\partial x^2} - v_x \frac{\partial C}{\partial x} = \frac{\partial C}{\partial t} \tag{18.1}$$

for the conditions

$$C(0, t) = C_0 \quad \text{and} \quad C(x, 0) = 0$$

The first boundary condition states that at $x = 0$, for all time t, the concentration is C_0 (that is, a continuous source). The second states that at all points x at time $t = 0$, the concentration is zero. Just at the start of the test, the concentration of tracer in the water in the column is zero. An exact solution (Eq. 18.1) for these boundary conditions is

$$\frac{C(x, t)}{C_0} = \left(\frac{1}{2}\right)\left(\text{erfc}\left[\frac{(x - vt)}{2(Dt)^{1/2}} \right] \right.$$
$$\left. + \exp\left(\frac{vx}{D}\right) \text{erfc}\left[\frac{(x + vt)}{2(Dt)^{1/2}} \right] \right) \tag{18.2}$$

where D is the coefficient of hydrodynamic dispersion, erfc is the complementary error function, and v is the linear ground-water velocity, which in this case is identical to the velocity of the tracer.

The second term in Eq. 18.2 is usually small in many practical situations so it will be ignored in future discussions. Later, we examine conditions when this simplification causes errors. Overall, the simplicity of this approach versus absolute accuracy is a reasonable trade-off in many applications. Second, the quantity $(Dt)^{1/2}$ can be written as $(\alpha_x vt)^{1/2}$ provided we ignore the diffusive contribution to hydrodynamic dispersion. In this abbreviated form, the solution becomes

$$\frac{C(x, t)}{C_0} = \left(\frac{1}{2}\right)\left(\text{erfc}\left[\frac{(x - vt)}{2(\alpha_x vt)^{1/2}} \right] \right) \tag{18.3}$$

Further discussions will focus only on the abbreviated form of Eq. 18.2.

The complementary error function occurs frequently in solutions to the advection–dispersion equation. Like the various well functions in hydraulic testing, it is a well-tabulated function so that graphs or tables can provide numerical values. It is convenient to express this complementary error function as

$$\text{erfc}\,(\beta) \qquad \text{where} \qquad \beta = \frac{x - vt}{2(\alpha_x vt)^{1/2}}$$

where β is the argument of the complementary error function. Thus, as with all well-tabulated functions, if we know β, we know erfc (β); and if we know erfc (β), we likewise know β.

Values for erfc (β) and erf (β) (or error function of β) for various values of the argument β are shown on Table

Table 18.1 Values of erf (β) and erfc (β) for Positive Values of β

β	erf (β)	erfc (β)
0	0	1.0
0.05	0.056372	0.943628
0.1	0.112463	0.887537
0.15	0.167996	0.832004
0.2	0.222703	0.777297
0.25	0.276326	0.723674
0.3	0.328627	0.671373
0.35	0.379382	0.620618
0.4	0.428392	0.571608
0.45	0.475482	0.524518
0.5	0.520500	0.479500
0.55	0.563323	0.436677
0.6	0.603856	0.396144
0.65	0.642029	0.357971
0.7	0.677801	0.322199
0.75	0.711156	0.288844
0.8	0.742101	0.257899
0.85	0.770668	0.229332
0.9	0.796908	0.203092
0.95	0.820891	0.179109
1.0	0.842701	0.157299
1.1	0.880205	0.119795
1.2	0.910314	0.089686
1.3	0.934008	0.065992
1.4	0.952285	0.047715
1.5	0.966105	0.033895
1.6	0.976348	0.023652
1.7	0.983790	0.016210
1.8	0.989091	0.010909
1.9	0.992790	0.007210
2.0	0.995322	0.004678
2.1	0.997021	0.002979
2.2	0.998137	0.001863
2.3	0.998857	0.001143
2.4	0.999311	0.000689
2.5	0.999593	0.000407
2.6	0.999764	0.000236
2.7	0.999866	0.000134
2.8	0.999925	0.000075
2.9	0.999959	0.000041
3.0	0.999978	0.000022

18.1 and are given graphically in Figure 18.3. Note on the figure that erf (β) ranges from -1 to $+1$ whereas erfc (β) ranges from 0 to $+2$. Further, erfc (β) takes on numbers greater than one only for negative values of the argument. This feature is not evident on Table 18.1. For negative values of the argument β, erfc $(-\beta)$ may be determined as

$$\text{erfc}\,(-\beta) = 1 + \text{erf}\,(\beta)$$

Other useful relationships include

$$\text{erf}\,(-\beta) = -\text{erf}\,(\beta) \qquad \text{and} \qquad \text{erfc}\,(\beta) = 1 - \text{erf}\,(\beta)$$

Now let us examine the physical meaning of the argument of erfc starting first with the numerator $x - vt$, where both x and vt are lengths. This expression simply identifies the point of observation x with respect to the position of the advective front vt. The point of observation can be in one of three possible places: at the advective front, in front of the advective front, or behind the advective front. At the advective front where $x = vt$, the argument is zero and erfc $(0) = 1$ so that from Eq. 18.3, $C = 0.5C_0$. This result is observed on the breakthrough curves of Figure 18.2. Beyond the advective front where $x \gg vt$, the argument is positive and as it approaches infinity, erfc $(\infty) = 0$ so that from Eq. 18.3, $C = 0$. In other words, the observation point x is far in front of the zone of dispersion so that mixing has not yet occurred and we are observing the concentration of the original fluid. However, in a practical sense, a positive argument equal to 2 or 3 suggests that erfc (β) is on the order of 10^{-3} to 10^{-5} (Table 18.1). Hence, for all practical purposes, $C = 0$ at positive values of β greater than 2.

If the point of observation is behind the advective front, the argument β is negative, and as it approaches negative infinity, erfc $(-\infty)$ equals 2 (Figure 18.3). This means that $C = C_0$, Eq. 18.3. However, a negative argument on the order of (-2) suggests that erfc (-2) is about 1.999. Thus, for all practical purposes, $C = C_0$ at small negative values of β. The denominator of the argument has the units of length and may be regarded as the longitudinal spreading length or a measure of the spread of the mass around the advective front. The larger

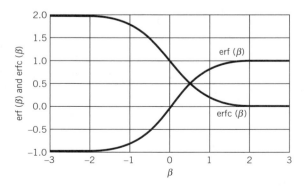

Figure 18.3 erf (β) and erfc (β) plotted versus β.

the denominator, the greater the spread about $C/C_0 = 0.5$.

Thus, the simple expression Eq. 18.3 tells us something about longitudinal dispersion. First, from a modeling perspective, longitudinal dispersion merely moves mass ahead of the advective front in a given flow tube. It contributes nothing to the lateral movement of material out of the tube. Second, as the distance of the advective front from the source increases with increasing time, the spread becomes larger around the advective front. Lastly, at some position $x \ll vt$, the concentration is maintained at the original source concentration C_0. The Ogata–Banks solution is not a field equation but a laboratory one because no matter how far the point of observation is from a continuous source, the equation predicts that the maximum concentration will eventually be equal to the source concentration. This condition is unrealistic, but such is the nature of models with only advection and longitudinal dispersion as the operative processes.

The example that follows illustrates one of the ways that analytical solutions can be used in this case to estimate a longitudinal dispersivity from breakthrough data.

Example 18.1

A nonsorbing species is sent through a column 30 cm in length at a velocity of 1×10^{-2} cm/s. C/C_0 ratios of 0.42 and 0.573 are noted at 46.6 and 53.3 min, respectively, after the test started. What is the longitudinal dispersivity?

Using the first breakthrough concentration

$$C/C_0 = \frac{1}{2}\,\text{erfc}\left[\frac{x - vt}{2(\alpha_x vt)^{1/2}}\right]$$

$$0.42 = \frac{1}{2}\,\text{erfc}\left[\frac{30 - 28}{2(\alpha_x 28)^{1/2}}\right]$$

$$0.84 = \text{erfc}\,(\beta)$$

Now solve for α_x

$$\beta = 0.14$$

$$0.14 = \left[\frac{30 - 28}{2(\alpha_x 28)^{1/2}}\right]$$

$$\alpha_x = 1.8\ \text{cm}$$

The second calculation is essentially the same

$$0.573 = \frac{1}{2}\,\text{erfc}\left[\frac{30 - 32}{2(\alpha_x 32)^{1/2}}\right]$$

$$1.146 = \text{erfc}\,(-\beta) = 1 + \text{erf}\,(\beta)$$

$$\beta = 0.13 \quad \text{and} \quad \alpha_x = 1.8\ \text{cm}$$

The Retardation Equation

For the case of mass transport accompanied by linear sorption, the governing equation is, as we discussed earlier,

$$\frac{D_x \partial^2 C}{R_f \partial x^2} - \frac{v_x \partial C}{R_f \partial x} = \frac{\partial C}{\partial t} \tag{18.4}$$

where

$$R_f = 1 + \left(\frac{1 - n}{n}\right)\rho_s K_d \tag{18.5}$$

The effect of sorption is to decrease the value of the transport parameters D and v. The solution to this equation for the same boundary conditions used with the Ogata–Banks equation (1961) is

$$\frac{C}{C_0} = \left(\frac{1}{2}\right)\text{erfc}\left[\frac{(R_f x - v_w t)}{2(\alpha_x v_w t R_f)^{1/2}}\right] \tag{18.6}$$

If $R_f = 1$, the Ogata–Banks result is recovered exactly. Note that v_w is the velocity of the water, which in this case is not the same as the velocity of the species undergoing sorption.

The relationship between an unretarded species, as described by the Ogata–Banks equation (1961), and a retarded one, as described by Eq. 18.6, is shown in Figure 18.4. Essentially, the velocity of the contaminant becomes less than the velocity of the ground water. Mathematically, the following equation relates these two velocities:

$$v_c = \frac{v_w}{R_f} = \frac{v_w}{1 + \dfrac{(1 - n)}{n}\rho_s K_d} \tag{18.7}$$

Equation 18.7 is known as the retardation equation. It predicts the position of the front of a plume due to advective transport with sorption described by a simple linear isotherm. The ratio v_w/v_c describes how many times faster the ground water (or a nonsorbing tracer) is moving relative to the contaminant being sorbed. Note that when K_d is zero (no sorption), the velocities are the same. In normal application of Eq. 18.7, v_c and v_w can be measured in any convenient units. However, K_d is normally reported in mL/g and ρ_s in g/cm³ so that their product is mL/cm³. No further conversions are required because of

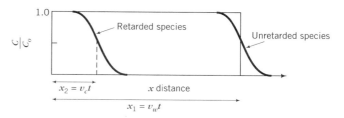

Figure 18.4 A comparison of concentration distributions for retarded and unretarded tracers.

the units chosen where there are 1000 cm³ per liter (or per 10³ mL). In applications, ρ_s may be taken as 2.65 g/cm³ and $(1 - n)/n$ is the inverse of the void ratio so that $(1 - n)e_s/n = \rho_s/e$, where e is the void ratio introduced in Chapter 2.

The retardation equation takes on a slightly different form if instead of defining a K_d value one uses

$$\frac{K_s \text{CEC}}{\tau} = K_d \qquad (18.8)$$

where K_s is the selectivity coefficient, CEC is the cation exchange capacity (meq/mass), and τ is the total competing cation concentration in solution (meq/mass). Thus for a problem involving binary exchange, the retardation equation is

$$\frac{v_w}{v_c} = 1 + \frac{\rho_s(1 - n)}{n\tau} K_s \text{CEC} \qquad (18.9)$$

The retardation equation has proven useful in evaluating problems of contamination from organic compounds and to a lesser extent radionuclides and trace metals. With rudimentary information about a site, one can begin to explain differences in the observed distributions of various contaminants and predict how contaminants might spread in the future. To illustrate how the equation could be applied in practice, consider the following example based on an analysis presented by Grisak and Jackson (1978) for the Whiteshell Nuclear Research Establishment (WNRE) in Canada.

Example 18.2

At WNRE, a sequence of glacial drift overlies Precambrian bedrock. The unit of interest in this problem is the basal sand, which overlies the bedrock and is confined on top by clay-loam till. The question to be answered is how fast a hypothetical contaminant such as ⁹⁰Sr might migrate in the basal sand should the primary containment fail. The seepage velocity through the sand is estimated to be 1.85 cm/day. Values of other necessary parameters are: $\rho_s(1 - n)$ is 1.9 g/cm³, CEC is 1.4 meq/100 g, τ is 4.8 meq/L, n is 0.35, and K_s is 1.3.

The velocity of ⁹⁰Sr can be determined by substituting the known parameters in consistent units in the retardation equation

$$v_{\text{Sr}-90} = \frac{v_w}{1 + \rho_s \dfrac{(1 - n)}{n\tau} \text{CEC} \cdot K_s}$$

$$v_{\text{Sr}-90} = \frac{1.85 \text{ cm/day}}{1 + \dfrac{1.9 \text{ g/cm}^3}{0.35 \cdot 0.0048 \text{ meq/mL}} 0.014 \text{ meq/g} \cdot 1.3}$$

$$= 0.086 \text{ cm/day}$$

Thus, the velocity of ⁹⁰Sr is about 5% of that of the ground water.

The simplicity of this calculation makes it useful for back-of-the-envelope type analyses of contamination problems. The method obviously breaks down when the physical setting is complex, or processes other than advection or sorption/exchange operate to control contaminant distributions.

Radioactive Decay, Biodegradation, and Hydrolysis

Let us now consider mass transport involving a first-order kinetic reaction. Mathematically, this problem is described by Eq. 14.16. For the same initial and boundary conditions of the Ogata–Banks equation (1961), the solution in one dimension is (Bear, 1979)

$$\frac{C}{C_0} = \left(\frac{1}{2}\right) \exp\left\{\left(\frac{x}{2\alpha_x}\right)\left[1 - \left(1 + \frac{4\lambda\alpha_x}{v}\right)^{1/2}\right]\right\}$$
$$\text{erfc}\left[\frac{x - vt(1 + 4\lambda\alpha_x/v)^{1/2}}{2(\alpha_x vt)^{1/2}}\right] \qquad (18.10)$$

where λ is the decay constant, or 0.693 divided by the half-life. Although this equation is not sufficiently complete for describing a field problem, it provides insight that does apply to field situations. First, if λ is zero, Eq. 18.10 reduces to the Ogata–Banks solution (1961). For this condition, the exponential term in the solution is one because $e^0 = 1$. Further, for a finite λ, the exponent will be raised to a negative number so that we immediately establish the appropriate limits for the exponential term

$$e^0 = 1 \qquad e^{-\infty} = 0$$

The important dimensionless group in the exponential term is $4\lambda\alpha_x/v$. As λ gets large with respect to the other variables in this dimensionless group, the exponential term approaches zero, and the concentration approaches zero. The material is reacting or decaying faster than it can be transported through the system. As the velocity becomes large with respect to the other variables, the dimensionless group approaches zero, and we approach the Ogata–Banks solution (1961). Transporting the mass faster than it can degrade or decay makes the decay ineffective. Competition thus exists between the transport and the kinetics, in particular the decay of a species and its velocity. As the longitudinal dispersivity α_x gets large, the dimensionless group of Eq. 18.10 gets large and decay again dominates. However, the dispersivity term occurs in two places in the exponential term of Eq. 18.10, and these apparently offset each other. In other words, any reasonable value of α_x can be used in the exponential term and the value of this term will generally change by less than 1% for a few-orders-in-magnitude range in longitudinal dispersivity. For example, for a nuclide with a half-life of 1000 yr, a velocity of 4 m/yr, a

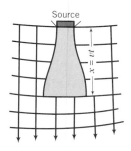

Figure 18.5 Advection accompanied by transverse dispersion.

distance of 5000 m, and α_x = to 30 m, 100 m, and 300 m, the exponential term equals 0.5, 0.505, and 0.51, respectively.

Equation 18.10 reflects a steady state when the argument of erfc approaches negative two, that is, the point of observation is behind the modified advective front $vt(1 + 4\lambda\alpha_x/v)^{1/2}$. This steady-state solution is

$$\frac{C}{C_0} = \exp\left\{\left(\frac{x}{2\alpha_x}\right)\left[1 - \left(1 + \frac{4\lambda\alpha_x}{v}\right)^{1/2}\right]\right\} \quad (18.11)$$

Although there appears to be a contradiction of terms when a steady state can be achieved for a decaying species, the condition can be realized if the source is continually renewed. As with the Ogata–Banks equation, the steady-state concentration is independent of the longitudinal dispersivity and in this case depends only on advection and decay. For all practical purposes, Eq. 18.11 describes the concentration of a species undergoing decay or degradation in a purely advective system.

There is one more thing to be learned from Eq. 18.11. The term v is velocity, but what velocity? The answer is the velocity of the contaminant. For an unretarded contaminant, the velocity of the contaminant is simply equal to the velocity of the water. For a retarded contaminant we can use the retarded velocity directly, or we may use v_w/R_f from Eq. 18.7. Thus, we note the importance of retardation in problems of decay or degradation in that

retarded velocities favor kinetics over transport. Slow-moving nuclides or organics have a greater opportunity to decay or degrade.

Transverse Dispersion

Figure 18.5 illustrates an advective model with transverse spreading (lateral dispersion). In the absence of longitudinal dispersion, no mass travels beyond the advective front. However, the mass is not restricted to a single flow tube but can spread in a direction transverse to the flow both laterally and vertically. Concentrations are everywhere less than the source concentration C_0, except at $x = 0$. Concentrations less than C_0 occur because the mass is equally distributed throughout the plume, but the volume it occupies increases with increasing distance from the source.

In all developments thus far, we specified a source concentration but no information concerning the geometry of the source. With transverse dispersion, something must be stated about the source geometry. Possible geometrical configurations include a point, a vertical line, or a planar area (Figure 18.6). The equations describing the parts of a plume controlled exclusively by transverse spreading for the point or line source are complicated. Thus, the equations that follow are limited to describing the maximum concentrations along the plane of symmetry, that is, along the x axis for y and $z = 0$ or

$$C_{\max} = C_0 \quad \text{(plug flow)}$$

$$C_{\max} = \frac{C_0 Q}{2v(\pi\alpha_y x)^{1/2}} \quad \text{(line source)} \quad (18.12)$$

$$C_{\max} = \frac{C_0 Q}{4x\pi v(\alpha_y\alpha_z)^{1/2}} \quad \text{(point source)}$$

where Q is a volumetric flow rate with units of L^2T^{-1} for the line source and L^3T^{-1} for the point source so that C_0Q has the units of $ML^{-1}T^{-1}$ and MT^{-1}, respectively. The line source result was developed by Wilson and Miller (1978) and the point source result by Hunt (1978).

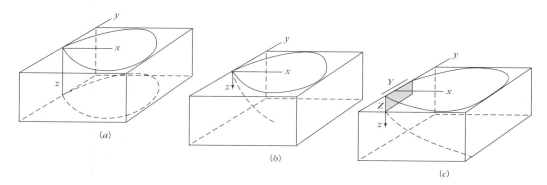

Figure 18.6 Geometrical considerations for a contaminant source. Shown in (*a*) is a vertical line source, in (*b*) a point source, and in (*c*) a finite plane source (from Huyakorn and others, 1987). Reprinted by permission of Ground Water. Copyright © 1987. All rights reserved.

Equations 18.12 clearly demonstrate the role of the transverse dispersivity in transport problems. For the line source, the larger α_y the greater the lateral extent of the plume and the smaller the concentrations. The point source result incorporates lateral α_y as well as vertical α_z spreading. The source flow rate $C_0 Q$ reflects the geometry of the source, expressed as mass per length per time for the line source and mass per time for the point source. Further, the concentrations are independent of time, and they approach infinity as x approaches zero. Thus, these equations are useful only for far field conditions where x is large.

The planar source is obviously the more practical one with lateral spreading in two directions and vertical spreading downward (Figure 18.6). Solutions for this source geometry have been developed by Domenico and Palciauskas (1982). For the lateral (y) component of spreading

$$\frac{C}{C_0} = \left(\frac{1}{2}\right)\left\{\text{erf}\left[\frac{(y+Y/2)}{2(\alpha_y x)^{1/2}}\right] - \text{erf}\left[\frac{(y-Y/2)}{2(\alpha_y x)^{1/2}}\right]\right\} \quad (18.13)$$

where the half source size $Y/2$ is part of the solution. The z component of spreading is

$$\frac{C}{C_0} = \left(\frac{1}{2}\right)\left\{\text{erf}\left[\frac{(z+Z)}{2(\alpha_z x)^{1/2}}\right] - \text{erf}\left[\frac{(z-Z)}{2(\alpha_z x)^{1/2}}\right]\right\} \quad (18.14)$$

where the full source size Z becomes part of the solution. For spreading directions in both y and z, we adhere to the product rule discussed earlier (Domenico and Palciauskas, 1982)

$$\frac{C}{C_0} = \left(\frac{1}{4}\right)\left\{\text{erf}\left[\frac{(y+Y/2)}{2(\alpha_y x)^{1/2}}\right] - \text{erf}\left[\frac{(y-Y/2)}{2(\alpha_y x)^{1/2}}\right]\right\}$$
$$\left\{\text{erf}\left[\frac{(z+Z)}{2(\alpha_z x)^{1/2}}\right] - \text{erf}\left[\frac{(z-Z)}{2(\alpha_z x)^{1/2}}\right]\right\} \quad (18.15)$$

For the plane of symmetry ($y = z = 0$)

$$C_{max} = C_0 \, \text{erf}\left[\frac{Y}{4(\alpha_y x)^{1/2}}\right] \text{erf}\left[\frac{Z}{2(\alpha_z x)^{1/2}}\right] \quad (18.16)$$

The source dimensions and transverse dispersivities control the maximum concentrations that are encountered in a steady-state plume. For example, as Y or Z is increased, or α_y and α_z decreased, the arguments of the error function approach +2, and the value of the error functions approach one; that is, spreading is restricted and the concentration is maintained near C_0 over small values of x.

Models for Multidimensional Transport

Multidimensional transport involves both longitudinal and transverse dispersion in addition to advection. The most complex form of the dispersion–advection equation that is amenable to an analytical solution includes three dispersive components, a constant advective veloc-

ity, and one kinetic term, as follows:

$$D_x\frac{\partial^2 C}{\partial x^2} + D_y\frac{\partial^2 C}{\partial y^2} + D_z\frac{\partial^2 C}{\partial z^2} - v_x\frac{\partial C}{\partial x} - \frac{r}{n} = \frac{\partial C}{\partial t} \quad (18.17)$$

where r is defined by some mathematical rate law.

Solutions to Eq. 18.17 provide the concentration distribution resulting from continuous or instantaneous sources. The following sections will examine these different source conditions in detail.

Continuous Sources

Models that include transverse spreading must incorporate information on the source geometry. Wilson and Miller (1978) provide a complete solution for a nonreacting species emanating from a line source, and Hunt (1978) presents a three-dimensional solution for a nonreacting species associated with a point source. However, because the finite planar source is the most realistic geometry for field problems, it has received the most attention. Two types of models are available. First there are finite source models that require numerical integration (for example, Cleary, 1978; Huyakorn and others, 1987). A second type of solution is given in a closed-form format, where no numerical integrations are necessary. Two models are available, one for advection and dispersion alone (Domenico and Robbins, 1985b), and another for mass transport together with a first-order reaction (Domenico, 1987).

The essence of the Domenico and Robbins model is apparent in Figure 18.7. Because of an assumed one-dimensional velocity, the plume shown in Figure 18.7d can be referred to as a plug-flow model with both longitudinal and transverse spreading (dispersion). The distribution of concentration (Figure 18.7d) has features common to both the longitudinal (Figure 18.7b) and the transverse spreading models (Figure 18.7c). There is a frontal zone of dispersion beyond the advective front caused exclusively by longitudinal dispersion and a zone of mass depletion behind the advective front. At some distance behind the advective front, depending largely on the value of the longitudinal dispersivity, the plume is at steady state. Because of transverse spreading, the steady-state concentrations are everywhere less than the source concentration C_0, except at $x = 0$. Results of the product rule discussed earlier capture the essential features of this plume for the source geometry of Figure 18.6c (Domenico and Robbins, 1985b)

$$\frac{C(x,y,z,t)}{C_0} = \left(\frac{1}{8}\right)\text{erfc}\left[\frac{(x-vt)}{2(\alpha_x vt)^{1/2}}\right]$$
$$\left\{\text{erf}\left[\frac{(y+Y/2)}{2(\alpha_y x)^{1/2}}\right] - \text{erf}\left[\frac{(y-Y/2)}{2(\alpha_y x)^{1/2}}\right]\right\}$$
$$\left\{\text{erf}\left[\frac{(z+Z)}{2(\alpha_z x)^{1/2}}\right] - \text{erf}\left[\frac{(z-Z)}{2(\alpha_z x)^{1/2}}\right]\right\}$$

$$(18.18)$$

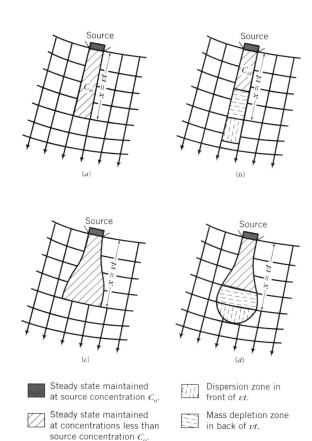

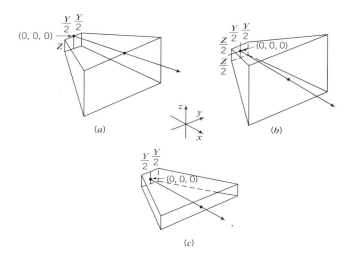

Figure 18.8 Idealized contaminant migration geometries for various transverse spreading directions (from Domenico and Robbins, 1985b). Reprinted by permission of Ground Water. Copyright © 1985. All rights reserved.

z (Figure 18.8b), the source dimension Z terms become $Z/2$. If z spreading is eliminated altogether (Figure 18.8c), the error functions containing the z terms in Eq. 18.18 are ignored, and $C_0/8$ becomes $C_0/4$. If the aquifer thickness is small, the plume may spread vertically and occupy the entire thickness (Figure 18.9). The length x' over which this spreading occurs can be approximated as

$$x' = \frac{(H - Z)^2}{\alpha_z} \qquad (18.20)$$

where H is the thickness of the aquifer (Domenico and Palciauskas, 1982). Hence, Eq. 18.18 is valid for all distances x equal to or less than $(H - Z)^2/\alpha_z$. For distances greater than x', the distance x in the denominator of the error function of the z term is replaced by x' (for example, Eq. 18.20). Thus, further spreading in z is prohibited for $x > x'$. Following is an example that illustrates how an analytical solution is used.

Steady state maintained at source concentration C_o.

Steady state maintained at concentrations less than source concentration C_o.

Dispersion zone in front of vt.

Mass depletion zone in back of vt.

Figure 18.7 Idealization of the dispersion process showing (a) plug flow, (b) longitudinal dispersion, (c) transverse dispersion, and (d) longitudinal and transverse dispersion (after Domenico, 1987). Reprinted from J. Hydrol., v. 91, by P. A. Domenico, An analytical model for multidimensional transport of a decaying contaminant species, p. 49–58, 1987 with kind permission from Elsevier Science, NL, Sara Burgerhartstraat 25, 1055 KV Amsterdam, The Netherlands.

For $x \ll vt$, the argument of erfc approaches negative two, and we recover the Domenico–Palciauskas (1982) steady-state model of Eq. 18.15. For the plane of symmetry $y = z = 0$

$$\frac{C(x, 0, 0, t)}{C_0} = \left(\frac{1}{2}\right) \text{erfc} \left[\frac{(x - vt)}{2(\alpha_x vt)^{1/2}} \right]$$
$$\left\{ \text{erf} \left[\frac{Y}{4(\alpha_y x)^{1/2}} \right] \text{erf} \left[\frac{Z}{2(\alpha_z x)^{1/2}} \right] \right\} \qquad (18.19)$$

This form also shows that as α_y and α_z approach zero, we recover the Ogata–Banks (1961) Eq. 18.3.

As with all of the transport equations, the velocity v is the contaminant velocity. For a retarded species, the contaminant velocity is used directly or v_w/R_f is substituted directly for v in Eq. 18.19. There is one other feature about this result that helps to accommodate various field conditions. As developed, Eq. 18.18 applies to a source geometry with two transverse and one vertical spreading direction (Figure 18.8a). For two spreading directions in

Example 18.3

Drums of diethyl ether (de) and carbon tetrachloride (ct) were buried in a sand aquifer 15 years ago. Calculate the concentrations of each contaminant along the plane of symmetry of the plume at the point ($x = 225$ m, $y = 0$ m, $z = 0$ m) at time 15 years (4.73×10^8 s). The linear velocity of the ground water is 1×10^{-6} m/s. The

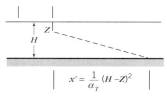

Figure 18.9 Vertical spreading of a contaminant limited by a low-permeability boundary.

retardation factor for de is 1.5 and for ct 27.4. The source concentration for de is 1×10^4 µg/L and for ct is 5×10^2 µg/L. The source size in Y is 25 m and in Z is 5 m. The estimated dispersivities are $\alpha_x = 1.0$ m, $\alpha_y = 0.1$ m, and $\alpha_z = 0.01$ m. Assume that the plume can spread in two z directions and two y directions so that

$$C(x, 0, 0, t) = \left(\frac{C_0}{2}\right) \text{erfc} \left[\frac{(x - vt)}{2(\alpha_x vt)^{1/2}}\right]$$

$$\left\{\text{erf}\left[\frac{Y}{4(\alpha_y x)^{1/2}}\right] \text{erf}\left[\frac{Z}{4(\alpha_z x)^{1/2}}\right]\right\}$$

For de,

$$C = \frac{1 \times 10^4}{2} \text{erfc}\left[\frac{(225 - 312)}{2(1 \times 312)^{1/2}}\right] \text{erf}\left[\frac{25}{4(0.1 \times 225)^{1/2}}\right]$$

$$\text{erf}\left[\frac{5}{4(0.01 \times 225)^{1/2}}\right]$$

$$= 5 \times 10^3 \text{ erfc}(-2.46) \text{ erf}(1.3) \text{ erf}(0.83)$$

$$= 5 \times 10^3 (2)(0.93)(0.759) = 7 \times 10^3 \text{ µg/L}$$

Because the argument of the complementary error function has a negative value greater than -2, the plume is at steady state at this point and the calculated concentration is a maximum value.

For ct,

$$C = \frac{5 \times 10^2}{2} \text{erfc}\left[\frac{(225 - 17.2)}{2(1 \times 17.2)^{1/2}}\right] \text{erf}\left[\frac{25}{4(0.1 \times 225)^{1/2}}\right]$$

$$\text{erf}\left[\frac{5}{4(0.01 \times 225)^{1/2}}\right]$$

Because erfc $[(225 - 17.2)/2(1 \times 17.2)^{1/2}] = \text{erfc}$ $(25) \approx 0$, it follows that carbon tetrachloride has not yet reached this point because of greater retardation. The maximum concentration that ct will eventually attain is

$$C_{\text{max}} = (5 \times 10^2)(0.93)(0.759) = 3.5 \times 10^2 \text{ µg/L}$$

Domenico (1987) uses similar arguments to incorporate decay or degradation in the following solution for multidimensional transport:

$$\frac{C(x, y, z, t)}{C_0} = \left(\frac{1}{8}\right) \exp\left\{\left(\frac{x}{2\alpha_x}\right)\left[1 - \left(1 + \frac{4\lambda\alpha_x}{v}\right)^{1/2}\right]\right\}$$

$$\text{erfc}\left[\frac{x - vt(1 + 4\lambda\alpha_x/v)^{1/2}}{2(\alpha_x vt)^{1/2}}\right]$$

$$\left\{\text{erf}\left[\frac{y + Y/2)}{2(\alpha_y x)^{1/2}}\right] - \text{erf}\left[\frac{(y - Y/2)}{2(\alpha_y x)^{1/2}}\right]\right\}$$

$$\left\{\text{erf}\left[\frac{(z + Z)}{2(\alpha_z x)^{1/2}}\right] - \text{erf}\left[\frac{(z - Z)}{2(\alpha_z x)^{1/2}}\right]\right\}$$

(18.21)

which reduces to the Domenico–Robbins (1985b) result when $\lambda = 0$. The concentration along the plane of symmetry is

$$\frac{C(x, 0, 0, t)}{C_0} = \left(\frac{1}{2}\right) \exp\left\{\left(\frac{x}{2\alpha_x}\right)\left[1 - \left(1 + \frac{4\lambda\alpha_x}{v}\right)^{1/2}\right]\right\}$$

$$\text{erfc}\left[\frac{x - vt(1 + 4\lambda\alpha_x/v)^{1/2}}{2(\alpha_x vt)^{1/2}}\right] \quad (18.22)$$

$$\left\{\text{erf}\left[\frac{Y}{4(\alpha_y x)^{1/2}}\right] \text{erf}\left[\frac{Z}{2(\alpha_z x)^{1/2}}\right]\right\}$$

As α_y and α_z approach zero, the result of Bear (1979) is recovered Eq. (18.10). Further, a steady-state form may be obtained when the argument of the complementary error function approaches negative two.

Virtually everything that has been stated for the Domenico–Robbins (1985b) result applies here also. Thus, Eq. 18.21 can be modified to incorporate different spreading geometries as well as to account for limitations in z spreading. Further, the velocity v in these equations is the contaminant velocity so that retardation is already incorporated or can be easily accommodated by substituting $v = v_w/R_f$.

The models discussed above have been applied to field problems by Fryar and Domenico (1989) and Ala and Domenico (1992). Their main use is in determining the transport parameters from concentration distributions observed in the field.

Numerical Integration of an Analytic Solution

A benefit of the mathematical procedure that we have just described is the inherent simplicity in the way components of the overall solution can be put together and the ability to manipulate these expressions to form a powerful method for parameter estimation (Domenico and Robbins, 1985b). Recall that the simplification came because we worked with a simplified form (Eq. 18.3) of the exact solution. This simplification, however, is not without some cost, which is reflected in some error of approximation.

No rigorous analytical solution to the multidimensional mass transport equation can avoid an extremely complicated numerical integration. Several of these solutions are available in the literature, including Wexler (1992), Leij and Dane (1990), and Leij and others (1991). One of these solutions (Leij and others, 1991) was used by Leslie Smith to test the accuracy of the approximate analytical solution developed in this chapter. Figure 18.10 shows that the approximate solution is less accurate close to the origin, and for relatively large values of dispersivity.

The Instantaneous Point Source Model

The accidental spill, frequently referred to as an instantaneous or pulse-type problem, represents another poten-

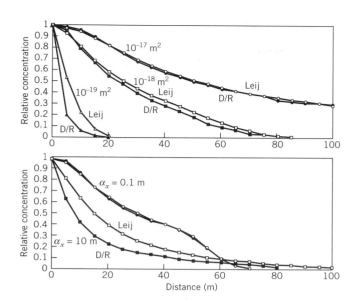

Figure 18.10 Comparison of the Domenico and Robbins (1985b) approximate solution (D/R) with an exact solution (Leij and others, 1991). (Unpublished analysis, kindly provided by Leslie Smith.)

tial contamination problem. Models for two types are available: the parallelepiped finite source model of Hunt (1978) and the point source model of Baetslé (1969). The point source of Baetslé (1969) is the more useful one because the geometry of a spill more closely resembles a point, at least when viewed from the far field, than a parallel piped. Baetslé's model is

$$C(x, y, z, t) = \left[\frac{C_0 V_0}{8(\pi t)^{3/2}(D_x D_y D_z)^{1/2}} \right]$$

$$\exp \left[-\frac{(x - vt)^2}{4D_x t} - \frac{y^2}{4D_y t} - \frac{z^2}{4D_z t} - \lambda t \right]$$

(18.23)

where C_0 is the original concentration; V_0 is the original volume so that the product $C_0 V_0$ is the mass involved in the spill; D_x, D_y, D_z are the coefficients of hydrodynamic dispersion; v is the velocity of the contaminant; x, y, and z are space coordinates; t is time; and λ is the disintegration constant for a decaying substance. For a nondecaying substance, the term λt is ignored. Again, this solution can be obtained readily as the product of three one-dimensional solutions.

With an idealized three-dimensional point source spill, spreading occurs in the direction of flow, and the peak or maximum concentration occurs at the center of the "cloud," that is, where $y = z = 0$ and $x = vt$ (Figure 18.11)

$$C_{\max} = \frac{C_0 V_0 e^{-\lambda t}}{8(\pi t)^{3/2}(D_x D_y D_z)^{1/2}}$$

(18.24)

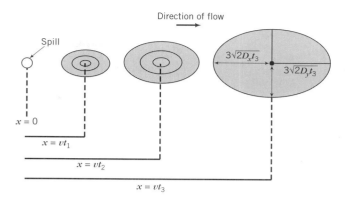

Figure 18.11 Plan view of the plume developed from an instantaneous point source at three different times.

The dimensions of the cloud, assuming it actually started as a point, are

$$\sigma_x = (2D_x t)^{1/2}; \qquad \sigma_y = (2D_y t)^{1/2}; \qquad \sigma_z = (2D_z t)^{1/2}$$

where σ is the standard deviation so that $3\sigma_x$, $3\sigma_y$, and $3\sigma_z$ represent three spreading lengths within which about 99.7% of the mass is contained.

Example 18.4 (after Baetslé, 1969)

A leak in a storage tank for radioactive waste results in an accidental release of 1000 ci of 10 years' cooled fission products, along with tritium. The waste contains 400 ci Cs-137 ($t_{1/2}$ 33 yr), 400 ci Sr-90 ($t_{1/2}$ 28 yr), 100 ci Pm-147 ($t_{1/2}$ 2.7 yr), and 100 ci H-3 ($t_{1/2}$ 12.26 yr). Assume that a river exists 100 m from the spill. Determine how much time is required for each constituent to reach the river and what the maximum concentrations will be. The linear velocity of the ground water is 10 cm/day, or 1.16 × 10⁻⁴ cm/s. Other data are given in Table E18.4-1.

(a) Time to reach a point at 100 m

From Eq. 18.7 where $R_f x = v_w t$, let x be the distance traveled by the nuclide. For tritium,

$$t = \frac{R_f x}{v_w} = (1) \cdot \frac{1 \times 10^4}{1.16 \times 10^{-4}} = 0.864 \times 10^8 \text{ s} = 2.75 \text{ yr}$$

Similar calculations show that Cs-147 will take 112 yr, Sr-90 6.9 yr, and Pm-147 1120 yr.

(b) Peak Concentration at 100 m

The half-life for tritium is 12.26 yr so that $e^{-\lambda t} = e^{-0.693}$

Table E18.4-1

Nuclide	K_d (mL/g)	R_f	D_x cm²/s	$D_y = D_z$
Cs-137	10	47.6	10^{-4}	10^{-3}
Sr-90	0.6	2.885	10^{-3}	10^{-5}
Pm-147	100	476.	10^{-5}	10^{-5}
H-3	0	1.	10^{-5}	10^{-5}

Table 18.2 FORTRAN Code for Evaluating the Analytical Solution for One-Dimensional Transport with Retardation

```
C      CODE FOR EVALUATING ONE-DIMENSIONAL DISPERSION
C         WITH RETARDATION
C
       DIMENSION C(50)
C
C      READ NECESSARY INPUT DATA
C
       OPEN(5,FILE='IN.DAT')
       OPEN(6,FILE='OUT.DAT')
       READ(5,100) NX
       WRITE(6,100) NX
       READ(5,110) CO,ALFX,VX,RF,TYR,DELX
       WRITE(6,110) CO,ALFX,VX,RF,TYR,DELX
C
          .
          .
          .
       (complete listing provided on disk as rone.for
       in the directory trans).
```

$t/(12.26 \times 365 \times 86,400)$ because $\lambda = \ln 2/\text{half-life}$ and $\ln 2 = 0.693$. The value of time for the calculation comes from part (a).

$$C_{max} = \frac{100\, e^{(-0.693 \times 0.864 \times 10E8)/(12.26 \times 365 \times 86,400)}}{8(3.14 \times 0.864 \times 10^8)^{3/2} \times 10^{-7.5}}$$

$$C_{max} = 7.6 \times 10^{-5}\ \text{ci/mL}$$

The rest of the results are Cs-137 1.08×10^{-7}; Sr-90 0.608×10^{-5}; and Pm-147 negligibly small 10^{-125} ci/mL.

18.2 Programming the Analytical Solutions for Computers

The closed-form analytical solutions can be applied directly to contamination problems. However, an alternative and more efficient approach is to construct codes to perform these calculations. Almost any type of computer can be used and the necessary codes can be constructed with little difficulty.

Most codes of this type carry out three operations: reading values of the transport parameters; solving for concentration at specified times and locations; and writing the final results. We can illustrate these steps in a practical way by developing a code for evaluating the analytical solution to Eq. 18.1 for one-dimensional mass transport subject to simple sorption. This example is a demonstration that is not applicable to field problems. According to the solution Eq. 18.3, suitably modified to account for sorption, three transport parameters v, α_x, and R_f, and the concentration at the upstream boundary (C_0) need to be specified. Coding begins (Table 18.2) by reading and writing these values and defining the time

and positions where concentrations are to be calculated. Input data are read from the file IN.DAT and output data are written to the file OUT.DAT. The analytical solution is evaluated within a DO loop for various distances from the source. The function subroutine ERFC evaluates the error functions. Finally, we write the calculated concentrations to the specified output file. The following hypothetical example illustrates how this code is used.

Example 18.5

An organic contaminant is disposed of continuously in a long narrow trench that fully penetrates a shallow, semi-infinite aquifer (see Figure 13.1a). For this aquifer $v = 2.31 \times 10^{-6}$ m/s, $\alpha_x = 4.3$ m, $R_f = 3.0$. The contaminant concentration at the source (C_0) remains constant with time at 1.0 mg/L. Calculate the contaminant concentration after 3 years at 10-m intervals from the source.

This problem is solved with the code listed in Table 18.2 and included on the disk as rone.for (see Appendix B). The data set is shown in Table E18.5-1.

Table E18.5-1

18					
1.0	4.3	.00000231	3.	3.	10.

For consistency in units, years in the input are converted to seconds in the code. The calculated concentration of contaminants at the end of 3 years down the system at 10-m intervals beginning with $x = 10$ m is shown in Table E18.5-2.

Table E18.5-2

3. YEARS					
0.99	0.98	0.96	0.91	0.82	0.70
0.55	0.39	0.25	0.14	0.07	0.03
0.01	0.00	0.00	0.00	0.00	0.00

Modeling is one of those activities that must be practiced to develop necessary coding skills, the ability to formulate a problem in terms of boundary conditions and data, and the ability to use a computer. Let us consider another, more complex problem.

Example 18.6

For the problem in Example 18.3, develop a computer code to predict the concentration of diethyl ether (de) along a horizontal plane through the middle of source (that is, $z = 0$ m).

There are five steps necessary to solve this problem: (1) formulate the problem mathematically and select an appropriate analytical solution, (2) establish what input parameters are required, (3) construct the computer code, (4) run the code using known or estimated information about the site, and (5) interpret the results.

Step 1

The information provided for this problem points to three-dimensional spreading of a plume in a one-dimensional velocity field. Diethyl ether with fixed initial concentration of 10,000 μg/L enters the ground water from a source of finite size. This problem can be solved using Eq. 18.18. This form of the transport equation accounts for dispersion in three directions, advection in one direction, and retardation due to sorption. The downstream side of the disposal site acts as the finite source with an area YZ and specified nonzero concentrations of each contaminant. The initial condition is that the concentration of either diethyl ether (C_{de}) is zero within the semi-infinite domain.

Step 2

The analytical solution requires values for v_x, α_x, α_y, α_z, C_0, $Z/2$, $Y/2$, and the location of points where we intend to calculate concentrations. This code has a similar scheme for data entry as the previous one except for requiring additional dispersivities and the dimensions of the source. Concentrations are calculated for a grid of points on a horizontal plane passing through the middle of the source at $z = 0$.

Step 3

Construct the code (included as rthree.for and rthree.exe on the disk in the directory trans.). A grid provides a convenient way to define the concentration of points on the plane through the center of the plume. NX and NY are the number of points in the x and y directions. DELX and DELY are the spacings between points.

x-dir: NX = 10 y-dir: NY = 7

DELX = 75.0 m DELY = 15.0 m

Step 4

Run the code using the following set of input data (on disk as in 3.dat).

The resulting set of input data is shown in Table E18.6-1. All units are consistent in meters and seconds. Because R is dimensionless we can use another set of internally

Table E18.6-1

10	7					
10000.		1.	0.1	.01	1.0E-06	1.5
15.	75.	15.	0.0			
5.	25.					

Table E18.6-2

15. YEARS						
0.00	0.00	2486.12	9575.76	2486.12	0.00	0.00
0.00	5.95	2757.84	8319.54	2757.84	5.95	0.00
0.00	34.59	2700.07	7137.75	2700.07	34.59	0.00
0.07	60.30	1886.02	4514.41	1886.02	60.30	0.07
0.00	1.21	21.63	47.73	21.63	1.21	0.00
0.00	0.00	0.00	0.00	0.00	0.00	0.00
0.00	0.00	0.00	0.00	0.00	0.00	0.00
0.00	0.00	0.00	0.00	0.00	0.00	0.00
0.00	0.00	0.00	0.00	0.00	0.00	0.00
0.00	0.00	0.00	0.00	0.00	0.00	0.00

consistent units. The simulated distribution of de is shown in Table E18.6-2.

18.3 Numerical Approaches

The numerical approaches are a family of computer-based techniques for solving the contaminant transport equations. They approximate forms of the advection–dispersion equation as a system of algebraic equations, or alternatively simulating transport through the spread of a large number of moving reference particles. Whatever the procedure used, it invariably has to be coded for solution on a high-speed computer.

Numerical approaches deal easily with variability in the flow and transport parameters (for example, hydraulic conductivity, porosity, dispersivity, cation exchange capacity, etc.). This flexibility in representing parameters facilitates modeling of layering or other, more complex geometries in two and three dimensions. Thus, one can simulate the complex plume shapes that often develop in natural systems. The analytical approaches cannot account for variability in transport parameters. Thus, numerical approaches are readily adapted to site-specific problems, which makes them particularly useful in practice.

Other important features of the numerical approaches are the flexibility in implementing complex boundary conditions and in accounting for a variety of important mass transport processes. Codes are now available to model (1) radionuclide chains in which radioisotopes and their daughters decay, each with a variable half-life, (2) equilibrium reactions such as precipitation, dissolution, and complexation, and (3) competitive sorption or ion exchange. Developing the increased sophistication to deal with reactions is necessary to accurately model contaminant groups such as trace metals or radionuclides.

A Generalized Modeling Approach

The modeling process can be examined independently of a particular solution technique. The starting place in nearly all cases is with one of the various forms of the advection–dispersion equation. Because of the power of the numerical approaches, the transport equation is commonly formulated in two or three dimensions. Although our emphasis here is with mass transport involving a single contaminant, modeling approaches have evolved to the point where a coupled set of differential equations can be formulated for problems involving several reacting species.

One of the first steps in developing a computer model is to subdivide the region in terms of cells or elements. This process makes it possible to account for varying parameter values within the domain. Furthermore, subdi-

viding the region also serves to define the nodes at which concentrations are calculated. The way in which nodes are defined and how the domain is subdivided (for example, squares, rectangles, triangles, stream tube segments, etc.) depends on the specific numerical technique. The Otis Air Base case study presented later shows how one goes about subdividing a region and assigning boundary conditions.

There is really no difference in the boundary conditions and transport parameters required in the numerical approaches as compared to the analytical approaches. Complicating the situation with numerical models is the need to provide boundary conditions at a large number of node points or nodal blocks (areas around the nodes). For example, values of concentration or loading rates defining various boundary conditions now need to be specified for all nodes located along the boundary of the domain. Initial conditions and transport parameters are specified for all nodes except in some cases for nodes with constant concentrations.

It is no real problem to assign transport parameters to nodal areas, other than perhaps not knowing what the values are. Velocity values are an important exception because they are more difficult to specify. In most real systems with pumping and injection wells and a variable hydraulic conductivity field, velocities can be extremely variable. Both the magnitude and the direction of flow can change in space and time. In terms of modeling, measured or guessed velocity values are not adequate. Continuity considerations in all numerical solutions require a smooth and accurate representation of the velocity field.

This kind of field can be obtained only by simulation with a flow model. Velocity values are calculated by applying the Darcy equation with calculated values of hydraulic head and known parameters, or directly calculating velocity values at the nodes. Figure 18.12 illustrates how flow and mass transport models are used together to predict contaminant distributions. Input necessary for the flow model could include the hydraulic conductivity distribution, the water-table configuration, and other boundary conditions. The transport model is coupled to the flow model by the velocity terms as shown in Figure 18.12.

Flow in many cases is assumed to be independent of the mass transport. In other words, the concentration of contaminants does not influence the flow by changing the fluid density. In this situation, there is no requirement to solve the flow and transport equations simultaneously. For steady flow, the ground-water flow equation is solved once and a single velocity field applies for all time. If flow is transient, the velocity field must be calculated each time that the contaminant concentration is required. However, again, the flow and transport equations are solved separately.

Contaminant concentrations are sometimes so large that the presence of the spreading plume causes the flow

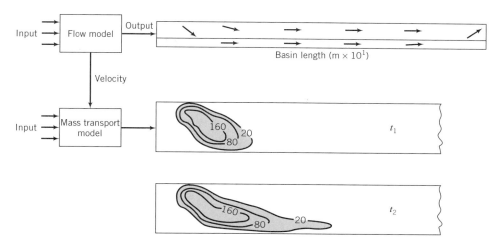

Figure 18.12 The relationship between ground-water flow and mass transport models in simulating the distribution of contaminants (from Schwartz and others, 1985). Reprinted with permission of Second Canadian/American Conference on Hydrogeology. Copyright © 1985. All rights reserved.

pattern to change. In these cases, the ground-water flow depends on the mass transport just as the mass transport depends on the ground-water flow. In Figure 18.12, the arrow joining the flow and transport models now will point in both directions. For this fully coupled situation, the flow and transport equations need to be solved simultaneously.

The Common Solution Techniques

The techniques for solving the advection–dispersion equation are extremely varied and would require almost a complete textbook to discuss them all in detail. Here, we present a brief summary of the most common methods mainly to introduce the pertinent literature and to create an awareness of the basic terminology. There are two general approaches available for solving mass transport problems. The direct solution techniques are the most common and involve a numerical solution of the advection–dispersion equation. Included in this group are the well-known finite-difference and finite-element techniques and the method of characteristics.

Research is continuing to develop new and improved approaches for solving mass transport equations. The continuing effort to develop numerical approaches is attempting to reduce the computational effort and to overcome numerical problems endemic to the conventional approaches. Examples of recent modeling approaches or significant refinements of existing methods include the principal direction method (Frind, 1982), the alternating direction Galerkin method (Daus and Frind, 1985), and the Laplace transform Galerkin technique (Sudicky, 1989).

The finite-difference methods have traditionally been applied to solve flow and transport equations. One of the most important implementations of the finite-difference approach is in the powerful code SWIFT (Dillon et al.,

1978) and its succeeding versions. This code has become a standard for use in studying the most complicated mass transport problems.

The essence of the finite-difference method is to replace the governing differential equation by a set of difference equations applicable to the system of nodes. The difference equations approximate the first- and second-order derivatives in the transport equation (that is, $\partial C/\partial x$ or $\partial^2 C/\partial x^2$) by concentration differences between node points. When each node in the network is considered, the result is a system of algebraic equations. This system can be solved with matrix algebra. The interested reader will readily discover a voluminous literature on the subject. An important feature of the finite-difference method is the relative ease in formulating the difference equations. Of all the numerical techniques available to solve differential equations, finite-difference techniques are easiest to understand from a conceptual point of view.

Finite-element approaches are common in solving mass transport equations. Although the approach was first developed in the petroleum industry, much of the subsequent refinements and applications were by hydrogeologists. The most important early studies included work by Pinder (1973) and Duguid and Reeves (1976) on the modeling of contaminant transport, and by Rubin and James (1973) on transport with multicomponent ion exchange. Various extensions of the technique examined density-dependent problems (see Huyakorn and Pinder, 1983).

Unlike the finite-difference methods, which involve solving the mass transport equation directly, the finite-element techniques deal with a mathematically equivalent, integral form of the mass transport equation. This integral form can be developed in using the method of weighted residuals or the variational method. Huyakorn and Pinder (1983) present a detailed discussion of these approaches and in particular the Galerkin procedure (a special case of the method of weighted residuals).

The method of characteristics is another useful approach for solving mass transport equations. It was first applied by Garder and others (1964). Subsequently, the method was used in hydrogeological applications to study saltwater intrusion (Pinder and Cooper, 1970), ground-water contamination (Reddell and Sunada, 1970; Pinder, 1973), and the role of various parameters in controlling the spread of contaminants (Schwartz, 1975). Konikow and Bredehoeft (1978) implemented this approach in a versatile two-dimensional computer code, which today is one of the most commonly used codes.

The method takes the advection–dispersion equation and breaks it down into a set of simpler differential equations. This formulation in effect provides a frame of reference that is moving with the linear ground-water velocity. Thus, the advection–dispersion equation, which is relatively difficult to solve, becomes a diffusion equation, which is simpler to solve.

The particle tracking method is different from the previous three numerical approaches. It does not solve the mass transport equation directly but simulates the spread of mass dissolved in water. This approach is formulated as a classical random-walk problem involving the motion of a swarm of reference particles. It has been applied in modeling atomic particle motion in nuclear reactors, the spread of air pollutants, and oil spills in oceans. Ahlstrom and others (1977) introduced the method, which subsequently was implemented in several computer codes (Schwartz and Crowe, 1980; Prickett and others, 1981).

The transport of contaminants is simulated by adding reference particles and moving them in a prescribed manner. By varying the number of particles added at the source during any one time step, it is possible to simulate complex loading functions. To account for advection, each particle moves in the direction of flow a distance that is determined by the product of the magnitude of the velocity and the size of the time step. With a small time step, this particle motion traces a pathline through the system. Dispersion is accounted for in the particle motion by adding a random component to the deterministic motion, which is a function of the dispersivities.

The mean concentration for each grid block is calculated as the sum of the mass carried by all of the particles located in a given block divided by the total volume of water in the block. By appropriate adjustments to the quantity of mass carried by each particle, first-order kinetic reactions or simple linear sorption can be simulated.

Adding Chemical Reactions

Going beyond problems involving simple reactions like linear sorption or first-order kinetic decay to those with several interacting solutes is much more difficult than any numerical approach considered so far. With a few exceptions, the possible approaches fall into two categories. With the "one-step" approach, a complete mass transport equation including all of the appropriate reactions is written for each solute species. The entire coupled set of algebraic equations is then solved simultaneously using either a finite-element or finite-difference method. This procedure is numerically complicated but is the most rigorous way to handle reactions in a transport framework.

Most of the more general implementations of this scheme have been for one-dimensional transport. Extending this approach to two-dimensional problems is a formidable task. One of the most sophisticated examples of the one-step procedure is a code developed by Miller and Benson (1983). Their model and the planned extensions account for one-dimensional advection and dispersion along with a host of equilibrium chemical reactions. Included in the suite of reactions are complexation in the aqueous phase, exchange of both ions and complexes, and mineral precipitation. Other examples of these models include Willis and Rubin (1987) and Ortoleva and others (1987a,b).

The "two-step" procedure separates the physical and chemical processes. A solution to the advection–dispersion equation provides an initial estimate of concentration. A second step corrects these concentrations to account for the partitioning of mass due to a suite of chemical reactions. Operationally, the two-step procedures require iteration between the steps until some specified convergence criterion is met. Iteration assures that the calculated concentrations are solutions to both the advection–dispersion equation and the material balance equations describing the chemical system. This method is simpler computationally although less rigorous than the one-step procedure. For this reason, the approach is amenable to two- and three-dimensional problems. It also provides a way of grafting sophisticated geochemical codes like PHREEQE, MINTEQ, or MINEQL to transport models.

An example in the development of this approach is described by Cederberg (1985). The transport equations are solved in two dimensions using a Galerkin finite-element procedure and the algebraic equations for the chemical reactions using a Newton–Raphson method. Chemical processes include complexation, as well as complex sorption, which considered the effects of surface ionization and complexation at the solid–solution interface. Other examples of two-step models come from work by Narasimhan and others (1986) and Liu and Narasimhan (1989).

18.4 Case Study in the Application of a Numerical Model

A model study of contamination at Otis Air Base on Cape Cod, Massachusetts (LeBlanc, 1984b), illustrates some of the steps in using a mass transport model. At Otis Air Base, approximately 1740 m³/day of treated wastewater has been disposed in the subsurface since 1936. The disposal unit is an unconfined sand and gravel aquifer approximately 35 m thick. A zone of contamination has developed that is approximately 915 m long, 23 m thick, and 3.35 km long. This site is being studied by the U.S. Geological Survey as part of the Toxic-Waste Ground-Water Contamination Program (LeBlanc, 1984a; Franks, 1987; Ragone, ed., 1988).

The treated sewage effluent contains above background concentrations of Na^+, Cl^-, ammonium, nitrate, phosphate, detergents, and several different volatile organic compounds. Boron, one of the trace metals in the effluent, was selected by LeBlanc (1984b) for a detailed model study because (1) its concentration has remained relatively constant in the effluent at 500 μg/L, (2) it has a relatively low background level in the native ground water (30 μg/L), and (3) it tends not to react chemically during transport. The objective in modeling was to guide the collection of data from the site and to test hypotheses concerning the character of contaminant migration.

The first step in developing the flow and boron transport models was to define the region of interest and to establish boundary conditions for flow. LeBlanc (1984b) used the following guidelines, which should apply generally to most studies. The domain had to be large enough to include all of the existing plume (Figure 18.13) as well as providing room for future spreading. The side and bottom boundaries down-gradient of the disposal area are ponds, rivers, and saltwater bays. These features are a good choice for boundaries because in the absence of detailed hydraulic head data, they provide the best places to estimate boundary conditions. In effect by assuming ground-water discharges at these locations, boundaries for ground-water flow can be estimated in terms of constant head nodes or leakage fluxes. North of Coonamessett and Johns Ponds, there are no natural hydrogeologic boundaries. In this area, flow lines formed the side boundaries (Figure 18.13). The flow lines are imaginary boundaries, located so as not to intersect the contaminant plume. The top or northern boundary was arbitrarily defined by the 60-ft equipotential line. This boundary could have been placed anywhere north of the site with the proviso again that contaminants from the site could not intersect this boundary.

The Konikow and Bredehoeft (1978) model requires that the site be subdivided into a regular finite difference grid. In this example, the grid consisted of 40 rows and 36 columns. The nodal blocks are square, with dimensions chosen so that there is not an unmanageably large number of cells (an upper limit might be 50 or so cells in the row/column directions). However, the blocks are sufficiently small to ensure that the plume is not localized in just a few cells. The plume in this case has a width of about seven cells. Figure 18.14 shows how the model

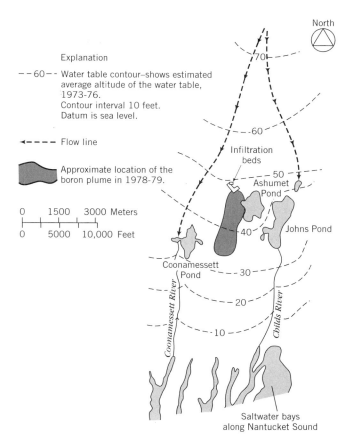

Figure 18.13 Map of important hydrologic features at the site in relation to the boron plume in 1978–79 (modified from LeBlanc, 1984b).

replicates the region shape and how the boundary conditions for flow are included.

LeBlanc modeled the pattern of ground-water flow first. Because the density and viscosity of the contaminated ground water are nearly the same as the uncontaminated water, the distribution of hydraulic head and hence the velocity field are unaffected by the migration of the plume. The flow equation therefore can be solved first independently of the mass transport equation. Observations at the site indicate further that the gradients in hydraulic head do not change significantly with time. Thus ground-water flow is steady, which means that the ground-water velocities need to be calculated only once.

Figure 18.15 compares the observed configuration of the water table with the "best fit" from a series of simulations. Such trials are designed to calibrate the model—a process of selecting the set of parameters that produces the best simulation of a known history. The calibration here took initial estimates of model parameters (for example, hydraulic conductivity, recharge rates) and adjusted them by trial and error until the model successfully reproduced the observed configuration of the water table. Because of the lack of data, no attempt was made to

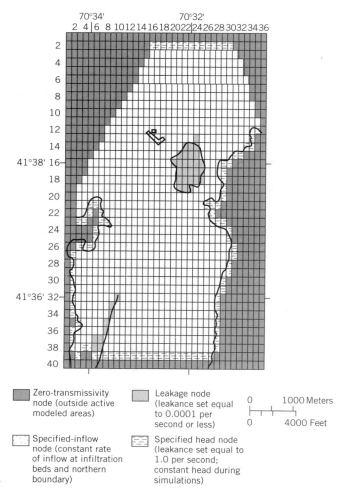

70°34'   70°32'

2 4 | 6 8 1012141618202224262830323436

▓ Zero-transmissivity node (outside active modeled areas) ▒ Leakage node (leakance set equal to 0.0001 per second or less)

0 ___ 1000 Meters

0 ___ 4000 Feet

Specified-inflow node (constant rate of inflow at infiltration beds and northern boundary) ▤ Specified head node (leakance set equal to 1.0 per second; constant head during simulations)

Figure 18.14 Grid system used to model both ground-water flow and mass transport. For ground-water flow modeling, the combination of zero transmissivity cells, leakance nodes, and specified head nodes define the region shape and bounary conditions (from LeBlanc, 1984b).

reproduce spatial variability in hydraulic conductivity or recharge rates. Nevertheless, the simulated and observed results compare well (Figure 18.15). The model as represented by the parameters and boundary conditions is not unique. Another set of different parameters and boundary conditions could give the same or a better fit. By keeping the hydraulic conductivity close to the measured values, the simulated hydraulic heads and resulting velocity field should be a reasonable representation of the flow system.

The only information required for the transport simulation with a conservative species like boron is the longitudinal and transverse dispersivities, plus the loading function at the source. In the absence of actual dispersivity data, values were selected from the literature for similar types of geologic materials. Here, this uncertainty in

choosing a value is not a problem because analysis shows that with active recharge, simulated concentrations are not sensitive to the particular choice of dispersivity value. A lack of data may not be so easily dealt with in every model study. Contaminant loading was approximated by fixing the boron concentration at 500 μg/L at the four cells representing the infiltration beds.

Tests with the model attempted to reproduce the historical spread of contaminants over a 40-year period between 1940 and 1978–79. Compared in Figures 18.16a and b are the simulated and the observed distributions of boron after 40 years. As LeBlanc (1984b) points out, the simulated path of the plume agrees reasonably well with that observed in 1978–79, particularly at concentrations above 50 μg/L. However, at lower concentrations, the center line of the simulated plume was located east

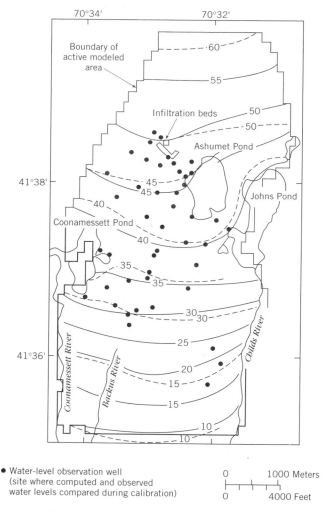

• Water-level observation well (site where computed and observed water levels compared during calibration)

0 ___ 1000 Meters

0 ___ 4000 Feet

Figure 18.15 A comparison between the observed (dashed lines) and simulated (solid lines) water table elevations for November 1979 (from LeBlanc, 1984b).

of the observed center line. The simulated plume is also somewhat longer and wider than the observed plume.

LeBlanc (1984b) explained the difference between observed and simulated plumes as follows. The observed plume could in fact be larger than is represented in Figure 18.16b because of uncertainty in establishing the edges of the plume at concentrations below 100 μg/L. Inaccuracies in the simulated flow field could have existed, producing somewhat more divergent flow than actually oc-

curs. This problem could be related to the complex interaction between ground waters and surface waters. Finally, the model could contain undetected numerical dispersion.

This study went on to evaluate the suitability of management options for the contamination. With reasonable confidence in the ability of this model to predict boron distributions, LeBlanc (1984b) examined the consequences of operating the site in the future as it is now or eliminating it.

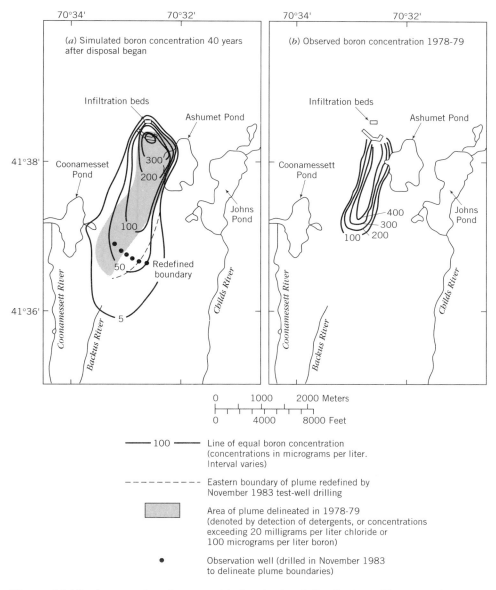

Figure 18.16 A comparison between (a) the simulated distribution of boron in the aquifer with (b) that observed in 1978–79. Also shown on (a) are indications of the plume size provided by Cl⁻ and detergents (from LeBlanc, 1984b).

Problems

1. In a one-dimensional column, a retarded species has moved a certain distance when $v_w t$ of a nonretarded species is taken as 28 cm and $\alpha_x = 2$ cm. The retardation factor is two. How far behind the advective front of the nonretarded species is the advective front of the retarded species?

2. In a one-dimensional column test, C/C_0 is noted to be on the order of 0.65. This means:

_____ a. The advective front has already broken through.

_____ b. The advective front has not broken through.

_____ c. The argument of the erfc is negative.

_____ d. The argument of the erfc is positive.

_____ e. a and c are correct.

_____ f. b and c are correct.

3. The erfc of negative beta is 1.146. What is the value of beta?

4. Match the breakthrough curves given on the diagram with the proper description given below.

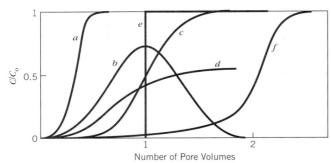

Problem 4. A pore volume in a column is the volume of water that will completely fill all of the voids in the column.

_____ This curve was obtained for continuous injection in a one-dimensional laboratory column where the constituents in the displacing fluid were being sorbed by the medium.

_____ This curve was obtained for continuous injection in a one-dimensional column with such a low velocity that plug flow (no dispersion) was observed.

_____ This curve was observed for a pulse (instantaneous) type source with both longitudinal and transverse dispersion.

_____ This curve was obtained for continuous injection in a one-dimensional (no transverse dispersion) laboratory column.

_____ This curve was obtained for continuous injection with both longitudinal and transverse dispersion.

5. An experiment with a step function input of nonreactive tracer is conducted. The velocity of the tracer through the column is about 36 cm/hr. The length of the flow tube is 30 cm. The $C/C_0 = 0.25$ point is arrived at the end of the column in approximately 0.7 hr. At time equal to 0.992 hr, $C/C_0 = 0.75$, and at time 0.833 hr., $C/C_0 = 0.5$.

a. Estimate the dispersion coefficient in cm²/hr.

b. Estimate the dispersivity in cm.

c. Can you use the $C/C_0 = 0.5$ at $t = 0.833$ hr in the preceding calculations? Why?

6. In a plan view of a contaminant plume, you notice that chloride has moved approximately 1000 m whereas cadmium has moved about 100 m. Assuming that both constituents were released from the source at the same time, find the distribution coefficient for cadmium if the porosity is 0.2 and the density ρ_s is 2.65 g/cm³.

7. Answer the following questions with respect to the Ogata–Banks equation.

$$\frac{C}{C_0} = \frac{1}{2}\,\text{erfc}\,\frac{x - vt}{2(D_x t)^{1/2}} = \frac{1}{2}\,\text{erfc}\,\frac{x - vt}{2(\alpha_x vt)^{1/2}}$$

a. The maximum observed concentration will be equal to _____.

b. This maximum will occur at some point x where x is _____.

c. At $x = vt$, the concentration C is equal to _____.

d. The concentration approaches zero at some point x where x is _____.
Prove or disprove each of the following.

e. The larger α_x, the more extensive the spread ahead of the advective front.

f. The larger α_x, the farther behind the center of mass the fluid takes on its maximum concentration.

g. The larger α_x, the longer the time required to reach steady state at the column exit.

h. When determining α_x in the laboratory, it is advisable to use the C/C_0 data at the column exit when $x = vt$.

i. The larger α_x, the lower the concentration at some point just ahead of the advective front.

8. A spill from a high-level storage tank releases the radionuclides shown in the table.

a. What constituent will be first to arrive at the property boundary?

b. What constituent will be last to arrive at the property boundary?

c. What constituent will have the smallest maximum concentration at the property boundary?

d. What constituent will have the largest maximum concentration at the property boundary?

e. What constituent will have the smallest spread around the center of mass in the direction of flow?

f. What constituent will have the largest spread around the center of mass in the direction of flow?

Nuclide	Activity (ci)	Half Life (yr)	K_d (mL/g)	D_x (cm²/s)	D_y (cm²/s)	D_z (cm²/s)
A	400	33	8	10^{-4}	10^{-5}	10^{-6}
B	400	28	52	10^{-3}	10^{-5}	10^{-7}
C	400	2.7	40	10^{-5}	10^{-5}	10^{-5}
D	400	20	0.1	10^{-6}	10^{-5}	10^{-6}

g. What constituent will have the same spread around the center of mass in all directions, that is, in the flow direction, transverse to the flow direction, and vertically as well?

9. A hazardous waste facility is to be constructed in a given area. The state requires a buffer zone between the waste trenches and the property boundary. Over 500 different constituents are to be included in the waste, some with reasonably large distribution coefficients, others with a likely distribution coefficient of zero. A few years after construction, a plume is first noted by observation wells at the waste boundary. The maximum concentration in the plume occurs in a zone of about 6 m thick and extends the entire length of the waste boundary or about 60 m. Because of low-permeability bounding materials, the plume is capable of spreading only in Y but not in Z.

Calculate the maximum concentration that might be expected (in terms of C_0, the plume concentration at the waste boundary) once the plume arrives at the buffer zone, which is 150 m from the waste boundary. The transverse dispersion coefficient is 1×10^{-3} m²/s and the linear velocity (flow through the pores) is 1×10^{-3} m/s.

10. Ground water travels with a linear velocity of 15 cm/day. What will the transport velocity be for an organic contaminant having a K_d of 6.6 mL/g in a medium with a porosity of 0.35 and a solids density (ρ_s) of 2.65 g/cm³?

11. Following is a site map and hypothetical plume. At the F-area and H-area seepage basins marked on the figure, waste water containing a variety of contaminants was disposed of over a 13-year period into an unconfined aquifer. Your objective is to model the transport of a contaminant at the F-area that is moderately sorbed and that is lost via a first-order kinetic reaction. Formulate this problem for a numerical solution by (1) defining a local region of interest and appropriate boundary conditions for flow and transport, (2) selecting appropriate forms of the flow and transport equations, and (3) indicating what parameter values are required for a simulation.

12. A plume of bromoform dissolved in ground water originates from two storage ponds with a total length of 150 m. The ponds penetrate the unconfined aquifer to a depth of 5 m. The total saturated thickness of the aquifer is 15 m. Use the computer code that was developed in Example 18.6 to solve this problem. Note that it will be necessary to modify the code to consider only one direction of z spreading (that is, Eq. 18.18). Predict the concentration distribution along the water table after 5110 days, assuming a linear ground-water velocity of 1.6×10^{-3}; dispersivities in x, y, and z of 5.0 m, 0.5 m, and 0.1 m, respectively; $C_0 = 4000$ mg/L; $f_{oc} = 0.005$; $\log K_{ow} = 2.38$; porosity 0.30; and a solids density of 2.65 g/cm³.

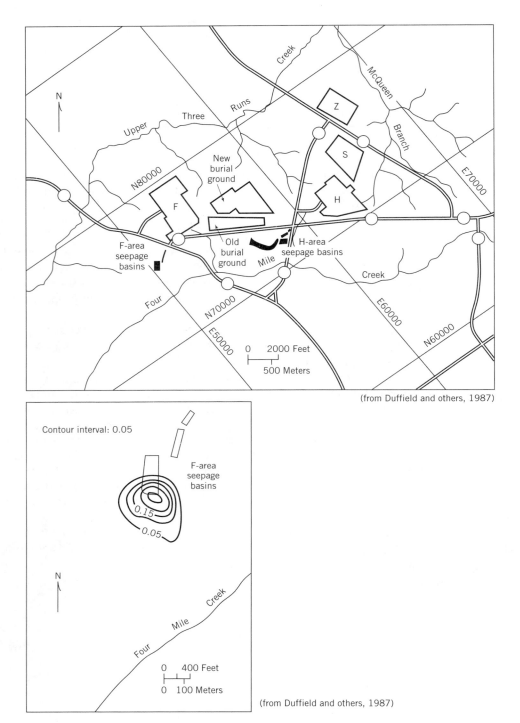

(from Duffield and others, 1987)

(from Duffield and others, 1987)

Problem 11

Chapter 19

Multiphase Fluid Systems

19.1 Basic Concepts
19.2 LNAPLs and DNAPLs
19.3 Partitioning
19.4 Fate of Organics in the Unsaturated Zone
19.5 Fate of Organics in the Saturated Zone
19.6 Air-Permeability Testing
19.7 Recognizing DNAPL Sites

Some contamination problems involve two or more fluids. Examples include air, water, and organic liquids in the unsaturated zone, or organic liquids and water in an aquifer. Organic liquids that have densities greater than water are referred to as DNAPLs (dense nonaqueous phase liquids). These compounds provide a significant challenge in both investigating sites and cleaning them up. Nonaqueous phase liquids that have densities less than water are called LNAPLs (light nonaqueous phase liquids). Commonly, contamination by LNAPLs involves spills of fuels like gasoline and jet fuel.

Problems involving NAPLs are more complex than dissolved plumes. To understand the main factors controlling the occurrence and flow of organic liquids will require several new basic concepts. We will take up these issues in the following section. Having developed these basic principles, the remainder of the chapter examines concepts and models affecting multiphase flow.

19.1 Basic Concepts

Saturation and Wettability

Saturation describes the relative abundance of fluid in a porous medium as the volume of the ith fluid per unit void volume. For a representative elemental volume,

$$S_i = V_i/V_{voids} \qquad (19.1)$$

where V_i is the volume of the ith fluid. In a multicomponent system, the sum of all the saturations is equal to one.

Wettability is the tendency for one fluid to be attracted to a surface in preference to another. According to Demond and Roberts (1987), the only direct measure of wetting is the contact angle. The idea of the contact angle can be illustrated by an experiment in which a surface such as a piece of glass is immersed in a reference fluid such as water and a drop of the test fluid is placed on the surface. The contact angle is established by measuring the tangent to the droplet from a point where all three components are in contact (that is, surface, reference liquid, and test liquid; Figure 19.1). A contact angle <90° (Figure 19.1) implies that the test liquid is wetting. A contact angle >90° implies that the reference liquid is wetting and that the test liquid is nonwetting.

Wettability is unique for given types of solids and fluids. However, a few generalizations hold:

- Water is always the wetting fluid with respect to oil or air on rock-forming minerals (Albertsen and others, 1986)

- Oil is a wetting fluid when combined with air, but a nonwetting fluid when combined with water (Albertsen and others, 1986)

393

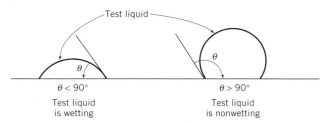

Figure 19.1 Definition of wetting on the basis of the contact angle (modified from Demond and Roberts, Water Resources Bulletin, v. 23, p. 617–628, 1987). Reproduced by permission of Amer. Water Resources Association.

- Oil is the wetting fluid on organic matter (for example, peat or humus) in relation to either water or air (Albertsen and others, 1986)
- The wetting character of organic contaminants remains uncertain.

Interfacial Tension and Capillary Forces

When a liquid is in contact with another substance (liquid or solid), there is a free interfacial energy between them, due largely to the difference between the inward attraction of that molecule in the interior of each phase. A surface that contains such free energy would like to contract, so that interfacial tensional forces are established. The tensional nature of these forces can be compared with those set up in a stretched membrane. Interfacial tension is defined as the work required to separate a unit area of one substance from that of another and is expressed as a force per unit length. Values range from zero for completely miscible liquids to 72 dynes/cm (0.072 N/m), which is the value for water at 25°C (Lyman and others, 1982). Values for most DNAPLs range between 15 and 50 dynes/cm (0.015 to 0.050 N/m) (Cohen and Mercer, 1993). Interfacial tension permits nonwetting DNAPLs to form globules in water-saturated materials.

As mentioned in Chapter 1, capillarity results from a combination of interfacial tension and the ability of certain liquids to wet the surfaces with which they come into contact. Interfacial tension is frequently determined by measuring the height a liquid will rise in a capillary tube. When the contact angle between a liquid and the wall is very small, as it is for water, the relationship between interfacial tension and capillary rise is given as

$$\sigma = \frac{\rho h r g}{2} \qquad (19.2)$$

where σ is the interfacial tension, h is the height of capillary rise, r is the radius of the tube, ρ is the liquid density, and g is the acceleration due to gravity.

The pressure discontinuity across any curved interface separating two immiscible fluids is referred to as the

capillary pressure, P_c, and is given as

$$P_c = P_{nw} - P_w \qquad (19.3)$$

where P_{nw} is the pressure of the nonwetting fluid (NAPL), and P_w is the pressure of the wetting fluid. For circular openings with a perfectly wetting fluid

$$P_c = \frac{2\sigma}{r} \qquad (19.4)$$

Thus, capillary pressure is directly proportional to the surface tension and inversely proportional to the radius of curvature.

Capillary pressure is a measure of the tendency of a porous medium to imbibe the wetting phase or to repel the nonwetting phase (Bear, 1972). As it is difficult to push a nonwetting fluid into a pore filled with wetting fluid, capillary pressure can also be thought of as the pressure required to move a particle of nonwetting fluid into a pore filled with a wetting fluid. Thus, as small pores provide resistance to entry due to capillarity, nonwetting NAPLs will tend to move through the coarser, more permeable zones of a heterogeneous medium. This behavior traps globules of water. In the unsaturated zone, capillary pressure (expressed as a negative pressure head) is referred to as suction.

Imbibition and Drainage

Imbibition and drainage are the dynamic processes by which fluids displace one another. Imbibition is the displacement of the nonwetting fluid by the wetting fluid, and drainage is just the opposite (Albertsen and others, 1986). Water being added to a dry soil or oil being gradually displaced by water from a water-wet oil reservoir are examples of imbibition. The entry of a nonwetting organic liquid into a water-wet aquifer is drainage.

Capillary pressure curves are frequently given as a function of the degree of saturation with respect to the wetting and nonwetting phases. In each case, the phase initially saturating the porous medium is slowly displaced by the other. As shown in Figure 19.2, the curves are frequently "L" shaped. The drainage or drying curve starts off at a wetting-fluid saturation of 100% but at some finite value of capillary pressure. For the displacement to take place, this value must be exceeded. This threshold value is called the displacement or entry pressure, which would be required for a DNAPL to enter a water-wet aquifer. As noted, capillary pressure becomes more negative as the wetting saturation decreases. With this increased suction, the displacement process eventually stops. The saturation value at which it stops is known as the residual wetting saturation and is designated S_{wr}. Here, the capillary pressure tends to infinity. Table 19.1 shows some estimates of displacement pressures calculated by Hubbert (1953).

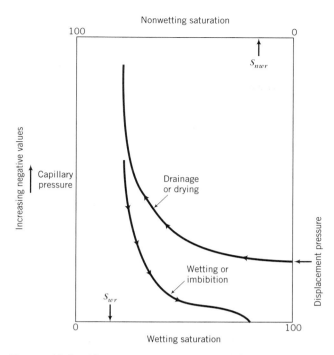

Figure 19.2 Observations: wetting or imbibition curve.

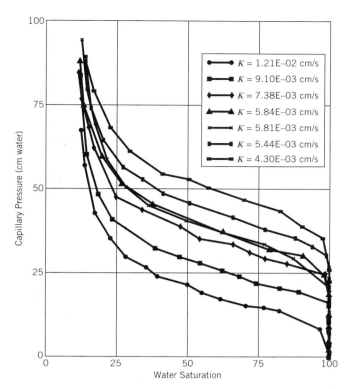

Figure 19.3 Tetrachloroethene–water drainage $P_c(S_w)$ curves determined for seven sands of varying hydraulic conductivity (from Kueper and Frind, Water Resources Res., v. 27, p. 1049–1058, 1991. Copyright © by American Geophysical Union).

The imbibition or wetting curve depicts initial saturation of a nonwetting phase that is slowly displaced by a wetting phase. It does not follow the same pathway as the drainage curve due to hysteresis. In this displacement, the capillary pressure becomes less negative as the wetting saturation increases. At zero capillary pressure, a point of residual nonwetting saturation is reached, designated as S_{nwr} on the figure. At this point, there is no further displacement of the nonwetting phase.

Kueper and Frind (1991) show capillary pressure–water saturation curves for a tetrachloroethylene–water system for sands of differing permeability (Figure 19.3). The entry pressure decreases with increasing permeability of the sands. Given such curves at a contaminated site, one can assess the potential for DNAPL to enter underlying fine-grained material (Kueper and McWhorter, 1991). The first step is to determine the saturation of the soil immediately above the fine-grained layer and obtain the capillary pressure from the curve. This capillary pressure can be compared to a measured or estimated entry pressure for the fine-grained layer.

Relative Permeability

The concept of relative permeability accounts for the tendency for fluids to interfere with one another as they flow (Demond and Roberts, 1987). This idea can be explained in relation to a Darcy's law expression for multifluid flow. For one-dimensional flow in a homogeneous medium, Darcy's equation written in terms of pres-

Table 19.1 Threshold Entry (Displacement) Pressures in Sediments with Various Grain Sizes Based on a ($\sigma \cos \phi$) Value of 25 dynes/cm (0.025 N/m) and Equation 4-5

Medium	Mean Grain Size Diameter (mm)	Threshold Entry Capillary Pressure (Pa)	Equivalent Capillary Rise of Water (cm)	Equivalent Capillary Rise of Water (ft)
Clay	<0.0039	>100,000	>1000	>33
Silt	0.0039–0.063	6300–100,000	65–1000	2.1–33
Sand	0.063–2.0	200–6300	2.1–65	0.06–2.1
Coarse sand	2.0–4.0	100–200	1.0–2.1	0.03–0.06

From Hubbert (1953).

sure gradients is (Muskat and Meres, 1936)

$$q_i = -\frac{k_i}{\mu_i}(\nabla P_i - \rho_i g \nabla b) \qquad (19.5)$$

where q_i is the flow of the ith fluid per unit area of the medium, k_i is the effective permeability of the medium to the ith fluid, μ is viscosity, P is pressure, ρ is density, g is the gravitational acceleration constant, and b is elevation. The relative permeability to the ith fluid is

$$k_{ri} = \frac{k_i}{k} \qquad (19.6)$$

where k_i as before is the effective permeability and k is the intrinsic permeability. Substitution of Eq. 19.6 into Eq. 19.5 yields the general form of Darcy's equation for multifluid flow (Demond and Roberts, 1987)

$$q_i = -\frac{kk_{ri}}{\mu_i}(\nabla P_i - \rho_i g \nabla b) \qquad (19.7)$$

In multifluid systems, k_{ri} ranges between zero and one. The product kk_{ri} represents the reduction in the intrinsic permeability because two or more liquids are present in the system. With a single-fluid system, k_{ri} is one and Eq. 19.7 reduces to the more familiar form of Darcy's equation. Exactly how k_{ri} varies between zero and one is a complex function of the relative saturation, whether the fluid is wetting or nonwetting with respect to the solids, and whether the system is undergoing imbibition or drainage. The role of relative saturation is shown in Figure 19.4. Note that at 100% saturation with NAPL, the relative permeability is one but decreases to zero as the water content within the pores increases. Similarly, at 100% saturation with water, the relative permeability is one.

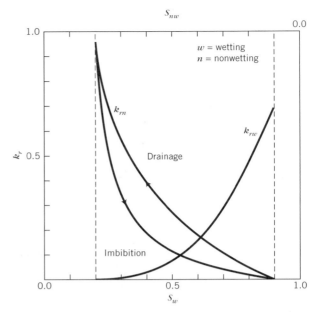

Figure 19.5   Typical relative permeability of curves (from Demond and Roberts, Water Resources Bulletin, v. 23, p. 617–628, 1987. Reproduced by permission of Amer. Water Resources Association).

Figure 19.5 shows typical curves of relative permeability as a function of the saturation for wetting and nonwetting fluids. Fixing the relative saturation of the wetting fluid (for example, $S_w = 0.8$) implicitly defines the relative saturation of the nonwetting fluid (for example, $S_n = 0.2$). For these relative saturations, $k_{rw} = 0.47$ and $k_{rn} = 0.04$ (for the imbibition curve, Figure 19.5).

There are several features common to most relative permeability curves (Demond and Roberts, 1987). First, when both fluids are present, the relative permeabilities rarely sum to one. For our example saturations, the sum is about 0.5. Second, as shown in the figure, the relative permeabilities of both the wetting and nonwetting fluids approach zero at finite saturations. In other words, some quantity of either wetting or nonwetting fluid in the pore system cannot move below some saturation threshold. In Figure 19.5, the relative permeability of the wetting fluid becomes zero at $S_w = 0.2$ and $k_{rn} = 0$ at $S_{nw} = 0.1$. These saturations are characteristic parameters known as residual saturations. The pattern of residual saturation shown by the example (that is, $S_{rw} > S_{rn}$) is observed frequently. A fluid at residual saturation is not capable of flow because at the low levels of saturation the fluid is not connected across the network of pores.

Residual saturation of the wetting fluid is sometimes called pendular saturation (Albertsen et al., 1986). The fluid is held by capillary forces in the narrowest parts of the pore space. Figure 19.6a provides an example of pendular saturation with water as the wetting fluid at residual saturation and air as the nonwetting fluid. Resid-

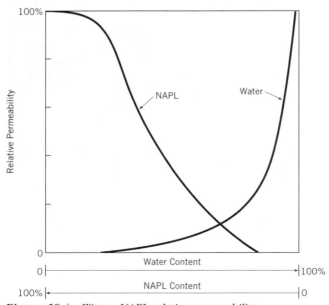

Figure 19.4   Water–NAPL relative permeability.

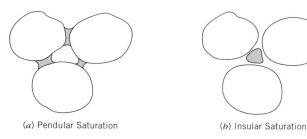

(a) Pendular Saturation   (b) Insular Saturation

Figure 19.6 Examples of different types of fluid saturation. In (a) the wetting fluid is at residual saturation; in (b) the nonwetting fluid is at residual saturation.

ual saturation of the nonwetting fluid is sometimes called insular saturation. This saturation is illustrated in Figure 19.6b, where a small quantity of organic liquid is at residual saturation in a water-wet aquifer. The fluid occurs as an isolated blob in the center of the pore.

A common situation is for water, air, and a contaminant such as oil to be found together in the unsaturated zone. Water is the wetting fluid, and air is the nonwetting fluid. Oil has intermediate wetting properties. It is nonwetting with respect to water but wetting with respect to air (Wilson and Conrad, 1984). The oil at residual saturation is caught as blobs and pendular rings, trapped between the water in the small parts of pores (Wilson and Conrad, 1984).

Because a contaminant such as oil behaves differently in the unsaturated versus the saturated zone, the residual saturations differ. For uniform glass beads packed in a column, the residual saturation is two to five times larger in the saturated zone, depending on bead size (Table 19.2).

Another feature of relative permeability curves is that for comparable saturations $k_{rn} > k_{rw}$. For the curves we show, at $S_w = 0.4$ and $S_n = 0.6$, the value of k_{rn} is about 0.1 (for imbibition) with $k_{rw} = 0.04$. The reason is that the wetting fluid occupies the smaller pores that contribute least to flow. The nonwetting fluid occupies the larger pores that contribute most to flow.

The last feature of note about the relative permeability curves is the hysteretic character of the nonwetting fluid. Many times this behavior is not evident because in petro-

leum studies authors typically present only imbibition curves for water displacing oil (Demond and Roberts, 1987). Hysteresis is evident because the nonwetting fluid occupies a different pore network during imbibition than during drainage.

The shape of the relative permeability curves depends on several variables such as (1) intrinsic permeability, (2) pore-size distribution, (3) viscosity ratio, (4) interfacial tension, and (5) wettability. Space is not sufficient here for us to discuss these factors. Readers can refer to an excellent article by Demond and Roberts (1987) and the text by Dullien (1979).

The volumetric retention capacity R_c has been introduced by de Pastrovich and others (1979) as a measure of the potential of the unsaturated zone to trap NAPL

$$R_c = 1000 \, S_r n \qquad (19.8)$$

where R_c is liters of residual NAPL per cubic meter of media, n is porosity, and S_r is the residual saturation of the NAPL. Retention capacity values and residual saturations in the unsaturated zone generally increase with decreasing moisture content, effective porosity, and intrinsic permeability. Residual saturation values in the unsaturated zone generally range between 0.1 and 0.2 (Cohen and Mercer, 1993). In the saturated zone, where NAPL is usually the nonwetting fluid, residual saturation values range from about 0.1 to 0.5.

Solubility and Effective Solubility

Solubility refers to the quantity of mass that will go into solution per unit volume of solution, such as mg/L, and is frequently taken as the maximum concentration that can be achieved in a contaminated body. Table 19.3 shows that solubilities can range over several orders of magnitude. However, concentrations are seldom, if ever, as high as the solubilities would indicate. Several factors are responsible for this, including heterogeneous flow behavior, complex kinetics in the dissolution process, and mixing of waters of different concentrations. An effective solubility for a given component can be approximated by multiplying the mole fraction of the chemical by its pure phase aqueous solubility, that is

Table 19.2 Residual Saturation Data for a Hydrocarbon in a Uniform Glass Bead Column for Three Fluid (Vadose Zone) and Two Fluid (Saturated Zone) Systems

Mean Bead Diameter (mm)	k (darcy)	Residual Saturation (%)	
		Vadose Zone	Saturated Zone
0.655	147	3	14
0.327	85	5	14
0.167	22	8	14

Modified from Wilson and Conrad (1984). Reprinted by permission of Petroleum Hydrocarbons and Organic Chemicals in Ground Water—Prevention, Detection, and Restoration. Copyright © 1984. All rights reserved.

Table 19.3 **Examples of Solubility Limits**

Constituent	Solubility Limit (mg/L)
Benzene	820–1800
Bromoform	3010–3190
Carbon tetrachloride	785
Chlorobenzene	448–500
Chlorodibromomethane	—
Chloroethane	5740
2-Chloroethylvinyl ether[a]	15,000
Chloroform	8200
Dichlorobromomethane	—
1,1,-Dichloroethane	5500
1,2-Dichloroethane	8690
1,1-Dichloroethylene	400
1,2-Dichloropropane	2700
1,3-Dichloropropylene	2700–2800
Ethylbenzene	152
Methyl bromide	900
Methyl chloride	6450–7250
1,1,2,2-Tetrachloroethane	2900
Tetrachloroethylene	150–200
Toluene	534
1,2-*trans*-Dichloroethylene	600
1,1,1-Trichloroethane	480–4400
1,1,2-Trichloroethane	4500
Trichloroethylene	1100
Vinyl chloride	1.1–60
Phenol (total)	—
Cyanide (total)	0.0054
2,4,6-Trichlorophenol	—
Arsenic	2.1E-06
Iron	0.0098
Copper	0.1054
Lead	20.58
Zinc	2.063E-07
Mercury	2.307E-21
Nickel	—
Barium	0.0563
Cadmium	—
Chromium	2.610E-04
Manganese	N/A
Silver	3.46E-09
Selenium	Soluble

[a] Unstable in water.

Solubility limits for inorganics are calculated using EQ3—Pitzer.

$$S_i^e = X_i S_i \qquad (19.9)$$

where S_i^e is the theoretical upper-limit dissolved-phase concentration of a constituent in equilibrium with ground water, X_i is the mole fraction of a component i in a DNAPL mix, and S_i is the pure phase solubility of component i. Thus, if the mole fraction of 1,1-dichloroethane is 0.1, the effective solubility is (Table 19.3)

$$5500 \times 0.1 = 550 \text{ mg/L}$$

As will be demonstrated later in the chapter, Raoult's law in combination with the ideal gas law will be used to calculate the mole fraction.

19.2 LNAPLs and DNAPLs

Consider a spill of LNAPLs or DNAPLs on the ground surface. With time, free product percolates downward through the unsaturated zone toward the water table. The most important process influencing downward movement of the free product is gravity-driven flow. Assume that in the unsaturated zone, water on the solid grains of the soil is the wetting phase and that the nonaqueous phase liquid (NAPL) is the wetting phase with respect to air on the water film. Because water wets the solids, the NAPL does not displace the water from the surface but moves from pore to pore once saturation exceeds the residual saturation. The NAPL displaces pore water that is not strongly held and soil gas.

Several important factors control the flow of NAPLs. In the case of a noncontinuous source (Figure 19.7), the volume of free product gradually decreases because some of the liquid is trapped in each pore at residual saturation. Thus, if the spill is relatively small, downward percolation in the unsaturated zone will stop once the total spill volume is at residual saturation. Another pulse of NAPLs is necessary to move the product downward. The main threat to ground water from such small spills is the opportunity for continuing dissolution of the NAPL by infiltration, or vapor phase migration in the vadose zone.

The NAPL in Figure 19.7 also tends to spread horizontally as it moves downward. This spreading is due to capillary forces, which operate together with gravity forces to control migration. The presence of layers of varying hydraulic conductivity also influences NAPL spread (Farmer, 1983). Even a relatively thin, low-permeability unit (Figure 19.8) will inhibit downward percolation and force the free product to move laterally. If such

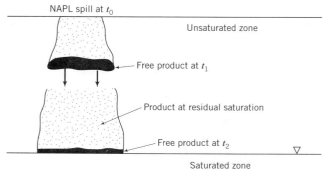

Figure 19.7 The downward percolation of a NAPL in the unsaturated zone. As the contaminant moves downward, the quantity of mobile fluid decreases and an increasing quantity is trapped as residual saturation.

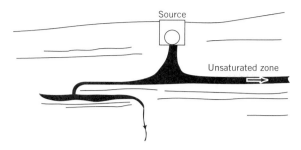

Figure 19.8 The presence of zones of low hydraulic conductivity within the unsaturated zone can cause NAPLs to mound and spread laterally. In this case, some of the contaminant is able to move around the ends of discontinuous units (modified from Farmer, 1983).

a layer is continuous, the NAPL will spread only within the unsaturated zone. If the layer is discontinuous, the NAPL will eventually spill over and continue to move downward toward the water table.

Conceptual Models for the Occurrence of LNAPLs

Chapter 17 described a conceptual model for LNAPL flow in the subsurface (Figure 17.1*a*). As LNAPL flowed downward toward the water table, it would encounter the capillary fringe. Because of increasing water saturations, relative permeability to the LNAPL declines, and buoyancy forces become important. Thus, less dense fluids (especially gasoline, kerosene, and other fuels) will float at the water table.

Classical models held that the free product would accumulate along the top of the capillary fringe (Figure 19.9). Once some critical thickness was achieved, NAPLs would flow down-gradient just above the water table. This conceptual model for the behavior of LNAPLs was useful because it provided a simple explanation about "true" versus "apparent" thickness of NAPLs at the capillary fringe (Figure 19.9). As we will see, this concept has outlived its usefulness. However, given that reference is

still made to true versus apparent thickness, we will describe this conceptual model briefly.

Assume that LNAPL occurs as a thin zone of saturation near the top of the capillary fringe (Figure 19.9). Under this condition, LNAPL product would flow into the well and fill the casing from the air–NAPL interface to the water table. Thus, the apparent thickness reflects the thickness of the capillary fringe in addition to the true product thickness. If the product is thick, it could also push down the water level in the well. The ratio of apparent to true product thickness is considered to range from two to about four, depending on site conditions.

More recent studies (Farr and others, 1990; Lenhard and Parker, 1990; Huntley and others, 1992) have shown this conceptual model to be erroneous. Instead of occurring as a discrete layer at the top of the capillary fringe, NAPL will be distributed throughout and even above the capillary fringe. Water saturations within the capillary fringe usually remain high, and NAPL saturations are relatively low. This pattern of saturation means that the actual volume of NAPL in the capillary fringe is relatively small (Huntley and others, 1992). The observation well, however, will still provide the impression of a much greater volume of NAPL present than is likely there.

Figure 19.10 from Huntley and others (1992) illustrates this alternative conceptualization. The LNAPL distributions actually observed in cores collected from a fine-grained unit are different than expected from the well measurement. Furthermore, the NAPL saturations at this site are quite low, indicating a lack of mobility. Thus, the more realistic model for NAPL saturation would provide for much less mobility than is suggested for high NAPL saturation within a single layer (Huntley and others, 1992).

In recent years, there has been interest in using alterna-

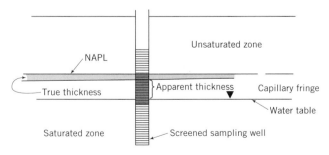

Figure 19.9 The concept of true and apparent thickness for LNAPLs. Product near the top of the capillary fringe moves down to the free surface in the well, making the apparent thickness much greater than the true thickness.

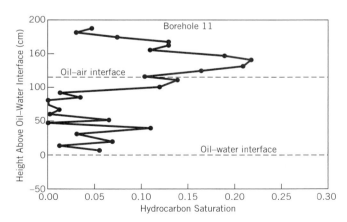

Figure 19.10 Measured hydrocarbon saturation profile compared to LNAPL distributions obtained from an observation well (from Huntley and others, 1992). Reprinted by permission of Conference on Petroleum Hydrocarbons and Organic Chemicals in Ground Water—Prevention, Detection, and Restoration. Copyright © 1992. All rights reserved.

tive approaches to obtain information on LNAPL distributions in the subsurface. The most promising approaches involve well tests (Gruszczenski, 1987; Hughes and others, 1988; and Sullivan and others, 1988) or dielectric well logging (Keech, 1988).

The flow of NAPLs also depends on the way the spill occurs (Farmer, 1983). Releasing a relatively large volume of contaminants over a relatively short time causes rapid downward and lateral migration (Figure 19.11*a*). Spreading is maximized and a relatively large volume of residual contamination remains in what Farmer (1983) refers to as the descent cone (Figure 19.11*a*). With slow leakage over a long time, the contaminant moves along the most permeable pathways (Figure 19.11*b*). These pathways can be a single channel or a more complex array of smaller channels, arranged in a dendritic pattern. Both the extent of lateral spreading and the volume of product held at residual saturation are considerably less than with a large volume spill. Overall, more of the mobile liquid will reach the water table from slow leakage.

The way LNAPLs interact with the capillary fringe depends on the rate at which product is supplied (Farmer, 1983). A slow rate of supply has little effect on the capillary fringe or the configuration of the water table (Figure 19.11*a*). Alternatively, a large volume of fluid, reaching the capillary fringe over a relatively short period of time, collapses the fringe and depresses the water table (Figure 19.11*b*). The extent of depression depends on the quan-

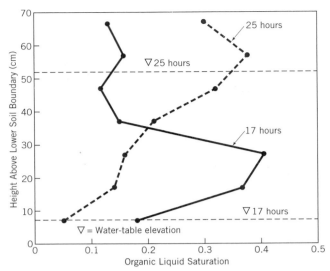

Figure 19.12 Results from a laboratory experiment showing how NAPL distribution changes in response to water-table rise (from Lenhard and others, 1993). Reprinted from J. Contam. Hydrol., v. 12, by R. J. Lenhard, T. G. Johnson, and J. C. Parker, Environmental observations of non-aqueous-phase liquid subsurface movement, p. 79–101, 1993, by kind permission from Elsevier Science, NL, Sara Burgerhartstraat 25, 1055 KV Amsterdam, The Netherlands.

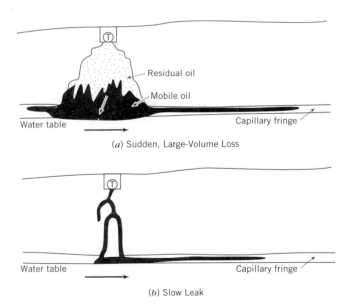

(a) Sudden, Large-Volume Loss

(b) Slow Leak

Figure 19.11 Pattern of flow is determined by the nature of the spill. In (*a*), a sudden, large-volume loss results in maximum spreading, large residual saturation, and collapse of the capillary fringe. In (*b*) a slow leak causes the product to follow a set of channels, a large volume of product to reach the water table, and minimal disturbance of the capillary fringe (modified from Farmer, 1983).

tity of product and its density. Even with this loading, the fluid occurs mainly within the capillary fringe (Figure 19.11). For spills that reach the capillary fringe, spreading continues until the total spill is at residual saturation.

Another feature of a ground-water system that influences LNAPL distributions is water-table fluctuations. The complexities in LNAPL saturation outlined above and the lack of mobility may conspire to trap LNAPLs below a rapidly rising water table. Lenhard and others (1993) present experimental results showing the effect of water-table rise over an LNAPL zone. Figure 19.12 shows how even in a relatively simple system a rising water table can trap significant LNAPL. Lenhard and others (1993) estimate that approximately one-half of the product present before the water table rose remained trapped. Even larger volumes may end up trapped if the water table rises into low-permeability units. Continuing water-table fluctuations over a long time will redistribute the free product across a larger zone.

The possibility for LNAPL to occur below the water table is illustrated by detailed saturation data developed for a single transect at a crude-oil spill near Bemidji, Minnesota (Hess and others, 1992b). The distribution of LNAPL at this site appears to be influenced by sediment heterogeneities and water-table fluctuations. In particular, Figure 19.13 illustrates how LNAPL may be trapped due to water-table fluctuations. Possibly, high-hydraulic-

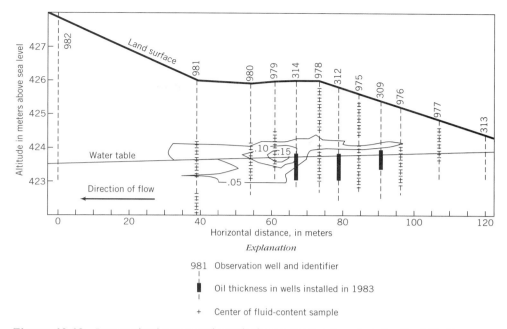

Figure 19.13 Longitudinal transect through the southern oil pool at the Bemidji site. Shown on the figure is the location of core samples, the oil thickness in wells, and the oil content as cm³/cm³ (from Hess and others, 1992). Reprinted from J. Contam. Hydrol., v. 10, by K. M. Hess, W. N. Herlkelrath, and H. I. Essaid, "Determination of subsurface fluid contents at a crude-oil spill," p. 75–96, 1992 with kind permission from Elsevier Science, NL, Sara Burgerhartstraat 25, 1055 KV Amsterdam, The Netherlands.

conductivity pathways in the vicinity of the water table induced the downward flow of LNAPLs (Hess and others, 1992b).

Occurrence of DNAPLs in Ground Water

For a simple geologic setting, DNAPLs have the potential to move downward to the base of the aquifer. Downward-moving DNAPLs displace water because they have a specific gravity much greater than that of water. DNAPL accumulating on low-permeability units will move downhill following the topography of the boundary (Figure 19.14). This flow in many cases will be in a direction that is different from that of the ground water. Spreading continues until the spill is at residual saturation. Within both the saturated and the unsaturated flow systems, these zones of residual saturation are a source of dissolved contamination as long as DNAPL remains. As was the case with LNAPLs, this simple conceptual model disguises the complex patterns of saturation in zones of "free product."

Not surprisingly, few sites are as simple as that depicted in Figure 19.14. Heterogeneities of all kinds may occur within the saturated ground-water system to produce complex patterns of NAPL distribution. Figure 19.15 illustrates how subtle variations in permeability and the attitude of heterogeneities can lead to complex distributions of DNAPL "pools". The figure also illustrates com-

plexities added by fracturing. Thus, in complex geologic settings, predicting migration pathways for DNAPLs can be relatively difficult.

Secondary Contamination Due to NAPLs

NAPLs in ground water can serve as important sources of secondary contamination. Problems develop when organic contaminants present as the free or residually saturated products partition into the soil gas through volatil-

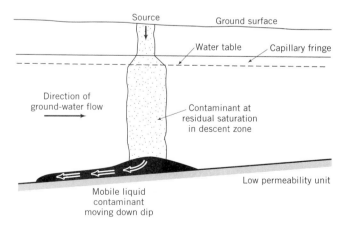

Figure 19.14 Pattern of DNAPL flow. In this case, the liquid moves along the bottom of the aquifer in a direction that is opposite to that of the ground water.

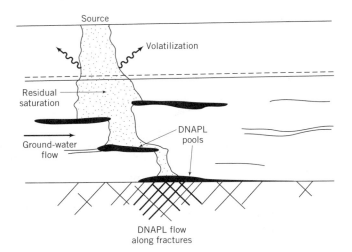

Figure 19.15 Complex distribution of DNAPLs in the vicinity of a surface source. Heterogeneities in hydraulic conductivity cause pooling. At depth, DNAPL flows into the bedrock along fractures.

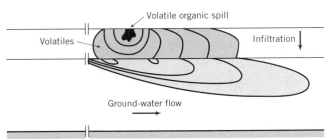

Figure 19.16 NAPLs in the vadose zone as a source of ground-water contamination. Volatilization of organic compounds leads to vapor phase migration and dissolution into the ground water (modified from Mendoza and McAlary, 1990). Reprinted by permission of Ground Water. Copyright © 1990. All rights reserved.

ization or into the ground water through dissolution. Figure 19.16 illustrates how even a small volume spill of a volatile organic liquid in the unsaturated zone can produce a plume of dissolved contaminants more significant than the original spill (Mendoza and McAlary, 1990). Volatiles spreading laterally and downward away from the source can partition into available soil moisture. Through continued infiltration, this water ultimately contaminates the aquifer.

Mendoza and Frind (1990a,b) simulated the vapor transport of volatile organic solvents in the unsaturated zone. Their model considered (1) transport due to diffusion and advection related to density gradients and vapor mass generation at the source, and (2) attenuation related to dissolution into soil moisture. Of particular importance is density-driven advection caused by density gradients that may develop with dense (that is, relative to soil gas)

chlorinated solvent vapors moving downward and away from a spill (Mendoza and Frind, 1990a). What is surprising about this transport is the speed and extent to which volatiles are able to spread through the soil gas system. Simulation results presented by Mendoza and Frind (1990a) showed vapor migration of up to several 10s of meters in a few weeks. This tendency for volatile NAPLs to contaminate soil gas will be enhanced as the spill grows in size and spreads along the capillary fringe.

NAPLs in the subsurface also have the opportunity to dissolve into mobile water that moves through the zone of contamination. Figure 19.17 illustrates the plume of dissolved contaminants that could develop in conjunction with a free LNAPL moving along the capillary fringe. Figure 19.18 shows the plume from DNAPLs at residual saturation within the cone of descent and from the free product. Experimental studies by Anderson and others (1987) and work they review show that water moving through DNAPLs at residual saturation requires only 10 cm or so of contact distance before saturation is reached in the ground water. Although the solubilities

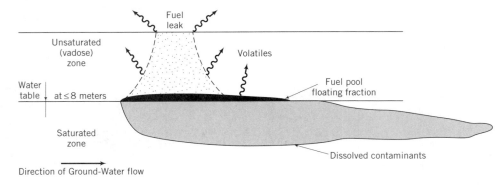

Figure 19.17 LNAPL dissolves into flowing ground water to form a contaminant plume (from Walther and others, 1986). Reprinted by permission of Conference on Petroleum Hydrocarbons and Organic Chemicals in Ground Water—Prevention, Detection, and Restoration. Copyright © 1986. All rights reserved.

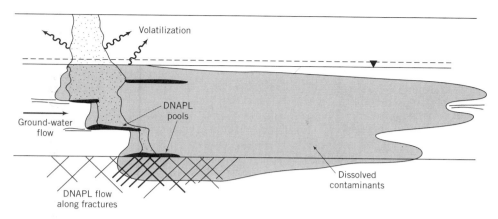

Figure 19.18 DNAPL dissolves into flowing ground water from residually saturated free product and pools to form a contaminant plume.

of volatile organic compounds (for example, benzene, carbon tetrachloride, trichloroethane) in water can be relatively low, the quantity of dissolved contaminant in water is often several orders of magnitude greater than the current standards permit (Anderson and others, 1987). Thus, these secondary sources of contamination are important in a regulatory sense.

Conceptual Models and Quantitative Methods

Several points have been brought out in the discussions above, largely with regard to the movement of NAPLs in the subsurface, heterogeneity, and residual saturations that hinder this movement, NAPLs as a source of secondary contamination due to dissolution and volatilization, and entry pressures leading to displacement. These have been discussed largely from a qualitative point of view because they are complex processes. However, if some assumptions can be made, some of these observations may be quantified as long as we keep in mind that a complex heterogeneous situation is being replaced by a geologic medium that is assumed to be the simplest type conceivable. Researchers at the Waterloo Centre for Groundwater Research have developed a series of illustrative conceptual models that illustrate these processes in differing geologic environments. Cohen and Mercer (1993) address these methods within the framework of conceptual models that lead to a series of questions to be addressed. We follow their lead, in part, by posing a given situation and presenting some simple equations that describe some of the more pertinent aspects of the situation. Because of space limitations, these equations are summarized in Table 19.4.

Example 19.1

In all the problems below, assume an interfacial tension of 0.040 N/m, a wetting contact angle of 32°, an effective

porosity of 25%, a residual saturation of 0.1, a DNAPL density of 1500 kg/m³, and a gravitational constant of 9.8 m/s².

(a) Consider a spill volume of approximately 0.3 m³. The volume of aquifer containing residual is

$$V_{res} = \frac{V_{spill}}{n_e S_{res}} = \frac{0.3 \text{ m}^3}{0.25 \times 0.1} = 12 \text{ m}^3$$

(b) Assuming the spill took place over a 2 m × 2 m area, the depth containing residual is

$$L = \frac{V_{res}}{\text{area}} = \frac{12 \text{ m}^3}{4} = 3 \text{ m}$$

(c) If this spill took place over this same area containing a 500-micron fracture (5 × 10⁻⁴ m), the depth of residual in the fracture is

$$L = \frac{V_{spill}}{S_{res} \text{ frac-thick} \cdot \text{frac-length}}$$

$$= \frac{0.3 \text{ m}^3}{0.1 \times 5 \times 10^{-4} \text{ m} \times 2 \text{ m}}$$

$$= 3000 \text{ m}$$

(d) What is the height of a DNAPL pool supported by a low-permeability layer with an average pore size of 10⁻³ mm (10⁻⁶ m)?

$$H_D = \frac{2\sigma \cos \phi}{r_{host} g (\rho_n - \rho_w)}$$

$$= \frac{2 \times 0.04 \text{ N/m} \times 0.848}{10^{-6} \text{ m} \times 9.8 \text{ m/s}^2 \times (1500 - 1000) \text{ kg/m}^3}$$

$$= 13.8 \times \text{N s}^2/\text{kg} = \text{m}$$

(Note: A Newton is a force and is equal to mass × acceleration)

(e) A DNAPL pool is 2 m in width, 2 m in height, and 10 m long and has an average concentration of 100

Table 19.4 Quantitative Methods (In the equations below, σ is the interfacial tension, ϕ is the wetting contact angle, r is a pore radius, ρ is a density of NAPL n or water w, g is the acceleration due to gravity.)

1. Volume of aquifer containing residual

$$V_{res} = \frac{V_{spill}}{nS_{res}}$$

2. The depth L of the residual given above

$$L = \frac{V_{res}}{\text{Area } A \text{ of spill}}$$

3. Depth of residual in a fracture

$$L = \frac{V_{spill}}{S_{res}\, \text{frac-thick} \cdot \text{frac-length}}$$

4. Height of a DNAPL pool that can be supported by low-permeability lens

$$H_D = \frac{2\sigma \cos\phi}{r_{host}\, g(\rho_n - \rho_w)}$$

5. Height of DNAPL pool that can be supported by a fractured capillary barrier

If residual DNAPL overlies the DNAPL in the saturated zone, replace r by the aperture b in the equation above

6. Thickness of DNAPL on the capillary fringe to cause DNAPL to enter the saturated zone

$$H_D = \frac{2\sigma \cos\phi}{rg\rho_n}$$

7. Critical DNAPL thickness for entry into a fracture at the top of the saturated zone

Replace r by the aperture size b in the above equation

8. The time required to dissolve a DNAPL pool in the saturated zone, where M is mass, v is the average ground-water velocity, C_w is the dissolved DNAPL concentration, and A is the cross-sectional area of DNAPL through which ground water flows

$$t = \frac{M}{vn_eC_wA}$$

g/m³ (mass = 400 g). Ground water with a DNAPL concentration of 10 g/m³ is moving through this pool at a velocity of 4×10^{-6} m/s. The time required to dissolve this pool is

$$t = \frac{M}{vn_eC_wA} = \frac{4000 \text{ g}}{4 \times 10^{-6} \text{ m/s} \times 0.25 \times 10 \text{ g/m}^3 \times 4 \text{ m}^2}$$
$$= 1 \times 10^8 \text{ s} = 3 \text{ yr}$$

A Case Study of Gasoline Leakage

The first case study (from O'Connor and Bouckhout, 1983) concerns a spill of gasoline from underground storage tanks in an urban area. This case is typical of the large number of spills involving petroleum products. More than 75,000 liters of gasoline leaked from buried underground storage tanks at Station A and possibly others nearby (Figure 19.19). The spill was first discovered when fumes leaked into sanitary sewers and the basements of houses. In the western half of the study area, from 4 to 8 m of gravel and sand overlie till (Figure 19.20). To the east, the gravel is thinner and capped by a thin silt or fine-sand layer.

Gasoline occurs at the water table beneath and east of Station A, but apparently bypasses areas where the till is at elevations above the water table (Figure 19.20). Thus, the spill, while generally moving down the gradient of the water table, is strongly influenced by the irregular topography of the till surface. The shaded areas (Figure 19.21) illustrate places where the surface of the till is above the water table. The arrows are pathways of migration where the water table is located in gravel. This complex pattern of flow explains why gasoline vapors were detected in only some of the buildings and why it was difficult to establish whether other stations were leaking. It also added to the difficulty in designing the collection system.

Hyde Park Landfill Case Study

The second case study, taken from Faust (1985) and Cohen and others (1987), examines the complex migration of both DNAPLs and associated aqueous phase liquids (APLs) from the Hyde Park Landfill at Niagara Falls, New York (Figure 19.22). Disposal of chemical wastes began at the site in 1954 by the then Hooker Chemical

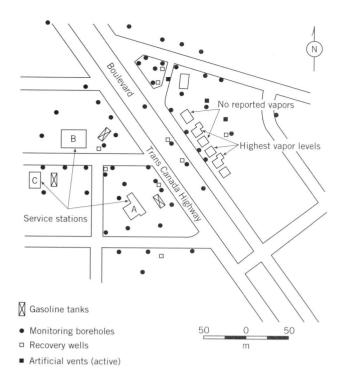

Gasoline tanks

● **Monitoring boreholes**

□ **Recovery wells**

■ **Artificial vents (active)**

Figure 19.19 Site plan showing the location of buried tanks, basements in which gasoline fumes were detected, the location of monitoring and recovery wells, and vents (from O'Connor and Bouckhout, 1983).

and Plastics Corporation. Ultimately, Hyde Park Landfill received approximately 80,000 tons of chemical wastes in both solid and liquid form. The wastes consist generally of chlorinated benzenes (21%), chlorinated toluenes (16%), chlorofluorotoluenes (11%), hexachlorocyclopentadiene (C-56) (75%), trichlorophenol (4%), chlorinated organic acids, pesticides, metal chlorides, and various other chemicals. Some of the contaminants, particularly chloroform, polychlorinated biphenyls, and mirex, are of concern because of their toxicity and persistence.

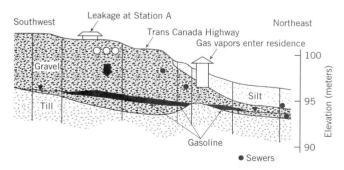

Figure 19.20 Cross section in the direction of spreading showing the general geology of the site and the pattern of gasoline spreading. The gasoline is moving down-gradient along the water table, moving around the till highs (modified from O'Connor and Bouckhout, 1983).

The landfill also contains an estimated 0.6 to 1.6 tons of 2,3,7,8-tetrachlorodibenzo-*p*-dioxin (TCDD), which is a potent carcinogen.

Ten meters or less of overburden overlies bedrock. Overburden consists of silt, clay, silty clay till, and a zone of sand. The uppermost bedrock units (Figure 19.22) include the Lockport Dolomite, Rochester Shale, and Irondequoit and Renales limestones. The Lockport Dolomite ranges in thickness from approximately 40 m immediately southeast of the landfill to approximately 20 m at the Niagara Gorge. Bedrock dips southward.

Ground water flows in a northwesterly direction toward the Niagara River in both overburden and bedrock. The hydraulic head is higher in the overburden than the Lockport Dolomite, providing a downward component of flow. The fractured upper 3 to 10 m of the Lockport Dolomite provide a highly permeable pathway for ground-water flow. Deeper parts of the Lockport are fractured but less permeable. The Rochester Shale is considered to be a low-permeability barrier to downward flow. Ground water discharges into the Niagara River along the gorge.

Figures 19.23*a* and *b* illustrate the extent of contamination by DNAPLs and dissolved compounds in the overburden and Lockport Dolomite. By far the greatest mass of contaminants is represented by the DNAPLs in the Lockport Dolomite. Dense liquids from the landfill are localized within the Lockport Dolomite to the top of the Rochester Shale. Lateral spreading of both dissolved contaminants and DNAPLs within the overburden has been relatively limited. Much of the contamination in the overburden (Figure 19.23*a*) is thought to be related to surface runoff more than lateral subsurface transport.

NAPLs in the landfill and those in the Lockport Dolomite are the source for dissolved contaminants that have been transported more than 750 m in the Lockport Dolomite. Contaminants discharged into the Niagara River northwest of the site. However, the DNAPLs in the Lockport Dolomite are not moving in the same direction as the ground water and the dissolved contaminants. These dense liquids are moving southward following the regional dip of the bedrock. This complex pattern of spreading helps to explain the differences in the DNAPL distribution and plumes of dissolved contaminants. Heterogeneity due to fracture development within the upper and lower parts of the Lockport Dolomite also results in localized variability in the contaminant distributions.

19.3 *Partitioning*

We have already learned some things about partitioning in Chapter 12. Included among the partition coefficients discussed there are Henry's dimensionless constant, the octonal-water partition coefficient, the soil-adsorption coefficient, and the distribution coefficient. These coef-

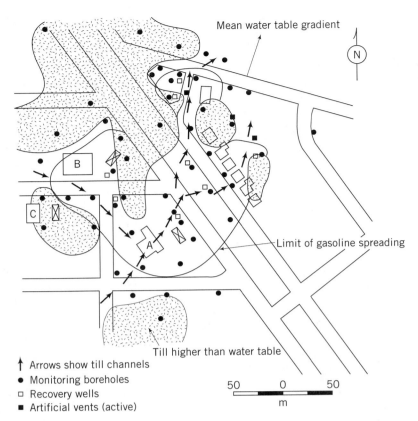

↑ Arrows show till channels
● Monitoring boreholes
□ Recovery wells
■ Artificial vents (active)

50   0   50
m

Figure 19.21 The pattern of gasoline migration in relation to the gradient on the water table. In general, the gasoline moves through the gravel, avoiding areas (hatched) where the till is at a higher elevation than the water table (from O'Connor and Bouckhout, 1983).

ficients tell us something about how organics partition between phases, that is, solid–liquid or liquid–gas partitioning. Henry's dimensionless constant gives an understanding of partitioning between the gas and liquid phases and is useful for problems involving a gas component. The distribution coefficient or the soil adsorption coefficient is useful for understanding partitioning between liquids and solids.

Figure 19.24 sums up the pertinent aspects of partitioning in the subsurface. Four phases can be present in the unsaturated zone: air, water, NAPL, and solid (soil). In the saturated zone, the phases include water, NAPL, and solid. As noted, partitioning in all cases is reversible. Partitioning to air depends on volatility and partitioning to water is controlled by solubility, whereas partitioning to soil depends on sorptivity. The important role of partitioning will be brought out in the next few sections.

19.4 Fate of Organics in the Unsaturated Zone

Figure 19.25 schematically illustrates the active processes in the unsaturated zone. NAPL can move as a separate phase, whereas advection of the water phase is enhanced by infiltrating rain water. The gas phase can also be in transport, driven by diffusion and density-driven advection. In this case, there is vaporization at the residual source. As the vapor migrates, it partitions to soil moisture and is transported both to the ground water and to the atmosphere. As noted, partitioning to air, water, and soil are dominant processes.

Several complicating factors are not shown on this diagram. First is the fact that we are dealing with a multicomponent mixture in a transient system that is undergoing advection as well as diffusive transport. The mass transfer process may be equilibrium or kinetically controlled in a varying temperature environment. Last, soil water content may increase with depth or be irregularly distributed.

Volatilization

Some of the concepts of significance to volatilization were discussed in Chapter 12. Henry's constant, in particular, is important in this discussion, as is Raoult's law. The higher Henry's constant, the more likely the com-

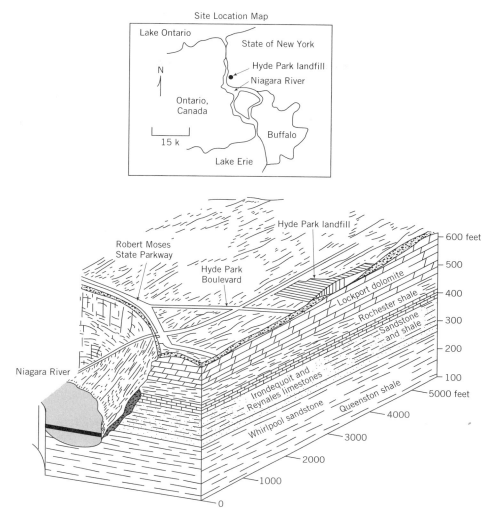

Figure 19.22 Location and geologic setting of the Hyde Park landfill (modified from Faust, 1985).

pound will partition to the gas phase. Among the group of compounds termed volatile organic compounds (VOC) are the chlorinated solvents and chlorinated aliphatic and aromatic hydrocarbons. The aromatic hydrocarbons include the water-soluble components of gasoline, benzene, toluene, ethylbenzene, and xylene (BTEX). The partitioning of these volatile organic chemicals in the unsaturated zone depends not only on volatility and solubility but on soil moisture content, the presence of organic matter, and temperature. Adsorption of vapor increases with decreasing soil moisture content and increasing organic content. Volatilization increases with increasing temperature.

With the movement of a volatile NAPL through the soil zone, some will remain as an immobile residual, providing the potential for long-term contamination by infiltrating ground water. Organic substances released to

the soil gas may migrate in response to a diffusive process, may be absorbed and therefore retarded by the solid particles, may be dissolved in ground water, or may escape to the atmosphere. Density-driven gas flow may result in contamination of an underlying ground-water body (Falta and others, 1989). The nature of the problem is shown in Figure 19.25.

As mentioned in Chapter 12, vapor pressure represents a compound's tendency to evaporate and is essentially a measure of the solubility of an organic solvent in gas. Thus obtaining a measure of a compound's propensity to volatilize requires Henry's law for partitioning of dissolved organic solutions from water to air and Raoult's law for partitioning between the pure solvent (NAPL) and air (see Problem 3, Chapter 12). On the other hand, transport of the vapor takes place by both an advective and diffusive process.

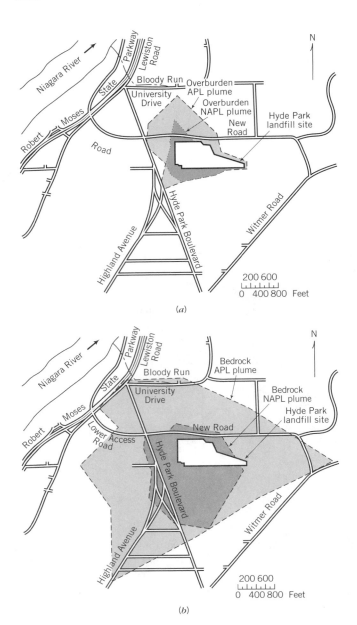

Figure 19.23 NAPL and dissolved contaminant plumes in (*a*) the overburden and (*b*) the Lockport Dolomite (modified from Faust, 1985).

Gas Transport by Diffusion

The complicating factors that arise from gas transport in the unsaturated zone are due to the fact that the system generally consists of a multicomponent mixture of differing density in a transient state, subject to temperature variations and reactions that may be equilibrium or kinetically controlled. Additionally, the transport may take place by both advection and diffusion. In this section, we simplify matters by considering diffusive transport of a single component subject to retardation. The conserva-

Zone	Phases	Partitioning
Unsaturated	Air	NAPL ↔ air
	Water	NAPL ↔ water
	NAPL	Water ↔ air
	Soil	Water ↔ soil
		NAPL ↔ soil
Saturated	Water	NAPL ↔ water
	NAPL	Water ↔ soil
	Soil	NAPL ↔ soil

Volatility	NAPL ↔ air	
	Water ↔ air	
Solubility	NAPL ↔ water	
Sorptivity	NAPL ↔ soil	
	Water ↔ soil	

Figure 19.24 Partitioning between phases.

tion statement required here was given already in Chapter 14, modified somewhat for the presence of gas in a partially saturated media

$$-\operatorname{div} \mathbf{J}_D = \frac{\partial(C_g n_g)}{\partial t} \qquad (19.10)$$

where $\mathbf{J}_D$ is a diffusive flux, C_g is the gas concentration, n_g is the part of the total porosity occupied by the gas phase, and t is time. This equation states that the mass outflow rate per unit volume equals the time rate of change of mass within the unit volume. For a one-dimensional system,

$$\frac{\partial J_D}{\partial x} = \frac{\partial(C_g n_g)}{\partial t} \qquad (19.11)$$

The diffusive flux J_D in one dimension is given by a modified form of Fick's law

$$J_D = -D_g^* \frac{\partial(C_g n_g)}{\partial x} \qquad (19.12)$$

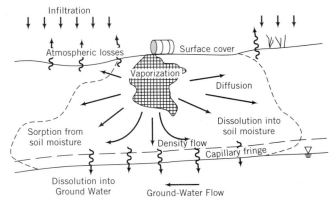

Figure 19.25 Mechanisms of contaminant transport in the unsaturated zone (from Mendoza and Frind, Water Resources Res., v. 26 no. 3, 1990. Copyright © by American Geophysical Union).

where D_g^* is an effective diffusion coefficient. Combining Eqs. 19.11 and 19.12 and considering the porosity n_g constant gives

$$D_g^* \frac{\partial^2 C_g}{\partial x^2} = \frac{\partial C_g}{\partial t} \qquad (19.13)$$

If gas transport by diffusion is subject to retardation, this becomes (see developments leading to Eq. 14.27)

$$D_g^* \frac{\partial^2 C_g}{\partial x^2} = \frac{\partial C_g}{\partial t} R_d \qquad (19.14)$$

where R_d is a retardation factor.

Equation 19.13 is frequently used in simulations where the source is given a concentration of unity and the relative concentrations are determined spatially for various times for both covered and uncovered ground. More realistic simulations have incorporated both advective and diffusive transport (Mendoza and Frind, 1990a,b), density-driven gas movement (Falta and others, 1989), and water–gas partitioning (Sleep and Sykes, 1989; Baehr, 1987).

Transport time to the saturated zone is strongly dependent on water saturation, giving rise to both water transport and air transport regimes. Part of this can be explained by the manner in which the gas-diffusion coefficient is treated. As mentioned, the diffusion coefficient in the unsaturated zone is an effective diffusion coefficient where the diffusion coefficient in air D_g is modified to account for the effects of tortuosity

$$D_g^* = \tau D_g \qquad (19.15)$$

where τ is frequently taken from the empirical development of Millington and Quirk (1961)

$$\tau = \frac{n_g^{7/3}}{n^2} \qquad (19.16)$$

The diffusion coefficient in free air varies for different constituents and can be as high as 1×10^{-1} cm²/s for methylene chloride to 8.11×10^{-2} cm²/s for trichloroethene. In most cases it ranges between 7 and 9×10^{-2} cm²/s. Thus, the diffusion coefficient decreases markedly with increased saturation, that is, small values of n_g. This suggests the presence of diffusion "barriers," such as perched water, low-permeability material, and the capillary fringe (Figure 19.26). However, with large saturations, the contaminant can partition to the water, adding to the dissolved phase. Additionally, lowering the water table in a pump-and-treat system will transform a fully saturated region into a partially saturated one, significantly increasing the diffusion coefficient.

The time for diffusive transport to the saturated zone is also sensitive to partitioning of vapor to the solid and aqueous phases. Retardation due to partitioning to the solid phase can be expressed as

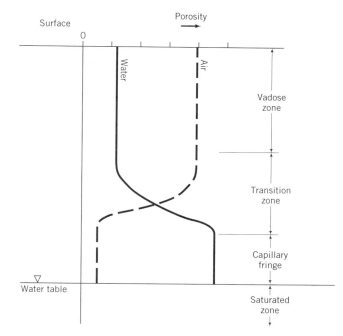

Figure 19.26 Water–air distribution in the unsaturated zone.

$$R_d = 1 + \frac{\rho_b K_d}{n_g H} \qquad (19.17)$$

where ρ_b is the bulk density of the solids, K_d is the distribution coefficient, taken as the product $K_{oc} f_{oc}$ (Chapter 12), and H is Henry's dimensionless constant. Retardation due to equilibrium partitioning to the aqueous phase is given as

$$R_d = 1 + \frac{n_w}{n_g H} \qquad (19.18)$$

where n_w is the part of the porosity containing water. The combined form represents retardation during diffusion transport

$$R_d = 1 + \frac{\rho_b K_d}{n_g H} + \frac{n_w}{n_g H} \qquad (19.19)$$

Of the variables in Eq. 19.19, the distribution coefficient can vary over a few orders of magnitude, depending on f_{oc}. Note also that large values of Henry's constant, which reflect a propensity to partition to the gas form, impede retardation of gas diffusion transport.

Equilibrium Calculations of Mass Distribution

As mentioned in Section 19.3, four phases can be present in the unsaturated zone: air, water (dissolved), NAPL, and solid (soil). Petroleum hydrocarbons, for example, can exist in the subsurface as a mobile free product, immobile residual, vapor, dissolved in ground water, and sorbed onto the solid particles. The distribution of mass among these phases will depend on the three key processes described in Figure 19.24, volatility, solubility, and sorptivity, as well as the hydrogeologic conditions at a particular site. Jury and others (1990) present methods that allow calculations for the distribution of mass in the subsurface. Many of the parameters required in these calculations can be found in handbooks of physical and chemical properties of organics (molecular weight, density, vapor pressure, solubility, adsorption coefficient), whereas other parameters require field data (porosity, residual water content, residual NAPL, mass density of the minerals making up the soil, normally 2.65 g/cm^3 for most sandy soils, and temperature). For the calculations below, consider a soil with a porosity n of 30%, a residual water content n_{rw} of 0.1, residual NAPL n_{nr} of 0.05, a gas content of n_g of $n - (n_{rw} + n_{nr}) = 0.15$, organic content f_{oc} of 0.004, mass density ρ_s of 2.65 g/cm^3, and a temperature T of 293°K. For trichloroethene, the following properties are noted: molecular weight M_w of 130 g, density of ρ_n of 1.48 g/cm^3, a partial vapor pressure Pv of 58 mm Hg, solubility of 1100 mg/L, and a soil adsorption coefficient K_{oc} of 139 mL/g.

The mass M_a contained in the aqueous phase is determined as the product of residual water saturation and solubility

$$M_a = n_{wr} \times S = \frac{0.1 \times 1100 \text{ mg/L} \times 10^3 \text{ L/m}^3}{1 \times 10^3 \text{ mg/g}} = 110 \text{ g/m}^3$$

$$(19.20)$$

The solubility is used here whereas for a mixture, a better choice might be the effective solubility (Eq. 19.9).

The mass contained in the sorbed phase is expressed

$$M_s = \rho_b C_s = 1.85 \, C_s \qquad (19.21)$$

where C_s is the mass sorbed on the solids (mg/g). It will be recalled that

$$K_d = K_{oc} f_{oc} = C_s/C_w \text{ (mL/g)} \qquad (19.22)$$

where C_w is the concentration in solution (mg/mL) so that

$$C_s = K_{oc} f_{oc} C_w \qquad (19.23)$$

If a pure phase is present, C_w is replaced by the solubility S (or by the effective solubility for a mixture). Thus,

$$C_s = K_{oc} f_{oc} S = \frac{0.004 \times 139 \text{ mL/g} \times 110 \text{ mg/L}}{1 \times 10^3 \text{ mL/L} \times 1 \times 10^3 \text{ mg/g}}$$

$$= 6.12 \times 10^{-4} \qquad (19.24)$$

Equation 19.21 then becomes

$$M_s = \rho_b C_s = 1.85 \text{ g/cm}^3 \times 6.12 \times 10^{-4} \times 1 \times 10^6 \text{ cm}^3/\text{m}^3$$

$$= 1132 \text{ g/m}^3 \qquad (19.25)$$

The mass in the NAPL phase is given as the product of the residual NAPL and the density of the NAPL

$$M_N = n_{nr} \times \rho_n = 0.05 \times 1.48 \text{ g/cm}^3 \times 1 \times 10^6 \text{ cm}^3/\text{m}^3$$

$$= 74,000 \text{ g/m}^3 \qquad (19.26)$$

The mass in the gas phase is simply the number of moles N times the molecular weight

$$M_g = N \times M_w \qquad (19.27)$$

where, from the equation of state for an ideal gas

$$N = \frac{PV}{RT} \qquad (19.28)$$

where P is the gas pressure, V is the gas volume, R is the gas constant, and T is temperature. Units are critical here. As N is the number of moles and temperature T is expressed in degrees Kelvin, the gas constant R can take on various units depending on the chosen units for pressure and volume

$$R = 0.08315 \text{ bar liter mol}^{-1}\,^{\circ}\text{K}^{-1}$$

$$= 8.315 \text{ Pa m}^3 \text{ mol}^{-1}\,^{\circ}\text{K}^{-1}$$

$$= 8.315 \times 10^{-5} \text{ bar m}^3 \text{ mol}^{-1}\,^{\circ}\text{K}^{-1} \qquad (19.29)$$

$$= 8.315 \times 10^6 \text{ mPa m}^3 \text{ mol}^{-1}\,^{\circ}\text{K}^{-1}$$

Some common conversions include 1 atmosphere equals 760 mm Hg and 1 bar equals 0.987 atmospheres.

If we wish to choose pressure in atmospheres and volume in m^3, the number of moles per cubic meter is determined from Eq. 19.28

$$N = \frac{(58/760) \text{ atm} \times 0.15 \times 1 \text{ m}^3 \times 10^3 \text{ L/m}^3}{0.08315 \text{ bar L mole}^{-1}\,^{\circ}\text{K}^{-1} \times 293^{\circ}\text{K} \times 0.987 \text{ atm bar}^{-1}}$$

$$= 0.4 \text{ moles (per m}^3) \qquad (19.30)$$

The mass in the gas phase, from Eq. 19.27, is

$$M_g = N \times M_w = 0.4 \text{ m}^{-1} \times 130 \text{ g} = 52 \text{ g/m}^3 \qquad (19.31)$$

To summarize, then, in grams per cubic meter

$$M_a = 110$$

$$M_s = 1132$$

$$M_n = 74,000$$

$$M_g = 52$$

Two points are worth noting here. The drinking water standard for this constituent is about 5 μg/L (5×10^{-3} g/m^3). Dividing the mass in grams by 5×10^{-3} gives the volume of water in m^3 that could be contaminated by the mass in each phase. The second point is that most of the mass resides in the NAPL residual and, to a lesser extent, soil phase, two of the more difficult phases to remediate. Vapor extraction or soil water suction techniques may be reasonably successful in dealing with the water and vapor phases, but this hardly seems worthwhile in view of the amount that would remain.

Example 19.2   Calculations of mass distribution in the unsaturated zone

This example is a little more elaborate than the preceding discussion. Consider a total porosity of 0.2, residual NAPL of 0.01, a residual water content of 0.1, and a gas content of $0.2 - (0.1 + 0.01) = 0.09$. The compound is benzene with a molecular weight of 78.11, a density of 0.88 g/cm^3, a vapor pressure of 0.1 atm for the pure organic solvent, and a solubility of 1780 mg/L. In the unsaturated zone, the gas concentration and the concentration in the sorbed phase vary, but we will take their average values at 503 μg/L and 100 μg/kg, respectively.

Mass of VOC in Gas Phase (Benzene)

$$M_g = \text{volumetric gas phase content}$$
$$\times \text{ concentration in gas phase}$$
$$= n_g \times C_g = 0.09 \times 503 \ (\mu\text{g/L})$$
$$\times 10^3 \ (1/\text{m}^3) = 45.27 \ (\text{mg/m}^3)$$

Mass of VOC in Aqueous Phase

$$M_a = \text{residual water saturation} \times \text{effective solubility}$$
$$= n_{rw} \times S_i^e$$

The effective solubility is determined by multiplying the pure-phase solubility of the organic constituent by the mole fraction of that component in the mixture. Raoult's law can be used to calculate the mole fraction

$$X = \frac{P_{org}}{P_{org}^0}$$

where X is the mole fraction of the component of a mixture, P_{org} is the partial vapor pressure of the organic in the gas phase, and P_{org}^0 is the vapor pressure of the pure organic solvent. The partial vapor pressure of the organic can be obtained from the ideal gas law

$$P_{org} = \frac{NRT}{V}$$

where N is the number of moles of gas, R is the proportional gas constant (0.08315 bar $\times$ liter mole^{-1} $\times$ °K^{-1}), T is the absolute Kelvin temperature, and V is the volume of gas. This can be restated as

$$P_{org} = \frac{C_g RT}{m}$$

where m is the molecular weight.

The partial pressure of benzene is determined

$$\frac{\left[\begin{array}{c} 503 \ \mu\text{g L}^{-1} \times 10^{-6} \ \text{g} \ \mu\text{g}^{-1} \times 0.0.08315 \ \text{bar} \\ \text{L} \times 293°\text{K} \times 0.987 \ \text{atm bar}^{-1} \ \text{mole}^{-1°}\text{K}^{-1} \end{array} \right]}{78.11 \ \text{g mole}^{-1}}$$

which equals 1.5×10^{-5} atm
The mole fraction of benzene is

$$\frac{1.5 \times 10^{-5} \ \text{atm}}{0.1 \ \text{atm}} = 1.5 \times 10^{-4}$$

The mass of benzene in the aqueous phase is

$$M_a = n_{rw} \times S \times X = 0.1 \times 1780 \ \text{mg/L} \times 1.5$$
$$\times 10^{-4} \times 10^3 \ \text{L/m}^3 = 27 \ \text{mg/m}^3$$

Mass of VOC in Sorbed Phase

Mass VOC in sorbed phase = bulk density $\times$ concentration in sorbed phase = $\rho_b \times C_s$, where ρ_b is the bulk density (1.41 g/cm^3) and C_s is the concentration in the sorbed phase. The mass of VOC in the sorbed phase is

$$M_s = 1.41 \ \text{g/cm}^3 \times 10^{-3} \ \text{kg/g} \times 100 \ \mu\text{g/kg}$$
$$\times 10^6 \ \text{cm}^3/\text{m}^3 = 141 \ \text{mg/m}^3$$

Mass of VOC in the NAPL Phase

Mass of VOC in NAPL phase = density of the NAPL
$$\times \text{ residual NAPL} = \rho_n \times n_{nr}$$

where ρ_n is the density of the NAPL and n_{nr} is the residual NAPL content. The mass of VOC in the NAPL state is

$$M_N = 0.88 \ \text{g/cm}^3 \times 0.01 \times 10^6 \ \text{cm}^3/\text{m}^3 = 8800 \ \text{g/m}^3$$

19.5   Fate of Organics in the Saturated Zone

Organic constituents in the saturated zone can occur in the dissolved, NAPL, and solid phases. The dissolved component may be the result of infiltration through a residual NAPL in the unsaturated zone, dissolution of residual DNAPL in the saturated zone, or direct infiltration of contaminated recharge. Movement of the dissolved contaminant coincides with the direction of ground-water movement (unless the density contrasts are quite large). DNAPL movement is far more complex. In the capillary fringe, DNAPL will tend to spread until the gravitational pressure at the base of the DNAPL exceeds the displacement or entry threshold pressure. At this point the DNAPL will displace the water and continue a downward vertical movement until it reaches a capillary barrier. The accumulated DNAPL will move in the direction of the slope of the barrier. The height of a DNAPL pool that can be supported by a low-permeability capillary barrier was given in Table 19.4. If the low-permeability lens has fractures, DNAPL migration into the fracture will commence, the potential for which increases in direct proportion to overlying DNAPL thickness and DNAPL–water density contrasts (Table 19.4). Downward migration through a capillary barrier is aided, not only by fractures with large apertures, but by root holes, stratigraphic windows, unsealed geotechnical boreholes, poorly constructed monitoring wells, and old uncased or abandoned water-supply wells.

As a consequence of the above process, DNAPL in the saturated zone will often be present in pools supported by low-permeability barriers, pools in large-aperture fractures surrounded by capillary barriers, and trapped as discontinuous globules and ganglia. If the source is finite and the fractures more or less continuous, residual saturation can be achieved, immobilizing the DNAPL.

Two- and three-phase flow simulation has long been the province of petroleum engineers interested in the flow of hydrocarbons to wells in reservoirs. In more recent years, multiphase codes have been developed for NAPL migration (see Abriola, 1988, for a review of multiphase codes). It is generally conceded, however, the immense data base required for simulation renders the use of these models impractical if the ultimate goal is prediction. They have, nonetheless, found some practical use in process analysis for uncalibrated situations.

Equilibrium Calculations of Mass Distribution

The total mass per unit volume in the saturated zone may be expressed as

$$\text{Total mass} = \text{amount in solution} + \text{amount sorbed}$$
$$+ \text{amount in NAPL} \quad (19.32)$$

or

$$\text{Total mass} = C_w(n - n_{nr}) + C_s\rho_b + n_{nr}\rho_n \quad (19.33)$$

where all terms have been previously defined. Recognizing that

$$K_d = \frac{C_s}{C_w} \quad (19.34)$$

Eq. 19.33 becomes

$$\text{Total mass} = C_w(n - n_{nr}) + K_dC_w\rho_b + n_{nr}\rho_n \quad (19.35)$$

Calculations are then carried out in much the same manner as for the unsaturated zone.

Example 19.3

Consider an absence of NAPL in the saturated zone for trichloroethylene, and calculate the percentage of mass in the sorbed and solution phases. The properties of the organic and soil mediums are as given for the calculations in the unsaturated zone; $n = 30\%$, $K_d = K_{oc}f_{oc} = 0.56$

mL/g, and $\rho_s = 2.65$ g/cm^3. Equation 19.35 becomes

$$\text{Total mass} = C_w \times 0.3 + 0.56 \times C_w \times 1.85$$
$$= 0.3\,C_w + 1.036\,C_w = 1.336\,C_w$$

$$\frac{\text{Mass in sorbed phase}}{\text{Total mass}} = \frac{1.036\,C_w}{1.336\,C_w} = 77\%$$

that is, 77% in sorbed phase, 23% in solution.

19.6 Air-Permeability Testing

A vapor-extraction system typically consists of a blower connected to a series of wells completed in the unsaturated zone. The performance of a venting system will depend largely on the chemical composition of the contaminant, vapor flow rates through the unsaturated zone, and the flow path of carrier vapors relative to the location of the contaminants (Johnson and others, 1990). Typically if the permeability to air is less than one darcy (10^{-12} m^2), flow rates may be too low to achieve successful removal in reasonable time frames (Cohen and Mercer, 1993). Consequently air-permeability tests are frequently required in the predesign studies.

Figure 20.26 shows a typical setup for an air-permeability test. The objective with the test is to remove air from the extraction well and make measurements of the transient subsurface pressure distribution. Thus, the procedures are quite similar to those employed in well hydraulics. It is important here to test the interval that will be vented during the actual operation. For shallow settings, pressure in a vapor monitoring probe is described mathematically (Johnson and others, 1990) as

$$P' = \frac{Q}{4\pi m(k/\mu)}\left[0.5772 - \ln\left(\frac{r^2n_g\mu}{4kP_a}\right) + \ln t\right] \quad (19.36)$$

where Q is a constant volumetric gas flow rate, P' is a gauge pressure measured at some distance r and time t, m is the thickness of the unit, r is the distance from the vapor extraction well, n_g is the air-filled porosity, t is time, μ is the viscosity of air (1.8×10^{-4} g/cm-s), P_a is the ambient atmospheric pressure (1 atm = 1.013×10^6 g/cm-s), and k is the permeability to air. Note that the term k/μ simplifies $k\rho g/\mu$, where ρg is assumed equal to one.

Johnson and others (1990) suggest two procedures to analyze test results. In all cases, Eq. 19.36 indicates that a plot of P' versus $\ln t$ would provide a straight line with slope A and y intercept B where

$$A = \frac{Q}{4\pi m(k/\mu)} \quad (19.37)$$

$$B = \frac{Q}{4\pi m(k/\mu)}\left[-0.5772 - \ln\left(\frac{r^2n_g\mu}{4kP_a}\right)\right] \quad (19.38)$$

If Q and m are known, Eq. 19.37 is used directly to calculate the air permeability. This calculation is illustrated with Example 20.6. If Q and m are unknown

$$k = \left(\frac{r^2 n_g \mu}{4P_a}\right) \exp\left(\frac{B}{A} + 0.5772\right) \quad (19.39)$$

During the air-permeability tests, at least one pore volume of vapor should be removed. This assures that all the free vapor in the pores will be removed prior to setting up the vapor extraction system. Normally, extraction rates will be initially very high, diminishing considerably as the total withdrawal approaches one pore volume. These latter lower rates of extraction are somewhat indicative of how the system will actually perform during the remediation.

The amount of vapor V_p contained in one pore volume is

$$V_p = n_g \pi R^2 H \quad (19.40)$$

where R is the radius of the contaminated zone and H is its thickness. The time to remove one pore volume can be determined by dividing the quantity by the extraction rate Q.

19.7 Recognizing DNAPL Sites

Often, success in cleaning up a contaminated site depends on recognizing whether NAPLs are present or not. This task is not as simple as it might appear, especially with DNAPLs. At many DNAPL sites, one also may be limited in the number of wells that can be installed because of the danger of creating permeable vertical pathways that could facilitate further product migration. A catalog of direct and indirect approaches for recognizing potential DNAPL sites has been developed mainly through efforts of researchers at the Waterloo Centre for Groundwater Research. Following below are the most commonly used criteria to establish whether pure-phase organic compounds are present.

1. *Site use and history,* as compiled during a site investigation, provide useful indirect evidence of the possibility that DNAPLs could be discovered. Clearly, if records show that DNAPLs have been used at a site, or the type of industrial processes indicate an application for solvents, then these compounds could show up as contaminants. A following section will illustrate how this approach has been systematized for site studies.

2. *Ground-water zone sampling for free product* involves sampling of monitoring wells to detect free product. DNAPLs, present, say, in pools, can be discovered when product is sufficiently mobile to flow into and to accumulate in monitoring wells. According to U.S. EPA (1992), the presence of DNAPLs in monitoring wells can be confirmed by NAPL–

water interface probes, inspection of the pumped fluid or samples collected by transparent bottom-loading bailers, or depth-discrete samplers.

3. *Anomalously high dissolved concentrations* are often observed at depth in the vicinity of a possible DNAPL source. A useful yardstick for determining whether the measured concentration of an organic contaminant is anomalous requires an estimate of the effective solubility. Recall from Section 19.1 that for a mixture of organic compounds, the effective solubility is the maximum concentration of a particular compound that one would find at equilibrium. The effective solubility is determined using Eq. 19.9 or $S_i^e = X_i S_i$.

   One may infer the presence of DNAPLs if the dissolved phase concentration is greater than 1% of S_i^e (U.S. EPA, 1992). Thus, for example, if the S_i^e for TCE was 110 mg/L, the presence of DNAPLs would be suggested by dissolved concentrations greater than 1.1 mg/L.

4. *Soil sampling* is another direct approach for NAPL identification. Soil samples collected from above and below the water table can be examined for the presence of NAPLs. Cohen and Mercer (1993) discuss several approaches to NAPL detection at a site including OVA analysis, visual detection, UV fluorescence analysis, and various shake or dye-shake tests. They point to UV fluorescence tests (for fluorescent NAPLs) and hydrophobic dye methods as being most effective. The rule of thumb is that the presence of NAPLs is indicated when laboratory analysis exceed 10,000 mg/kg (>1% of soil mass) (Cohen and Mercer, 1993).

   Another approach involves taking the measured concentration of the organic compound in the saturated soil (μg/g) and determining what the theoretical concentration would be if the contaminant were present as a dissolved phase (Feenstra and others, 1991). The calculation, however, needs to account for the possibility of sorption onto the solid phase. The theoretical pore-phase concentration is given as (Feenstra and others, 1991)

$$C_w = \frac{M\rho_b}{K_d \rho_b + n_w} \quad (19.41)$$

where C_w is the theoretical concentration in the pore water, M is the total mass of the contaminant measured in the soil sample (μg/g dry weight), K_d is the distribution coefficient (cm^3/g), n_w is the water-filled porosity (volume fraction), and ρ_b is the bulk density (g/cm^3). One compares the theoretical concentration (C_w) to the effective solubility (S_i^e). If $C_w > S_i^e$, then DNAPLs are likely present. If $C_w < S_i^e$, then DNAPLs are likely not present. Following is an example that illustrates this calculation.

Example 19.4

A solvent spill is comprised of 75% TCE (mole fraction) and 25% PCE. Soil samples from the site contained TCE concentrations of 3000, 300, and 30 mg/kg. The organic carbon content of the soil is thought to be relatively high. Using values for the unknowns given below, determine which of these three sampling sites is likely to have TCE present as DNAPL.

$$S_i^e = (0.75)(1100 \text{ mg/L}) = 825 \text{ mg/L}$$

$$K_{ow} = 200 \text{ and log } K_{ow} = 2.3$$

$$\log K_{oc} = 2.3 - 0.21 = 2.09$$

$$K_{oc} = 124 \text{ mL/g}; \quad \rho_b = 2.0 \text{ g/mL}$$

$$f_{oc} = 0.03 \text{ mg/mg}; \quad \theta_w = 0.30$$

$$K_d = 124 \cdot 0.03 = 3.72 \text{ mL/g}$$

$$C_w = \frac{C_t \cdot 2.0}{[3.72 \cdot 2.0 + 0.30]} = 0.26 C_t$$

(1) $C_t = 3000$ and $C_w = 780$, which is about the same as S_i^e, possible DNAPL site

(2) $C_t = 300$ and $C_w = 78$, which is much less than S_i^e, not DNAPL site

(3) $C_t = 30$ and $C_w = 7.8$, which is much less than S_i^e, not DNAPL site

Systemic Screening Procedure

The U.S. EPA has developed a systematic approach for identifying the possibility of DNAPL contamination at a site. It is based on the screening criteria and calculations outlined above. The approach combines historical information about site use (Figure 19.27) with site-specific information (Figure 19.28) to form a summary evaluation (Figure 19.29). This evaluation provides three categories including: (I) DNAPLs are confirmed or the potential is high, (II) there is a moderate potential for DNAPLs, or (III) there is a low potential for DNAPLs at the site.

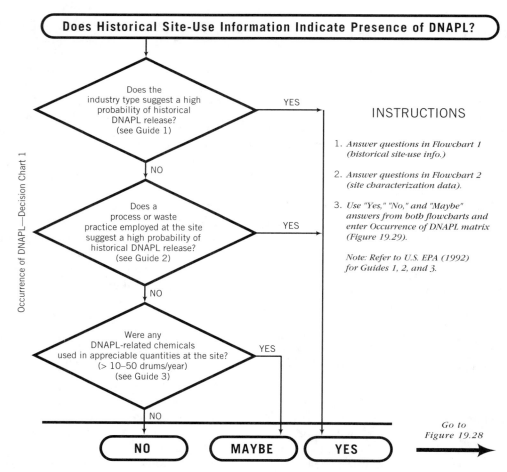

Figure 19.27 Assessment of the potential for DNAPLs to be present based on information on how the site has been used (modified from U.S. EPA, 1992).

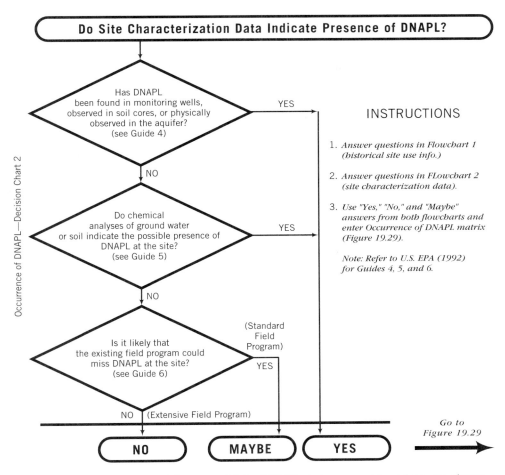

Figure 19.28 Assessment of the potential for DNAPLs to be present based on site characterization data (modified from U.S. EPA, 1992).

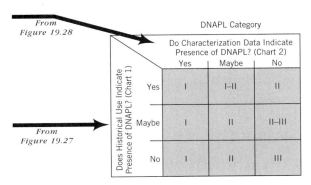

Figure 19.29 Summary evaluation of the potential for DNAPL occurrence (from U.S. EPA, 1992).

Problems

1. There are 100 g/m³ of some organic constituent in an aquifer. The solubility of the constituent is 1000 μg/L and the drinking water standard is 5 μg/L. Calculate the potential volume of water that can be rendered undrinkable.

2. Two immiscible fluids are present in the subsurface. The saturation of the wetting fluid is 0.8. From Figure 19.5, for the imbibition curve determine

a. the relative permeability with respect to the wetting fluid
b. the relative permeability with respect to the nonwetting fluid
c. at what saturation the relative permeability of the wetting fluid equals zero.
d. at what saturation the relative permeability of the nonwetting fluid equals zero.

3. Given a spill of 0.1 m³ of product, 35% porosity, and a residual NAPL saturation of 5%, calculate the volume of aquifer containing the residual.

4. Assuming a 10 × 10 cm cross-sectional flow path, calculate the depth of the residual.

5. Consider an aquifer contaminated with an aqueous phase and a sorbed phase. The organic content f_{oc} is 0.0005, the porosity is 35%, bulk density is 1.6 g/cm³, and K_{oc} is 126 mL/g. For a unit volume of aquifer, calculate the percentage of mass in the solution (aqueous) phase and the percentage of mass in the sorbed phase.

6. Consider an organic compound with the following properties: molecular weight 153 g/mol, density 1.59 g/cm³, partial vapor pressure in the gas phase 90 mm Hg, solubility 785 mg/L. Consider the following properties for a soil contaminated with this contaminant. Porosity 0.3, residual water content 0.1, residual DNAPL 0.05, gas phase content 0.15, solid-phase organic content (f_{oc}) 0.001, bulk density 1.5 g/cm³, K_{oc} 430 mL/g, temperature 20°C (293°K). Using the actual solubility (785 mg/L) in the calculations, determine the mass in the NAPL phase, the gas phase, the aqueous phase, and the sorbed phase for a bulk volume of 1 m³. Note: one atmosphere = 760 mm Hg.

7. An organic solvent has a vapor pressure of 0.1 atmospheres. It has a solubility of 1500 mg/L. Within a contaminated soil, a gas concentration of 600 μg/L is noted. The soil has a residual water content of 0.12. Calculate the effective solubility and the mass of this constituent in the aqueous phase. Assume a molecular weight of 100 g/mole.

8. Benzene has a Henry's law constant of 5.5×10^{-3} atm-m³/mole. From the data presented in Example 19.2, calculate the retardation of benzene due to partitioning to the aqueous phase. If the concentration of the water was 200 μg/L and the mass sorbed on the solids was 400 μg/kg, calculate the retardation due to partitioning to the solid phase.

9. On diagrams (a) and (b) shown in the figure for this problem, illustrate the contaminant distribution for the organic compounds with the specified chemical properties. Consider all important spreading mechanisms.

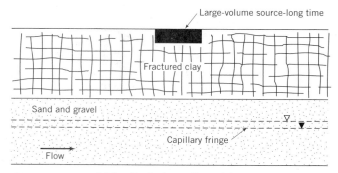

(*a*) Gasoline Leak over Fissured Clay
Two soluble and volatile components of concern in gasoline
benzene—log K_{ow} = 2.04: solubility 1780 mg/L
ethylbenzene—log K_{ow} = 3.15: solubility 140 mg/L

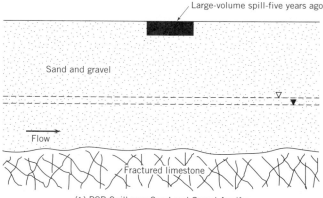

(*b*) PCB Spill over Sand and Gravel Aquifer
PCBs slightly soluble in water; log K_{ow} = 4 to 6; not volatile; specific gravity = 1.5

Problem 9

Chapter 20

Remediation: Overview and Removal Options

20.1   Containment
20.2   Management Options
20.3   Overview of Methods for Contaminant Removal
20.4   Pump and Treat
20.5   Interceptor Systems for NAPL Recovery
20.6   Soil-Vapor Extraction
20.7   Air Sparging
20.8   Case Studies in Site Remediation

Thus far, we have examined the causes of ground-water contamination, how contaminants are transported, and how to characterize transport in terms of key parameters. What is left for this and the following chapter are questions of how to correct a contamination problem. There are generally four main alternatives for dealing with contamination problems: (1) containing of the contaminants in place, (2) attenuating the possible hazard by institutional controls, (3) removing contaminants from the ground altogether, and (4) treating the contaminants in situ. This chapter will consider the first three in varying detail. In situ methods for contaminant treatment, especially methods for bioremediation, will be discussed in Chapter 21. Table 20.1 summarizes the remedial technologies that we will discuss in this and the following chapter.

20.1   Containment

The containment options are control measures that keep the contaminants in the ground but prevent further spread. They are often standard geotechnical approaches that have been used historically to control ground-water seepage but have been adapted to problems of contaminant control.

Slurry Walls

Slurry walls are low-permeability barriers emplaced in the subsurface. They confine contaminants by either surrounding the entire spill (Figure 20.1) or by removing the potential for flow through the source. With the trench method, a trench is dug through a bentonite slurry to the desired depth (Figure 20.2). The wall is solidified either by incorporating excavated material with the bentonite or by adding cement to the bentonite slurry. Walls are from about 0.5 to 2 m wide and can be installed to depths of about 50 m.

Walls also can be constructed using various specialized equipment. With the vibrating-beam method, a steel plate is forced into the ground. As this plate is gradually removed, bentonite is injected through the plate to fill the space that was created. The slurry walls created in this way are narrower in width than a traditional wall and emplaced at a shallower depth (Knox and others, 1984). Because the walls are thin and placed in segments, the barrier may leak.

A California company has developed the S.M.W.™ cut-off wall technique, which is an in situ mixing technology to repair highly permeable zones that sometimes develop in conventionally installed slurry-trench walls. This approach uses three hollow-stem augers both to drill and

Table 20.1 **Tabulation of the Remedial Alternatives Discussed Here and in the Following Chapter**

Containment	Contaminant Removal
Slurry walls	Excavation and ex situ
Sheet pile walls	treatment
Grouting	Pump and treat
Surface seals and surface	Interceptor systems for NAPL
drainage	recovery
Hydrodynamic controls	Soil-vapor extraction
Stabilization and solidification	Air sparging
Management Options	**In Situ Destruction**
(See Table 20.2)	Biodegradation
	Reactive barriers

to inject slurry. Paddles on the augers mix the bentonite slurry with the native soils. Walls can be installed or repaired to depths of about 60 m.

Sheet Pile Walls

Sheet pile cutoff walls are constructed by driving interlocking piles into the ground to form a continuous barrier (Rogoshewski and others, 1983). Most commonly, steel piles are used. The joints between conventional piles usually leak until fines migrate and fill the joints between piles. A new technique, however, has been developed at the University of Waterloo for sealing the joints in steel piles. Field tests at Canadian Forces Base Borden have shown sealable-joint pile walls to be extremely effective in controlling contaminant migration.

Grouting

Grout walls can also control the migration of contaminants. A grout wall or curtain is constructed by injecting special liquids under pressure into the ground. The grouting material moves away from the zone of injection, gels or solidifies, and ultimately reduces the permeability of the geologic material (Rogoshewski and others, 1983).

The distance of grout penetration varies from site to site but can be relatively small, requiring closely spaced injection holes (for example, 1.5 m). In practice, holes are drilled in two or three staggered rows to ensure a more or less continuous barrier. Typical grouting fluids include cement, bentonite, or specialty fluids like silicate or lignochrome grouts (Rogoshewski and others, 1983).

Grouting techniques are most useful for sealing fractured rocks, where it is not possible to construct a low-permeability barrier in any other way. The high cost and potential contaminant/grout interactions, however, limit the applicability. In addition, most grouts can be injected only in materials with sand-sized and larger grain sizes (Knox and others, 1984).

Surface Seals and Surface Drainage

Most waste containment approaches require surface controls to work effectively. Surface seals and drains are used together to control infiltration moving downward through the contaminant source. The reduction of infil-

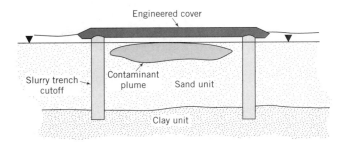

Figure 20.1 Example of a dissolved contaminant plume contained by slurry walls (modified from Knox and others, 1984).

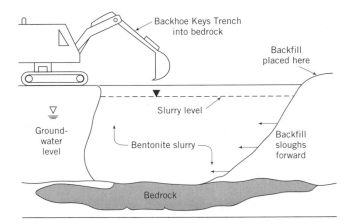

Figure 20.2 Schematic of the preparation of a slurry wall (from U.S. EPA, 1985).

tration passing through contaminated source materials decreases the volume of dissolved contaminants that may be generated. In addition, covering and controlling the drainage will reduce the possibility of exposure and erosion of contaminated materials, as well as the rate of transfer of volatiles or other hazardous materials (for example, asbestos) into the atmosphere.

Surface seals are a common element in many different types of waste-management problems. Given the relative ease of constructing a surface seal, a variety of materials can be used. Examples of cover materials include compacted clay and silty clay; mixtures of natural soils and stabilizers such as cement; bitumen or fly ash; bentonite layered with other natural materials; sprayed bituminous membranes; synthetic membranes [e.g., high-density polyethylene (HDPE) or polyvinyl chloride (PVC)] again used in combination with other materials; and finally all sorts of waste materials such as furnace slag, incinerator residue, fly ash, and clinkers (Rogoshewski and others, 1983). Figure 20.3 shows examples of these surface liners used at hazardous waste sites in North America.

The technology for sealing sites is proven. The main source of uncertainty in the long-term functioning of a

seal is the quality of ongoing management. A seal can be damaged by erosion from runoff; cracking due to desiccation, frost effects, or subsidence; punctures created by burrowing animals or plant roots; and deterioration of synthetic membranes by tearing or exposure to sunlight. Thus, seals need to be inspected and repaired on an ongoing basis.

The surface of seals are graded to provide lateral drainage to ditches. Water is moved off the surface seal rapidly to minimize infiltration, while also minimizing erosion. General specifications for the design of a landscape in terms of slope lengths and gradients are provided by Rogoshewski and others (1983). They also present guidance in designing, constructing, and maintaining the different elements of a network including drainage ditches, diversions (a combined ditch and dike constructed across a slope), grassed waterways, and drainage benches (terraces built across long slopes to reduce the steepness).

Hydrodynamic Controls

Pumping or injection of water can control the spread of contaminants by controlling hydraulic or pressure gradients. These approaches isolate the contaminants without removing them. For example, lowering the water table may prevent discharge of contaminants to rivers or lakes and reduce the rate of contaminant generation by desaturating the waste. Similarly, a plume of dissolved contaminants can be confined in a potentiometric low created by an appropriate combination of pumping and injection wells (Rogoshewski and others, 1983).

The design approach is a straightforward extension of well-hydraulics concepts developed previously. The containment strategy will work in most cases given sufficient care in designing and monitoring the operation of the system. Of the techniques considered so far, this one requires the greatest ongoing supervision. Wells and pumps need to be kept properly serviced and pumping rates need to be adjusted to respond to changing groundwater conditions. For example, above-average recharge may raise water levels and destroy confinement.

Another example of hydrodynamic control is the use of positive or negative differential pressure systems to control vapor migration (O'Connor and others, 1984a). With some hydrocarbon spills, the free product is not as much of a problem as the rapid movement of vapors. Even when the free product has been removed, the vapors may continue to be a problem at a site (O'Connor and others, 1984a). Vapors migrating into buildings can pose a health threat when breathed for long periods and if accumulated to sufficient concentrations can explode. Adding or removing air provides a way to control contaminant spread in the vapor phase. Figure 20.4a shows how to keep volatiles out of a structure using both positive- and negative-pressure systems (O'Connor and others, 1984a).

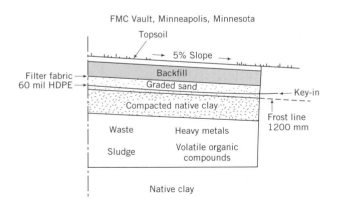

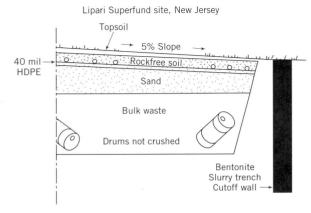

Figure 20.3 Examples of engineered covers (from Mclelwain and Reades, 1985). Reprinted by permission of Second Canadian/American Conference on Hydrogeology. Copyright © 1985. All rights reserved.

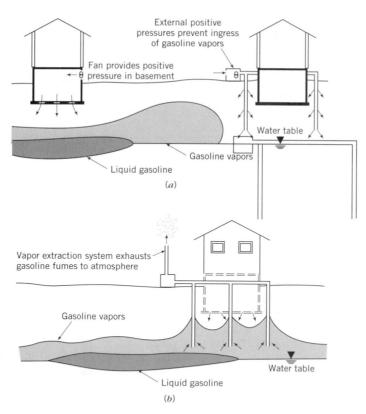

External positive pressures prevent ingress of gasoline vapors

Fan provides positive pressure in basement

θ

θ

Water table

Gasoline vapors

Liquid gasoline

(a)

Vapor extraction system exhausts gasoline fumes to atmosphere

Gasoline vapors

Liquid gasoline

Water table

(b)

Figure 20.4 Control of subsurface gasoline vapors using (*a*) positive and (*b*) negative differential pressure systems (from O'Connor and others, 1984a). Reprinted by permission of Conference on Petroleum Hydrocarbons and Organic Chemicals in Ground Water—Prevention, Detection and Restoration. Copyright © 1984. All rights reserved.

Stabilization and Solidification

Solidification and stabilization techniques involve treating contaminated soil (commonly lagoon or pond sludges) to improve the physical characteristics of the material (for example, so that it might be moved) and to reduce the leachability and mobility of the contaminants. Stabilization attempts to bind the contaminants to the solid, effectively reducing their solubilities. By making a more robust solid out of the contaminated material, additional benefits in material performance can be achieved by reducing the permeability.

Materials are usually processed in place with mixing augers or digging tools adapted to mix the additives with the contaminated soils. Common additives include (1) cement, (2) lime spiked with fly ash or sodium silicate, (3) asphalt or bitumen, or (4) various organic polymers. In some applications, combinations of additives are used together. By adjusting the quantity of additives, one can provide for permanent solidification at the site (monolithic treatment) or for material appropriate for transport after treatment (loose structure treatment). Test results indicate that the hydraulic conductivity of soils was re-

duced from 1.8×10^{-2} cm/s to 4.2×10^{-7} cm/s following treatment with a cement–clay combination. The method appears to be most effective in the stabilization of metals and organic contaminants like PCBs.

20.2   *Management Options*

The management options (Table 20.2) are institutional actions to avoid health-risks problems or to shut down a source of contamination without removing the contamination from the ground. These alternatives are straightforward and will not be examined in detail.

20.3   *Overview of Methods for Contaminant Removal*

Contaminant removal is usually an integral part of most practical strategies for site remediation. Four approaches are commonly used including excavation, pumping, interceptor systems, and soil venting.

Table 20.2 **Management Alternatives for Dealing with Ground-Water Contamination**

Management options: Management options are usually applied either to prevent further contamination or to protect potential exposure points from contaminated ground water. These methods thus focus on sources and exposure points rather than on the contaminants per se. The methods also tend to be institutionally based rather than technology based.

1. *Limit/terminate aquifer use:* Limits access or exposure of receptors to contaminated ground water.
2. *Develop alternative water supply:* Involves the substitution of contaminated ground water with alternative supplies (for example, surface-water diversions and/or storage, desalination, and new wells).
3. *Purchase alternative water supply:* Includes bottled water and water imports.
4. *Source removal:* Involves the physical removal of the source of contamination and includes measures to eliminate, remove, or otherwise terminate source activities; could also include modification of a source's features (for example, operations, location, or product) to reduce, eliminate, or otherwise prevent contamination.
5. *Monitoring:* Involves an active evaluation program with a "wait and see" orientation.
6. *Health advisories:* Involves the issuance of notifications about ground-water contamination to potential receptors.
7. *Accept increased risk:* Involves the decision to accept increased risk; is usually a "no action" alternative.

Modified from Office of Technology Assessment (1984).

Excavation and Ex Situ Treatment

Excavation is a straightforward corrective action. In its simplest form, it is an exercise in digging and trucking. The greatest problems are health risks associated with potential human exposure to vapors and particles, costs involved, and finding an appropriate place to dispose of the contaminated soils. A variation on the method, ex situ treatment, involves removing the contaminants from the soils and replacing the treated material at the site. Examples of these ex situ treatment techniques include soil washing, biomounding, low-temperature thermal desorption, and high-temperature incineration. Given our interest here in the hydrogeologic aspects of site remediation, we will not discuss these approaches further.

Pump and Treat

Pump-and-treat schemes remove NAPLs and aqueous-phase contaminants from the ground using wells. This technique has been successful in a variety of applications and site conditions. Because of its importance, we will discuss issues of both design and operation of pump-and-treat systems in detail in Section 20.4.

In developing pumping alternatives, there is always the problem of what to do with the contaminants removed from the ground. Of necessity, on-site treatment is required before injecting the water to the subsurface or releasing it to surface-water bodies. Extensive discussions of physical and chemical–biological treatment methods are given in Knox and others (1984).

Interceptor Systems

Interceptor systems involve the use of drains, trenches, and lined trenches to collect contaminants close to the water table. Knox and others (1984) define a drain as a line of buried perforated pipe or drainage tiles. A trench is an excavation that is either open or backfilled with gravel to promote wall stability. Typically, interceptor systems are installed close to the ground surface and 1 or 2 m below the water table. In the case of dissolved contaminants, the drains or trenches operate as an infinite line of wells and are thus efficient at removing shallow contamination. We will examine this approach in detail in Section 20.5, particularly as it relates to the recovery of LNAPLs.

Soil-Vapor Extraction

Soil-vapor extraction (SVE) removes volatile organic compounds from the unsaturated zone by passing large volumes of air through the zone of contamination. Air flow through the spill is maintained by vacuum pumping one or more wells (Figure 20.5). Surface treatment is usually required to remove contaminants from the air before release to the atmosphere (Rogoshewski and others, 1983).

SVE is the leading method for cleaning up fuel spills and industrial solvents (that is, BETX compounds, and VOCs) from unsaturated settings. It provides one of the few alternatives to soil removal in the unsaturated zone where pure organic liquids often occur at residual saturation. The method is able to treat relatively large volumes

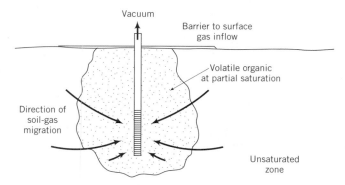

Figure 20.5 Use of a vacuum system for removing volatile organic compounds from the unsaturated zone (from Schwartz, 1988). Reprinted from J. Hydrol, v. 100, by F. W. Schwartz, Contaminant hydrogeology—dollars and sense, p. 453–470, 1988 with kind permission from Elsevier Science, NL, Sara Burgerhartstraat 25, 1055 KV Amsterdam, The Netherlands.

of material with minimal disruption at the surface. Gas stations or airports with fuel spills can be operated in a normal manner as the cleanup proceeds. SVE will be discussed in detail in Section 20.6.

20.4 Pump and Treat

Pump-and-treat schemes are used to clean up (1) dissolved contaminants present as plumes, (2) NAPLs, or (3) sites with both types of contamination. Systems consist of three major components. Recovery wells remove contaminated water or NAPLs from the ground. Treatment equipment on the surface separates produced water and NAPLs and removes dissolved contaminants that remain in the process water. Monitoring wells establish how well the system is meeting design and cleanup objectives.

Pump-and-treat systems are designed to effect either (1) substantially complete removal of contaminants from the subsurface or (2) hydraulic control of a plume without significant mass removal. With mass removal in mind, the system cleans up the contamination within some specified time. With hydraulic control systems, the system is run in perpetuity to keep contamination from spreading. If the recovery wells are ever shut down, the contamination problem will redevelop because most of the contaminant mass remains in the ground.

The Problem of Pump and Treat

When used as intended, pump-and-treat schemes are a cost-effective alternative for cleaning up plumes of dissolved contaminants. They have received bad press in recent years because projects that started out as mass-removal schemes actually functioned as hydraulic control systems. Sites were not being cleaned up as expected,

much to the chagrin of those paying the bills and the regulators.

Figure 20.6 depicts two idealized monitoring records for a well being used to recover dissolved contaminants. First, consider the reduction in concentration in an ideal well that is shown as the "theoretical removal" curve. The concentration eventually falls to zero once the well removes a volume of water approximately equivalent to the volume of contaminated water (Figure 20.6). Mathematically, the time required for a perfect pump-and-treat system to clean up a plume is given as,

$$t_c = \frac{V_T}{\Sigma Q_i} \qquad (20.1)$$

where V_T is the total volume of contaminated water and Q_i is the pumping rate from each of i recovery wells. The following example illustrates how this equation is used.

Example 20.1 (adapted from Mercer and others, 1990)

A ground-water plume has an areal extent of approximately 41,490 m² and is being cleaned up by a three-well system, with each well being pumped at 0.13 m³/min. Assume that the plume is 16.7 m thick in an aquifer having a porosity of 0.3. Ideally, how long will it take to clean up this plume?

$$t_c = \frac{41,490 \times 16.7 \times 0.3}{3 \times 0.13} = 5.33 \times 10^5 \text{ min} = 1 \text{ yr}$$

Pump-and-treat systems often exhibit tailing, exemplified by the second curve in Figure 20.6. With tailing, the concentration of contaminants declines but never reaches zero, even after two or more plume volumes are pumped.

There are three main causes of tailing: the presence of NAPLs, the effects of sorption, and heterogeneity in hydraulic conductivity (Keely, 1989). Figure 20.7 illustrates a common situation where a pump-and-treat system

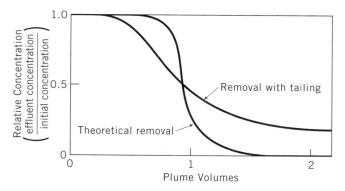

Figure 20.6 Effects of tailing on number of plume volumes that need to be removed by the pump-and-treat system (modified from Mercer and others, 1990; and Keeley, 1989).

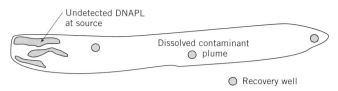

Figure 20.7 Conceptualization of a pump-and-treat system for DNAPL recovery being operated without knowledge of the existence of residual and pooled DNAPL in the vicinity of the source.

is being used to clean up a dissolved organic contaminant in a large plume. In the vicinity of the source, the pure DNAPL remains undetected, and the bulk of contaminant resides in the DNAPL phase. A relatively small proportion of the total contaminant mass is dissolved in the plume. Thus, the pump-and-treat system would only remove a small proportion of the total contaminant mass and is only effective in controlling further spreading. If the recovery wells were shut off, the plume would form again.

This hypothetical example illustrates why it is so important to establish whether NAPLs occur at a site. When NAPLs are present, the source must be isolated using barriers or wells if a pump-and-treat system is to be effective. A variety of other source-intensive strategies could then be pursued to remove the NAPLs.

The possibility for contaminants to be sorbed to the porous medium also reduces the efficiency of pump-and-treat systems. Sorption effectively reduces the velocity of contaminant transport relative to water. Thus, if a contaminant is strongly sorbed, many "plume" volumes need to be pumped before the sorbed mass is fully removed. A simple modification of Eq. 20.1 provides an estimate of the cleanup time when the contaminant is sorbed, or

$$t_c = \frac{R_f V_T}{\sum Q_i} \qquad (20.2)$$

where R_f is the retardation factor for the contaminant (defined as Eq. 18.7). Let us return to Example 20.1 and consider a case with a sorbing contaminant.

Example 20.2

Assume that the contaminant in Example 20.1 sorbs with a K_d of 0.2 mL/g. Assume a solid density of aquifer materials is 2.65 g/cm³. Determine by how much the cleanup time would be increased.

The retardation factor from Eq. 18.7 is given as

$$R_f = 1 + \left(\frac{1-n}{n}\right)\rho_s K_d = 1 + \left(\frac{1-0.3}{0.3}\right)2.65 \cdot 0.2$$

$$= 2.23$$

Thus, the cleanup would now take 2.23 years instead of 1 year.

The efficiency of pump-and-treat systems will also be reduced in highly heterogeneous media. Over long times, dissolved contaminants will diffuse from high-permeability zones into low-permeability zones (Figure 20.8). Once cleanup begins, wells will rapidly remove dissolved mass from the zones of higher permeability. However, contaminants within low-permeability zones will be transported back out only by the slow process of diffusion. Water pumped from the wells will contain low levels of contamination for relatively long times.

In summary, pump-and-treat systems for dissolved contaminants work best when the rate of cleanup is controlled by advection. If rate-limiting processes become controlling, such as DNAPL dissolution rates or diffusion rates, then pump-and-treat systems will not be efficient at mass removal.

Technical Considerations with Injection–Recovery Systems

Contaminants dissolved in water are relatively easy to deal with in terms of well design. Because water containing dissolved contaminants is being pumped, water well technology is usually appropriate. Thus, ideas presented in Chapter 6 on constructing water wells are directly transferable. Some care is required in selecting casing materials and pumps that will not fail due to chemical interaction with the contaminant.

The cleanup of NAPLs at the water table requires that the collection device accomplish two things: (1) create a cone of depression on the water table causing the product to flow toward the well, and (2) remove free product moving into the well. Wells designed for this purpose are usually one of three types. The single-pump system with one recovery well (Figure 20.9) is the simplest and least expensive (Blake and Fryberger, 1983). The pump is positioned in the well so that as it cycles on

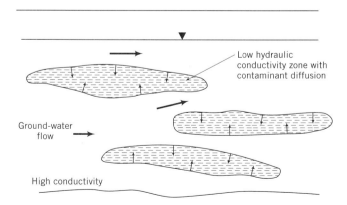

Figure 20.8 Idealization of difficulties in contaminant recovery in a heterogeneous aquifer with advective flushing through the high hydraulic conductivity zones and diffusion from the low hydraulic conductivity zones.

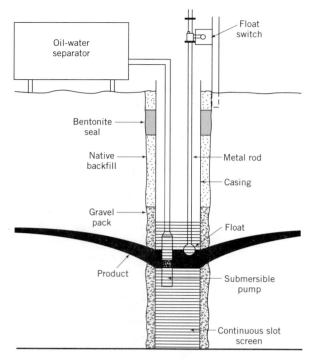

Figure 20.9 Example of a one-well, one-pump system for recovering NAPLs (from Blake and Lewis, 1982). Reprinted by permission of Second National Symposium on Aquifer Restoration and Ground Water Monitoring. Copyright © 1982. All rights reserved.

and off, both water and product are recovered. Pumping depresses the water table and removes the floating product. The biggest problem is that a high-speed pump emulsifies the oil in water. This mixture requires treatment, which at the minimum involves separating the oil and water. However, the water once separated contains soluble hydrocarbons that could require further treatment (Blake and Fryberger, 1983). Piston pumps or other noncentrifugal pumps can reduce the treatment because less fluid mixing is involved.

Other pumping schemes avoid surface treatment by preventing the oil and water from mixing during pumping. Figure 20.10 provides two examples of two-well, two-pump systems in which the deep water well screened below the spill lowers the water table and the shallow well collects the free product (Blake and Fryberger, 1983). However, installing two wells and two pumps pushes up the cost per installation. The one-well, two-pump design is a compromise design that is less expensive and avoids the large-volume treatment problem of the one-well, one-pump systems. Because two pumps and operational monitoring equipment are required in the same well, casing diameters need to be larger than either of the other two types of systems. Two variations on the one-well, two-pump design are shown in Figure 20.11.

The operation of recovery systems must be carefully monitored. For example, with the one-well, one-pump

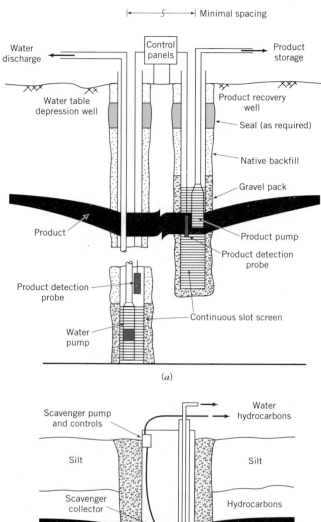

(a)

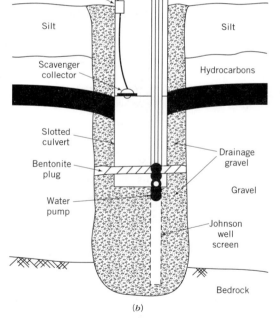

(b)

Figure 20.10 Examples of a two-well, two-pump system for the recovery of NAPLs. Panel (a) is from Blake and Lewis (1982) and is reprinted by permission of Second National Symposium on Aquifer Restoration and Ground Water Monitoring. Copyright © 1982. Panel (b) is modified from O'Connor and others (1984b) and is reprinted by permission of Conference on Petroleum Hydrocarbons and Organic Chemicals in Ground Water—Prevention, Detection and Restoration. Copyright © 1984. All rights reserved.

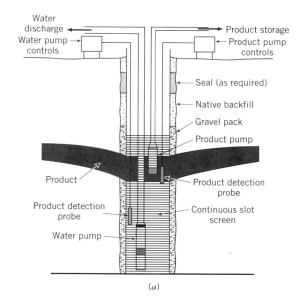

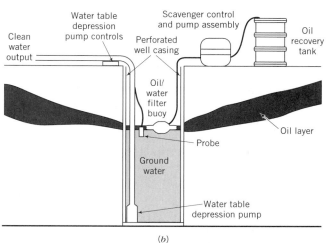

Figure 20.11 Examples of a one-well, two-pump system for the recovery of NAPLs. Panel (*a*) is from Blake and Lewis (1982) and is reprinted by permission of Second National Symposium on Aquifer Restoration and Ground Water Monitoring. Copyright © 1982. Panel (*b*) is modified from Yaniga and Mulry (1984) and is reprinted by permission of Conference on Petroleum Hydrocarbons and Organic Chemicals in Ground Water—Prevention, Detection and Restoration. Copyright© 1984. All rights reserved.

system, the pump must be placed at the proper depth and the float switch adjusted so that product is being pumped toward the end of the pumping cycle. With the two-pump systems, the water well must not be over-pumped to enable product to be pumped through the water pump (Blake and Fryberger, 1983). Failure of the product pump in a one-well, two-pump system could also cause product to accumulate in the well and eventually to reach the intake of the water pump. Part of the job of the control system is to shut off the water pump when free product is detected by sensors above the pump.

Blake and Fryberger (1983) provide additional guidance on operational and safety details that need to be considered, especially in selecting the proper equipment and working with potentially explosive liquids.

Research on the technical aspects of emplacing wells for the collection of DNAPLs is extremely limited. Ferry and others (1986) studied how the efficiency of removing DNAPLs from low-permeability bedrock depended on the pumping system. One pumping scheme involved simply removing the accumulated DNAPL at the bottom of the well with an air-activated purge pump. This approach of collecting DNAPL through gravity drainage into the well was not successful. Once the fluid in the well was removed, no other inflow occurred up to a day afterward. In their second test, a submersible pump installed higher in the well was used to produce a significant cone of depression. Further, they monitored inflow of DNAPL into the bottom of the well during pumping. In one case, after about 1140 min of pumping and a final drawdown of about 15 m, the level of DNAPL increased by about 1 m.

In general, experience suggests that one-pump and two-pump systems have a role to play in DNAPL recovery. However, in some settings, the one-pump system may not be the best choice or may not even work. At first glance, the tendency for DNAPL and water to move in different directions suggests that moving large volumes of water may not be helpful. However, creating a situation for a "natural water drive" can be helpful in moving DNAPL into a well.

Methods for Designing Pump-and-Treat Systems

The design of a pump-and-treat system attempts to meet the desired cleanup objective at some minimal cost. Cost is minimized by installing the smallest number of recovery wells that are pumped at some minimal rate. Thus, the design strategy is to minimize the overall cost of installing and operating the wells and treating the water at the surface. In practice, there are generally four basic design approaches: expanding pilot-scale systems, capture-zone type curves, trial-and-error simulations using models, and simulation-optimization techniques.

Expanding Pilot-Scale Systems
This approach involves starting recovery operations with a pilot system and monitoring how the system performs. Evaluation of the monitoring data provides a basis for expanding the pilot by selecting places where the next well will do the most good. In the case of NAPLs, this well could be placed where (1) thick zones of product are not being reduced by existing pumping or (2) the product is moving in a direction that cannot be tolerated, for example, toward wells, rivers, or buildings. For dissolved contaminants, the pumping wells could be placed where the plume continues to grow, or does not appear

to be shrinking in size. Injection wells are located to provide barriers to flow or to increase the velocity of flow to collection wells.

Systems designed in this manner rely on experience and a cautious, incremental expansion of the network to ensure that the remedy is effective and economical. This approach is used commonly with spills involving NAPLs. Although modeling approaches are available to optimize the design, a lack of available data usually limits the applicability of these techniques. The situation with respect to dissolved contaminants is better with possibilities of using several different analytical and numerical procedures.

Capture Zones

Most of the quantitative design approaches use the concept of a capture zone. A capture zone is the area surrounding a well that supplies water to that well. In effect, it is the zone of contribution to the well. Figure 20.12 illustrates the idealized shape of the capture zone for a well in a homogeneous aquifer. The capture zone, indicated by the shaded area, extends from the vicinity of the well far up-gradient. All contaminants in the capture zone should be recovered by the well (Figure 20.12).

The shape the capture zone develops results from the combined effects of drawdown caused by the well and uniform regional flow. In most cases, the full extent of a capture zone (that is, the steady-state capture zone) is rarely shown (Figure 20.13). Most often, a capture zone is described in a time context, such as the one-year capture zone or the five-year capture zone (Figure 20.13). The water in the volume enclosed by the time-related capture zone will be recovered by the well over that specified time. The width of the capture zone depends on the pumping rate. Pump the well harder and the capture zone gets wider.

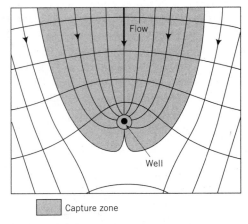

Figure 20.12 The effects of pumping superimposed on a regional flow system creates a capture zone (modified from Gorelick, 1987). Reprinted by permission of Solving Ground Water Problems with Models. Copyright © 1987. All rights reserved.

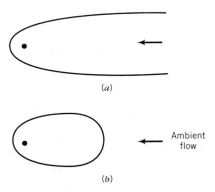

Figure 20.13 Illustration of (a) a steady-state and (b) a time-related capture zone associated with a single recovery well (from Blandford and Huyakorn, 1991).

Analytical Approaches to Defining Capture Zones

Javandel and Tsang (1986) describe an analytical approach to the design of contaminant recovery systems. The approach provides a way to minimize the number of injection or recovery wells, the pumping rates, and the distance between wells. The optimal scheme provides a capture zone that is slightly larger than the plume being collected.

We know from well hydraulics theory that a well in an aquifer can be pumped only at some maximum rate determined by the available drawdown. In terms of contaminant removal, pumping only one well at its maximum rate may not provide a capture zone that is large enough. The solution is to keep adding wells until the pumping capacity is sufficient to create a capture zone of the appropriate size. However, once more than one well is added, the spacing must be such to avoid contaminants passing between the two wells. Thus, with more than one well, pumping rates and optimum spacings between wells are determined.

Javandel and Tsang (1986) use complex potential theory as the basis for a simple graphical procedure to design a well system capable of creating a capture zone of the appropriate size. To develop a design requires the type curves for one to four wells (Figure 20.14), and values for two parameters—B, the aquifer thickness (assumed to be constant), and U, the specific discharge or Darcy velocity (also assumed constant)—for the regional flow system. The method involves the following five steps (Javandel and Tsang, 1986).

1. A map of the contaminant plume is constructed at the same scale as the type curves. The margin of the plume should be indicated along with the direction of regional ground-water flow.

2. The type curve for one well is superimposed on the plume, keeping the x axis parallel to the direction of regional ground-water flow and along the midline of the plume. Approximately equal proportions of

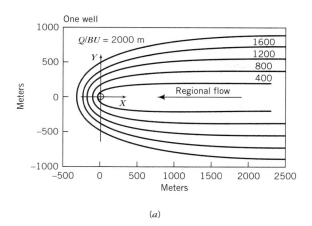

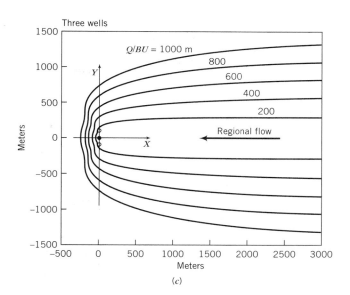

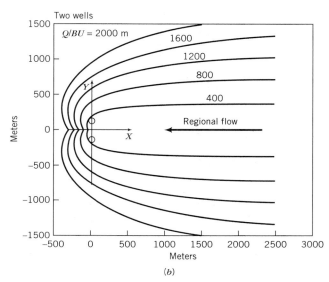

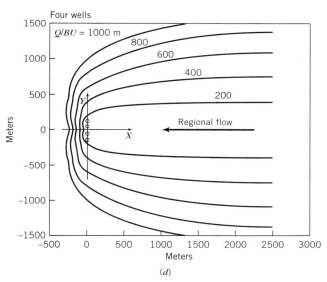

Figure 20.14   Capture-zone type curves for one, two, three, and four wells (from Javandel and Tsang, 1986). Reprinted by permission of Ground Water. Copyright © 1986. All rights reserved.

Figure 20.14   Continued

the plume should lie on each side of the *x* axis. The well on the type curve will be at the downstream end of the plume. The type curve is adjusted so that the plume is enclosed by a single *Q/BU* curve.

3. The single well pumping rate (*Q*) is calculated using the known values of aquifer thickness (*B*) and the Darcy velocity for regional flow (*U*) along with the value of *Q/BU* indicated on the type curve (*TCV*) with the equation

$$Q = B \cdot U \cdot TCV \qquad (20.3)$$

4. Next, check whether the well can support the calculated pumping rate over the estimated time required for the cleanup. If the production rate is feasible, one well is required with pumping rate *Q*. If the required production is not feasible due to a lack of available drawdown, additional wells need to be added (go to step 5).

5. Repeat steps 2, 3, and 4 using the two, three, or four well-type curves in that order, until a pumping rate is determined for each well that the aquifer can support. The optimum spacing between wells is calculated using the following formulas:

two wells: $\dfrac{Q}{\pi BU}$

three wells: $\dfrac{1.26Q}{\pi BU}$

four wells: $\dfrac{1.2Q}{\pi BU}$

Be sure to account for the interference among the pumped wells in the drawdown calculations. The wells are always located symmetrically around the x axis, as the type curves show.

Reinjecting the treated water produced by the wells will accelerate the rate of aquifer cleanup. The procedure is the same as that just discussed except the type curves are reversed and the wells are injecting instead of pumping. The injection wells should be moved slightly upstream of the calculated location to avoid causing parts of the plume to follow a long flow path.

Javandel and Tsang's rule of thumb is to place a well or line of wells half the distance between the theoretical location and the tail of the plume. The following example taken from Javandel and Tsang (1986) illustrates how the technique is used.

Example 20.3

Shown in Figure 20.15 is a plume of trichloroethylene (TCE) present in a shallow confined aquifer having a thickness of 10 m, a hydraulic conductivity of 10^{-4} m/s, an effective porosity of 0.2, and a storativity of 3×10^{-5}. The hydraulic gradient for the regional flow system is 0.002, and the available drawdown for wells in the aquifer is 7 m. Given this information, design an optimum collection system.

Values of B and U are required for the calculation. B is given as 10 m, but U needs to be calculated from the Darcy equation:

$$U = K \,\text{grad}(b) \quad \text{or} \quad U = 10^{-4} \times (0.002)$$
$$= 2 \times 10^{-7} \text{ m/s}$$

Superposition of the type curve for one well on the plume (Figure 20.16a) provides a Q/BU curve of about 2500. Using this number and the values of B and U, the single-well pumping rate is

$$Q = B \cdot U \cdot TCV \quad \text{or} \quad 10 \times (2 \times 10^{-7}) \times 2500$$
$$= 5 \times 10^{-3} \text{ m}^3/\text{s}$$

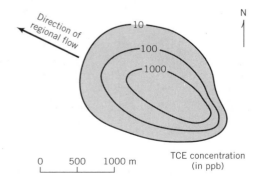

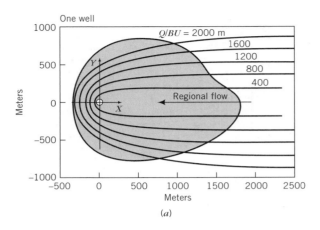

(a)

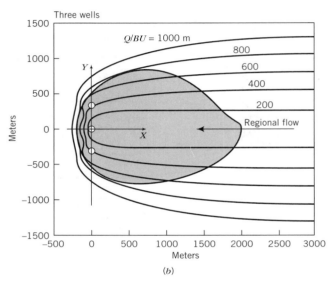

(b)

Check whether this pumping rate can be supported for the aquifer using the Cooper–Jacob (1946) equation to calculate drawdown at the well. Assume $r = 0.2$ m and the pumping period of one year

$$s = \frac{2.3Q}{4\pi T} \log \frac{2.25Tt}{r^2 S} \tag{20.4}$$

where $Q = 5 \times 10^{-3}$ m^3/s, $T = KB = 10^{-3}$ m^2/s, $t = 1$ yr or 3.15×10^7 s, $r = 0.2$ m, $S = 3 \times 10^{-5}$.

The pumping period represents some preselected planning horizon for the cleanup. Substitution of the known values into Eq. 20.4 gives a drawdown of 9.85 m. Even without accounting for well loss, the calculated drawdown exceeds the 7 m available. Thus, a multiwell system is necessary.

Superimposing the plume on the two-well type curve provides a Q/BU value of 1200, which in turn gives a Q for each of the two wells of $10 \times (2 \times 10^{-7}) \times 1200$ or

2.4×10^{-3} m³/s. The optimum distance between wells is $Q/(\pi BU)$ or $2.4 \times 10^{-3}/[\pi \times 10 \times (2 \times 10^{-7})] = 382$ m. Again, we check the predicted drawdown at each well after one year against the available 7 m. Because of symmetry, the drawdown in each well is the same. The total drawdown at one of the wells includes the contribution of that well pumping plus the second one 382 m away or

$$s = \frac{2.3Q}{4\pi T} \left[\log \frac{2.25Tt}{r_w^2 S} + \log \frac{2.25Tt}{r^2 S} \right]$$

The calculated drawdown is 6.57 m, which is less than the available drawdown. However, well loss should be considered, which makes the two-well scheme unacceptable. Moving to three wells, Q/BU is 800 (Figure 20.16*b*), which translates into a single pumping rate of 1.6×10^{-3} m³/s for each well. Carrying out the drawdown calculation for three wells located $1.26Q/(\pi BU)$ or 320 m apart provides an estimate of 5.7 m for the center well, which is comfortably less than the available drawdown. Thus, three wells are required 320 m apart. Each would be pumped at 1.6×10^{-3} m³/s.

This method is extremely useful, but it cannot be applied to every design problem encountered in practice. There are assumptions of constant transmissivity, fully penetrating wells, no recharge, and isotropic permeability built into calculation. We will explore the implications of the simplifications in a later section.

Model-Based Approaches for the Design of Recovery Systems

For most real problems, mathematical models are used for designing well systems for contaminant recovery. The simplest models are based on analytical theory similar to Javandel and Tsang (1986), but use numerical schemes to define pathlines. An example is the WHPA code distributed by the U.S. Environmental Protection Agency (Blandford and Huyakorn, 1991). Numerical models, such as MODFLOW and MODPATH, are also used for design. The mathematical approaches rely on the modeler's experience and a series of model runs to come up with a reasonable design. These designs are not optimized in the formal mathematical sense, but the modeling procedures are straightforward.

Larson and others (1987) apply a three-dimensional ground-water flow model with a second program that calculates ground-water pathlines. A finite-difference model is used to solve for the steady-state distribution of hydraulic head on a three-dimensional nodal network. These values of hydraulic head are input to the second program along with known values of hydraulic conductivity and porosity. Pathlines are constructed by particle tracking. Given a well pumping in a three-dimensional system, these pathlines can be used to define the capture

zone for the well. The following example from Larson and others (1987) will illustrate this final step.

Assume that a partially penetrating well is being pumped at 21.5 gpm in an aquifer having a $K_b = 100$ ft/day, $K_v = 1.00$ ft/day, and a recharge rate of 4 in./yr. By examining the pathlines in plan view and cross sections, the capture zone can be readily defined (Figure 20.17). Interestingly, the lateral extent of the capture zone is larger than predicted by the Javandel and Tsang analysis, while the vertical size is somewhat less (Figure 20.17). The partially penetrating well, the recharge, and anisotropism are complexities that violate the basic assumptions made in the development of the capture-zone type curve procedure.

Simulation–Optimization Techniques

A promising approach to the design of contaminant recovery systems involves the combination of ground-water simulation with mathematical optimization (Gorelick and others, 1984; Atwood and Gorelick, 1985; Lefkoff and Gorelick, 1986; Gorelick, 1987). This procedure combines the power of numerical flow modeling to account for the complexities of the hydrogeologic setting and the capabilities of formal optimization techniques to provide the best design subject to imposed constraints. Discussion of this sophisticated modeling approach is beyond the scope of this book. This and similar ap-

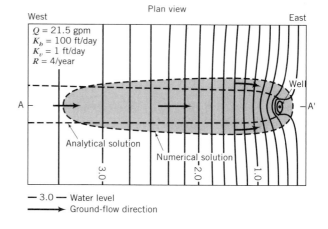

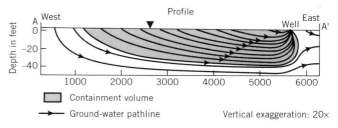

Figure 20.17 Capture zone calculated using a three-dimensional ground-water model (from Larson and others, 1987). Reprinted by permission of Solving Ground Water Problems with Models. Copyright © 1987. All rights reserved.

proaches have promise of being able to overcome the inherent limitations of the analytical approaches in dealing with complex systems and the non-uniqueness of the trial-and-error procedures.

Issues in the Design of Capture Zones

The design of well systems to provide a given capture zone is carried out using average values of hydraulic conductivity, porosity, and the like. The natural heterogeneity of geologic media is not usually considered because at most sites there are insufficient data to define the spatial structure in various hydraulic parameters. However, analyses (Gorelick, 1987) have shown that not considering the heterogeneous structure of an aquifer can lead to an underdesigned recovery system.

Gorelick (1987) generated a series of heterogeneous random transmissivity fields with a mean ln T of -3.22, a standard deviation of 1.0, and a correlation length of 10 m. An example is shown in Figure 20.18 along with a series of nodes that represent possible locations for recovery wells. Capturing the plume shown in Figure 20.18 with an optimum system would require three wells pumping at a combined rate of 110 L/s. This pumping rate compares to an ideal combined rate of 96 L/s for a homogeneous version of the same aquifer (that is, no spatial variation in transmissivity). Other examples of the heterogeneous aquifer required from 2 to 58% higher pumping rates as compared to the homogeneous case. Importantly, the local variability in transmissivity played

an important role in determining where the individual wells were placed.

This analysis assumed that one knew exactly what the transmissivity field looked like. When the variability in the transmissivity field is unknown, which is the usual case, even more pumping capacity is required to create a robust design for any type of heterogeneity. Thus, ignorance about the nature of the heterogeneity translates into greater pumping rates (Gorelick, 1987).

Another factor impacting the shape of the capture zone for a given well is the extent to which the well penetrates the aquifer. Most of the simple approaches assume a fully penetrating well irrespective of the aquifer thickness. Wells are rarely designed in this manner, especially for thick aquifers. Bair and Lahm (1996) conducted an analysis of the effect of partially penetrating wells on the shape of capture zones. They showed that neglecting the effects of partial penetration can lead to substantial errors in assessing the abilities of an individual well to capture a plume.

Under a given flow regime, screening only a fraction of the aquifer can lead to flow under the well. This effect is shown in Figure 20.19. Decreasing the percentage of well penetration (80% to 40% to 10%) eventually leads to the loss of containment of the plume (see Figure 20.19c). A less significant but still important effect of partial penetration is to increase the width of the capture zone. The need to consider the effects of partial penetration become more important as the hydraulic gradient increases and the aquifer becomes increasingly anisotropic (that is, $K_x \gg K_z$) (Bair and Lahm, 1996).

20.5 Interceptor Systems for NAPL Recovery

Interceptor systems are useful in recovering NAPLs at shallow depths. Figure 20.20a,b illustrates two designs for open trench systems. In the first, a skimmer pump selectively removes product entering the trench. Because little water is removed, a barrier needs to be installed to prevent the product from reentering the subsurface on the down-gradient side of the trench (Figure 20.20a). The second design is similar except that a greater volume of water is removed to contain the product hydraulically in the trench (Figure 20.20b). This design avoids the impermeable barrier and speeds up product recovery because the flow gradient to the trench is increased (Blake and Fryberger, 1983). However, a greater volume of contaminated water must be dealt with at the surface.

An alternative design involves closed trenches or drains (Figure 20.21). Product moves through gravel or drain pipes in the trench laterally toward collection sumps. Product is removed from the sumps by pumping. Again, hydraulic control keeps the product from simply passing through the trench (Figure 20.21c,d). Treatment of water

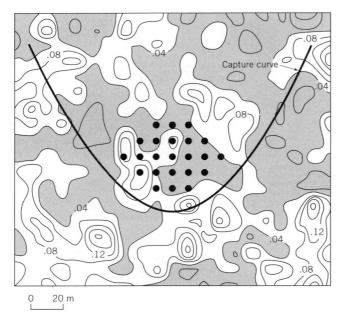

Figure 20.18 Simulated transmissivity map for capture zone analysis. The closed circles represent possible locations for one to five recovery wells (modified from Gorelick, 1987). Reprinted with permission of Solving Ground Water Problems with Models. Copyright © 1987. All rights reserved.

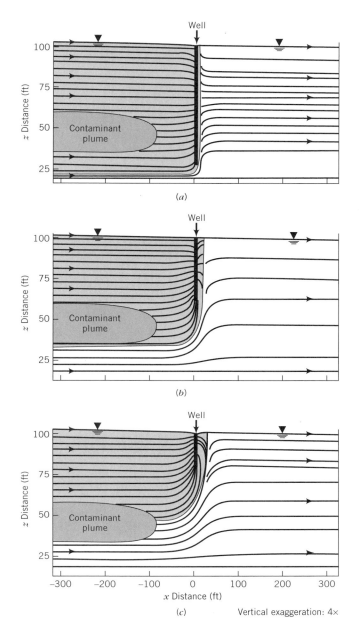

Figure 20.19 Cross sections in the direction of flow that illustrate how partial penetrations of (*a*) 80%, (*b*) 40%, and (*c*) 10% influence the vertical extent of the capture zone. In these simulations, $K_H = K_V$ with a regional flow gradient of 0.006 (from Bair and Lahm, 1996). Reprinted by permission of Ground Water. Copyright © 1996. All rights reserved.

from the sumps again would be required. The main attractions of interceptor systems are their overall simplicity, availability of construction materials, and speed of installation (Blake and Fryberger, 1983).

Unlike drainage systems for farmland, interceptor systems are not commonly installed as a large areal network to collect product everywhere. More often, they are installed across the nose of a migrating spill or at critical points along the perimeter of the spill to prevent further

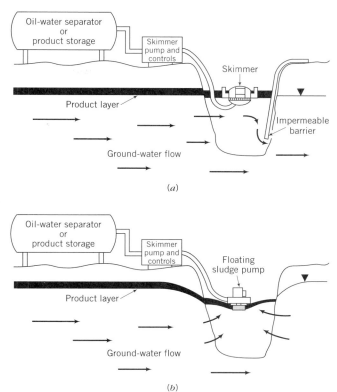

Figure 20.20 Examples of interceptor trenches used to control the spread of NAPLs. In (*a*) product is recovered using a skimmer pump with an impermeable barrier to control product leakage. In (*b*) product and water are recovered with a floating sludge pump. Potential product leakage is controlled hydrodynamically (from Blake and Fryberger, 1983).

spreading or inflow to surface water bodies. Given the complexity of multicomponent flow, there are no quantitative approaches available for designing a system. The main considerations in design are to ensure that the trench is long enough to avoid the product migrating around the end and deep enough to provide the necessary hydraulic control and to avoid the water table falling beneath the bottom of the trench during dry periods.

20.6 *Soil-Vapor Extraction*

Soil-vapor extraction (SVE) has emerged as the key technology for cleaning up volatile organic compounds in the unsaturated zone. Most often, SVE is used with NAPLs. However, it also works with dissolved contaminants occurring in soil water or sorbed to solids. SVE uses vacuum blowers in combination with soil-gas extraction wells to create an air flow through the zone of contamination (Figure 20.22). The contaminants volatilize into this air stream at concentrations dictated by their vapor pressure. VOCs carried to the well are removed. The air stream is cleaned up at the surface by carbon adsorption or some

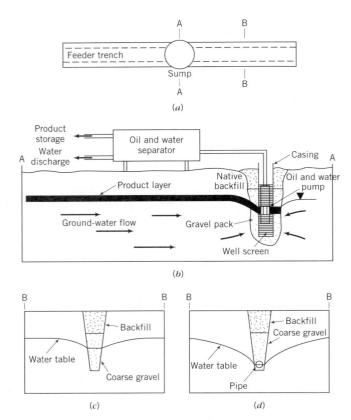

(a)

(b)

(c)   (d)

Figure 20.21 Schematic representation of two interceptor systems. Water and contaminants move along the feeder trench to a sump (*a*). A pumping system moves the fluids to the surface for treatment (*b*). The feeder trenches can be filled with gravel (*c*) or actual drains (*d*) from Blake and Fryberger, 1983).

type of thermal treatment (incineration or catalytic oxidation).

Components of an SVE System

An SVE system consists of extraction wells connected to the treatment system and the vacuum blower. Typically, the extraction wells are one of two types (Figure 20.23). The simplest design is a single screened well installed

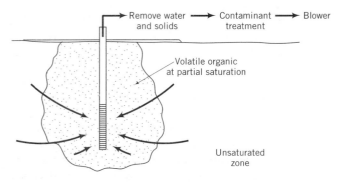

Figure 20.22 Schematic of how an SVE system operates.

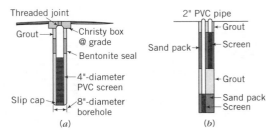

(a)   (b)

Figure 20.23 Examples of extraction wells used for SVE systems. Panel (*a*) is a single well completion (from U.S. EPA, 1991). Panel (*b*) is a multiwell completion.

in an auger hole (Figure 20.23*a*). The multilevel design involves two or more screened sections completed in a single auger hole (20.23*b*). The choice of design usually depends on the vertical extent of contamination. If the spill has a limited vertical extent, then the first design is more appropriate. If the spill is several tens of feet thick or zoned, then the multilevel approach is preferable. Good practice dictates that the well screens be kept less than 2 to 3 m to maintain air flow through the contaminated zones. When screens are longer than 5 m, airflow may be channeled along a few preferential pathways and bypass much of the contaminated soil.

Monitoring wells are installed to provide vapor-sampling points away from the extraction wells. Normally, vapor-monitoring wells (Figure 20.24) are permanently installed, as opposed to, say, soil-gas probes, which are inserted and removed from the ground. Specialized equipment such as a Geoprobe® sampling rig can install a permanent sampling point. More complex multilevel wells can be installed using an auger rig in cases where the contamination is thick.

Whether installing vapor extraction or monitoring wells, care must be taken to seal the annulus above screened sections with bentonite. Any short circuiting

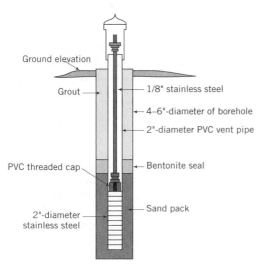

Figure 20.24 Example of a vadose-zone monitoring well (from U.S. EPA, 1991).

of airflow along the casing will reduce the airflow through the zone of contamination and the overall efficiency of the system. This caution applies especially to the conversion of existing wells. In many projects, one might be tempted to convert water-table monitoring wells to keep drilling costs down. Often the seals in these wells may be inappropriate for vapor extraction wells.

Surface treatment of the vapor stream usually begins by removal of small solids and liquids. This treatment is usually accomplished by a "knock out" drum or vapor–liquid separator. The next treatment step deals with the contaminated vapor. Vapor combustion units incinerate the contaminants by burning. These units are most efficient when vapor concentrations are high—on the order of a few percent by volume (Johnson and others, 1990). They are less economical at lower concentrations because a supplemental fuel needs to be added. Another treatment technique involves passing the vapor over a catalytic oxidation bed. These catalytic oxidation units work best with vapor concentrations less than 8000 ppm (Johnson and others, 1990). With carbon treatment, vapor passes through a column of activated charcoal, which preferentially sorbs organic compounds. This treatment procedure is most economical when vapor concentrations are very low (Johnson and others, 1990).

The most economical way of dealing with vapors is to simply release them into the air using what is called a diffuser stack. This approach obviously does not treat the air stream and, thus, has health and safety implications. This approach is not commonly acceptable to regulators.

When Can SVE Systems Be Used?

The two main factors determining when SVE systems can be used are the chemical character of the contaminant and the airflow rate through the unsaturated zone. These two factors are reflected in the theoretical contaminant removal rate per day at a site (R_{est}) given as

$$R_{est} = C_{org}Q \qquad (20.5)$$

where C_{org} is the concentration of the contaminant in the gas phase and Q is the gas flow rate. C_{org} depends on the vapor pressure for the contaminant, and Q depends mainly on the permeability of the medium.

For SVE schemes to work, the organic compound or mixture of organic compounds in the subsurface must be somewhat volatile. In other words, the rate of volatilization into the moving air phase must be sufficiently large to reduce the total contaminant mass over a meaningful time frame. A rule of thumb for considering SVE for a given contaminant is a vapor pressure greater than 0.015 atm (14 mm Hg at 20°C). Table 20.3 lists commonly occurring contaminants that are likely candidates for SVE.

By far, most SVE systems are used for cleaning up fuel spills. However, fuels include a spectrum of different organic compounds, and each type can respond differently to SVE systems. Gasoline contains a relatively large proportion of volatile compounds. Thus, gasoline spills are most responsive to SVE systems. The middle distillates, such as jet fuel and diesel, contain a relatively large proportion of less volatile compounds. Thus, SVE is slow to clean up these kinds of spills.

When an SVE system is first turned on, it is effective in removing the most volatile components of the given fuel. For example, in the case of gasoline, the more volatile compounds, such as butane and benzene, are removed in preference to the heavier and less volatile compounds. As the mass of contaminants in the soil decreases, the remaining compounds are enriched in the less volatile compounds (Johnson and others, 1990). Thus, with time the residual product will be harder and

Table 20.3 **List of Contaminants That Can Be Cleaned Up Using SVE**

Volatiles:	methanol
benzene	acetone
toluene	pyridine
xylenes	tetrahydrofuran
ethylbenzene	carbon tetrachloride
hexane	trichloroethane (TCA)
chloroform	
methylene chloride	**Semivolatiles:**
tetrachloroethylene (PCE)	chlorobenzene
trichloroethylene (TCE)	dichlorobenzene (DCB)
dichloroethylene (DCE)	trichloropropane
ethyl acetate	
cyclohexane	**Hydrocarbons:**
methyl ethyl ketone (MEK)	gasoline
methyl isobutyl ketone (MIBK)	jet fuel
	diesel
	kerosene
	heavy naphthas

From U.S. EPA (1989b).

harder to remove by SVE. This tendency for residual fuel to remain limits the practical level of cleanup that can be achieved at a site by SVE (Johnson and others, 1990).

Estimating Removal Rates

Application of Eq. 20.5 first requires an estimate of the concentration of the organic compound in the vapor phase or C_{org}. This calculation was illustrated with Example 12.3. For a simple mixture, the only information that needs to be provided is P_{org}, the vapor pressure of the pure organic solvent, the formula weight of the contaminant, and the mole fraction of the organic solvent.

The airflow rate (Q) is determined by the permeability of the porous medium and the vapor extraction rate. Given that constraints exist on the ultimate vacuum that can be achieved in a well (and hence the extraction rate), permeability is the more important variable. Generally, SVE systems work best in materials, such as sand and gravel, that are highly permeable. U.S. EPA (1991) provides the following guidance in relating the potential success of an SVE system to soil texture:

Stop and Evaluate Carefully—clay, silty clay, silty clay loam

Some Difficulty—sandy clay, clay loam

Good—sandy clay loam, silt loam

Very Good—sand, loamy sand, sandy loam, loam

Creating appropriate vapor flow rates through rocks or fine-grained sediments requires a natural fracture network connected to wells by hydraulically induced fractures. Increasing water saturations reduces the permeability of the medium to gas flow because of the effects of multiphase flow.

Johnson and others (1990) provide a detailed methodology to examine whether an SVE system is likely to work. They exploit the analogy between the flow of gas and water to develop the gas flow rate per unit thickness of well screen (Q/H) as

$$\frac{Q}{H} = \pi \frac{k}{\mu} P_w \frac{(1 - (P_{atm}/P_w)^2)}{\ln(R_w/R_i)} \quad (20.7)$$

where H is screen length (cm), k is the soil permeability to airflow (cm²), μ is the viscosity of air (1.8×10^{-4} g/cm-s), P_w is the absolute pressure at the extraction well (g/cm-s²), P_{atm} is the absolute ambient pressure or 1.01×10^6 g/cm-s² (1 atm), R_w is the radius of the vapor extraction well, and R_i is the radius of influence of the vapor extraction well. If k can be estimated, the only unknown in Eq. 20.6 is the radius of influence. Values of R_i can be estimated from a range of about 9 to 30 m. Because of the form of the equation, airflow rates are not particularly sensitive to R_i.

Example 20.4

A soil-gas extraction well is completed in a medium sand having an estimated permeability of 1×10^{-7} cm² (that is, 10 darcy). Assume that the radius of influence of this well is 12 m (1200 cm), the radius of the well is 5.1 cm, the vacuum pressure in the well is 0.90×10^6 g/cm-s² (0.9 atm), and other pertinent parameters are given above. Calculate Q/H for this well.

Substitute known values in Eq. 20.6 to give

$$\frac{Q}{H} = 3.14 \frac{10^{-7}}{1.8 \times 10^{-4}} 0.9$$

$$\times 10^6 \frac{(1 - (1.0 \times 10^6/0.9 \times 10^6)^2)}{\ln(5.1/1200)}$$

$$\frac{Q}{H} = 1.74 \times 10^{-3} \times 0.9 \times 10^6 \frac{(1 - 1.23)}{-5.46}$$

$$\frac{Q}{H} = 6.6 \times 10^1 \text{ cm}^3/\text{cm-s} = 0.4 \text{ m}^3/\text{m-min}$$

To make this calculation easier, Johnson and others (1990) have provided the nomogram shown in Figure 20.25. It provides estimates of Q/H in standard volumetric units Q^*/H, which are equal to $Q/H(P_w/P_{atm})$. The nomogram assumes $R_w = 5.1$ cm (2 in.) and $R_i = 12$ m. For other conditions, the following simple corrections apply

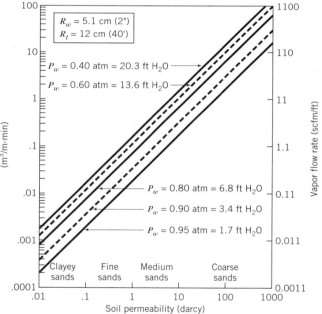

[ft H₂O] denote vacuums expressed as equivalent water column heights

Figure 20.25 Predicted steady-state flow rates (per unit well screen length) for a range of soil permeabilities and applied vacuums (from Johnson and others, 1990). Reprinted by permission of Ground Water Monitoring Review. Copyright © 1990. All rights reserved.

$R_w = 5.1$ cm (2 in.) $R_i = 7.6$ m (25 ft)

multiply Q^*/H by 1.09

$R_w = 5.1$ cm (2 in.) $R_i = 23$ m (75 ft)

multiply Q^*/H by 0.90

$R_w = 7.6$ cm (3 in.) $R_i = 12$ m (40 ft)

multiply Q^*/H by 1.08

$R_w = 10$ cm (4 in.) $R_i = 12$ m (40 ft)

multiply Q^*/H by 1.15

$R_w = 10$ cm (4 in.) $R_i = 7.6$ m (25 ft)

multiply Q^*/H by 1.27

Removal Rate Calculations

As we saw with Eq. 20.5, the contaminant removal rate can be calculated from a knowledge of C_{org} and the airflow rate Q. The main design issue is whether an SVE scheme can recover the contaminants within a realistic time frame. We will illustrate this idea with the following calculation.

Example 20.5

Within the radius of influence of a single extraction well, assume there is 5000 kg of chlorobenzene. The equilibrium concentration of this contaminant in air is 55 mg/L, and the well is capable of providing an airflow of 280 L/min (10 standard cubic feet per minute). Estimate the cleanup time using this single well.

From Eq. 20.5, the recovery rate is calculated

$R_{est} = 55 \times 280 = 1.54 \times 10^4$ mg/min   or   22 kg/day

Given the initial mass of contaminants, it can be determined that

$$t_{cleanup} = \text{total mass/recovery rate}$$

$$t_{cleanup} = 5000/22 = 227 \text{ days}$$

This relatively short cleanup time makes the SVE scheme attractive.

Field Estimates of Soil Permeability

Air-permeability testing in the field is an integral part of any site investigation where SVE is being considered. The most common test involves removing soil gas at a constant rate from an extraction well and monitoring pressure declines in adjacent observation wells (Figure 20.26). Usually, a minimum of three monitoring wells is used. These wells are located at distances of approximately one to three times the depth to the water table away from the extraction well (Peargin and Mohr, 1994). The maximum distance is determined by the radius of influence of the well, commonly 9 to 30 m. In lower permeability settings, more wells are advisable with some

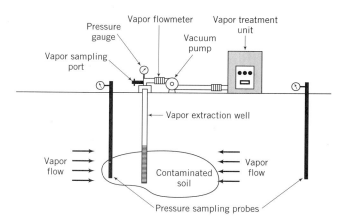

Figure 20.26 Schematic of an air-permeability test run in the field (from Johnson and others, 1990). Reprinted by permission of Ground Water Monitoring Review. Copyright © 1990. All rights reserved.

of the wells located close to the extraction well (Peargin and Mohr, 1994).

Section 19.6 showed how well hydraulics theory could be modified to interpret the results of air-permeability tests. Here, we illustrate how field data can be interpreted using the Cooper–Jacob method. Begin by plotting drawdown versus the log of time. With a constant rate of withdrawal in a confined and infinite aquifer, the data should plot as a straight line. For the gas-flow problem, the measured pressure decline in the wells should follow a straight line.

Example 20.6 (adapted from Massmann, 1989)

An air-permeability test was carried out in January 1986 at the Midway Landfill. Air pressure decline versus time for one monitoring well (as equivalent centimeters of water) is shown in Table 20.4. Given that the test zone is 25 m thick, the observation well is 38 m away from the extraction well, the air removal rate of 105 L/s, and the vapor density of the air is 0.001 g/cm³, calculate the hydraulic conductivity to gas and the permeability of the medium.

The first step is to plot the drawdown in centimeters of water versus the log of time (Figure 20.27). From this graph, the drawdown per log cycle is 8.2 cm water or 8.2/0.001 = 8200 cm of gas. Rearrange the Cooper–Jacob equation and substitute the data, or

$$K = \frac{2.3Q}{4\pi m\,\Delta s} = \frac{2.3 \times 105 \times 10^3}{4 \times 3.14 \times 2500 \times 8200} = 0.0009 \text{ cm/s}$$

Permeability is calculated as

$$k = \frac{K\mu}{\rho g} = \frac{.0009 \times 1.45 \times 10^{-4}}{.001 \times 980}$$

$$= 1.33 \times 10^{-7} \text{ cm}^2 = 13.3 \text{ darcys}$$

Table 20.4 **Drawdowns Determined from Field Data**

Time (min)	Drawdown (cm water)
1	0.0
2	0.0
3	0.02
5	0.15
6	0.25
7	0.43
9	0.74
11	1.04
14	1.52
18	2.16
21	2.60
26	3.22
31	3.81
41	4.82
46	5.23
61	6.35
101	8.00
166	9.14
306	10.16

Modified from Massmann (1989).

The design and execution of air-permeability tests is not without problems. Peargin and Mohr (1994) studied a data base of 80 SVE pilot tests to identify common mechanical or procedural problems. Three problems occur commonly: (1) submerged extraction well screens, (2) insufficient well seals, and (3) vacuum monitoring points either too few or too far away. The screen can be submerged when the bottom of the screen is set too close to the water table. When the extraction well is turned on, the water table will mound at the extraction well due to the decline in pressure. Problems with the integrity of the seals commonly lead to short-circuiting of the airflow. Seals should be constructed from materials

like a bentonite slurry or bentonite–cement grouts and be longer than 1 m. Commonly, the design of tests often places monitoring wells inappropriately—at the margin of the zone affected by the test. To be most effective, monitoring wells should be located in the zones with large expected airflows.

Heterogeneity and the Efficiency of SVE Systems

The efficiency of an SVE system is affected considerably by heterogeneity both in the distribution of the contaminant and the permeability of the medium. All of the analyses presented so far have assumed implicitly that the contaminants are distributed homogeneously throughout the region of active airflow. However, the airflow often does not effectively contact the contaminants in the system. Johnson and others (1990) provide conceptual models of contaminant distributions, which provide slower-than-expected contaminant recovery (Figure 20.28).

In the first case, a significant fraction of the airflow is not passing through the spill because of the well location. The recovery rate of the contaminant for this case be-

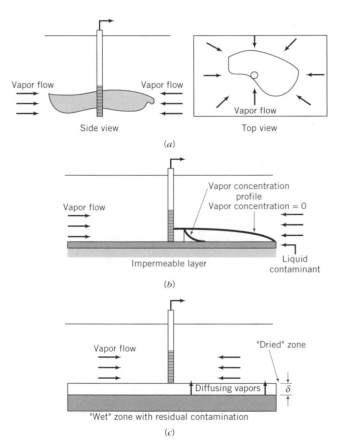

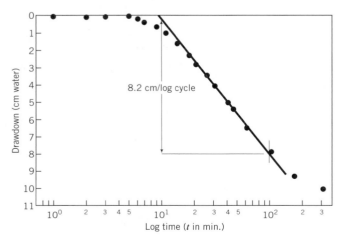

Figure 20.27 Graph of drawdown versus log time.

Figure 20.28 Three examples illustrating situations when the contaminant recovery rate may be less than expected (from Johnson and others, 1990). Reprinted by permission of Ground Water Monitoring Review. Copyright © 1990. All rights reserved.

comes

$$R_{est} = (1 - \phi)C_{org}Q \qquad (20.7)$$

where ϕ is the fraction of the total airflow that is moving through uncontaminated soil. For the case illustrated in Figure 20.28*a*, this fraction would be about 0.3. Effectively, the recovery rate is 70% of what one might expect.

Figure 20.28*b* illustrates a case where the NAPL has pooled and reached relatively high saturations. The air flows parallel to the surface of the pool but not through the contaminated soil. A sample calculation provided by Johnson and others (1990) shows that the removal rate might be less than 10% of the maximum rate. Figure 20.28*c* is a case where the airflow is mainly through an adjacent layer separated from the zone of contamination by a so-called dried zone. Contaminant removal is diffusion controlled, and thus, inefficient.

Recovery will also be inefficient when a few highly permeable units attract most of the airflow through the system. These units may clean up very quickly but adjacent lower permeability zones will not. According to Peargin and Mohr (1994), airflow in low-permeability materials is largely controlled by preferential pathways. Cleanup levels and times are directly related to the density and regularity of the spacing of these pathways.

20.7 *Air Sparging*

Air sparging can recover volatile organic compounds, semivolatiles, chlorinated solvents, and some petroleum compounds (C_3–C_{10}) from the saturated ground-water system. Air is injected into the zone of contamination. It flows through channels and forms an in situ, air stripping system, as volatile contaminants partition into the airflow (Figure 20.29). The contaminant-laden air arriving at the water table is collected by a conventional SVE system. Besides stripping volatiles from the zone of contamination, the addition of oxygen may promote biodegradation. However, many VOCs are not biodegradable under oxic conditions.

Sparging wells are constructed using either PVC or metal casings. The well screen and filter pack must be carefully sealed with bentonite or neat cement grout to avoid air leaks. Individual wells are outfitted at the surface with a pressure gage and regulator to facilitate the monitoring and operation of the well. Check valves are typically included at the well head to avoid the backflow of water when the compressor is shut off. Air is supplied to the wells from an oil-free compressor through metal pipes or rubber hoses. These robust materials can accommodate the heat generated as the air is compressed.

Airflow and Channeling

Early work depicted the flow of air through an aquifer as small bubbles. However, airflow mostly occurs in chan-

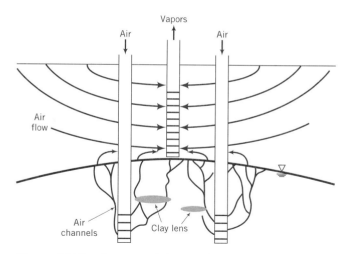

Figure 20.29 Conceptual design of an air sparging system (from Newell and others, 1995).

nels as depicted in Figure 20.29. Ahlfeld and others (1994) discuss the mechanics of channel formation once a sparging well is turned on. In essence, a "pocket" of pressurized air will begin following the path of least resistance, which consists of media having the lowest air entry pressures or smallest capillary pressure. Thus, channels form preferentially in media having the largest pore size, and by inference, the highest permeability (Ahlfeld and others, 1994). Airflows will bypass lower permeability units. Once air channels are established, they will be maintained as long as the air-entry pressure is maintained.

Contaminants are removed because dissolved VOCs partition into the adjacent air phase. With time, the contaminant concentrations decline adjacent to the air channel. The resulting concentration gradient in the aqueous phase promotes diffusion of contaminants toward the air channels. The magnitude of interphase transfer is proportional to the area of the interface that exists between the air and water systems (Ahlfeld and others, 1994). The greatest mass transfer would be promoted by many small channels, evenly distributed as opposed to a few large channels spaced far apart.

Ahlfeld and others (1994) discuss calculations that illustrate the importance of maintaining small diffusion pathways. For example, the time to reduce TCE concentrations from 5000 to 5 μg/L by flowing a column of air with a one-inch radius through soil columns having radii of 20 in., 5 in., and 2 in. was 78 years, 1.7 years, and 27 days, respectively. The smaller the cleanup radius around a given channel, the more rapidly the contaminant concentration is reduced.

Air sparging works effectively with airflow rates in a range of 3 to 10 standard cubic feet per minute per sparge point. The actual pressure required to produce these rates depends in part on the depth of the well below the water table or the static head above the sparge point.

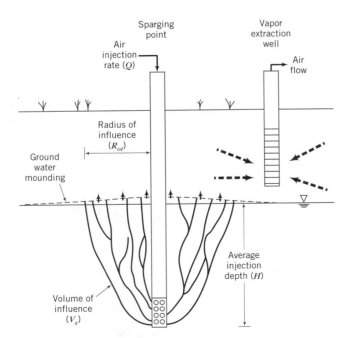

Figure 20.30 Schematic diagram of an air-sparging and vapor-extraction system showing the radius of influence and volume of influence (modified from Sellers and others, 1992). Reprinted by permission of Conference on Petroleum Hydrocarbons and Organic Chemicals in Ground Water— Prevention, Detection and Restoration. Copyright © 1992. All rights reserved.

Deeper wells require higher injection pressures. These airflow rates are promoted by sands and gravels. This airflow also must find its way to the unsaturated zone. The presence of low-permeability layers above the well intake may promote the formation of gas pockets and the lateral spreading of contaminants. Thus, there are significant geologic limitations as to where the method can be used.

Designing Air-Sparging Systems

The main variable determining the number of sparging wells is the radius of influence for an individual well. As

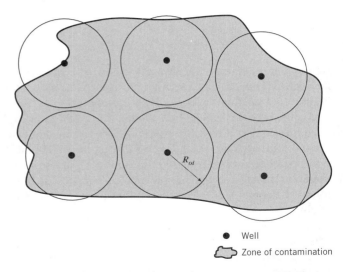

● Well
◌ Zone of contamination

Figure 20.31 Sizing an air-sparging system to match the extent of contamination.

Figure 20.30 illustrates, the radius of influence (R_{oi}) is defined in relation to the volume of influence, which is the zone around the well that experiences the airflow. Thus, the wells should provide a total area of influence that corresponds to the area of the contaminated zone (Figure 20.31).

A variety of measurements are made in the field to interpret the radius of influence, including

- changes of pressure in the saturated and unsaturated zones
- concentration of VOCs in the unsaturated zone
- the extent of water-table mounds, and
- tracer-gas responses.

Lundegard (1994) used a highly instrumented test involving a single sparge well in a simple geologic setting to examine these interpretive methods critically. Interestingly, he found that the common approaches listed above overestimated the radius of influence by a factor of two to eight. He suggests that in more complex settings it will be even more difficult to use conventional approaches.

20.8 Case Studies in Site Remediation

Four case studies have been selected to illustrate the typical range of problems encountered in the cleanup of sites. The first case is typical of literally thousands of spills of gasoline and oil. The next two case studies are larger and much more serious problems resulting from the disposal of toxic organic chemicals at uncontrolled sites. The Gilson Road site in New Hampshire is important historically because it was the subject of the first cooperative agreement signed under the Superfund law

in the United States. The Hyde Park site is famous in its own right as one of the set of large hazardous-waste disposal sites located in and around Niagara Falls, New York. The last case study examines a demonstration of SVE technology for solvents (mainly TCE) at an industrial site in Massachusetts.

Oil Spill: Calgary, Alberta

The case of a large spill of petroleum products in floodplain sediments is related to an asphalt plant and terminal at Calgary, Alberta (O'Connor and others, 1984b). The spill first came to light when petroleum products discharged at the surface in the area marked by "Early oil seeps" (Figure 20.32). After plant personnel constructed an interceptor ditch along the southern plant boundary (Figure 20.32), it was thought the problem was solved. However, at this early stage, the full extent of the problem was not known. Oil discharged subsequently into lagoons of the Inglewood Bird Sanctuary.

The need to control seepage in the bird sanctuary in advance of returning geese required emergency action. An interceptor drain was installed across the zone of seepage (Figure 20.32). A horizontal box culvert and gravel buried in the trench provided lateral drainage of the oil to a series of sumps. Migration of oil past the interceptor system was eliminated by adding an impermeable barrier to the downstream wall of the culvert. Cutoff walls added to both ends of the interceptor system deflected oil into the box culvert system. This action eliminated oil seepage into the north lagoon.

The detailed site investigation revealed that oil had collected on the water table over a relatively large area and was migrating toward the bird sanctuary, following the local gradient of the water table. The contaminants were spreading in a sand and gravel unit approximately 10 m thick.

Because of the size of the spill, the interceptor system was not sufficient to contain and recover all of the oil. A two-well, two-pump extraction system was installed in the bird sanctuary (refer to Figure 20.10b). The water wells produced up to 2.5×10^{-2} m^3/s and created the necessary potential low to move the oil into the product-collection wells (Figure 20.33). Clean water pumped from the water wells was discharged into the interceptor ditches constructed earlier at the site.

This collection system was run for five years and its performance reviewed in 1983. A lack of well mainte-

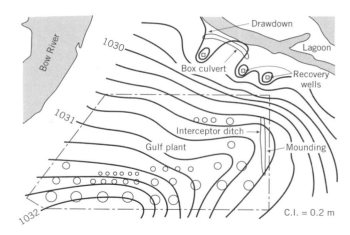

Figure 20.33 Elevation of the water table (meters) follow-. ing startup of the recovery wells, autumn 1979 (from O'Connor and others, 1984b). Reprinted by permission of Conference on Petroleum Hydrocarbons and Organic Chemicals in Ground Water—Prevention, Detection and Restoration. Copyright © 1984. All rights reserved.

nance contributed to reduced product recovery. Nevertheless, during this period of operation, approximately 1,300,000 L of oil were collected. A significant quantity of product remained in the subsurface, especially under the plant site and away from the area of most intense remedial activity. The increasing thickness of product at the plant site pointed to the continued leakage of oil at the source. Based on the performance of the system to 1983, about six more years would be required to collect the remaining free product.

Gilson Road: Nasbua, New Hampshire

The Gilson Road Superfund site in New Hampshire is important historically because it was the subject of the first cooperative agreement signed under the Superfund law in the United States. Chemical sludge and organic wastes in drums were placed together with refuse in a sand and gravel pit. In addition, some 4,000,000 L of liquid chemical wastes were discharged into the ground at a nearby trench (Figure 20.34: Ayres and others, 1983). The most hazardous contaminants included heavy metals and volatile and extractable organic compounds. Spreading of contaminants was facilitated by the presence of 6 to 35 m of permeable sand and gravel, which overlie till and fractured bedrock. Initial studies in 1980 defined a contaminant plume more than 450 m long and up to 33 m thick (Ayres and others, 1983). By December 1980, some contaminants from the front of the plume had begun to discharge into Lyle Reed Brook, located at the western edge of the study area. Following completion of the site investigation in 1981, a feasibility study established a strategy for remedial action. The possible alternatives included combined hydrologic containment and

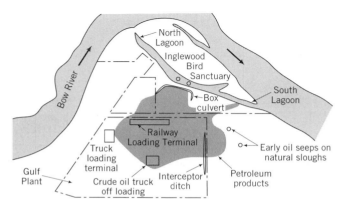

Figure 20.32 Site of an oil spill at Calgary, Alberta (from O'Connor and others, 1984b). Reprinted by permission of Conference on Petroleum Hydrocarbons and Organic Chemicals in Ground Water—Prevention, Detection and Restoration. Copyright © 1984. All rights reserved.

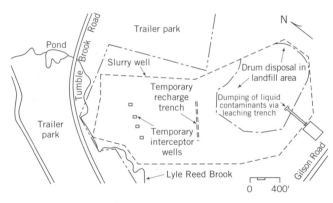

Figure 20.34 Layout of the Gilson Road hazardous waste-disposal site at Nashua, New Hampshire, showing the landfill and liquid waste-disposal areas as well as facilities for controlling contaminant migration (modified from Ayres and others, 1983). Reprinted by permission of Third National Symposium on Aquifer Restoration, and Ground Water Monitoring. Copyright © 1983.

ground-water treatment as well as total removal (Ayres and others, 1983). The overall strategy that was adopted involved isolating the most contaminated parts of the plume using a cutoff wall and surface seals, as well as a withdrawal–treatment system to remove the contaminants from the ground water.

Before the remedy could be implemented in 1982, monitoring showed that a highly contaminated portion of the plume was about to discharge into the brook. To delay the arrival of this front, an emergency recirculation system was designed to remove contaminated ground water from four wells at the nose of the plume (Figure 20.34) and to return it without treatment using a temporary recharge trench. A two-dimensional computer model was developed and calibrated against field data to assist in the design of this system (Ozbiligin and Powers, 1984). Subsequent monitoring indicated that pumping the wells at a total of 0.25 m³/min was successful in temporarily retarding the advance of the plume (Ozbiligin and Powers, 1984).

The conventional cutoff wall ranged in thickness from 0.5 to 1.25 m (Ayres and others, 1983; Schulz and others, 1984). It surrounded an area of about 80,000 m² to depths ranging from 10 to 33 m. The performance of the wall was evaluated using various mathematical models. The quantity of contaminated water passing through the site was reduced from 284 m³/day to 114 m³/day during seasons with low rainfall (Ozbiligin and Powers, 1984). Containment was being lost not through the grout wall but through fractures and fracture zones in the bedrock (Ozbiligin and Powers, 1984).

The literature we examined discussed the design of the system for collecting, treating, and recirculating water but not its implementation. Computer modeling played an important role in the design of the well system. In addition to simply collecting water for treatment at

the surface, the system was to provide further containment for contaminants. With this design, ground water only flows into the area surrounded by the grout wall. This control could be affected by pumping a total of 1.14 m³/min, treating 0.19 m³/min to drinking water standards for discharge to Lyle Reed Brook, and returning 0.95 m³/min of partially treated water back to the ground via trenches.

Hyde Park Landfill: Niagara Falls, New York

Coming out of the Settlement Agreement for Hyde Park Landfill in 1982 was a provision to evaluate the available data and to develop a plan for remedial action. The remedial measures finally agreed upon were reviewed in detail by Faust (1985) and are described here as an illustration of the variety of factors that must be considered when dealing with a complex site and a variety of different contaminants.

The overall strategy for remediation involved three parts: (1) source control, (2) control and collection of dissolved and aqueous phase contaminants from the overburden, and (3) control and collection of contaminants from the Lockport Dolomite. The complexity of the geologic setting, and uncertainty about how well certain remedial strategies might work, necessitated a phased approach that involved using some prototype systems along with extensive data collection. Proceeding in this way avoids "the situation where remedial measures are installed and later to be found ineffective without any simple recourse to correct the deficiencies" (Faust, 1985).

The main reason for source control was to reduce the quantity of contaminants moving out of the landfill. A surface seal will reduce infiltration from precipitation and the flux of water and contaminants. Collection wells will be used to remove fluids from the actual landfill itself. Initial designs are for this to be a prototype system involving two wells.

Remedial action in the overburden will (1) contain the lateral spread of contaminants and (2) provide a way of collecting them. The measure that is proposed is a subsurface drain around the area where free organic liquids are known to occur. Contaminants not collected by the drain will move down into bedrock, where other collection systems are in place. These bedrock systems include an injection–withdrawal scheme for DNAPLs (Figure 20.35) and a withdrawal scheme for collecting dissolved contaminants close to the Niagara Gorge. The system proposed for DNAPLs is analogous to a secondary recovery scheme for oil (Faust, 1985). The advantages of the injection–withdrawal scheme as compared to pumping alone include: (1) it avoids dewatering of the upper part of the Lockport Dolomite, which is contaminated and requires ground water to circulate for cleanup to occur, and (2) larger volumes of water can be circulated through contaminated parts of the plume. Both two-

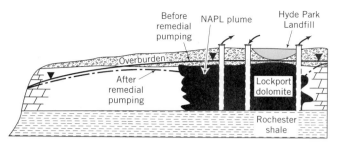

Figure 20.35 Conceptual diagram of the injection/withdrawal system for removing DNAPLs from the Lockport Dolomite at the Hyde Park Landfill (from Faust, 1985).

and three-dimensional flow modeling was conducted to evaluate the effectiveness of the prototype system, which involves four collection and two injection wells. Pumping rates of 20 gallons per minute for each withdrawal well and 10 gallons per minute for each injection well should provide a capture zone larger than the DNAPL plume. After further study, an operational system will be put in place. The second withdrawal system will collect dissolved contaminants in the Lockport Dolomite. These contaminants are sufficiently far from the site that they will be in general unaffected by the cone of depression created by the first DNAPL removal wells. Flow, however, will eventually cause these contaminants to discharge along the Niagara Gorge. The plan is to install a line of interceptor wells across the plume west of the site.

In the remedial system at Hyde Park Landfill, monitoring is an integral part of ongoing operations. Monitoring will establish whether a particular remedial alternative has met specific performance goals, provide criteria for logically expanding the prototype systems, and safeguard the public against accidental exposure to the contaminants.

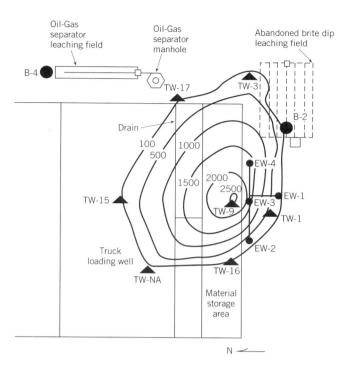

Figure 20.37 Concentration (ppm) of contaminants in the soil at the Groveland site (from U.S. EPA, 1989a).

Groveland Wells Site, Massachusetts

The last case study involves Terra Vac, Inc.'s in situ vacuum extraction process, which was demonstrated as part of the Superfund Innovative Technology Evaluation (SITE) Program (U.S. EPA, 1989c). The site selected for the demonstration was contaminated by a leaking storage tank and improper disposal practices. Approximately 3,000 to 30,000 lb of TCE were spilled.

A generalized geologic cross section at the site (Figure

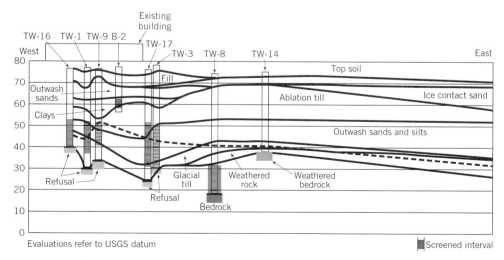

Figure 20.36 Generalized east-west cross section through the zone of contamination (modified from EPA, 1989).

Table 20.5 **Distribution of TCE as a Function of Depth Before and After Treatment**

Depth (ft)	Permeability (cm/s)	TCE Conc. ppm	
		Pre	Post
0-2	10^{-4}	2.94	ND
2-4	10^{-4}	29.90	ND
4-6	10^{-5}	260.0	39
6-8	10^{-5}	303.0	9
8-10	10^{-4}	351.0	ND
10-12	10^{-4}	195.0	ND
12-14	10^{-4}	3.14	2.3
14-16	10^{-8}	ND	ND
16-18	10^{-8}	ND	ND
18-20	10^{-8}	ND	ND
20-22	10^{-4}	ND	ND
22-24	10^{-3}	6.71	ND

Modified from U.S. EPA (1989c).

20.36) shows shallow outwash sand underlain by silty clay and clay till. A second silt and sand unit occurs below these units. Most of the contamination is in the shallow unit where TCE concentrations reach a maximum of approximately 2500 mg/kg. Relatively minor contamination was detected in the deeper sand unit as well. Figure 20.37 shows the areal distribution of contaminants and the location of the extraction wells (EW 1–EW 4) used to demonstrate the SVE system.

A multiwell design was utilized with two 2-in. wells completed in each borehole in both the sand zones. The system was tested for 59 days and removed a total of 590 kg of contaminants. Table 20.5 compares observed soil concentrations of TCE in the vicinity of EW 4 before and after the test. At other wells, the reduction in TCE concentrations in the soil was less—ranging from 9 to 30%. Overall, this test showed the potential of SVE to clean up VOCs.

Problems

1. The capture-zone type curve for a single well is superimposed on an outline of a dissolved contaminant plume as shown in the figure for this problem.

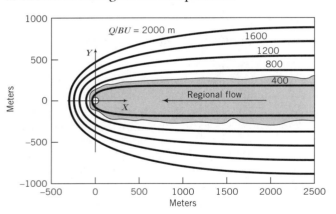

Problem 1

a. Given that the thickness of the aquifer is 10 m and the Darcy flow velocity is 1×10^{-6} m/s, calculate the single-well pumping rate.

b. For an aquifer transmissivity of 0.005 m²/s, storativity of 1×10^{-4}, and available drawdown of 10 m, determine whether the calculated production rate is feasible.

2. A soil-vapor extraction system is proposed to clean up pure chlorobenzene spilled at a site.

a. Knowing that the formula weight of chlorobenzene is 113 g/mol and its vapor pressure 0.012 atm at 20°C, estimate the vapor-phase concentration (as mg/L) in equilibrium with the pure liquid.

b. Estimate the rate of removal of chlorobenzene from the ground assuming an airflow rate of 140 L/min

and equilibrium between the phases. Express your answer in both mg/min and kg/day.

c. Compare your answer in (b) with that obtained by the nomogram.

3. In Chapter 19 through discussion, we described the case of a gasoline spill in gravel along the Trans Canada Highway. Develop two feasible strategies in qualitative terms for remediating the site. Be sure to explain the positive and negative factors that would influence the choice of a particular strategy.

4. An air-permeability test was conducted using an extraction rate of 75 L/s. The drawdowns shown below were measured in an observation well located 5 m away. Given that the vapor density is 0.001 g/cm³ and the zone is 2.5 m thick, calculate the hydraulic conductivity to gas.

Time (min)	Drawdown (cm water)
10.	69.2
100.	108.4
1000.	147.6

5. An underground storage tank at an industrial facility leaked approximately 90,000 L of gasoline over a 5-year period into a shallow sand aquifer. An aggressive program of product recovery collected approximately 35,000 L of free product.

a. Explain what potential problems would occur by simply leaving the residually saturated fraction of the spill in the ground.

b. What are the feasible alternatives for removing this volume of the gasoline?

Chapter 21

In Situ Destruction and Risk Assessment

21.1   Intrinsic Bioremediation
21.2   Bioventing and Bioslurping
21.3   Abiotic Chemical Destruction
21.4   Risk Assessment
21.5   Fernald Case Study

Techniques for degrading, detoxifying, or immobilizing contaminants in situ rely on biological and chemical processes to get the job done. With biodegradation, bacteria mineralize or transform contaminants using organic contaminants as an energy source. Under favorable conditions, biodegradation produces harmless compounds like water and $CO_2(g)$ from complex organic molecules. One new alternative, intrinsic bioremediation, evaluates the natural capabilities of a hydrogeologic system to biodegrade contaminants without active treatment. The challenges with this approach are to demonstrate that cleanup will be achieved and that the monitoring scheme is appropriate.

Other common approaches to bioremediation are more proactive. They are aimed at enhancing the growth potential of existing populations of microorganisms through the addition of nutrients like nitrogen and phosphorous, as well as electron acceptors like oxygen in the case of aerobic bacteria. There also has been promising work on introducing specially engineered bacteria to metabolize hard-to-treat contaminants like the chlorinated solvents. All of these approaches will be discussed in the following sections.

Chemical degradation methods treat contaminants in situ through the addition of treatment agents. In this chapter, we will be concerned especially with techniques that break down contaminants to less harmful constituents. To a lesser extent, we will consider techniques that chemically transform contaminants to an immobile phase.

The chemical schemes are specific to particular types of compounds. For example, destruction of PCE and TCE can be achieved through reductive dechlorination, catalyzed by zero-valent iron. Similarly, TCE can be abiotically transformed using oxidizing agents like potassium permanganate or ozone. Overall, the chemical technologies are less well developed than the biological approaches, with demonstration projects just beginning to emerge.

21.1   Intrinsic Bioremediation

Intrinsic bioremediation relies on the natural system to affect the in situ degradation of contaminants without engineering efforts to stimulate the natural populations. This approach is particularly well suited for sites with a low risk of impacting sensitive off-site receptors. Many believe that intrinsic bioremediation is a no-action alternative. However, pursuing this strategy requires (1) an investigative effort to demonstrate that the natural system is capable of eliminating the specific contaminants, (2) a detailed analysis to demonstrate that significant health

443

risks do not exist, or adjacent properties or aquifers would not be impacted, and (3) a long-term monitoring effort to verify that the expected cleanup is occurring (NRC, 1993).

Intrinsic bioremediation is mainly used to clean up petroleum spills. The tendency for petroleum hydrocarbons to biodegrade is reasonably well understood, and a variety of site studies have documented the capabilities of natural systems to degrade contaminants. Observations made at the site of a pipeline break at Bemidji, Minnesota, are particularly instructive. Crude oil contaminated a shallow unconfined aquifer. Monitoring of the site (Baedecker and others, 1989) showed that dissolved compounds had migrated initially some 200 m. However, since 1987, the plume has not advanced. Additional evidence for intrinsic bioremediation comes from a detailed evaluation of chemical components in the plume.

It is difficult to determine that biodegradation is occurring at a site, and to quantify the rate of degradation in fate and transport analyses. Dispersion and sorption lead to concentration reductions, but not to an overall reduction in the contaminant mass. Buscheck and others (1993) describe a method for using monitored well data to quantify patterns of contaminant attenuation as a function of either time or distance. The method attempts to fit simple first-order kinetic models to concentration data from one or more monitoring sites.

The concentration versus time analysis establishes the rate of concentration reduction over some period of record at a single monitoring well. Mathematically, a first-order model provides the concentration as a function of the initial concentration and decay rate as

$$C(t) = C_0 e^{-kt} \qquad (21.1)$$

where $C(t)$ is the contaminant concentration at any time > 0, C_0 is the initial concentration at $t = 0$, and k is the decay rate constant (t^{-1}). Such a decrease in contaminant concentration could be observed at or near the fringe of a shrinking plume. In wells closer to the source, an exponentially decreasing concentration could suggest significant source decay (Buscheck and others, 1993).

The concentration versus distance model uses data from several monitoring wells, collected at the same time along a pathline. By recasting time as distance over velocity, Eq. 21.1 can be rewritten as

$$C(x) = C_0 e^{-k\frac{x}{v}} \qquad (21.2)$$

where $C(x)$ is concentration as a function of distance x, and v is the linear ground-water velocity (no retardation assumed). This model conceptualizes the plume being at steady state. In other words, dissolved contaminants being advected and biodegraded will achieve a steady-state concentration distribution (Buscheck and others, 1993).

Fitting an exponential model to monitoring well data should yield an estimate of the decay rate from Eq. 21.1, or k/v from Eq. 21.2. If an estimate of v is available, then the latter equation can also provide an estimate of k. However, evidence of a temporal or spatial reduction in concentration by itself is not firm evidence that biodegradation is operative. Independent confirmation should be provided by examining various parameters such as dissolved oxygen or other electron acceptors, E_H measurements, or the enumeration of microbial populations attached to the porous medium (Buscheck and others, 1993).

This theory was applied to a storage facility at Fairfax, Virginia (Buscheck and others, 1993). Benzene data are plotted in Figure 21.1 for wells located along a pathline at distances approximately 0, 25, 175, and 505 ft from the contaminant source. The best-fit line through the data provides a k/v estimate of -0.011. With a velocity estimate of 0.05 ft/day, the decay rate for benzene is 5.5×10^{-4} day^{-1} or 0.055% day^{-1}. Toluene, ethylbenzene, and xylene values are 0.045, 0.045, and 0.04% day^{-1}, respectively. An analysis of the indicator parameters provides corroborating evidence of biological activity.

Table 21.1 is a summary of data on a variety of sites studied by Buscheck and others (1993). Overall, most of the decay rates fall in a range of from 0.1 to 0.46% day^{-1}. These rates represent half-lives of approximately 150 to 693 days, which is comparable to half-lives due to biodegradation (Buscheck and others, 1993).

There are data available to indicate that chlorinated aliphatic hydrocarbons may biodegrade naturally under some conditions (Major and others, 1991; McCarty and Wilson, 1992). One recent study by Semprini and others (1995) examined solvent contamination in a sand aquifer near St. Joseph, Michigan. The study found that TCE,

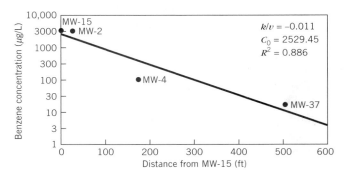

Figure 21.1 Change in benzene concentration along a pathline at a site in Fairfax, Virginia (modified from Buscheck and others, 1993). Reprinted by permission of Conference on Petroleum Hydrocarbons and Organic Chemicals in Ground Water—Prevention, Detection and Restoration. Copyright © 1993. All rights reserved.

Table 21.1 Summary of Estimated Intrinsic
Biodegradation Rates

Site	Compound	k (% per day)
Campbell, CA	Benzene	1.20
	Ethylbenzene	0.67
	Xylene	1.12
	Benzene	0.42
Palo Alto, CA	Benzene	0.30
Virginia Beach, VA	PCE	0.46
	TCE	0.30
Montrose County, CO	Benzene	0.42
Provo, UT	Benzene	0.23
San Jose, CA	Benzene	0.16
	Benzene	0.10
Chemical facility	Toluene	0.39
	PCE	0.34
	TCE	0.26

which had been present in the aquifer for more than a decade, had degraded to *cis*-1,2-dichloroethene (*c*-DCE), vinyl chloride (VC), and ethene. The distribution of these compounds was correlated with sulfate and methane concentrations. The transformation of TCE to DCE is associated with sulfidogenic conditions, while reductions to VC and ethene require methanogenic conditions (Semprini and others, 1995). Given the relatively slow reaction rates involved and the incomplete nature of the reductive dechlorination reactions, it is unlikely that a case can be made for applying intrinsic bioremediation schemes to problems with chlorinated solvents. However, these studies do provide some hope that systems contaminated by solvents that are not being treated could have some capacity for intrinsic remediation.

21.2 Bioventing and Bioslurping

Bioremediation schemes promote the development of natural bacterial populations to transform organic contaminants into less hazardous compounds. Unlike other cleanup technologies, biodegradation can reduce contamination to extremely low levels. For example, cleaning up a gasoline spill by pumping leaves behind a relatively significant proportion of the spill. This fraction is the gasoline held at residual saturation in the pores. If left in place, this remaining contamination will continue to be a source for dissolved compounds.

A variety of schemes have been proposed to utilize natural bacteria to destroy contaminants. In this respect, both bioventing and bioslurping have a proven track record for cleaning up fuel spills. They facilitate the stimu-

lation of an existing population of bacteria in aerobic situations. Other research has focused on methods that deal with anaerobic or co-metabolic systems, which are required for the breakdown of chlorinated solvents like PCE and TCE.

Both bioventing and bioslurping are used primarily to clean up fuels from the unsaturated zone and the vicinity of the water table. They rely on air from the atmosphere to provide oxygen necessary for biodegradation. In some respects, bioventing is similar to SVE methods in that a vacuum-extraction well moves air through the spill. However, unlike SVE systems that strive to maximize the volatilization of low-molecular-weight compounds, bioventing systems are operated to maximize the in situ biodegradation of compounds. In other words, SVE systems involve volatilization with minor biodegradation, and bioventing systems involve biodegradation with minor volatilization.

A system for in situ bioventing, thus, is the same as that in Figure 20.5. The main difference, however, is the airflow rate. In SVE systems, rates are high to maximize phase partitioning of the contaminants. In bioventing systems, rates are much lower because the system needs to supply oxygen to bacteria only in the zone of contamination. In general, the rates of oxygen utilization by bacteria are relatively small so that airflow rates can be kept small.

Bioventing destroys contaminants in the subsurface. Thus, air removed by the extraction well should contain relatively low vapor-phase concentrations of contaminants and may not require treatment at the surface. In large projects, reducing the costs for long-term treatment can dramatically affect the overall economics.

Because bioventing systems rely on bacterial destruction, they can be used to clean up relatively nonvolatile fuels. For example, middle petroleum distillates, like diesel fuel, contain a smaller proportion of volatile compounds compared to gasoline. Thus, SVE systems are relatively inefficient. These compounds, however, are commonly biodegradable and will respond to bioventing.

Bioslurping is an approach that combines two remedial technologies: free product recovery and bioventing. Free-phase LNAPLs are recovered using vacuum-enhanced pumping. The vacuum system also contributes to bioventing in the unsaturated zone. A bioslurper well consists of an exterior casing and well screen that is installed across the water table (Figure 21.2). A one-inch PVC drop tube enters the well through a vacuum-tight seal and extends to the ground-water–product interface (Kittel and others, 1994). When free product is present in the well, the vacuum system removes it. However, much of the time the system pumps air. Equipment on the surface separates water and free product and treats contaminants in the air stream (Kittel and others, 1994).

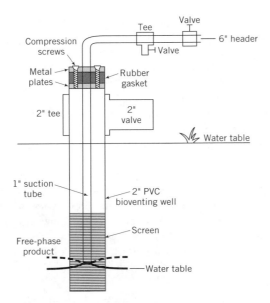

Figure 21.2 Schematic diagram of a bioslurper well design (from Kittel and others, 1994). Reprinted by permission of Conference on Petroleum Hydrocarbons and Organic Chemicals in Ground Water—Prevention, Detection and Restoration. Copyright © 1994. All rights reserved.

Applicability of the Technology to Contaminant Groups

The essence of a biotransformation reaction is the oxidation of an organic compound by an electron acceptor such as $O_2(g)$, NO_3^-, SO_4^{2-}, or CO_2 in a reaction whose rate is controlled by bacteria. To date, the most common application of biodegradation schemes is for cleaning up petroleum hydrocarbons. The monoaromatic and aliphatic constituents in these hydrocarbons are mineralized in aerobic reactions as follows:

$$C_6H_6 + \tfrac{15}{2}O_2 \rightarrow 6CO_2 + 3H_2O \qquad (21.3)$$

Representative of petroleum hydrocarbons and derivatives are gasoline, fuel oil, polycyclic aromatic hydrocarbons (PAHs), creosote, ethers, alcohols, ketones, and esters (NRC, 1993). The greatest problem in degrading these compounds comes with complex molecules with a strong tendency to sorb onto aquifer solids.

Another class of potentially treatable organic contaminants is the halogenated aliphatic compounds (Lee and others, 1987). Given their abundance and relative mobility, schemes to treat these compounds are under active research. Certain highly chlorinated contaminants, such as PCE, are resistant to degradation by aerobic microbes (NRC, 1993). However, they may be degraded by anaerobic organisms. Thus, many of the conventional approaches for treating petroleum spills are ineffective for these compounds.

The feasibility for degrading halogenated aliphatic compounds in aerobic settings improves as the degree of halogenation decreases (NRC, 1993). Typically, degradation is achieved using a mixed culture of bacteria, which grows by oxidizing methane (usually added) and co-metabolizes the halogenated aliphatic compounds (Wilson and Wilson, 1985; Fogel and others, 1987).

Another important class of halogenated organic contaminants is the halogenated aromatic compounds. These compounds contain one or more halogen-bearing benzene rings (NRC, 1993). Examples of these compounds include chlorinated benzenes, polychlorinated biphenyls (PCBs), and pentachlorophenol. Highly chlorinated compounds do not appear to degrade aerobically, although they are biodegradable to some extent by anaerobes. Again, as the degree of halogenation is reduced, compounds are susceptible to aerobic degradation (NRC, 1993).

Because the methods for dealing with compounds degraded under anaerobic settings are under development, the remainder of the discussion in this section will be concerned mostly with the remediation of petroleum spills.

Requirements for Success with Bioventing Systems

For aerobic bioventing methods to succeed, reasonably large volumes of air, containing oxygen, must be moved into the subsurface. The air permeability of a medium depends primarily on the type of geologic material present and/or the extent of fracturing. Soil moisture is another important variable affecting air permeability. High soil-moisture levels can effectively reduce the air permeability of the medium due to multiphase effects. In a performance analysis of bioventing at 103 test locations, Miller and others (1993) found that bioventing was infeasible at only three sites due to a combination of high water table, high moisture contents, and fine-grained soils.

One absolute requirement to make a bioventing system work is the presence of a microbial population that is capable of degrading the organic contaminant. For many applications, where the susceptibility of the contaminant to degrade is well known (for example, gasoline), this requirement is usually met. For other less well known contaminants, laboratory treatability testing (as outlined in Chapter 13) and field testing are necessary to demonstrate biodegradability. A field procedure for treatability testing is outlined in the following section.

Miller and others (1993) identified soil pH, nutrient level, and temperature as other factors possibly important in affecting biodegradation rates. Of this list, temperature appears to be most important. Biodegradation rates at cold sites are lower than at warmer sites. However, even in cold soils (that is, 3°C), biological activity is measurable

(Miller and others, 1993). As a rule of thumb, biological activity appears to double for every 10°C, following the van Hoff–Arrhenius equation. Nutrient levels appear to be somewhat important. However, preliminary experiments in the field find no beneficial effects of adding nutrients.

In Situ Respiration Testing

Kittel and others (1993) describe a field test to determine the efficacy of bioventing. The in situ respiration test uses observed oxygen utilization rates to estimate in situ biodegradation rates. These estimates establish whether bacteria can biodegrade the contaminant in question. The test involves (1) flooding the unsaturated zone with air, containing oxygen and an inert tracer (helium), (2) cutting off the source of oxygen, and (3) quantifying how fast bacteria utilize oxygen.

Running one of these tests involves the following steps:

- site evaluation and test-design with the help of a soil gas survey,

- installation of instrumentation,

- test setup and monitoring, and finally,

- data reduction and rate calculations (Kittel and others, 1993).

The first step is to decide on locations for the monitoring wells. Most wells are installed in zones of contamination. One or more background wells are placed away from sites of contamination. Kittel and others (1993) recommend that well sites be selected by a soil-gas survey with concentration measurements for CO_2, O_2, and hydrocarbons. For shallow contamination, sampling is conducted with hand-driven monitoring points (Figure 21.3). Soil gas can be sampled at depths greater than about 6 m with hydraulically driven probes (Kittel and others, 1993). Figure 21.3 illustrates how the different gas detectors are connected to the system to make the various measurements. Kittel and others (1993) list equipment manufacturers.

Using the soil-gas measurements, an obviously contaminated site can be selected. At such a site, O_2 concentrations would fall in a range of 0 to 2%, CO_2 concentrations in a range from 5 to 20%, and hydrocarbon concentrations > 10,000 mg/L, depending on the contaminant. The control well(s) would be located at an uncontaminated site with O_2, about 15 to 20%, CO_2 about 1 to 5%, and hydrocarbons < 100 mg/L.

The next step is to install the various monitoring wells. Kittel and others (1993) recommend using permanently installed gas-monitoring wells. Normally, one would collect data from four or so wells in the zone of contamination and one or two background wells. The test begins by injecting air with 1 to 2% helium into the background and monitoring points (Figure 21.4). Gas-flow rates of 60 to 100 ft³/hr are maintained for 24 hr. The helium is used

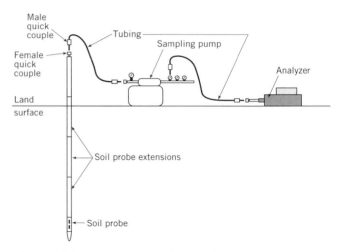

Figure 21.3 Equipment used to conduct a soil-gas sampling survey (from Kittel and others, 1993). Reprinted by permission of Conference on Petroleum Hydrocarbons and Organic Chemicals in Ground Water—Prevention, Detection and Restoration. Copyright © 1993. All rights reserved.

to determine whether O_2 concentrations are influenced by other processes like diffusion.

After the air injection stops, the monitoring wells are sampled on a regular basis using a soil-gas monitoring pump with appropriate analyzers for O_2, CO_2, helium, and total hydrocarbons. Gas-flow rates are kept small—2 to 4 ft³/hr—to avoid drawing in air from the atmosphere. With time, O_2 levels decline and CO_2 levels increase as oxygen-enhanced biodegradation occurs. The test is halted when O_2 levels reach about 5%, or after 5 days of sampling (Kittel and others, 1993).

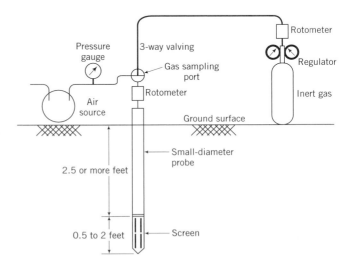

Figure 21.4 Equipment used to flood the porous medium with air and an inert tracer in preparation for an in situ respiration test (from Kittel and others, 1993). Reprinted by permission of Conference on Petroleum Hydrocarbons and Organic Chemicals in Ground Water—Prevention, Detection and Restoration. Copyright © 1993. All rights reserved.

The resulting data are interpreted graphically to provide O_2 utilization rates. Kittel and others (1993) prefer to work with O_2 rather than CO_2, because fewer factors influence O_2 concentration. The percent change in O_2 with time and an assumed stoichiometric relationship for the reaction of hydrocarbon and oxygen is used in the following equation to estimate biodegradation rates

$$K_B = -K_O A D_O \frac{C}{100} \qquad (21.4)$$

where K_B is the biodegradation rate (mg hydrocarbon/ kg soil day), K_O is the measured oxygen utilization rate (percent per day), A is the volume of air per m^3 of soil (L/kg), D_O is the density of oxygen gas (mg/L), and C is the mass ratio of hydrocarbons to oxygen required for mineralization. This latter parameter comes from assuming that the hydrocarbon reacts like hexane as

$$C_6H_{14} + 9.5O_2 \rightarrow 6CO_2 + 7H_2O$$

One mole of hexane is 86 g and 9.5 moles of oxygen is 304 g, which yields a hydrocarbon/oxygen ratio of 1/3.5.

The background measurements provide a basis for correcting the observed O_2 utilization rates for other processes. For example, O_2 can be utilized in the oxidation of iron. Similarly, helium behavior in all of the wells is used to indicate possible short-circuiting of flow. The following example illustrates the calculation.

Example 21.1

Kittel and others (1993) report results from an in situ respiration test at Fallon Naval Air Station (NAS) in Nevada. Table 21.2 lists pretest measurements along with 11 other measurements of O_2 and CO_2 concentrations. Assuming a porosity of 0.3; a soil bulk density of 1440 kg/m^3; an oxygen density of 1330 mg/L (varies with tem-

Table 21.2 **Summary of Monitoring Data for O_2 and CO_2 at Fallon NAS**

| Time (hr) | Percent | |
	O_2	CO_2
−23.5	0.05	20.4
0	20.9	0.05
2.5	20.3	0.08
5.25	19.8	0.10
8.75	18.7	0.13
13.25	18.1	0.16
22.75	15.3	0.14
27.0	15.2	0.22
32.5	13.8	0.14
37.0	12.9	0.23
46.0	11.2	0.22
49.5	10.6	0.16

From Kittel and others (1993).

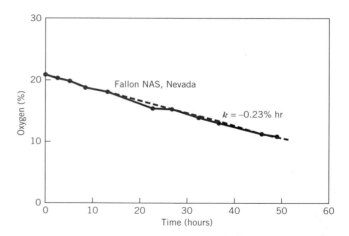

Figure 21.5 Results of an in situ respiration test at Fallon NAS (modified from Kittel and others, 1993). Reprinted by permission of Conference on Petroleum Hydrocarbons and Organic Chemicals in Ground Water—Prevention, Detection and Restoration. Copyright © 1993. All rights reserved.

perature, altitude, and atmospheric pressure); a hydrocarbon-to-oxygen ratio for hexane of 1/3.5; and no other processes affecting gas concentrations, calculate the biodegradation rate.

A plot of the data (Figure 21.5) yields an oxygen utilization rate of −0.23%/hr. The value of A is determined from the porosity and soil bulk density as 300/1440 or 0.21 L/kg.

$$K_B = -K_O \frac{(0.21)(1330)(1/3.5)}{100} = 0.8 K_O$$

Expressing the utilization rate for O_2 in days gives K_O a value of 5.52%/day (−0.23%/hr × 24 hr/day). Thus, the biodegradation rate is

$$(0.8)(-5.52) \quad \text{or} \quad -4.4 \text{ mg/kg day}$$

In situ respiration tests have been conducted at a variety of Air Force sites. Data summarized by Miller and others (1993) show O_2 utilization rates commonly in a range from 3 to 12% per day (Figure 21.6). These rates produce hydrocarbon biodegradation rates of approximately 1.5 to 7.5 mg fuel/kilogram of soil each day. With levels of hydrocarbon contamination (TRPH) in the order of 2000 to 3000 mg hydrocarbons/kg soil, it will require about 1 to 5 years to effect a cleanup.

Progress in Solvent Bioremediation

The bioremediation schemes in this chapter have been targeted almost exclusively to fuel spills. While there is an obvious need for biodegradation schemes in cleaning up solvents (that is, halogenated aliphatic compounds), there are important technical limitations. These com-

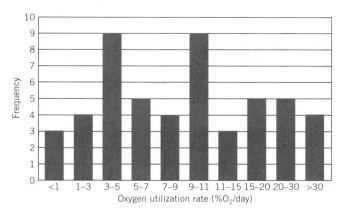

Figure 21.6 Survey of results from a large number of in situ respiration tests (from Miller and others, 1993). Reprinted by permission of Conference on Petroleum Hydrocarbons and Organic Chemicals in Ground Water—Prevention, Detection and Restoration. Copyright © 1993. All rights reserved.

pounds are resistant to biodegradation in aerobic environments because of a lack of appropriate degradative enzymes (Chapelle, 1993). In fact, few if any bacteria can use halogenated aliphatic compounds as a primary energy source (Chapelle, 1993). Thus, solvents in the soil and the ground water are extremely persistent.

Experiments (for example, Wilson and Wilson, 1985) have shown that compounds like TCE can be biodegraded under aerobic conditions. Essentially, methanogenic bacteria are capable of co-metabolizing aliphatic compounds. These bacteria can be stimulated in aerobic settings through the addition of methane gas. The potential of this approach was validated in a three-year field investigation at the Moffett NAS in California (Roberts and others, 1990; Semprini and others, 1990). A variety of in situ tests showed the possibilities of degrading solvents by stimulating naturally occurring populations of methanotrophic bacteria.

A full-scale demonstration of methane injection to promote the biodegradation of solvents was undertaken by the U.S. Department of Energy (DOE). The results of this study are presented in a large number of DOE/OTD reports. The site for the demonstration was the M Area at the Savannah River Site, Aiken, South Carolina. At the M Area, a leaking process water line resulted in contamination by TCE, PCE, and TCA. DOE estimates that between 260,000 and 450,000 lb of organic contaminants are present in the ground water, with additional mass sorbed (U.S. DOE, 1995). Prior to the implementation of the in situ bioremediation scheme, concentrations of TCE and PCE in the water ranged from 10 to 1031 μg/L and 3 to 124 μg/L, respectively. The sorbed concentration of TCE and PCE on the sediments ranged from 0.67 to 6.29 mg/kg and 0.44 to 1.05 mg/kg, respectively (U.S. DOE, 1995). The zone of contamination has an areal

extent of about 1200 acres and is approximately 150 feet thick.

The demonstration involves two horizontally drilled wells, installed above and below the water table (Figure 21.7). A variety of operational modes were tested including;

* SVE in the vadose zone (20 days)

* air sparging through the deeper well, and SVE in the shallow well (33 days),

* methane and nutrient addition to the air-sparging well (Figure 21.7) as,

    * 1% methane addition (107 days)

    * 4% methane addition (79 days)

    * pulsed 4% methane addition (94 days)

    * addition of gaseous nutrients in the form of nitrous oxide, triethyl phosphate in combination with 4% methane in pulses (94 days).

Over the 384 days of the test, approximately 17,000 lb of solvents were removed. The SVE system removed about 12,096 lb and biodegradation led to the mineralization of 4838 lb. The assessment indicated that before stimulation by methane, the microbial populations were low and under nutrient stress. The addition of methane stimulated the growth of methanotrophs, which were effective in the biodegradation of TCE. Apparently, contaminants were mineralized without the formation of intermediates. PCE degradation was accomplished in anaerobic pockets created within the larger aerobic system. The pulsed operation of the system, thus, may have promoted both aerobic and anaerobic degradation simultaneously (U.S. DOE, 1995).

21.3 *Abiotic Chemical Destruction*

Remedial alternatives are now being studied that involve the in situ abiotic destruction of aqueous contaminants.

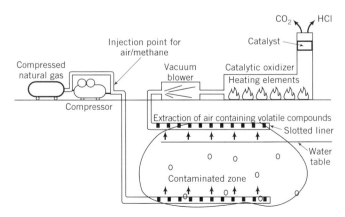

Figure 21.7 Schematic of the in situ bioremediation system demonstrated at the Savannah River Site (modified from U.S. DOE, 1995).

An exciting new approach involves using reduction reactions to transform certain chlorinated solvents. Kriegman-King and Reinhard (1992) showed how carbon tetrachloride could be made to transform to chloroform and CS_2 in the presence of pyrite, sulfide, biotite, or vermiculite. More recently, Gillham and O'Hannesin (1994) were successful in using zero-valent iron (iron filings) as an "enhancing agent" to promote the dehalogenation of chlorinated methanes, ethanes, and ethenes. They found that all of the chlorinated compounds that they tested, with the exception of dichloromethane, were transformed at relatively rapid rates. Reactions generally followed pseudo-first-order rate laws with half-lives ranging from 0.25 to 432 hr (Gillham and O'Hannesin, 1994). Overall, their experiments indicated that the presence of iron enhanced transformation rates 3 to 13 orders of magnitude.

Gillham and O'Hannesin (1994) concluded that the reaction is electrochemical in nature and involves the oxidation of iron and the reductive dechlorination of the organic compounds. Following is their suggested overall reaction describing this process:

$$Fe^0 + 3H_2O + X\text{-}Cl \rightarrow 2Fe^{2+} + 3OH^- + H_2 + X\text{-}H + Cl^-$$

Somewhat less advanced is work that examines the potential of oxidizing organic solvents using powerful oxidizing agents like hydrogen peroxide (H_2O_2) and potassium permanganate ($KMnO_4$). These reactions are particularly attractive because they appear capable of mineralizing the compound to CO_2 and water.

Reactive Barrier Systems

The relatively short half-lives for abiotic transformations involving chlorinated organic contaminants have prompted the development of creative strategies for engineering the systems. The reactive barrier system is one such example. In its simplest form, the reactive barrier is a chemically active zone created in the subsurface to facilitate the transformation or mineralization of chlorinated solvents. Ground water passes through the barrier, leaving the contaminant to be transformed within the barrier (Figure 21.8). There is the hope that the barriers would operate as a passive remedial system. In other words, once the barrier is constructed, it would work for years without extensive maintenance. At the time of writing, Gillham and co-workers have several field trials under way to evaluate the performance of zero-valent iron barrier systems under field conditions.

Other possibilities exist for the treatment of contaminants in reactive barrier systems. For example, the barrier could contain dense populations of microorganisms to biodegrade organic compounds. Work has been under way at Lawrence Livermore National Laboratory to develop this concept. Xu and Schwartz (1994) describe the results of experiments that show how lead and potentially other metals could be immobilized in a reactive

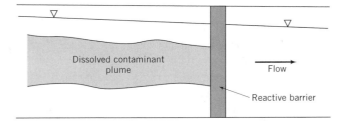

Figure 21.8 Schematic representation of a reactive barrier system. Ground water passes through the barrier, while contaminants are removed in situ at the barrier.

barrier. In this application, apatites placed in the wall provide small quantities of phosphate that react with dissolved lead to produce lead phosphate minerals. The resulting precipitates have an extremely low solubility.

Funnel-and-Gate Systems

Consideration of reactive barrier technologies often leads to the question of how one would actually install one of these systems. Simply digging out a piece of aquifer and replacing it with a reactive medium may be feasible for shallow small spills. However, this approach would be infeasible for plumes that are wide or located deep in the ground. The funnel-and-gate system described by Starr and Cherry (1994) overcomes these limitations. Their method uses low-conductivity cutoff walls (the funnel) to focus flow through a gap (the gate), which contains the chemically active zone.

As Figure 21.9 illustrates, the funnel-and-gate system can be deployed in a variety of ways. The choice of the design is dictated by the types of contaminants involved and the types of contaminants present. If the plume is wide, several gates could be included in the system. If a variety of contaminants are involved, different reactor systems could be stacked together (Figure 21.9b). With LNAPLs, it is likely that a hanging gate would suffice (Figure 21.9c).

The design of the funnel-and-gate system needs to be closely tied to rate of the reaction acting to destroy the contaminant (Starr and Cherry, 1994). A sufficiently long residence time in the reactive portion of the system needs to be provided to reduce the contaminant concentrations appropriately. Figure 21.10 displays potential concentration reductions that are possible for given residence times and reaction half-lives. The residence times are design variables determined by the length of the cutoff walls and the extent to which the cutoff walls surround the zone of contamination (Starr and Cherry, 1994).

21.4 Risk Assessment

One of the realities of the site-remediation business is that most contaminated sites will never be cleaned up. Overall, the cost of cleaning up sites far exceeds the

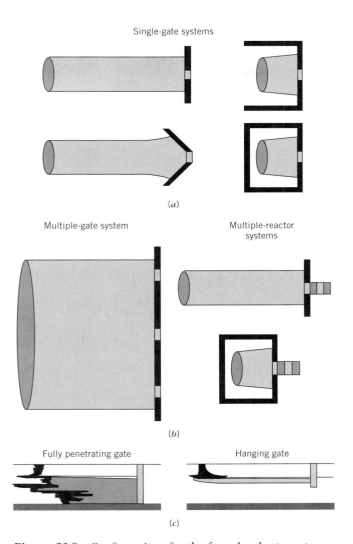

Single-gate systems

(a)

Multiple-gate system

Multiple-reactor systems

(b)

Fully penetrating gate

Hanging gate

(c)

Figure 21.9 Configurations for the funnel-and-gate system (from Starr and Cherry, 1994). Reprinted by permission of Ground Water. Copyright © 1994. All rights reserved.

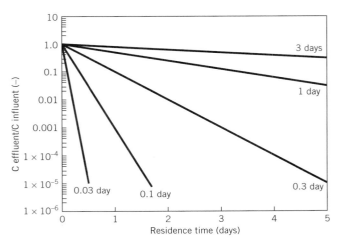

Figure 21.10 Effect of residence time and reaction half-life on concentration reductions achievable in the gate (from Starr and Cherry, 1994). Reprinted by permission of Ground Water. Copyright © 1994. All rights reserved.

dollars available for cleanup, especially for orphan sites. If cleanups are limited by the availability of funding, then it makes sense to develop criteria to select those sites that pose the greatest "risk" to public health. Eventually, if a site is to be cleaned up, the remedy should reduce the health risk to some appropriately small level. It is in these contexts that risk assessment has emerged as a significant activity involved with the study of contaminated sites. Risk assessment is broadly defined as the process of characterizing the adverse health effects of human exposures to environmental hazards (NAS, 1983). Beyond evaluating contaminated sites, risk assessment provides a means to assess the impact of waste-disposal schemes, like landfills or nuclear-waste repositories. Though risk assessments are used most commonly to examine risk of human health, the approaches can be applied more generally to other environmental receptors (for example, birds and fish). The U.S. Environmental Protection Agency (EPA) (1989b,c) outlined the basic framework for evaluating hazardous-waste sites to assess potential health and environmental impacts. The framework may be summarized in the following steps: (1) data collection and data evaluation; (2) exposure assessment; (3) toxicity assessment; (4) public health risk characterization; and (5) environmental risk assessment.

Data Collection and Data Evaluation

The objective of data collection and data evaluation is to provide basic information for risk assessment. This process includes reviewing available site information, identifying potential human exposure, defining further data-collection needs (background sampling, modeling, and other sampling needs) and strategy, and actual data collection. This process will also combine data available from site investigations to evaluate analytical methods, chemicals, and extent of contamination, and to develop a set of chemical data and information for use in the risk assessment.

Exposure Assessment

Throughout this book, we have touched on the health implications of contamination, but never really emphasized the ways in which contaminants could move from ground water into humans. In part, our particular focus on geologic systems is less of an oversight and more a recognition that the main role of a ground-water hydrologist is to provide background data for a risk assessment. However, to do this job properly requires basic knowledge of what this assessment is trying to accomplish.

The concept of an exposure pathway is fundamental to a risk assessment. An exposure pathway is the route that contaminants follow from the point of release to some point in the human body where the dose is delivered. A reader progressing systematically through the book should have intimate knowledge about how con-

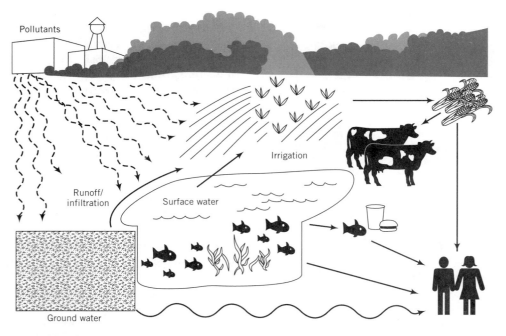

Figure 21.11 Conceptual model of a generalized water pathway from a contaminant source to humans (from FERMCO, 1994).

taminants move from a source along the ground-water pathway (that is, geosphere transport). This pathway is but one of several "water-based" pathways available to contaminants to move away from a source. Further, as Figure 21.11 illustrates, the pathway to humans wends its way through the biosphere as parts of complex plant and animal food chains. The relationships among these pathways are usually captured in a flow chart. Figure 21.12 exemplifies the general environmental pathways for radionuclides.

A risk assessment requires an estimate of existing contaminant concentrations and sometimes the likely future

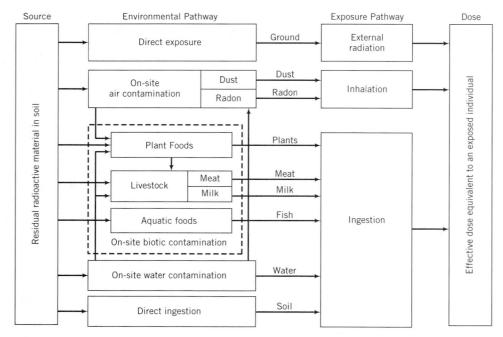

Figure 21.12 Schematic representation of pathways by which humans could receive a dose from radionuclides in soil (from Yu and others, 1993).

concentrations related to each of the exposure pathways (for example, dust radon, meat, milk, etc.; Figure 21.12). These concentrations are usually unavailable, even for the analysis of present conditions. Models, then, provide predictions of biosphere migration and accumulation. As an example, we will consider the code RESRAD (Yu and others, 1993) that calculates risk and doses related to radioactive contaminants. The plant food pathway (Figure 21.12) considers more specific transfer of radionuclides due to (1) root uptake of crops grown in contaminated soil, (2) foliar uptake from contaminated dust deposited on the plant, (3) root uptake of contaminated irrigation water, and (4) foliar uptake from contaminated irrigation water.

These general models are customized to represent details of the contaminated site (for example, whether surface water is available for irrigation), and the lifestyles of individuals living in the vicinity of the site. Thus, assessments tend to be quite site specific. To provide some basis for comparisons among sites, investigators rely on reference scenarios. One of the most well known is the "born loser" scenario. This scenario presents the case for a subsistence farmer who spends his or her entire life on a farm located at the site boundary. This farmer drinks and bathes in contaminated water, eats contaminated food grown on the site, and is exposed to contaminated dust. This scenario represents the case for the maximally exposed individual, although this individual would not be found at most sites.

Multimedia models are commonly used to track the complexities of contaminant fate and transport within the biosphere. These models represent environmental media, water, soil, or plants as homogeneous compartments and estimate contaminant fluxes among compartments or accumulation within compartments (Renner, 1995). The models are formulated based on conservation of mass principles accompanied by appropriate source/sink adjustments related to chemical and biological processes. Examples of such models include CalTOX, Chem-CAN, HAZCHEM, and SimpleBOX (Cowan and others, 1995).

The biosphere analysis provides an estimate of contaminant concentrations available in environmental media. This information establishes intake or exposure levels for each relevant environmental medium, plus a total exposure. The intake equation for air, water, milk, and food has the following form

$$TI = EI_1 + EI_2 + \cdots + EI_n \qquad (21.5)$$

where TI is the total intake of contaminants with units mg/day, and EI is the estimated daily intake for each of the environmental media or other exposures (for example, dermal). EI values for eating and drinking are determined by multiplying C_i, the contaminant concentration in the ith environmental medium, by R_i, the rate of intake.

Values of C_i have units such as, mg/m³ in air, mg/L in water, or mg/kg in food. Values of R_i have units like L/day in the case of water. In some cases, TI in Eq. 21.5 is multiplied by a dimensionless fraction (1 or less) that represents the fraction of the year or the fraction of one's lifetime that a person spends on the site. The total intake (TI) in this latter case becomes the average daily exposure rate expected during an average lifetime. For comparative purposes, values of TI can be normalized to one kg of body weight (divide by body weight, BW). Thus, intake or exposure levels have units like mg/day or mg/kg-BW day.

Example 21.2

An individual consumes contaminated water, milk, and beef containing 3 mg/L, 0.003 mg/L, and 9.5×10^{-8} mg/kg of the contaminant, respectively. Calculate the exposure level to a receptor who weighs 70 kg and consumes 1.5 L/day of water, 0.5 L/day of milk, and 0.2 kg/day of beef.

$$TI = EI_{water} + EI_{milk} + EI_{beef}$$
$$= C_{water}R_{water} + C_{milk}R_{milk} + C_{beef}R_{beef}$$
$$TI = \frac{(3.0 \times 1.5) + (0.003 \times 0.5) + (9.5 \times 10^{-8} \times 0.2)}{70.0}$$
$$= 0.064 \text{ mg/kg-BW day}$$

The greatest intake comes from drinking contaminated water.

Toxicity Assessment

The toxicity assessment evaluates toxicological effects and carcinogenic potentials of chemical and radioactive contaminants. The objectives of the toxicological assessment are to assess the toxicity of the potential contaminants for the site considered and to provide the toxicity values (reference doses for noncarcinogenic effects; slope factors for carcinogenic effects) for risk assessment. Reference Dose (RfD) is an estimate (with an uncertainty typically spanning an order of magnitude) of daily exposure (mg/kg day) to the general human population (including sensitive subgroups) that is likely to be without an appreciable risk of deleterious effects during a lifetime of exposure. Slope Factor (SF) (or called risk factor, RF) is the slope of the dose–response curve in the low-dose region. Source materials for developing toxicological profiles include the EPA's Integrated Risk Information System (IRIS), the U.S. EPA Health Effects Assessment Summary Tables (HEAST), the Agency for Toxic Substance Disease Registry Toxicological Profiles (the Department of Health and Human Services), the *Handbook on the Toxicology of Metals*, and the peer-reviewed scientific literature.

Health-Risk Assessment

The health impact of a contaminant depends not on the intake to the human body but on the dose that the body receives. For example, some contaminants may pass through the body without accumulating in various organs. One of the simplest ways of determining the dose is to multiply the intake by a factor, called the bioavailability factor. This factor represents the fraction of the daily intake of the contaminant that is retained to produce the adverse health effect. The total dose of a contaminant (*TD*) is obtained as

$$TD = EI_1 \cdot B_1 + EI_2 \cdot B_2 + \cdot \cdot \cdot + EI_n \cdot B_n \quad (21.7)$$

where B_i is the bioavailability factor. Values of B_i are available from the medical literature. *TD* values have units of mass/time and are usually normalized by dividing by the body weight (for example, mg/kg-BW day).

Before discussing how these dose calculations are used, let us consider dose thresholds. The existence of a dose threshold implies that if the contaminant intake is below some threshold, there is no health effect. When thresholds do not exist, even a very small dose may have health implications. Cancer risks are typically modeled in this manner given that even one molecule of a contaminant may impact DNA and lead to the growth of a tumor (Reichard and others, 1990).

One of the most straightforward approaches to screening health risks where thresholds are applicable is to compare the total daily dose (*TD* values) to reference standards. The hazard quotient (*HQ*) for the human receptor and the particular contaminant is calculated as

$$HQ = \frac{TD}{RfD} \quad (21.8)$$

with an $HQ > 1$ indicating some risk to the receptor.

This same approach is followed with radionuclides, where the dose is established depending on the type of radiation and the location of the radiation (that is, inside or outside of the body). Thus, the dose will vary depending on the body organs in which the radionuclides accumulate. Dose then can be expressed as an "organ" dose or some "effective" dose that involves a weighted average of the doses to various organs. The weights are proportional to the total risk associated with the irradiation of those organs (Yu and others, 1993).

Dose conversion factors provide a relatively straightforward way of estimating a dose from some intake quantity. These factors are derived from detailed modeling of the decay rates, daughter in-growths, and the metabolism of radionuclides in the body. The dose to a particular organ from a given nuclide would be calculated as

$$TD = EI_1 \cdot DCF_1 + EI_2 \cdot DCF_2 + \cdot \cdot \cdot + EI_n \cdot DCF_n \quad (21.9)$$

where *TD* is dose with units mrem/yr, EI_i as before is $C_i \cdot R_i$ (now for example, curies kg^{-1} × kg yr^{-1}), and

Table 21.3 **Conversion Factors for Common Units of Radionuclide Activity and Dose**

Multiply	By	To Obtain
pCi	10^{-12}	Ci
pCi	0.037	Becquerel (Bq)
mrem	0.001	rem
mrem	10^{-5}	Sieverts (Sv)

DCF_i is the dose conversion factor with units of (mrem/curie) that converts nuclide activity into dose in mrems. The dose, mrem (milli Roentgen Equivalent Man), accounts for the relative biological damage due to the type of radiation involved (FERMCO, 1994). Table 21.3 summarizes the equivalences between commonly used measures of radionuclide activity and dose.

Standards for radionuclides are often presented in terms of dose. For example, in the United States, DOE Order 5400.5 (U.S. DOE, 1990) provides guidelines for radiation exposure to the public based on internationally accepted standards. Figure 21.13 shows how the dose limits are applied to air and ground-water pathways at the DOE site at Fernald, Ohio. The 100 mrem/yr dose accounts for all pathways and is the weighted sum of the dose in various organs.

Although intake or dose-based standards are intended to be protective of public health, they may not capture

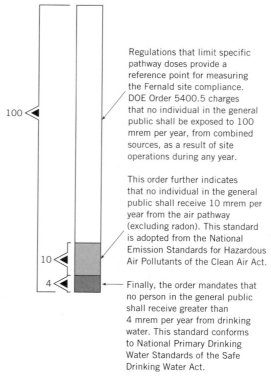

Regulations that limit specific pathway doses provide a reference point for measuring the Fernald site compliance. DOE Order 5400.5 charges that no individual in the general public shall be exposed to 100 mrem per year, from combined sources, as a result of site operations during any year.

This order further indicates that no individual in the general public shall receive 10 mrem per year from the air pathway (excluding radon). This standard is adopted from the National Emission Standards for Hazardous Air Pollutants of the Clean Air Act.

Finally, the order mandates that no person in the general public shall receive greater than 4 mrem per year from drinking water. This standard conforms to National Primary Drinking Water Standards of the Safe Drinking Water Act.

Figure 21.13 Example of dose limits applied to Department of Energy sites, in particular the site at Fernald, Ohio (from FERMCO, 1994).

health risks like cancer where there is no chronic threshold. A major problem in assembling the dose–response data from laboratory experiments is that health effects of chemicals in animals are collected when large doses are applied. These data need to be extrapolated to the low doses that are more common in contaminant settings (Reichard and others, 1990).

Let us briefly consider how lifetime cancer risks are calculated in the case of exposure to some radionuclide. Commonly risk is presented as *individual lifetime risk*, the estimated increase in probability that an individual will experience a health effect over a lifetime, due to exposure (Hamilton and others, 1993). As before, the intake (EI) of the nuclide pCi/yr is multiplied by the dose conversion factor in mrem/pCi to obtain a dose. These effective doses are summed over an assumed lifetime of 70 years as $E_{70}L$. The individual lifetime risks for cancer (ILR) are determined as

$$ILR = E_{70}L \cdot RF \qquad (21.10)$$

where $E_{70}L$ is the sum of effective doses over a 70-year lifetime (mrem), and RF is the risk factor (deaths/mrem) (Hamilton and others, 1993). These risk factors are derived from internationally accepted standards.

Hamilton and others (1993) explain in simple terms how risk data can be interpreted. Assuming a U.S. population of 240 million, a 10^{-6} (one in a million) excess risk would mean that 240 people would die prematurely sometime within their approximate 70-year life span. The resulting excess 3.4 deaths per year would compare to 2.1 million deaths each year from all causes. The U.S. EPA standard for excess individual cancer risks is 1×10^{-4}, or one in 10,000.

Types of Risk Assessments

Risk assessments can vary in complexity, scope, data needs, and ultimately the confidence in the overall result, as represented in Figure 21.14 (Hamilton and others, 1993). Categorization studies provide a simple way to identify from a collection of contaminated sites those that are of greatest potential concern. For example, with the passage of the Comprehensive Environmental Response, Compensation and Liability Act (CERCLA; that is, the Superfund Law), the U.S. EPA was required to develop priorities for undertaking remedial action and possibly removal actions. The Hazard Ranking System (HRS) was implemented in 1982 as a scoring system to determine whether or not a given hazardous site should be placed on the National Priority List for further study and remediation. The original HRS evaluated three migration pathways (ground water, surface water, and air), as well as threats from direct contact, fire, and explosions.

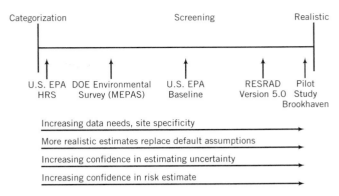

Figure 21.14 Spectrum of approaches to risk assessment (modified from Hamilton and others, 1993).

The system was revised in 1992 to consider relative risks to human health by scoring exposure along four pathways. Sites scoring above 28.5 may be proposed for the National Priority List.

The HRS evaluation is quite detailed, but makes use of only basic information about the site. The ground-water migration pathway examines the likelihood for releases to the aquifer, the characteristics of the waste (such as toxicity, mobility, volume), and impact (present and future) on various "targets" (that is, wells, populations, resources). The HRS, however, does not provide a quantitative assessment of potential risks associated with contamination at a given site (Hamilton and others, 1993).

Figure 21.14 identifies "screening" approaches, which provide increased complexity and accuracy in the risk assessment. Examples of these are the Multimedia Environmental Exposure System (MEPAS) model (Droppo and others, 1990) and baseline risk assessments performed at Superfund sites as part of Remedial Investigation/Feasibility Studies (RI/FS) (U.S. EPA, 1989a).

The MEPAS model is a quantitative risk assessment model, designed to assist in the investigation of DOE sites. It consists of pathway models coupled to a set of exposure models that form the basis for an assessment of health effects associated with the site and relative rankings among environmental problems. Although this model can be used in a detailed and site-specific manner, the risk assessment is still quite simplified (Hamilton and others, 1993). Baseline risk assessments are also considered "screening-level" assessment because of conservative assumptions that are embedded in the analyses and simplified flow and transport models (Hamilton and others, 1993).

Moving to the right in Figure 21.14, the analyses become more rigorous and incorporate more of the detail of the actual site. The RESRAD code was developed at Argonne National Laboratory to implement DOE guidelines for residual radioactive materials in soils. The model considers the critical pathways for human exposure, such as direct exposure, internal doses due to inhalation, and internal doses due to eating or drinking contaminated materials.

The Brookhaven Pilot Study exemplifies a detailed and quantitatively rigorous analysis. This kind of assessment might be carried out at a site that poses significant health concerns. Unlike the screening assessments, which likely contain conservative assumptions and safety factors, the realistic approach employs reasonable parameter estimates and realistic exposure pathways (Hamilton and others, 1993). Flow and transport models that are part of such a modeling effort are tied strongly to an extensive data base of information about the site.

Environmental Risk Assessment

The objective of environmental assessment is to determine if the contaminants detected at a particular site have the potential to adversely affect the existing biological community (plants, wildlife, and livestock) at or surrounding the site. The U.S. EPA guideline (1989c) may be used qualitatively for this purpose.

21.5 Fernald Case Study

The Fernald Case illustrates the application of two different health-risk assessments. Fernald is a DOE facility that produced uranium metal for defense activities. Production at the site ceased in 1989, and now activities are targeted toward cleanup and closure (FERMCO, 1994). The site is located along the Great Miami River, approximately 27 km northwest of Cincinnati, Ohio. A considerable volume of uranium metal waste (low-level radioactivity) is contained in six waste pits. The site also has two silos that contain wastes from refining pitchblende ore. These silos leak radon gas to the atmosphere. A cross section (Figure 21.15) shows how the site is situated on a mainly low-permeability, glacial-drift unit. This unit protected the underlying Great Miami Aquifer from contamination to a significant extent. However, surface-water drainage from the site toward the Miami River (for example, along Paddys Run; Figure 21.16) enables contaminants to enter the aquifer where the drift unit is absent. The so-called South Plume (Figure 21.16) contains uranium concentrations as high as 27 to 190 pCi/L (Hamilton and others, 1993) with a relatively large zone with concentrations above 13.5 pCi/L (20 μg/L).

FERMCO (1994) describes the results of a risk assessment designed to evaluate whether the site met the DOE guidelines in Figure 21.14. The analysis examined both air and water pathways (Table 21.4) using the CAP-88 computer code.

Shown in Table 21.5 is the calculated dose in 1993 to some maximally exposed individual, determined mainly by where this person lives. As indicated, the greatest dose would come from consumption of contaminated well water. Comparison with applicable guidelines shows that the total dose of 1 mrem is far less than the U.S. DOE guideline. Similar dose calculations made annually since 1989 show that restoration activities have reduced the annual dose to the maximally exposed individual from about 19 mrem in 1989 to about 1 mrem at present (FERMCO, 1994).

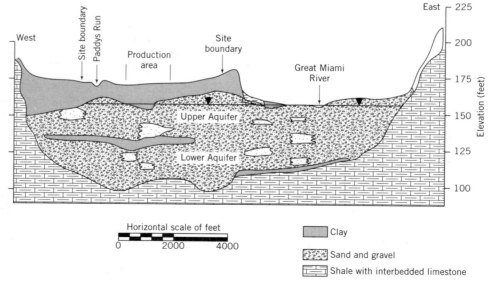

Figure 21.15 Map showing the location of the South Plume. Contamination occurred due to infiltration of water from Paddys Run and the Outfall Ditch, where the overburden is absent (modified from FERMCO, 1994).

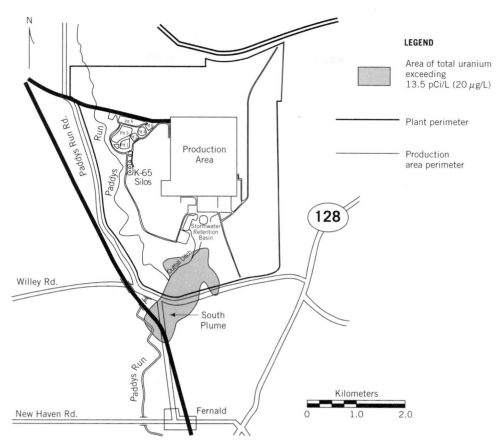

Figure 21.16 East–west geologic cross section at the Fernald site. The regional water table occurs in the Great Miami aquifer. Locally, the overburden is saturated (modified from FERMCO, 1994).

Detailed Risk Assessment

Researchers from Brookhaven National Laboratory and Lawrence Livermore National Laboratory used the Fernald site to demonstrate an approach to nonconservative health-based risk assessments (Hamilton and others, 1993). Their approach emphasizes the use of site-specific data, multidimensional transport codes, and the latest scientific information describing dose–response relationships. Their analysis examined the implications of radon emissions from the K-65 silos and uranium contamination associated with the South Plume. Our discussion here focuses on the plume.

The main toxic effect for both long- and short-term exposure is related to the liver, where uranium tends to accumulate. However, it has been generally difficult to characterize the extent of kidney damage. In the absence of detailed epidemiological or other data, conservative dose–response relationships are often used (Hamilton and others, 1993). The study of Hamilton and others modeled uranium transfer within the body with a Monte Carlo scheme to account for uncertainty of the parame-

ters. The threshold for kidney damage was determined to be in a range of 0.1 to 1 μg-U/g-kidney (Hamilton and others, 1993).

Uranium also may contribute to the development of human cancers. The main problem in dealing with health risks is that the environmental exposures are relatively low. To develop realistic risk numbers, Hamilton and others (1993) examined to what extent uranium accumulated in bones over a lifetime to facilitate estimating dose from intake rates. The estimate of the risk factor came from assessments of the results of international studies. The risk factor from low-dose radiation was estimated to be 0.05 deaths/Sievert.

The analysis was designed to predict the exposure and risks 70 years into the future. Estimates of uranium concentrations in the ground water for 70 years were obtained using the flow and transport code SWIFT III with two conditions of source loading. The first assumed that the source of uranium to the South Plume was removed after 10 years. The second assumed that loading continued to occur at rates comparable to the modeling period. Details on the grid design and parameter selection

Table 21.4 **Summary of Pathways for Human Exposure to Radioactive Contamination from the Fernald Site**

Pathway	Description
Air	
Airborne emissions	Dust, water mists, considers release points, wind speed, direction, and population distribution
Foodstuffs	Dose from eating foodstuffs grown within about 5 km of the site—due to uranium deposited in the soil
Direct radiation	Doses from gamma and X-rays due to radioactive waste stored in K-65 silos, and thorium elsewhere
Water	
Well water	Dose provided by the consumption of water containing uranium in areas south and east of the plant
Fish from Great Miami River	Various water pathways let uranium enter the river. Dose is provided by eating 4.5 kg of fish each year.

are provided by Hamilton and others (1993). The model was coupled to the risk assessment in the sense that individual concentration estimates for each node could be converted to dose and cancer risks. In effect, Hamilton and others (1993) were able to describe health risks in detail as a function of location and time.

The risk assessment considered three exposure pathways for contaminated water: direct ingestion in water, irrigation of crops, and watering of dairy cattle. The multimedia analysis led to estimates of uranium concentration in soil and vegetables and in home-produced food and milk. For the liver assessment, the uranium intake in food from other sources was also included. These concentrations were converted to intake using the equations in

Table 21.6 (from Hamilton and others, 1993). Note that separate equations are written for toxic intakes and cancer-causing intakes. The unit for intake in this latter case is Becquerel (Bq), which is a unit comparable to pCi.

The total uranium dose intake from the sources above are given as

$$TI_c \text{ (Bq/day)} = EI_{wc} + EI_{mc} + EI_{vc} \qquad (21.11)$$

$$TI_t \text{ (}\mu\text{g/day)} = EI_{wt} + EI_{mt} + EI_{vt} + EI_{ot} \qquad (21.12)$$

The problems at the end of the chapter will explore a few of the calculations with these equations in more detail.

The next steps in the toxicity analysis involve converting the intakes to concentrations of uranium in the kidney and comparing these results to the effect threshold discussed previously. Hamilton and others (1993) propose the following equation to estimate the concentration of uranium in the kidney from some chronic intake

$$A = TI_t \cdot (f_1/\ln 2) \cdot \alpha T \qquad (21.13)$$

where A is the uranium mass in the liver (μg U), TI_t is the uranium intake rate (μg/day), f_1 is the fractional transfer from gut to blood, α is the fractional transfer of blood to kidney compartment 1, and T is the biological half-life of uranium in kidney compartment 1 (days). The parameters f_1, α, and T were assigned mean values of 1.03, .063, and 11.1 days, respectively. When A is divided by the total kidney mass, one has a uranium concentration (μg-U/g-kidney) that can be compared to toxic threshold developed previously. Given that model predictions of uranium concentrations in ground water are available for 70 years, one could carry out such a calculation to assess future risks as well as present risks.

Example 21.3

A person drinks 1.2 L/day of water containing a uranium concentration of 20 μg/L. Assuming a kidney mass of

Table 21.5 **Estimates of the 1993 Dose to the Maximally Exposed Individual**

Pathway	Dose Attributable to the Site		Applicable Guideline
Air			
Estimated 1993 emissions	0.016	mrem	10 mrem/air
Foodstuffs grown in Fernald area	0.01	mrem	100 mrem/all pathways
Direct radiation	0.0	mrem	100 mrem/all pathways
Liquid			
Well water in the Fernald area	0.7	mrem	4 mrem/drinking water
Fish from Great Miami River	0.01	mrem	100 mrem/all pathways
Maximally exposed individual	**~1.0**	**mrem**	**100 mrem/all pathways**

From FERMCO (1994).

Table 21.6 **Summary of Equations Used to Calculate Intake Due to Ground-Water Exposure**

Pathway	Equation and Parameters
Ingestion in water	C_{wc} (Bq/L) = C_{wt} (μg/L) $\cdot$ 0.68 (pCi/μg) $\cdot$ 0.037 (Bq/pCi)
	EI_{wc} (Bq/day) = C_{wc} (Bq/L) $\cdot$ R_w (L/day)
	EI_{wt} (μg/day) = C_{wt} (μg/L) $\cdot$ R_w (L/day)
	C_{wc} = uranium concentration expressed as activity (*c* cancer)
	C_{wt} = uranium concentration in water (*t* toxic)
	EI_{wc} = daily intake for calculating cancer risk
	EI_{wt} = daily intake for calculating risk of liver damage
	R_w = rate of water intake
Ingestion in home-produced milk	C_{mc} (Bq/L) = C_{mt} (mg/L) $\cdot$ 0.68 (pCi/μg) $\cdot$ 1000 (μg/mg) $\cdot$ 0.037 (Bq/pCi)
	EI_{mc} (Bq/day) = C_{mc} (Bq/L) $\cdot$ R_m (L/day)
	EI_{mt} (μg/day) = C_{mt} (mg/L) $\cdot$ 1000 (μg/mg) $\cdot$ R_m (L/day)
	C_{mc} = uranium concentration in milk expressed as activity
	C_{mt} = uranium concentration in milk
	EI_{mc} = daily intake for calculating cancer risk
	EI_{mt} = daily intake for calculating risk of liver damage
	R_m = rate of milk intake
Ingestion in home-grown fruit and vegetables	C_{vc} (Bq/kg) = C_{vt} (mg/kg) $\cdot$ 0.68 (pCi/μg) $\cdot$ 1000 (μg/mg) $\cdot$ 0.037 (Bq/pCi)
	EI_{vc} (Bq/day) = $FH \cdot C_{vc}$ (Bq/kg) $\cdot$ R_v (kg/day)
	EI_{vt} (μg/day) = $FH \cdot C_{vt}$ (mg/kg) $\cdot$ 1000 (μg/mg) $\cdot$ R_v (kg/day)
	C_{vc} = uranium concentration in fruit and vegetables as activity
	C_{vt} = uranium concentration in fruit and vegetables
	EI_{vc} = daily intake for calculating cancer risk
	EI_{vt} = daily intake for calculating risk of liver damage
	R_v = rate of fruit and vegetable intake
	FH = fraction of fruit and vegetables that are homegrown
Ingestion of food from other sources	EI_{ot} (μg/day) = $FN \cdot R_o$ (μg/day)
	EI_{ot} = daily intake for calculating risk of liver damage
	R_o = rate of food intake from other sources
	FN = fraction of fruit and vegetables that are not homegrown (1 − *FH*)

750 g, no other uranium intakes, and the parameters given for uranium behavior in the kidney (Hamilton et al., 1993), assess whether this intake could be toxic.

$$EI_{wt}\ (\mu g/day) = C_{wt}\ (\mu g/L) \cdot R_w\ (L/day)$$

$$EI_{wt} = 20.0 \times 1.2 = 24\ \mu g/day$$

$$A = TI_t \cdot (f_1/\ln 2) \cdot \alpha T$$

$$A = 24.0 \times (1.03/0.693) \times .063 \times 11.1$$

$$= 24.9\ \mu g\text{-}U$$

Expressing concentration per gram of kidney is 24.9/750 or 0.033 μg-U/g-kidney. This concentration is less than the threshold for toxicity (0.1 to 1.0 μg-U/g-kidney) and so the intake is not toxic.

At Fernald, Hamilton and others (1993) predicted no toxic effects due to the South Plume. They estimated that uranium concentrations needed to be of the order of 500 μg/L to be capable of producing liver concentrations at the threshold. The main dose came from drinking contaminated water, rather than from milk or food intakes.

The cancer risk requires integrating intakes over a period of time. In effect, a dose estimate is obtained for intakes over each of 70 years. The sum of these effective doses ($E_{70}L$ [Sv]) is multiplied by the risk factor (deaths/ Sv) to estimate the individual lifetime risks for cancer mortality. The analysis showed that all cancer risks were small. The future risk for using water from wells south of the facility was 2.2 × 10^{-5}.

Overall, the results of the two risk assessments for Fernald yielded comparable results. The health risk for individuals in the vicinity of the site is small.

Problems

1. An in situ respiration test yielded an oxygen utilization rate of -0.166%/hr. Assuming a porosity of 0.30, a soil bulk density of 1440 kg/m^3, an oxygen density of 1330 mg/L, and hexane as the reference hydrocarbon, determine the biodegradation rate.

2. A 70-kg rancher living close to a contaminated site is exposed to a contaminant through (1) the consumption of homegrown meat, and (2) the incidental ingestion of dust. Assuming the following information:

For Beef	For Dust
$C_b = 9.5 \times 10^{-8}$ mg/kg	$C_d = 0.024$ mg/kg
$R_b = 0.18$ kg/day	$R_d = 28$ mg/day
$f_b = 0.25$	$f_d = 0.17$

Note f_b multiplies $C_b R_b$, accounting for the fact that the rancher was exposed for only a fraction of his life, and not all the beef was homegrown. f_d is used similarly, reflecting that not all dust is contaminated and the source was present for a fraction of one person's lifetime.

a. Use Eq. 21.5 to estimate the total contaminant intake (as mg/day and mg/kg-BW day) expressed over an average lifetime (that is, multiply by f, which includes this adjustment).

b. Assuming a bioavailability factor of 0.64, estimate the daily dose as mg/kg-BW day.

c. The chronic oral exposure limit (RfD) for this contaminant is 1×10^{-4} mg/kg-BW day. Estimate the hazard quotient, and determine whether the contaminant poses a health risk.

3. A residential well in the vicinity of a hypothetical site was found to contain uranium at a concentration of 100 μg/L. Estimate the risk of this well causing liver damage. Assume that exposure comes from drinking contaminated water and milk and eating contaminated food. The following information is provided in the table.

a. Calculate the intake of uranium from each of the environmental media (that is, EI_{wt}, EI_{mt}, and EI_{vt}).

b. Calculate the total intake of uranium (TI_t).

c. Using the data concerning the behavior of uranium in the body, estimate the concentration in the kidney with a mass of 500 g.

d. Is this dose toxic?

Environmental Medium	Concentration	Intake Rate	Fraction Home Produced
Water	100 μg/L	1.2 L/day	na
Milk	0.006 mg/L	0.25 L/day	1.
Food	0.30 mg/kg	0.36 kg/day	0.22

Answers to Problems

Chapter 1

1. $k = 1.18 \times 10^{-7}$ s^{-1}
2. $k = 2.22 \times 10^{-7}$ s^{-1}
   Recharge $= 1870 \times 10^6$ ft^3
3. Recession length 7.7 months
4. Change in storage $= 6.561 \times 10^6$ m^3

Chapter 3

1. $q = 1.89 \times 10^{-8}$ m/s
2. Effective porosity $= 0.1$, $v = 1.9 \times 10^{-7}$ m/s
   Effective porosity $= 0.0001$, $v = 1.9 \times 10^{-4}$ m/s
7. Total head: 300 m, 205 m, 110 m, 15 m
   Pressure head: 200 m, 190 m, 50 m, 8 m
   Elevation head: 100 m, 15 m, 60 m, 7 m
8. $K = 3.28 \times 10^{-6}$ ft/s, $K = 2.12$ gal/day ft^2,
   $k = 1 \times 10^{-9}$ cm^2, $k = 1.04 \times 10^{-1}$ darcys
11. $K_x = 3.7 \times 10^{-7}$ m/s, $K_z = 2.7 \times 10^{-8}$ m/s
12. $q_x = 1 \times 10^{-4}$ cm/s
16. $q_x = 5 \times 10^{-8}$ cm/s, downward flow
17. **a.** Inflow both aquifers, $Q = 1 \times 10^{-4}$ m^3/s
    b. Outflow aquifer A, $Q = 1.5 \times 10^{-4}$ m^3/s
       Outflow aquifer B, $Q = 5 \times 10^{-5}$ m^3/s
    c. Outflow area A, 1.5×10^{-1}
       Outflow area B, 0.5×10^{-1}

Chapter 4

2. **a.** 300 m^3/m
   b. $S = 8.4 \times 10^{-5}$
   c. 3.6×10^{-10} m^2/N
4. **a.** 6250 lb/ft^2, 4368 lb/ft^2, 1882 lb/ft^2
   b. 4998 lb/ft^2, 4368 lb/ft^2, 630 lb/ft^2
   d. 30 ft
5. 378 m^2/day

Chapter 5

(Answers subject to interpolation of tabulated values.)

3. **a.** At the canal 7.77 m, 5.25 m, 4.2 m, 17.9 m
      At the divide 21 m, 21 m, 20 m, 17.9 m
   b. $q' = 1379$ m^2/yr at 10 years, $q' = 250$ m^2/yr at 100 years

5. $Q = 0.247$ ft^3/s
6. Inland distance 4125 ft with seaward flow reduced 8250 ft

Chapter 6

(Answers subject to interpolation of tabulated values.)

1. $T = 1850$ m^2/day
2. $T = 520$ m^2/day, $S = 1.8 \times 10^{-4}$
3. 1 year 49 m, 3 years 51 m
4. $Q = 427$ m^3/day
5. $Q = 2249$ m^3/day
6. Zero drawdown at stream, 3.7 m at midpoint
7. Approximately 39%
11. **a.** Approximately 6.8 m
12. **a.** 49,543 m
    b. 2.28 m
13. In order of blanks and selections
    100, 6, 40, 4, add, 30, 2, add
14. 594 m
15. Approximately 27 m

Chapter 8

1. **a.** Elastic compression $= 0.042$ ft
   b. Inelastic compression $= 2$ ft
   c. Bulk modulus $= 3 \times 10^7$ lb/ft^2
3. $S = 2.2 \times 10^{-4}$
7. 19% at 2000 m, 60% at 10,000 m
8. **a.** 150 m/my
   b. 300 m/my
   d. 15
11. About 40° above ambient
13. Change in total stress 450 kbars, change in fluid pressure 54 kbars, change in effective stress 396 kbars

Chapter 9

1. 1.9 and 0.8 cal m^{-1} s^{-1} °C^{-1} for the saturated sandstone
3. $q = 2.5 \times 10^{-3}$ cm/s
4. **a.** $k = 7 \times 10^{-9}$ cm^2

Chapter 10

1. Fractured 2.85 yr; unfractured 2850 yr
2. Approximately 130 years
3. 4×10^{-6} cm²/s
5. Advection 0.1 mg/m²s; diffusion 0.4×10^{-5} mg/m²s
6. **b.** 2×10^{-2} cm
7. 28.2 m, 8.9 m, and 2.8 m in x, y, z, respectively
8. Approximately 0.25 m
9. 10 m
10. **a.** 18 m²
    b. 0.21 m

Chapter 11

1. **a.** 2.3×10^{-3} M; 0.83×10^{-3} M; 5.5×10^{-3} M; 0.25×10^{-3} M; 4.7 meq/L; 0.83 meq/L; 5.47 meq/L 0.25 meq/L
   b. 0.012 M
   c. 0.66 and 0.90
3. SI = 0.03 slightly oversaturated
4. Cation/anion balance—0.99; good analysis

Chapter 12

2. **a.** 1.7×10^{-2} atm
   b. Strongly oversaturated $- IAP/K = 9.2$
3. **a.** 2×10^{-5} atm
   b. 0.1 atm
4. **a.** $n = 0.75$; $K = 56$
   b. $K_d = 0.60$ mL/g
5. **a.** (i) 3480, (ii) 355, (iii) 5.2×10^5
   b. (i) 34.8, (ii) 3.6, (iii) 5180
6. 4.16, 3.36, 4.57
7. **a.** 5, 1, 6, 2, 3, 4
   b. 4, 3, 2, 6, 1, 5
8. pe = 4.65; E_H = 0.27 V
9. $Fe^{3+} = 9.1 \times 10^{-7}$; $Fe^{2+} = 9.1 \times 10^{-6}$
10. 13.8 μg/L
12. $\delta^{18}O = -18^0/_{00}$; $\delta D = -135^0/_{00}$ SMOW

Chapter 13

1. 13.9 days
2. 0.092 day^{-1}, 7.5 day
3. **a.** Acetic acid, and 1,1-DCE
   b. Acetic acid

Chapter 15

7. 2850 years old
8. 24,300 years old
9. About 1950, 1963, 1970, 1976

Chapter 18

1. 14 cm
3. 0.13
5. **a.** 36 cm²/hr
   b. 1.0 m

6. 0.85 mL/g
9. $C = 0.91C_0$
10. 0.45 cm/day
12. Input data:

```
10      7
4000.00      5.00      0.50     0.10     0.00     5.97
     14.00    200.00     50.00     0.00
      5.00    150.00
```

Output Data:
14 Years

0.00	88.02	2195.20	2283.22	2195.20	88.02	0.00
0.15	179.12	1516.28	1695.10	1516.28	179.12	0.15
1.55	216.39	1191.31	1404.60	1191.31	216.39	1.55
4.92	231.56	997.65	1219.36	997.65	231.56	4.92
9.33	226.28	828.12	1035.77	828.12	226.28	9.33
6.75	104.52	339.73	430.82	339.73	104.52	6.75
0.49	5.48	16.27	20.78	16.27	5.48	0.49
0.00	0.01	0.04	0.05	0.04	0.01	0.00
0.00	0.00	0.00	0.00	0.00	0.00	0.00
0.00	0.00	0.00	0.00	0.00	0.00	0.00

Chapter 19

1. 2×10^4 m³
2. **a.** $k_{rw} = 0.47$
   b. $k_{rn} = 0.04$
   c. $S_w = 0.2$
   d. $S_{nw} = 0.1$
3. Volume of aquifer containing residual = 5.71 m³
4. Depth of residual = 571 m
5. 22.4% sorbed, 77.6% in solution
6. NAPL phase 79,500 g/m³, gas phase 113.2 g/m³
7. Effective solubility = 2.2×10^{-3} g/L, Mass in aqueous phase = 0.26 g/m³
8. Retardation due to aqueous phase = 5.83, solid phase = 137.2

Chapter 20

1. **a.** 6×10^{-3} m³/s
   b. Feasible—drawdown about 2.5 m after 1 year
2. **a.** 56 mg/L
   b. 10.8 kg/day
   c. Agrees
4. 1.4×10^{-3} cm/s

Chapter 21

1. -3.1 mg hydrocarbon/kg soil day
2. **a.** $EI_b = 4.3 \times 10^{-9}$ mg/day or 6.1×10^{-11} mg/kg-BW/day
   $EI_d = 1.1 \times 10^{-7}$ mg/day or 1.6×10^{-9} mg/kg-BW/day
   $TI_d = 1.1 \times 10^{-7}$ mg/day or 1.6×10^{-9} mg/kg-BW/day
   b. $TD = 1.1 \times 10^{-9}$ mg/kg-BW/day
   c. $HQ = 1.1 \times 10^{-5}$
3. **a.** $EI_{wt} = 120$ μg/day
   $Ei_{mt} = 1.5$ μg/day
   $Ei_{vt} = 23.7$ μg/day
   b. $TI_t = 145$ μg/day
   c. $A = 0.30$ μg/g
   d. Possibly, at threshold

Appendix A

Derivation of the Flow Equation in a Deforming Medium

In the derivation of the flow equation in Chapter 4 it was stated that Eq. 4.7 was inadequate in that both the fluids and the solids are in motion in deforming media, requiring two conservation statements, one for the fluids and one for the solids. These conservation statements are

$$-\text{div}(\rho_w n \mathbf{v}_w) = \frac{\partial(\rho_w n)}{\partial t} \quad \text{(A1.1a)}$$

$$-\text{div}[\rho_s(1 - n)\mathbf{v}_s] = \frac{\partial[\rho_s(1 - n)]}{\partial t} \quad \text{(A1.1b)}$$

where $\mathbf{v}_w$ is the fluid velocity, $\mathbf{v}_s$ is the solid velocity, and ρ_s is the density of the solids. In this form $\rho_w n$ is the total mass of fluids per unit volume and $\rho_s(1 - n)$ is the total mass of solids per unit volume. In this development $\mathbf{v}_w$ and $\mathbf{v}_s$ are the average velocities of the fluids and the solids with respect to a stationary coordinate system. The Darcy flux q is described relative to the solid matrix

$$\mathbf{q} = n(\mathbf{v}_w - \mathbf{v}_s) \quad \text{(A1.2)}$$

Now we must introduce the idea of a material derivative that follows the motion of the solid phase, that is, $Dn/Dt = \partial n/\partial t + \mathbf{v}_s \cdot \nabla n$, which we notice becomes equal to the partial derivative when the last term is small. Introducing Eq. A1.2 into Eq. A1.1a and assuming further that the solid grains are incompressible, $\rho_s = $ constant, Eq. A1.1 becomes

$$-\text{div}(\rho_w \mathbf{q}) - \rho_w n\, \text{div}(\mathbf{v}_s) = \frac{D(\rho_w n)}{Dt} \quad \text{(A1.3a)}$$

$$(1 - n)\, \text{div}(\mathbf{v}_s) = \frac{Dn}{Dt} \quad \text{(A1.3b)}$$

Eliminating the divergence of the solids between Eqs. A1.3 gives the final form for the conservation of fluid and solid mass (Palciauskas and Domenico, 1980)

$$-\left(\frac{1}{\rho_w n}\right)\text{div}(\rho_w \mathbf{q}) = \frac{1}{\rho_w}\frac{D\rho_w}{Dt} + \frac{1}{n(1 - n)}\frac{Dn}{Dt} \quad \text{(A1.4)}$$

If the fractional compression is small, the material derivatives are replaced by partial derivatives

$$-\left(\frac{1}{\rho_w}\right)\text{div}(\rho_w \mathbf{q}) = \frac{n}{\rho_w}\frac{\partial \rho_w}{\partial t} + \frac{1}{(1 - n)}\frac{\partial n}{\partial t} \quad \text{(A1.5)}$$

Now we may use the relationships discussed in Chapter 4 to get the correct form of the diffusion equation. First, it is assumed that the density does not vary spatially so that it can be taken out as a constant on the left-hand side of Eq. A1.5. Substituting Darcy's law in the first term of Eq. A1.5 gives us, as described previously

$$-\text{div } \mathbf{q} = K\nabla^2 b \quad \text{(A1.6)}$$

The first term on the right-hand side of Eq. A1.5 becomes, by way of Eq. 4.20,

$$\frac{n}{\rho_w}\frac{\partial \rho_w}{\partial t} = n\beta_w\frac{\partial P}{\partial t} = n\rho_w g\beta_w\frac{\partial b}{\partial t} \quad \text{(A1.7)}$$

In the last term of Eq. A1.5 we recognize that $dn/(1 - n)$ is merely the change in pore volume per unit volume dV_p/V_T, where, in accordance with our definition of compressibility (Eq. 4.26) and effective stress, $dV_p/V_T = -\beta_p d\overline{\sigma} = \beta_p dP = \rho_w g\beta_p db$. Thus, the last term of Eq. A1.5 becomes

$$\frac{1}{1 - n}\frac{\partial n}{\partial t} = \rho_w g\beta_p\frac{\partial b}{\partial t} \quad \text{(A1.8)}$$

Collecting Eqs. A1.6, A1.7, and A1.8 gives the diffusion equation

$$\nabla^2 b = \frac{\rho_w g(\beta_p + n\beta_w)}{K}\frac{\partial b}{\partial t} = \frac{S_s}{K}\frac{\partial b}{\partial t} \quad \text{(A1.9)}$$

Appendix B

About the Computer Disk

A computer disk is provided with the book. It contains the well hydraulics package WELLz and a number of other code segments provided with the book. The standard high density disk should be readable on most DOS- or Windows-based personal computers. To help organize the information on the disk, the various code segments are installed in separate subdirectories. The contents of each subdirectory are listed below.

a:\woost

In Chapter 7, we briefly described the application of MODFLOW to the Wooster study site described by Springer and Bair (1992). This subdirectory contains the MODFLOW data files necessary to run this program including

woo.bas, woo.bcf, woo.ghb, woo.oc, woo.rch, woo.riv, woo.sip, woo.wel.

This data set was tested using MODFLOW/EM (version 9412) from Maximal Engineering Software, Inc. Two other files are provided in the data set. The first is woo.mfn, which contains the list of MODFLOW codes used by MODFLOW/EM. The second is woo.nam, which is an MFI/EM preprocessor name file containing a list of all files used by the MFI Preprocessor and the I-numbers assigned to each. Copy these files to your hard drive.

a:\trans

Chapter 18 contains two simple codes for simulating one- and three-dimensional mass transport. The code listings are supplied along with the sample dataset. In the interests of conserving space, only the three-dimensional code is provided as an executable (that is, exe file). We have used various versions of Lahey compilers to compile these codes.

rone.for	—source FORTRAN code for the one-dimensional model, Table 18.2
in.dat	—example data set used in Example 18.5
rone.exe	—needs to be created by compiling and linking rone.exe
rthree.for	—source code for the three-dimensional code
in3.dat	—example data set used in Example 18.6
rthree.exe	—executable code

Copy these files onto your hard drive. Once installed, the code is run as follows

1. Click on the START button on the Windows 95 task bar or choose the File menu on Windows 3.1x
2. Choose the RUN command
3. Type c:\trans\rthree.exe
4. Close the C: window that opens
5. Read/print the output (out3.dat) using any standard word processor

Normally, one can make changes to the input file (in3.dat) using a word processor. However, be careful to preserve the input format as shown in the book. Also when you have made the changes, save the modified file (in3.dat) in **ASCII format**. A FORTRAN code will not be able to read the standard files created by a word processor.

a:\wellz

This directory contains the Windows version of WELLz. It has been developed and tested with Windows 95. It runs with Windows 3.1x, although on some older machines it can be somewhat unstable.

The code installs automatically. Choose the RUN command (under the START button on Windows 95, or under File menu on Windows 3.1x). Type 'a:\setup' or 'b:\setup' as appropriate. The code operates as a normal Windows program. Details on operating the code are provided in Chapter 6. Here we present a few additional details.

Files on the Disk

Several data files are provided on the disk. Each time a user selects File–New, the data set "test" is loaded into the Windows. This file is edited by the user for the problem at hand. When you select File–Open, note that test is there as "test.wzf". There are two other files present on the disk "ex6_10.wzf" and "ex6_11.wzf" that are the data files to run Examples 6.10 and 6.11. As you create new data sets, these are saved as with the "wzf" extensions.

Adding and Labeling Contour Lines

The code lets you plot data points and label the contour lines for drawdown. For example, run "test" by proceeding through the various data screens in order. The plot illustrates a pattern of circular drawdown with no labels.

1. To determine what contour lines are shown:
   —click 'Options' and select 'Contour values'
   —you can either keep the default set of contours or provide your own set of values by editing the 'Contour values dialog' box
   —accept the values by clicking 'OK' and the plot will reflect your choice of contours

2. To label contours or selected points:
   —click 'Options' and select 'Label drawdown-values'
   —click the mouse to add labels at appropriate points

3. To get rid of contour labels or other labels:
   If you save a data set and later reopen it and edit parameters, the labels on the contour line will remain. The old labels can be removed in one of two ways. With the contour plot in the Window
   —click 'File' and select 'New' this returns you to the first window, whereupon you run through the set of Windows and obtain a new plot, or
   —click 'Editing' and proceed to the last Window

4. Plotting precautions:
   For the plot to be prepared correctly, the region to be plotted must have coordinate locations greater than zero. Stated another way, the coordinate axis (0, 0) should always be at the bottom left of the contour map. It is possible to use negative well coordinates, but the contour plot cannot include these wells.

Appendix C

Table of Atomic Weights

Element	Symbol	Atomic Number	Atomic Weight	Element	Symbol	Atomic Number	Atomic Weight
Actinium	Ac	89		Hydrogen	H	1	1.01
Aluminum	Al	13	26.98	Indium	In	49	114.82
Americium	Am	95		Iodine	I	53	126.90
Antimony	Sb	51	121.75	Iridium	Ir	77	192.20
Argon	Ar	18	39.95	Iron	Fe	26	55.85
Arsenic	As	33	74.92	Krypton	Kr	36	83.80
Astatine	At	85		Lanthanum	La	57	138.91
Barium	Ba	56	137.34	Lead	Pb	82	207.19
Berkelium	Bk	97		Lithium	Li	3	6.94
Beryllium	Be	4	9.01	Lutetium	Lu	71	174.97
Bismuth	Bi	83	208.98	Magnesium	Mg	12	24.31
Boron	B	5	10.81	Manganese	Mn	25	54.94
Bromine	Br	35	79.91	Mendelevium	Md	101	
Cadmium	Cd	48	112.30	Mercury	Hg	80	200.59
Calcium	Ca	20	40.08	Molybdenum	Mo	42	95.94
Californium	Cf	98		Neodymium	Nd	60	144.24
Carbon	C	6	12.01	Neon	Ne	10	20.18
Cerium	Ce	58	140.12	Neptunium	Np	93	
Cesium	Cs	55	132.90	Nickel	Ni	28	58.71
Chlorine	Cl	17	35.45	Niobium	Nb	41	92.91
Chromium	Cr	24	52.00	Nitrogen	N	7	14.01
Cobalt	Co	27	58.93	Nobelium	No	102	
Copper	Cu	29	63.54	Osmium	Os	76	190.20
Curium	Cm	96		Oxygen	O	8	16.00
Dysprosium	Dy	66	162.50	Palladium	Pd	46	106.40
Einsteinium	Es	99		Phosphorus	P	15	30.97
Erbium	Er	68	167.26	Platinum	Pt	78	195.09
Europium	Eu	63	151.96	Plutonium	Pu	94	
Fermium	Fm	100		Polonium	Po	84	
Fluorine	F	9	19.00	Potassium	K	19	39.10
Francium	Fr	87		Praseodymium	Pr	59	140.91
Gadolinium	Gd	64	157.25	Promethium	Pm	61	
Gallium	Ga	31	69.72	Protactinium	Pa	91	
Germanium	Ge	32	72.59	Radium	Ra	88	
Gold	Au	79	196.97	Radon	Rn	86	
Hafnium	Hf	72	178.49	Rhenium	Re	75	186.20
Helium	He	2	4.00	Rhodium	Rh	45	102.90
Holmium	Ho	67	164.93	Rubidium	Rb	37	85.47

Element	Symbol	Atomic Number	Atomic Weight	Element	Symbol	Atomic Number	Atomic Weight
Ruthenium	Ru	44	101.07	Thallium	Tl	81	204.37
Samarium	Sm	62	150.35	Thorium	Th	90	232.04
Scandium	Sc	21	44.96	Thulium	Tm	69	168.93
Selenium	Se	34	78.96	Tin	Sn	50	118.69
Silicon	Si	14	28.09	Titanium	Ti	22	47.90
Silver	Ag	47	107.87	Tungsten	W	74	183.85
Sodium	Na	11	22.99	Uranium	U	92	238.03
Strontium	Sr	38	87.62	Vanadium	V	23	50.94
Sulfur	S	16	32.06	Xenon	Xe	54	131.30
Tantalum	Ta	73	180.95	Ytterbium	Yb	70	173.04
Technetium	Tc	43		Yttrium	Y	39	88.90
Tellurium	Te	52	127.60	Zinc	Zn	30	65.37
Terbium	Tb	65	158.92	Zirconium	Zr	40	91.22

References

Abriola, L. M. 1988. Multiphase flow and transport models for organic chemicals: A review and assessment. Electric Power Research Inst., Palo Alto, Calif.

Adams, E. E., and L. W. Gelhar. 1992. Field study of dispersion in a heterogeneous aquifer, 2. Spatial moments analysis. Water Resources Res., v. 28, no. 12, p. 3293-3307.

Ahlfeld, D. P., A. Dahmani, and W. Ji. 1994. A conceptual model of field behavior of air sparging, and its implications for application. Ground Water Mon. Rev., Fall, p. 132-139.

Ahlstrom, S. W., H. P. Foote, R. C. Arnett, C. R. Cole, and R. J. Serne. 1977. Multicomponent Mass Transport. Model: Theory and Numerical Implementation (discrete-parcel-random-walk version). Battelle Northwest Laboratories, BNWL—2127, Richland, Wash.

Ala, N. K., and P. A. Domenico. 1992. Inverse analytical techniques applied to coincident contaminant distributions at Otis Air Force Base, Massachusetts. Ground Water, v. 30, p. 212-218.

Albertsen, M., and an ad hoc Task Force. 1986. Beurteilung und Behandlung von Mineralölschadensfallen im Hinblick auf den Grundwasserschutz, Jeil 1, Die wissenschaftlichen Grundlagen zum Verständnis des Verhaltens von Mineralöl im Untergrund. Federal Office of the Environment, LTw S no. 20, 178 p.

Alexander, M. 1994. Biodegradation and bioremediation: Academic Press, San Diego, Calif., 302p.

Alpay, O. A. 1972. A practical approach to defining reservoir heterogeneity: J. Petrol. Technol., v. 20, no. 7, p. 841-848.

American Chemical Society. 1980. Guidelines for data acquisition and data quality evaluation in environmental chemistry. Anal. Chem., v. 52, p. 2242-2249.

Amundson, N. R. 1950. Mathematics of adsorption in beds, II. J. Phys. Colloid. Chem., v. 54, p. 812-820.

Anderson, G. M., and R. W. Macqueen. 1982. Ore deposit models—Mississippi Valley type lead zinc deposits. Geoscience Canada, v. 9, p. 108-117.

Anderson, M. P. 1984. Movement of contaminants in groundwater: groundwater transport-advection and dispersion. Groundwater Contamination, NRC Studies in Geophysics, National Academy Press, Washington, D.C., p. 37-45.

Anderson, M. P., and J. A. Munter. 1981. Seasonal reversals of groundwater flow around lakes and the relevance to stagnation points and lake budgets. Water Resources Res., v. 17, p. 1139-1150.

Anderson, M. P., and W. W. Woessner. 1992. Applied Groundwater Modeling. Academic Press, San Diego, Calif., 381 p.

Anderson, M. R., R. L. Johnson, and J. F. Pankow. 1987. The dissolution of dense non-aqueous phase liquid (DNAPL) from a saturated porous medium. Proc. of Petroleum Hydrocarbons and Organic Chemicals in Ground Water: Prevention, Detection and Restoration, National Ground Water Assoc., Columbus, Ohio, p. 409-428.

Anderson, R. E., and R. L. Laney. 1975. The influence of late Cenozoic stratigraphy on distribution of impoundment related seismicity at Lake Mead, Nevada, Arizona. U.S. Geol. Survey J. Res., v. 3, p. 337-343.

Aris, R., and N. R. Amundson. 1973. Mathematical Methods in Chemical Engineering, II. Prentice-Hall, Englewood Cliffs, N.J., 369 p.

Arizona Department of Water Resources. 1994. Riparian Protection Program Legislative Report. Arizona Department of Water Resources, Phoenix, Ariz., p. 209-280.

Armstrong, J. M., W. Korreck, L. E. Leach, R. M. Powell, S. V. Vandergrift, and J. T. Wilson. 1988. Bioremediation of a fuel spill: Evaluation of techniques for preliminary site characterization. Proc. of Petroleum Hydrocarbons and Organic Chemicals in Ground Water: Prevention, Detection and Restoration. National Ground Water Assoc., Columbus, Ohio, p. 931-944.

Athy, L. F. 1930. Density, porosity, and compaction of sedimentary rocks. Bull. Am. Assoc. Petrol. Geol. v. 14, p. 1-24.

Atkinson, T. C. 1977. Diffuse flow and conduit flow in limestone terrain in the Mendip Hills, Somerset (Great Britain). J. Hydrol., v. 35, p. 93-110.

Atwater, G. I., and E. E. Miller. 1965. The effect of decreases in porosity with depth on future development of oil and gas reserves in South Louisiana. Paper presented at AAPG meeting, New Orleans.

Atwood, D. F., and S. M. Gorelick. 1985. Hydraulic gradient

control for groundwater contaminant removal. J. Hydrol., v. 76, p. 85–106.

Ayres, J. E., D. C. Lager, and M. J. Barvenik. 1983. The first EPA Superfund cutoff wall: Design and specifications. In D. M. Nielsen (ed.), Proc. of the Third National Symposium on Aquifer Restoration and Ground Water Monitoring. National Water Well Assoc., Dublin, Ohio, p. 13–22.

Back, W. 1960. Origin of hydrochemical facies of groundwater in the Atlantic Coastal Plain. Proc. Intern. Geol. Cong., Copenhagen, p. 87–95.

Back, W. 1961. Techniques for mapping of hydrochemical facies. U.S. Geol. Surv. Prof. Paper 424-D, p. 380–382.

Back, W., and R. A. Freeze. 1983. Chemical Hydrogeology. Benchmark papers in Geology. Hutchinson Ross Pub. Co., Stroudsburg, Pa.

Baedecker, M. J., D. I. Siegal, P. E. Bennett, and I. M. Cozzarelli. 1989. The fate and effects of crude oil in a shallow aquifer: 1. The distribution of chemical species and geochemical facies. U.S. Geol. Surv., Water-Resources Investigations Report 88-4220.

Baehr, A. L. 1987. Selective transport of hydrocarbons in the unsaturated zone due to aqueous and vapor phase partitioning. Water Resources Res., v. 23, p. 1926–1938.

Baetslé, L. H. 1969. Migration of radionuclides in porous media. In A. M. F. Duhamel (ed.), Progress in Nuclear Energy Series XII, Health Physics, Pergamon Press, Elmsford, N.Y., p. 707–730.

Bahr, J. M., and J. Rubin. 1987. Direct comparison of kinetic and local equilibrium formulations for solute transport affected by surface reactions. Water Resources Res., v. 23, p. 438–452.

Bailey, J. E., and D. F. Ollis. 1987. Biochemical Engineering: New York, McGraw-Hill, 753 p.

Bair, E. S., and T. D. Lahm. 1996. Variations in capture-zone geometry of a partially penetrating pumping well in an unconfined aquifer: Ground Water, v. 34, no. 5, p. 842–852.

Balkwill, D. L., J. K. Fredrickson, and J. M. Thomas. 1989. Vertical and horizontal variations in the physiological diversity of the aerobic chemoheterotrophic bacteria microflora in deep southeast Coastal Plain subsurface sediments. Appl. Environ. Microbiol., v. 55, p. 1058–1065.

Ball, J. W., E. A. Jenne, J. M. Burchard, and A. H. Truesdell. 1976. Sampling and preservation techniques for water with a section on gas collection. Proc. First Workshop on Sampling Geothermal Effluents, EPA-600/9-76-0111, p. 219–234.

Banks, H. O. 1953. Utilization of underground storage reservoirs. Trans. Am. Soc. Civil Engrs., v. 118, p. 220–234.

Banks, H. O., and R. C. Rictor. 1953. Sea water intrusion into groundwater basins bordering the California coast and inland bays. Trans. Am. Geophys. Union, v. 34, p. 575–582.

Barari, A., and L. Hedges. 1985. Movement of water in glacial till. Memoires, Hydrogeology of Rocks of Low Permeability. Inter. Assoc. Sci. Hydrol., Tucson, Ariz., p. 129–134.

Barker, J. A., and J. H. Black. 1983. Slug tests in fissured aquifers. Water Resources Res., v. 19, p. 1558–1564.

Barker, J. F., G. C. Patrick, and D. Major. 1987. Natural attenuation of aromatic hydrocarbons in a shallow sand aquifer. Ground Water Mon. Rev., p. 64–71.

Barker, C. 1972. Aquathermal pressuring—role of temperature in development of abnormal pressure zones. Am. Assoc. Petrol. Geol. Bull. 56, p. 2068–2071.

Bastin, E. S. (ed.). 1939. Contributions to the knowledge of the lead-zinc deposits of the Mississippi Valley Region. Geol. Soc. Am. Special Paper, v. 24, 16 p.

Bathhurst, R. G. C. 1971. Carbonate sediments and their diagenesis. In Developments in Sedimentology, no. 12, Elsevier, New York.

Baumann, P. 1963. Theoretical and practical aspects of well recharge. Trans. Am. Soc. Civil Engrs., v. 128, p. 739–764.

Baumann, P. 1965. Technical development in groundwater recharge. In V. T. Chow (ed.), Advances in Hydroscience, v. 2, p. 209–279.

Bear, J., and Y. Bachmat. 1967. A generalized theory on hydrodynamic dispersion in porous media: Intern. Assoc. Sci. Hydrol., Symp. Haifa, Publ. 72, p. 7–16.

Bear, J. 1972. Dynamics of Fluids in Porous Media. Elsevier, New York, 764 p.

Bear, J. 1979. Hydraulics of Groundwater. McGraw-Hill, New York.

Beck, A. E., G. Garven, and L. Stegena. 1989. Hydrogeologic regimes and their subsurface thermal effects. Geophysical Monogram 47, IUGG, v. 2.

Behnke J., and W. Bianchi. 1965. Pressure distributions in layered sand columns during transient and steady state flows. Water Resources Res., v. 1, p. 557–562.

Bekins, B. A., E. M. Godsy, and D. F. Goerlitz. 1993. Modeling steady-state methanogenic degradation of phenols in groundwater. J. Contam. Hydrology, v. 14, p. 279–294.

Bell, M. L., and A. Nur. 1978. Strength changes due to reservoir induced pore pressure and stresses and application to Lake Oroville. J. Geophys. Res., v. 83, p. 4469–4483.

Bengtsson, G. 1985. Microcosm for ground water research: Ground Water Quality, C. H. Ward, W. Giger, and P. L. McCarty (eds.), John Wiley & Sons, New York, p. 330–341.

Bennett, R. R., and R. R. Meyer. 1952. Geology and groundwater resources of the Baltimore area. Maryland Board Nat. Resources, Dept. Geology, Mines, Water Resources Bull. 4.

Bennion, D. W., and J. C. Griffiths. 1966. A stochastic model for predicting variations in reservoir rock properties. Trans. Am. Inst. Mining Met. Engrs., v. 237, no. 2, p. 9–16.

Bentley, H. W., F. M. Phillips, S. N. Davis, M. A. Habermehl, P. L. Airey, G. E. Calf, D. Elmore, H. E. Grove, and T. Torgersen. 1986. Chlorine 36 dating of very old groundwater: 1. The great artesian basin, Australia. Water Resources Res., v. 22, no. 13, p. 1991–2001.

Berner, R. A. 1974. E. D. Goldberg (ed.), In the Sea, v. 5, p. 427–450. John Wiley & Sons, New York.

Berner, R. A. 1980. Early Diagenesis. Princeton University Press, Princeton, N. J., 241 p.

Berner, R. A., and J. W. Morse. 1974. Dissolution kinetics of calcium carbonate in sea water: IV. Theory of calcite dissolution. Am. J. Sci., v. 274, p. 108–134.

Betcher, R. N. 1977. Temperature distribution in deep groundwater flow systems—a finite element model. Master's Thesis. Univ. of Waterloo, Ontario, Canada.

Bethke, C. M. 1985. A numerical model of compaction driven groundwater flow and heat transfer and its application to the paleohydrology of intracratonic sedimentary basins. J. Geophys. Res., v. 90, p. 6817–6828.

Bethke, C. M. 1986. Hydrologic constraints on the genesis of the upper Mississippi Valley Mineral District from Illinois Basin Brines. Econ. Geol., v. 81, p. 233–249.

Bethke, C. M. 1988. Origin of long range brine migration across the North American craton (abst.). EOS Trans. Am. Geophys. Union, v. 69, p. 470.

Bethke, C., and S. Marshak. 1990. Brine migrations across North America—the plate tectonics of ground water. Ann. Rev. Earth Planet. Sci., v. 18, p. 287–315.

Bevin, K., and P. Germann. 1982. Macropores and water in soils. Water Resources Res., v. 18, p. 1311–1325.

Biot, M. A. 1941. General theory of three dimensional consolidation. Appl. Physics, v. 12, p. 155–164.

Biot, M. A., and D. G. Willis. 1957. The elastic coefficients of the theory of consolidation. J. Appl. Mech., v. 24, p. 594–601.

Blake S. B., and R. W. Lewis. 1982. Underground oil recovery. Second National Symposium on Aquifer Restoration and Ground Water Monitoring, Proc., D. M. Nielsen (ed.); National Water Well Assoc., Dublin, Ohio, p. 69–75.

Blake S. B., and J. S. Fryberger. 1983. Containment and recovery of refined hydrocarbons from groundwater. Proc. Seminar on Groundwater and Petroleum Hydrocarbons Protection, Detection and Restoration. Petroleum Assoc. for Conservation of the Canadian Environment, Ottawa, p. 4-1–4-47.

Blanchard, F. G., and P. Byerly. 1935. A study of a well gage as a seismograph. Bull. Seismol. Soc. Am., v. 25, p. 313–321.

Blanchard, P. E., and J. Sharp. 1985. Possible free convection in thick Gulf Coast sandstone sequences. Southwest Section of the Am. Assoc. Petrol. Geol. Trans., p. 6–12.

Blandford, T. N., and P. S. Huyakorn. 1991. WHPA Well Head Protection Area Delineation Code. Unpublished report to U.S. Environmental Protection Agency, code available from Intl. Ground Water Modeling Center, Golden, Colo.

Blatt, H. 1979. Diagenetic processes in sandstone. In P. A. Scholle and P. R. Schluger (eds.), Aspects of Diagenesis. Soc. Econ. Paleontol. and Mineral Special Pub. 26, p. 141–157.

Bodvarsson, G. 1970. Confined fluids as strain meters. J. Geophys. Res., v. 75, p. 2711–2718.

Boggs, J. M., and others. 1992. Field study of dispersion in a heterogeneous aquifer. 1, Overview and site description: Water Resources Res., v. 28, no. 12, p. 3281–3291.

Boggs, S. 1987. Principles of Sedimentology and Stratigraphy. Merrill Pub. Co., Columbus, Ohio.

Bogomolov, V., and others. 1978. The principles of paleohydrological reconstruction of groundwater formation. Internat. Geol. Congress, 24th Sec II. p. 205–226.

Bolt, G. H., and P. H. Groenvelt. 1969. Coupling phenomena as a possible cause for non-Darcian behavior of water in soil. Bull. Intern. Assoc. Sci. Hydrol. 14, p. 17–26.

Bonham, L. C. 1980. Migration of hyrdrocarbons in compacting basins. In W. H. Roberts and R. J. Cordell (eds.), Problems of Petroleum Migration. Am. Assoc. Petrol. Geol. Studies in Geology, v. 10, p. 69–88, Tulsa, Okla.

Borden, R. C., and P. B. Bedient. 1986. Transport of dissolved hydrocarbons influenced by oxygen-limited biodegradation. 1, Theoretical development: Water Resources Res., v. 22, no. 13, p. 1973–1982.

Born, S. M., S. A. Smith, and D. A. Stephenson. 1974. The hydrologic regime of glacial terrain lakes with management and planning applications. An Inland Lake Renewal and Management Demonstration Project Report. Upper Great Lakes Regional Comm.

Born, S. M., S. A. Smith, and D. A. Stephenson. 1979. The hydrologic regime of glacial terrain lakes. J. Hydrol., v. 43, p. 7–44.

Bosworth, W. 1981. Strain induced partial dissolution of halite. Tectonophysics, v. 78, p. 509–525.

Boulton, N. S. 1954. The drawdown of the water table under nonsteady conditions near a pumped well in an unconfined formation. Proc. Inst. Civil Engrs., v. 3, p. 564–579.

Boulton, N. S. 1955. Unsteady radial flow to a pumped well allowing for delayed yield from storage. Proc. Gen. Assembly, Rome Intern. Assoc. Sci. Hydrol. Publ., v. 37, p. 472–477.

Boulton, N. S. 1963. Analysis of data from nonequilibrium pumping test allowing for delayed yield from storage. Proc. Inst. Civil Engrs., v. 26, p. 469–482.

Boulton, N. S., and S. V. Streltsova. 1977. Unsteady flow to a pumped well in a fissured water bearing formation. J. Hydrol., v. 35, p. 257–269.

Bouwer, E. J., and P. H. McCarty. 1984. Modeling of trace organics biotransformation in the subsurface. Ground Water, v. 22, no. 4, p. 433–440.

Bouwer, H. 1985. Renovating waste water with groundwater recharge. In E. T. Smerdon and W. R. Jordan (eds.), Issues in Groundwater Management, Texas A&M Univ., College Station, p. 331–346.

Box, G. E. P., and G. W. Jenkins. 1976. Time Series Analysis: Forecasting and Control. Holden-Day Inc., San Francisco, 575 p.

Boyd, F. M., and C. W. Kreitler. 1986. Hydrogeology of a Gypsum Playa. Northern Salt Basin, Texas. Bureau of Economic Geology, The Univ. of Texas at Austin, Report of Investigations No. 158, 37 p.

Brace, W. F. 1980. Permeability of crystalline rocks and argellaceous rocks. Intern. J. Rock Mech. Min. Sci. Geomechanics, v. 17, p. 241–251.

Brace, W. F., B. W. Paulding, and C. Sholz. 1966. Dilatency in the fracture of crystalline rocks. J. Geophys. Res., v. 71, p. 3939–3953.

Brace, W. F., J. B. Walsh, and W. T. Fangos. 1968. Permeability of granite under high pressure. J. Geophys. Res., v. 63, p. 2225–2236.

Bradbury, K. R. 1984. Major ion and isotope geochemistry of groundwater in clayey till, northwestern Wisconsin, U.S.A. In B. Hitchon and E. I. Wallick (eds.), First Canadian/American Conference on Hydrogeology. National Water Well Assoc., Dublin, Ohio, p. 284–289.

Bradley, J. S. 1975. Abnormal formation pressure. Am. Assoc. Petrol. Geol. Bull. 59, p. 957–973.

Brashears, M. L. 1946. Artificial recharge on Long Island, New York. Econ. Geol., v. 41, p. 503–516.

Bredehoeft, J. D. 1965. The drill stem test: The petroleum industry's deep-well pumping test. Ground Water, v. 3, p. 31–36.

Bredehoeft, J. D. 1967. Response of well-aquifer systems to earth tides. J. Geophys. Res., v. 72, p. 3075–3087.

Bredehoeft J., and B. Hanshaw. 1968. On the maintenance of anomalous fluid pressures: I. Thick sedimentary sequences. Geol. Soc. J. Am. Bull. 79, p. 1097–1106.

Bredehoeft, J. D., and L. F. Konikow. 1993. Editorial—Groundwater models: validate or invalidate. Ground Water, v. 31, no. 2, p. 178–179.

Bredehoeft, J., and I. S. Papadopulos. 1965. Rates of vertical groundwater movement estimated from the earth's thermal profile. Water Resources Res., v. 1, p. 325–328.

Bredehoeft, J., and R. Young. 1970. The temporal allocation of groundwater—A simulation approach. Water Resources Res., v. 16, p. 3–21.

Breen, K. J., A. L. Kontis, G. L. Rowe, and R. J. Haefner. 1995. Simulated ground-water flow and sources of water in the Killbuck Creek Valley near Wooster, Wayne County, Ohio. U.S. Geological Survey, Water-Res. Investigations Report 94-4131.

Brooks, R. H., and A. T. Corey. 1966. Properties of porous media affecting fluid flow. J. Irrigation and Drainage Div., Proc. Am. Soc. Civil Eng., v. 92, IR2, p. 61–88.

Brown, R. F., and T. L. Lambert. 1963. Reconnaissance of ground water resources in the Mississippi Plateau region, Kentucky. U.S. Geological Survey Water Supply Paper 1603, 58 p.

Brown, R. H., and G. G. Parker. 1945. Salt water encroachment in limestone of Silver Bluff, Miami, Florida. Econ. Geol., v. 40, p. 235–262.

Brown, J. A. 1937. Discussion on "Effect of a sea level canal on the groundwater level of Florida." Econ. Geol., v. 32, p. 589–599.

Browne, P. R. L. 1978. Hydrothermal alteration in active geothermal fields. In F. A. Donath, F. G. Stehli, and G. Wetherill (eds.), Ann. Rev. Earth Planet Sci., v. 6, p. 229–250.

Bryan, K. 1925. Geology of reservoir and dam sites. U.S. Geol. Survey Water Supply Papers, 597A.

Buddemeier, R. W., and J. R. Hunt. 1988. Transport of colloidal contaminants in groundwater: Radionuclide migration at the Nevada Test Site. Appl. Geochem., v. 3, p. 535–548.

Buras, N. 1966. Dynamic programming in water resources development. Advances in Hydroscience, V. T. Chow (ed.), vol. 3: New York, Academic Press, p. 367–412.

Bureau of Reclamation. 1960. Design of Small Dams. U.S. Gov. Printing Office, Washington, D.C.

Burns, W. A. 1969. New single well test for determining vertical permeability. Trans. Am. Inst. Mining Engrs., v. 246, p. 743–752.

Burst, J. F. 1969. Diagenesis of Gulf Coast clayey sediments and its possible relation to petroleum migration. Am. Assoc. Petrol. Geol. Bull. 53, p. 73–93.

Buscheck, T. E., P. E. Kirk, T. O'Reilly, and R. G. Nelson. 1993. Evaluation of intrinsic bioremediation at field sites: Proc. of Petroleum Hydrocarbons and Organic Chemicals in Ground Water—Prevention, Detection and Restoration. National Ground Water Assoc., Columbus, Ohio, p. 367–381.

Busenberg, E., and L. N. Plummer. 1992. Use of chlorofluoromethanes (CCl_3F and CCl_2F_2) as hydrologic tracers and age-dating tools: Example, the alluvium and terrace system of Central Oklahoma. Water Resources Res., v. 28, p. 2257–2283.

Busenberg, E., E. P. Weeks, L. N. Plummer, and R. C. Bartholomay. 1993. Age dating ground water by use of chlorofluorocarbons (CCl_3F and CCl_2F_2), and distribution of chlorofluorocarbons in the unsaturated zone, Snake River Plain Aquifer, Idaho National Engineering Laboratory, Idaho: U.S. Geol. Surv. Water Resour. Invest., 93-4054, 47 pp.

Butler, R. D. 1984. Hydrogeology of the Dakota aquifer system, Williston Basin, North Dakota. In D. G. Jorgensen and D. C. Signor (eds.), Proc. of the Geohydrology Dakota Aquifer Symposium, p. 12–23. Water Well Journal Pub. Co., Worthington, Ohio.

Byerlee, J. D. 1967. Frictional characteristics of granite under high confining pressure. J. Geophys. Res., v. 72, p. 2629–2648.

Cady, R. C. 1941. Effect upon groundwater levels of proposed surface water storage in Flatland Lake, Montana. U.S. Geol. Survey Water Supply Paper 849 b, p. 51–81.

California Dept. Water Resources. 1958. Sea water intrusion in California. Calif. Dept. Water Res. Bull. 63, Sacramento.

California Dept. Water Resources. 1980. Groundwater basins in California. Calif. Dept. Water Res. Bull. 118–180, Sacramento.

Cant, D. J. 1982. Fluvial facies models and their application. In P. A. Scholle and D. Spearing (eds.), Sandstone Depositional Environments. Am. Assoc. Petrol. Geol. Mem. 31, p. 115–138.

Carder, D. S. 1970. Reservoir loading and local earthquakes. In W. M. Adams (ed.), Engineering Seismology (Eng. Geol. Case Histories #8), Geol. Soc. Am.

Cardwell, W. T., and R. L. Parsons. 1945. Average permeabilities of heterogeneous oil sands. Trans. Am. Inst. Min. Met. Engrs. v. 169, p. 34–42.

Carlson, W. D. 1983. The polymorphs of $CaCO_3$ and the aragonite-calcite transformation. In R. J. Reeder (ed.), Carbonates: Mineralogy and Chemistry. Reviews in Mineralogy, v. 11, Mineral. Soc. Am., p. 191–225.

Carlston, C. W., L. L. Thatcher, and E. C. Rhodehamel. 1960. Tritium as a hydrologic tool, the Wharton Tract Study. Int. Assoc. Sci. Hydrol., Pub. 52, p. 503–512.

Carothers, W. W., and Y. K. Kharaka. 1978. Aliphatic acid anions in oil field water—implications for origin of natural gas. Am. Assoc. Petrol. Geol. Bull. 62, p. 2441–2453.

Carr, P. A., and G. S. van der Kamp. 1969. Determining aquifer characteristics by the tidal method. Water Resources Res., v. 5, p. 1023–1031.

Carrillo, N. 1948. Influence of artisan wells in the sinking of Mexico City. Proc. 2nd Intern. Conf. Soil Mech. and Found. Eng., v. 7, p. 156–159.

Carslaw, H., and J. Jeager. 1959. Conduction of Heat in Solids. Clarendon Press, Oxford.

Cartwright, K. 1970. Groundwater discharge in the Illinois Basin as suggested by temperature anomalies. Water Resources Res., v. 6, p. 912–918.

Cartwright, K. 1973. The effect of shallow groundwater flow systems on rock and soil temperatures. Ph.D. Thesis, Geology Dept. Univ. of Illinois at Urbana.

Cartwright, K., and M. R. McComas. 1968. Geophysical surveys in the vicinity of sanitary landfills in Northern Illinois. Ground Water, v. 16, no. 5, p. 23–30.

Cathels, L. M. 1977. An analysis of cooling of intrusives by groundwater convection which includes boiling. Econ. Geol., v. 72, p. 804–826.

Cathels, L. M., and A. T. Smith. 1983. Thermal constraints on the formation of Mississippi Valley type lead-zinc deposits and their implications for episodic basin dewatering and deposit genesis. Econ. Geol., v. 78, p. 983–1002.

Cederberg, G. A. 1985. TRANQL: A ground-water mass-transport and equilibrium chemistry model for multicomponent systems. Ph.D. dissertation, Stanford Univ., Stanford, Calif., 117 p.

Cedergren H. R. 1967. Seepage, Drainage, and Flow Nets. John Wiley & Sons, New York.

Chamberlin, T. C. 1885. The requisite and qualifying conditions of artesian wells. U.S. Geol. Survey 5th Ann. Rept. p. 131–173.

Champ, D. R., J. Gulens, and R. E. Jackson. 1979. Oxidation-reduction sequences in ground-water flow systems. Can. J. Earth Sci., v. 16, no. 1, p. 12–23.

Champ, D. R., J. L. Young, D. E. Robertson, and K. H. Abel. 1984. Chemical speciation of long-lived radionuclides in a shallow groundwater flow system. Water Pollut. Res. J. Can., v. 19, p. 35–54.

Chapelle, F. H. 1993. Ground-Water Microbiology and Geochemistry. John Wiley & Sons, New York, 424 p.

Chapelle, F. H., J. L. Zelibor, Jr., D. J. Grimes, and L. L. Knobel. 1987. Bacteria in deep coastal plain sediments of Maryland: A possible source of CO_2 to groundwater. Water Resources Res., v. 23, no. 8, p. 1625–1632.

Chapman, R. E. 1980. Mechanical versus thermal causes of abnormally high pore pressure in shales. Am. Assoc. Petrol. Geol. Bull. 64, p. 2179–2183.

Chapman, R. E. 1981. Geology and Water. Martinus Nijhoff/Dr. W. Junk Pub., Boston.

Charlax, E., E. Guyon, and N. Rivier. 1984. A criterion for percolation threshold in a random array of plates. Solid State Communications, v. 20, p. 999–1002.

Chebotarev, I. I. 1955. Metamorphism of natural water in the crust of weathering. Geochim. Cosmochim. Acta, v. 8, p. 22–48, 137–170, 198–212.

Cherry, J. A. 1983. Piezometers and other permanently-installed devices for groundwater quality monitoring. Proc. Seminar on Groundwater and Petroleum Hydrocarbons Protection, Detection, Restoration. Petroleum Assoc. for Conservation of the Canadian Environment, Ottawa, p. IV-1–IV-39.

Chia, Y. P. 1979. Digital simulation of compaction in sedimentary sequences. Ph.D. thesis, Geology Dept., Univ. of Illinois, Urbana.

Chiang, C. Y., K. R. Loos, R. A. Klopp, and M. C. Beltz. 1989. A real-time determination of geological/chemical properties of an aquifer by penetration testing. Proc. NWWA/API Conference Petroleum Hydrocarbons and Organic Chemicals in Ground Water—Prevention, Detection and Restoration. National Water Well Association, Dublin, Ohio, p. 175–189.

Chiou, C. T., D. W. Schmedding, and M. Manes. 1982. Partitioning of organic compounds in octanol-water systems. Environ. Sci. Technol., v. 16, p. 4–10.

Chun, R. Y. D., L. R. Mitchell, and K. W. Mido. 1964. Groundwater management for the nation's future—optimum conjunctive operation of groundwater basins. J. Hydrauls. Div. Am. Soc. Civil Engrs. HY4, v. 90, p. 79–105.

Claasen, H. C. 1982. Guidelines and Technologies for Obtaining Water Samples that Accurately Represent the Water Chemistry of an Aquifer. U.S. Geol. Surv. Open-File Report 82-1024, 49 p.

Cleary, R. W. 1978. Report on 208 Long Island pollution study. Water Resources Program, Princeton University. Princeton, N.J.

Cochran, G. 1968. Optimization of conjunctive use of groundwater and surface water for urban supply. M.S. Thesis in Civil Eng., Univ. Nevada at Reno.

Cochran, J. R., and K. E. Dalton. 1995. Using high-density magnetic and electromagnetic data for waste site characterization. Proc. of the Symposium on the Application of Geophysics to Engineering and Environmental Problems, Environmental and Engineering Geophysical Society, Englewood Colo., p. 117–127.

Cohen, P., O. Frank, and B. Foxworthy. 1968. An atlas of Long Island's water resources. New York State Water Res. Comm., Bull. GW62.

Cohen, R. M., and J. W. Mercer. 1993. DNAPL Site Evaluation. CRC Press, 2000 Corporate Blvd., Boca Raton, Fla.

Cohen, R. M., R. R. Rambold, C. F. Faust, J. O. Rumbaugh, III, and J. R. Bridge. 1987. Investigation and hydraulic containment of chemical migration: Four landfills in Niagara Falls. Civil Eng. Pract., Spring, 1987, p. 33–58.

Coleman, J. M., and D. B. Prior. 1982. Deltaic environments of deposition. In P. A. Scholle and D. Spearing (eds.), Sandstone Depositional Environments. Am. Assoc. Petrol. Geol. Mem. 31, p. 139–178.

Collins, R. E. 1961. Flow of Fluids Through Porous Materials. Reinhold, New York.

Collins, W. D. 1923. Graphical representation of water analysis. Ind. Eng. Chem., v. 15, p. 394.

Collins, W. D. 1925. Temperature of water available for individual use in the United States. U.S. Geol. Surv. Water Supply Papers 520-F, p. 97–104.

Combarnous, M. A., and S. A. Bories. 1975. Hydrothermal convection in saturated porous media. In V. T. Chow (ed.), Advances in Hydroscience. Academic Press, New York, v. 10, p. 231–307.

Conkling, H. 1946. Utilization of groundwater storage in stream system development. Trans. Am. Soc. Civil Engrs., v. 111, p. 275–305.

Cooper, H. H., J. D. Bredehoeft, I. S. Papadopulos. 1967. Response of a finite diameter well to an instantaneous charge of water. Water Resources Res., v. 3, p. 263–269.

Cooper, H. H., and C. E. Jacob. 1946. A generalized graphical method for evaluating formation constants and summarizing well field history. Trans. Am. Geophys. Union, v. 27, p. 526–534.

Cooper, H. H., and others. 1965. The response of well aquifer systems to seismic waves. J. Geophys. Res., v. 70, p. 3915–3926.

Cooper, H. H., and M. I. Rorabaugh. 1963. Groundwater movements and bank storage due to flood stages in surface streams. U.S. Geol. Survey Water Supply Paper 1536-J, p. 343–366.

Corey, J. D., D. R. Nielsen, and J. W. Biggar. 1963. Miscible displacement in saturated and unsaturated sandstone. Soil Sci. Soc. Am. Proc., v. 27, p. 258–262.

Coustau, H., and others. 1975. Classification hydrodynamique des basins sedimentaires utilisation combinée avec d'autres méthodes pour rationaliser l'exploration dans des bassins non produotifs. Proc. 9th World Petrol. Congr. Tokyo. Section II, p. 105–119.

Cowan, C. E., and others. 1995. The Multimedia Fate Model: A Vital Tool for Predicting the Fate of Chemicals. SETAC Press, Pensacola, Fla.

Cozzareli, I. M., R. P. Eganhouse, and M. J. Baedecker. 1988. The fate and effects of crude oil in a shallow aquifer: II. Evidence of anaerobic degradation of monoaromatic hydrocarbons. U.S. Geol. Survey Toxic Waste Ground-Water Contamination Program, U.S. Geol. Surv. Water-Resources Investigations Rept., 88-4220, p. 21–33.

Craig, H. 1961. Isotopic variations in meteoric water. Science, v. 133, p. 1702–1703.

Craun, G. F. 1984. Health aspects of groundwater pollution.

In G. Bitton and C. P. Gerba (eds.), Groundwater Pollution Microbiology. John Wiley & Sons, New York, p. 135–179.

Croff, A. G., T. F. Lomenick, R. S. Lowrie, and S. H. Stow. 1985. Evaluation of five sedimentary rocks other than salt for high level waste repository siting purposes. Oak Ridge Nat. Lab. ORNL/CF-85/2/V2. Oak Ridge, Tenn.

Cuevas, J. A. 1936. Foundation conditions in Mexico City. Proc. Intern. Conf. Soil Mech., v. 3, Cambridge, Mass.

Dagan, G. 1967. A method of determining the permeability and effective porosity of unconfined anisotropic aquifers. Water Resources Res. v. 3, p. 1059–1071.

Dagan, G. 1982. Stochastic modeling of groundwater flow by unconditional and conditional probabilities. 2. The solute transport. Water Resources Res., v. 18, no. 4, p. 835–848.

Dagan, G. 1984. Solute transport in heterogeneous porous formations. J. Fluid Mech., v. 145, p. 151–177.

Dagan, G. 1989. Flow and Transport in Porous Formations. Springer-Verlag, New York, 465 p.

Dagan, G., and J. Bear. 1968. Solving the problem of local interface upconing in a coastal aquifer by the method of small perturbations. J. Hydrol. Res., v. 6, p. 15–44.

Daines, S. R. 1982. Aquathermal pressuring and geopressure evaluation. Am. Assoc. Petrol. Geol. Bull. 66, p. 931–939.

Daniels, J. J. 1989. Technical review: ground penetrating radar. In Proc. (2nd) Symposium on the Application to Engineering and Environmental Problems, Soc. Engineering and Mineral Exploration Geophysicists, Golden, Colo., p. 62–142.

Darcy, H. P. G. 1856. Les fontaines publiques de la Ville de Dijon. Victor Dalmont, Paris.

Darton, N. H. 1909. Geology and underground waters of South Dakota. U.S. Geol. Surv. Water Supply Papers 227.

Daus, A. D., and E. O. Frind. 1985. An alternating direction Galerkin technique for simulation of contaminant transport in complex groundwater systems. Water Resources Res., v. 21, no. 5, p. 653–664.

Davidson, M. R. 1984. A green-ampt model of infiltration in a cracked soil. Water Resources Res., v. 20, p. 1685–1690.

Davidson, M. R. 1985. Numerical calculations of saturated-unsaturated infiltration in a cracked soil. Water Resources Res., v. 21, p. 709–714.

Davis, G. H., and others. 1959. Groundwater conditions and storage capacity in the San Joaquin Valley, California: U.S. Geol. Survey Water Supply Papers 1469.

Davis, G. H., J. B. Small, and H. B. Counts. 1963. Land subsidence related to decline of artesian pressure in the Ocala limestone at Savannah, Georgia. Eng. Geol. Case Histories 4, Geol. Soc. Am., p. 1–8.

Davis, J. C. 1986. Statistics and data analysis in geology: John Wiley & Sons, New York, 646 p.

Davis, S. N. 1969. In R. J. M. DeWiest (ed.), Porosity and Permeability of Natural Materials in Flow through Porous Materials. Academic Press, New York, p. 54–89.

Davis, S. N., and R. J. M. De Wiest. 1966. Hydrogeology. John Wiley & Sons, New York, 463 p.

de Swaan, A. O. 1976. Analytical solutions for determining naturally fractured reservoir properties by well testing. Soc. Petrol. Eng. J., v. 16, p. 117–122.

de Marsily, G. 1985. Flow and transport in fracture rock. Memoires. Hydrogeology of Rocks of Low Permeability. Intern. Assoc. Hydrogeol., p. 267–277, Tucson, Ariz.

de Pastrovich, T. L., Y. Baradat, R. Barthel, A. Chiavelli, and D. R. Fussell. 1979. Protection of ground water form oil pollution. CONCAWE, The Hague.

dePastrovich, T. L., and others. 1979. Protection of groundwater from oil pollution: CONCAWE, The Hague, 61 p.

DeBoer, R. B. 1977. On the thermodynamics of pressure solution—interaction between chemical and mechanical forces. Geochim. Cosmochim. Acta, v. 41, p. 249–256.

Deere, D. U., and F. D. Patton. 1967. Effect of pore pressure on the stability of slopes. Geol. Soc. Am. Soc. Civil Engrs. Symp., New Orleans, La.

Deju, R. A. 1973. A worldwide look at the occurrence of high fluid pressures in petroliferous basins. Special publication of Gulf Research and Development Co.

Delhomme, J. P. 1979. Spatial variability and uncertainty in groundwater flow parameters: A geostatistical approach. Water Resources Res., v. 15, no. 2, p. 269–280.

Demond, A. H., and P. V. Roberts. 1987. An examination of relative permeability relations for two-phase flow in porous media. Water Resources Bull. 23, no. 4, p. 617–628.

Dept. of Energy. 1980. Statement of position of the United States Department of Energy, proposed rule making on the storage and disposal of nuclear waste. Washington, D.C.

Dept. of Energy. 1986. Environmental assessment. Reference repository location, Hanford Site, Washington. DOE/RW-0070.

Dept. of Energy. 1988. Consultation draft site characterization plan, Yucca Mountain site, Nevada. US DOE OCRWM, I-III, Washington, D.C.

Desaulniers, D. E., J. A. Cherry, and P. Fritz. 1981. Origin, age, and movement of pore water in argillareous quaternary deposits at four sites in southwestern Ontario. J. Hydrol., v. 50, p. 231–257.

DeSitter, L. U. 1956. Structural Geology. McGraw-Hill, New York.

Dewers, T., and P. Ortoleva. 1988. The role of geochemical self-organization in the migration and trapping of hydrocarbons. Appl. Geochem., v. 3, p. 287–316.

Dickinson, G. 1953. Geologic aspects of abnormal reservoir pressures in Gulf Coast Louisiana. Am. Assoc. Petrol. Geol. Bull. 37, p. 410–432.

Dienes, J. K. 1982. Permeability, percolation, and statistical crack mechanisms. In Issues in Rock Mechanics, Proc. 22nd Symp. on Rock Mechanics, Berkeley, Calif.

Dillon, R. T., R. B. Lantz, and S. B. Pahwa. 1978. Risk Methodology for Geologic Disposal of Radioactive Waste. The Sandia Waste Isolation and Flow (SWIFT) Model. NUREG/CR-0424, Nuclear Regulatory Comm., Washington, D.C.

Domenico, P. A. 1972. Concepts and Models in Groundwater Hydrology. McGraw-Hill, New York.

Domenico, P. A. 1977. Transport phenomena in chemical rate processes in sediments. In F. A. Donath, F. G. Stehli, and G. Wetherill (eds.), Ann Rev. Earth Planet Sci., v. 5, p. 287–317.

Domenico, P. A. 1983. Determination of bulk rock properties from groundwater level fluctuations. Bull. Assoc. Eng. Geol. 20, p. 283–287.

Domenico, P. A. 1987. An analytical model for multidimensional transport of a decaying contaminant species. J. Hydrol., v. 91, p. 49–58.

Domenico, P. A., and M. D. Mifflin. 1965. Water from low

permeability sediments and land subsidence. Water Resources Res., v. 4, p. 563–576.

Domenico, P. A., M. D. Mifflin, and A. L. Mindling. 1966. Geologic controls on land subsidence in Las Vegas Valley. Proc. Ann. Eng. Geol., Soils Eng. Symp., Moscow, Idaho, p. 113–121.

Domenico, P. A., and V. V. Palciauskas. 1973. Theoretical analysis of forced convective heat transfer in regional groundwater flow. Geol. Soc. Am. Bull. 84, p. 3803–3814.

Domenico, P. A., and V. V. Palciauskas. 1979. Thermal expansion of fluids and fracture initiation in compacting sediments. Geol. Soc. Am. Bull. Part II, 90, p. 953–979.

Domenico, P. A., and V. V. Palciauskas. 1982. Alternative boundaries in solid waste management. Ground Water, v. 20, p. 303–311.

Domenico, P. A., and V. V. Palciauskas. 1988. The generation and dissipation of abnormal fluid pressures in active deposition environments. In W. Back, J. S. Rosenshein, and P. R. Seaber (eds.), The Geology of North America, Hydrogeology, v. 0–2, p. 435–445.

Domenico, P. A., and G. A. Robbins. 1984. A dispersion scale effect in model calibrations and field tracer experiments. J. Hydrol., v. 7, p. 121–132.

Domenico, P. A., and G. A. Robbins. 1985a. The displacement of connate water from aquifers, Geol. Soc. Am. Bull. 96, p. 328–335.

Domenico, P. A., and G. A. Robbins. 1985b. A new method of contaminant plume analysis. Ground Water, v. 23, p. 476–485.

Donaldson, I. G. 1962. Temperature gradients in the upper layers of the earth's crust due to convective water flows. J. Geophys. Res., v. 67, p. 3449–3459.

Downey, J. S. 1984. Geology and hydrology of the Madison limestone and associated rocks in parts of Montana, Nebraska, North Dakota, South Dakota, and Wyoming. U.S. Geol. Surv. Prof. Paper 1273-G, 152 p.

Dozy, J. J. 1970. A geologic model for genesis of lead zinc ores of the Mississippi Valley. U.S.A., Inst. Min. Metal. Trans., v. 79, p. B163–B170.

Drever, J. I. 1982. The Geochemistry of Natural Waters. Prentice-Hall, Englewood Cliffs, N.J., 388 p.

Dreybrodt, W. 1990. The role of dissolution kinetics in the development of karst aquifers in limestone: A model simulation of karst evolution. J. Geol., v. 98, p. 639–655.

Driscoll, F. G. 1986. Groundwater and Wells. Johnson Division, St. Paul, Minn.

Droppo, J. G., Jr., J. W. Buck, D. L. Strenge, and M. R. Siegel. 1990. Analysis of Health Impact Inputs to the U.S. Department of Energy's Risk Information System. Pacific Northwest Laboratory, PNL-7432, Richland, Wash.

Duffield, G. M., D. R. Buss, D. E. Stephenson, and J. W. Mercer. 1987. A grid refinement approach to flow and transport modeling of a proposed groundwater corrective action at the Savannah River Plant, Aiken, South Carolina. Solving Ground Water Problems with Models, Proc., Vol. 2: National Water Well Assoc., Dublin, Ohio, p. 1087–1120.

Duguid, J. O., and M. Reeves. 1976. Material transport through porous media: A finite element Galerkin model. Oak Ridge National Laboratory, ORNL-4928, Oak Ridge, Tenn.

Dullien, F. A. L. 1979. Porous Media: Fluid Transport and Pore Structure. Academic Press, New York, 396 p.

Dunbar, C. O., and K. M. Waage. 1969. Historical Geology. John Wiley & Sons, New York.

Dunham, K. C. 1970. Mineralization by deep formation waters: a review. Inst. Min. Metall. Trans., v. 79, p. B127–B136.

Dunkle S. A., L. N. Plummer, E. Busenberg, P. J. Phillips, J. M. Denver, P. A. Hamilton, R. L. Michel, and T. B. Coplen. 1993. Chlorofluorohydrocarbons (CCl_3F and CCl_2F_2) as dating tools and hydrologic tracers in shallow groundwater of the Delmarva Peninsula, Atlantic Coastal Plain, United States. Water Resources Res., v. 29, p. 3837–3860.

Durant, N. D., L. P. Wilson, and E. J. Bouwer. 1995. Microcosm studies of subsurface PAH-degrading bacteria from a former manufactured gas plant. J. Contam. Hydrol., v. 17, p. 213–237.

Eakin, T. A. 1964. Groundwater appraisal of Coyote Spring and Kane Spring Valleys and Muddy River Springs area, Lincoln and Clark Counties, Nevada. Groundwater Resources Reconnaissance Series Rept. 25, U.S. Geol. Survey, Carson City, Nev.

Eakin, T. A. 1966. A regional interbasin groundwater system in the White River area, southeastern Nevada. Water Resources Res., v. 2, no. 2, p. 251–271.

Eakin, T. A., and others. 1976. Summary appraisals of the Nation's groundwater resources: Great Basin. U.S. Geol. Survey Prof. Paper 813-G.

Earlougher, R. C. 1977. Pressure buildup testing. Advances in Well Test Analysis. Monograph 5, Soc. Petrol. Engrs. Am. Inst. Mech. Engrs., New York.

Eaton, J. P., and K. J. Takasaki. 1959. Seismological interpretation of earthquake induced water level fluctuations in wells. Bull. Seismol. Soc. Am., v. 49, p. 227–245.

Edge, R. W., and K. Cordry. 1989. The Hydropunch™: An in situ sampling tool for collecting ground water from unconsolidated sediments. Ground Water Mon. Rev., v. 9, no. 3, p. 177–183.

Eliott, T. 1978. Deltas. In H. G. Reading (ed.), Sedimentary Environments and Facies. Blackwell, Oxford, p. 97–142.

Elkins, J. W., T. M. Thompson, T. H. Swanson, J. H. Butler, B. D. Hall, S. O. Cummings, D. A. Fisher, A. G. Raffo. 1993. Decrease in growth rates of atmospheric chlorofluorocarbons 11 and 12: Nature, 364, 780–783.

Elzeftawy, A., R. S. Mansell, and H. M. Selim. 1976. Distribution of water and herbicide in Lakeland Sand during initial stages of infiltration. Soil Sci., v. 122, p. 297–307.

Endoe, H. K. 1984. Mechanical transport in two dimensional networks of fractures. Ph.D. Dissertation. Univ. of California, Berkeley.

Energy Research and Development Administration (ERDA). 1976. A bibliography of geothermal resources—exploration and exploitation. Washington, D.C.

Ernest, L. F. 1969. Groundwater flow in the Netherlands delta area and its influence on the salt balance of the future Lake Zeeland. J. Hydrol., v. 8, p. 137–172.

Evans, D. 1966. The Denver area earthquakes and the Rocky Mountain Arsenal Disposal Well. Mtn. Geol., v. 3, p. 23–36.

Ewers, R. O. 1982. Cavern development in the dimensions of length and breadth. Unpublished Ph.D. Thesis, McMaster University, Hamilton, Canada, 398 p.

Fabryka-Martin, J., D. O. Whittemore, S. N. Davis, P. W. Kubik, and P. Sharma. 1991. Geochemistry of halogens in the Milk River aquifer, Alberta, Canada: Applied Geochemistry, v. 6, pp. 447–464.

Fabryka-Martin, J., S. J. Wightman, W. J. Murphy, M. P. Wickham, M. W. Cafee, G. J. Nimz, J. R. Southon, and P. Sharma. 1993. Distribution of chlorine-36 in the unsaturated zone at Yucca Mountain: an indicator of fast transport paths. Proc. Focus '93: Site Characterization and Model Validation, Am. Nuclear Soc., La Grange Park, Ill., p. 58–68.

Falta, R. W., I. Javandel, K. Pruess, and P. A. Witherspoon. 1989. Density drive flow of gas in the unsaturated zone due to evaporation of volatile organic chemicals. Water Resources Res., v. 25, p. 2159–2169.

Farmer, V. E. 1983. Behaviour of petroleum contaminants in an underground environment. Proc. Seminar on Groundwater and Petroleum Hydrocarbons, Protection, Detection, Restoration. Petroleum Assoc. for Conservation of the Canadian Environment, Ottawa, II-1–II-16.

Farr, A. M., R. J. Houghtalen, and D. B. McWhorter. 1990. Volume estimation of light nonaqueous phase liquids in porous media. Ground Water, v. 28, p. 48–56.

Farvolden, R. N. 1964. Geologic controls on groundwater storage and base flow. J. Hydrol., v. 1, p. 219–250.

Faust, C. R. 1985. Affidavits in Civil Action No. 79-989. United States District Court for the Western District of New York, 92 p.

Feenstra, S., D. M. Mackay, and J. A. Cherry. 1991. Presence of residual NAPL based on organic chemical concentrations in soil samples. Ground Water Mon. Rev., v. 11, p. 128–136.

FERMCO, Fernald Environmental Restoration Management Corporation. 1994. Fernald 1993 Site Environmental Report. FEMP-2342, 137 p.

Ferris, J. G. 1951. Cyclic fluctuations of water levels as a basis for determining aquifer transmissibility. Int. Assoc. Sci. Hydrol. Pub., v. 33, p. 148–155.

Ferris, J. G. 1959. Groundwater. In C. O. Wisler and E. F. Brater (eds.), Hydrology. John Wiley & Sons, New York.

Ferris, J. G., and others. 1962. Theory of aquifer tests. U.S. Geol. Surv. Water Supply Papers 1536-E., p. 69–174.

Ferry, J. P., P. J. Dougherty, J. B. Moser, and R. M. Schuller. 1986. Occurrence and recovery of a DNAPL in a low-yielding bedrock aquifer. Proc. of the NWWA/API Conference on Petroleum Hydrocarbons and Organic Chemicals in Ground Water—Prevention, Detection and Restoration. National Water Well Assoc., Dublin, Ohio, p. 722–735.

Fletcher, J. B., and L. R. Sykes. 1977. Earthquakes related to hydraulic mining and natural seismic activity in western New York State. J. Geophys. Res., v. 82, p. 3767–3780.

Fogel, S., M. Findlay, A. Moore, and M. Leahy. 1987. Biodegradation of chlorinated chemicals in groundwater by methane oxidizing bacteria. Proc. of the NWWA/API Conference on Petroleum Hydrocarbons and Organic Chemicals in Ground Water—Prevention, Detection and Restoration. National Water Well Assoc., Dublin, Ohio, p. 167–185.

Fogg, G. E., and C. W. Kreitler. 1982. Ground-water Hydraulics and Hydrochemical Facies in Eocene Aquifers of the East Texas Basin. Bureau of Economic Geology, The Univ. of Texas at Austin, Report of Investigations No. 127, 75 p.

Folk, R. L., and L. S. Land. 1975. Mg/Ca ratio and salinity: two controls over crystallization of dolomite. Am. Assoc. Petrol. Geol. Bull. 59, p. 60–68.

Fontes, J. C., and J. M. Garnier. 1979. Determination of the initial ^{14}C activity of the total dissolved carbon: A review of the existing models and a new approach: Water Resources Res., v. 15, no. 2, p. 399–413.

Fontes, J. C. 1980. Environmental isotopes in groundwater hydrology, Chapter 3. In P. Fritz and J. C. Fontes (eds.), Handbook of Environmental Isotope Geochemistry, v. 1. Elsevier, Amsterdam, p. 75–140.

Ford, D. C., and P. Williams. 1989. Karst geomorphology and hydrology. Unwin Hyman, London, 601 p.

Forster, C., and L. Smith. 1988. Groundwater flow systems in mountainous terrain: 2. Controlling factors. Water Resources Res., v. 24, p. 1011–1023.

Forster, C., and L. Smith. 1989. The influence of groundwater flow on thermal regimes in mountainous terrain. J. Geophys. Res., v. 94, p. 9439–9451.

Forstner, U., and T. W. Wittman. 1981. Metal Pollution in the Aquatic Environment. Springer-Verlag, Berlin, Heidelberg, 486 p.

Foster, M. D. 1950. The origin of high sodium bicarbonate waters in the Atlantic and Gulf Coastal Plains. Geochim. Cosmochim. Acta, v. 1, p. 33–48.

Frank-Kamenetskii, P. A. 1955. Diffusion and Heat Transfer in Chemical Kinetics, trans. by J. P. Appleton (1969): New York, Plenum Press.

Franks, B. J. (ed.). 1987. U.S. Geological Survey Program on Toxic Waste—Ground-Water Contamination: Proc. of the Third Technical Meeting, Pensacola, Florida, March 23–27, 1987. U.S. Geol. Surv. Open File Report 87-109.

Fredrickson, J. K., F. J. Brockman, R. J. Hicks, and B. A. Denovan. 1990. Biodegradation of nitrogen-containing aromatic compounds in deep subsurface sediments. In C. B. Fliermans and T. C. Hazen (eds.), Proc. of the First International Symposium on Microbiology of the Deep Subsurface. Westinghouse Savannah River Company, p. 6-27 to 6-44.

Freeze, R. A. 1969. The mechanism of natural groundwater recharge and discharge; I: one dimensional vertical, unsteady unsaturated flow above a recharging or discharging groundwater flow system. Water Resources Res., v. 5, p. 153–171.

Freeze, R. A. 1971. Three dimensional transient saturated-unsaturated flow in a groundwater basin. Water Resources Res., v. 7, p. 347–366.

Freeze, R. A., and J. A. Cherry. 1979. Groundwater. Prentice-Hall, Inc. Englewood Cliffs, N.J., 604 p.

Freeze, R. A., J. Masmann, L. Smith, T. Sperling, and B. James. 1990. Hydrogeologic decision analysis: 1. A framework. Ground Water, v. 28, no. 5, p. 738–766.

Freeze, R. A., and P. A., Witherspoon. 1966. Theoretical analysis of regional groundwater flow. I: Analytical and numerical solutions to the mathematical model. Water Resources Res., v. 2, p. 641–656.

Freeze, R. A., and P. A. Witherspoon. 1967. Theoretical analysis of regional groundwater flow. II: Effect of water table configuration and subsurface permeability variations. Water Resources Res., v. 3, p. 623–634.

Freyberg, D. L. 1986. A natural gradient experiment on solute transport in a sand aquifer. 2. Spatial moments and the advection and dispersion of nonreactive tracers. Water Resources Res., v. 22, no. 13, p. 2031–2046.

Freyberg, D. L. 1988. An exercise in ground-water model calibration and prediction: Ground Water 26(3), pp. 350–360.

Fried, J. J. 1975. Groundwater Pollution. Elsevier, Amsterdam, 330 p.

Fried, J. J., and M. A. Combarnous. 1971. Dispersion in porous media. Adv. Hydrosci., v. 7, p. 169–282.

Friedman, L. C., and D. E. Erdman. 1982. Quality assurance practices for the chemical and biological analyses of water and fluvial sediments. U.S. Geol. Surv. TWRI, Book 5, Chapter A6, 181 p.

Frind, E. O. 1982. The principal direction technique a new approach to ground-water contaminant transport modeling. Fourth International Conference on Finite Elements in Water Resources, Proc.: New York, Springer-Verlag, p. 13-25–13-42.

Frind, E. O., and D. Germain. 1986. Simulation of contaminant plumes with large dispersive contrast: evaluation of alternating direction Galerkin models. Water Resources Res., v. 22, no. 13, p. 1857–1873.

Fritz, P., and J. C. Fontes. 1980. Introduction. In P. Fritz and J. C. Fontes (eds.), Handbook of Environmental Isotope Geochemistry, v. 1. Elsevier, Amsterdam, p. 1–19.

Frind, E. O. 1982. The principal direction technique: a new approach to groundwater contaminant transport modeling. In Proceedings, Fourth International Conference on Finite Elements in Water Resources. Springer-Verlag, New York, p. 13-25–13-42.

Fryar, A. E., and P. A. Domenico. 1989. Analytical inverse modeling of regional scale tritium waste migration. J. Contam. Hydrol., v. 4, p. 113–125.

Gabrysch, R. K., and C. W. Bonnet. 1975. Land surface subsidence in the Houston-Galveston region, Texas. Texas Water Dev. Board Rept. 188.

Gale, J. E. 1975. A numerical field and laboratory study of flow in rocks with deformable fractures. Ph.D. dissertation, Univ. of California, Berkeley.

Gale, J. E., A. Rouleau, and L. C. Atkinson. 1985. Hydraulic properties of fractures. Memoires. Hydrogeology of Rocks of Low Permeability. Inter. Assoc. Sci. Hydrol., Tucson, Ariz., p. 1–11.

Galloway, W. E. 1978. Uranium mineralization in a coastal-plain fluvial aquifer system; Catahoula Formation, Texas. Econ. Geol., v. 73, p. 1656–1676.

Galloway, W. E., and D. K. Hobday. 1983. Terrigenous Clastic Depositional Systems: Application to Petroleum, Coal and Uranium Exploration. Springer-Verlag, New York.

Galloway, W. E., and W. R. Kaiser. 1980. Catahoula Formation of the Texas coastal plain: origin, geochemical evolution, and characteristics of uranium deposits. Bureau of Economic Geology, The Univ. Texas, Austin, Report Invest., no. 100.

Gambolati, G., and R. A. Freeze. 1973. Mathematical simulation of the subsidence of Venice. I. Theory. Water Resources Res., v. 9, p. 721–733.

Gambolati, G., P. Gatto, and R. A. Freeze. 1974a. Mathematical simulation of the subsidence of Venice. 2. Results. Water Resources Res., v. 10, p. 563–577.

Gambolati, G., P. Gatto, and R. A. Freeze. 1974b. Predictive simulation of the subsidence of Venice. Science, v. 183, p. 849–851.

Gangi, A. F. 1978. Variation of the whole and fractional porous rock permeability with confining pressure. Int. J. Rock Mech., v. 15, p. 249–257.

Garabedian, S. P., and D. R. LeBlanc. 1991. Overview of research at the Cape Cod Site: Field and laboratory studies of hydrologic, chemical, and microbial processes affecting transport in a sewage-contaminated sand and gravel aquifer. U.S. Geol. Survey Toxic Waste Ground-water Contamination Program, U.S. Geol. Surv., Water-Resoures Invest. Report, 91-4034, p. 1–9.

Garder, A. O., A. L. Peaceman, and R. Pozzi. 1964. Numerical calculation of multi-dimensional displacement by the method of characteristics. Soc. Petrol. Eng. J., v. 4, no. 1, p. 26–38.

Garrels, R. M. 1960. Mineral Equilibrium at Low Temperature and Pressure. Harper, New York.

Garrels, R. M., and F. T. MacKenzie. 1967. Origin of the chemical compositions of some springs and lakes. In R. F. Gould (ed.), Equilibrium Concepts in Natural Water Systems. Am. Chem. Soc., Adv. Chem. Ser. 67, Washington, D.C., p. 222–242.

Garrison, A. W., L. H. Keith, and W. H. Shackelford. 1977. Occurrence, registry, and classification of organic pollutants in water, with development of a master scheme for their analysis. In O. Hutzinger, I. H. Van Lelyveld, and B. C. J. Zoeteman (eds.), Aquatic Pollutants: Transformation and Biological Effects. Pergamon Press, Oxford, p. 39–68.

Garven, G. 1988. Paleohydrology of oil migration in the Alberta Basin (abst). EOS, Trans. Am. Geophys. Union, v. 69, p. 360.

Garven, G. 1995. Continental scale ground water flow and geologic processes. Ann. Rev. Earth Planet. Sci., v. 23, p. 89–117.

Garven, G., and R. A. Freeze. 1984a. Theoretical analysis of the role of groundwater flow in the genesis of stratabound ore deposits. I. Mathematical and numerical model. Am. J. Sci., v. 284, p. 1085–1124.

Garven, G., and R. A. Freeze. 1984b. Theoretical analysis of the role of groundwater flow in the genesis of stratabound ore deposits. II. Quantitative results. Am. J. Sci., v. 284, p. 1125–1174.

Gary, S. K., and D. R. Kassoy. 1981. Convective heat and mass transfer in hydrothermal systems. In L. Rybach and L. J. P. Muffler (eds.), Geothermal Systems, Principles and Case Histories. John Wiley & Sons, New York, p. 37–76.

Gat, J. R. 1980. The isotopes of hydrogen and oxygen in precipitation. Chapter 1. In P. Fritz and J. C. Fontes (eds.), Handbook of Environmental Isotope Geochemistry, v. 1. Elsevier, Amsterdam, p. 21–47.

Gelhar, L. W., A. L. Gutjahr, and R. L. Naff. 1979. Stochastic analysis of macrodispersion in a stratified aquifer: Water Resources Res., v. 15, no. 6, p. 1387–1397.

Gelhar, L. W., and C. L. Axness. 1983. Three-dimensional stochastic analysis of macrodispersion in aquifers. Water Resources Res., v. 19, no. 1, p. 161–180.

Gelhar, L. W., A. Mantoglou, C. Welty, and K. R. Rehfeldt. 1985. A review of field-scale physical solute transport processes in saturated and unsaturated porous media. Electric Power Research Institute EPRI EA-4190 Project 2485-5, 116 p.

Gelhar, L. W., C. Welty, and K. R. Rehfeldt. 1992. A critical review of data on field-scale dispersion in aquifers. Water Resources Res., v. 28, no. 7, p. 1955–1974.

Ghiorse, W. C., and J. T. Wilson. 1988. Microbial ecology of the terrestrial subsurface. Adv. Appl. Microbiol., v. 33, p. 107–172.

Ghyben, W. B. 1899. Notes in verband met Voorgenomen Put boring Nabji Amsterdam, Tijdschr. Koninhitk, Inst. Ingrs., The Hague.

Gibson, R. E. 1958. The progress of consolidation in a clay layer increasing in thickness with time. Geotechnique, v. 8, p. 171–182.

Gibb, J. P., and M. J. Barcelona. 1984. Sampling for organic contaminants in ground water. Ground Water, v. 76, no. 5, p. 48–51.

Gibb, J. P., R. M. Schuller, and R. A. Griffin. 1981. Procedures for the Collection of Representative Water Quality Data from Monitoring Wells. Cooperative Ground Water Report. Illinois State Water Survey and U.S. Environ. Protection Agency, 60 p.

Gibson, R. E. 1958. The progress of consolidation in a clay layer increasing in thickness with time: Geotechnique, v. 8, p. 171–182.

Gilkeson, R. H., P. C. Heigold, and D. E. Laymon. 1986. Practical application of theoretical models to magnetometer surveys of hazardous waste disposal sites—A case study. Ground Water Mon. Rev., v. 6, 54–61.

Gillham, R. W., and S. F. O'Hannesin. 1994. Enhanced degradation of halogenated aliphatics by zero-valent iron. Ground Water, v. 32, no. 6, p. 958–967.

Gilluly, J., and U. S. Grant. 1949. Subsidence in Long Beach Harbor Area, California. Bull. Geol. Soc. Am. v. 60, p. 461–521.

Glasstone, S., and D. Lewis. 1960. Elements of Physical Chemistry. D. Van Nostrand Company, Princeton, N.J., 758 p.

Glover, R. E. 1964. The pattern of freshwater flowing in a coastal aquifer. In Sea Water in Coastal Aquifers. U.S. Geol. Survey Water Supply Paper 1613-C, p. 32–35.

Goerlitz, D. F., and E. Brown. 1972. Methods for analysis of organic substances in water. U.S. Geol. Surv. TWRI, Book 5, Chapter A3, 40 p.

Goguel, J. 1976. Geothermics (trans. by A. Rite). McGraw-Hill, New York.

Goode, D. J., and C. A. Appel. 1992. Finite-difference interblock transmissivity for unconfined aquifers and for aquifers having smoothly varying transmissivity. U.S. Geol. Surv., Water-Resources Investigations Rept., 92–4124, 79 p.

Gordan, F. R. 1970. Water level changes preceding the Neckering Western Australia earthquake of October 14, 1968. Bull. Seismol. Soc. Am. v. 60, p. 1739–1740.

Gorelick, S. M. 1987. Sensitivity analysis of optimal ground water contaminant capture curves. Solving Ground Water Problems with Models. National Water Well Assoc., Dublin, Ohio, p. 133–146.

Gorelick, S. M., C. I. Voss, P. Gill, W. Murray, M. Saunders, and M. Wright. 1984. Aquifer reclamation design: The use of contaminant transport simulation combined with nonlinear programming. Water Resources Res., v. 20, no. 4, p. 415–427.

Gounaris, V., P. R. Anderson, and T. M. Holsen. 1993. Characteristics and environmental significance of colloids in a landfill leachate. Environ. Sci. Technol., v. 27, p. 1381–1387.

Graton, L. C., and H. J. Fraser. 1935. Systematic packing of spheres with particular relation to porosity and permeability and experimental study of the porosity and permeability of clastic sediments. Geology, v. 43, p. 785–909.

Green, H. W. 1984. "Pressure solution" creep: some causes and mechanisms. J. Geophys. Res., v. 89, p. 4313–4318.

Green, D. H., and H. F. Wang. 1986. Fluid pressure response to undrained compression in saturated sedimentary rock. Geophysics, v. 51, p. 948–956.

Green, D. W., and R. H. Perry. 1961. Heat transfer with a fluid flowing through porous media. Chem. Eng. Prog. Symp. Ser. 57, p. 61–68.

Greenhouse, J. P., and D. D. Slaine. 1983. The use of reconaissance electrical methods to map contaminant migration. Ground Water Mon. Rev., v. 3, p. 47–49.

Greenkorn, R. A. 1983. Flow Phenomena in Porous Media: Fundamentals and Applications in Petroleum, Water and Food Production. Marcel Dekker, New York, 550 p.

Greenkorn, R. A., and D. P. Kessler. 1972. Transfer Operations. McGraw-Hill, New York, 548 p.

Gretener, P. E. 1972. Thoughts on overthrust faulting in a layered sequence. Bull. Canadian Petrol. Geol. v. 20, p. 583–607.

Gretener, P. E. 1981. Pore pressure: fundamentals, general ramifications, and implications for structural geology (revised). Education Course Note Series 4, Am. Assoc. Petrol. Geol. Tulsa, Okla.

Griffin, R. A., and W. R. Roy. 1985. Interaction of Organic Solvents with Saturated Soil-water Systems. Environmental Institute for Waste Management Studies, Univ. of Alabama, Open File Report No. 3, 86 p.

Grisak, G. E., and R. E. Jackson. 1978. An appraisal of the Hydrogeological Processes Involved in Shallow Subsurface Radioactive Waste Management in Canadian Terrain. Scientific Series No. 84, Inland Waters Directorate, Environment Canada, Ottawa, 194 p.

Grisak, G. E., and J. F. Pickens. 1980. Solute transport through fractured media, 1. The effect of matrix diffusion. Water Resources Res., v. 16, no. 4, p. 719–730.

Grove, D. B. 1971. U.S. Geological Survey tracer study, Armargosa Desert, Nye County, Nevada. II. An analysis of the flow field of a discharging-recharging pair of wells. U.S. Geol. Surv. Rep. USGS—474-99, 56 p.

Grove, D. B., and W. A. Beetem. 1971. Porosity and dispersion constant calculations for a fractured carbonate aquifer using the two well tracer method. Water Resources Res., v. 7, p. 128–134.

Gruszczenski, T. S. 1987. Determination of a realistic estimate of the actual formation product thickness using monitor wells: a field bailout test. Proc. of the NWWA/API Conference on Petroleum Hydrocarbons and Organic Chemicals in Ground Water—Prevention, Detection and Restoration. National Water Well Assoc. Dublin, Ohio, p. 235–253.

Gschwend, P. M., and M. D. Reynolds. 1987. Monodisperse ferrous phosphate colloids in an anoxic groundwater plume. J. Contam. Hydrol., v. 1, p. 309–327.

Gupta, M. L., and B. K. Rastogi. 1970. Dams and earthquakes. Dev. Engr. Geol., v. 11.

Gupta, M. L., B. K. Rastogi, and H. Narain. 1972. Common features of the reservoir associated seismic activities. Bull. Seismol. Soc. Am. v. 6, p. 481–492.

Gupta, M. L., and B. S. Suknija. 1974. Preliminary studies of some geothermal areas in India: Geothermics, v. 3, p. 105–112.

Guth, P. L., K. V. Hodges, and J. H. Willemin. 1982. Limitations on the role of pore pressure in gravity sliding. Bull. Geol. Soc. Am. v. 93, p. 606–612.

Gutjahr, A. L., L. W. Gelhar, A. A. Bakr, and J. R. McMillan. 1978. Stochastic analysis of spatial variability in subsurface flow: 2. Evaluation and application. Water Resources Res., v. 14, p. 953–959.

Güven, O., R. W. Falta, F. J. Molz, and J. G. Melville. 1986. A simplified analysis of two-well tracer tests in stratified aquifers. Ground Water, v. 24, no. 1, p. 63–71.

Hamilton, L. D., A. F. Meinhold, S. L. Baxter, S. Holtzman, S. C. Morris, R. Pardi, M. D. Rowe, and C. Sun. 1993. Pilot Study

Risk Assessment at the Fernald Environmental Management Project (FEMP). Brookhaven National Laboratory, BNL-48777, 168 p.

Handin, J., and others. 1963. Experimental deformation of sedimentary rocks under confining pressure: pore pressure tests. Am. Assoc. Petrol. Geol. Bull. 47, p. 717–755.

Hansch, C., and A. Leo. 1979. Substituent constants for correlation analysis in chemistry and biology. John Wiley & Sons, New York, 339 p.

Hanshaw, B., W. Back, and R. Deike. 1971. A geochemical hypothesis for dolomitization by groundwater. Econ. Geol., v. 66, p. 710–724.

Hanshaw, B., and E. Zen. 1965. Osmotic equilibrium and overthrust faulting. Geol. Soc. Am. Bull. 76, p. 1379–1386.

Hantush, M. S. 1956. Analysis of data from pumping tests in leaky aquifers. Trans. Am. Geophys. Union, v. 37, p. 702–714.

Hantush, M. S. 1957. Nonsteady radial flow to a well partially penetrating an infinite leaky aquifer. Proc. Iraqi Sci. Soc., v. 1, p. 10–19.

Hantush, M. S. 1960. Modification of the theory of leaky aquifers. J. Geophys. Res., v. 65, p. 3713–3725.

Hantush, M. S. 1961. Aquifer tests on partially penetrating wells. Proc. Am. Soc. Civil Engrs., v. 87, p. 171–195.

Hantush, M. S. 1964. Hydraulics of wells. In V. T. Chow (ed.), Advances in Hydroscience. Academic Press, New York, p. 281–432.

Hantush, M. S. 1966a. Wells in homogeneous anisotropic aquifers. Water Resources Res., v. 2, p. 273–279.

Hantush, M. S. 1966b. Analysis of data from pumping test in anisotropic aquifers. J. Geophys. Res., v. 71, p. 421–426.

Hantush, M. S., and C. E. Jacob. 1955. Nonsteady radial flow in an infinite leaky aquifer. Trans. Am. Geophys. Union, v. 36, p. 95–100.

Hantush, M. S., and R. G. Thomas. 1966. A method for analyzing a drawdown test in anisotropic aquifers. Water Resources Res., v. 2, p. 281–285.

Hardie, L. A. 1991. On the significance of evaporites. Ann. Rev. Earth Planet. Sci., v. 19, p. 131–168.

Hardie, L. A., and H. P. Eugster. 1970. The evolution of closed-basin brines. Mineral. Soc. Am. Spec. Publ., v. 3, p. 273–290.

Harris, J. C. 1982. Rate of hydrolysis. In W. J. Lyman, W. F. Reehl, and D. H. Rosenblatt (eds.), Handbook of Chemical Property Estimation Methods. McGraw-Hill, New York, 977 p.

Harris, J. F., G. L. Taylor, and J. L. Walper. 1960. Relation of deformational fracture in sedimentary rocks to regional and local structure. Am. Assoc. Petrol. Geol. Bull. 44, p. 1853–1873.

Harvey, R. W., and S. P. Garabedian. 1991. Use of colloid filtration theory in modeling movement of bacteria through a contaminated sandy aquifer. Environ. Sci. Technol., v. 25, no. 1, p. 178–185.

Harvey, R. W., and L. George. 1986. Bacterial distribution and transport in a plume of sewage-contaminated ground water. U.S. Geol. Survey Toxic Waste Ground-water Contamination Program, U.S. Geol. Surv. Open File Rept., 86-481. p. B25–B26.

Harvey, R. W., L. H. George, R. L. Smith, and D. R. LeBlanc. 1989. Transport of microspheres and indigenous bacteria through a sandy aquifer: Results of natural- and forced-gradient tracer experiments. Environ. Sci. Technol., v. 23, no. 1, p. 51–56.

Harvey, R. W., R. L. Smith, and L. George. 1984. Microbial distribution and heterotropic uptake in a sewage plume. Movement and Fate of Solutes in a Plume of Sewage-Contaminated Ground Water, Cape Cod, Massachusetts: U.S. Geol. Surv. Toxic Waste Ground-Water Contamination Program, U.S. Geol. Surv. Open-File Rept. 84-475, p. 139–152.

Hashin, Z., and S. Shtrikman. 1962. A variation approach to the theory of effective magnetic permeability of multiphase materials. J. Appl. Phys., v. 33, p. 3125–3131.

Hassett, J. J., W. L. Banwart, and R. A. Griffin. 1983. Correlation of compound properties with sorption characteristics of nonpolar compounds by soils and sediments: Concepts and limitations. In C. W. Francis and S. I. Auerbach (eds.), Environment and Solid Wastes: Characterization, Treatment and Disposal. Butterworth Publishers, Chap. 15, p. 161–178.

Hastings, T. 1986. A theoretical approach for assessing the role of rock and fluid properties in the development of abnormal pressure. M.S. thesis, Texas A&M Univ.

Hayes, J. B. 1977. Sandstone diagenesis, the hole truth. In P. A. Scholle and R. R. Schluger (eds.), Aspects of Diagenesis. Soc. Econ. Paleontol. and Mineral Spec. Pub. 26, p. 127–139.

Healy, J. W., and others. 1968. The Denver earthquakes. Science, v. 161, p. 1301–1310.

Heard, H. C., and W. W. Rubey. 1966. Tectonic implications of gypsum dehydration. Geol. Soc. Am. Bull. 77, p. 741–760.

Heath, R. C. 1982. Classification of groundwater systems of the United States. Ground Water, v. 20, p. 393–401.

Heath, R. C. 1984. Groundwater regions of the United States. U.S. Geol. Survey Water Supply Paper 2242.

Helfferich, F. 1966. In J. A. Marinsky (ed.), Ion Exchange, A Series of Advances. Marcel Dekker, New York p. 65–100.

Helm, D. C. 1975. One dimensional simulation of aquifer compaction system near Pixley, California. Water Resources Res., v. 11, p. 465–478.

Hem, J. D. 1959. Study and interpretation of the chemical characteristics of natural water. U.S. Geol. Survey Water Supply Papers, 1473.

Hendry, M. J., J. A. Cherry, and E. I. Wallick. 1986. Origin and distribution of sulfate in a fractured till in southern Alberta, Canada. Water Resources Res., v. 22, no. 1, p. 45–61.

Hendry, M. J., and F. W. Schwartz. 1988. An alternative view on the origin of chemical and isotopic patterns in groundwater from the Milk River aquifer. Water Resources Res., v. 24, no. 10, p. 1747–1764.

Hendry, M. J., and F. W. Schwartz. 1990. The chemical evolution of ground water in the Milk River aquifer, Canada. Ground Water, v. 28, no. 2, p. 253–261.

Herrin, H., and T. Goforth. 1975. Environmental problems associated with power production from geopressured reservoirs. Proc. 1st Geopress. Geotherm Conf., p. 311–320.

Herzberg B. 1901. Die Wasserversovgung einiger Nordseebaser. J. Gasbeleucht und wasserversov, v. 44, p. 815–819.

Hess, K. M., W. N. Herkelrath, and H. I. Essaid. 1992. Determination of subsurface fluid contents at a crude-oil spill site. J. Contam. Hydrol., v. 10, p. 75–96.

Hess, K. M., S. H. Wolf, and M. A. Celia. 1992. Large-scale natural gradient tracer test in sand and gravel, Cape Cod, Massachusetts, 3. Hydraulic conductivity variability and calculated macrodispersivities. Water Resources Res., v. 28, no. 8, p. 2011–2027.

Hewitt, T. A. 1986. Fractal distributions of reservoir heterogeneity and their influence on fluid transport. Proc. 61st Annual Tech. Conf. Soc. Petrol. Eng., New Orleans, Paper SPE-15386.

Heyl, A. V. and others. 1970. Guidebook to the Upper Mississippi Valley basic-metal district. Univ. of Wisconsin Geol. and Nat. History Survey, Inf. Circ. No. 16, 49 p., Madison.

Hickey, J. J. 1984. Field testing of the hypothesis of Darcian flow through a carbonate aquifer. Ground Water, v. 22, p. 544–547.

Hill, M. C. 1990. Preconditioned Conjugate-Gradient 2 (PCG2), A computer program for solving ground-water flow equations. U.S. Geol. Surv., Water-Resources Investigations Report, 90-4048, 43 p.

Hill, M. C. 1992. A computer program (MODFLOWP) for estimating parameters of a transient, three-dimensional ground-water flow model using nonlinear regression. U.S. Geol. Surv., Open-File Report 91-484.

Hillel, D. 1971. Soil and Water: Physical Principles and Processes. Academic Press, New York.

Hirasaka, G. J. 1974. Pulse tests and other early transient pressure analysis for in-situ estimation of vertical permeability. Trans. Am. Inst. Mining Engr., v. 257, p. 75–90.

Hoefner, M. L., and H. S. Fogler. 1988. Pore evolution and channel formation during flow and reaction in porous media. AIChE J., v. 34, p. 45–54.

Hoek, E., and J. W. Bray. 1981. Rock Slope Engineering. Inst. of Min. and Met., London.

Hornberger, G. M., J. Ebert, and I. Remson. 1970. Numerical solution of the Boussinesq equation for aquifer-stream interaction. Water Resources Res., v. 6, p. 601–608.

Hornberger, G. M., I. Remson, and A. A. Fungaroli. 1969. Numeric studies of a composite soil moisture groundwater system. Water Resources Res., v. 5, p. 797–802.

Horner, D. R. 1951. Pressure buildup in wells. Proc. 3rd World Petrol. Congress. Sect. II. E. J. Brill, Leiden, Holland. p. 503–521.

Horton, C. W., and F. T. Rogers. 1945. Convection currents in a porous medium. J. Appl. Phys., v. 16, p. 367–370.

Horton, R. E. 1933. The role of infiltration in the hydrologic cycle. Trans. Am. Geophys. Union, v. 14, p. 446–460.

Horton, R. E. 1940. Approach toward a physical interpretation of infiltration capacity. Soil Sci. Soc. Am. Proc., v. 5, p. 339–417.

Howard, A. D., and C. G. Groves. 1995. Early development of karst systems: 2. Turbulent flow. Water Resources Res., v. 31, p. 19–26.

Hsieh, P. A., and J. D. Bredehoeft. 1981. A reservoir analysis of the Denver earthquakes. J. Geophys. Res., v. 86, p. 903–920.

Hsieh, P. A., and J. R. Freckleton. 1993. Documentation of a computer program to simulate horizontal-flow barriers using the U.S. Geological Survey's modular three-dimensional finite-difference ground-water flow model. U.S. Geol. Surv., Open-File Report 92-477, 32 p.

Hsieh, P. A., and S. P. Neuman. 1985. Field determination of the three dimensional hydraulic conductivity tensor of anisotropic media. I: Theory. Water Resources Res., v. 21, p. 1655–1665.

Hsieh, P. A., and others. 1985. Field determination of the three dimensional hydraulic conductivity tensor of anisotropic media: II. Methodology and application to fractured rocks. Water Resources Res., v. 21, p. 1667–1676.

Hsu, K. J. 1977. Studies of Ventura field, California: II. Lithology,

compaction, and permeability of sands. Am. Assoc. Petrol. Geol. Bull. 61, p. 169–191.

Hubbert, M. K. 1940. The theory of groundwater motion. J. Geol., v. 48, p. 785–944.

Hubbert, M. K. 1953. Entrapment of petroleum under hydrodynamic conditions. Bull. Am. Assoc. Petrol. Geol. 37, p. 1944–2026.

Hubbert, M. K. 1956. Darcy's law and the field equations of the flow of underground fluids. Trans. Am. Inst. Min. Met. Engrs., v. 207, p. 222–239.

Hubbert, M. K., and W. W. Rubey. 1959. Role of fluid pressure in mechanics of overthrust faulting. Part I: Mechanics of fluid filled porous solids and its application to overthrust faulting. Bull. Geol. Soc. Am. 70, p. 115–166.

Hubbert, M. K., and D. G. Willis. 1957. Mechanics of hydraulic fracture. Am. Inst. Min. Engr. Trans., v. 210, p. 153–168.

Hughes, J. P., C. R. Sullivan, and R. E. Zinner. 1988. Two techniques for determining the true hydrocarbon thickness in an unconfined sandy aquifer. Proc. of the NWWA/API Conference on Petroleum Hydrocarbons and Organic Chemicals in Ground Water—Prevention, Detection and Restoration. National Water Well Assoc., Dublin, Ohio, p. 291–314.

Huitt, J. L. 1956. Fluid flow in simulated fractures. Am. Inst. Chem. Eng., v. 2, p. 259.

Hull, R. W., Y. K. Kharaka, A. S. Maest, and T. L. Fries. 1984. Sampling and analysis of subsurface water. In B. Hitchon and E. I. Wallick (eds.), First Canadian/American Conference on Hydrogeology. National Water Well Assoc., Dublin, Ohio, p. 117–126.

Hunt, B. 1978. Dispersive source in uniform groundwater flow. J. Hydrol. Div., Proc. Amer. Soc. Civil Eng., v. 104, p. 75–85.

Hunt, B. W. 1978. Dispersive sources in uniform groundwater flow. J. Hydraulics. Div., ASCE, v. 99, no. HYi, p. 13–21.

Hunter, J. A., S. E. Pullan, R. A. Burns, R. M. Gagne, and R. L. Good. 1984. Shallow seismic reflection mapping of the overburden-bedrock interface with the engineering seismograph—some simple techniques. Geophysics, v. 49, p. 1381–1385.

Hunter, J. A., S. E. Pullan, R. A. Burns, R. M. Gagne, and R. L. Good. 1988. Applications of a shallow seismic reflection method to groundwater and engineering studies. In Proc. of Exploration '87: Third Decennial International Conference on Geophysical and Geochemical Exploration for Minerals and Groundwater, Ontario Geological Survey, Special Volume, p. 704–715.

Huntley, D., R. N. Hawk, and H. P. Corley. 1992. Non-aqueous phase hydrocarbon saturations and mobility in a fine-grained, poorly consolidated sandstone. Proc. of the 1992 Petroleum Hydrocarbons and Organic Chemicals in Ground Water: Prevention, Detection, and Restoration. National Ground Water Assoc., Columbus, Ohio, p. 223–237.

Huyakorn, P. S., P. F. Andersen, O. Güven, and F. J. Molz. 1986. A curvilinear finite element model for simulating two-well tracer tests and transport in stratified aquifers. Water Resources Res., v. 22, no. 5, p. 663–678.

Huyakorn, P., and others. 1987. A three dimensional analytical method for predicting leachate migration. Ground Water, v. 25, p. 588–598.

Huyakorn, P. S., and G. F. Pinder. 1983. Computational Methods in Subsurface Flow. Academic Press, New York, 473 p.

Hvorslev, M. J. 1951. Time lag and soil permeability in ground-water observations. U.S. Army Corps of Engr. Waterway Exp. Stat. Bull. 36, Vicksburg, Miss.

Ibrahim, H. A., and W. Brutsaert. 1965. Inflow hydrographs from large unconfined aquifers. J. Irr. Drain. Div., Proc. Am. Soc. Civil Engrs., v. 91, p. 21–38.

Ingerson, E., and J. F. Pearson, Jr. 1964. Estimation of age and rate of motion of ground-water by the ^{14}C method. Recent Research in the Fields of Hydrosphere, Atmosphere and Nuclear Chemistry. Maruzen, Tokyo, p. 263–283.

Ivanovich, M., K. Frohlich, and M. J. Hendry. 1991. Dating very old groundwater, Milk River Aquifer, Alberta, Canada. Appl. Geochem., v. 6, no. 4, p. 367–472.

Jackson, R. E., and J. Inch. 1980. Hydrogeochemical processes affecting the migration of radionuclides in a fluvial sand aquifer at the Chalk River Nuclear Laboratories. Environment Canada, National Hydrol. Research Instit., Paper No. 7, Science Series, Ottawa, 58 p.

Jackson, R. E., and R. J. Patterson. 1982. Interpretation of pH and E_H trends in a fluvial-sand aquifer system. Water Resources Res., v. 18, no. 4, p. 1255–1268.

Jackson, R. E., R. J. Patterson, B. W. Graham, J. Bahr, D. Belanger, J. Lockwood, and M. Priddle. 1985. Contaminant Hydrogeology of Toxic Organic Chemicals at a Disposal Site, Gloucester, Ontario: 1. Chemical Concepts and Site Assessment. Environment Canada, National Hydrol. Research Instit., Paper No. 23, Ottawa, 114 p.

Jackson, S. A, and Beales, F. W. 1967. An aspect of sedimentary basin evolution: The concentration of Mississippi Valley type ores during late stages of diagenesis. Canadian Petrol. Geol. Bull. 15, p. 383–437.

Jacob, C. E. 1939. Fluctuations in artesian pressure produced by passing railroad trains as shown in a well on Long Island, New York. Trans. Am. Geophys. Union, v. 20, p. 666–674.

Jacob, C. E. 1940. On the flow of water in an elastic artesian aquifer. Trans. Am. Geophys. Union, v. 22, p. 574–586.

Jacob, C. E. 1941. Notes on the elasticity of the Lloyd Sand on Long Island, New York. Trans. Am. Geophys. Union, v. 22, p. 783–787.

Jacob, C. E. 1950. Flow of groundwater. In H. Rouse (ed.), Engineering Hydraulics. John Wiley & Sons, New York, p. 321–386.

Jacquin, C., and M. Poulet. 1970. Study of hydrodynamic pattern in a sedimentary basin subject to subsidence. Soc. Petrol. Engrs. Am. Inst. Mining, Metal. Petrol. Engs. Paper no. SPE2988, 10 p.

Jacquin, C., and M. Poulet. 1973. Essas de Restitution des conditions hydrodynamiques regnant dans un bassin sedimentarire au coors de son evolution. Revue de L'Institut Francais du Petrole, v. 28, p. 269–297.

Jaeger, J. C., and N. G. W. Cook. 1969. Fundamentals of Rock Mechanics. Methuen Press, London.

Jamieson, G. R., and R. A. Freeze. 1983. Determining hydraulic conductivity distribution in a mountainous area using mathematical modeling. Ground Water, v. 21, p. 168–177.

Javandel, I., and C. Tsang. 1986. Capture-zone type curves: a tool for aquifer cleanup. Ground Water, v. 24, no. 5, p. 616–625.

Johnson, A. I. 1948. Groundwater recharge on Long Island. J. Am. Water Works Assoc., v. 49, p. 1159–1166.

Johnson, A. I. 1967. Specific yield-compilation of specific yields for various materials. U.S. Geol. Survey Water Supply Paper 1662-D, 74 p.

Johnson, A. I., and D. A. Morris. 1962. Physical and hydrologic properties of water bearing deposits from core holes in the Las Banos–Kettleman City area, California. U.S. Geol. Survey Open. File Rept. Denver, Colo.

Johnson, K. S. 1981. Dissolution of salt on the east flank of the Permian Basin in the southwestern U.S.A. In W. Back and R. Létolle (eds.), J. Hydrol. Special Issue, Symposium on Geochemistry of Groundwater, v. 54, p. 75–94.

Johnson, P. C., C. C. Stanley, M. W. Kemblowski, D. L. Byers, and J. B. Colthart. 1990. A practical approach to the design, operation, and monitoring of in situ soil-venting systems. Ground Water Mon. Rev., Spring, p. 159–178.

Johnson, R. L., S. M. Brillante, L. M. Isabelle, J. E. Houck, and J. F. Pankow. 1985. Migration of chlorophenolic compounds at the chemical waste disposal site at Alkali Lake, Oregon. 2. Contaminant distributions, transport and retardation. Ground Water, v. 23, no. 5, p. 652–665.

Jones, J. W., and others. 1985. Field and theoretical investigation of fractured crystalline rock near Oracle, Southern Arizona. NUREG/CR-3736. Nuclear Reg. Comm.

Jordan T. E., D. G. Leask, D. Slain, I. MacLeod, and T. M. Dobush. 1991. The use of high resolution electromagnetic methods for reconaissance mapping of buried wastes. Proc. of the Fifth National Outdoor Action Conference on Aquifer Restoration: Ground Water Monitoring and Geophysical Methods, p. 849–862.

Jouanna, P. 1972. In situ permeability tests under applied stresses. Proc. Symp. Percolation through Fissured Rock.

Jury, W. A., D. Russo, G. Streile, and H. El Abd. 1990. Evaluation of volatilization by organic chemicals residing below the soil surface. Water Resources Res., v. 26, p. 13–20.

Kaida, Y., M. Matsubara, R. Ghose, and T. Kanemori. 1995. Very shallow seismic reflection profiling using portable vibrator. Proc. of the Symposium on the Application of Geophysics to Engineering and Environmental Problems, Environ. and Engr. Geophys. Soc., Englewood, Colo., p. 601–607.

Karickhoff, S. W. 1981. Semi-empirical estimation of sorption of hydrophobic pollutants on natural sediments and soils: Chemosphere, v. 10, no. 8, p. 833–846.

Karickhoff, S. W. 1984. Organic pollutant sorption in aquatic systems. J. Hydraul. Engr. (ASCE), v. 110, no. 6, p. 707–735.

Karickhoff, S. W. 1985. Pollutant sorption in environmental systems. In W. B. Neely and G. E. Blau (eds.), Environmental Exposure form Chemicals, Volume I. CRC Press Inc., Boca Raton, Fla., p. 49–62.

Karickhoff, S. W., D. S. Brown, and T. A. Scott. 1979. Sorption of hydrophobic pollutants on natural sediments. Water Res., v. 13, p. 241–248.

Kasten, P. R., L. Lapidus, and N. R. Amundson. 1952. Mathematics of adsorption in beds, V. Effect of intraparticle diffusion in flow systems in fixed beds. J. Phys. Chem., v. 56, p. 683–688.

Kazmann, R. G. 1956. Safe yield in groundwater development, reality or illusion. Proc. Am. Soc. Civil Engrs., v. 82.

Keech, D. A. 1988. Hydrocarbon thickness on groundwater by dielectric well logging. Proc. of the NWWA/API Conference on Petroleum Hydrocarbons and Organic Chemicals in Ground Water—Prevention, Detection and Restoration. National Water Well Assoc., Dublin, Ohio, p. 275–289.

Keeley, J. F. 1989. Performance Evaluations of Pump-and-Treat Remediations. U.S. Environmental Protection Agency, EPA/540/4-89/005, 19 p.

Kelly, F. X., K. J. Dapsis, and D. A. Lauffenburger. 1988. Effect of bacterial chemotaxis on dynamics of microbial competition. Microb. Ecol., v. 16, p. 115-131.

Kelly, V. 1985. Field Determination of Dispersivity of Comingling Plumes: Unpublished M.S. Theses, Texas A&M University, College Station.

Keswick, B. H., and C. P. Gerba. 1980. Viruses in groundwater. Environ. Sci. Technol., v. 14, p. 1290-1297.

Kharaka, Y. K., and I. Barnes. 1973. SOLMNEQ: Solution-Mineral Equilibrium Computations. National Technol. Info. Serv. Technol. Rept. PB214-899, 82 p.

Kiersch, G. 1973. The Vaiont Reservoir Disaster. In R. Tank (ed.), Focus on Environmental Geology. Oxford Univ. Press, New York.

Killey, R. W. D., J. O. McHugh, D. R. Champ, E. L. Cooper, and J. L. Young. 1984. Subsurface cobalt-60 migration from a low-level waste disposal site. Environ. Sci. Technol., v. 18, p. 148-157.

Kilty, K., and D. S. Chapman. 1980. Convective heat transfer in selected geological situations. Ground Water, v. 18, p. 386-394.

Kimball, B. A. 1984. Ground water age determinations, Piceance Creek Basin, Colorado. In B. Hitchon and E. I. Wallick (eds.), First Canadian/American Conference on Hydrogeology. National Water Well Assoc., Dublin, Ohio, p. 267-283.

Kimmel, G. E., and O. C. Braids. 1980. Leachate plumes in ground water from Babylon and Islip landfills, Long Island, New York. U.S. Geol. Surv., Prof. Paper 1085, 38 p.

King, F. H. 1899. Principles and conditions of the movements of groundwater. U.S. Geol. Survey 19th Ann. Rept., pt. 2, p. 59-294.

Kirkham, D., and W. L. Powers. 1972. Advanced Soil Physics. Wiley-Interscience, New York.

Kissen, I. G. 1978. The principal distinctive features of the hydrodynamic regime of intensive earth crust downwarping areas. In Hydrogeology of Great Sedimentary Basins: Pub. of the Intern. Assoc. Sci. Hydrol., no. 120, p. 178-185.

Kittel, J. A., R. E. Hinchee, R. Hoeppel, and R. Miller. 1994. Bioslurping—Vacuum-enhanced free-product recovery coupled with bioventing: A case study. Proc. of Petroleum Hydrocarbons and Organic Chemicals in Ground Water—Prevention, Detection and Restoration. National Ground Water Assoc., Columbus, Ohio, p. 255-270.

Kittel, J. A., R. E. Hinchee, R. D. Miller, C. Vogel, and R. Hoeppel. 1993. In situ respiration testing: A field treatability test for bioventing. Proc. of Petroleum Hydrocarbons and Organic Chemicals in Ground Water—Prevention, Detection and Restoration. National Ground Water Assoc., Columbus, Ohio, p. 351-366.

Klotz, I. M. 1946. The adsorption wave. Chem. Rev., v. 39, p. 241-268.

Klotz, I. M. 1950. Chemical Thermodynamics. Prentice-Hall, Englewood Cliffs, N.J., 369 p.

Knapp, R. B., and J. E. Knight. 1977. Differential thermal expansion of pore fluids: fracture propagation and microearthquake production in hot pluton environments. J. Geophys. Res., v. 82, p. 2515-2522.

Knox, R. D., L. W. Canter, D. F. Kincannon, E. L. Stover, and C. H. Ward. 1984. State-of-the-art of aquifer restoration. National Center for Groundwater Research, University of Oklahoma, Oklahoma State University and Rice University, Norman, Okla., National Technical Information Service, 371 p.

Kohout, F. A. 1960. Flow pattern of fresh water and salt water in the Biscayne aquifer of the Miami area, Florida: Intern. Assoc. Sci. Hydrol., Comm. Subter. Waters, publ. 52, p. 440-448.

Kohout, F. A. 1961. Case history of salt water encroachment caused by a storm sewer in Miami. J. Am. Water Works Assoc., v. 53, p. 1406-1416.

Kohout, F. A., and H. Klein. 1967. Effect of pulse recharge on the zone of diffusion in the Biscayne aquifer. Intern. Assoc. Sci. Hydrol. Symp. Haifa, Israel, pub. 72, p. 252-270.

Kölbel-Boelke, J., E. M. Anderson, and A. Nehrkorn. 1988. Microbial communities in saturated groundwater environment. II. Diversity of bacterial communities in a Pleistocene sand aquifer and their *in vitro* activities. Microb. Ecol., v. 16, p. 31-48.

Koltermann, C. E., and S. M. Gorelick. 1992. Paleoclimate signature in terrestrial flood deposits. Science, v. 256, p. 1775-1782.

Konikow, L. F., and J. D. Bredehoeft. 1978. Computer model of two-dimensional solute transport and dispersion in ground water. U.S. Geol. Surv. TWRI, Book 7, 1 Chap. C2, Reston, Va., 40 p.

Koons, P. O. 1988. The evolution of the fluid flow regime during continental collison (abst). EOS. Trans. Am. Geophys. Union, v. 69, p. 361.

Kreitler, C. W. 1977. Fault control of subsidence, Houston, Texas. Ground Water, v. 3, p. 203-214.

Kriegman-King, M. R., and M. Reinhard. 1992. Transformation of carbon tetrachloride in the presence of sulfide, biotite, and vermiculite. Environ. Sci. Technol., v. 26, p. 2198-2206.

Krupp, H. K., S. W. Biggar, and D. R. Nielsen. 1972. Relative flow rates of salt in water in soil. Soil Sci. Soc. Am. Proc., v. 36, p. 412-417.

Kueper, B. H., and E. O. Frind. 1991. Two phase flow in heterogeneous porous media, 2. Model development. Water Resources Res., v. 27, p. 1049-1058.

Kueper, B. H., and D. B. McWorter. 1991. The behavior of dense, nonaqueous phase liquids in fractured clay and rock. Ground Water, v. 29, p. 716-728.

Kunii, D., and J. M. Smith. 1961. Heat transfer characteristics of porous rocks, II. Thermal conductivities of unconsolidated particles with flowing fluids. J. Am. Inst. Chem. Eng., v. 7, p. 29-31.

Lachenbruch, A. H., and J. H. Sass. 1977. Heat flow in the United States. In J. Heacock (ed.), The Earth's Crust. Monogr. Ser. Am. Geophys. Union, Washington, D.C., p. 626-675.

Lai, S. H., and J. J. Jurinak. 1971. Numerical approximation of cation exchange in miscible displacement through soil columns. Soil Sci. Soc. Am. Proc., v. 35, p. 894-899.

Land, L. S. 1973. Holocene meteoric dolomitization of Pleistocene limestone in North Jamaica. Sedimentology, v. 20, p. 411-422.

Land, L. S., F. T. MacKenzie, and S. J. Gould. 1967. Pleistocene history of Bermuda. Geol. Soc. Am. Bull. 78, p. 993-1006.

Land, L. S., and D. Prezbindowski. 1981. The origin and evolution of saline formation waters; lower Cretaceous carbonates, South-Central Texas, U.S.A. J. Hydrol., v. 54, p. 51-74.

Landes, K. K., and others. 1960. Petroleum resources in basement rock. Am. Assoc. Petrol. Geol. Bull. 44, p. 1682-1691.

Langmuir, D. 1971. The geochemistry of some carbonate

ground waters in central Pennsylvania. Geochim. Cosmochim. Acta, v. 35, p. 1023–1045.

Langmuir, D. 1978. Uranium solution–mineral equilibria at low temperatures with applications to sedimentary ore deposits. Geochim. Cosmochim. Acta, v. 42, p. 547–569.

Langmuir, D., and J. Mahoney. 1984. Chemical equilibrium and kinetics of geochemical processes in ground water studies. In B. Hitchon and E. I. Wallick (eds.), First Canadian/American Conference on Hydrogeology. National Water Well Assoc., Dublin, Ohio, p. 69–95.

Lapidus, L., and N. R. Amundson. 1952. Mathematics of adsorption in beds. VI. the effect of longitudinal diffusion in ion exchange and chromatographic columns. J. Phys. Chem., v. 56, p. 984–988.

Lapwood, E. R. 1948. Convection of a fluid in a porous medium. Cambridge Phil. Soc. Proc., v. 44, p. 508–521.

Larson, S. P., C. B. Andrews, M. D. Howland, and D. T. Feinstein. 1987. A three-dimensional modeling analysis of ground water pumping schemes for containment of shallow ground water contamination. Solving Ground Water Problems with Models. National Water Well Assoc., Dublin, Ohio, p. 517–531.

Larson, S. P., M. S. McBride, and R. J. Wolf. 1976. Digital models of a glacial outwash aquifer in the Pearl-Sallie Lakes area, West-Central Minnesota. Water Res. Invest., U.S. Geol. Survey, p. 40–75.

Lattman, L. A., and R. R. Parizek. 1964. Relationship between fracture traces and the occurrence of groundwater in carbonate rocks. J. Hydrol., v. 2, p. 73–91.

Law, J. 1944. A statistical approach to the interstitial heterogeneity of sand reservoirs. Trans. Am. Inst. Min. Met. Engr., v. 155, p. 202–222.

Lawrence Berkeley Laboratory, 1984. Panel report on coupled thermo-mechanical-hydro chemical process associated with a nuclear waste repository. Prepared for office of Nuclear Regulatory Commission, Rept. LBL-18250.

Leake, S. A., and D. E. Prudic. 1991. Documentation of a Computer Program to Simulate Aquifer-System Compaction Using the Modular Finite-Difference Ground-Water Flow Model: Techniques of Water-Resources Investigations, U.S. Geological Survey, 68 p.

LeBlanc, D. R. (ed.) 1984a. Movement and Fate of Solutes in a Plume of Sewage-Contaminated Ground Water, Cape Cod, Massachusetts. U.S. Geological Survey Toxic Waste Ground-Water Contamination Program. U.S. Geol. Surv. Open-File Rept., 84-475, 175 p.

LeBlanc, D. R. 1984b. Digital modeling of solute transport in a plume of sewage-contaminated ground water. Movement and Fate of Solutes in a Plume of Sewage-Contaminated Ground Water, Cape Cod, Massachusetts: U.S. Geological Survey Toxic Waste Ground-Water Contamination Program. U.S. Geol. Surv. Open-File Rept., p. 11–45.

LeBlanc, D. R., S. P. Garabedian, K. M. Hess, L. W. Gelhar, R. D. Quadri, K. G. Stollenwerk, and W. W. Wood. 1991. Large-scale natural gradient tracer test in sand and gravel, Cape Cod, Massachusetts, 1, Experimental design and observed tracer movement: Water Resour. Res., 27(5), p. 895–910.

Lee, C. H. 1915. The determination of safe yield of underground reservoirs of the closed basin type. Trans. Am. Soc. Civil Engrs., v. 78, p. 148–151.

Lee, M. D., V. W. Jamison, and R. L. Raymond. 1987. Applicability of in-situ bioreclamation as a remedial action alternative.

Proc. of the NWWA/API Conference on Petroleum Hydrocarbons and Organic Chemicals in Ground Water—Prevention, Detection and Restoration. National Water Well Assoc., Dublin, Ohio, p. 167–185.

Lefkoff, L. J., and S. M. Gorelick. 1986. Design and cost analysis of rapid aquifer restoration systems using flow simulation and quadratic programming. Ground Water, v. 24, no. 6, p. 777–790.

LeGrand, H. E. 1954. Geology and groundwater in the Statesville area, North Carolina. North Carolina Dept. Conser. Dev., Div. Mineral Resources, Bull. 68.

Leij, F. J., and J. H. Dane. 1990. Analytical solutions of the one-dimensional advection equation and two- or three-dimensional dispersion equation. Water Resources Res., v. 26, no. 7, p. 1475–1482.

Leij, F. J., T. H. Skaggs, and M. T. van Genuchten. 1991. Analytical solutions for solute transport in three-dimensional semi-infinite porous media. Water Resources Res., v. 27, no. 10, p. 2719–2733.

Lenhard, R. J., T. G. Johnson, and J. C. Parker. 1993. Experimental observations of non-aqueous-phase liquid subsurface movement. J. Contam. Hydrol., v. 12, p. 79–101.

Lenhard, R. J., and J. C. Parker. 1990. Estimation of free hydrocarbon volume from fluid levels in monitoring wells. Ground Water, v. 28, p. 57–67.

Leonards, G. A. 1962. Engineering properties of soils. In G. A. Leonards (ed.), Foundation Engineering. McGraw-Hill, New York, p. 66–240.

Levine, S. N., and W. C. Ghiorse. 1990. Analysis of environmental factors affecting abundance and distribution of bacteria, fungi and protozoa in subsurface sediments of the Upper Atlantic Coastal Plain, USA. In C. B. Fliermans and T. C. Hazen (eds.), Proc. of the First International Symposium on Microbiology of the Deep Subsurface. Westinghouse Savannah River Company, p. 5-31–5-45.

Li, Y.-H., and S. Gregory. 1974. Diffusion of ions in sea water and in deep-sea sediments: Geochim. Cosmochim. Acta., v. 38, p. 703–714.

Lico, M. S., Y. K. Kharaka, W. W. Carothers, and V. A. Wright. 1982. Methods for Collection and Analysis of Geopressured, Geothermal and Oil Field Water. U.S. Geol. Surv., Water Supply Paper 2194, 21 p.

Lindberg, R. D., and D. D. Runnells. 1984. Ground water redox reactions: An analysis of equilibrium state applied to E_H measurements and geochemical modeling. Science, v. 225, p. 925–927.

Linsley, R. K., M. A. Kohler, and J. L. H. Paulhus. 1958. Hydrology for Engineers. McGraw-Hill, New York.

Liu, C. W., and T. N. Narasimhan. 1989. Redox-controlled multiple-species reactive chemical transport, 1. Model development. Water Resources Res., v. 25, no. 5, p. 869–882.

Lohman, S. W. 1961. Compression of elastic aquifers. U.S. Geol. Survey Prof. Paper 424-B, p. 47–48.

Long, J. C. S., and others. 1982. Porous media equivalents for networks of discontinuous fractures. Water Resources Res., v. 18, p. 645–658.

Longman, M. W. 1982. Carbonate diagenesis as a control on stratigraphic traps. Am. Assoc. Petrol. Geol. Eduction Course Notes 21.

Loo, W. W., K. Frantz, and G. R. Holzhausen. 1984. The application of telemetry to large scale horizontal anisotropic permeability determination by surface tiltmeter survey. Ground Water Mon. Rev, v. 4, p. 124–130.

Lovering, T. S., and H. D. Goode. 1963. Measuring geothermal gradients in drill holes less than 60 feet deep, East Tintic District, Utah. U.S. Geol. Survey Bull. 1172.

Lundegard, P. D. 1994. Actual versus apparent radius of influence—an air sparging pilot test in a sandy aquifer. In Proc. Petroleum Hydrocarbons and Organic Chemicals in Ground Water—Prevention, Detection and Restoration. National Ground Water Assoc., Columbus, Ohio, p. 191–206.

Lusczynski, N. J., and W. V. Swarzenski. 1966. Salt water encroachment in Southern Nassau and Southeastern Queens Counties, Long Island, New York. U.S. Geol. Survey Water Paper 1613-F.

Lyman, W. J., W. F. Reehl, and D. D. Rosenblatt. 1982. Handbook of chemical property estimation methods. Environmental Behavior of Organic Compounds. McGraw-Hill, New York.

Lyngkilde, J., and T. H. Christensen. 1992. Fate of organic contaminants in the redox zones of a landfill leachate pollution plume (Vejen, Denmark). J. Contam. Hydrol., v. 10, p. 291–307.

Mackay, D. M., D. L. Freyberg, and P. V. Roberts. 1986. A natural gradient experiment on solute transport in a sand aquifer. 1. Approach and overview of plume movement. Water Resources Res., v. 22, no. 13, p. 2017–2029.

Mackay, D. M., and P. J. Leinonen. 1975. Rate of evaporation of low-solubility contaminants from water bodies to atmosphere. Environ. Sci. Technol., v. 9, no. 8, p. 1178–1180.

Mackay, D. M., P. V. Roberts, and J. A. Cherry. 1985. Transport of organic contaminants in groundwater. Environ. Sci. Technol., v. 19, no. 5, p. 384–392.

Mackay, D. M., and T. M. Vogel. 1985. Ground water contamination by organic chemicals: Uncertainties in assessing impact. Second Canadian/American Conference on Hydrogeology, B. Hitchon and M. Trudell (eds.): Dublin, Ohio, National Water Well Assoc., p. 50–59.

Macumber, P. G. 1984. Hydrochemical processes in the regional ground water discharge zones of the Murray Basin, Southeastern Australia. First Canadian/American Conference on Hydrogeology. B. Hitchon and E. I. Wallick (eds.): Dublin, Ohio, National Water Well Assoc., p. 47–63.

Magara, K. 1975. Importance of aquathermal pressuring effect in Gulf Coast. Am. Assoc. Petrol. Geol. Bull. 59, p. 2037–2045.

Magara, K. 1978. Compaction and Fluid Migration: Practical Petroleum Geology. Elsevier Science, Amsterdam.

Major, D. W., E. W. Hodgins, and B. J. Butler. 1991. Field and laboratory evidence of in situ biotransformation of tetrachloroethene to ethene and ethane at a chemical transfer facility in North Toronto. In R. E. Hinchee and R. G. Olfenbuttel (eds.), On-Site Bioreclamation Processes for Xenobiotic and Hydrocarbon Treatment. Butterworth, Stoneham, Mass., p. 113–133.

Malley, M. J., W. W. Bath, and L. H. Bongers. 1985. A case history: Surface static collection and analysis of chlorinated hydrocarbons from contaminated ground water. Proc. of the NWWA/API Conference on Petroleum Hydrocarbons and Organic Chemicals in Ground Water—Prevention, Detection and Restoration. National Water Well Assoc., Dublin, Ohio, p. 276–290.

Malmberg, G. T. 1960. An analysis of the hydrology of the

Las Vegas groundwater basin, Nevada, U.S. Geol. Sur. Open File Rept.

Manheim, F. T., and J. F. Bischoff. 1969. Geochemistry of pore waters from Shell Oil Company drill holes on the continental slope of the northern Gulf of Mexico. In E. E. Angino and G. K. Billings (eds.), Chem. Geol. Special Issue, Geochemistry of Subsurface Brines, v. 4, p. 63–82.

Manheim, F. T., and C. K. Paull. 1981. Patterns of groundwater salinity changes in a deep continental–oceanic transect off the southeastern Atlantic coast of the U.S.A. In W. Back and R. Létolle (eds.), J. Hydrol. Special Issue, Symposium on Geochemistry of Groundwater, v. 54, p. 95–106.

Manov, G. C., R. C. Bates, W. J. Hamer, and S. F. Acree. 1943. Values of the constants in the Debye-Huckel equation for activity coefficients: Jour. Amer. Chem. Soc., v. 65, p. 1765–1767.

Mantoglou, A., and J. L. Wilson. 1982. The turning bands method for simulation of random fields using line generation by a spectral method: Water Resour. Res., 18(5), 1379–1394.

Massmann, J. W. 1989. Applying groundwater flow models in vapor extraction system design. J. Environ. Eng., v. 115, no. 1, p. 129–149.

Matheron, G. 1967. Eléments pour une théorie des milieux poreux. Masson et Cie, Paris.

Matthess, G. 1982. The Properties of Groundwater: New York, John Wiley & Sons, 406 p.

———, and A. Pekdeger. 1985. Survival and transport of pathogenic bacteria and viruses in ground water. Ground Water Quality, eds. C. H. Ward, W. Giger, and P. L. McCarty: New York, John Wiley & Sons, p. 472–482.

Matthews, C. S., and D. G. Russel. 1967. Pressure buildup analysis. In Pressure Buildup and Flow Test in Wells. Monograph 1. Soc. Petrol. Engrs., Am. Inst. Mech. Engrs., New York.

Maxwell, J. C. 1964. Influence of depth, temperature and geologic age on porosity of quartzose sandstone. Bull. Am. Assoc. Petrol. Geol. 48, p. 697–709.

McBride, M. S., and H. O. Pfannkuch. 1975. The distribution of seepage within lake beds. J. Res., U.S. Geol. Surv., v. 3, p. 505–512.

McCarthy, J. F., and J. M. Zachara. 1989. Subsurface transport of contaminants. Environ. Sci. Technol., v. 23, no. 5, p. 496–502.

McCarty, P. L., M. Reinhard, and B. E. Rittmann. 1981. Trace organics in groundwater. Environ. Sci. Technol., v. 15, no. 1, p. 40–51.

McCarty, P. L., B. E. Rittman, and E. J. Bouwer. 1984. Microbiological processes affecting chemical transformations in groundwater. In G. Bitton and C. P. Gerba (eds.), Groundwater Pollution Microbiology, John Wiley & Sons, New York, p. 89–115.

McCarty, P. L., and J. T. Wilson. 1992. Natural anaerobic treatment of a TCE plume, St. Joseph Michigan, NPL site. In Bioremediation of Hazardous Wastes. U.S. Environmental Protection Agency, EPA/600R-92/16, p. 68–72.

McDonald, M. G., and A. W. Harbaugh. 1988. A modular three-dimensional finite-difference ground-water flow model. U.S. Geol. Surv., Techniques of Water-Resources Investigations 06-A1, 576 p.

McDonald, M. G., A. W. Harbaugh, B. R. Orr, and D. J. Ackerman. 1991. A method of converting no-flow cells to variable-head cells for the U.S. Geological Survey Modular Finite-Difference

Ground-Water Flow Model. U.S. Geological Survey, Open-File Report, 91-536, 99 p.

McDowell-Boyer, L. M., J. R. Hunt, and N. Sitar. 1986. Particle transport through porous media. Water Resources Res., v. 22, no. 13, p. 1901–1921.

McGarr A., and N. C. Gay. 1978. State of stress in the earth's crust. Ann. Rev. of Earth Planet. Sci., v. 6, p. 405–436.

McGowen, J. H., and C. G. Groat. 1971. Van Horn Sandstone, West Texas: An alluvial fan model for mineral exploration. Texas Bur. Econ. Geol. Rept. Inv., v. 72, Austin.

Mclelwain, T. A., and Reades, D. W., 1985, Status Report: The use of engineered covers at waste disposal sites. Second Annual Canadian/American Conference on Hydrology, Proc.: Dublin, Ohio, National Water Well Assoc., p. 176–182.

McNabb, J. F., and G. Mallard. 1984. Microbial sampling in the assessment of groundwater pollution. In G. Bitton and C. P. Gerba (eds.), Groundwater Pollution Microbiology. John Wiley & Sons, New York, p. 235–260.

Mead, D. W. 1919. Hydrology. McGraw-Hill, New York.

Meinzer, O. E. 1917. Geology and water resources of Big Smokey, Clayton, and Alkali Springs Valleys, Nevada. U.S. Geol. Survey Water Supply Papers 423.

Meinzer, O. E. 1922. Map of pleistocene lakes of the Basin and Range Province and its significance. Bull. Geol. Soc. Am. 33, p. 541–552.

Meinzer, O. E. 1923. The occurrence of groundwater in the United States with a discussion of principles. U.S. Geol. Surv. Water Supply Papers, v. 489.

Meinzer, O. E. 1923. Outline of groundwater in hydrology with definitions. U.S. Geol. Surv. Water Supply Papers, v. 494.

Meinzer, O. E. 1927. Plants as indicators of groundwater. U.S. Geol. Survey Water Supply Papers, v. 577.

Meinzer, O. E. 1928. Compressibility and elasticity of artesian aquifers. Econ. Geol., v. 23, p. 263–291.

Meinzer, O. E. 1932. Outline of methods for estimating ground-water supplies. U.S. Geol. Survey Water Supply Papers 638-C, p. 94–144.

Meinzer, O. E. 1939. Groundwater in the United States. U.S. Geol. Survey Water Supply Papers 836-D, p. 157–232.

Meinzer, O. E. 1942 (ed.). Hydrology. McGraw-Hill, New York.

Meinzer, O. E., and H. H. Hard. 1925. The artesian water supply of the Dakota Sandstone in North Dakota, with special reference to the Edgeley Quadrangle. U.S. Geol. Survey Water Supply Papers 520-E, p. 73–95.

Meinzer, O. E., and N. D. Stearns. 1928. A study of groundwater in the Pomperaug Basin, Conn., with special reference to intake and discharge. U.S. Geol. Survey Water Supply Papers v. 597, p. 73–146.

Melnyk, T. W., F. B. Walton, and H. L. Johnson. 1983. High level waste glass field burial tests at CRNL: The effect of geochemical kinetics on the release and migration of fission products in a sandy aquifer. Rep. AECL-6836, Atom. Energy, Can. Ltd. Chalk River, Ontario.

Mendoza, C. A., and E. O. Frind. 1990a. Advective–dispersive transport of dense organic vapors in the unsaturated zone. 1. Model development. Water Resources Res., v. 26, no. 3, p. 379–387.

Mendoza, C. A., and E. O. Frind. 1990b. Advective–dispersive transport of dense organic vapors in the unsaturated zone. 2. Sensitivity analysis. Water Resources Res., v. 26, no. 3, p. 388–398.

Mendoza, C. A., and T. A. McAlary. 1990. Modeling of ground-water contamination caused by organic solvent vapors. Ground Water, v. 28, no. 2, p. 199–206.

Mercer, J. W., G. F. Pinder, and I. G. Donaldson. 1975. A Galerkin finite element analysis of the hydrothermal system at Wairakei, New Zealand. J. Geophys. Res., v. 80, p. 2608–2621.

Mercer, J. W., D. C. Skipp, and D. Griffin. 1990. Basics of pump-and-treat remediation technology. U.S. Environ. Protection Agency, EPA-600/8-90/003, 31 p.

Meyboom, P. 1960. Geology and groundwater resources of the Milk River Sandstone in southern Alberta. Research Council of Albert Memoir 2, Edmonton.

Meyboom, P. 1961. Estimating groundwater recharge from stream hydrographs. J. Geophys. Res., v. 66, p. 1203–1214.

Meyboom, P. 1962. Patterns of groundwater flow in the Prairie environment. Can. Hydrol. Symp., 3rd, p. 5–33.

Meyboom, P. 1966a. Unsteady groundwater flow near a willow ring in hummocky moraine. J. Hydrol., v. 4, p. 38–62.

Meyboom, P. 1966b. Groundwater studies in the Assiniboine River drainage basin, pt. 1. The evaluation of a flow system in southcentral Saskatchewan. Can. Dept. Mines Tech. Survey. Geol. Surv. Can. Bull. 139.

Meyboom, P. 1967. Mass transfer studies to determine the groundwater regime of permanent lakes in hummocky moraine of western Canada. J. Hydrol., v. 5, p. 117–142.

Mifflin, M. D. 1968. Delineation of groundwater flow systems in Nevada. Desert Research Inst. Tech. Rept. Ser. H-W, no. 4, Reno.

Mikels, F. C. 1952. Report on hydrogeological survey for city of Zion, Illinois. Ranney Method Water Supplies, Inc., Columbus, Ohio.

Miller, C. W., and L. V. Benson. 1983. Simulation of solute transport in a chemically reactive heterogeneous system: model development and application. Water Resources Res., v. 19, no. 2, p. 381–391.

Miller, R. N., D. C. Downey, V. A. Carmen, R. E. Hinchee, and A. Leeson. 1993. A summary of bioventing performance at multiple Air Force sites. Proc. of Petroleum Hydrocarbons and Organic Chemicals in Ground Water—Prevention, Detection and Restoration. National Ground Water Assoc., Columbus, Ohio, p. 397–411.

Millington, R. J. and J. P. Quirk. 1961. Permeability of porous solids. Trans., Faraday Soc., v. 57, p. 1200–1207.

Minear, R. A., and L. H. Keith (eds.). 1984. Water Analysis, v. III, Organic Species. Academic Press, Orlando, Fla., 456 p.

Mines, B. S., J. L. Davidson, D. Bloomquist, and T. B. Stauffer. 1993. Sampling of VOCs with the BAT Ground Water Sampling System. Ground Water Mon. Remediation, v. 13, no. 1, p. 115–120.

Mitsdarffer, A. R. 1985. Hydrodynamics of the Mission Canyon formation in the Billings Nose area, North Dakota. M. S. thesis, Geology Dept., Texas A&M University, College Station.

Moltz, F. J., A. D. Parr, and P. F. Anderson. 1981. Thermal energy storage in a confined aquifer, second cycle. Water Resources Res., v. 17, p. 641–645.

Moltz, F. J., M. A. Widdowson, and L. D. Benefield. 1986. Simulation of microbial growth dynamics coupled to nutrient and oxygen transport in porous media: Water Resources Res., v. 22, no. 8, p. 1207–1216.

Moody, J. B. 1982. Radionuclide Migration/Retardation: Research and Development Technology Status Report. Office of Nuclear Waste Isolation, Battelle Memorial Ins., ONWI-321, 61 p.

Mook, W. G. 1980. Carbon-14 in hydrogeological studies. Chapter 2. In P. Fritz and J. C. Fontes (eds.), Handbook of Environmental Isotope Geochemistry, v. 1. Elsevier, Amsterdam, p. 49–74.

Moran, S. R., G. H. Groenwold, and J. A. Cherry. 1978a. Geologic, Hydrologic, and Geochemical. Concepts and Techniques in Overburden Characterization for Mined-land Reclamation. North Dakota Geological Survey, Report of Investigation no. 63, 152 p.

Moran, S. R., J. A. Cherry, P. Fritz, W. M. Peterson, M. H. Somerville, S. A. Stancel, and J. H. Ulmer, 1978b, Geology, Groundwater Hydrology, and Hydrochemistry of a Proposed Surface Mine and Lignite Gasification Plant Site Near Dunn Center, North Dakota: North Dakota Geological Survey, Report of Investigation 61, 263 p.

Morel, F. M. M. 1983. Principles of Aquatic Chemistry. John Wiley & Sons, New York, 446 p.

Morel, F. M. M., and J. G. Hering. 1993. Principles and Applications of Aquatic Chemistry. John Wiley & Sons, New York, 588 p.

Morgan, P. V., and others. 1981. A groundwater convective model for Rio Grande Rift geothermal systems. Geotherm. Resour. Council Trans., v. 5, p. 193–196.

Morganstern, N. R. 1970. The influence of groundwater on stability. Proc. 1st Intern. Conf. on Stability in Open Pit Mining. Vancouver, British Columbia.

Morrison, R. D., and P. E. Brewer, 1981. Airlift samples for zone-of-saturation monitoring: Ground Water Monitoring Review, v. 1, no. 1, p. 52–55.

Muskat, M. 1937. The Flow of Homogeneous Fluids Through Porous Media. McGraw-Hill, New York.

Muskat, M., and M. W. Meres. 1936. The flow of heterogeneous fluids through porous media. Physics, v. 7, p. 346–363.

Narasimhan, T. N., A. F. White, and T. Tokunaga. 1986. Groundwater contamination from an inactive uranium mill tailings pile. 2. Application of a dynamic mixing model. Water Resources Res., v. 22, no. 13, p. 1820–1834.

Nathenson, M., and L. J. P. Muffler. 1975. Geothermal resources in hydrothermal convection systems and conduction dominated areas. In Assessment of Geothermal Resources of the United States—1975, U.S. Geol. Surv. Circ. v. 726, p. 104–121.

National Academy of Science. 1983. Risk Assessment in the Federal Government: Managing the Process. U.S. Government Printing Office, Washington, D.C.

Nativ, R., and others. 1995. Water recharge and solute transport through the vadose zone of fractured chalk under desert conditions. Water Resources Res., v. 31, p. 253–261.

Neely, W. B. 1985. Hydrolysis. In W. B. Neely and G. E. Blau (eds.), Environmental Exposure from Chemicals, v. I. CRC Press Inc., Boca Raton, Fla., p. 157–173.

Neretnieks, I. 1985. Transport in fractured rocks. Proc. of Hydrogeology of Rocks of Low Permeability. Tucson Arizona International Assoc. of Hydrogeologists, p. 306.

Neuman, S. P. 1972. Theory of flow in unconfined aquifers considering delayed response of the water table. Water Resources Res., v. 8, p. 1031–1045.

Neuman, S. P. 1974. Effect of partial penetration on flow in unconfined aquifers considering delayed response of the water table. Water Resources Res., v. 9, p. 1102–1103.

Neuman, S. P. 1975. Analysis of pumping test data from anisotropic unconfined aquifers considering delayed gravity reponse. Water Resources Res., v. 11, p. 329–342.

Neuman, S. P., and P. A. Witherspoon. 1969a. Theory of flow in confined two aquifer system. Water Resources Res., v. 5, p. 803–816.

Neuman, S. P., and P. A. Witherspoon. 1969b. Applicability of current theories of flow in leaky aquifers. Water Resources Res., v. 5, p. 817–829.

Neuman, S. P., and P. A. Witherspoon. 1972. Field determination of hydraulic properties of leaky multiple aquifer systems. Water Resources Res., v. 8, p. 1284–1298.

Neuzil, C. E. 1986. Groundwater in low permeability environments. Water Resources Res., v. 22, p. 1163–1195.

Neuzil, C. E., and D. W. Pollock. 1983. Erosional unloading and fluid pressures in hydraulically tight rocks. J. Geol., v. 9, p. 179–193.

Newell, C. J., S. D. Acree, R. D. Ross, and S. G. Huling, 1995. Light nonaqueous phase liquids: U.S. Environmental Protection Agency, EPA/540/S-95/500, 28 p.

Nield, D. A. 1968. Onset of thermohaline convection in a porous medium. Water Resources Res., v. 4, p. 1553–1560.

Nielsen, P. H., H. Bjarnadottir, P. L. Winter, and T. H. Christensen. 1995. In situ and laboratory studies on the fate of specific organic compounds in an anaerobic landfill leachate plume. 2. Fate of aromatic and chlorinated aliphatic compounds. J. Contam. Hydrol., v. 20, p. 51–66.

Nielsen, P. H., and T. H. Christensen. 1994. Variability of biological degradation of aromatic hydrocarbons in an aerobic aquifer determined by laboratory batch experiments. J. Contam. Hydrol., v. 15, p. 305–320.

Nielsen, P. H., and T. H. Christensen, 1994. Variability of biological degradation of phenolic hydrocarbons in an aerobic aquifer determined by laboratory batch experiments: J. Contam. Hydrol., v. 17, p. 55–67.

Nobel, E. A. 1963. Formation of ore deposits by water of compaction. Econ. Geol., v. 58, p. 1145–1156.

Nordstrom, D. K., and J. L. Munoz. 1986. Geochemical Thermodynamics. Blackwell Scientific Publications. Palo Alto, Calif., 477 p.

Norton, D. 1978. Sourcelines, source regions, and pathlines in hydrothermal systems related to cooling plutons. Econ. Geol., v. 73, p. 21–28.

Norton, D., and R. Knapp. 1977. Transport phenomena in hydrothermal systems: Nature of porosity. Am. J. Sci., v. 27, p. 913–936.

NRC, National Research Council. 1993. In Situ Bioremediation: When Does It Work. National Academy Press, Washington, D.C., 207 p.

Nur, A. 1972. Dilatancy, pore fluids and premonitory variations of ts/tp travel times. Bull. Seismol. Soc. Am. v. 62, p. 1217–1222.

Nyer, E., G. Boettcher, and B. Morello. 1991. Using the properties of organic compounds to help design a treatment system. Ground Water Mon. Rev., v. 11, no. 4, p. 115–120.

O'Connor, M. J., J. G. Agar, and R. D. King. 1984a. Practical experience in the management of hydrocarbon vapors in the subsurface. Proc. of the NWWA/API Conference on Petroleum Hydrocarbons and Organic Chemicals in Ground Water—Prevention, Detection and Restoration. National Water Well Assoc., Dublin, Ohio, p. 519–533.

O'Connor, M. J., and L. W. Bouckhout. 1983. Gasoline spills in urban areas: a comparison of two case histories. Proc. Seminar on Groundwater and Petroleum Hydrocarbons—Protection, Detection, Restoration. Petroleum Assoc. for Conservation of the Canadian Environment, Ottawa, p. VIII-1–VIII-34.

O'Connor, M. J., A. M. Wofford, and S. K. Ray. 1984b. Recovery of subsurface hydrocarbons at an asphalt plant: Results of a five year monitoring program. Proc. of the NWWA/API Conference on Petroleum Hydrocarbons and Organic Chemicals in Ground Water—Prevention, Detection and Restoration National Water Well Assoc., Dublin, Ohio, p. 359–376.

Office of Technology Assessment. 1982. Summary: Managing commercial high level radioactive work. U.S. Congress, Washington, D.C.

Office of Technology Assessment, 1984. Protecting the nation's groundwater from contamination. Office of Technology Assessment, OTA-0-233: Washington, D.C., OTA, 244 pp.

Ogata, A., and R. B. Banks. 1961. A solution of the differential equation of longitudinal dispersion in porous media. U.S. Geol. Surv. Prof. Paper 411-A.

Ohle, E. L. 1959. Some consideration in determining the origin of ore deposits of the Mississippi Valley Type. Part I, Econ. Geol., v. 54, p. 769–789.

Oliver, J. 1986. Fluids expelled tectonically from orogenic belts: Their role in hydrocarbon migration and other geologic phenomena. Geology, v. 14, p. 99–102.

Oreskes, N., K. Shrader-Frechette, and K. Belitz. 1994. Verification, validation, and confirmation of numerical models in earth sciences. Science, v. 263., p. 641–646.

Ortoleva, P., J. Chadam, E. Merino, and A. Sen. 1987b. Geochemical self-organization. II: The reactive-infiltration instability. Am. J. Sci., v. 287, p. 1008–1040.

Ortoleva, P., E. Merino, C. Moore, and J. Chadam. 1987a. Geochemical self-organization. I: Reaction-transport feedbacks and modeling approach. Am. J. Sci., v. 287, p. 979–1007.

Ozbiligin, M. M., and M. A. Powers. 1984. Hydrodynamic isolation in hazardous waste containment. Proc. of the Fourth National Symposium and Exposition on Aquifer Restoration & Groundwater Monitoring. National Water Well Assoc., Dublin, Ohio, p. 44–49.

Paige, S. 1936. Effect of a sea-level canal on the groundwater level of Florida. Econ. Geol., v. 31, p. 537–570.

Paige, S. 1938. Effect of a sea-level canal on the groundwater level of Florida—A reply. Econ. Geol., v. 33, p. 647–665.

Palciauskas, V. V., and P. A. Domenico. 1976. Solution chemistry, mass transport, and the approach to chemical equilibrium in porous carbonate rocks and sediments. Geol. Soc. Am. Bull. 87, p. 207–214.

Palciauskas, V. V., and P. A. Domenico. 1980. Microfracture development in compacting sediments: Relation to hydrocarbon maturation. Am. Assoc. Petrol. Geol. Bull. 64, p. 927–937.

Palciauskas, V. V., and P. A. Domenico. 1982. Characterization of drained and undrained response of thermally loaded repository rocks. Water Resources Res., v. 18, p. 281–290.

Palciauskas, V. V., and P. A. Domenico. 1989. Fluid pressure in deforming porous rocks. Water Resources Res., v. 25, p. 203–213.

Palmer, A. N. 1991. Origin and morphology of limestone caves. Geol. Soc. Am. Bull. 103, p. 1–21.

Palmer, C. D., and J. A. Cherry. 1984. Geochemical evolution of groundwater in sequences of sedimentary rocks. J. Hydrol., v. 75, p. 27–65.

Pankow, J. F., R. L. Johnson, J. E. Houck, S. M. Brillante, and W. J. Bryan. 1984. Migration of chlorophenolic compounds at the chemical waste disposal site at Alkali Lake, Oregon. 1. Site description and ground-water flow. Ground Water, v. 22, no. 5, p. 593–601.

Papadopulos, I. S. 1965. Nonsteady flow of a well in an infinite anisotropic aquifer. Proc. Dubrovnik Symp. on Hydrol. of Fractured Rock, Inter. Assoc. Sci. Hydrol., p. 21–31.

Papadopulos, I. S., J. D. Bredehoeft, and H. H. Cooper. 1973. On the analysis of slug test data. Water Resources Res., v. 9, p. 1087–1089.

Papadopulos, I. S., and H. H. Cooper. 1967. Drawdown in a well of large diameter. Water Resources Res., v. 3, p. 241–244.

Papadopulos, S. S., and others. 1975. Assessment of onshore geopressured–geothermal sources of the United States. Assessment of Geothermal Resources in the United States—1975. U.S. Geol. Surv. Circ. 726, p. 125–142.

Parker, G. G., and V. T. Springfield. 1950. Effects of earthquakes, rains, tides, winds, and atmospheric pressure changes on the water in geologic formations of southern Florida. Econ. Geol., v. 45, p. 441–460.

Parkhurst, D. L., D. C. Thorstenson, and L. N. Plummer. 1980. PHREEQE—A Computer Program for Geochemical Calculation. U.S. Geol. Surv. Water Resources Invest. Report. 80-96, 210 p.

Parsons, M. L. 1970. Groundwater thermal regime in a glacial complex. Water Resources Res., v. 6, p. 1701–1720.

Patton, F. D., and A. J. Hendron, Jr. 1974. General report on mass movements. Proc. 2nd Intern. Congress, Intern. Assoc. Eng. Geol., San Paulo, Brazil, v. 2, p. V-GR.1–V-GR.57.

Peargin, T. R., and D. H. Mohr. 1994. Field criteria for SVE pilot tests to evaluate data quality and estimate remediation feasibility. Proc. of Petroleum Hydrocarbons and Organic Chemicals in Ground Water—Prevention, Detection and Restoration. National Ground Water Assoc., Columbus, Ohio, p. 337–350.

Pearson, F. J., Jr., and B. B. Hanshaw. 1970. Sources of dissolved carbonate species in groundwater and their effects on carbon-14 dating. Isotope Hydrology. International Atomic Energy Agency, Vienna, p. 271–286.

Perkins, T. K., and O. C. Johnston. 1963. A review of diffusion and dispersion in porous media. J. Soc. Petrol. Eng., v. 3, p. 70–83.

Peter, K. L. 1984. Hydrochemistry of lower Cretaceous sandstone aquifers, Northern Great Plains. In D. G. Jorgenson and D. C. Signor (eds.), Proc. of the Geohydrology Dakota Aquifer Symposium, Water Well Journal Publ., Worthington, Ohio, p. 163–174.

Peters, R. R., and E. A. Klavetter. 1988. A continuum model for water movement in an unsaturated fractured rock mass. Water Resources Res., v. 24, p. 416–430.

Pfannkuch, H. O. 1962. Contribution à l'étude des déplacement de fluides miscible dans un milieu poreux. Rev. Inst. Fr. Petrol., v. 18, no. 2, p. 215–270.

Phillips, F. M., H. W. Bentley, S. N. Davis, D. Elmore, and G. Swanick. 1986. Chlorine 36 dating of very old groundwater. 2. Milk River Aquifer, Alberta. Water Resources Res., v. 22, no. 13, p. 2003–2016.

Phillips, F. M., J. L. Wilson, and J. M. Davis. 1989. Statistical analysis of hydraulic conductivity distributions: A quantitative geological approach. In F. J. Molz, J. C. Melville, and O. Guven (eds.), Proc. Conf. on New Field Techniques for Quantifying the Physical and Chemical Properties of Heterogeneous Aquifers. National Ground Water Assoc., Dublin, Ohio, p. 19–31.

Pickens, J. F., R. E. Jackson, K. J. Inch, and W. F. Merritt. 1981. Measurement of distribution coefficients using a radial injection dual-tracer test. Water Resources Res., v. 17, no. 3, p. 529–544.

Pinder, G. F. 1973. A Galerkin finite-element simulation of groundwater contamination on Long Island, New York. Water Resources Res., v. 9, no. 6, p. 1657–1664.

Pinder, G. 1979. State of the art review of geothermal reservoir engineering. Rept. LBL-9093, Lawrence Berkeley Lab., Berkeley, Calif.

Pinder, G. F., and H. H. Cooper, Jr. 1970. A numerical technique for calculating the transient position of the saltwater front. Water Resources Res., v. 6, p. 875–882.

Piper, A. M., 1944, A graphic procedure in the geochemical interpretation of water analysis: Trans. Amer. Geophys. Union, v. 25, p. 914–923.

Pitzer, K. S., and J. J. Kim. 1974. Thermodynamics of electrolytes: 4. Activity and osmotic coefficients for mixed electrolytes. J. Am. Chem. Soc., v. 96, p. 5701–5707.

Plummer, L. N. 1984. Geochemical modeling: A comparison of forward and inverse methods. In B. Hitchon and E. I. Wallick (eds.), First Canadian/American Conference on Hydrogeology. National Water Well Assoc., Dublin, Ohio, p. 149–177.

Plummer, L. N., B. F. Jones, and A. H. Truesdell. 1976. WATEQF—A Fortran IV version of WATEQ, A Computer Program for Calculating Chemical Equilibrium of Natural Waters. U.S. Geol. Surv. Water Resources Invest. 76-13, 61 p.

Plummer, L. N., and T. M. L. Wigley. 1976. The dissolution of calcite in CO_2-saturated solutions at 25°C and 1 atmosphere total pressure. Geochim. Cosmochim. Acta, v. 40, p. 191–202.

Poland, J. F. 1961. The coefficient of storage in a region of major subsidence caused by compaction of an aquifer. U.S. Geol. Prof. Paper 424-B, p. 52–54.

Poland, J. F., and G. H. Davis. 1956. Subsidence of the land surface in Tulare–Wasco (Delano) and Los Banos–Kettleman City area, San Joaquin Valley, California. Trans. Am. Geophys. Union, v. 37, p. 287–296.

Poland, J. F., and others. 1975. Land subsidence in the San Joaquin Valley, California as of 1972. U.S. Geol. Survey Prof. Paper 437-H.

Pollock, D. W. 1989. Documentation of computer programs to compute and display pathlines using results from the U.S. Geological Survey modular three-dimensional finite-difference ground-water flow model. U.S. Geol. Surv., Open-File Report 89-381, 188 p.

Powers, M. C. 1967. Fluid-release mechanisms in compacting marine mudrocks and their importance in oil exploration. Am. Assoc. Petrol. Geol. Bull. 51, p. 1240–1254.

Pratts, M. 1966. The effect of horizontal flow on thermally induced convection currents in porous mediums. J. Geophys. Res., v. 71, p. 4835–4838.

Pratts, M. 1970. A method for determining the net vertical permeability near a well from in-situ measurements. Trans. Am. Inst. Mining Engrs., v. 249, p. 637–643.

Price, M. 1994. A method for assessing the extent of fissuring in double porosity aquifers, using data from packer tests. IAHS Publ. No. 222, p. 271–278.

Prichett, W.C. 1980. Physical properties of shales and possible origin of high pressures. Soc. Petrol. Eng. J., v. 20, p. 341–348.

Prickett, T. A. 1965. Type curve solution to aquifer tests under water table conditions. Ground Water, v. 3, p. 5–14.

Prickett, T. A., T. G. Naymik, and C. G. Lonquist. 1981. A Random-Walk Solute Transport Model for Selected Groundwater Quality Evaluations. Illinois State Water Surv. Bull. 65, Champaign, Ill., 103 p.

Prudic, D. E. 1989. Documentation of a computer program to simulate stream–aquifer relations using a modular, finite-difference, ground-water flow model. U.S. Geol. Surv., Open-File Rept. 88-729, 113 p.

Puls, R. W., and R. M. Powell. 1992. Transport of inorganic colloids through natural aquifer material: Implications for contaminant transport. Environ. Sci. Technol., v. 26, no. 3, p. 614–621.

Pytkowicz, R. P. 1983. Equilibria, Nonequilibria & Natural Waters, Vol. II. John Wiley & Sons, New York, 353 p.

Quinlan, J. F. 1990. Special problems of ground-water monitoring in karst terranes. In D. M. Nielson and A. I. Johnson (eds.), Ground Water and Vadose Zone Monitorings. Am. Soc. Testing and Materials, Philadelphia, p. 275–304.

Quinlan, J. F., and R. O. Ewers. 1985. Ground water flow in limestone terranes: Strategy, rationale and procedure for reliable, efficient monitoring of ground water quality in Karst areas. Proc. Fifth National Symposium and Exposition on Aquifer Restoration and Ground Water Monitoring. National Water Well Assoc., Dublin, Ohio, p. 197–234.

Quinlan, J. F., and J. A. Ray. 1981. Groundwater basins in the Mammoth Cave Region, Kentucky. Occ. Publ. No. 1, Friends of the Karst, Mammoth Cave.

Quinlan, J. F., and D. R. Rowe. 1977. Hydrology and water quality in the central Kentucky karst: Phase I. Univ. Kentucky Water Resources Res. Institute, Research Report No. 101, 93 p.

Ragone, S. E. (ed.). 1988. U.S. Geological Survey Program on Toxic Waste—Ground Water Contamination. Proc. of the Second Technical Meeting, Cape Cod, Massachusetts, October 21–25, 1985. U.S. Geol. Surv. Open-File Rept. 86-481.

Rainwater, F. H., and L. L. Thatcher. 1960. Methods for Collection and Analysis of Water Samples. U.S. Geol. Surv. Water Supply Paper 1454, 301 p.

Raleigh, C. B. 1971. Earthquakes and fluid injection. Am. Assoc. Petrol. Geol. Memoirs, no. 18, p. 273–279.

Raleigh, C. B., J. H. Healy, and J. D. Bredehoeft. 1972. Faulting and crustal stress at Rangely, Colorado. Geophys. Union Monog. Series 16, p. 275–284.

Raleigh, C. B., J. H. Healy, and J. D. Bredehoeft. 1976. An experiment in earthquake control at Rangeley, Colorado. Science, v. 191, p. 1230–1236.

Randall, J. H., and T. R. Schultz. 1976. Chlorofluorocarbons as hydrologic tracers: A new technology. Hydrol. Water Res. Ariz. Southwest, v. 6, p. 189–195.

Ranganathan, V., and J. S. Hanor. 1987. A numerical model for the formation of saline waters due to diffusion of dissolved NaCl in subsiding sedimentary basins with evaporites. J. Hydrol., v. 92, p. 97–120.

Rasmussen, T. C., J. H. Blanford, and P. J. Sheets. 1989. Characterization of unsaturated fractured tuff at the Apache Leap Site: comparison of laboratory/field, matrix/fracture and water/air data. Paper presented at Nuclear Waste Isolation in the Unsaturated Zone, joint sponsorship by the American Nuclear Soc. and the Geological Soc. of America, Las Vegas, Nev.

Raven, K., and K. S. Novakowski. 1984. Field investigation of the solute-transport properties of fractures in monzonitic gneiss. International Symposium on Groundwater Resources Utilization and Contaminant Hydrogeology, Vol. II. Atomic Energy of Canada Ltd., Pinawa, Manitoba, p. 507–516.

Reardon, E. J., and P. Fritz, 1978, Computer modelling of ground water ^{13}C and ^{14}C isotope compositions: J. Hydrol., v. 36, p. 201–224.

Reddell, D. L., and D. K. Sunada. 1970. Numerical Simulation of Dispersion in Groundwater Aquifers. Colorado State Univ., Hydrol. Paper 41, 79 p.

Regan, L. J., and A. W. Hughes. 1949. Fractured reservoirs of Santa Maria District, California. Bull. Am. Assoc. Petrol. Geol. 33, p. 32–51.

Rege, S. D., and H. S. Folger. 1989. Competition among flow, dissolution and precipitation in porous media. AIChE J., v. 35, p. 1177–1185.

Rehfeldt, K. R., and others. 1992. Field study of dispersion in a heterogeneous aquifer. 1, Geostatistical analysis of hydraulic conductivity: Water Resources Res., vol. 28, no. 12, p. 3309–3324.

Reichard, E., C. Cranor, R. Raucher, and G. Zapponi. 1990. Groundwater contamination risk assessment: A guide to understanding and managing uncertainties. Intl. Assoc. Hydrol. Sci., IAHS Press, Publication 196, 204 p.

Reinson, C. E. 1984. Barrier Island and associated strand plain systems. In R. G. Walker (ed.), Facies Models. GeoSciences Canada Reprint Series 1, p. 119–140.

Reisinger, H. J., D. R. Burris, L. R. Cessar, and G. D. McCleary. 1987. Factors affecting the utility of soil vapor assessment data. Proc. First Outdoor Action Conference on Aquifer Restoration, Ground Water Monitoring and Geophysical Methods. National Water Well Assoc., Dublin, Ohio, p. 425–435.

Relyea, J. F. 1982. Theoretical and experimental considerations for the use of the column method for determining retardation factors. Radioactiv. Manage. Nucl. Fuel Cycle, v. 3, p. 151–166.

Renner, J. L., D. E. White, and D. L. Williams. 1975. Hydrothermal convection systems. In Assessment of Geothermal Resources of the United States—1975. U.S. Geol. Surv. Circ. 726, p. 47–56.

Renner, R. 1995. Predicting chemical risks with multimedia fate models. Environ. Sci. Technol., v. 29, no. 12, p. 556–559.

Richards, L. A. 1931. Capillary conduction of liquids through porous mediums. Physics, v. 1, p. 318–333.

Rittmann, B. E., and P. L. McCarty. 1980. Model of steady-state biofilm kinetics. Biotech. Bioeng., v. 22, p. 2343–2357.

Rittmann, B. E., and P. L. McCarty. 1980. Evaluation of steady-state biofilm kinetics: Biotech. Bioeng., v. 22, p. 2359–2373.

Rittmann, B. E. 1993. The significance of biofilms in porous media. Water Resources Res., v. 29, no. 7, p. 2195–2202.

Robbins, G. A. 1983. Determining Dispersion Parameters to Predict Groundwater Contamination. Ph.D. dissertation, Texas A&M Univ., College Station, 226 p.

Roberts, P. V., G. D. Hopkins, D. M. MacKay, and L. Semprini.

1990. A field evaluation of in-situ biodegradation of chlorinated ethenes: Part 1. Methodology and field site characterization. Ground Water, v. 28, no. 4, p. 591–604.

Roberts, P. V., L. Semprini, G. D. Hopkins, D. Grbic-Galic, P. L. McCarty, and M. Reinhard. 1989. In-situ Restoration of Chlorinated Aliphatics by Methanogenic Bacteria. U.S. Environ. Protection Agency, EPA/600/2-89/033, 214 p.

Robin, M. J. L., A. L. Gutjahr, E. A. Sudicky, and J. L. Wilson. 1993. Cross-correlated random field generation with direct Fourier transform method. Water Resources Res., v. 29, no. 7, p. 2385–2397.

Robinson, G. D. 1985. Structure of pre-Cenozoic rocks in the vicinity of Yucca Mountain, Nye County, Nevada-A potential nuclear waste disposal site: U.S. Geol. Survey Bull., p. 1647.

Robinson, P. C. 1982. Connectivity of fracture systems: A percolation theory approach. Theor. Phys. Div., AERE Harwell, HL82/960.

Robinson, T. W. 1939. Earth tides shown by fluctuations of water levels in wells in New Mexico and Iowa. Trans. Am. Geophys. Union, v. 20, p. 656–666.

Rogoshewski, P., H. Bryson, and K. Wagner. 1983. Remedial Action Technology for Waste Disposal Sites. Pollution Technology Review No. 101. Noyes Data Corp., Park Ridge, N.J., 498 p.

Rojstaczer, S., and J. D. Bredehoeft. 1988. Groundwater and fault strength. In J. J. Back, J. Rosenshein, and P. R. Seaber (eds.), Hydrogeology, The Geology of North America. Geol. Soc. Am. Pub., p. 447–460.

Romm, E. S. 1966. Flow characteristics of fractured rocks (in Russian). Nedra, Moscow.

Roth, J. P. 1969. Earthquakes and reservoir loading. Proc. 4th World Conf. on Earthquake Engineering, Santiago, Chile.

Royster, D. L. 1979. Landslide remedial measures. Bull. Assoc. Eng. Geol. v. 16, p. 301–352.

Rubin, J. 1968. Theoretical analysis of two dimensional transient flow of water in unsaturated and partly saturated soils. Soil Sci. Am. Proc. v. 32, p. 607–615.

Rubin, J., and R. V. James. 1973. Dispersion affected transport of reacting solutes in saturated porous media: Galerkin method applied to equilibrium controlled exchange in unidirectional steady water flow. Water Resources Res., v. 9, no. 5, p. 1332–1356.

Russell, W. L. 1972. Pressure-depth relations in Appalachian region. Am. Assoc. Petrol. Geol. Bull. 56, p. 528–536.

Rust, B. R., and E. H. Koster. 1984. Coarse alluvial deposits. In R. G. Walker (ed.), Facies Models. GeoSciences Canada Reprint Series 1.

Rutter, E. H. 1983. Pressure solution in nature, theory and experiment. J. Geol. Soc., v. 140, p. 725–740.

Sadovsky, M. A., and others, 1969, The process preceding strong earthquakes in some regions of Middle Asia: Tectonophysics, v. 23, p. 247–255.

Sass, J. W., and H. Lachenbrach. 1982. Preliminary interpretation of thermal data from the Nevada Test Site. U.S. Geol. Surv. Open File Rept. 82-973, Denver.

Sass, J. W., and others. 1976. A new heat flow contour map of the conterminous United States. U.S. Geol. Surv. Open File Rept. 76-756, Denver, Colo.

Sauter, M. 1992. Assessment of hydraulic conductivity in a karst aquifer at local and regional scale. In J. F. Quinlan and A. Stanley (eds.), Hydrology, Ecology, Monitoring, and Management of Karst Terranes Conference. Proc. National Ground Water Assoc., Dublin, Ohio, p. 39-57.

Sauty, J. P., and others. 1982a. Sensible energy storage in aquifers. I. Theoretical study. Water Resources Res., v. 18, p. 245-252.

Sauty, J. P., and others. 1982b. Sensible energy storage in aquifers. II. Field experiments and comparison with theoretical results. Water Resources Res., v. 18, p. 253-265.

Savin, S. M. 1980. Oxygen and hydrogen isotope effects in low-temperature mineral-water interactions. Chapter 8. In P. Fritz and J. C. Fontes (eds.), Handbook of Environmental Isotope Geochemistry, v. 1. Elsevier, New York, p. 283-327.

Scalf, M. R., J. F. McNabb, W. J. Dunlap, R. L. Cosby, and J. Fryberger. 1981. Manual of Ground Water Sampling Procedures. U.S. Environ. Protection Agency, Reports EPA 660/2-81-160, 93 p.

Schiff, L. 1955. The status of water spreading for groundwater replenishment. Trans. Am. Geophys. Union, v. 36, p. 1009-1020.

Schleicher, D. 1975. A model for earthquakes near Palisades Reservoir, Southeast Idaho. U.S. Geol. Survey J. Res., v. 3, p. 393-400.

Schmidt, V., and D. A. McDonald. 1977. The role of secondary porosity in the course of sandstone diagenesis. In P. A. Scholle and P. R. Schluger (ed.), Aspects of Diagenesis. Soc. Econ. Paleontol. and Mineral. Spec. Pub. 26, p. 175-225.

Schmorak, S. 1967. Saltwater encroachment in the Coastal Plain of Israel. Inter. Assoc. Sci. Hydrol. Symp. Haifa, Pub. 72, p. 305-318.

Schmorak, S., and A. Mercado. 1969. Upconing of freshwater-seawater interface below pumping wells. Water Resources Res., v. 5, p. 1290-1311.

Schneider, R. 1964. Relation of temperature distribution to groundwater movement in carbonate rocks of Central Israel. Geol. Soc. Am. Bull. 75, p. 209-216.

Scholz, C. H., L. R. Sykes, and Y. P. Aggerwal. 1973. Earthquake prediction. A physical basis. Science, v. 181, p. 803-809.

Schuler, R. W., V. P. Stallings, J. M. Smith. 1952. Heat and mass transfer in fixed bed reactors. Chem. Eng. Prog. Symp. Ser. 48, p. 19-30.

Schulz, D., M. Barvenik, and J. Ayres. 1984. Design of soil-bentonite backfill mix for the first Environmental Protection Agency Superfund Cutoff wall. Proc. of the Fourth National Symposium and exposition on Aquifer Restoration & Groundwater Monitoring. National Water Well Assoc., Dublin, Ohio, p. 8-17.

Schwartz, F. W. 1975. On radioactive waste management: An analysis of the parameters controlling subsurface contaminant transfer. J. Hydrol., v. 27, p. 51-71.

Schwartz, F. W. 1988. Contaminant hydrogeology—dollars and sense: J. Hydrol., v. 100, p. 453-470.

Schwartz, F. W., and A. S. Crowe. 1980. A Deterministic-Probabilistic Model for Contaminant Transport. NUREG/CR-1609, Nuclear Regulatory Comm., Washington, D.C., 158 p.

Schwartz, F. W., and D. Gallup. 1978. Some factors controlling the major ion chemistry of small lakes: Examples from the Prairie Parkland of Canada. Hydrobiologia, v. 58, no. 1, p. 65-81.

Schwartz, F. W., G. L. McClymont, and L. Smith, 1985. On the role or mass transport modeling. Canadian/American Confer-

ence on Hydrogeology, Proc.: National Water Well Assoc., Dublin, Ohio, p. 2-12.

Schwartz, F. W., and K. Muehlenbachs. 1979. Isotope and ion geochemistry of groundwaters in the Milk River aquifer, Alberta. Water Resources Res., v. 15, no. 2, p. 259-268.

Schwarzenbach, R. P., and W. Giger, 1985. Behavior and fate of halogenated hydrocarbons in ground water. Ground Water Quality, C. H. Ward, W. Giger, and P. L. McCarty (eds.): New York, Wiley-Interscience, p. 446-471.

Schwarzenbach, R. P., P. M. Gschwend, and D. M. Imboden. 1993. Environmental Organic Chemistry. John Wiley & Sons, New York, 681 p.

Schwarzenbach, R. P., and J. Westall. 1981. Transport of nonpolar organic compounds from surface water to groundwater. Laboratory studies. Environ. Sci. Technol., v. 15, p. 1300-1367.

Schwille, F. W. 1985. Migration of organic fluids immiscible with water in the unsaturated and saturated zones. Second Canadian/American Conference on Hydrogeology, B. Hitchon and M. Trudell (eds.): National Water Well Assoc., Dublin, Ohio, p. 31-35.

Schwille, F., 1988, Dense chlorinated solvents in porous and fractural media: Lewis Publishers, Chelsea, Mich., 146 p.

Scoffin, T. 1987. An Introduction to Carbonate Sediments and Rocks. Chapman and Hall, New York.

Scott, R. F. 1963. Principles of Soil Mechanics. Addison-Wesley, Reading, Mass.

Seaburn, G. E. 1970. Preliminary analysis of rate of movement of storm runoff through the zone of aeration beneath a recharge basin on Long Island. New York. U.S. Geol. Surv. Prof. Paper 700-B, p. 196-198.

Segol, G., and G. F. Pinder. 1976. Transient simulation of saltwater intrusion in southeastern Florida. Water Resources Res., v. 12, p. 65-70.

Sellers, K. L., and R. P. Schreiber. 1992. Air sparging model for predicting groundwater cleanup rate: in Proceedings Petroleum Hydrocarbons and Organic Chemicals in Ground Water: Prevention, Detection and Restoration, National Ground Water Association, Columbus, Ohio, p. 365-376.

Semprini, L., P. K. Kitanidis, D. H. Kampbell, and J. T. Wilson. 1995. Anaerobic transformation of chlorinated aliphatic hydrocarbons in a sand aquifer based on spatial chemical distributions. Water Resources Res., v. 31, no. 4, p. 1051-1062.

Semprini, L., P. V. Roberts, G. D. Hopkins, and P. L. McCarty. 1990. A field evaluation of in-situ biodegradation of chlorinated ethenes: Part 2, Results of biostimulation and biotransformation experiments: Ground Water, v. 28, no. 5, p. 715-727.

Senger, R. K., and G. E. Fogg. 1987. Regional underpressuring in deep brine aquifers, Palo Duro Basin, Texas: 1. Effects of hydrostratigraphy and topography. Water Resources Res., v. 23, p. 1481-1493.

Senger, R. K., G. E. Fogg, and C. Kreitler. 1987. Effect of hydrostratigraphy and basin development on hydrodynamics of the Palo Duro Basin, Texas. Rept. of Invest. 165, Bureau Econ. Geol., Austin, Tex.

Senger, R. K., C. Kreitler, and G. Fogg. 1987. Regional underpressuring in deep brine aquifers, Palo Duro Basin, Texas: 2. The effect of Cenozoic development. Water Resources Res., v. 23, p. 1494-1504.

Senn, R. B., and M. S. Johnson. 1985. Interpretation of gas

chromatography data as a tool in subsurface hydrocarbon investigations. In Petroleum Hydrocarbons and Organic Chemicals in Ground Water—Prevention, Detection and Restoration. National Ground Water Assoc., Columbus, Ohio, p. 331-357.

Serafin, J. L., and A. del Campo. 1965. Interstitial pressure on rock foundations of dams. J. Soil Mech. Found. Div., Am. Assoc. Civil Engrs., v. 91, p. 65-85.

Shante, V. K. S., and S. Kirkpatrick. 1971. An introduction to percolation theory. Adv. Phys., v. 20, p. 352-357.

Sharp, J. M. 1978. Energy and momentum transport model of the Ouachita Basin and its possible impact on formation of economic mineral deposits. Econ. Geol., v. 73, p. 1057-1068.

Sharp, J. M. 1983. Permeability control on aquathermal pressure. Am. Assoc. Petrol. Geol. Bull. 67, p. 2057-2061.

Sharp, J. M., and P. A. Domenico. 1976. Energy transport in thick sequences of compacting sediment. Geol. Soc. Am. Bull. 87, p. 390-400.

Sharp, J. M., and J. R. Kyle. 1988. The role of groundwater processes in the formation of ore deposits. In W. Back, J. Rosenshein, and P. Seaber (eds.), The Geology of North America, Hydrogeology. Geol. Soc. Am., v. 0-2, p. 461-483.

Sharp, J. C., and Y. N. T. Maini. 1972. In W. Wittke (ed.), Fundamental considerations on the hydraulic characteristics of joints in rock in percolation through fractured rock. Intern. Soc. Rock Mech., Stuttgart.

Shi, Y., and C. Y. Wang. 1986. Pore pressure generation in sedimentary basins: Overloading versus aquathermal. J. Geophys. Res., v. 91, p. 2153-2162.

Signor, D. C. 1985. Groundwater sampling during artificial recharge: equipment, techniques, and data analyses. In T. Asano (ed.), Artificial Recharge of Groundwater. Butterworth, Stoneham, Mass., p. 151-202.

Signor, D. C., and others. 1970. Annotated bibliography on artificial recharge of groundwater, 1955-1967. U.S. Geol. Survey Water Supply Paper 1990.

Simon, R. B. 1969. Seismicity of Colorado. Science, v. 165, p. 897-899.

Singer, E., and R. H. Wilhelm, 1950. Heat transfer in packed beds: Analytical solution and design methods: Chem. Eng. Prog., v. 46, p. 343-357.

Slaine, D. D., P. E. Pehme, J. A. Hunter, S. E. Pullan, and J. P. Greenhouse. 1990. Mapping overburden stratigraphy at a proposed hazardous waste facility using shallow seismic reflection methods. In S. H. Ward (ed.), Special Volume on Environmental Physics, Soc. Explor. Geophysics, Tulsa, Okla.

Slattery, J. C. 1972. Momentum, Energy, and Mass Transfer in Continua. McGraw-Hill, New York.

Sleep, B. E., and J. F. Sykes. 1989. Modeling the transport of volatile organics in variably saturated media. Water Resources Res., v. 25, p. 81-92.

Slichter, C. S. 1899. Theoretical investigation of the motion of groundwater. U.S. Geol. Survey Annual Rept. 19 (part 2), p. 295-384.

Smith, L., and D. S. Chapman. 1983. On the thermal effects of groundwater flow: I. Regional scale systems. J. Geophys. Res., v. 88, p. 593-608.

Smith, L., and F. W. Schwartz. 1981. Mass transport. 2. Two-dimensional simulations. Water Resources Res., v. 17, no. 2, p. 351-369.

Smith, R. E., and D. A. Woolhiser. 1971. Overland flow on an infiltrating surface. Water Resources Res., v. 7, p. 899-913.

Smith, R. L., R. W. Harvey, J. H. Duff, and D. R. LeBlanc. 1987. Importance of close interval vertical sampling in delineating chemical and microbiological gradients in ground-water studies. U.S. Geol. Surv. Open-File Rept. 87-109, p. B33-B35.

Smolenski, W. J., and J. M. Sulflita. 1987. Biodegradation of cresol isomers in anoxic aquifers. Appl. Environ. Microbiol., v. 53, p. 710-716.

Smolley, M., and J. C. Kappmeyer. 1991. Cone penetrometer tests and Hydropunch® sampling: A screening technique for plume definition. Ground Water Mon. Rev., v. 11, no. 2, p. 101-106.

Smoluchowski, M. S. 1909. Some remarks on the mechanics of overthrusting. Geol. Magazine, v. 6, p. 1133-1138.

Smyth, J. R. 1982. Zeolite stability constraints on radioactive waste isolation in zeolite bearing rocks. J. Geol., v. 90, p. 195-201.

Snow, D. T. 1968. Rock fracture spacings, openings, and porosities. J. Soil Mech., Found. Div., Proc. Am. Soc. Civil Engrs., v. 94, p. 73-91.

Snow, D. T. 1972. Geodynamics of seismic reservoirs. Proc. Symp. on Percolation through Fissured Rock, Stuttgart, T2-J.

Solomon, D. K., R. J. Poreda, P. G. Cook, and A. Hunt. 1995. Site characterization using ^{3}H/^{3}He ground-water ages, Cape Cod, Mass. Ground Water, v. 33, no. 6, p. 988-996.

Sorey, M. L. 1971. Measurement of vertical groundwater velocity from temperature profiles in wells. Water Resources Res., v. 7, p. 963-970.

Sorey, M. L. 1978. Numerical modeling of liquid geothermal systems. U.S. Geol. Surv. Prof. Papers 1044-D.

Sposito, G. 1984. The Surface Chemistry of Soils. Oxford University Press, New York, 234 p.

Springer, A. E. 1990. An evaluation of wellfield-protection area delineation methods as applied to municipal wells in the stratified-drift aquifer at Wooster, Ohio. M.S. thesis, The Ohio State University, 167 p.

Springer, A. E., and E. S. Bair. 1992. Comparison of methods used to delineate capture zones of wells: 2. Stratified-drift buried-valley aquifer. Ground Water, v. 30, no. 6, p. 908-917.

Stallman, R. W. 1952. Nonequilibrium type curves for two well systems. U.S. Geol. Surv. Groundwater Notes, Open File Rept. 3.

Starr, R. C., and J. A. Cherry. 1994. In situ remediation of contaminated ground water: The funnel-and-gate system. Ground Water, v. 32, no. 3, p. 465-476.

Stearns, D. W. 1964. Macrofracture patterns on Tetron anticline, Northwest Montana (abst). Am. Geophys. Union Trans. v. 45, p. 107-108.

Stearns, D. W. 1967. Certain aspects of fracture in naturally deformed rock. In R. E. Riecker (ed.), National Science Foundation Advanced Science Seminar in Rock Mechanics. Air Force Cambridge Research Lab Spec. Rept., Bedford, Mass., p. 97-118.

Stearns, D. W., and M. Friedman. 1972. Reservoirs in fractured rock. In Fracture Controlled Production. Am. Assoc. Petrol. Geol Reprint Ser. 21, p. 174-198.

Steefel, C. I., and A. C. Lasaga. 1988. The space-time evolution of dissolution patterns: Permeability change due to coupled flow and reaction. Submitted to Am. Chem. Soc. Symp. Set.

Stein, R., and F. W. Schwartz. 1990. On the origin of saline soils at Blackspring Ridge, Alberta, Canada: J. Hydrol., v. 117, p. 99-131.

Stewart, M. T., and M. C. Gay. 1986. Evaluation of transient electromagnetic soundings for deep detection of conductive fluids. Ground Water, v. 24, p. 351–356.

Stiff, H. A., Jr. 1951. The interpretation of chemical water analysis by means of patterns. J. Petrol. Technol., v. 3, no. 10, p. 15–17.

Stoiber, R. 1941. Movement of mineralizing solutions in the Picher Field, Oklahoma-Kansas. Econ. Geol., v. 46, p. 800–812.

Stollenwerk, K. G., 1991. Simulation of molybdate sorption with diffuse layer surface-complexation model. U.S. Geol. Surv. Water Resources Investigations Report, 91-4034, p. 47–52.

Stoodley, P., D. de Beer, and Z. Lewandowski. 1994. Liquid flow in biofilm systems. Appl. Environ. Microbiol., Aug., p. 2711–2716.

Streltsova, T. D. 1972. Unsteady radial flow in an unconfined aquifer. Water Resources Res., v. 8, p. 1059–1066.

Streltsova, T. D., and K. Rushton. 1973. Water table drawdown due to a pumped well. Water Resources Res., v. 9, p. 236–242.

Stringfield, V. T., and H. E. LeGrande. 1966. Hydrology of limestone terranes. Geol. Soc. Am. Spec. Paper 3.

Strom, E. W. 1993. A model of the large hydraulic gradient at Yucca Mountain, Nevada Test Site, based on hydraulic conductivity contrasts between Cenozoic and Paleozoic rocks. M.S. thesis, Texas A&M Univ., Dept. Geol. and Geop., College Station, Tex.

Stumm, W., and J. J. Morgan. 1981. Aquatic Chemistry, 2nd ed. John Wiley & Sons, New York, 780 p.

Sudicky, E. A. 1986. A natural gradient experiment on solute transport in a sand aquifer: Spatial variability of hydraulic conductivity and its role in the dispersion process. Water Resources Res., v. 22, no. 13, p. 2069–2082.

Sudicky, E. A. 1989. The Laplace transform Galerkin technique: A time-continuous finite element theory and application to mass transport in groundwater. Water Resources Res., v. 25, no. 8, p. 1833–1846.

Sullivan, C. R., R. E. Zinner, and J. P. Hughes. 1988. The occurrence of hydrocarbon on an unconfined aquifer and implications for the liquid recovery. Proc. of the NWWA/API Conference Petroleum Hydrocarbons and Organic Chemicals in Ground Water—Prevention, Detection and Restoration. National Water Well Assoc., Dublin, Ohio, p. 135–156.

Surdam, R. C., S. W. Boese, and L. J. Crossey. 1984. The chemistry of secondary porosity. In D. A. McDonald and R. C. Surdam (eds.), Clastic Diagenesis, AAPG Memoir 37, p. 127–149.

Swenson, F. A. 1968. New theory of recharge to the artesian basin of the Dakotas. Geol. Soc. Am. Bull. 79, p. 163–182.

Szymanski, J. 1989. Conceptual considerations of the Yucca Mountain groundwater system with special emphasis on the adequacy of this system to accommodate a high level nuclear waste repository. Dept. of Energy Rept. (unpublished). Nevada operations office, Las Vegas.

Tamers, M. A. 1967. Surface-water infiltration and groundwater movement in arid zones of Venezuela. Isotopes in Hydrology. Vienna, Intl. Atomic Energy Agency, p. 339–351.

———, 1975, Variability of radiocarbon dates on groundwater. Geophys. Surv., v. 2, p. 217–239.

Tang, D. H., E. O. Frind, and E. A. Sudicky. 1981. Contaminant transport in fractured porous media: Analytical solution for a single fracture. Water Resources Res., v. 17, no. 3, p. 555–564.

Taylor, T. W., and M. C. Serafini. 1988. Screened auger sampling: The technique and two case studies. Ground Water Mon. Rev., v. 8, no. 3, p. 145–152.

Terjesen, S. G., O. Erga, G. Thorsen, and A. Ve. 1961. Phase boundary processes as rate determining steps in reactions between solids and liquids. The inhibitory action of metal ions on the formation of calcium bicarbonate by the reaction of calcite with aqueous carbon dioxide. Chem. Eng. Sci., v. 14, p. 277–289.

Terzaghi, K. 1950. Mechanism of landslides. In Berkeley, Vol., Application of geology to engineering practice. Geol. Soc. Am. Publ., p. 83–123.

Terzaghi, K., and R. B. Peck. 1948. Soil Mechanics in Engineering Practice. John Wiley & Sons, New York.

Theis, C. V. 1935. The relation between the lowering of the piezometric surface and rate and duration of discharge of a well using groundwater storage. Trans. Am. Geophys. Union, v. 2, p. 519–524.

Theis, C. V. 1940. The source of water derived from wells—essential factors controlling the response of an aquifer to development. Civil Eng., Am. Soc. Civil Eng., p. 277–280.

Theis, C. V., R. H. Brown, and R. R. Meyers. 1963. Estimating the transmissibility of aquifers from the specific capacity of wells. In Methods of Determining Permeability, Transmissibility, and Drawdown. U.S. Geol. Survey Water Supply Papers 1536-I.

Thiem, G. 1906. Hydrologische Methode. Gebhardt, Leipzig.

Thomas, H. C., 1944. Heterogeneous ion exchange in a flowing system: J. Amer. Chem. Soc., v. 66, p. 1664–1666.

Thomas, H. E. 1951. The Conservation of Groundwater. McGraw-Hill, New York.

Thompson, C., G. McMechan, R. Szerbiak, and N. Gaynor. 1995. 3-D GPR imaging of complex stratigraphy within the Ferron Sandstone, Castle Valley, Utah. Proc. of the Symposium on the Application of Geophysics to Engineering and Environmental Problems, Environmental and Engineering. Geophys. Soc., Englewood, Colo., p. 435–443.

Thompson, D. G., O. E. Meinzer, and V. T. Stringfield. 1938. Discussion on "Effect of a sea level canal on the groundwater level of Florida." Econ. Geol., v. 33, p. 87–107.

Thompson, G. M. 1976. Trichloromethane: A new hydrologic tool for tracing and dating groundwater. Ph.D. dissertation, Dept. of Geology, Indiana University, Bloomington, 93 p.

Thompson, G. M., and J. M. Hayes. 1979. Trichlorofluoromethane in groundwater: A possible tracer and indicator of groundwater age. Water Resources Res., v. 15, p. 546–554.

Thornthwaite, C. W. 1948. An approach toward a rational classification of climate. Geograph. Rev., v. 38, p. 55–94.

Thurman, E. M. 1985. Organic Geochemistry of Natural Waters. Martinus Nijhoff/Dr. W. Junk Publishers, Dordrecht, The Netherlands, 497 p.

Todd, D. K. 1959a. Annotated bibliography on artificial recharge of groundwater through 1954: U.S. Geol. Surv. Water Supply Papers, 1477.

Todd, D. K. 1959b. Groundwater Hydrology, 1st ed. John Wiley & Sons, New York.

Todd, D. K. 1980. Groundwater Hydrology. 2nd ed. John Wiley & Sons, New York, 535 p.

Tokyo Inst. Civ. Eng. 1975. Subsidence as of 1974. Tokyo Metrop. Govt. Annual Rept. (in Japanese).

Tortensson, B. A., and A. M. Petsonk. 1988. A hermetically isolated sampling method for ground-water investigations. In A. G. Collins and A. I. Johnson (eds.), Ground-Water Contamination: Field Methods. ASTM STP 963, Am. Soc. for Testing and Materials, Philadelphia, Pa., p. 274–289.

Tóth, J. 1962. A theory of groundwater motion in small drainage basins in Central Alberta. J. Geophys. Res., v. 67, p. 4375–4387.

Tóth, J. 1963. A theoretical analysis of groundwater flow in small drainage basins. J. Geophys. Res., v. 68, p. 4795–4812.

Tóth, J. 1966. Mapping and interpretation of field phenomena for groundwater reconnaissance in a prairie environment, Alberta, Canada. Int. Assoc. Sci. Hydrol., 11 Année, no. 2, p. 1–49.

Tóth, J. 1980. Cross formational gravity flow of groundwater—a mechanism of the transport and accumulation of petroleum. In W. H. Roberts and R. J. Cordell (eds.), Problems of Petroleum Migration. Am. Assoc. Petrol. Geol. Studies in Geol., no. 10, Tulsa, Okla., p. 121–167.

Tóth, J. 1988. Ground water and hydrocarbon migration. In W. Back, J. S. Rosenshein, and P. R. Seaber (eds.), The Geology of North America, v. O-2. Geol. Soc. Am., Boulder, Colo., p. 485–501.

Tóth J., and T. Corbett. 1986. Post-Paleocene evolution of regional groundwater flow systems and their relation to petroleum accumulations, Taber area, southern Alberta. Bull. Canadian Petrol. Geol. v. 34, no. 3, p. 339–363.

Truesdell, A. H., and J. R. Hulston. 1980. Isotopic evidence on environments of geothermal systems. Chapter 5. In P. Fritz and J. C. Fontes (eds.), Handbook of Environmental Isotope Geochemistry, v. 1. Elsevier, Amsterdam, p. 179–226.

Tsang, C. F., T. Buscheck, and C. Doughty. 1981. Aquifer thermal storage. A numerical simulation of Auburn University Field experiments. Water Resources Res., v. 17, p. 647–658.

ULARA Watermaster. 1994. Watermaster service in the upper Los Angeles River Area, Los Angeles County. ULARA Watermaster Report, May 1994.

United Nations. 1976. Proc. 2nd U.N. Symposium on Development and Use of Geothermal Resources, 3 Vols. San Francisco: United Nations.

U.S. Environmental Protection Agency (EPA). 1985. Remedial Action at Waste Disposal Sites. EPA/625/6-85/006 (NTIS PB87-201034).

U.S. EPA. 1989a. Risk Assessment Guidance for Superfund. Vol. I. Human Health Evaluation Manual (Part A), Interim Final. U.S. Environmental Protection Agency, Office of Emergency and Remedial Response, EPA/540/1-89/002, Washington, D.C.

U.S. EPA. 1989b. Terra Vac In Situ System, Applications Analysis Report. Environmental Protection Agency, EPA/540/A5-89/003, 52 p.

U.S. EPA. 1989c. Technology Evaluation Report: Site Program Demonstration Test Terra Vac In Situ Vacuum Extraction System Groveland, Mass. Environmental Protection Agency, EPA/540/5-89/003a, 97 p.

U.S. EPA. 1991. Site characterization for subsurface remediation seminar publication, EPA/625/4-91/026, 259 p.

U.S. EPA. 1992. Estimating potential for occurrence of DNAPL at superfund sites. Publ. 9355.4-07FS, 9 p.

U.S. EPA. 1993. Subsurface characterization and monitoring techniques: A desk reference guide. EPA/625/RR-93/003a, Vol. I.

U.S. Department of Energy. 1990. Radiation Protection for the Public and the Environment. U.S. Department of Energy Order 5400, v. 5, February 8.

U.S. Department of Energy. 1995. In Situ Bioremediation Using Horizontal Wells. Innovative Technology Summary Report, U.S. DOE, DOE/EM-0270.

United States Dept. of the Navy, Bureau of Yards and Docks. 1961. Design manual, Soil mechanics, Foundations, and Earth Structures. DM-7 Chap. 4.

United States Geological Survey Annual Report 19. 1899. U.S. Geol. Surv. Ann. Rept.

United States Geological Survey. 1970. The national atlas of the United States of America.

Vaccaro, J. 1986. Columbia plateau basalt regional aquifer system study. From Ren Jen Sun (ed.), Regional aquifer system analysis program of the U.S. Geological Survey summary of projects, 1978-84, U.S. Geol. Sur. Circ. 1002.

Valocchi, A. J. 1985. Validity of the local equilibrium assumption for modeling sorbing solute transport through homogeneous soils. Water Resources Res., v. 21, p. 808–820.

Van Golf-Racht, T. D. 1982. Fundamentals of Fractured Reservoir Engineering. Elsevier, New York.

Van Genuchten, M. Th. 1981. Non-equilibrium transport parameters from miscible displacement experiments. U.S. Dept. Agric. U.S. Salinity Lab., Res. Rept. No. 119, 88 p.

Van der Kamp, G. 1972. Tidal fluctuations in a confined aquifer extending under the sea. Environ. Canada reprint 242, p. 101-106, Ottawa.

Van der Kamp, G., and S. Bachu, 1989. Use of dimensional analysis in the study of thermal effects of various hydrogeological regimes, in Hydrogeologic regimes and their subsurface thermal effects, Beck and others (eds.): Geophysical Monograph 47, IUGG, v. 2.

Van der Kamp, G., and J. E. Gale. 1983. Theory of earth tide and barometric effects in porous formations with compressible grains. Water Resources Res., v. 19, p. 538–544.

Van Everdingen, R. O. 1972. Observed changes in groundwater regime caused by the creation of Lake Diefenbaker, Saskatchewan. Can. Dept. of the Environ., Inland Waters Branch Tech. Bull. 59.

Van Orstrand, C. E. 1934. Temperature gradients. In Problems of Petrol. Geol. Am. Assoc. Petrol. Geol., Tulsa, Okla., p. 989–1021.

Verma, R. D., and W. Brutsaert. 1970. Unconfined aquifer seepage by capillary flow theory. J. Hydraul. Div., Proc. Am. Assoc. Civil Engrs., v. 96, p. 1331–1344.

Verma, R. D., and W. Brutsaert. 1971. Unsteady free surface groundwater seepage. J. Hydraul. Div., Proc. Am. Assoc. Civil Engrs., v. 97, p. 1213–1229.

Vernon, R. O. 1969. The geology and hydrology associated with a zone of high permeability (Boulder Zone) in Florida. Soc. Min. Engr. Preprint 69-AG12, 24 p.

Verschueren, K. 1983. Handbook of Environmental Data on Organic Chemicals, 2nd ed. Van Nostrand Reinhold, New York, 1310 p.

Vogel, J. C. 1967. Investigation of groundwater flow with radiocarbon. Isotopes in Hydrology. Intl. Atomic Energy Agency, Vienna, Austria, p. 355–368.

Vogel, J. C. 1970. Carbon-14 dating of groundwater. Isotope Hydrology. Intl. Atomic Energy Agency, Vienna, Austria, p. 235–237.

Vogel, T. M., and P. L. McCarty. 1987. Abiotic and biotic transformations of 1,1,1-trichloroethane under methanogenic conditions. Environ. Sci. Technol., v. 21, p. 1208–1213.

Vorhis, R. R. 1955. Interpretation of hydrologic data resulting from earthquakes. Geologischer Rundschar, v. 43, p. 47–52.

Walder, J., and A. Nur. 1984. Porosity reduction and crustal pore pressure development. J. Geophys. Res., v. 89, p. 11539–11548.

Walker, R. G., and D. J. Cant. 1984. Sandy fluvial systems. In R. Walker (ed.), Facies Models. GeoSciences Canada Reprint Series 1, p. 71–89.

Wallace, R. E. 1974. Goals, strategy, and tasks of the earthquake hazard reduction program. U.S. Geol. Survey Circular 701.

Wallick, E. I., H. R. Krouse, and A. Shakur. 1984. Environmental isotopes: Principles and applications in ground water geochemical studies in Alberta, Canada. In B. Hitchon and E. J. Wallick (eds.), First Canadian/American Conference on Hydrogeology. National Water Well Assoc., Dublin, Ohio, p. 249–266.

Wallis, P. M., H. B. N. Hynes, and S. A. Telang. 1981. The importance of groundwater in the transportation of allochthonous dissolved organic matter to the streams draining a small mountain basin. Hydrobiologia, v. 79, p. 77–90.

Walsh, J. B. 1978. The effect of fractures on compressibility, resistivity, and permeability. EOS, no. 59, p. 1193.

Walsh, J. B. 1981. Effect of the pore pressure and confining pressure on fracture permeability. Int. J. Rock Mech. Min. Soc. and Geomech., v. 18, p. 429–435.

Walther, E. G., A. M. Pitchford, and G. R. Olhoeft. 1986. A strategy for detecting subsurface organic contaminants. Proc. of the Petroleum Hydrocarbons and Organic Chemicals in Ground Water: Prevention, Detection and Restoration: Dublin, Ohio, National Water Well Assoc., p. 357–381.

Walton, W. C. 1962. Selected analytical methods for well and aquifer evaluation. Illinois State Water Survey, Bull. 49, 81 p.

Walton, W. C. 1970. Groundwater Resource Evaluation. McGraw-Hill, New York.

Wang, J. S. Y., and T. N. Narasimhan. 1985. Hydrologic mechanisms governing fluid flow in a partially saturated fractured porous medium. Water Resources Res., v. 21, p. 1861–1874.

Wang, J. S. Y., and others. 1978. Transient flow in tight fractures. Proc. Invitational Well Testing Symp., Univ. of California, Lawrence Berkeley Lab., Berkeley, Calif., p. 103–116.

Warren, J. E., and H. S. Price. 1961. Flow in heterogeneous porous media. Soc. Petrol. Eng. J., v. 1, p. 153–169.

Watts, E. V. 1948. Some aspects of high pressure in the D7 zone of the Ventura Avenue Field. Am. Inst. Mining Engrs. Petrol. Div., v. 174, p. 191–200.

Way, S. C., and C. R. McKee. 1982. In-situ determination of three dimensional aquifer permeabilities. Ground Water, v. 20, p. 594–603.

Weeks, E. P. 1969. Determining the ratio of horizontal to vertical permeability by aquifer test analysis. Water Resources Res., v. 5, p. 196–214.

Wenzel, L. K. 1942. Methods for determining permeability of water bearing materials with special reference to discharging well methods. U.S. Geol. Surv. Water Supply Papers 887.

Werner, P. W., and D. Noren. 1951. Progressive waves in non-artesian aquifers. Trans. Am. Geophys. Union, v. 32, p. 238–244.

Wershaw, R. L., M. J. Fishman, R. R. Gabbe, and L. E. Lowe. 1983. Methods for the determination of organic substances in water and fluvial sediments. U.S. Geol. Surv. TWRI, Book 5, Chapter A3, 173 p.

Wesson, R. L. 1981. Interpretation of changes in water level accompanying fault creep and implications for earthquake prediction. J. Geophys. Res., v. 86, p. 9259–9267.

Wexler, E. J. 1992. Analytical solutions for one-, two-, and three-dimensional solute transport in ground-water systems with uniform flow. Techniques of Water-Resources Investigations of the U.S. Geol. Surv., Book 3, Chapter B7, 190 p.

Weyl, P. K. 1959. Pressure solution and the force of crystalization. A phenomenological theory. J. Geophys. Res., v. 64, p. 2001–2025.

Wheatcraft, S. W., and S. W. Tyler. 1988. An explanation of scale-dependent dispersivity in heterogeneous aquifers using concepts of fractal geometry. Water Resources Res., v. 24, p. 566–578.

Whitaker, S. 1967. Diffusion and dispersion in porous media. Am. Inst. Chem. Eng. J., v. 13, p. 420–427.

White, D. E., and D. L. Williams. 1975. Summary and Conclusion. In Assessment of Geothermal Resources of the United States—1975. U.S. Geol. Surv. Circ. 726, p. 147–155.

White, D. E. 1968. Environments of generation of some basic-metal ore deposits. Econ. Geol., v. 63, p. 301–335.

White, W. B., and E. L. White. 1987. Ordered and stochastic arrangements within regional sinkhole populations. In B. F. Beck and W. L. Wilson (eds.), Karst Hydrogeology: Engineering and Environmental Applications. A.A. Balkema Publishers, Accord, Mass., p. 85–90.

Whittman, S. G. 1985. Use of soil gas sampling techniques for assessment of ground water contamination. Proc. of the NWWA/API Conference on Petroleum Hydrocarbons and Organic Chemicals in Ground Water—Prevention, Detection and Restoration. National Water Well Assoc., Dublin, Ohio, p. 291–309.

Wierenga, P. J., L. W. Gelhar, C. S. Simmons, G. W. Gee, and T. J. Nicholson. 1986. Validation of Stochastic Flow and Transport Models for Unsaturated Soil: A Comprehensive Field Study. U.S. Nuclear Regulatory Commission, NUREG/CR-4622.

Wigley, T. M. L., L. N. Plummer, and F. J. Pearson, Jr. 1978. Mass transfer and carbon isotope evolution in natural water systems. Geochim. Cosmochim. Acta, v. 42, p. 1117–1139.

Willis, C., and J. Rubin. 1987. Transport of reacting solutes subject to a moving dissolution boundary: Numerical methods and solutions. Water Resources Res., v. 23, no. 8, p. 1561–1574.

Wilson, E. B., Jr. 1952. An Introduction to Scientific Research. McGraw-Hill, New York, 373 p.

Wilson, G., and H. Grace. 1942. The settlement of London due to underdrainage of the London clay. J. Inst. Civil Engr. (London), v. 19, p. 100–127.

Wilson, J., and M.J. Noonan. 1984. Microbial activity in model aquifer systems. In G. Bitton and C. P. Gerba (eds.), Groundwater Pollution Microbiology, John Wiley & Sons, New York, p. 117–133.

Wilson, J. L., and S. H. Conrad. 1984. Is physical displacement of residual hydrocarbons a realistic possibility in aquifer restoration? Proc. of the NWWA/API Conference on Petroleum Hydrocarbons and Organic Chemicals in Ground Water—Prevention, Detection and Restoration. National Water Well Assoc., Dublin, Ohio, p. 274–298.

Wilson, J. L., and C. Jordan. 1983. Middle shelf. In P. A. Scholle, D. G. Bebout, and C. H. Moore (eds.), Carbonate Depositional Environments. Am. Assoc. Petrol. Geol. Mem. 33, p. 297–344.

Wilson, J. L., and P. J. Miller. 1978. Two dimensional plume in uniform groundwater flow. J. Hydraul. Div., Proc. Am. Assoc. Civil Eng., v. 104, p. 503–514.

Wilson, J. T., and B. H. Wilson. 1985. Biotransformation of trichloroethylene in soil. Appl. Environ. Microbiol., v. 49, p. 242–243.

Wilson, J. T., M. J. Noonan, and J. F. McNabb. 1985. Biodegradation of contaminants in the subsurface: Ground Water Quality, C. H. Ward, W. Giger, and P. L. McCarty (eds.), John Wiley & Sons, New York, p. 483–492.

Winograd, I. J. 1962. Interbasin movement of groundwater at the Nevada Test Site, Nevada. U.S. Geol. Surv. Prof. Papers 450C, p. 108–111.

Winograd, I. J., and W. Thordarson. 1975. Hydrogeologic and hydrochemical framework, South Central Great Basin, with special reference to the Nevada Test Site. U.S. Geol. Surv. Prof. Papers 712-C.

Winslow, A. G., and L. A. Wood. 1959. Relation of land subsidence to groundwater withdrawals in the upper Gulf Coastal region of Texas. Am. Inst. Min. Engrs. Min. Div., Min. Eng., p. 1030–1034.

Winter, T. C. 1976. Numerical simulation analysis of the interaction of lakes and groundwaters. U.S. Geol. Surv. Prof. Papers, 1001.

Winter, T. C. 1978. Numerical simulation of steady state three dimensional groundwater flow near lakes. Water Resources Res., v. 14, p. 245–254.

Witherspoon, P. A., and J. E. Gale. 1977. Mechanical and hydraulic properties of rocks related to induced seismicity. Eng. Geol., v. 11, p. 23–55.

Witherspoon, P. A., and others. 1980. Validity of cubic law for fluid flow in a deformable rock fracture. Water Resources Res., v. 16, p. 1016–1024.

Wolery, T. J. 1979. Calculation of chemical equilibrium between aqueous solutions and minerals: The EQ3/EQ6 software package. UCRL-52658, Univ. of California, Lawrence Livermore Laboratory, Livermore, Calif., 41 p.

Wolff, R. G. 1970. Field and laboratory determination of the hydraulic diffusivity of a confining bed. Water Resources Res., v. 6, p. 194–203.

Wolff, R. G., and S. S. Papadopulos. 1972. Determination of the hydraulic diffusivity of a heterogeneous confining bed. Water Resources Res., v. 8, p. 1051–1058.

Wood, J. R., and T. A. Hewett. 1982. Fluid convection and mass transfer in porous sandstone—a theoretical model. Geochim. Cosmochim. Acta, v. 46, p. 1707–1713.

Wood, W. W. 1976. Guidelines for the collection and field analysis of ground water samples for selected unstable constituents. U.S. Geol. Surv. TWRI, Book 1, Chapter D2, 24 p.

Woodbury, A. D., and L. Smith. 1985. On the thermal effects of three dimensional groundwater flow. J. Geophy. Res., v. 90. p. 759–767.

Woodbury, A. D., and E. A. Sudicky. 1991. The geostatistical characteristics of the Borden aquifer. Water Resources Res., v. 27, no. 4, p. 533–546.

Woodward and Clyde, Consultants. 1979. Study of reservoir induced seismicity, Reston, Virginia. U.S. Geol. Surv. technical report contract 14-08-0001-16809.

Worthington, S., and D. Ford. 1995. Borehole tests for megascale channeling in carbonate aquifers. Proc. XXVI Congress of the International Assoc. of Hydrogeologists, Edmonton, Alberta.

Xu, Y., and F. W. Schwartz. 1994. Lead immobilization by hydroxyapatite in aqueous solutions. J. Contam. Hydrol., v. 15, no. 3, p. 187–206.

Yagi, S., and D. Kunii. 1957. Studies on effective thermal conductivities in packed beds. Am. Inst. Chem. Eng. J., v. 3, p. 373–381.

Yaniga, P. M., and J. Mulry, 1984. Accelerated aquifer restoration: In-situ applied techniques for enhanced free product recovery/absorbed hydrocarbon reduction via bioreclamation. NWWA/API Conference on Petroleum Hydrocarbons and Organic Chemicals in Ground Water—Prevention, Detection and Restoration, Proc.: Dublin, Ohio, National Water Well Assoc., p. 421–440.

Yoder, H. S. 1955. Role of water in metamorphism. Geol. Soc. Am. Spec. Paper 62, p. 505–524.

Yong, R. N. 1985. Interaction of clay and industrial waste: A summary review. In B. Hitchon and M. Trudell (eds.), Second Canadian/American Conference on Hydrogeology. National Water Well Assoc., Dublin, Ohio, p. 13–25.

Young, R., and J. Bredehoeft. 1972. Digital computer simulation for solving management problems of conjunctive groundwater and surface water systems. Water Resources Res., v. 8, p. 533–556.

Yu, C., A. J. Zielen, J.-J. Cheng, Y. C. Yuan, L. G. Jones, D. J. LePoire, Y. Y. Wang, C. O. Loureiro, E. Gnanapragasam, E. Faillance, A. Wallo III, W. A. Williams, and H. Peterson. 1993. Manual for Implementing Residual Radioactive Material Guidelines Using RESRAD, Version 5.0. Argonne National Laboratory, ANL/EAD/LD-2, Argonne, Ill.

Zachara, J., 1990, Hydrogeology in relation to microorganisms: Proceedings of the First International Symposium on Microbiology of the deep subsurface, C. B. Fliermans and T. C. Hazen (eds.), Westinghouse Savannah River Company, p. 5-3 to 5-13.

Zalasiewicz, J. A., S. J. Mathers, and J. D. Cornwell. 1985. The application of ground conductivity measurements to geological mapping. Q. J. Eng. Geol. London, v. 18, p. 139–148.

Zaporozec, A. 1972. Graphical interpretation of water quality data: Ground Water, v. 10, p. 32–43.

Zemo, D. A., T. E. Graf, and J. E. Bruya. 1993. The importance and benefit of fingerprint characterization in site investigation and remediation focusing on petroleum hydrocarbons. Proc. of the Petroleum Hydrocarbons and Organic Chemicals in Ground Water—Prevention, Detection and Restoration. National Ground Water Assoc., Columbus, Ohio, p. 39–54.

Zenger, D. H. 1972. Dolomitization and uniformitarianism. J. Geol. Educ., v. 20, p. 107–124.

Zheng, C., and Bennett, G. D. 1995. Applied contaminant transport modeling. Intern. Thomson Pub. Co., New York. 440 p.

Zimmerman, R. W., W. H. Somerton, and M. S. King. 1986. Compressibility of porous rocks. J. Geophys. Res., v. 91, p. 12,765–12,777.

Zoback, M. D., and J. D. Byerlee. 1976. A note on the deformation behavior and permeability of crushed granite. Inter. J. Rock Mech., Min. Sci., and Geomech., v. 13, p. 291–294.

Index

Limited Use License Agreement

This is the John Wiley and Sons, Inc. (Wiley) limited use License Agreement, which governs your use of any Wiley proprietary software products (Licensed Program) and User Manual (s) delivered with it.

Your use of the Licensed Program indicates your acceptance of the terms and conditions of this Agreement. If you do not accept or agree with them, you must return the Licensed Program unused within 30 days of receipt or, if purchased, within 30 days, as evidenced by a copy of your receipt, in which case, the purchase price will be fully refunded.

License: Wiley hereby grants you, and you accept, a non-exclusive and non-transferable license, to use the Licensed Program and User Manual (s) on the following terms and conditions only:

a. The Licensed Program and User Manual(s) are for your personal use only.
b. You may use the Licensed Program on a single computer, or on its temporary replacement, or on a subsequent computer only.
c. The Licensed Program may be copied to a single computer hard drive for playing.
d. A backup copy or copies may be made only as provided by the User Manual(s), except as expressly permitted by this Agreement.
e. You may not use the Licensed Program on more than one computer system, make or distribute unauthorized copies of the Licensed Program or User Manual(s), create by decompilation or otherwise the source code of the Licensed Program or use, copy, modify, or transfer the Licensed Program, in whole or in part, or User Manual(s), except as expressly permitted by this Agreement.
    If you transfer possession of any copy or modification of the Licensed Program to any third party, your license is automatically terminated. Such termination shall be in addition to and not in lieu of any equitable, civil, or other remedies available to Wiley.

Term: This License Agreement is effective until terminated. You may terminate it at any time by destroying the Licensed Program and User Manual together with all copies made (with or without authorization).
    This Agreement will also terminate upon the conditions discussed elsewhere in this Agreement, or if you fail to comply with any term or condition of this Agreement. Upon such termination, you agree to destroy the Licensed Program, User Manual (s), and any copies made (with or without authorization) of either.

Wiley's Rights: You acknowledge that all rights (including without limitation, copyrights, patents and trade secrets) in the Licensed Program (including without limitation, the structure, sequence, organization, flow, logic, source code, object code and all means and forms of operation of the Licensed Program) are the sole and exclusive property of Wiley. By accepting this Agreement, you do not become the owner of the Licensed Program, but you do have the right to use it in accordance with the provisions of this Agreement. You agree to protect the Licensed Program from unauthorized use, reproduction, or distribution. You further acknowledge that the Licensed Program contains valuable trade secrets and confidential information belonging to Wiley. You may not disclose any component of the Licensed Program, whether or not in machine readable form, except as expressly provided in this Agreement.

WARRANTY: TO THE ORIGINAL LICENSEE ONLY, WILEY WARRANTS THAT THE MEDIA ON WHICH THE LICENSED PROGRAM IS FURNISHED ARE FREE FROM DEFECTS IN THE MATERIAL AND WORKMANSHIP UNDER NORMAL USE FOR A PERIOD OF NINETY (90) DAYS FROM THE DATE OF PURCHASE OR RECEIPT AS EVIDENCED BY A COPY OF YOUR RECEIPT. IF DURING THE 90 DAY PERIOD, A DEFECT IN ANY MEDIA OCCURS, YOU MAY RETURN IT. WILEY WILL REPLACE THE DEFECTIVE MEDIA WITHOUT CHARGE TO YOU. YOUR SOLE AND EXCLUSIVE REMEDY IN THE EVENT OF A DEFECT IS EXPRESSLY LIMITED TO REPLACEMENT OF THE DEFECTIVE MEDIA AT NO ADDITIONAL CHARGE. THIS WARRANTY DOES NOT APPLY TO DAMAGE OR DEFECTS DUE TO IMPROPER USE OR NEGLIGENCE.
    THIS LIMITED WARRANTY IS IN LIEU OF ALL OTHER WARRANTIES, EXPRESSED OR IMPLIED, INCLUDING, WITHOUT LIMITATION, ANY WARRANTIES OF MERCHANTABILITY OR FITNESS FOR A PARTICULAR PURPOSE.
    EXCEPT AS SPECIFIED ABOVE, THE LICENSED PROGRAM AND USER MANUAL(S) ARE FURNISHED BY WILEY ON AN "AS IS" BASIS AND WITHOUT WARRANTY AS TO THE PERFORMANCE OR RESULTS YOU MAY OBTAIN BY USING THE LICENSED PROGRAM AND USER MANUAL(S). THE ENTIRE RISK AS TO THE RESULTS OR PERFORMANCE, AND THE COST OF ALL NECESSARY SERVICING, REPAIR, OR CORRECTION OF THE LICENSED PROGRAM AND USER MANUAL(S) IS ASSUMED BY YOU.
    IN NO EVENT WILL WILEY OR THE AUTHOR, BE LIABLE TO YOU FOR ANY DAMAGES, INCLUDING LOST PROFITS, LOST SAVINGS, OR OTHER INCIDENTAL OR CONSEQUENTIAL DAMAGES ARISING OUT OF THE USE OR INABILITY TO USE THE LICENSED PROGRAM OR USER MANUAL(S), EVEN IF WILEY OR AN AUTHORIZED WILEY DEALER HAS BEEN ADVISED OF THE POSSIBILITY OF SUCH DAMAGES.

General: This Limited Warranty gives you specific legal rights. You may have others by operation of law which varies from state to state. If any of the provisions of this Agreement are invalid under any applicable statute or rule of law, they are to that extent deemed omitted.
    This Agreement represents the entire agreement between us and supersedes any proposals or prior Agreements, oral or written, and any other communication between us relating to the subject matter of this Agreement.
    This Agreement will be governed and construed as if wholly entered into and performed within the State of New York. You acknowledge that you have read this Agreement, and agree to be bound by its terms and conditions.